Interdisciplinary Biotechnological Advances

This series is an authoritative resource that explains recent advances and emerging directions in biotechnology, reflecting the forefront of research clearly and reliably, without excessive hype. Each volume is written by authors with excellent reputations and acknowledged expertise in the topic under discussion. The volumes span the entire field from an interdisciplinary perspective, covering everything from biotechnology principles and methods to applications in areas including genetic engineering, transgenic plants and animals, environmental problems, genomics, proteomics, diagnosis of disease, gene therapy, and biomedicine. The significance of these applications for the achievement of UN Sustainable Development Goals is highlighted. The series will be highly relevant for Master's and PhD students in Biotechnology, Nanochemistry, Biochemical Engineering, and Microbiology, medical students, academic and industrial researchers, agricultural scientists, farmers, clinicians, industry personnel, and entrepreneurs.

Santa Ram Joshi • Pradeep Verma
Subhro Banerjee
Editors

Mineral Transformation and Bioremediation by Geo-Microbes

Springer

Editors
Santa Ram Joshi
Department of Biotechnology and Bioinformatics
North Eastern Hill University
Shillong, Meghalaya, India

Pradeep Verma
Department of Microbiology
Central University of Rajasthan
Kishangarh, Rajasthan, India

Subhro Banerjee
Department of Zoology
SRM University Sikkim
Gangtok, Sikkim, India

ISSN 2730-7069 ISSN 2730-7077 (electronic)
Interdisciplinary Biotechnological Advances
ISBN 978-981-96-3032-5 ISBN 978-981-96-3033-2 (eBook)
https://doi.org/10.1007/978-981-96-3033-2

This Springer imprint is published by the registered company Springer Nature Singapore Pte Ltd.
The registered company address is: 152 Beach Road, #21-01/04 Gateway East, Singapore 189721, Singapore

Preface

Until recently, geomicrobiology has been a least concerned branch of biology. However, it has now revolutionized the field of biology worldwide. Needless to say that the interest in geomicrobiology is garnering up with each passing day. Geomicrobiology is the study of the microbial role in geochemical and geological processes that have carved the Earth and continue to function to this day. Microbes play a prominent part in generating, recycling, sequestering, and neutralizing a gamut of chemical elements and substances in nature through biogeochemical cycles that cover the deep lithosphere, hydrosphere, and atmosphere.

This book comprises 22 chapters, covering a wide range of topics, such as general geomicrobiology, and the applications of geo-microbes in diverse fields, including the advanced field of astrobiology. This book provides an exhaustive account of the diversity of geo-microbes, their mechanisms of action in bioremediation, and the transformation of minerals. Additionally, the search for extraterrestrial life in celestial bodies and its correlation with microbial life in subterranean environments are described in detail. Microbial ecology, specifically in mineral ore deposits, coal mines, oil spills, and salt marshes, and respective interactions are also discussed. The current need for enhanced bioremediation of toxic environments and the transformation of minerals into less noxious ones are also covered in this edited volume. With the growing interest in searching for life on other planets, particularly on Mars, a separate chapter has been dedicated to possible models for extraterrestrial life, catering to the interest of researchers.

Although there are a number of research monographs and advanced books on geomicrobiology, there has been no up-to-date book suitable for students and researchers. This book fills a significant gap in the field and we are pleased to see it published. Research scientists and technicians already working in geomicrobiology will also benefit from its practical perspective. In addition to geomicrobiologists, this book will be of interest to personnel from various multidisciplinary domains, including environmental scientists, organic chemists, astrophysicists, and chemical engineers. The authors are pleased to offer this modern, attractive, and accessible book, which will be equally beneficial to practicing researchers and students.

The authors have received encouragement, help, and suggestions from their well-wishers, colleagues, and friends while compiling this book. We gratefully acknowledge the advice and recommendations provided by our fellow workers and experts. In the preparation of this book, we are immensely thankful to the contributors of the chapters for their kind cooperation. Our sincere thanks are also due to our Production Editor (books), Ms. Nandhini Natchatra of Springer, for her guidance and timely interventions in the production of this volume.

We welcome any constructive suggestions from our readers.

Shillong, Meghalaya, India Santa Ram Joshi
Kishangarh, Rajasthan, India Pradeep Verma
Gangtok, Sikkim, India Subhro Banerjee

Contents

Geomicrobiological Interactions in the Frozen World: Potential for Exobiological Processes

Khashti Dasila, Dheeraj Shootha, and Anita Pandey

Contents

Abstract The study of microorganism-mineral interactions in different geographical environments is known as geomicrobiology. It is concerned with the involvement of microbes in several elemental cycles, mineral transformation, and geochemical processes. Geomicrobiology combines environmental science, geology, chemistry, and microbiology that improve the insights into Earth's systems. Geomicrobiological interactions reveal the fascinating resilient and adaptablity

K. Dasila
Department of Forestry and Environmental Science, School of Earth and Environmental Science, Uttarakhand Open University, Haldwani, Uttarakhand, India
e-mail: kdasila@uou.ac.in

D. Shootha
School of Applied and Biotechnology, Shoolini University of Biotechnology and Management Science, Bajhol Solan, Himachal Pradesh, India

A. Pandey (✉)
Department of Biotechnology, Graphic Era (Deemed to be University), Dehradun, Uttarakhand, India
e-mail: anita.pandey@geu.ac.in

S. R. Joshi et al. (eds.), *Mineral Transformation and Bioremediation by Geo-Microbes*, Interdisciplinary Biotechnological Advances,
https://doi.org/10.1007/978-981-96-3033-2_1

of life in extreme environments offering insights into potential exobiological processes on icy worlds beyond the Earth. Extremely low temperatures, limited nutrition availability, high pressure, and low liquid water content are characteristics of frozen environments, including Antarctic ice sheets, Arctic permafrost, and glacier ecosystems. Having harsh conditions, a variety of microbes have adapted to survive and exert significant influence in shaping such extreme environments by active participation in various biogeochemical cycles. The possible contribution of microorganisms in chemical weathering is by the production of organic acids that dissolve minerals and release nutrients essential to their survival. The microbial-assisted biogeochemical activity may potentially result in the emergence of new minerals, altering the landscape over time. The interactions between microorganisms and their extreme habitat (frozen) not only demonstrate the adaptability of life but also raise the possibility that life exists in comparable conditions on other planets. A highly fascinating aspect of these studies is the potential for finding life on cold planets and moons like Europa and Enceladus. It is known that many celestial bodies contain seas beneath their frozen crusts, which might provide habitats for life. Geomicrobiological interactions on the Earth in such extreme conditions make it probable that analogous processes may be occurring on these frozen moons.

Keywords Geomicrobiology · Exobiological processes · Frozen environments · Biogeochemical cycle · Celestial bodies

Abbreviations

%	Percentage
°C	Degree celsius
atm	Atmosphere
C	Carbon
$CaCO_3$	Calcium carbonate
CO_2	Carbon dioxide
FTIR	Fourier-transform Infrared spectroscopy
Ga	Gigaannum
H_2	Hydrogen
K	Kelvin
m	Metre
pH	Power of hydrogen
SEM	Scanning electron microscopy
SSU rRNA	Small subunit ribosomal RNA
TEM	Transmission electron microscopy

1 Introduction

Geomicrobiology is an multidisciplinary field that explains the intricate relationship that exists between geological events and microorganisms with practical implications of microorganisms in various geological and environmental phenomena on the Earth (Hallsworth et al. 2021). The key components of geomicrobiology are to understand the physicochemical characteristics of the Earth's surface and subsurface are geology and geomorphology, respectively. Metagenomics, which became the powerful tool available to study the collective genome of microbial communities, has emerged as a potent tool and revealed the enormous diversity and metabolic potential of microbes inhabiting different geological areas (Valdivia-Granda 2008; Mocali and Benedetti 2010; Nesme et al. 2016). Soil harbor the utmost microbial diversity on Earth, and soil microbiology is an important area of study within geomicrobiology (Mocali and Benedetti 2010). Numerous essential ecosystem services, such as nutrient cycling, the breakdown of organic matter, and the stimulation of plant growth, are mediated by soil microbes (Comerford et al. 2013). Geomicrobiology explores the functional microbial diversity in terms of soil formation and mineral weathering, investigation of microbial life in harsh environments, analysis of microbially influenced geological processes, development of biotechnological applications including bioremediation, and sustainable resource extraction (Choi et al. 2016).

Frozen environments on the Earth like Arctic and Antarctic regions have been explored by researchers due to their unique and crucial role in the global ecosystem (Thomas and Dieckmann 2002; Pikuta and Hoover 2003; Quillfeldt 2010; Guven and Demirci 2012; Clarke et al. 2013; Elliott et al. 2017). These frozen regions are characterized by vast expanses of ice. Limited sunlight and extreme temperatures, which harbor a varied range of extremophilic organisms, have evolved with remarkable adaptations to thrive in the harsh conditions (Thomas and Dieckmann 2002; Correa and Abreu 2020). The frozen environment motivates researchers to explore due to its significance that is discussed as follows:

1. High possibilities of valuable biotechnological products, such as antifreeze proteins, enzymes, and biomolecules with industrial and medical applications (Martínez-Rosales and Castro-Sowinski 2011; Martorell et al. 2017; Correa and Abreu 2020; Lima et al. 2022).
2. The psychrophilic and psychrotrophic microorganisms inhabiting frozen environment can provide valuable insight on the adaptive mechanism in extreme conditions and limits of life, which could lead to implications in search of extraterrestrial life (Nichols et al. 1999; Pikuta and Hoover 2003; Correa and Abreu 2020; Lima et al. 2022).
3. The study of frozen habitats/environments is essential to comprehending the effect of global climate system (Goordial 2021).
4. The study of frozen environments has implications to understand the broader planetary systems and the possibilities of life on other icy worlds (Pikuta and Hoover 2003; Rampelotto 2013; Correa and Abreu 2020; Goordial 2021).

2 Cryosphere: Frozen Environments

Frozen environments are known as the cryosphere. It is the regions of the Earth characterized by presence of permafrost/ice and sub-zero temperatures. Such unique habitats are distributed in polar and alpine regions and support a various array of microbial world that has adapted to flourish in extreme conditions (Pikuta and Hoover 2003; Goordial 2021). Universally, over 80% of the Earth's biosphere is eternally cold with major regions located in the polar ice caps (Arctic and Antarctic regions) and high-altitude regions (glaciers and permafrost) (Garcia-Lopez and Cid 2017). The frozen environments of our planet, a remarkable array of physical, chemical, and biological characteristics, are crucial for understanding climate dynamics and ecological interactions (Table 1).

Polar ice caps are mainly composed of various layers of snow and ice forming compact structures (Burgess and Sharp 2004) and are characterized by extreme temperatures, often ranging around −20 °C to −65 °C (Britannica 2016). Such extreme environment contributes significantly to a high solar albedo effect, reflecting substantial radiation and influencing global climate patterns (Flanner et al. 2007). Moreover, the composition of these ice caps includes fresh water, glacial melt containing salts and nutrients into the surrounding environments and influencing the biogeochemistry (Golden et al. 1998). Ice caps sustain various microbial communities, including psychrophilic archaea and bacteria, which contribute significantly to nutrient cycling in these extreme conditions (Pielou 2008; Abirami et al. 2021).

Glaciers play an important role in frozen ecosystems, which flow under the influence of gravity and are affected by various factors like temperature and pressure. The internal dynamics of glaciers allow them to erode bedrock while transporting essential nutrients downstream (Bockheim 2007). The crystalline ice structure of glaciers varies with pressure and age, affecting its strength and melting point (Petrenko and Whitworth 1999). The glaciers harbor unique ecosystems for biological dynamics that predominantly composed of specific microbial phyla, such as Bacteroidetes, Cyanobacteria, and Proteobacteria (Waldrop et al. 2023). These microorganisms flourish during melting periods of glaciers, which subsequently support invertebrate communities that benefit from nutrients deposited due to the glacial erosion (Bagshaw et al. 2012).

Permafrost, the permanently frozen state, exhibits significant inconsistency soil composition and thaw depth, which is severely influenced by local environmental conditions (Zhang 2005). The permafrost soil is often rich in organic materials along with ice, affecting land stability and hydrology (Williams and Smith 1989). Permafrost contributed significantly to carbon sinks as it stores vast amounts of organic carbon, which may be released as greenhouse gases in the atmosphere once thawed (Schuur et al. 2015). Therefore, such potential for carbon release poses a significant concern with context to global climate change. The diverse microbial communities in permafrost with predominant phyla include Proteobacteria and Actinobacteria, which play a crucial role in absorbing organic matter (Jansson and Taş 2014; Mackelprang et al. 2017). The microbial diversity is affected by thawing

Table 1 Physiochemical and biological characteristics of frozen environments

Sr no.	Frozen environments	Parameters	Characteristics
Physical characteristics			
1.	Polar ice caps	Structure	Their structure is typically composed of layers of compacted snow and ice (Burgess and Sharp 2004)
		Temperature	Extremely low temperatures, often averaging −20 °C to −65 °C (Britannica 2016)
		Albedo effect	High albedo effect reflects significant solar radiation, impacting the global climate patterns (Flanner et al. 2007)
2.	Glaciers	Movement and dynamics	The movement of glaciers like sliding and internal deformation can be influenced by temperature and pressure (Benn and Evans 2014)
		Ice structure	The crystals of glacial ice can vary and depends on the age and pressure, affecting its strength and melting point (Petrenko and Whitworth 1999)
3.	Permafrost	Thaw depth	The permanently frozen state is known as permafrost. It is composed with active compact layers that thaw seasonally. The thickness and compactness can vary significantly depending on location (Zhang 2005)
		Soil composition	It contains ice-rich sediments and organic materials that may affect land stability and hydrology (Williams and Smith 1989)
Chemical characteristics			
1.	Polar ice caps	Salinity	Ice caps are fresh ice but glacial melt can introduce salts and nutrients into the surrounding marine environment (Golden et al. 1998)
2.	Glaciers	Nutrient transport	Glaciers can act as channel for nutrients as they bedrock erode and downstream transport of minerals occurs (Bockheim 2007)
3.	Permafrost	Carbon storage	It acts as major carbon sink and store organic carbon, which can release as greenhouse gases upon thawing (Schuur et al. 2015)
Biological characteristics			
1.	Polar ice caps	Microbial activities	The diversity in archaeal, bacterial, and eukaryotic microbial and viral communities that establish the important part of the polar ecosystems (Abirami et al. 2021). Microbial communities exist, including psychrophiles that thrive in extreme cold, contributing to nutrient cycling (Pielou 2008)
2.	Glaciers	Biological composition and ecosystems	The dominant microbial phyla in glaciers are Proteobacteria (48.8%), Bacteroidetes (29.1%), and Cyanobacteria (16.3%). Glacial seawater microorganisms belong to phyla Bacteroidetes (40.3%), followed by Actinobacteria (31.7%) and Proteobacteria (25.4%). Other phyla such as Firmicutes, Gemmatimonadetes, Nitrospirae, Chloroflexi, Planctomycetes, Verrucomicrobia, and Chlamydiae are least abundant (Waldrop et al. 2023). Glacier melting supports the growth of unique microbial and invertebrate ecosystems (Bagshaw et al. 2012)

(continued)

Table 1 (continued)

Sr no.	Frozen environments	Parameters	Characteristics
3.	Permafrost	Microbial community and activity	Dominant members of the permafrost community include Actinobacteria followed by Proteobacteria, whereas Bacteroidetes, Chloroflexi, and Firmicutes can vary substantially with different physicochemical factors and age (Jansson and Taş 2014; Mackelprang et al. 2017). Permafrost hosts diverse microbial communities that can metabolize organic matter. Thawing increases microbial activity leading to nutrient release (Jansson and Taş 2014)

process; as it thaws, the microbial activity increases significantly, leading to nutrients release that can further influence the nutrient status of surrounding ecosystems (Jansson and Taş 2014).

3 Microbial Existence in Frozen Environments

The Earth's icy biosphere, which includes a wide variety of frozen habitats, from polar regions to high-altitude glaciers, harbors a remarkable diversity of microbial life. These psychrophilic microorganisms have developed notable adaptations to sustain themselves in harsh conditions like sub-zero temperatures, high levels of ultraviolet radiation, and limited nutrient availability (Priscu and Christner 2003). The study of frozen microbial communities is essential to understand the evolutionary and ecological processes, particularly with the resilience toward extreme conditions that shape the cryosphere. Modern emerging tools and techniques significantly expand the understanding of abundance and functional activity of microbial communities of extreme environments (Margesin and Collins 2019). Glaciers and permafrost are the most extensively studied frozen habitat of the cryosphere and reveal the microbial diversity with representatives from all three domains of life—Archaea, Bacteria, and Eukarya (Margesin and Collins 2019). These microorganisms have a potential role in biogeochemical processes like nutrient cycling and overall functioning of their respective frozen ecosystems.

Microbial communities in frozen environment exhibit remarkable resilience and adaptations (Fig. 1). These adaptations include the production of antifreeze proteins, accumulation of compatible solutes, maintenance of membrane fluidity and development of cold-active enzymes, and the ability to maintain metabolic activity at subzero temperature cryosphere (Priscu and Christner 2003; Margesin and Collins 2019). The study of extremophiles is important to explore the diversity, ecology and their potential applications in biotechnology and astrobiology (Margesin and Miteva 2011; Goordial 2021). Furthermore, the biogeography of these microbial communities is highly influenced by factors such as moisture,

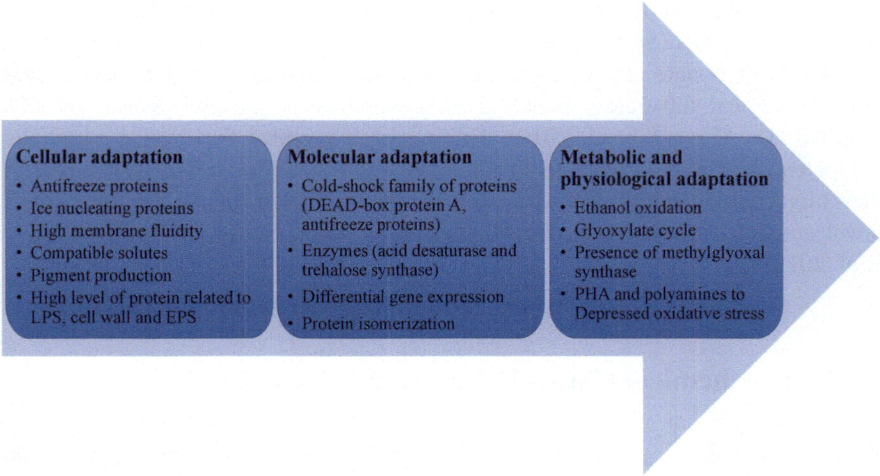

Fig. 1 Adaptation strategies of microorganisms in frozen environment

temperature, and the nutrient availability including organic and inorganic substrates with diverse microbial community compositions reported from different frozen habitats.

4 Geomicrobiological Interactions

Geomicrobiology discovers the complex relationships between microorganisms and their geological environments where they interact and are influenced by the physical and chemical characteristics of the Earth's crust. One of the important aspects of geomicrobiology is the microbial-induced precipitation of calcium carbonate ($CaCO_3$). In this process, the bacteria and fungi facilitate and exhibit the potential applications in soil stabilization, construction materials, and environmental remediation (De Muynck et al. 2010; Seifan et al. 2017; Chen et al. 2022). Studies reported that some bacterial species, such as *Arthrobacter*, *Bacillus*, *Lysinibacillus*, *Rhododcoccus*, and *Sporosarcina*, produce calcium carbonate (Seifan et al. 2016). These microorganisms induce the production of $CaCO_3$ through various metabolic pathways like urea hydrolysis, anaerobic respiration, and anoxygenic and oxygenic photosynthesis. The microbial-mediated precipitation of $CaCO_3$ is used in binding soil particles, immobilizing heavy metals, and sealing cracks (De Muynck et al. 2010; Ortega-Villamagua et al. 2020). The microscale visualization techniques, such as atomic force microscopy and scanning electron microscopy, are employed by researchers to harness and understand the morphology and distribution of precipitated mineral crystals (Wang et al. 2019). Besides these techniques, the advanced spectroscopic methods, such as Raman spectroscopy and X-ray diffraction, have been used to characterize the mineralogical composition and crystalline structure of the $CaCO_3$ deposits (Seifan et al. 2017; Wang et al. 2019). In addition to the study of calcium carbonate precipitation, the interactions between microorganisms and other geological substrates, including sulfides, silicates, and phosphates, are also reported (De Muynck et al. 2010). This research revealed the diverse metabolic competences of microorganisms to solubilize and transform various minerals with their biotechnological implications, such as bioremediation and the development of novel construction materials (De Muynck et al. 2010; Seifan et al. 2017; Wang et al. 2019).

5 Biogeochemical Cycles in Frozen Environments

Microorganisms play an essential role in governing of the biogeochemical cycles on the Earth. Progressions in analytical and molecular techniques have significantly expanded the understanding of the microbial diversity and its functional capabilities within cryogenic environments. The importance of considering microbial ecology in frozen environment plays a crucial role in regulating the global climate systems. Cryophilic microorganisms may influence the reflectivity of ice surfaces, subsequently affecting the albedo and energy balance of the Earth (Goordial 2021). Furthermore, the thawing of permafrost due to climate change has the potential to release greenhouse gases is the concern subject of study the permafrost microbiology for predicting and mitigating the global warming impacts (Margesin and Collins 2019). The evolving technologies like high-throughput sequencing, in situ microscopy, and stable isotope probing expand the understating on the complex microbial-mediated biogeochemical cycle in cryosphere (Priscu and Christner 2003; Margesin and Collins 2019; Goordial 2021). The microbial-mediated biogeochemical cycles in frozen environment are presented in Fig. 2. The cryophilic microorganisms regulate the carbon sequestration and release of locked carbon in permafrost-frozen soil (Margesin and Collins 2019). Similarly, microbial communities in glacial and permafrost habitat significantly transform the atmospheric nitrogen by the processes of fixation, ammonification, nitrification, and denitrification. Chemolithotrophic microorganisms are responsible for the sulfur cycle in glacial and subglacial habitats and oxidize the reduced sulfur compounds, influencing the speciation and mobility of sulfur in these habitats (Boyd et al. 2014). The insoluble organic phosphorus is solubilized and mineralized by psychrophilic microbes, making it available for other organisms in nutrient-limited polar and alpine environments (Margesin and Miteva 2011; Margesin and Collins 2019). The climate change significantly impacts the frozen biome of the Earth; therefore, understating the potential and role of microbial communities in such extreme habitat is crucial to predict and mitigate the effect of global climate change.

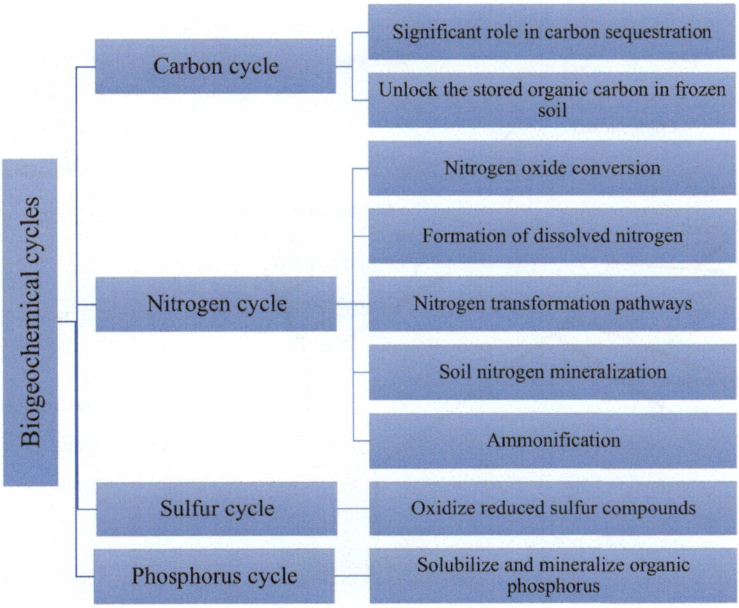

Fig. 2 Microbial-associated biogeochemical processes in frozen environment

6 Potential of Exobiological Processes

Emerging advancements in technologies allowed the researchers to discover the abundance, functional aspects, and intricate lifestyles of these cryophilic (cold-loving) microbes, providing valuable insights into the life limit on the Earth and the potential for microbial existence on other celestial bodies (Margesin and Collins 2019; New et al. 2021).

Exobiology is defined as the relationship between life and the physiochemical system of the Earth. It is highly interdisciplinary effort of subdisciplines, including geochemistry, paleontology, sedimentology, microbiology, molecular biology, and oceanography (Martin and Johnson 2012). Some fundamental questions that have been posed for many generations are answered by astrobiology: How did life start and develop? Is there life elsewhere in the universe, and what is fate of life on Earth and beyond? As a result, astrobiology includes research on the possibility of life existing on planets other than our solar system, as well as field and laboratory studies on the origins, early evolutionary life, and the possibility that life may adapt to new environments both on Earth and in space (Des Marais et al. 2003). Molecular biology has utterly changed our understanding of the biosphere, particularly with respect to the role of microbial organisms in the total diversity of life on Earth (Pace 1997). Environmental genomics recently discovered a planet inhabited by an extraordinary variety of microbial entities, and identified species were classified within one of the three life domains (Woese 1987; Woese et al. 1990;

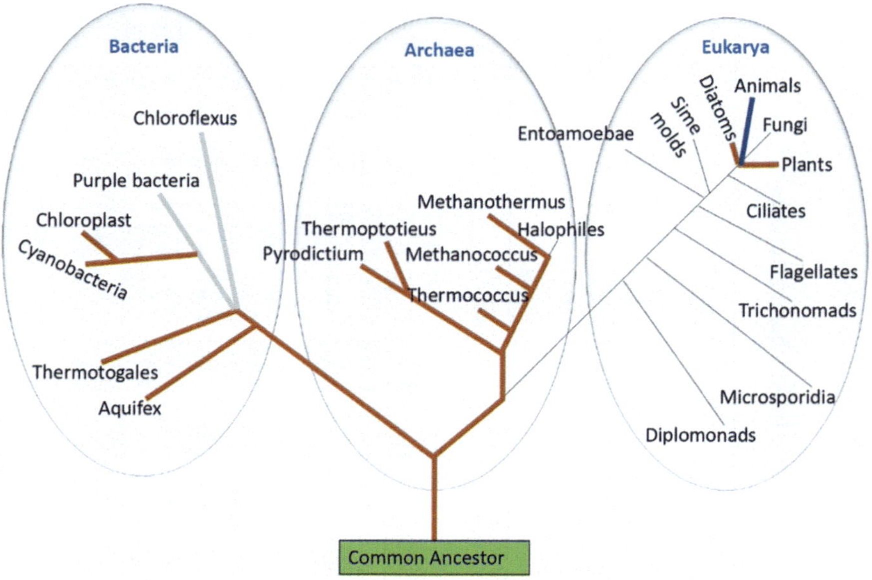

Fig. 3 Three domains of life constructed through comparative analyses of 16s rRNA gene sequences found in prokaryotes. The orange branches within Bacteria and Archaea domains indicate hyperthermophiles. The gray and black branches denote oxygenic and anoxygenic photosynthetic organisms, respectively. The orange branch in all the domains is attributed to endosymbiotic events involving cyanobacteria that produce oxygen

Cavalier-Smith 2004). The domains Archaea, Bacteria, and Eukarya represent a fundamental classification of life (Fig. 3). The domains Bacteria and Archaea are solely composed of microbial communities and encompass less complex organisms that lack a defined nucleus or membrane-bound organelles. The continuous discovery of novel species has accumulated as biologists have expanded their investigations into the intricate ecological system of our planet; this attempt has significantly increased the diversity of physiology, genotypes, metabolic, and phyla classifications (Woese 1987).

6.1 Terrestrial Analogues

The dominance of silicate minerals on terrestrial planets and moons may support complicated organic chemistry. Variation in physiochemical properties like temperature, pH, salinity, pressure, and radiation levels are capable to support prokaryotic and eukaryotic life on the Earth (Rothschild and Mancinelli 2001). In the present scenario, the most effective approach for examining the habitability of moons and other planets needs the development of tools and technology for their exploration through terrestrial analogue studies. However, they have drawbacks like no analogue site or replica of a different planet or moon along with a lack of established standards measures to follow them (Soare et al. 2001). For example, the statistics and photos confirm that Mars contains ice volcanoes, caps, volcanic rocks, and impact craters, all may be potential locations for habitable niches. The volcanoes, basalts, impact craters, and ice-covered areas on the Earth are considered first-order analogues due to the significant difference between the ecosystems of the two planets.

6.2 *Terrestrial Planets*

6.2.1 Mars

Mars have similarity with the Earth planet in many ways and thought to have once offered a considerably wetter and warmer environment that was suitable for the emergence and continuation of life. However, Mars have experienced global environmental alteration in the past and its age assumed roughly around 700–900 million years of Martian history (Table 2), thus reflecting the planetary similarity between the two planets. The surface water was abundant, and the temperature was globally cold (Fairén et al. 2009). The atmosphere on Mars receives around 43% of the radiation as Earth that does receive. However, CO_2-rich thin atmosphere minimally reduces this radiation, leading to high surface fluxes (>200 nm) of radiation. Surface oxidants degrade organic carbon, explaining the negative results of the Viking missions in 1976 (Biemann et al. 1977). The basalt weathering produces

Table 2 Geological evolution of Mars in different climatic era and its reported terrestrial analogues (Wordsworth 2016; Kite 2019)

Climate era	Geologic time	Absolute age (Ga)	Analogues
Hyperarid Mars	Late Amazonian	0.6–0.3 to present	Atacama Desert, Chile
	Middle Amazonian	2.1–1.4 to 0.6–0.3	
	Early Amazonian	3.1–1.4 to 0.6–0.3	
Snowball Mars	Late Hesperian	3.6 to 3.1–2.9	Axel Heiberg Island Beacon Valley
	Early Hesperian	3.7 to 3.6	
Cold and wet Mars	Late Noachian	3.82 to 3.7	Pilbara Kalgoorlie Rio Tinto Canadian Arctic
	Middle Noachian	3.92 to 3.82	
	Early Noachian	>3.95	

Ga Gigaannum

phyllosilicates, and orbiter spectroscopic imagery of Noachian surface of Mars has shown Al-rich and Fe/Mg phyllosilicate minerals (Bibring et al. 2007; Mustard et al. 2008; Carter et al. 2010). The phyllosilicate deposits are significant targets for researching their habitability potential and the search for life on Mars because of their age, wide spatial distribution, known propensity to accumulate and preserve organic compounds, and the geochemical environments in which they formed (Kahle et al. 2004; Wiseman and Püttmann 2006). The degradation of organic material and the importance of oxidation processes on Mars are still being debated (Davila et al. 2008; Bonaccorsi et al. 2010).

6.2.2 Venus

Venus has the highest surface temperature due to its close proximity to the Sun and the strong greenhouse effect created by its thick 90-bar atmosphere of more than 96% CO_2. Therefore, thick carbon dioxide atmosphere is responsible for the complete evaporation of any water, rendering the planet uninhabitable Walker 1975; Kasting 1988; Hasimoto et al. 2008). The surface temperature is approximately 460 °C, eliminating the possibility of stable organic molecules or liquid water. However, there is speculation regarding the possibility of supercritical water in the subsurface (Schulze-Makuch and Irwin 2002). However, a possibly sustainable zone exists far above the surface where liquid water is available and temperatures range from freezing to 120 °C. Thus, an aerial biosphere could be supported by the cloud deck, mainly the lower and middle in high-elevation habitable zones (Cockell 1999). These clouds provide persistent water droplets, but they are quite acidic due to dissolved H_2SO_4, and photosynthesis or sulfate reduction can provide metabolic energy sources (Schulze-Makuch et al. 2004). Hyperthermophilic acidophiles, such as *Acidianus infernus*, which thrive optimally at a temperature of 88 °C under highly acidic (pH 0.5) conditions, represent significant analogue organisms for Venus (Segerer et al. 1986). Additionally, viable bacterial populations have been detected at substantially higher altitudes, surpassing the tropopause boundary within the stratosphere, facilitated by sampling conducted by means of stratospheric balloon flights and high-altitude research aircraft (Harris et al. 2002; Wainwright et al. 2003; Smith et al. 2010).

6.2.3 Icy Moons

The terrestrial planets currently exhibit conditions that are not particularly conducive to habitability on their surfaces. However, Mars may harbor conserved biosignatures on its surface resources or may possess dynamic communities situated in its subsurface regions, while Venus may provide habitable settings at elevated altitudes within its atmosphere. The most promising icy satellites orbiting Jupiter and Saturn, particularly Enceladus, Europa, and Titan, have been demonstrated to be geologically dynamic entities that possess a myriad of potentially habitable environments (Table 3). Europa is perpetually subjected to ionizing radiation due to its position within the magnetosphere of Jupiter, as noted by Baumstark-Khan and Facius (2002), while the surface temperatures fluctuate between 86 and 132 K, which is significantly below the minimum threshold requisite for microbial proliferation (Spencer et al. 1999). Consequently, the ice does not constitute a viable habitat capable of sustaining any presently recognized terrestrial organisms; nonetheless, it may furnish adequate shielding from radiation, thereby facilitating the preservation of organic materials and any biotic entities contained within it (Gilichinsky et al. 1993). There exists a significant amount of indirect evidence that substantiates the hypothesis regarding the existence of a subsurface brine ocean under the icy exterior of Europa (Pappalardo 2010), alongside propositions indicating that its chemical makeup may include hydrated minerals (Carlson et al. 1999). Titan is the only moon in the solar system having atmosphere like Earth. Its surface exhibiting many dynamic processes like nitrogen-rich atmosphere as a sign of dendritic networks resembling fluvial systems (Tomasko et al. 2005), seas and lakes (Hayes et al. 2008; Paillou et al. 2008) composed of methane and ethane along with heavier hydrocarbons, dissolved nitriles, and/or atmospheric gases (Mitri et al. 2007). The amalgamation of free H_2 in proximity to the surface engenders a highly dynamic ecological context. The environmental conditions on Titan fulfill the essential prerequisites for life: integral thermodynamic disequilibrium, a copious carbon inventory, and a fluidic milieu (Lorenz et al. 2008).

Table 3 The analogue sites for Europa and Titan on the Earth

Location	Features	References
Analogue sites for Europa		
Permafrost, Multiple Sites (Arctic regions of Alaska, Canada, Greenland, and Siberia)	• Sustain the Earth's 20% land surface with frozen sediments that harbor different microbial communities	Gilichinsky et al. (1993)
	• Preserves ancient life upto three million years old and remains viable	
	• The stable ice act as preservation medium for life over geological time; biosignature and survival of microbes	
The Borup Fiord Pass (glaciers), Canada	• Perennial subsurface alkaline spring waters	Grasby et al. (2003), Lorenz et al. (2011)
	• Deposits of gypsum, calcite, and elemental sulfur on glacial ice	
	• Microbial-mediated sulfur (sulfur-reducing bacteria)	
	• Analogue chemical reactions on the icy surface of Europa, sulfur-based spectral signatures ecosystem	

(continued)

Table 3 (continued)

Location	Features	References
The Mariana Trench, Pacific Ocean	• The deepest ocean basin on earth with lowest temperature 1 °C and 1100 bars pressure	Sharma et al. (2002), Glud et al. (2013)
	• Bacteria can tolerate up to 16,800 bars pressures	
	• Chemolithotrophs can survive on the ocean surface	
Analogue sites for Titan		
Pitch Lake, Trinidad and Tobago	• A natural asphalt lake, analogue of possibility of habitability of liquid hydrocarbon lakes	Schulze-Makuch et al. (2011)
Alaskan Oil Fields, USA	• Continental petroleum reservoir upto 1670 m depth	L'Haridon et al. (1995)
	• The temperatures and pressure range 65–70 °C and 50–160 atm, respectively	
	• Occurrence of hyperthermophilic archaea and indigenous thermophilic bacteria.	
	• Obligate anaerobic nitrite- and sulfur- oxidizing bacteria	
	• Strict anaerobic metabolisms with archaeal-mediated methane metabolism	
	• Analogue for life in petroleum reservoirs, deep subsurface environments, and microbial communities	
Rancho La Brea Tar Pits, California, USA	• Natural asphalt-soil seepages of recalcitrant petroleum hydrocarbons from the Pleistocene	Kim and Crowley (2007)
	• Soil-seepage range from high to low salinities and acidic to alkaline	
	• Presence of water in pockets and pore spaces	
	• Dominance of purple sulfur bacteria and halophilic archaea microbial communities	
	• Microbial communities analogue of natural asphalt; microbial preservation of fossils and biosignatures	

7 Tools and Technologies Used in Geomicrobiological Research

The technological advancement in the field of geomicrobiology resulted in scaling up the research and has given an advanced view of the working on microbes (Fig. 4).

1. **Metagenomics:** This method simply entails the direct examination of genetic material from samples collected from the environment. Researchers can analyze whole groups of microbes in their natural environments without isolating them.
2. **Lipid Biomarkers:** The primary lipid constituents of microorganisms are phospholipid fatty acids. Such information obtained can be on biomass and community structure, and even on metabolic activities of microbes.
3. **Molecular Marker:** The most widely used molecular markers, namely small subunit ribosomal RNA (SSU rRNA), for the identification and detection of relatedness among unexplored phylogenetic diversity lie in various environmental settings.
4. **Fourier-transform infrared spectroscopy (FTIR):** This method is used to analyze the absorption and emission of radiation of molecular vibrations in a sample. This technique is also useful to detect the protonation and deprotonation behaviors of bacteria as a response to changes in pH, sorption of contaminants onto minerals/bacteria, and further monitoring the mineral precipitation or degradation of contaminants in a system of interest.
5. **Microscopy**: Geomicrobiological samples procured from natural ecosystems or controlled laboratory experiments frequently encompass microbial cells, primary minerals, and/or (biogenic) secondary minerals manifesting as extracellular precipitates on cellular surfaces. High-resolution images can be made using transmission electron microscopy (TEM) and scanning electron microscopy (SEM).

   (a) **Transmission electron microscopy (TEM):** It allows scientists to research tiny living things by describing their shape and materials utilized and their interaction with rocks and minerals (Shuster et al. 2019).

   (b) **Scanning electron microscopy (SEM):** It characterizes having good resolving features at high magnification, thereby offering exceptional two-dimensional and three-dimensional structural characterization of living cells and geological materials. This technique often serves as the principal means of visual exploration of geo-microbial interaction and interpreting biogeochemical processes.

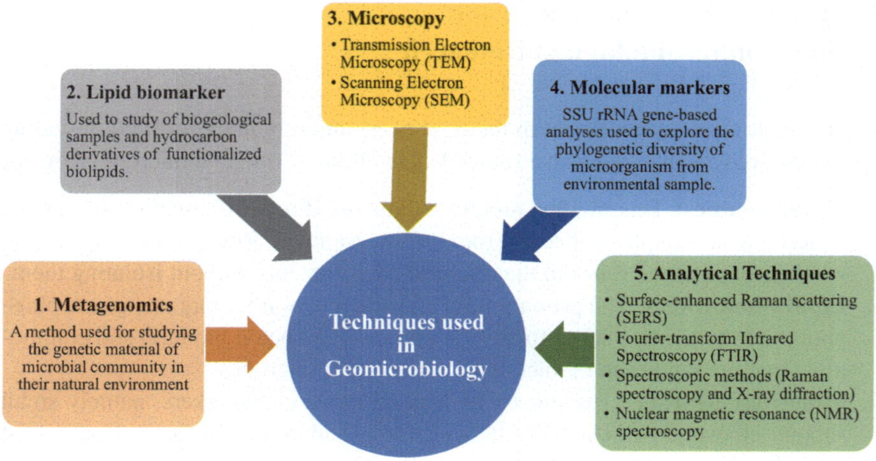

Fig. 4 Analytical and molecular techniques used in geomicrobiology for characterization of microbial community

8 Significance of Astrobiology in Extraterrestrial Life

Astrobiology is the complex field devoted to explore the evolution and allocation of life in the universe, although its significance encompasses beyond the scientific investigations, reforming human perception for the cosmos. One of the interesting aspects of astrobiology is discerning the extraterrestrial life. Astrobiologists recognized the importance of microbial extremophiles, as they can serve as models for different life patterns that may exist in the extreme conditions found on other planetary bodies (Hoover and Rozanov 2002). For example, microorganisms discovered from subglacial lakes of Antarctica may provide clues about the possibilities of life on the moon of Jupiter, i.e., Europa, which is believed to harbor a subsurface ocean. Similarly, the study of cryophilic microorganisms and their capability to thrive in freezing temperatures hints at the possibility of finding life on the frozen landscape of Mars or icy moons of Saturn and Jupiter (Pikuta and Hoover 2003; Rampelotto 2013; Goordial 2021).

Various research findings profoundly augmented the possibility that life may exist in the other cosmos (Table 4). The high-priced possibilities of life existed are being validated by researchers. Bacterial paleontology, an emerging field, has provided key evidence on the morphology and chemical characteristics of microfossils, allowing the detection of biosignatures on other planetary surfaces. The advances studied in geomicrobiology have also revealed microbial-mediated biochemical processes contributing significantly to global biogeochemical cycles and subsequently in the evolution of the atmosphere and surface composition of the Earth. The astrobiological study of Mars and the icy satellites indicates the outer solar

Table 4 Geomicrobiological processes and their implications for extraterrestrial life

Geomicrobiological processes	Insights into extraterrestrial life	References
Microbial extremophiles	Models for life in harsh environments on other planets and moons	Hoover and Rozanov (2002)
Subsurface biosphere	Potential for life in the subsurface of Mars and icy satellites	Farmer (2013)
Bacterial paleontology	Recognition of biosignatures on other planetary surfaces	Hoover and Rozanov (2002)

system has benefited greatly from the subsurface biosphere, which is fueled by inorganic chemical energy and able to withstand harsh environmental conditions (Marais and Walter 1999; Hoover and Rozanov 2002; Farmer 2013). By exploring the survival and adaptation strategies of extremophiles on Earth, astrobiologists may predict valuable information into the possibilities for life to exist in the harsh and often inhospitable environments of other planetary bodies, such as the icy moons of Saturn and Jupiter. These analogues can inform the development of more targeted and effective search strategies, increasing the likelihood of detecting potential biosignatures or even existing extraterrestrial life.

9 Conclusion

Geomicrobiology is an intricate multidisciplinary field that deals with the interaction between microorganisms and geological processes on the Earth. It focuses on the microbial-assisted environmental phenomenon, including soil formation and nutrient recycling in different habitat conditions. Unique ecosystem, like frozen environment from polar region to high-altitude areas, hold special interest due to its unique geomorphological characteristics as well as extremophilic microorganisms. These regions are characterized by extreme temperatures and limited sunlight is critical for studying microbial diversity under harsh conditions. Scientific investigations in these areas aim to reveal and explore biotechnological applications and understandings into the limits of life and climate change impacts with implications for astrobiology.

The cryosphere, covering glaciers and polar ice caps, contributed significantly in global climate dynamics. It harbors diverse microbial communities such as psychrophilic and psychrotrophic microorganisms. These microbial communities influence biogeochemical cycles by facilitating nutrient mineralization, solubilization, and transformations. The role of microbial communities in frozen habitats can be understood as permafrost thawing can release greenhouse gases and contribute significantly to climate regulation. Geomicrobiological interactions in frozen environments include microbial-induced mineral precipitation, particularly calcium carbonate

($CaCO_3$), which can be utilized in improving soil structure, stabilization, environmental remediation by reducing the permeability and improving soil strength. The improvement in analytical techniques is a must to explore the intricate relationships between microorganisms and geological substrates. The geomicrobiology of frozen environments also provides critical insights on the effects of climate change on the Earth and the potential for life on other icy celestial bodies. This interdisciplinary field continues to evolve, enhancing our understanding of microbial resilience and adaptation in extreme environments.

The study of astrobiology emphasizes on terrestrial planets and moons, particularly composed of silicate minerals that may support complex organic chemistry. Various environmental aspects, such as temperature, pH, and radiation, can foster life and are helpful in the development of terrestrial analogues key for exploring habitability on other celestial bodies. However, these analogues have limitations since no site can fully replicate on another planet.

Mars is often considered the most Earth-like planet, having once had a wetter and warmer environment that may have been favorable for life. However, it has since undergone significant environmental changes. The presence of minerals like phyllosilicate formed in wet environments indicates that certain regions on Mars may harbor conserved biosignatures. Similarly, Venus having extreme temperatures and a thick CO_2-rich atmosphere also indicates a hostile environment for life. However, its upper atmosphere may provide a habitable zone, where unique microbial life could exist in acidic cloud layers.

The geological dynamic environment of icy moons like Enceladus, Europa, and Titan are significant targets for astrobiologists. It is believed that Europa has a subsurface ocean protected by an icy shell, whereas Titan has a compact atmosphere and liquid methane lake on the surface, creating conditions that may support life.

Progressions in techniques and tools used in geomicrobiology, like metagenomics and microscopy, significantly contributed to understand the microbial life in frozen habitats, assisting the search for extraterrestrial life. Microbial extremophiles serve as model systems for discovering the possibility of life on other planets and underscoring their significance on the Earth to apprise astrobiological exploration.

Competing Interests All the authors declare that they have no competing interests.

References

Abirami B, Manigundan K, Radhakrishnan M, Gopikrishnan V, Bhaskar PV, Shanmugasundaram T, Dastager SG (2021) Diversity and bioprospecting potentials of antarctic (polar) microbes, 1st edn. CRC Press, Boca Raton, pp 349–365

Bagshaw EA, Stibal M, Anesio AM, Bellas C, Tranter M, Telling J, Wadham JL (2012) Glacier surface habitats. In Life at extremes: environments, organisms and strategies for survival (pp. 155–175). Wallingford UK: CABI

Baumstark-Khan C, Facius R (2002) Life under conditions of ionizing radiation. In: Astrobiology: the quest for the conditions of life. Springer Berlin, Heidelberg, pp 261–284
Benn D, Evans DJ (2014) Glaciers and glaciation, 2nd edn. Routledge
Bibring JP, Arvidson RE, Gendrin A, Gondet B, Langevin Y, Le Mouelic S, Mangold N, Morris RV, Mustard JF, Poulet F, Quantin C (2007) Coupled ferric oxides and sulfates on the Martian surface. Science 317(5842):1206–1210
Biemann K, Oro JIIIPT, Toulmin P III, Orgel LE, Nier AO, Anderson DM, Simmonds PG, Flory D, Diaz AV, Rushneck DR, Biller JE (1977) The search for organic substances and inorganic volatile compounds in the surface of Mars. J Geophys Res 82(28):4641–4658
Bockheim JG (2007) Importance of cryoturbation in redistributing organic carbon in permafrost-affected soils. Soil Sci Soc Am J 71(4):1335–1342
Bonaccorsi R, McKay CP, Chen B (2010) Biomass and habitability potential of clay minerals-and iron-rich environments: testing novel analogs for Mars Science Laboratory landing sites candidates. Philos Mag 90(17–18):2309–2327
Boyd ES, Hamilton TL, Havig JR, Skidmore ML, Shock EL (2014) Chemolithotrophic primary production in a subglacial ecosystem. Appl Environ Microbiol 80(19):6146–6153
Britannica, E., 2016. Noet edition. Chicago, IL: Encyclopædia Britannica. https://www.britannica.com/science/snow-and-ice-climate. Accessed 26 Sept 2024
Burgess DO, Sharp MJ (2004) Recent changes in areal extent of the Devon ice cap, Nunavut, Canada. Arct Antarct Alp Res 36(2):261–271
Carlson RW, Johnson RE, Anderson MS (1999) Sulfuric acid on Europa and the radiolytic sulfur cycle. Science 286(5437):97–99
Carter J, Poulet F, Bibring JP, Murchie S (2010) Detection of hydrated silicates in crustal outcrops in the northern plains of Mars. Science 328(5986):1682–1686
Cavalier-Smith T (2004) Only six kingdoms of life. Proc Biol Sci 271(1545):1251–1262
Chen J, Liu B, Zhong M, Jing C, Guo B (2022) Research status and development of microbial induced calcium carbonate mineralization technology. PLoS One 17(7):0271761
Choi A, Cho H, Kim SH, Thamdrup B, Lee S, Hyun JH (2016) Rates of N2 production and diversity and abundance of functional genes associated with denitrification and anaerobic ammonium oxidation in the sediment of the Amundsen Sea Polynya, Antarctica. Deep-Sea Res II Top Stud Oceanogr 123:113–125
Clarke A, Morris GJ, Fonseca F, Murray BJ, Acton E, Price HC (2013) A low temperature limit for life on Earth. PLoS One 8(6):66207
Cockell CS (1999) Life on Venus. Planet Space Sci 47(12):1487–1501
Comerford NB, Franzluebbers AJ, Stromberger ME, Morris L, Markewitz D, Moore R (2013) Assessment and evaluation of soil ecosystem services. Soil Horizons 54(3):1–14
Correa T, Abreu F (2020) Antarctic microorganisms as sources of biotechnological products. In: Salwan R, Sharma V (eds) Physiological and biotechnological aspects of extremophiles. Academic Press, pp 269–284
Davila AF, Fairén AG, Gago-Duport L, Stoker C, Amils R, Bonaccorsi R, Zavaleta J, Lim D, Schulze-Makuch D, McKay CP (2008) Subsurface formation of oxidants on Mars and implications for the preservation of organic biosignatures. Earth Planet Sci Lett 272(1–2):456–463
De Muynck W, De Belie N, Verstraete W (2010) Microbial carbonate precipitation in construction materials: a review. Ecol Eng 36(2):118–136
Des Marais DJ, Allamandola LJ, Benner SA, Boss AP, Deamer D, Falkowski PG, Farmer JD, Hedges SB, Jakosky BM, Knoll AH, Liskowsky DR (2003) The NASA astrobiology roadmap. Astrobiology 3(2):219–235
Elliott GD, Wang S, Fuller BJ (2017) Cryoprotectants: a review of the actions and applications of cryoprotective solutes that modulate cell recovery from ultra-low temperatures. Cryobiology 76:74–91
Fairén AG, Davila AF, Gago-Duport L, Amils R, McKay CP (2009) Stability against freezing of aqueous solutions on early Mars. Nature 459(7245):401–404

Farmer JD (2013) Role of geobiology in the astrobiological exploration of the solar system. Geol Soc Am Spec Pap 500:567–589
Flanner MG, Zender CS, Randerson JT, Rasch PJ (2007) Present-day climate forcing and response from black carbon in snow. J Geophys Res Atmos 112(D11):345–349
Garcia-Lopez E, Cid C (2017) Glaciers and ice sheets as analog environments of potentially habitable icy worlds. Front Microbiol 8:1407
Gilichinsky DA, Soina VS, Petrova MA (1993) Cryoprotective properties of water in the Earth cryolithosphere and its role in exobiology. Orig Life Evol Biosph 23(1):65–75
Glud RN, Wenzhöfer F, Middelboe M, Oguri K, Turnewitsch R, Canfield DE, Kitazato H (2013) High rates of microbial carbon turnover in sediments in the deepest oceanic trench on Earth. Nat Geosci 6(4):284–288
Golden KM, Ackley SF, Lytle VI (1998) The percolation phase transition in sea ice. Science 2825397:2238–2241
Goordial J (2021) Cryomicrobial ecology: still much to learn about life left out in the cold. Msystems 6(5):10–1128
Grasby SE, Allen CC, Longazo TG, Lisle JT, Griffin DW, Beauchamp B (2003) Supraglacial sulfur springs and associated biological activity in the Canadian High Arctic—signs of life beneath the ice. Astrobiology 3(3):583–596
Guven S, Demirci U (2012) Integrating nanoscale technologies with cryogenics: a step towards improved biopreservation. Nanomedicine 7(12):1787–1789
Hallsworth JE, Mancinelli RL, Conley CA, Dallas TD, Rinaldi T, Davila AF, Benison KC, Rapoport A, Cavalazzi B, Selbmann L, Changela H (2021) Astrobiology of life on Earth. Environ Microbiol 23(7):3335–3344
Harris MJ, Wickramasinghe NC, Lloyd D, Narlikar JV, Rajaratnam P, Turner MP, Al-Mufti S, Wallis MK, Ramadurai S, Hoyle F (2002) Detection of living cells in stratospheric samples. In: Instruments, methods, and missions for astrobiology IV, SPIE 4495. SPIE, pp 192–198
Hasimoto GL, Roos-Serote M, Sugita S, et al (2008) Felsic highland crust on Venus suggested by Galileo Near-Infrared Mapping Spectrometer data. J Geophys Res Planets 113. https//doi.org/10.1029/2008je003134
Hayes A, Aharonson O, Callahan P, Elachi C, Gim Y, Kirk R, Lewis K, Lopes R, Lorenz R, Lunine J, Mitchell K (2008) Hydrocarbon lakes on Titan: distribution and interaction with a porous regolith. Geophys Res Lett 35(9):L09204
Hoover RB, Rozanov AY (2002) Astrobiology: traces of life in the cosmos. In: International conference on X-ray and neutron capillary optics, SPIE 4765. SPIE, pp 1–19
Jansson JK, Taş N (2014) The microbial ecology of permafrost. Nat Rev Microbiol 12(6):414–425
Kahle M, Kleber M, Jahn R (2004) Retention of dissolved organic matter by phyllosilicate and soil clay fractions in relation to mineral properties. Org Geochem 35(3):269–276
Kasting JF (1988) Runaway and moist greenhouse atmospheres and the evolution of Earth and Venus. Icarus 74(3):472–494
Kim JS, Crowley DE (2007) Microbial diversity in natural asphalts of the Rancho La Brea Tar Pits. Appl Environ Microbiol 73(14):4579–4591
Kite ES (2019) Geologic constraints on early Mars climate. Space Sci Rev 215:1–47
L'Haridon S, Reysenbacht AL, Glenat P, Prieur D, Jeanthon C (1995) Hot subterranean biosphere in a continental oil reservoir. Nature 377(6546):223–224
Lima IG, Bispo JR, Agostinho AY, Queiroz ACD, Moreira MSA, Passarini MRZ, Oliveira VMD, Sette LD, Rosa LH, Duarte AWF (2022) Antarctic environments as a source of bacterial and fungal therapeutic enzymes. An Acad Bras Cienc 94:20210452
Lorenz RD, Mitchell KL, Kirk RL, Hayes AG, Aharonson O, Zebker HA, Paillou P, Radebaugh J, Lunine JI, Janssen MA, Wall SD (2008) Titan's inventory of organic surface materials. Geophys Res Lett 35(2):L02206

Lorenz RD, Gleeson D, Prieto-Ballesteros O, Gomez F, Hand K, Bulat S (2011) Analog environments for a Europa lander mission. Adv Space Res 48(4):689–696
Mackelprang R, Burkert A, Haw M, Mahendrarajah T, Conaway CH, Douglas TA, Waldrop MP (2017) Microbial survival strategies in ancient permafrost: insights from metagenomics. ISME J 11(10):2305–2318
Marais DD, Walter MR (1999) Astrobiology: exploring the origins, evolution, and distribution of life in the universe. Annu Rev Ecol Syst 30(1):397–420
Margesin R, Collins T (2019) Microbial ecology of the cryosphere (glacial and permafrost habitats): current knowledge. Appl Microbiol Biotechnol 103:2537–2549
Margesin R, Miteva V (2011) Diversity and ecology of psychrophilic microorganisms. Res Microbiol 162(3):346–361
Martin YE, Johnson EA (2012) Biogeosciences survey: studying interactions of the biosphere with the lithosphere, hydrosphere and atmosphere. Prog Phys Geogr 36(6):833–852
Martínez-Rosales C, Castro-Sowinski S (2011) Antarctic bacterial isolates that produce cold-active extracellular proteases at low temperature but are active and stable at high temperature. Polar Res 30(1):7123
Martorell MM, Ruberto LAM, Fernández PM, Castellanos de Figueroa LI, Mac Cormack WP (2017) Bioprospection of cold-adapted yeasts with biotechnological potential from Antarctica. J Basic Microbiol 57(6):504–516
Mitri G, Showman AP, Lunine JI, Lorenz RD (2007) Hydrocarbon lakes on Titan. Icarus 186(2):385–394
Mustard JF, Murchie SL, Pelkey SM, Ehlmann BL, Milliken RE, Grant JA, Bibring JP, Poulet F, Bishop J, Dobrea EN, Roach L (2008) Hydrated silicate minerals on Mars observed by the Mars Reconnaissance Orbiter CRISM instrument. Nature 454(7202):305–309
Mocali S, Benedetti A (2010) Exploring research frontiers in microbiology: the challenge of metagenomics in soil microbiology. Res Microbiol 161(6):497–505
Nesme J, Achouak W, Agathos SN, Bailey M, Baldrian P, Brunel D, Frostegård Å, Heulin T, Jansson JK, Jurkevitch E, Kruus KL (2016) Back to the future of soil metagenomics. Front Microbiol 7:73
New JS, Kazemi B, Spathis V, Price MC, Mathies RA, Butterworth AL (2021) Quantitative evaluation of the feasibility of sampling the ice plumes at enceladus for biomarkers of extraterrestrial life. Proc Natl Acad Sci 118(37):2106197118
Nichols D, Bowman J, Sanderson K, Nichols CM, Lewis T, McMeekin T, Nichols PD (1999) Developments with Antarctic microorganisms: culture collections, bioactivity screening, taxonomy, PUFA production and cold-adapted enzymes. Curr Opin Biotechnol 10(3):240–246
Ortega-Villamagua E, Gudiño-Gomezjurado M, Palma-Cando A (2020) Microbiologically induced carbonate precipitation in the restoration and conservation of cultural heritage materials. Molecules 25(23):5499
Pace NR (1997) A molecular view of microbial diversity and the biosphere. Science 276(5313):734–740
Paillou P, Lunine J, Ruffié G, Encrenaz P, Wall S, Lorenz R, Janssen M (2008) Microwave dielectric constant of Titan-relevant materials. Geophys Res Lett 35(18):L18202
Pappalardo RT (2010) Seeking Europa's ocean. Proc Int Astronomical Union 6(S269):101–114
Petrenko VF, Whitworth RW (1999) Physics of ice. Oxford University Press
Pielou EC (2008) After the ice age: the return of life to glaciated North America. University of Chicago Press
Pikuta EV, Hoover RB (2003) Psychrophiles and astrobiology: microbial life of frozen worlds. In: Instruments, methods, and missions for astrobiology VI, SPIE 4939. SPIE, pp 103–116
Priscu JC, Christner BC (2003) Earth's icy biosphere. In: Microbial diversity and bioprospecting. Wiley, pp 130–145

Quillfeldt JA (2010) Astrobiologia: água e vida no sistema solar e além. Caderno Brasileiro de Ensino de Física 27:685–697

Rampelotto PH (2013) Extremophiles and extreme environments. Life 3(3):482–485

Rothschild LJ, Mancinelli RL (2001) Life in extreme environments. Nature 409(6823):1092–1101

Schulze-Makuch D, Irwin LN (2002) Energy cycling and hypothetical organisms in Europa's ocean. Astrobiology 2(1):105–121

Schulze-Makuch D, Grinspoon DH, Abbas O, Irwin LN, Bullock MA (2004) A sulfur-based survival strategy for putative phototrophic life in the Venusian atmosphere. Astrobiology 4(1):11–18

Schulze-Makuch D, Haque S, de Sousa Antonio MR, Ali D, Hosein R, Song YC, Yang J, Zaikova E, Beckles DM, Guinan E, Lehto HJ (2011) Microbial life in a liquid asphalt desert. Astrobiology 11(3):241–258

Schuur EA, McGuire AD, Schädel C, Grosse G, Harden JW, Hayes DJ, Hugelius G, Koven CD, Kuhry P, Lawrence DM, Natali SM (2015) Climate change and the permafrost carbon feedback. Nature 520(7546):171–179

Segerer A, Neuner A, Kristjansson JK, Stetter KO (1986) Acidianus infernus gen. nov., sp. nov., and Acidianus brierleyi comb. nov.: facultatively aerobic, extremely acidophilic thermophilic sulfur-metabolizing archaebacteria. Int J Syst Evol Microbiol 36(4):559–564

Seifan M, Samani AK, Berenjian A (2016) Bioconcrete: next generation of self-healing concrete. Appl Microbiol Biotechnol 100:2591–2602

Seifan M, Khajeh Samani A, Hewitt S, Berenjian A (2017) The effect of cell immobilization by calcium alginate on bacterially induced calcium carbonate precipitation. Fermentation 3(4):57

Sharma A, Scott JH, Cody GD, Fogel ML, Hazen RM, Hemley RJ, Huntress WT (2002) Microbial activity at gigapascal pressures. Science 295(5559):1514–1516

Shuster J, Reith F, Southam G (2019) Applications of transmission electron microscopy in geomicrobiology. Cambridge University Press, Cambridge, pp 166–186

Smith DJ, Griffin DW, Schuerger AC (2010) Stratospheric microbiology at 20 km over the Pacific Ocean. Aerobiologia 26:35–46

Soare R, Pollard W, Green D (2001) Deductive model proposed for evaluating terrestrial analogues. EOS Trans Am Geophys Union 82(43):501–501

Spencer JR, Tamppari LK, Martin TZ, Travis LD (1999) Temperatures on Europa from Galileo photopolarimeter-radiometer: nighttime thermal anomalies. Science 284(5419):1514–1516

Thomas DN, Dieckmann GS (2002) Antarctic sea ice—a habitat for extremophiles. Science 2955555:641–644

Tomasko MG, Archinal B, Becker T, Bézard B, Bushroe M, Combes M, Cook D, Coustenis A, De Bergh C, Dafoe LE, Doose L (2005) Rain, winds and haze during the Huygens probe's descent to Titan's surface. Nature 4387069:765–778

Valdivia-Granda W (2008) The next meta-challenge for bioinformatics. Bioinformation 2(8):358

Wainwright M, Wickramasinghe NC, Narlikar JV, Rajaratnam P (2003) Microorganisms cultured from stratospheric air samples obtained at 41 km. FEMS Microbiol Lett 218(1):161–165

Waldrop MP, Chabot CL, Liebner S, Holm S, Snyder MW, Dillon M, Dudgeon SR, Douglas TA, Leewis MC, Walter Anthony KM, McFarland JW (2023) Permafrost microbial communities and functional genes are structured by latitudinal and soil geochemical gradients. ISME J 17(8):1224–1235

Walker JC (1975) Evolution of the atmosphere of Venus. J Atmos Sci 32(6):1248–1256

Wang Y, Soga K, DeJong JT, Kabla AJ (2019) Microscale visualization of microbial-induced calcium carbonate precipitation processes. J Geotech Geoenviron 145(9):04019045

Williams PJ, Smith M (1989) The frozen earth: fundamentals of permafrost physics. Polar Geogr 13(1):1–24

Wiseman CLS, Püttmann W (2006) Interactions between mineral phases in the preservation of soil organic matter. Geoderma 134(1–2):109–118

Woese CR (1987) Bacterial evolution. Microbiol Rev 51(2):221–271
Woese CR, Kandler O, Wheelis ML (1990) Towards a natural system of organisms: proposal for the domains archaea, bacteria, and eucarya. Proc Natl Acad Sci 87(12):4576–4579
Wordsworth RD (2016) The climate of early Mars. Annu Rev Earth Planet Sci 44(1):381–408
Zhang Y (2005) Influence of climate change on permafrost and its implications for the environment. Environ Res Lett 2(3):29–47

Geomicrobes in Rehabilitation of Saturated Lands

Arigo R. W. Sangma and Santa Ram Joshi

Contents

Abstract Saturated lands are among the planet's most productive ecosystems, providing crucial habitat for various animal and plant species and an abundance of additional ecosystem services, such as carbon sequestration, flood control, and water filtration. However, wetlands are also highly susceptible to deterioration because of pollution, habitat loss, and climate change. Utilizing geomicrobes, or microorganisms that live beneath the surface, can aid in the ecological restoration of damaged saturated land ecosystems. These microorganisms mediate biogeochemical

A. R. W. Sangma · S. R. Joshi (✉)
Microbiology Laboratory, Department of Biotechnology and Bioinformatics, North-Eastern Hill University, Shillong, Meghalaya, India

S. R. Joshi et al. (eds.), *Mineral Transformation and Bioremediation by Geo-Microbes*, Interdisciplinary Biotechnological Advances,
https://doi.org/10.1007/978-981-96-3033-2_2

processes, such as the cycling of nutrients, the breakdown of organic matter, and the degradation of contaminants. Harnessing the inherent capabilities of these bacteria can hasten the restoration of wetland ecosystems by increasing fertilizer availability, strengthening soil structure, and deteriorating contaminants. Using geomicrobes can be a powerful tool for recovering and restoring wetland ecosystems that have been damaged or degraded. By leveraging the natural processes mediated by these microorganisms, the restoration of crucial wetland habitats and functions can be accelerated.

Geomicrobes provide an invaluable resource for wetland restoration projects since they are essential to the biogeochemical processes, soil stability, and pollutant degradation of wetland ecosystems. Wetland restoration initiatives can enhance habitat quality, hasten ecosystem recovery, and promote long-term ecological stability by utilizing the ecological roles of geomicrobes. Wetland conservation and recovery initiatives could greatly benefit from the application of geomicrobial techniques; however, to optimize the efficacy of these strategies, more investigation is required to truly understand the complex interactions among wetland plants, geomicrobes, and environmental conditions.

Keywords Wetlands · Geomicrobes · Restoration · Biogeochemical cycles · Ecosystem

Abbreviations

Al	Aluminium
AMF	Arbuscular mycorrhizal fungi
BOD	Biological oxygen demand
C	Carbon
CO_2	Carbon dioxide
COD	Chemical oxygen demand
CW	Constructed wetlands
Fe	Iron
Mn (IV)	Manganese
MNC	Multi-nutrient cycling
N	Nitrogen
N_2	Nitrogen gas
N_2O	Nitrous oxide
NOM	Natural organic matter
OM	Organic matter
P	Phosphorus
PAH	Polyhydroxyalkanoate
pH	Potential of hydrogen
RW	Restored wetlands
S	Sulfur
SRB	Sulfate-reducing bacteria

1 Introduction

Global climate change is recognized as a threat to both species' survival and natural systems' integrity. Researchers from all over the world are examining how ecosystems and water management are affected by climate change. Future attempts to manage and restore saturated lands like marshes, swamps, bogs, etc. clubbed as wetlands will be more difficult due to climate change (Erwin 2009). The loss of wetlands worldwide has decreased ecosystem services and biodiversity. The decreases in wetlands have prompted numerous landowners to consider restoring them; however, measuring the outcomes of these endeavors remains uncertain due to the diverse methods required to assess ecosystem services (Tomscha et al. 2021). Though they comprise only 3% of all soils worldwide, wetlands and peatlands are responsible for 21% of the soil's organic carbon (C) stock. More than 10% of wetlands worldwide have been cleared for farming due to their abundant nutrients, and more drainage is anticipated to meet rising agricultural demands (Anthony 2021).

Saturated land restoration success has long been challenging to assess and occasionally controversial because the term is ambiguous and can mean different things to different people and contexts. The ecological characteristics of the wetland systems and their proportional contribution to ecosystem services were evaluated using constructed wetlands (CW) that were developed to employ surface, subsurface, or hybrid treatment systems to handle a broad range of wastewaters and restored wetlands (RW) that were usually positioned and planned to have a stronger relationship with local ecosystems to create self-sustaining ecosystems through early successional management, revegetation, and construction (Spieles 2022). Although many attempts have been made to restore ecosystems to recover vital ecosystem services, there is still uncertainty regarding the extent to which the ecosystems' structure and functionality will be recovered due to these initiatives (Moreno-Mateos et al. 2012). To restore and improve the function of wetlands, a thorough grasp of the composition, roles, and environmental potential of microorganisms in CW sediments is crucial (Liang et al. 2023).

An ecosystem of wetlands serves as a major repository for microbial diversity and is essential to reducing greenhouse gas emissions. It is widely acknowledged that increased nitrogen (N) inputs from fossil fuel combustion and agriculture seriously threaten wetlands' ability to control greenhouse gas emissions and maintain biodiversity (Chandra et al. 2020). Microbes are essential to the biosphere's geoactive processes, especially when it comes to the formation of soil sediment and transformations of metals and minerals, bioweathering, decomposition, and element biotransformation along with biogeochemical cycling. Transformations of metals and minerals facilitate numerous geomicrobial processes, encompassing a diverse array of microbes, including prokaryotes, eukaryotes, and symbiotic relationships with higher organisms and among themselves (Gadd 2013).

2 Saturated Land Degradation and Ecological Impact

The health of saturated lands is continuously declining, sometimes to the point where it is doubtful that they will survive in the near future due to improper extraction of uses and misconceptions directed toward wetlands. The primary challenges to wetlands in the country include the expansion of conventional and contemporary agriculture, persistent land degradation, urbanization, industrialization, a lack of resources, ecological and natural issues, and insufficient institutional frameworks and policies (Gebresllassie et al. 2014). Nutrient enrichment, salinization, pesticide and heavy metal pollution, exotic plant invasion, loss of bordering vegetation, and altered hydrological regimes brought on by urbanization and agricultural practices are some of the clear and immediate causes. The underlying causes are a lack of knowledge about wetland hydrology and ecology on the part of the private sector and planning agencies, as well as a lack of coordination among the numerous agencies in charge of managing wetlands (Davis and Froend 1999). Wetlands were harmed by population growth, irrational land use, and unfavorable natural eco-environmental factors like increased soil evaporation, higher soil salinization, and decreased precipitation during the development of the petroleum sector and urban proliferation. Reduction of the area of coastal wetlands, contamination of the seawater environment, deterioration of the ecological environment, sharp decline in the wetlands' biodiversity, and other characteristics are indicative of degradation trends (Han et al. 2006). Since there hasn't been much human interference, the spatial pattern of wetlands is very natural; however, because of widespread irrigation by nearby residents, there has been ecological degradation. The primary determinants of the ecological pattern and the structural integrity of wetlands are the water availability and the geomorphological gradient. As the water level drops, marsh areas become more and more confined to lower-lying land; as a result, non-marsh plants replace typical marsh plants along the geomorphological gradient, expanding the marsh area. The physical features of the nearby water system have altered as a result of agricultural activity. The habitat of marshes has been negatively impacted by human activity's redistribution of regional water resources (Zhou et al. 2009). Because urbanization causes habitat destruction and fragmentation, it is one of the main causes of species loss and can have an impact on wetlands; there is a way to compare the state of wetlands across different land use categories. Cypress domes, which are solitary wetlands dominated by *Taxodium distichum* cypress, frequently found in urban settings were either destroyed or degraded, and nearly half of managed forests saw their land consumed by agriculture or urban development (McCauley et al. 2013). The hydrology of specific wetland ecosystems will be impacted by climate change primarily through modifications to global temperature and precipitation regimes (Erwin 2009).

3 Importance of Saturated Land Restoration Efforts

Saturated lands or wetlands are important to the regional ecosystem because they maintain the equilibrium of the local water supply, clean up the environment, and regulate the climate. For a wide variety of plants and animals, the wetland offers vital habitat. Wetlands are among the world's most productive ecosystems and serve as a link between land and water. Several common names for various kinds of wetlands include bog, marsh, and swamp. In addition to offering essential ecological services, wetlands control water regimes and serve as sources of biodiversity on all scales, including species, genetics, and ecosystems. For the community, wetlands are a valuable resource from an economic, scientific, cultural, and recreational standpoint (Amenu and Mamo 2018). The main ecological steps to restore degraded wetlands include introducing salt-tolerant flora to enhance species diversity and plant coverage, thereby augmenting the wetland's capacity for bioremediation and pollutant absorption. Additionally, safeguarding pristine plant communities and facilitating their regeneration through site preparation and improved environmental conditions are crucial. These efforts should culminate in a comprehensive land-use strategy that eliminates detrimental facilities accordingly. Despite their ecological significance, wetlands are often altered or reclaimed for financial and commercial purposes. However, they provide numerous goods and services that hold significant economic value for both nearby communities and people beyond the wetland area (Schuyt 2005). Alongside geological mapping and absolute dating, the recognition of macrofossils as indicators of limno-telmatic conditions can help trace the evolution of peat-forming wetlands spanning thousands of years (Wójcicki and Woskowicz-Ślęzak 2015). Preventing or minimizing additional stress that can impair wetlands' capacity to adapt to climate change is a crucial management tactic. The resilience of the wetland ecosystems in the delta can be improved by reducing pollution, preserving hydrology, and safeguarding the biological diversity and integrity of the wetland ecosystems. This will ensure that the ecosystems can continue to provide vital services even in the face of changing weather patterns (Wang et al. 2012).

4 Introduction to Geomicrobes

Combining geology and microbiology, geomicrobiology studies how microorganisms interact with their surroundings, including sedimentary rocks. In the last 10 years, this young and quickly expanding field has produced a drastically altered understanding of the variety and activity of microbial life on Earth. The study of geomicrobiology looks at the part that microorganisms play in various essential geological processes, both now and in the past. Microbes have a vital function in the ecosystem. These organisms carry out biogeochemical cycles, which are essential to the continuation of regular life on Earth. From an economic perspective, these microorganisms are crucial to the pharmaceutical and petroleum industries. In addition, natural products, a variety of industrially significant compounds isolated from these microorganisms, and enhanced agricultural productivity due to soil microorganisms have made this field popular as a focus for current and future researchers (Jain et al. 2010). The mineral formation, mineral degradation, organic and inorganic matter cycling, chemical and isotopic fractionation, and the genesis and degradation of fossil fuels are all influenced by geomicrobial activity. Weathering, bioleaching, and the formation and transformation of soil and sediment (diagenesis) are examples of phenomena related to microbial mineral degradation. The genesis and decomposition of fossil fuels, such as methane, peat, coal, and petroleum, are facilitated by microbes to differing degrees. Owing to their roles in biogeochemical cycles and environmental transformations, members of the Prokaryote domain, such as fungi and protozoa, are widely acknowledged as important geomicrobial agents. Conversely, members of the Eukaryote domain, such as bacteria (Eubacteria), Archaea, and certain Algae, also play significant roles (Ehrlich 2006).

In the biogeochemical cycle, numerous bacteria are significant players. Bacteria that perform nitrification and denitrification, as well as other bacteria that oxidize and reduce iron, sulfur, and manganese, are examples of significant physiological groups. A crucial step in the transformation of ammonia into an anionic nitrogen species is performed by nitrifying bacteria (Barns and Nierzwicki-Bauer 1997). Fungi are essential to certain geomicrobial transformation processes, such as the weathering of certain rocks, the early phases of the breakdown of complex natural polymers like cellulose and lignin that make up the organic matter in the soil's litter zone, and certain bioremediation procedures involving xenobiotics in soil and water (Ehrlich 2006). The predominant soil archaea showed distinct continental patterns and specialized in differentiating habitats between wetland and dryland, which were connected to soil pH and average annual temperature (Jiao et al. 2019).

5 Carbon Cycle and Saturated Lands

The carbon cycle is the most effective way to demonstrate the interconnection of organisms in various ecosystems. Methanogens utilize carbon dioxide as the final electron acceptor in anaerobic respiration, resulting in the build-up of methane in certain anaerobic settings through the production of reduced one-carbon compounds. These life forms are present in soil, wetlands, and aquatic settings, along with animal stomachs, sewage treatment plants, and landfills. They can also exist in anaerobic settings with organic molecules (Keenleyside 2019). Methane metabolism, carbon fixation, and carbon degradation are all parts of the carbon cycle (Meng et al. 2022). Wetland soils rich in carbon (C) are frequently drained for agricultural purposes because of their ability to maintain high net primary productivity. The agricultural needs of an expanding population are predicted to be met this century by increased drainage. Significant losses of soil carbon and concentrations of leftover soil minerals like iron (Fe) and aluminum (Al) can be caused by wetland drainage. The amount of mineral material present in residual soils after wetlands are drained can affect patterns in C concentrations after wetland restoration as well as C cycling on drained soils (Anthony 2021). The yearly carbon dioxide flux from fossil fuel emissions is dwarfed by ten times by the respiration flux of CO_2 released into the atmosphere as a result of soil microorganism activity. Thus, even slight modifications to the soil's carbon cycle could have a significant effect on the amount of CO_2 in the atmosphere (Gougoulias et al. 2014). Due to their ability to function as carbon "sources" or "sinks," wetlands have a significant impact on the global carbon cycle. Wetlands' source-sink capacity is controlled by biogeochemical processes mediated by microbes, which are also influenced by external factors (Bonetti et al. 2021).

6 Nitrogen Cycle and Saturated Lands

Nitrogen is present in various biological macromolecules like proteins and nucleic acids, but introducing nitrogen into living organisms can be difficult. Prokaryotes play a crucial role in the nitrogen cycle by transforming nitrogen into various forms to meet their needs, which in turn supports other organisms. The primary providers of nitrogen in the ecosystem are free-living, symbiotic diazotrophic bacteria and archaea. These organisms incorporate nitrogen into their macromolecules by utilizing a process known as nitrogen fixation, which involves an assimilatory reduction reaction. In aquatic and marine habitats, cyanobacteria play a crucial role as primary producers and are also key suppliers of fixed nitrogen. Microorganisms change organic nitrogen into nitrogen gas. Nitrification, denitrification, and ammonification are the three most recognized processes (Keenleyside 2019). The primary microorganisms responsible for catalyzing N_2 fixation in terrestrial settings are actinomycetes and symbiotic N_2-fixing bacteria that are linked to legume root nodules. These species include *Rhizobium* spp., *Bradyrhizobium* spp., and *Frankia* spp. A variety of free-living soil and regolith bacteria, including Bacillus, Azotobacter, and Clostridium species, can also fix atmospheric N_2 (Jones et al. 2003; Gutknecht et al. 2006; Reith et al. 2008).

7 Sulfur Cycle and Saturated Lands

Sulfide, thiosulfate, or polysulfide are produced in deep sediments when organic carbon is oxidized and then moves upward to be oxidized. This procedure involves intricate chemistry and lacks precise measurement. As sulfide is carried upward, it quickly reacts with Fe (III), Mn (IV), and oxygen to produce different sulfur compounds that mix with other substances, creating a complex system. Most of the time, the sediment's ability to oxidize ensures that sulfide is used up within the sediment by oxygen or other oxidizing agents such as nitrate or metals (Nealson and

Stahl 1997). The dissolved sulfate in Earths oceans is the biggest source of S for the biosphere. The atmosphere also contains sulfur, primarily from volcanic eruptions, microbial processes (like dissimilative sulfate reduction or the breakdown of organic matter), and human sources (like industrial and transportation discharges) (Reith et al. 2008). Due to its high reactivity, redox sensitivity, and microbial activity, sulfur is frequently found in CW influent streams. As such, it influences the activity of sulfur-oxidizing and sulfate-reducing bacteria in CW systems that are intended to treat wastewater from mining and other sources, as well as other beneficial and harmful processes in CWs.

8 Iron Cycle and Saturated Lands

Strong relationships between microbes and plants are evident in the biogeochemical cycling of nitrogen (N), sulfur (S), and iron (Fe) at different scales, ranging from the rhizosphere to the landscape and Fe cycling is closely related to S and P cycling. Microbial oxidation of oxygen in wetlands, caused by radial oxygen loss from roots, leads to acidification; on the other hand, reduction via alternative electron acceptors produces alkalinity, which modifies rhizosphere pH and, consequently, plant composition (Lamers et al. 2012). Redox reactions, which are brought on by the mineralization of organic matter, have an impact on the cycling of iron and sulfur in wetlands and other aquatic environments (Stumm and Sulzberger 1992). The wetland's natural organic matter (NOM) concentrations significantly mobilize ferric iron because ferric oxide oxidizes when NOM is present as some ferric oxide stabilizes as colloidal iron (Peiffer et al. 1999).

9 Symbiotic Relationships Between Geomicrobes and Saturated Land Vegetation

Lakes are home to a variety of functional groups of microorganisms that support complex microbial food webs that impact the generation and use of greenhouse gases and connect different biological cycles (such as the nitrogen, sulfur, and carbon cycles) through redox coupling (Ibrahim et al. 2023). Known as natural biofertilizers, arbuscular mycorrhizal fungi (AMF) are a type of soil microorganisms that exhibit symbiotic relationships with most plants (Mohan and Joshi 2024). Wetland plants benefit from AMF in terms of nutrient uptake, growth, resistance to salt stress, and resistance to pollutants; however, wetland plants, soil characteristics, water, and season may have an impact on AMF diversity, infection, spatial distribution, spore density, growth, and development (Yan JinZhuo et al. 2008). Wetlands' ecological significance is suggested by the presence of AMF in them. Depending on the plant attribute, such as the host plant's identity, the addition of phosphorus, and the availability of water in the soil where both symbionts develop, the beneficial AMF effect varied in magnitude (Ramírez-Viga et al. 2018). Because of their great yield, wetlands are vital ecosystems for both the economy and the environment. Wetland habitats contain a diverse and productive community of microorganisms that consistently transform nutrients from decaying plants into nitrogen, phosphorus, and other plant-friendly nutrients. The microorganisms consume the substances released by plant roots (Singh et al. 2020). In wetlands, submerged plants are crucial ecosystem engineers that enhance elemental cycling and self-purification (Wang et al. 2021). The profiles of the bacterial and fungal communities differed depending on the type of vegetation; in particular, a densely vegetated area promoted higher microbial abundance and a higher ratio of fungi to bacteria. Co-occurrence analysis showed that there were significant interactions between bacteria and fungi in vegetated zones of wetland ecosystems (Chen et al. 2019).

10 Geomicrobial Processes in Saturated Land Rehabilitation

10.1 Organic Matter Decomposition

Geomicrobes break down complex organic matter, releasing nutrients that can be utilized by plants and other organisms. The four processes that make up the decomposition of organic matter are mineralization, comminution, leaching, and photo-oxidation. Arthropods play a major role in mineralization and comminution. The three most common metrics used to quantify factors influencing decomposition include rate of respiration, proportion of litter input to standing crop, and speed of litter disappearance. Decomposition rate data can also be obtained using isotopic tracers (Schowalter 2016). One crucial aspect of how ecosystems react to global change is how soil organic matter (OM) breaks down when temperatures increase. The conflicting outcomes of field and lab studies, primarily focusing on easily degradable carbon with quick turnover rates, have not yet determined the impact of climate change on decomposition dynamics. Nevertheless, organic carbon comprises the largest portion of total soil carbon reserves, with turnover rates varying from decades to centuries (Conant et al. 2011). Plants with higher nitrogen or phosphorus content and lower levels of lignin or cellulose will decompose faster than other types of plant debris. The biochemical composition at the beginning usually has a significant impact. Location is also crucial in decomposition, typically, decay occurs faster in warm, humid, and nutrient-rich areas (Findlay 2021). Sulfur, nitrogen, and phosphorus are stored in soil organic matter. Very little inorganic nitrogen is present in soil; instead, most of the nitrogen is obtained from organic forms. Soil microbiomes produce extracellular hydrolytic enzymes that assist in breaking down organic matter through enzymatic processes. Arylsulfatase, amylase, chitinase, cellulase, dehydrogenase, urease, and phosphatase are vital soil enzymes originating from plants, animals, organic compounds, microorganisms, and soil participating in biochemical processes when breaking down organic material (Ganeshamurtily et al. 1995).

10.2 Contaminant Remediation

Some geomicrobes can degrade or immobilize pollutants, helping to purify wetland waters and sediments. Bioremediation is the term used for the utilization of biological systems to eliminate both organic and inorganic contaminants. Fungi and bacteria are essential organisms for recovering, immobilizing, or detoxifying metallic and radioactive pollutants. The catalytic and other properties of specific biominerals or metallic elements deposited by microbes in the form of nanoparticles, crystalline forms, or colloidal forms influence the creation of new biomaterials for technological and antimicrobial uses (Gadd 2010). Trace metal concentrations (zinc and copper) exhibited a positive correlation with SRB and a negative correlation with oxidizers and methane generators. Overall, the wetland system was effective in reducing methane emissions by gradually removing methane producers using Geobacter and SRB, as well as immobilizing copper and zinc through the anaerobic respiration of sulfate by SRB (Elhaj Baddar and Xu 2022). Using a constructed wetland with macrophytes to treat various wastewaters shows promise for significantly reducing BOD, COD, suspended particles, all solids, total nitrogen, and metals with high density. It can also be used to remediate xenobiotics, pesticides, and polyaromatic hydrocarbons. The rhizosphere of macrophytes such as Phragmites, Typha, Juncus, Spartina, and Scirpus serves as a lively and moving area for the microbial breakdown of organic pollutants and storage of inorganic pollutants, leading to the successful treatment of domestic, textile, and other effluents. This technology treats a variety of wastewaters, including those containing xenobiotics, pesticides, household, dairy, and wastewaters. Promoting plant-microbe interaction in designer rhizospheres with pigmentation, biostimulation, and genetic modifications can address limiting factors such as wetland loading limits, removal efficiency, biomass disposal, and seasonal growth variability (Bhatia and Goyal 2014). Constructed wetlands, or CWs, are an inexpensive eco-technological solution for managing industrial wastewater. The CWs are autonomous remediation systems that remove contaminants using biological (bioremediation and phytoremediation), chemical, and physical methods without needing an outside energy supply (Sharma and Malaviya 2022). Constructed wetlands (CWs), one of the technologies used in PAH remediation, offer several economical and environmentally beneficial benefits. These include advances in the understanding and optimization of PAH-removal mechanisms that concentrate on sediments, plants, and microorganisms. Important elements influencing the processes of PAH elimination in CWs include adsorption, uptake, phytovolatilization, and biodegradation in wetland plants (Zhao et al. 2021). Wetlands' physical and chemical characteristics offer numerous advantages for removing pollutants. For microbes involved in degradation, the extensive rhizosphere of wetland herbaceous shrub and tree species provides a fertile cultivation area. The majority of wetland soil and sediment areas exhibit redox conditions that enhance degradation pathways, requiring the presence of reducing conditions. Apart from planted poplar communities, the primary emphasis of wetland phytoremediation research has been on laboratory microcosm and mesocosm technologies. Extensive field studies have not focused on the remedial actions performed by natural wetland communities (Williams 2002).

10.3 Nutrient Cycling

Geomicrobes are involved in the cycling of essential nutrients like nitrogen, phosphorus, and sulfur, making them available for primary producers. Soil health has declined gradually, and a thriving soil system now relies on a complicated mix of physical, chemical, and biological elements. Healthy soils need to meet the requirements of both farmers and the community in order to maintain balance. Soil organic matter is essential for preserving soil health and quality, counteracting toxins, preventing pathogens, and supporting environmental sustainability (Sahu et al. 2017). Several microbes (fungi and bacteria) found in the ectomycorrhizosphere can stimulate one another to form a particular interface between the soil and the trees. Fungi have a major impact on soil, rock, and mineral environments, and they are known to cause geochemical alterations in the plant root zone. As a result, they play various roles in the cycling of nutrients that maintains the soil's quality for plant growth. The majority of ectomycorrhizal root tips are found in mineral horizons, where they take advantage of the nutrient-rich substrates and aid in their translocation and assimilation. The microbes living in the mycorrhizosphere can mobilize iron and dissolve other organic nutrients, such as organic phosphate (Koshila Ravi et al. 2019). The major nutrient cycles of silicon, phosphorus, and nitrogen are closely related to the ocean's carbon cycle. Hence, the marine nutrient biogeochemistry and the microbes responsible for key nutrient transformations will probably be greatly affected by the increasing presence of anthropogenic CO_2 in the ocean and the resulting acidification. According to the most reliable data, the nitrogen cycle may respond to elevated CO_2 levels by potentially boosting global N_2 fixation, reducing nitrification, and potentially enhancing denitrification (Hutchins et al. 2009). The level of the connections between soil microbial diversity and various nutrients cycling in different environments is closely linked to the equilibrium of positive and negative bacterial-fungal relationships. The most powerful non-biased method of microbial organization in farming soils indicates that transforming natural habitats into agricultural land, such as converting forests into fields for crops and wetlands, or grasslands into agricultural crops, could encourage random movement and distribution of microorganisms and strengthen the connection between biodiversity and MNCs (Jiao et al. 2021).

10.4 Greenhouse Gas Regulation in Saturated Lands

Certain geomicrobes can either produce or consume greenhouse gases like methane and carbon dioxide, influencing the overall carbon balance of saturated lands. There are many different uses for microorganisms, especially in the management and decrease of greenhouse gas emissions through nutrient recycling systems. It works well when producing or consuming these gases by helping it reduce environmental risks brought on by human activity and the natural world (Abatenh et al. 2018). Microorganisms, categorized into different functional groups, primarily produce and consume methane and nitrous oxide, which are potent greenhouse gases, linking their production and consumption in oxic and anoxic environments and impacting various other biogeochemically active molecules. Abiotic processes, such as nitrogen oxides reacting with metals, have a strong influence on microorganisms and play a major role in greenhouse gas emissions (Stein 2020). The production of biogas from manure, organic wastes, and landfills shows how methanization can produce renewable energy, thereby reducing dependency on fossil fuels. Moreover, the transformation of aerobic wastewater treatment into anaerobic nitrogen removal, like anaerobic ammonium oxidation, may lessen the release of nitrous oxide (N_2O) into the atmosphere, further mitigating greenhouse gas emissions. Methane oxidation can also aid in managing current methane emissions (Insam and Wett 2008). The depth of the water table strongly affects the makeup of the microbial community. During dry periods or when the water table is low, fungi decrease, while Gram-negative bacteria and actinobacteria increase. This shift appears to boost methanotrophy, leading to a temporary reduction in methane emissions, turning the wetland from a methane source into a methane sink. During these dry conditions, methane emissions from the wetland were greatly reduced, even showing negative flux values. Overall, wetland carbon dynamics and greenhouse gas emissions are closely linked to the rapid response of microbial communities and carbon metabolism to short-term hydrological changes (Zhang et al. 2022). Adding bait caused a significant decrease in the dissolved oxygen levels in the water above, while various sediment characteristics (like organic matter, sulfide, and ammonium) and the dissolved organic carbon in the water above showed contrasting changes. The bait input led to a significant decrease in the microbial community's richness and diversity, causing a shift in microbial compositions from aerobic, slow-growing, and oligotrophic (*Actinobacteriota*, *Chloroflexi*, and *Acidobacteriota*) to anaerobic, fast-growing, and copiotrophic (*Firmicutes* and *Bacteroidota*). The aquaculture ecosystem may be a major global hotspot for greenhouse gas emissions, as evidenced by the fact that bait input dramatically increased both the production of greenhouse gases and the potential for global warming. The physicochemical properties and the sediment microbial community responded rapidly to the bait input (Lin and Lin 2022). Increased nitrogen input into pristine wetlands alters the composition and activity of microorganisms, particularly in relation to dominant wetland plants. When nitrogen levels rise in incubations with *Juncus acutiflorus*, greenhouse gas emissions increase, shifting the system from a greenhouse gas sink to a source. This suggests that higher nitrogen input causes changes in microbial activity within the rhizosphere, impacting nitrogen cycling dynamics (Hester et al. 2018).

11 Geomicrobes in Specific Saturated Land Types

The breakdown of organic matter in freshwater wetlands is mediated by the competitive and cooperative interactions of respiration, fermentation, and methanogenic microbes. Saltwater intrusion disturbances have the potential to disrupt these interactions by enhancing the activity of sulfate-reducing bacteria (SRB) and intensifying their competition for electron donors with methanogens and syntrophic bacteria. By utilizing metabolic inhibitors to separate the activity of distinct microbial functional groups and to study the effects on biogeochemical and microbial responses, researchers conducted a simulation of saltwater intrusion in wetland soil microcosms (Berrier et al. 2022) and found that it is not sulfate, but rather other ions found in seawater that trigger ecological and biogeochemical reactions to seawater intrusion. These ions may also be responsible for elevated methane emissions in soils that have been artificially supplemented with seawater (Bueno de Mesquita et al. 2024). The relative abundance of methanogens and putative denitrifying methanotrophs decreased as they moved from freshwater to marine wetlands. Methanogens are outcompeted by sulfate reducers with higher salinity and greater sulfate availability, even with minimal salt exposure (Zhao et al. 2023).

Salinity caused a significant alteration in the composition of the P-cycling microbial community, leading to a distinct separation between the freshwater and brackish wetland populations (Hu et al. 2023). The microbial communities that live in hypersaline wetlands are essential to the biogeochemical cycles that guarantee ecosystem services. These communities are well-adapted to the environmental fluctuations brought on by flooding and desiccation events where the microbial community alters in response to water flooding conditions and provides important data for biomonitoring and management of these wetlands subject to irreversible changes in salinity and extreme aridity as a result of human activity (Bourhane et al. 2023).

12 Techniques for Enhancing Geomicrobial Activity in Rehabilitation

The capacity of a microbial population to degrade contaminants in the environment can be enhanced through bioaugmentation, which involves introducing specific microorganisms, or biostimulation, which entails supplying nutrients or electron acceptors to native microorganisms (El Fantroussi and Agathos 2005). As an environmentally friendly and cost-effective method, bioremediation—which involves bioaugmentation and/or biostimulation—has become the most popular method for cleaning up contaminated soil and water that contains organic pollutants and/or heavy metals (Tyagi et al. 2011). To restore the contaminated ecosystems, bioaugmentation, enhanced natural attenuation, and biostimulation have been key initiatives (Jumbo et al. 2024). Many factors, including the kind and amount of pollution, the time it takes to reach the target ecosystem, the biodiversity of microorganisms, pollutant bioavailability, and many more, affect how effective hydrocarbon bioremediation techniques are (Mohajeri et al. 2017). Native groups of microorganisms were found to break down pollutants from petroleum hydrocarbons in acidic wetlands more rapidly than other groups of microorganisms (Jumbo et al. 2023), and when compared to biostimulation alone, the combination of biostimulation and bioaugmentation showed greater and faster affectivity on oil depletion (Darmayati et al. 2015). The use of bioremediation was a successful strategy for treating endosulfan-contaminated constructed wetlands where the levels of endosulfan in plants and sediment decreased, with biodegradation likely being promoted by bioaugmentation and biostimulation using sucrose (Zhao et al. 2014).

13 Future Directions and Emerging Research in Geomicrobial Saturated Land Rehabilitation

The microorganisms found in coastal wetlands exhibit a great degree of biodiversity and spatial heterogeneity, and they are crucial to the upkeep of biogeochemical cycles. The swift advancement of molecular biology technologies has greatly expanded our comprehension of the microbial community and their possible roles. Still, there are many significant obstacles in CW research (Liang et al. 2023). Because of its straightforward principles, extensive application, and room for growth, microbial geotechnical engineering is a novel curing technology that is worthwhile investigating and utilizing where future directions could involve optimizing bacterial species to increase environmental adaptability, increasing urease persistence through increased urease activity, and enhancing the uniformity of mineralization through improved grouting techniques (Zhou et al. 2024). The value and development of organisms as essential elements of environmental engineering and habitat restoration projects are growing. The advantages of using organisms in engineering projects are becoming more widely acknowledged in a variety of fields, including sustainable urban drainage systems, urban building stone decay, coastal flood defences, and river management (Naylor 2005). A whole new realm of rhizosphere research is made possible by the rapid development of new, much more affordable molecular techniques. Basic and applied ecological concerns regarding plant-microbe communities that were previously impossible or very difficult to address can now be approached (Lamers et al. 2012). The majority of wetland ecosystem service restoration projects that are successful entail collaboration between several agencies, obtaining funding from unconventional sources, consulting scientists about ecosystem processes, and fostering positive working relationships between biologists, managers, and maintenance personnel (Smith et al. 2008).

14 Conclusion

The chapter highlights the vital role that geomicrobes play in the ecological restoration of wetlands, specifically focusing on their involvement in biogeochemical cycles such as carbon, nitrogen, sulfur, and phosphorus cycling. Ecosystem services and biodiversity are at risk due to wetland degradation brought on by climate change, agriculture, and urbanization. By dissolving pollutants and promoting plant growth, geomicrobes improve wetland restoration efforts through processes like bioremediation, nutrient recycling, and greenhouse gas regulation. To ensure sustainable ecosystems and advance restoration techniques, a multidisciplinary approach combining geomicrobiology, biotechnology, and environmental management is essential. To develop more effective restoration techniques, future research should prioritize optimizing microbial interactions.

References

Abatenh E, Gizaw B, Tsegaye Z, Tefera G (2018) Microbial function on climate change—a review. Open J Environ Biol 3:1–7. https://doi.org/10.17352/ojeb.000008

Amenu BT, Mamo GS (2018) Review on wetland ecosystem destruction. Int J Sci Res Civ Eng 2:5–15

Anthony TL (2021) Quantification and controls on soil carbon and greenhouse gas fluxes from managed wetland ecosystems. PhD dissertation, UC Berkeley

Barns SM, Nierzwicki-Bauer SA (1997) Microbial diversity in ocean, surface and subsurface environments. In: Geomicrobiology: interactions between microbes and minerals 35. Mineralogical Society of America, pp 35–79. https://doi.org/10.1515/9781501509247-004

Berrier DJ, Neubauer SC, Franklin RB (2022) Cooperative microbial interactions mediate community biogeochemical responses to saltwater intrusion in wetland soils. FEMS Microbiol Ecol 98:fiac019. https://doi.org/10.1093/femsec/fiac019

Bhatia M, Goyal, D (2014) Analyzing remediation potential of wastewater through wetland plants: A review. Environ Prog Sustain 33:9–27. https://doi.org/10.1002/ep.11822

Bonetti G, Trevathan-Tackett SM, Carnell PE, Treby S, Macreadie PI (2021) Local vegetation and hydroperiod influence spatial and temporal patterns of carbon and microbe response to wetland rehabilitation. Appl Soil Ecol 163:103917. https://doi.org/10.1016/j.apsoil.2021.103917

Bourhane Z, Cagnon C, Castañeda C, Rodríguez-Ochoa R, Álvaro-Fuentes J, Cravo-Laureau C, Duran R (2023) Vertical organization of microbial communities in Salineta hypersaline wetland, Spain. Front Microbiol 14:869907. https://doi.org/10.3389/fmicb.2023.869907

Bueno de Mesquita CP, Hartman WH, Ardón M, Tringe SG (2024) Disentangling the effects of sulfate and other seawater ions on microbial communities and greenhouse gas emissions in a coastal forested wetland. ISME Commun 4:ycae040. https://doi.org/10.1093/ismeco/ycae040

Chandra P, Enespa, Kumar M (2020) Contribution of microbes in the renovation of wetlands. In: Restoration of wetland ecosystem: a trajectory towards a sustainable environment. Springer, Singapore, pp 101–124. https://doi.org/10.1007/978-981-13-7665-8_8

Chen M, He S, Li J, Hu W, Ma Y, Wu L, Gang G (2019) Co-occurrence patterns between bacterial and fungal communities in response to a vegetation gradient in a freshwater wetland. Can J Microbiol 65:722–737. https://doi.org/10.1139/cjm-2019-0147

Conant RT, Ryan MG, Agren GI, Birge HE, Davidson EA, Eliasson PE, ... & Bradford MA (2011) Temperature and soil organic matter decomposition rates–synthesis of current knowledge and a way forward. Glob Change Biol 17:3392–3404. https://doi.org/10.1111/j.1365-2486.2011.02496.x

Darmayati Y, Sanusi HS, Prartono T, Santosa DA, Nuchsin R (2015) The effect of biostimulation and biostimulation-bioaugmentation on biodegradation of oil-pollution on sandy beaches using mesocosms. Int J Marine Sci 5:1–11. https://doi.org/10.5376/ijms.2015.05.0027

Davis JA, Froend R (1999) Loss and degradation of wetlands in southwestern Australia: underlying causes, consequences and solutions. Wetl Ecol Manag 7:13–23. https://doi.org/10.1023/A:1008400404021

Ehrlich HL (2006) Geomicrobiology: relative roles of bacteria and fungi as geomicrobial agents. In: Fungi in biogeochemical cycles. Cambridge University Press, Cambridge, pp 1–27

El Fantroussi S, Agathos SN (2005) Is bioaugmentation a feasible strategy for pollutant removal and site remediation? Curr Opin Microbiol 8:268–275. https://doi.org/10.1016/j.mib.2005.04.011

Elhaj Baddar Z, Xu X (2022) Evaluation of changes in the microbial community structure in the sediments of a constructed wetland over the years. Arch Microbiol 204:552. https://doi.org/10.1007/s00203-022-03157-5

Erwin KL (2009) Wetlands and global climate change: the role of wetland restoration in a changing world. Wetl Ecol Manag 17:71–84. https://doi.org/10.1007/s11273-008-9119-1

Findlay SE (2021) Organic matter decomposition. In Fundamentals of ecosystem science. Academic Press. pp 81–102

Gadd GM (2010) Metals, minerals and microbes: geomicrobiology and bioremediation. Microbiology 156:609–643. https://doi.org/10.1099/mic.0.037143-0

Gadd GM (2013) Microbial roles in mineral transformations and metal cycling in the Earth's critical zone. In: Xu J, Sparks D (eds) Molecular environmental soil science. Progress in soil science, pp 115–165. https://doi.org/10.1007/978-94-007-4177-5_6

Ganeshamurthy AN, Singh G, Singh NT (1995) Sulphur status and response of rice to sulphur on some soils of Andaman and Nicobar Islands. J India Soc Soil Sci 43:637–641

Gebresllassie H, Gashaw T, Mehari A (2014) Wetland degradation in Ethiopia: causes, consequences and remedies. J Environ Earth Sci 4:40–48

Gougoulias C, Clark JM, Shaw LJ (2014) The role of soil microbes in the global carbon cycle: tracking the below-ground microbial processing of plant-derived carbon for manipulating carbon dynamics in agricultural systems. J Sci Food Agric 94:2362–2371. https://doi.org/10.1002/jsfa.6577

Gutknecht JLM, Goodman RM, Balser TC (2006) Linking soil process and microbial ecology in freshwater wetland ecosystems. Plant Soil 289:17–34. https://doi.org/10.1007/s11104-006-9105-4

Han Q, Huang X, Shi P, Zhang Q (2006) Coastal wetland in South China: degradation trends, causes and protection countermeasures. Chin Sci Bull 51:121–128. https://doi.org/10.1007/s11434-006-9121-z

Hester ER, Harpenslager SF, van Diggelen JMH, Lamers LL, Jetten MSM, Lüke C, Lücker S, Welte CU (2018) Linking nitrogen load to the structure and function of wetland soil and rhizosphere microbial communities. Msystems 3:10–1128. https://doi.org/10.1128/msystems.00214-17

Hu M, Le Y, Sardans J, Yan R, Zhong Y, Sun D, Tong C, Peñuelas J (2023) Moderate salinity improves the availability of soil P by regulating P-cycling microbial communities in coastal wetlands. Glob Chang Biol 29:276–288. https://doi.org/10.1111/gcb.16465

Hutchins DA, Mulholland MR, Fu F (2009) Nutrient cycles and marine microbes in a CO_2-enriched. Oceans 22:128–145. https://doi.org/10.2307/24861030

Ibrahim HAH, Abdeltwab AM, Amer MS, Gad AA, El Sayed WMM (2023) Geomicrobiology of African lakes. In: Lakes of Africa. Elsevier, pp 195–246. https://doi.org/10.1016/B978-0-323-95527-0.00018-X

Insam H, Wett B (2008) Control of GHG emission at the microbial community level. Waste Manag 28:699–706. https://doi.org/10.1016/j.wasman.2007.09.036

Jain KS, Khan AA, Rai KM (2010) Geomicrobiology. CRC Press, London

Jiao S, Xu Y, Zhang J, Lu Y (2019) Environmental filtering drives distinct continental atlases of soil archaea between dryland and wetland agricultural ecosystems. Microbiome 7:15. https://doi.org/10.1186/s40168-019-0630-9

Jiao S, Peng Z, Qi J, Gao J, Wei G (2021) Linking bacterial-fungal relationships to microbial diversity and soil nutrient cycling. Msystems 6:10–1128. https://doi.org/10.1128/msystems.01052-20

Jones DL, Farrar J, Giller KE (2003) Review article associative nitrogen fixation and root exudation-what is theoretically possible in the rhizosphere? Symbiosis 35:19–38

Jumbo RB, Coulon F, Bortone I, Jing Y (2023) 5 Simultaneous use of biostimulation and bioaugmentation as optimised strategies for remediating oil impacted wetlands. Enhancing bioremediation efficiency of acidic wetlands contaminated with crude oil 183. Doctoral dissertation, Cranfield University

Jumbo RB, Atai E, Azuazu I, Bortone I, Coulon F, Jiang Y (2024) Sustainable low carbon and bioaugmentation strategies for bioremediation of oil-contaminated acidic wetlands. In: Ortega-Calvo JJ, Coulon F (eds) Soil remediation science and technology. Springer, Cham, pp 293–309. https://doi.org/10.1007/698_2024_1077

Keenleyside W (2019) Microbiology: Canadian Edition. Pressbooks, Toronto

Koshila Ravi R, Anusuya S, Balachandar M, Muthukumar T (2019) Microbial interactions in soil formation and nutrient cycling. In: Varma A, Choudhary D (eds) Mycorrhizosphere and pedogenesis. Springer, Singapore, pp 363–382. https://doi.org/10.1007/978-981-13-6480-8_21

Lamers LPM, Van Diggelen JMH, Op den Camp HJM, Visser EJW, Lucassen EC, Vile MA, Jetten MSM, Smolders AJP, Roelofs JGM (2012) Microbial transformations of nitrogen, sulfur, and iron dictate vegetation composition in wetlands: a review. Front Microbiol 3:156. https://doi.org/10.3389/fmicb.2012.00156

Liang S, Li H, Wu H, Yan B, Song A (2023) Microorganisms in coastal wetland sediments: a review on microbial community structure, functional gene, and environmental potential. Front Microbiol 14:1163896. https://doi.org/10.3389/fmicb.2023.1163896

Lin G, Lin X (2022) Bait input altered microbial community structure and increased greenhouse gases production in coastal wetland sediment. Water Res 218:118520. https://doi.org/10.1016/j.watres.2022.118520

McCauley LA, Jenkins DG, Quintana-Ascencio PF (2013) Isolated wetland loss and degradation over two decades in an increasingly urbanized landscape. Wetlands 33:117–127. https://doi.org/10.1007/s13157-012-0357-x

Meng S, Peng T, Liu X, Wang H, Huang T, Gu J-D, Hu Z (2022) Ecological role of bacteria involved in the biogeochemical cycles of mangroves based on functional genes detected through GeoChip 5.0. mSphere 7:e0093621. https://doi.org/10.1128/msphere.00936-21

Mohajeri L, Zahed MA, Abdul Aziz H, Hasnain Isa M (2017) Assessment of bioaugmentation and biostimulation efficiencies for petroleum contaminated sediments. Environ Energy Econ Res 1:8998. https://doi.org/10.22097/eeer.2017.46459

Mohan VK, Joshi SR (2024) Mycorrhizal fungi as Ecofriendly interventions for crop productivity. In: Industrial microbiology and biotechnology: an insight into current trends. Springer, pp 663–677. https://doi.org/10.1007/978-981-97-1912-9_26

Moreno-Mateos D, Power ME, Comín FA, Yockteng R (2012) Structural and functional loss in restored wetland ecosystems. PLoS Biol 10:e1001247. https://doi.org/10.1371/journal.pbio.1001247

Naylor LA (2005) The contributions of biogeomorphology to the emerging field of geobiology. Palaeogeogr Palaeoclimatol Palaeoecol 219:35–51. https://doi.org/10.1016/j.palaeo.2004.10.013

Nealson KH, Stahl DA (1997) Microorganisms and biogeochemical cycles: what can we learn from layered microbial communities? Rev Mineral 35:5–34. https://doi.org/10.1515/9781501509247-003

Peiffer S, Walton-Day K, Macalady DL (1999) The interaction of natural organic matter with iron in a wetland (Tennessee Park, Colorado) receiving acid mine drainage. Aquat Geochem 5:207–223. https://doi.org/10.1023/A:1009617925959

Ramírez-Viga TK, Aguilar R, Castillo-Argüero S, Chiappa-Carrara X, Guadarrama P, Ramos-Zapata J (2018) Wetland plant species improve performance when inoculated with arbuscular mycorrhizal fungi: a meta-analysis of experimental pot studies. Mycorrhiza 28:477–493. https://doi.org/10.1007/s00572-018-0839-7

Reith F, Dürr M, Welch S, Rogers SL, Scott K, Pain CF (2008) The geomicrobiology of the Regolith. CSIRO Press, Melbourne

Richard T, Conant Michael G, Ryan Göran I, Ågren Hannah E, Birge Eric A, Davidson Peter E, Eliasson Sarah E, Evans Serita D, Frey Christian P, Giardina Francesca M, Hopkins Riitta, Hyvönen Miko UF, Kirschbaum Jocelyn M, Lavallee Jens, Leifeld William J, Parton Jessica, Megan Steinweg Matthew D, Wallenstein JÅ, Martin Wetterstedt Mark A, Bradford (2011) Temperature and soil organic matter decomposition rates - synthesis of current knowledge and a way forward Global Change Biology 17(11):3392–3404. https://doi.org/10.1111/j.1365-2486.2011.02496.x

Sahu N, Vasu D, Sahu A, Lal N, Singh SK (2017) Strength of microbes in nutrient cycling: a key to soil health. Agriculturally important microbes for sustainable agriculture. Plant-Soil-Microbe Nexus 1:69–86. https://doi.org/10.1007/978-981-10-5589-8_4

Schowalter TD (2016) Decomposition and Pedogenesis. Insect Ecology: 477–510. https://doi.org/10.1016/b978-0-12-803033-2.00014-5

Schuyt KD (2005) Economic consequences of wetland degradation for local populations in Africa. Ecol Econ 53:177–190. https://doi.org/10.1016/j.ecolecon.2004.08.003

Sharma R, Malaviya P, (2022) Constructed wetlands for textile wastewater remediation: A review on concept, pollutant removal mechanisms, and integrated technologies for efficiency enhancement. Chemosphere 290:133358. https://doi.org/10.1016/j.chemosphere.2021.133358

Singh A, Singh PK, Wang W, Shrivastava AK (2020) Microbes biology: microbes in wetland and bioprospection of microbes. In: Restoration of wetland ecosystem: a trajectory towards a sustainable environment. Springer, pp 87–99. https://doi.org/10.1007/978-981-13-7665-8_7

Smith LM, Euliss NH, Wilcox DA, Brinson MM (2008) Application of a geomorphic and temporal perspective to wetland management in North America. Wetlands 28:563–577. https://doi.org/10.1672/07-155.1

Spieles DJ (2022) Wetland construction, restoration, and integration: a comparative review. Land 11:554. https://doi.org/10.3390/land11040554

Stein LY (2020) The long-term relationship between microbial metabolism and greenhouse gases. Trends Microbiol 28:500–511. https://doi.org/10.1016/j.tim.2020.01.006

Stumm W, Sulzberger B (1992) The cycling of iron in natural environments: considerations based on laboratory studies of heterogeneous redox processes. Geochim Cosmochim Acta 56:3233–3257. https://doi.org/10.1016/0016-7037(92)90301-X

Tomscha SA, Bentley S, Platzer E, Jackson B, de Roiste M, Hartley S, Norton K, Deslippe JR (2021) Multiple methods confirm wetland restoration improves ecosystem services. Ecosyst People 17:25–40. https://doi.org/10.1080/26395916.2020.1863266

Tyagi M, da Fonseca MMR, de Carvalho CCCR (2011) Bioaugmentation and biostimulation strategies to improve the effectiveness of bioremediation processes. Biodegradation 22:231–241. https://doi.org/10.1007/s10532-010-9394-4

Wang M, Qi S, Zhang X (2012) Wetland loss and degradation in the Yellow River Delta, Shandong Province of China. Environ Earth Sci 67:185–188. https://doi.org/10.1007/s12665-011-1491-0

Wang B, Zheng X, Zhang H, Yu X, Lian Y, Yang X, Yu H, Hu R, He Z, Xiao F, Yan Q (2021) Metagenomic insights into the effects of submerged plants on functional potential of microbial communities in wetland sediments. Marine Life Sci Technol 3:405–415. https://doi.org/10.1007/s42995-021-00100-3

Williams JB (2002) Phytoremediation in wetland ecosystems: progress, problems, and potential. Crit Rev Plant Sci 21:607–635. https://doi.org/10.1080/0735-260291044386

Wójcicki KJ, Woskowicz-Ślęzak B (2015) Anthropogenic causes of wetland loss and degradation in the lower Kłodnica valley (southern Poland). Environ Socio-Econ Stud 3:20–29. https://doi.org/10.1515/environ-2015-0070

Yan JinZhuo YJ, Wu FaSi WF, Feng HuYuan FH (2008) Review on the relationship between wetland plants and arbuscular mycorrhizal fungi (AMF). Acta Botan Boreali-Occiden Sin 28:836–842

Zeinah, Elhaj Baddar Xiaoyu, Xu (2022) Evaluation of changes in the microbial community structure in the sediments of a constructed wetland over the years Archives of Microbiology 204(9) https://doi.org/10.1007/s00203-022-03157-5

Zhang Y, Naafs BDA, Huang X, Song Q, Xue J, Wang R, Zhao M, Evershed RP, Pancost RD, Xie S (2022) Variations in wetland hydrology drive rapid changes in the microbial community, carbon metabolic activity, and greenhouse gas fluxes. Geochim Cosmochim Acta 317:269–285. https://doi.org/10.1016/j.gca.2021.11.014

Zhao C, Xie H, Mu Y, Xu X, Zhang J, Liu C, Liang S, Ngo HH, Guo W, Xu J, Wang Q (2014) Bioremediation of endosulfan in laboratory-scale constructed wetlands: effect of bioaugmentation and biostimulation. Environ Sci Pollut Res 21:12827–12835. https://doi.org/10.1007/s11356-014-3107-1

Zhao J, Chakrabarti S, Chambers R, Weisenhorn P, Travieso R, Stumpf S, Standen E, Briceno H, Troxler T, Gaiser E (2023) Year-around survey and manipulation experiments reveal differential sensitivities of soil prokaryotic and fungal communities to saltwater intrusion in Florida Everglades wetlands. Sci Total Environ 858:159865. https://doi.org/10.1016/j.scitotenv.2022.159865

Zhou D, Gong H, Wang Y, Khan S, Zhao K (2009) Driving forces for the marsh wetland degradation in the Honghe National Nature Reserve in Sanjiang plain, Northeast China. Environ Model Assess 14:101–111. https://doi.org/10.1007/s10666-007-9135-1

Zhao C, Xu J, Shang D, Zhang Y, Zhang J, Xie H,... & Wang Q (2021) Application of constructed wetlands in the PAH remediation of surface water: A review. Science of The Total Environment 780:146605. https://doi.org/10.1016/j.scitotenv.2021.146605

Zhou Y, Li Z, Zhang P, Wang Q, Pan W, Wang S, Xie X (2024) Research status, hot spots, difficulties and future development direction of microbial geoengineering. J Road Eng 4:234–255. https://doi.org/10.1016/j.jreng.2024.04.001

Geomicrobiology in Hot Spring Hydrothermal Systems

Yukti Oza, Swarali Bagade, and Avinash Sharma

Contents

Abstract Hot water springs generated by geothermally heated groundwater are of interest to microbial ecologists, as they demonstrate that life can thrive in temperatures where the survival of other organisms is impossible. The microbes surviving under such high-temperature conditions are known as thermotolerants, thermophiles, or hyperthermophiles, depending on the temperature range they can survive. Thermophiles and hyperthermophiles are particularly important for their ability to

Y. Oza · S. Bagade · A. Sharma (✉)
BRIC-National Centre for Cell Science, Pune, Maharashtra, India
e-mail: avinash@nccs.res.in

S. R. Joshi et al. (eds.), *Mineral Transformation and Bioremediation by Geo-Microbes*, Interdisciplinary Biotechnological Advances,
https://doi.org/10.1007/978-981-96-3033-2_3

produce enzymes that function optimally at high temperatures, making them ideal for industrial processes where conventional enzymes tend to denature and lose their activity under extreme thermal conditions. Understanding the intricate ecosystems where these microbes thrive is both fascinating and challenging. However, with advancements in high throughput sequencing technologies, it is now possible to explore the diversity and roles of these microorganisms in ecosystems with greater depth. These technologies provide valuable insights into microbial diversity, enabling the development of strategies for isolating and cultivating specific microbial groups. Additionally, whole genome sequencing offers the opportunity to identify novel genes and metabolic pathways in these organisms, paving the way for innovations in biotechnology.

Keywords Hot springs · High throughput sequencing · Thermophiles · Extremozymes · Microbial ecology · Biological processes

Abbreviations

Cu	Copper
DNA	Deoxyribose nucleic acid
Fe	Iron
Mn	Manganese
RNA	Ribonucleic acid
SO_4^{2-}	Sulfates
UNESCO	United Nations Educational, Scientific and Cultural Organization

1 Introduction

In the depths of our planet, where geological forces and heat arise from the molten core, lie some of the most intriguing and extreme environments known, the hydrothermal systems. These dynamic and geologically active ecosystems serve as a "yet to be revealed ocean" for understanding the interplay between geology, chemistry, and biology. These unique manifestations of hydrothermal activity provide unique niches for life to flourish under conditions of elevated temperatures, high pressures, and extreme chemical gradients. Microorganisms that thrive in such extreme environments are known as "extremophiles," the term first coined by R.D. Macelroy (1974).

Thermophiles (*thermo* means heat, and *phile* means loving), with their remarkable ability to thrive in extreme heat, have captured the interest of researchers seeking innovative biotechnological applications. These resilient organisms produce a wealth of heat-stable enzymes and proteins that are highly valuable for industrial processes. In the realm of biocatalysis, the optimization of enzyme properties, such as stability at high temperatures, substrate affinity, and tolerance to hypersalinity, has traditionally relied on tedious and time-consuming genetic engineering approaches often lacking clear rational design. However, an exciting alternative emerges through the utilization of "extremozymes," enzymes sourced from extremophilic microorganisms, which offer inherent adaptations, making them suitable for industrial applications. These extremozymes exhibit catalytic properties in non-aqueous environments, hypersaline conditions, under extreme pressures, across a wide pH spectrum (acidic to alkaline), and at temperatures soaring to 140 °C or plummeting towards the water's freezing point (Adams et al. 1995). From the production of biofuels to the development of eco-friendly detergents, thermophilic microbes offer enzymatic tools that can withstand high temperatures, harsh chemical conditions, and other challenges that conventional enzymes cannot endure.

The recent advancement of high-throughput sequencing has enabled us to dive deeper into the genetic makeup of the microorganisms. Through functional metagenomics, one can identify and isolate the essential functional genes from the metagenome and clone them into libraries (Mirete et al. 2016). These libraries can be expressed in the cultivable microorganisms and assayed for their expression of the desired gene of interest, and its products can be further analyzed (Fig. 1). Recent advancements in functional screening of metagenomic libraries sourced from high-temperature habitats, like hot springs, have yielded a wealth of thermostable enzymes with significant biotechnological potential.

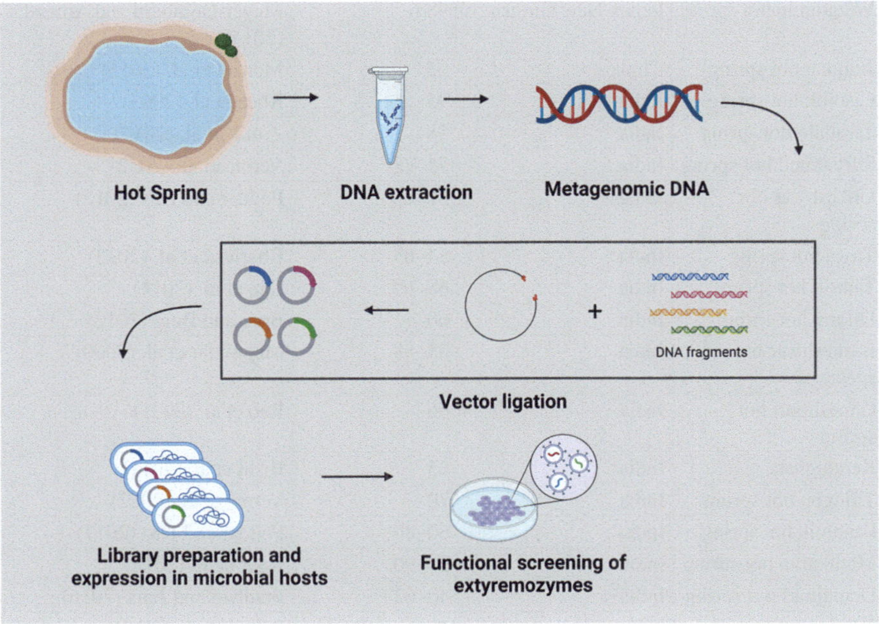

Fig. 1 Schematic diagram depicting the steps involved in preparation of cloning libraries for screening of extremozymes. (Created with BioRender.com)

2 Key Hot Springs Across the Globe

Hot springs are renowned not only for their thermal properties but also for their distinctive mineral content, which includes salts, sulfur compounds, and elements, which may have therapeutic effects (Debnath et al. 2019; Ranjit 2022; Tolessa 2020). These extreme environments host a plethora of microbial life, each adapted to their specific conditions, offering valuable insights and applications in biotechnology (Table 1). Traditionally, microbial diversity was explored using cultivation-based approach, which allowed the isolation and characterization, followed by the identification of these microorganisms. However, this method captured only a fraction of the microbial community. Recent advances in microbiology, particularly high-throughput sequencing techniques, such as metagenomics and amplicon sequencing, have revolutionized the study of hot spring microbiomes by providing a more comprehensive understanding of unculturable microbial populations. For instance, the Sungai Klah hot spring in Malaysia, hosting biofilm-producing

Table 1 Key hotsprings across the globe

Name of the hot springs	Country	Temperature (°C)	References
Octopus hot spring	United States of America	93	Moser et al. (2012)
Cistern hot spring	United States of America	95	Guidry and Chafetz (2003)
Waramung hot spring	Papua New Guinea	98.6	Meyer-Dombard and Amend (2014)
Jiuquan hot spring	China	65	Menzel et al. (2015)
Cisolok hot spring	Indonesia	95	Rhee et al. (2005)
Tarabalo hot spring	India	58	Zimik et al. (2017)
Suryakund hot spring	India	75–78	Verma et al. (2022)
Unkeshwar hot spring	India	50–60	Poddar and Das (2017)
Tuwa hot spring	India	54–65	Bharwad et al. (2023)
Tantoli hot spring	India	68–69	Jain et al. (2014)
Dirang hot spring	India	60	Bora and Bora (2012)
Barkeshwar hot spring	India	35–88	Majumdar et al. (2000)
Ganeshpuri hot spring	India	66	Rao et al. (2021)
Kendughata	India	65	Bisht et al. (2011)
Thingbu hot spring	India	70	Verma et al. (2022)
Panamik hot spring	India	60–80	Poddar and Das (2017)
Manikaran hot spring	India	50–90	Kaur et al. (2018)
Deulajhari hot spring	India	40–62	Pradhan and Jena (2016)
Jakrem hot spring	India	46	Panda et al. (2015)
Yumthang hot spring	India	43–63	Sahay et al. (2017)

bacteria, particularly *Chloroflexota* and *Armatimonadota*, has been identified through metagenomic sequencing. These microorganisms are proficient in breaking down plant litter components, showcasing metabolic diversity and producing glycosyl hydrolases, which have significant biotechnological applications (Liew et al. 2024). The microbial activity in these hot springs underscores their potential in industrial processes, such as biomass degradation and enzyme production. In South Africa, the hot springs of Limpopo Province are home to *Bacillus* species that thrive in high pH levels and temperatures ranging from 42 °C to 68 °C, which possess diverse enzymatic activities, including protease production and the decomposition of pollutants using phenol and amylase (Jardine et al. 2018). The enzymes may find their potential applications in waste management and bioremediation, highlighting the ecological and industrial significance of these hot springs. Furthermore, the imputed metagenomics applied on the samples of hot springs of Maharashtra, India, revealed genes that are able to metabolize carbon, methane, sulfur, and nitrogen (Kumar et al. 2022); the findings of which could be beneficial in mitigating climate change.

2.1 Octopus Hot Springs, Yellowstone National Park, USA

Octopus Hot Springs are situated in the Lower Geyser Basin in Yellowstone National Park, USA. They show a remarkable geothermal feature with temperatures ranging from 84 °C to 88 °C and a slightly alkaline pH of 8.1. This hot spring is an exceptional site for studying microbial ecology, as it hosts a diverse array of unique and newly discovered species, including biofilm-forming bacteria and archaea (Summons et al. 2005). Among these is the *Korarchaeota*, a newly identified branch of Archaea, providing insights into evolutionary relationships within this domain, and phototrophic iron bacteria capable of oxidation were revealed (Ferris et al. 2001).

2.2 Noboribetsu Hot Spring, Japan

Noboribetsu Hot Spring, located in Japan, is renowned for its geothermal activity and its potential for geothermal energy reservoirs. It is situated along a northwest to southeast axis and benefits from a deep-seated hydrology with estimated temperatures exceeding 270 °C, primarily due to the proximity of volcanic activity. Noboribetsu Hot Spring is home to some of the most extreme acidophilic archaea ever discovered, including *Picrophilus oshimae* and *Picrophilus torridus* (Kaushik et al. 2021). These archaea thrive in environments with an exceptionally low pH of 0.06, showcasing their extreme acidophilic nature. Additionally, these microorganisms exhibit polyextremophilic characteristics, allowing them to adapt to multiple extremes (Yadav et al. 2024). This adaptability is achieved through modifications in their lipid membrane fluidity, enabling them to survive and function in such harsh conditions where pH levels fluctuate dramatically. The unique combination of extreme temperature and low pH conditions at Noboribetsu Hot Spring provides a valuable natural laboratory for studying the mechanisms of microbial adaptation and survival.

2.3 Pamukkale Hot Springs, Turkey

Pamukkale, which means "Cotton Castle" in Turkish, is a breathtaking natural wonder known for its stunning terraces of white travertine which attracts from across the world. These terraces are formed by the flow of hot spring waters rich in calcium carbonate, deposited through a karst system composed of limestone (Simsek 2022). Recognized as a UNESCO World Heritage site, Pamukkale is not only a geological marvel but also a site of historical importance, with its use dating back to Roman times (Scoon 2021). Pamukkale boasts 17 active hot springs with temperatures soaring from 35 °C to 100 °C (Scoon 2021). These terraces provide a unique habitat for a diverse array of phototrophic microorganisms, including diatoms, *Chlorophyceae*, and *Cyanobacteria* species (Pentecost et al. 1997). These phototrophs contribute to the distinct coloration and ecological diversity of the travertine formations.

2.4 Chumathang Hot Spring, India

The Himalayan Geothermal Belt (HGB), initially reported by Tong and Zhang in 1981, is the greatest of geothermal areas in our Earth's arena (Hochstein and Regenauer-Lieb 1998; Das and Thakur 2021). Among these, the Chumathang hot spring lies to the north of the Indus Tsangpo Suture Zone (ITSZ) in Ladakh, India (Rawat et al. 2020). These thermal springs, arising from Earth's crust, reflect distinct aquatic habitats, giving rise to novel species in such a pristine environment where India has been thriving around a countless number of hot springs for many centuries. With the use of culture-independent approach, researchers have found out the abundance of *Firmicutes* and *Proteobacteria* in water bodies, while *Proteobacteria* and *Bacteroidetes* predominate the soil (Anu et al. 2024), where the results show promising potential for various biotechnological applications along with microbial interactions in these ecosystems.

2.5 *Ringigad Hot Spring, India*

Ringigad thermal spring is located 17 km from Joshimath towards Suraithota at an altitude of 1850 m above sea level (ASL) in the Chamoli district of Uttarakhand. The temperature range of this hot water spring is 82–89 °C. Luxurious algal mats cover its surroundings with some fern species as well. In spite of this place harboring many potentially important microorganisms, the Ringigad hot spring is less studied. However, few studies, which employed 16S rRNA gene sequencing, have found out a wide range of thermophiles from this thermal spring (Pandey et al. 2014). Reports have also shown that there are approximately 35 species of Cyanobacteria at this site (Bhardwaj et al. 2011; Singh et al. 2018). Hence, further detailed studies employing metagenomics might give us a deeper insight into the thermophiles, which serve a significant biotechnological potential.

2.6 *Soldhar Hot Spring, India*

The Soldhar Hot Springs, located in the Chamoli district of Uttarakhand at an altitude of 1900 m ASL, India, is a biodiversity hotspot due to its distinct characteristics, including a slightly alkaline pH, heavy metals like Cu, Fe, Mn, SO_4^{2-}, and low concentration of organic matter. The temperature goes upto 95 °C, and it boasts a wide range of thermophilic bacteria. Due to the state being prone to natural calamities, this hot spring and its associated microbial diversity is essential for conservation. With the use of culture independent method, studies have shown the dominance of Proteobacteria, Deinococcus-Thermus, Aquificae, Actinobacteria, and Firmicutes phyla (Sharma et al. 2014, 2017). Moreover, a study employing high-throughput sequencing of Soldhar Hot Springs revealed thermophiles with high temperature (upto 80 °C) and pH ranges (upto 14) (Pandey et al. 2014).

2.7 *Kamchatka Hot Springs, Russia*

Located within the volcanic caldera of the Kamchatka Peninsula, Russia, the Kamchatka Hot Springs are renowned for their unique and diverse microbial life. The extreme thermal conditions of these hot springs provide an ideal environment for thermophilic organisms, particularly prokaryotes, playing a significant role in the degradation of complex biopolymers (Kublanov et al. 2009). Hydrolytic experiments using multiple polymeric substances in the hot springs of Kamchatka have revealed significant microbial activity, highlighting the presence of thermophiles capable of breaking down complex biopolymers in hostile environments. Such enzymes hold immense importance for industrial applications, attracting biotechnologists worldwide (Kublanov et al. 2009). Key microbial species identified in

Kamchatka Hot Springs also include sulfur-oxidizing bacteria, such as *Sulfurihydrogenibium* and *Caldimicrobium*, which thrive in the sulfur-rich environments (Merkel et al. 2016).

2.8 Kotli Hot Springs, Pakistan

The Kotli geothermal springs, also known as Tatta Pani, are located adjacent to the Main Karakoram Thrust in the Himalayas of eastern Pakistan. These springs, with temperatures reaching up to 98 °C and pH levels ranging from 6.2 to 9.4, are renowned for their unique microbial diversity. Research indicates that at lower temperatures around 60 °C, the hot spring environment at Kotli supports the growth of the phylum Proteobacteria, which is associated with various extremophilic bacteria (Amin et al. 2017). Notable species identified include *Rhizobium etli* and *Rhizobium daejeonense*, both of which thrive under these intense conditions (Amin et al. 2017). Additionally, a novel Gram-positive thermophilic bacterium was discovered in these springs that showed 89% similarity to *Geobacillus debilis*, which exhibits the ability to assimilate ortho-nitrophenyl-β-D-galactopyranoside and hydrolyze gelatin at 65 °C (Zahoor et al. 2016). Furthermore, *Thermoactinomyces sacchari*, identified from this hot spring, was found to produce the enzyme amylase, which has significant industrial applications (Das and Thakur 2021).

2.9 Pamir Hot Springs, Kyrgyzstan

Situated in the majestic Pamir Mountains, the Pamir Hot Springs are located at the convergence of China, Pakistan, Afghanistan, and Kyrgyzstan. This geothermal site features a wealth of thermal and cold springs, with approximately 72 spots exhibiting thermal discharge along with the physicochemical characterization of sulfur and nitrogen (Breu and Hurni 2003). In an effort to explore the biodiversity of these springs, a bioindication methodology was implemented for the first time to interpret six mineral and thermal springs in Pamir. This metagenomics study revealed that diatoms, such as *Achnanthidium* sp., *Caloneis* sp., *Amphora* sp., *Achnanthes* sp., *Actinella* sp., and *Brachysira* sp. have adapted to temperate water habitats, performing photosynthesis for protein synthesis (Niyatbekov and Barinova 2018a, b). These algae communities are significant indicators of the unique ecological adaptations present in the Pamir Hot Springs. Early-stage research on the microbial diversity of these hot springs proposed that *Thermus ruber* and *Thermus flavus* are prevalent, indicating a close proximity in their thermal adaptation (La and Ig 1975).

2.10 *Lobios Hot Spring, Spain*

Located in the scenic Galician region of north-western Spain, the Lobios Hot Spring is a notable geothermal feature characterized by its high temperature and alkalinity. Situated near the town of Lobios along the Río Caldo River, this hot spring lies about 70 km from Ourense, Spain, and 6 km from Portugal. The Lobios Hot Spring, with temperatures exceeding 76 °C and a pH level over 8.2, represents one of the hottest and most alkaline springs in the region. It is part of the abundant mineral water springs originating from meteoric sources in Galicia. Rainwater infiltrates through fractures into the subsoil, undergoing chemical and thermal transformations before emerging as natural springs. This unique hydrothermal environment creates an ideal niche for extremophiles. Researchers have focused on the Lobios Hot Spring to uncover and characterize enzymes capable of functioning under high temperatures and alkaline pH. One notable discovery is the thermostable xylanase isolated from the soil sediment of the Lobios Hot Spring (Knapik et al. 2019). Xylanases are enzymes with broad applications in various industries, comprising food, paper and pulp, textile, and biofuel production industries (Collins et al. 2005). In a nutshell, the microbial diversity perceived in Lobios Hot Springs is probably a consequence of its distinctive physicochemical characteristics, wherein thermophilic, sulfur-utilizing, and heterotrophic microorganisms thrive, thereby enriching the intricate ecological interactions within these geothermal habitats (DeCastro et al. 2021).

3 Applications of Extremozymes in the Industries

Hot springs are rich in diverse microorganisms, primarily thermophiles thriving at temperatures from 45 °C to over 100 °C. These thermophiles are classified into moderate (ranging from 45 °C to 60 °C), extreme (ranging from 70 °C to 80 °C), and hyperthermophiles (above 80 °C) (Arbab et al. 2021). Thermophiles possess unique adaptations, such as heat-stable enzymes, specialized membrane compositions, and efficient DNA repair mechanisms, which are invaluable for biotechnological applications. Their enzymes can operate at temperatures above 70 °C, which is sufficiently high to kill almost all pathogenic bacteria. This reduces the risk of contamination, particularly in food processing, ensuring higher safety and quality standards. Also, at elevated temperatures, the viscosity of process fluids decreases, lowering shear forces, which in turn reduces costs for pumping, filtration, and centrifugation. This reduces water levels during processing, leading to more efficient operations (Bruins et al. 2001). Extremozymes maintain their catalytic activity in extreme pH environments and high-pressure conditions, making them suitable for applications in various chemical and pharmaceutical processes where traditional enzymes would denature or lose efficacy. Previous study demonstrated that the thermostability of proteins is closely related to the number of hydrogen bonds and

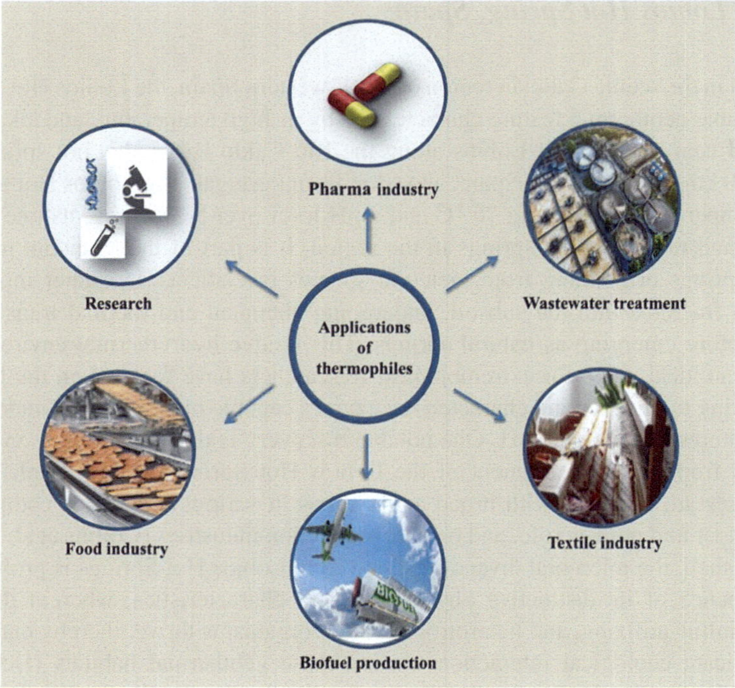

Fig. 2 Applications of thermophiles with significant biotechnological potential

fractional polar surface (Vogt et al. 1997). Increased hydrogen bonding density along with the surrounding water molecules contributes significantly to the stability of these proteins. This enhanced hydrogen bonding network stabilizes the protein structure, preventing denaturation at high temperatures. Disulfide bridges also play a crucial role in stabilizing proteins (Matsumura et al. 1989). Extremozymes inherently possess properties that typically require extensive engineering in other enzymes through directed evolution, truncation, DNA shuffling, gene fusion, or rational protein synthesis (Davids et al. 2013; Rizk et al. 2012; Nisha and Satyanarayana 2013). Their ability to function under harsh conditions makes them suitable for a variety of industrial applications, offering several significant advantages (Fig. 2, Table 2).

3.1 Discovery of Taq *Polymerase*

Thermophilic enzymes are invaluable in research due to their robustness and ability to function at high temperatures. The most common example is *Taq* polymerase, which is obtained from the thermophilic bacterium *Thermus aquaticus* commonly found in hot water springs. This enzyme's major contributuion in molecular biology

Table 2 Industrially important enzymes obtained from the thermophilic microorganisms

Enzyme	Source	Temperature (°C)	Industrial applications	References
Lipase	*Thermoanaerobacter thermohydrosulfuricans*	75	Stereo-specific reactions	Royter et al. (2009)
Chitinase	*Bacillus thurigiensis*	110	Chitin modification	Kuzu et al. (2012)
	Sulfolobus tokodaii	70	Food and health products	Staufenberger et al. (2012)
Endoxylanase	*Thermotoga petrophila*	95	Paper and Pulp industry	Haq et al. (2012)
α-Amylase	*Anoxybacillus* sp., *Bacillus* sp., *Halorubrum xinjiangense*	60–90	Pharmaceutics, food industry, textile industry, detergents	Liao et al. (2019), Mantiri et al. (2019), Suvd et al. (2001), Srivastava and Baruah (1986), Yuuki et al. (1985), Machius et al. (1998), Dobreva et al. (1994), Moshfegh et al. (2013)
Amylopullulanase	*Thermococcus kodakarensis*	100	Starch industry, for the production of maltose syrups	Han et al. (2013)
β-Glucosidase	*Thermofilum pendens*	90	Cellulose saccharification, glucoside hydrolysis, bioactive substrates liberation	Li et al. (2013)
	Thermotoga thermarum			Zhao et al. (2013)
Esterase	*Thermus thermophilus*	80	High resistance for chemical and physical denaturing agents	Fuciños et al. (2014)
Proteolysin	*Coprothermobacter proteolyticus*	80	Biocatalytic processes for organic synthesis	Toplak et al. (2013)
Thermopsin-like protease	*Sulfolobus solfataricus*	70	Signal-transduction pathways	Gogliettino et al. (2014)

(continued)

Table 2 (continued)

Enzyme	Source	Temperature (°C)	Industrial applications	References
Polygalaturonase	*Caldicellulosirutor bescii*	72	Food and textile industry.	Chen et al. (2014)
β-Mannosidase	*Pyrococcus furiosus*	90	Food and pharmaceutical applications	Park et al. (2011)
Xylanase	*Thermotoga thermarum*	90	Biomass utilization	Shi et al. (2013)

was enabling the development of the polymerase chain reaction (PCR) (Chien et al. 1976). *Taq* polymerases maintain their activity at elevated temperatures, making them ideal for molecular biology techniques that demand extreme conditions. Apart from *Thermus aquaticus*, similar polymerases have been isolated from other thermophilic organisms such as *T. littoralis*, *T. maritima*, *P. woesei*, and *P. furiosus*, expanding their applicability in PCR (Satyanarayana et al. 2004). The *Thermus* genus is particularly noteworthy for its biotechnological potential. Many members of this genus have been isolated to obtain enzymes crucial for both research and industrial applications. For instance, enzymes involved in DNA replication, DNA repair, and RNA maturation have been isolated from extreme thermophilic halotolerant *Thermus* strains from Japan (Henne et al. 2004).

3.2 Food and Beverage Processing

In the food industry, thermozymes play a crucial role in breaking down starch into sugars at high temperatures. This process not only improves efficiency but also increases yield. The history of starch-degrading enzymes dates back to 1833 when Anselme Payen, a French chemist, discovered diastase, the first such enzyme (Elleuche et al. 2014). The breakdown of starch to glucose requires temperatures up to 105 °C (Sarian et al. 2023), which requires enzymes capable of functioning under such extreme conditions. Key enzymes catalyzing starch hydrolysis include α- and β-amylases, glucoamylases, α-glucosidases, and pullulanases, which are prevalent in thermophilic microorganisms (Sunna et al. 1997; Nisha and Satyanarayana 2013). The most heat-stable enzymes, such as pullulanase and α-amylase, can function at temperatures up to 120 °C; examples include pullulanase from *Thermococcus kodakarensis KOD1* (Han et al. 2013) and α-amylase from *Methanococcus jannaschii* (Kim et al. 2001).

3.3 Biofuel Production

Biofuel production is a sustainable alternative to fossil fuels, which comprises the formation of hydrogen, methane, ethanol, butanol, and long-chain alkyl ester derivatives through fermentation and microbial metabolism (Samanta et al. 2022). Amylases and lipolytic enzymes are particularly preferred for biofuel production due to their efficiency in breaking down biomass. Promising thermozymes for biofuel production include α-amylase isolated from *Anoxybacillus rupiensis TS-4* and xylanase from *Anoxybacillus flavithermus TWXYL3* (Kikani et al. 2019; Ellis and Magnuson 2012). These enzymes facilitate the conversion of complex polysaccharides into fermentable sugars, which are then transformed into biofuels by microbial action.

3.4 Wastewater Treatment and Bioremediation

Thermophilic aerobic processes offer significant advantages in treating high-strength wastewaters due to their rapid biodegradation rates and minimal sludge production. These processes have been successfully applied to diverse industrial effluents, including those from pulp and paper production, livestock farming, and various other sources. Recent research, reported by Al-Rasheedi et al. (2022), underscores the efficacy of thermophilic bacteria in treating authentic wastewater, ensuring compliance with environmental discharge standards. The binding capacity to metals in *Geobacillus thermodenitrificans*, isolated from the Damodar River in India, was evaluated by employing synthetic metal solutions and industrial wastewater. This strain exhibited significant reduction in metal concentrations in the industrial wastewater under optimal pH conditions (Chatterjee et al. 2010). Also, they are prevalent in both natural geological settings and human-made thermal environments enriched with high amount of dissolved metals and exhibit distinctive cell wall structures, metabolic pathways, and enzymatic activities that play pivotal roles in metal-thermophile interactions (Mir et al. 2021). These microbial communities possess the dual capability of degrading organic and inorganic substrates while facilitating metal reduction processes. Notably, thermophilic bacteria demonstrate proficiency in reducing a wide array of metals, thereby offering potential applications in immobilizing toxic metals and radionuclides during the bioremediation of hot wastewater and disposal sites for radioactive wastes. This capability is particularly advantageous given the prolonged favorable temperature conditions conducive to thermophiles.

3.5 Pharma Industry

Pharma industries are involved in the research, discovery, production, and distribution of medicines and drugs. The processes involved require enzymes with stability and activity at high temperatures, extreme pH, and salinity. Thermophilic enzymes such as acylaminoacyl peptidase, amylases, and lipases are essential in the production of pharmaceuticals. The acylaminoacyl peptidase, isolated from an archaeon *Aeropyrum pernix* K1, is an extremely stable enzyme showing its optimum activity at 90 °C (Gao et al. 2003). Thus, they serve as important enzymes for synthesis of drug precursors.

3.6 Textile Industry

The textile industry includes processes such as fabric treatment, dyeing, and finishing. Extremozymes play a significant role by their ability to function effectively not only under harsh conditions but also in the presence of harsh chemicals, making them ideal for processes like bio-polishing, bio-scouring, and dye degradation. These enzymes help improve fabric quality, reduce energy consumption, and minimize the environmental impact by replacing harsh chemicals with more sustainable biological alternatives. De-inking in textile industries is carried out using cellulases or combinations of cellulases and xylanases (Ibarra et al. 2012; Saxena and Chauhan 2016). Cellulases and hemicellulases have been reported to remove the ink from the surface of the pulp by partial hydrolysis of carbohydrate molecules (Rahkamo et al. 1998). The use of extremozymes thus contributes to more efficient and eco-friendly textile production, further aligning with the industry's push toward sustainability.

4 Conclusion

Hot springs serve as a reserve for the thermophilic microorganisms, and with the recent advances in the high-throughput sequencing techniques, enzymes derived from hot spring microorganisms offer a wealth of biotechnological applications. Due to their stability and efficiency at high temperatures, their applications span multiple industries ranging from molecular biology to biofuel production and hold promise for future innovations in bioremediation, pharmaceuticals, and agriculture. Continued research and development in this field will undoubtedly lead to new and exciting biotechnological advancements. Moreover, the rapid development in the "-omics" field like metagenomics, proteomics, and metabolomics, clubbed with enzymes and molecular biology techniques, will soon provide biological catalysts with desired properties, leading to the development of a sustainable industry.

Competing Interests All the authors declare that they have no competing interests.

References

Adams MWW, Perler FB, Kelly RM (1995) Extremozymes: expanding the limits of biocatalysis. Nat Biotechnol 13(7):662–668

Al-Rasheedi S, Tawabini B, Nazal M, Khalil A (2022) Application of thermophilic bacteria for the treatment of municipal wastewaters. Arab J Sci Eng 47(6):6881–6888

Amin A, Ahmed I, Salam N, Kim B, Singh D, Zhi X, Xiao M, Li W (2017) Diversity and distribution of thermophilic bacteria in hot springs of Pakistan. Microb Ecol 74(1):116–127

Anu K, Kumari S, Choudhary G, Devi S (2024) Microbial diversity analysis of Chumathang geothermal spring, Ladakh, India. Braz J Microbiol 55(2):1545–1555

Arbab S, Ullah H, Khan MIU, Khattak MNK, Zhang J, Li K, Hassan IU (2021) Diversity and distribution of thermophilic microorganisms and their applications in biotechnology. J Basic Microbiol 62(2):95–108

Bhardwaj K, Tiwari S, Pandey A (2011) Conservation of thermophilic cyanobacterial diversity and physicochemical characteristics of thermal springs of Tapoban geothermal field, Himalayan province (India). Natl Acad Sci Lett 34:77–82

Bharwad K, Shekh S, Singh NK, Patel A, Joshi C (2023) Heterologous expression and biochemical characterization of novel multifunctional thermostable α-amylase from hot-spring metagenome. Int J Biol Macromol 242:124810

Bisht SS, Das NN, Tripathy NK (2011) Indian hot-water springs: a bird's eye view. J Energy Environ Carbon Credits 1(1):1–15

Bora L, Bora M (2012) Optimization of extracellular thermophilic highly alkaline lipase from thermophilic Bacillus sp isolated from Hotspring of Arunachal Pradesh, India. Braz J Microbiol 43(1):30–42

Breu TM, Hurni H (2003) The Tajik Pamirs: challenges of sustainable development in an isolated mountain region. Geographica Bernensia

Bruins ME, Janssen AEM, Boom RM (2001) Thermozymes and their applications: a review of recent literature and patents. Appl Biochem Biotechnol 90(2):155–186

Chatterjee S, Bhattacharjee I, Chandra G (2010) Biosorption of heavy metals from industrial waste water by *Geobacillus thermodenitrificans*. J Hazard Mater 175(1–3):117–125

Chen Y, Sun D, Zhou Y, Liu L, Han W, Zheng B, Wang Z, Zhang Z (2014) Cloning, expression and characterization of a novel thermophilic polygalacturonase from Caldicellulosiruptor bescii DSM 6725. Int J Mol Sci 15(4):5717–5729

Chien A, Edgar DB, Trela JM (1976) Deoxyribonucleic acid polymerase from the extreme thermophile Thermus aquaticus. J Bacteriol 127(3):1550–1557

Collins T, Gerday C, Feller G (2005) Xylanases, xylanase families and extremophilic xylanases. FEMS Microbiol Rev 29(1):3–23

Das S, Thakur N (2021) The microbial diversity of hot springs located in Himalayan geothermal belts (HGB). In: Extreme environments. CRC Press, pp 18–43

Davids T, Schmidt M, Böttcher D, Bornscheuer UT (2013) Strategies for the discovery and engineering of enzymes for biocatalysis. Curr Opin Chem Biol 17(2):215–220

Debnath T, Kujur RRA, Mitra R, Das SK (2019) Diversity of microbes in hot springs and their sustainable use. In: Microbial diversity in ecosystem sustainability and biotechnological applications. Springer, pp 159–186

DeCastro M, Escuder-Rodríguez J, Becerra M, Rodríguez-Belmonte E, González-Siso M (2021) Comparative metagenomic analysis of two hot springs from Ourense (Northwestern Spain) and others worldwide. Front Microbiol 12:769065

Dobreva E, Ivanova V, Emanuilova E (1994) Effect of temperature on some characteristics of the thermostable α-amylase from Bacillus licheniformis. World J Microbiol Biotechnol 10(5):547–550

Elleuche S, Schröder C, Sahm K, Antranikian G (2014) Extremozymes—biocatalysts with unique properties from extremophilic microorganisms. Curr Opin Biotechnol 29:116–123

Ellis JT, Magnuson TS (2012) Thermostable and alkalistable xylanases produced by the thermophilic bacterium Anoxybacillus flavithermus TWXYL3. ISRN Microbiol 2012:1–8
Ferris MJ, Nold SC, Santegoeds CM, Ward DM (2001) Examining bacterial population diversity within the octopus spring microbial mat community. In: Thermophiles biodiversity, ecology, and evolution. Springer, pp 51–64
Fuciños P, Atanes E, López-López O, Solaroli M, Cerdán ME, González-Siso MI, Pastrana L, Rúa ML (2014) Cloning, expression, purification and characterization of an oligomeric His-tagged thermophilic esterase from thermus thermophilus HB27. Process Biochem 49(6):927–935
Gao R, Feng Y, Ishikawa K, Ishida H, Ando S, Kosugi Y, Cao S (2003) Cloning, purification and properties of a hyperthermophilic esterase from archaeon Aeropyrum pernix K1. J Mol Catal B Enzym 24–25:1–8
Gogliettino M, Riccio A, Cocca E, Rossi M, Palmieri G, Balestrieri M (2014) A new pepstatin-insensitive thermopsin-like protease overproduced in peptide-rich cultures of Sulfolobus solfataricus. Int J Mol Sci 15(2):3204–3219
Guidry SA, Chafetz HS (2003) Anatomy of siliceous hot springs: examples from Yellowstone National Park, Wyoming, USA. Sediment Geol 157(1–2):71–106
Han T, Zeng F, Li Z, Liu L, Wei M, Guan Q, Liang X, Peng Z, Liu M, Qin J, Zhang S, Jia B (2013) Biochemical characterization of a recombinant pullulanase from Thermococcus kodakarensis KOD1. Lett Appl Microbiol 57(4):336–343
Haq NIU, Hussain Z, Khan MA, Muneer B, Afzal S, Majeed S, Akram F (2012) Kinetic and thermodynamic study of cloned thermostable endo-1,4-β-xylanase from Thermotoga petrophila in mesophilic host. Mol Biol Rep 39(7):7251–7261
Henne A, Brüggemann H, Raasch C, Wiezer A, Hartsch T, Liesegang H, Johann A, Lienard T, Gohl O, Martinez-Arias R, Jacobi C, Starkuviene V, Schlenczeck S, Dencker S, Huber R, Klenk H, Kramer W, Merkl R, Gottschalk G, Fritz H (2004) The genome sequence of the extreme thermophile Thermus thermophilus. Nat Biotechnol 22(5):547–553
Hochstein M, Regenauer-Lieb K (1998) Heat generation associated with collision of two plates: the Himalayan geothermal belt. J Volcanol Geotherm Res 83(1–2):75–92
Ibarra D, Monte MC, Blanco A, Martínez AT, Martínez MJ (2012) Enzymatic deinking of secondary fibers: cellulases/hemicellulases versus laccase-mediator system. J Ind Microbiol Biotechnol 39(1):1–9
Jain P, Reza HM, Pal S (2014) Molecular phylogenetic analysis of bacterial community and characterization of Cr(VI) reducers from the sediments of Tantloi hot spring, India. Aquatic Biosyst 10(1):7
Jardine J, Stoychev S, Mavumengwana V, Ubomba-Jaswa E (2018) Screening of potential bioremediation enzymes from hot spring bacteria using conventional plate assays and liquid chromatography—tandem mass spectrometry (Lc-Ms/Ms). J Environ Manag 223:787–796
Kaur R, Rajesh C, Sharma R, Boparai JK, Sharma PK (2018) Metagenomic investigation of bacterial diversity of hot spring soil from Manikaran, Himachal Pradesh, India. Ecol Genet Genomics 6:16–21
Kaushik S, Alatawi A, Djiwanti SR, Pande A, Skotti E, Soni V (2021) Potential of extremophiles for bioremediation. In: Microorganisms for sustainability. Springer, pp 293–328
Kikani BA, Kourien S, Rathod U (2019) Stability and thermodynamic attributes of starch hydrolyzing α-amylase of Anoxybacillus rupiensis TS-4. Stärke/Starch 72(1–2):1900105
Kim JW, Flowers LO, Whiteley M, Peeples TL (2001) Biochemical confirmation and characterization of the family-57-like α-amylase ofMethanococcus jannaschii. Folia Microbiol 46(6):467–473
Knapik K, Becerra M, González-Siso M (2019) Microbial diversity analysis and screening for novel xylanase enzymes from the sediment of the Lobios hot spring in Spain. Sci Rep 9(1):11195
Kublanov IV, Perevalova AA, Slobodkina GB, Lebedinsky AV, Bidzhieva SK, Kolganova TV, Kaliberda EN, Rumsh LD, Haertlé T, Bonch-Osmolovskaya EA (2009) Biodiversity of thermophilic prokaryotes with hydrolytic activities in Hot Springs of Uzon Caldera, Kamchatka (Russia). Appl Environ Microbiol 75(1):286–291

Kumar RM, Jani K, Parvathi JR, Thomas BM, Raja SSS, Pandey A, Sharma A (2022) Bacterial diversity of geochemically distinct hot springs located in Maharashtra, India. Arch Microbiol 204(1):110

Kuzu SB, Güvenmez HK, Denizci AA (2012) Production of a thermostable and alkaline chitinase by Bacillus thuringiensis subsp. kurstaki strain HBK-51. Biotechnol Res Int 2012:1–6

La E, Ig L (1975) Distribution of highly thermophilic, nonsporulating bacteria in the hot springs of Tadzhikistan. Mikrobiologiia 44(5):938–942

Li D, Li X, Dang W, Tran PL, Park S, Oh B, Hong W, Lee J, Park K (2013) Characterization and application of an acidophilic and thermostable β-glucosidase from thermofilum pendens. J Biosci Bioeng 115(5):490–496

Liao S, Liang G, Zhu J, Lu B, Peng L, Wang Q, Wei Y, Zhou G, Huang R (2019) Influence of calcium ions on the thermal characteristics of α-amylase from thermophilic Anoxybacillus sp. GXS-BL. Protein Peptide Lett 26(2):148–157

Liew KJ, Shahar S, Shamsir MS, Shaharuddin NB, Liang CH, Chan K, Pointing SB, Sani RK, Goh KM (2024) Integrating multi-platform assembly to recover MAGs from hot spring biofilms: insights into microbial diversity, biofilm formation, and carbohydrate degradation. Environ Microbiome 19(1):29

Macelroy RD (1974) Some comments on the evolution of extremophiles. Biosystems 6(1):74–75

Machius M, Declerck N, Huber R, Wiegand G (1998) Activation of Bacillus licheniformis α-amylase through a disorder→order transition of the substrate-binding site mediated by a calcium–sodium–calcium metal triad. Structure 6(3):281–292

Majumdar R, Majumdar N, Mukherjee A (2000) Geoelectric investigations in Bakreswar geothermal area, West Bengal, India. J Appl Geophys 45(3):187–202

Mantiri FR, Rumende RRH, Sudewi S (2019) Identification of α-amylase gene by PCR and activity of thermostable α-amylase from thermophilic Anoxybacillus thermarum isolated from Remboken hot spring in Minahasa, Indonesia. IOP Conf Ser Earth Environ Sci 217:012045

Matsumura M, Signor G, Matthews BW (1989) Substantial increase of protein stability by multiple disulphide bonds. Nature 342(6247):291–293

Menzel P, Gudbergsdóttir SR, Rike AG, Lin L, Zhang Q, Contursi P, Moracci M, Kristjansson JK, Bolduc B, Gavrilov S, Ravin N, Mardanov A, Bonch-Osmolovskaya E, Young M, Krogh A, Peng X (2015) Comparative metagenomics of eight geographically remote terrestrial hot springs. Microb Ecol 70(2):411–424

Merkel AY, Pimenov NV, Rusanov II, Slobodkin AI, Slobodkina GB, Tarnovetckii IY, Frolov EN, Dubin AV, Perevalova AA, Bonch-Osmolovskaya EA (2016) Microbial diversity and autotrophic activity in Kamchatka hot springs. Extremophiles 21(2):307–317

Meyer-Dombard DR, Amend JP (2014) Geochemistry and microbial ecology in alkaline hot springs of Ambitle Island, Papua New Guinea. Extremophiles 18(4):763–778

Mir MY, Hamid S, Rohela GK, Parray JA, Kamili AN (2021) Composting and bioremediation potential of thermophiles. In: Soil bioremediation: an approach towards sustainable technology. Wiley, pp 143–174

Mirete S, Morgante V, González-Pastor JE (2016) Functional metagenomics of extreme environments. Curr Opin Biotechnol 38:143–149

Moser MJ, DiFrancesco RA, Gowda K, Klingele AJ, Sugar DR, Stocki S, Mead DA, Schoenfeld TW (2012) Thermostable DNA polymerase from a viral metagenome is a potent RT-PCR enzyme. PLoS One 7(6):e38371

Moshfegh M, Shahverdi AR, Zarrini G, Faramarzi MA (2013) Biochemical characterization of an extracellular polyextremophilic α-amylase from the halophilic archaeon Halorubrum xinjiangense. Extremophiles 17(4):677–687

Nisha M, Satyanarayana T (2013) Recombinant bacterial amylopullulanases. Bioengineered 4(6):388–400

Niyatbekov T, Barinova S (2018a) Bioindication of water properties by algal communities in the Pamir High Mountain mineral and thermal springs. Transylvanian Rev Syst Ecol Res 20(3):1–30

Niyatbekov T, Barinova SS (2018b) Bioindication of aquatic habitats with diatom algae in the Pamir Mountains, Tajikistan. MOJ Ecol Environ Sci 3(3):117–120

Panda AK, Bisht SS, Kumar NS, De Mandal S (2015) Investigations on microbial diversity of Jakrem hot spring, Meghalaya, India using cultivation-independent approach. Genomics Data 4:156–157

Pandey A, Dhakar K, Sharma A, Priti P, Sati P, Kumar B (2014) Thermophilic bacteria that tolerate a wide temperature and pH range colonize the Soldhar (95 °C) and Ringigad (80 °C) hot springs of Uttarakhand, India. Ann Microbiol 65(2):809–816

Park SH, Park KH, Oh BC, Alli I, Lee BH (2011) Expression and characterization of an extremely thermostable β-glycosidase (mannosidase) from the hyperthermophilic archaeon Pyrococcus furiosus DSM3638. New Biotechnol 28(6):639–648

Pentecost A, Bayari S, Yesertener C (1997) Phototrophic microorganisms of the Pamukkale travertine, Turkey: their distribution and influence on travertine deposition. Geomicrobiol J 14(4):269–283

Poddar A, Das SK (2017) Microbiological studies of hot springs in India: a review. Arch Microbiol 200(1):1–18

Pradhan B, Jena R (2016) Spatial relationship between earthquakes, hot-springs and faults in Odisha, India. IOP Conference Series Earth and Environmental Science 37:012070

Rahkamo L, Siika-Aho M, Viikari L, Leppänen T, Buchert J (1998) Effects of cellulases and hemicellulase on the alkaline solubility of dissolving pulps. Holzforschung 52(6):630–634

Ranjit M (2022) Hot springs in Nepal. Springer eBooks

Rao MPN, Dong Z, Luo Z, Li M, Liu B, Guo S, Hozzein WN, Xiao M, Li W (2021) Physicochemical and microbial diversity analyses of Indian Hot Springs. Front Microbiol 12:627200

Rawat G, Bartarya SK, Singh B, Bhasin RK (2020) Geophysical characterization of Chumathang (Ladakh) hot spring. In: Springer geophysics. Springer, pp 363–376

Rhee J, Ahn D, Kim Y, Oh J (2005) New thermophilic and thermostable esterase with sequence similarity to the hormone-sensitive lipase family, cloned from a metagenomic library. Appl Environ Microbiol 71(2):817–825

Rizk M, Antranikian G, Elleuche S (2012) End-to-end gene fusions and their impact on the production of multifunctional biomass degrading enzymes. Biochem Biophys Res Commun 428(1):1–5

Royter M, Schmidt M, Elend C, Höbenreich H, Schäfer T, Bornscheuer UT, Antranikian G (2009) Thermostable lipases from the extreme thermophilic anaerobic bacteria Thermoanaerobacter thermohydrosulfuricus SOL1 and Caldanaerobacter subterraneus subsp. tengcongensis. Extremophiles 13(5):769–783

Sahay H, Yadav AN, Singh AK, Singh S, Kaushik R, Saxena AK (2017) Hot springs of Indian Himalayas: potential sources of microbial diversity and thermostable hydrolytic enzymes. 3 Biotech 7(2):118

Samanta D, Govil T, Saxena P, Thakur P, Narayanan A, Sani RK (2022) Extremozymes and their applications. In: Extremozymes and their industrial applications. Elsevier, pp 1–39

Sarian FD, Leemhuis H, Nurachman Z, Van Der Maarel MJEC, Dessy N (2023) Utilization of Indonesian root and tuber starches for glucose production by cold enzymatic hydrolysis. Biologia 78(7):1889–1897

Satyanarayana T, Noorwez S, Kumar S, Rao J, Ezhilvannan M, Kaur P (2004) Development of an ideal starch saccharification process using amylolytic enzymes from thermophiles. Biochem Soc Trans 32(2):276–278

Saxena A, Chauhan PS (2016) Role of various enzymes for deinking paper: a review. Crit Rev Biotechnol 37(5):598–612

Scoon RN (2021) Correction to: the GeoTraveller. Springer

Sharma A, Jani K, Shouche YS, Pandey A (2014) Microbial diversity of the Soldhar hot spring, India, assessed by analyzing 16S rRNA and protein-coding genes. Ann Microbiol 65(3):1323–1332

Sharma A, Paul D, Dhotre D, Jani K, Pandey A, Shouche YS (2017) Deep sequencing analysis of bacterial community structure of Soldhar hot spring, India. Microbiology 86(1):136–142
Shi H, Zhang Y, Li X, Huang Y, Wang L, Wang Y, Ding H, Wang F (2013) A novel highly thermostable xylanase stimulated by Ca^{2+} from Thermotoga thermarum: cloning, expression and characterization. Biotechnol Biofuels 6(1):26
Simsek Ş (2022) Karstic hot water aquifers in Turkey. In: Cave and karst systems of the world. Springer, pp 115–118
Singh Y, Gulati A, Singh D, Khattar J (2018) Cyanobacterial community structure in hot water springs of Indian North-Western Himalayas: a morphological, molecular and ecological approach. Algal Res 29:179–192
Srivastava R, Baruah JN (1986) Culture conditions for production of thermostable amylase by Bacillus stearothermophilus. Appl Environ Microbiol 52(1):179–184
Staufenberger T, Imhoff JF, Labes A (2012) First crenarchaeal chitinase found in Sulfolobus tokodaii. Microbiol Res 167(5):262–269
Summons R, Meyer-Dombard D, Bradley A, Dibbell A, Fredricks H, Hinrichs K-U, Jahnke L, Shock E, Amend J (2005) Molecular studies of filamentous and biofilm-forming hyperthermophilic communities in Yellowstone National Park. AGU fall meeting abstracts
Sunna A, Moracci M, Rossi M, Antranikian G (1997) Glycosyl hydrolases from hyperthermophiles. Extremophiles 1(1):2–13
Suvd D, Fujimoto Z, Takase K, Matsumura M, Mizuno H (2001) Crystal structure of Bacillus stearothermophilus alpha-amylase: possible factors determining the thermostability. J Biochem 129(3):461–468
Tolessa D (2020) Historical and anthropological study of natural hot springs: healing ways and its role to health tourism, the case of Oromia Special Zone, Amhara Regional State Ethopia. Am Res J Hist Cult 6(1):1–8
Toplak A, Wu B, Fusetti F, Quaedflieg PJLM, Janssen DB (2013) Proteolysin, a novel highly thermostable and cosolvent-compatible protease from the thermophilic bacterium coprothermobacter proteolyticus. Appl Environ Microbiol 79(18):5625–5632
Verma J, Sourirajan A, Dev K (2022) Bacterial diversity in 110 thermal hot springs of Indian Himalayan Region (IHR). 3 Biotech 12(9):238
Vogt G, Woell S, Argos P (1997) Protein thermal stability, hydrogen bonds, and ion pairs. J Mol Biol 269(4):631–643
Yadav G, Bharindwal S, Mehrotra A, Kumari S, Roohi N, Jobby R (2024) Implications of polyextremophiles in astrobiology research. In: Trends in biotechnology of polyextremophiles. Springer, pp 119–149
Yuuki T, Nomura T, Tezuka H, Tsuboi A, Yamagata H, Tsukagoshi N, Udaka S (1985) Complete nucleotide sequence of a gene coding for heat- and pH-stable α-amylase of Bacillus licheniformis: comparison of the amino acid sequences of three bacterial liquefying α-amylases deduced from the DNA sequences. J Biochem 98(5):1147–1156
Zahoor S, Javed MM, Babarl ME (2016) Characterization of a novel hydrolytic enzyme producing thermophilic bacterium isolated from the hot spring of Azad Kashmir-Pakistan. Braz Arch Biol Technol 59:e16150662
Zhao L, Xie J, Zhang X, Cao F, Pei J (2013) Overexpression and characterization of a glucose-tolerant β-glucosidase from Thermotoga thermarum DSM 5069T with high catalytic efficiency of ginsenoside Rb1 to Rd. J Mol Catal B Enzym 95:62–69
Zimik HV, Farooq SH, Prusty P (2017) Geochemical evaluation of thermal springs in Odisha, India. Environ Earth Sci 76(17):593

Geomicrobes in Transformation of Arsenic

Nirmala Akoijam and Santa Ram Joshi

Contents

Abstract Arsenic contamination of the environment (soil and water) has several causes and is a long-standing problem in many areas of the world. The anthropogenic additions of arsenic through activities like mining and rampant use of arsenic-containing chemical fertilizers in agricultural activities greatly outweigh that of the natural flux from rock, sediments, and so on. This, in part, causes its leaching into groundwater and agricultural soils. The health hazards (including cancers of skin, lungs) associated with arsenic contamination are numerous and mostly prove fatal. This metalloid makes its way to the human body more commonly than expected through the consumption of contaminated water and food, as is reported in countries like Bangladesh, Vietnam, and Germany. In the environ-

N. Akoijam · S. R. Joshi (✉)
Microbiology Laboratory, Department of Biotechnology and Bioinformatics, North-Eastern Hill University, Shillong, Meghalaya, India

S. R. Joshi et al. (eds.), *Mineral Transformation and Bioremediation by Geo-Microbes*, Interdisciplinary Biotechnological Advances,
https://doi.org/10.1007/978-981-96-3033-2_4

ment, geochemical cycling of arsenic is driven both by biotic and abiotic elements. Speciation and solubility of arsenic species in soil and water greatly depend on pH and redox potential, which are in turn influenced by the complex processes of biotic origins. This chapter focuses on the microbial role in arsenic speciation and transformation in both water and soil and the molecular machineries that are essential for these processes. We also discuss various existing remediation measures and future bioremediation strategies that can be investigated based on the natural microbial mechanisms.

Keywords Arsenic · Soil · Groundwater · Microorganisms · Microbes · Bioremediation

Abbreviations

As	Arsenic
AST	Arsinothricin
CAOs	Chemolithoautotrophic arsenite oxidizers
CBS domain	Cystathionine-β-synthase domain
DARPs	Dissimilatory arsenate-reducing prokaryotes
DMA	Dimethyl arsine
DMSO	Dimethylsulfoxide
GMOs	Genetically modified organisms
HAOs	Heterotrophic arsenite oxidizers
MMA	Monomethyl arsine
MSMA	Monosodium methanearsonate
NIPs	Nodulin 26-like intrinsic membrane proteins
TMA	Trimethyl arsine
USEPA	United States Environmental Protection Agency

1 Introduction

Arsenic (As) is a ubiquitous metalloid that naturally occurs in the Earth's crust, commonly associated with elements like sulfur and iron in mineral form. However, anthropogenic activities like mining and the use of synthetic agricultural inputs have mobilized arsenic, leading to its leaching into groundwater and soil (Mandal and Suzuki 2002). Arsenic is predominantly found in soil types like peaty, mudstone, phyllite, marine shale, and slate. Its natural distribution is also linked to global tectonic activity, with arsenic-affected regions mainly located in sedimentary basins near recently formed mountain systems, such as the Himalayas and their deltaic areas, including Bangladesh and West Bengal, India (Shaji et al. 2021). It is usually found in high concentrations in sulfide deposits in the form of sulfides, arsenides, and sulfosalts. Additionally, inorganic arsenic in groundwater has been reported in numerous countries, including Argentina, Cambodia, Chile, China, Mexico, Pakistan, the United States, Vietnam, and Germany (Shaji et al. 2021). Elevated levels of arsenic are also reported in some groundwaters of active geothermal areas and in some waters of high salt content (Grimmett and Mcintosh 1939; White et al. 1963).

Some environmentally common arsenic minerals include arsenolite and claudetite, which are oxides of arsenic. Iron-associated minerals include arsenopyrite, löllingite, and scorodite, while haidingerite, hörnesite, pharmacolite, picropharmacolite, and weilite are arsenate minerals associated with calcium and magnesium (Drahota and Filippi 2009). The leaching of arsenic from these minerals occurs when pH conditions deviate from neutral (pH7). Arsenic exhibits greater mobility in soil under acidic (pH4) and alkaline (pH10) conditions, with acidic conditions having a more pronounced effect (Shaw 2006).

Although arsenic is notorious for its toxicity, its historical use as a medicinal agent is noteworthy. For instance, the arsenical salvarsan, developed by Paul Ehrlich in the early 1900s, was used to treat syphilis (Paul et al. 2023). Similarly, other arsenicals were employed in treating diseases like trypanosomiasis, also known as sleeping sickness. Notably, melarsoprol, an arsenical, is still used for treating advanced stages of human African trypanosomiasis (Rodgers 2009).

The valence states of arsenic species significantly influence their toxicity profiles. Inorganic trivalent arsenites are more toxic than pentavalent arsenates due to their higher solubility, which leads to rapid absorption in the human body (Squibb and Fowler 1983). The most toxic form of arsenic is arsine gas, characterized by its garlic odor. Arsine is extremely flammable and can severely affect kidney tissues, with acute exposure potentially causing renal failure. Conversely, organic arsenicals, such as arsenobetaine found in marine organisms, are less toxic because their arsenic atoms are not readily available to bond with proteins and other crucial biomolecules (Sakurai 2002).

Geomicrobiology explores the intricate relationship that exists between microbial life and the lithosphere. Although microbes are ubiquitous in nature, the habitats that they prevail in are extraordinarily very specific. This has garnered significant attention to understand the biological and physical interactions that influence elemental cycling, locally and globally (Robbins and Konhauser 2021). Microbes that have evolved to withstand environmental conditions like extreme temperature, pressure, and pH are collectively known as extremophiles (Thakur et al. 2022). Based on their nutritional and metabolic needs, microbes transform inorganic and organic compounds, releasing wastes that act as substrates for other organisms, microbes, or otherwise. In this way, there is a continuous recycling of elements that shape the dynamics of microbial communities that further control the composition of various habitats across the entire planet (DasSarma et al. 2020).

In metal-rich environments, microbes play a role in precipitating soluble metals by forming insoluble but non-crystalline metal species as in the case of biomineralization of magnetite from iron (Raab and Feldmann 2003). In another instance, a bacterium, *Desulfotomaculum auripigmentum*, precipitates arsenic trisulfide by respirative reduction of arsenate to arsenite and sulphate to sulfide (Newman et al. 1997a, b). Due to their dynamic chemical properties, inorganic arsenic compounds interact with various living organisms in addition to microorganisms, including humans, plants, and fish. Some of these organisms can even utilize arsenic for essential biological processes. This chapter summarizes the interaction of arsenic with different living organisms, with a special focus on the role of bacterial mechanisms in the transformation and interconversion of arsenic species in the environment, as well as the numerous factors that influence them and possible bioremediation strategies that can be developed for alleviating arsenic contamination in the environment.

2 Chemistry of Arsenic in the Environment

Understanding arsenic chemistry in the environment is crucial for developing effective remediation and treatment technologies, preserving ecosystem integrity, enhancing regulatory policies, and advancing scientific knowledge. As arsenic primarily occurs in soil and groundwater, this section will explore its various forms and the biotic and abiotic factors that influence its interactions with other metals and living organisms.

2.1 Arsenic in Groundwater and Soil

The inorganic trivalent and pentavalent species of arsenic are the two most dominant arsenic species in both groundwater and soil due to their high solubility (Vaughan 1993). Factors such as pH and redox conditions influence the partitioning of arsenic between the solid and aqueous phase as well as interconversion between the two inorganic species, thereby influencing arsenic mobility and bioavailability in soil (Al-Abed et al. 2007; Chen et al. 2017b; Masscheleyn et al. 1991; Shaw 2006). In natural waters, hydrated ferric oxides provide a positive surface charge for the binding of negatively charged arsenic compounds (Bissen and Frimmel 2003).

In soil solutions, arsenic exists as oxyanionic acids, viz., arsenous acid (H_3AsO_3) contains arsenite, and arsenic acid (H_3AsO_4) contains arsenate. Arsenous acid is a weak acid with pKa values of 9.23, 12.13, and 13.4, while arsenic acid has pKa values of 2.22, 6.98, and 11.53 (Goldberg 2002). At high soil redox levels, arsenic solubility is low, and the predominant species is arsenate; in alkaline conditions, there is substantial reduction of arsenate to arsenite, which results in increased mobility of arsenic in solution. Since arsenic is immobilized in the complexes of iron oxyhydroxides, their solubility is highly dependent on the dissolution of the latter, which occurs at moderately reduced conditions (Ascar et al. 2008;

Masscheleyn et al. 1991; Signes-Pastor et al. 2007). In natural aquatic conditions (circumneutral pH and reducing conditions), the mobility of arsenite is higher than that of arsenate due to its uncharged state and subsequent poor adsorption on most mineral sediments (Gulens et al. 1973). In 1997, Manning et al. demonstrated that arsenite was fairly stable between pH 4 and pH 9, and its oxidation to arsenate occurred in solutions at or above pH 9.2 (Manning and Goldberg 1997). Another interesting observation was reported in several studies, whereby bicarbonate ions enhanced arsenic leaching into solution. This is a consequence of the formation of carbonate complexes on the surface of iron hydroxide, which substitutes arsenic from mineral and sediment surfaces and releases arsenic to groundwater (Anawar et al. 2004; Kim et al. 2000; Saalfield and Bostick 2010; Stachowicz et al. 2007).

Arsenates bind to aluminum hydroxides and aluminosilicate clay minerals more strongly than arsenites (Kumar et al. 2016; Xu et al. 1988). Similarly, minerals like birnessite containing manganese oxides have a high affinity for arsenic, and most of it is found in the pentavalent form as Mn (III/IV) being a good oxidizing agent that converts arsenite to arsenate readily (Manning et al. 2002). Several studies have also emphasized the role of soil organic matter (SOM) in the solid-solution phase partitioning of arsenic. Organic matter amendments such as biochar in soil enhance microbial reduction of iron (Fe) oxyhydroxide as well as pentavalent arsenate adsorbed on Fe oxides, resulting in the release of the trivalent inorganic species and increasing its bioavailability (Wang et al. 2017b). However, an interesting observation was made by Wang et al. (2024), wherein biochar-supported zero-valent iron (ZVI/BC) reduced the bioavailability of arsenite in soil. Removal of the SOM biochar in fact resulted in increased arsenate reduction due to the proliferation of arsenic- and iron-reducing bacteria, such as *Bacillus* and *Ammoniphilus*.

3 Implications of Arsenic in Living Systems

Arsenic poses a serious threat to living systems due to its toxic, carcinogenic, and disruptive effects on various biological processes. Understanding these implications is crucial for formulating effective public health strategies to reduce exposure and avert the associated adverse health effects.

1. **Arsenic bioaccumulation in mammals**

   In humans, arsenic induces oxidative stress, causing DNA damage and lipid peroxidation. The severity of these effects depends on the individual's antioxidant and nutrient levels. Arsenic inhibits selenoenzymes, such as thioredoxin reductase and glutathione peroxidase, which are special systems that alleviate stress in the placenta. Selenium deficiency leads to the buildup of arsenic in the brain and maternal liver in mice (Miyazaki et al. 2005).

   In a study by Lanz Jr. et al. (1950), it was observed that hemoglobin in rat blood retained most of the ingested arsenic, with minimal retention in plasma proteins. Nanoelectrospray mass spectrometry later revealed that most inorganic arsenic bound to the α chain of hemoglobin (Lu et al. 2004). However, this accumulation was specific to rats. In cattle, inorganic trivalent arsenic was rapidly excreted in urine, with only pentavalent species retained in tissues for a short duration (Buck 1970; Moxham and Coup 1968).

2. **Arsenic bioaccumulation in aquatic organisms**

   Marine organisms such as phytoplankton, microalgae, and bacteria can reduce arsenate to arsenite and excrete methylated forms that are much less toxic than inorganic arsenic. In contrast, higher marine organisms tend to accumulate organic forms of arsenic, such as arsenobetaine, in their tissues, with limited absorption of inorganic arsenic. A similar pattern is observed in freshwater environments, where lower trophic level organisms like cyanobacteria, phytoplankton, stoneflies, and snails exhibit higher concentrations of arsenic compared to higher trophic level organisms like fish (Burger et al. 2002; Byeon et al. 2021; Spehar et al. 1980; Zhang et al. 2013). Understanding the dynamics of arsenic flow within marine and freshwater organisms is essential for their potential use as bioindicators of water contamination.

3. **Arsenic bioaccumulation in plants**

   Inorganic arsenic species can penetrate the plant cuticle and enter the apoplast system. Trivalent arsenic, in particular, interacts with sulfhydryl enzymes in plants, causing wilting due to loss of turgor from weakened membrane integrity (Abbas et al. 2018). This phytotoxic property of arsenicals has been exploited in developing herbicides, such as monosodium methanearsonate (MSMA), which the United States Environmental Protection Agency (USEPA) permitted for use on sod farms, golf courses, and cotton fields. However, subsequent amendments have limited its use (EPA 2013).

   Arsenates and arsenites are often transported into plants via existing nutrient transporters. For example, the chemical similarity between arsenate and phosphate, and between arsenite and silicate, allows them to be easily transported from soil solutions across plant barriers (Bucher 2007; Ma and Yamaji 2006; Ma et al. 2007; Meharg et al. 1994). Arsenite transport is also mediated by aquaporin family proteins such as nodulin 26-like intrinsic membrane proteins (NIPs). In rice and *Arabidopsis thaliana*, NIPs like OsNIP and AtNIP have been identified as efficient arsenite influx proteins in root cells (Isayenkov and Maathuis 2008; Xu et al. 2015).

   Plants in the Pteridaceae family, particularly those in the genus *Pteris*, such as *Pteris vittata*, *P. cretica*, *P. umbrosa*, *P. longifolia*, *P. biaurita*, and *P. oshimensis*, are natural hyperaccumulators of arsenic (Ma et al. 2001; Srivastava et al. 2010; Wang et al. 2007; Zhao et al. 2002).

4 Geochemical Interactions of Arsenic with Metals

The geochemical interactions of arsenic with metals are multifaceted as they are influenced by several environmental factors, like pH, redox conditions, and microbial activity. These interactions are key to understanding the mobility and availability and to strategizing remediation measures for arsenic contamination. Arsenic is majorly found adsorbed on clay minerals and oxides of metals like iron (Fe), manganese (Mn), and aluminum (Al). The arsenate species bind to the oxides more readily than arsenite, which makes the latter more mobile in the environment (Panagiotaras and Nikolopoulos 2015). Studies have shown that the concentration of arsenic and other metals increases with decreasing particle size in soils from abandoned mine sites. This finding has significant implications for designing bioremediation measures, as the removal efficiencies of the contaminants are higher in coarse soil particles (Bari et al. 2020).

Microbial activity plays a key role in the biogeochemical cycling of arsenic. Microbes act on As-bearing Fe(III) oxides by reductive dissolution and enable the release of arsenic from sediments into groundwater. Figure 1 is an overview of the redox reactions that take place on the surface of arsenic-containing minerals when acted upon by microbes.

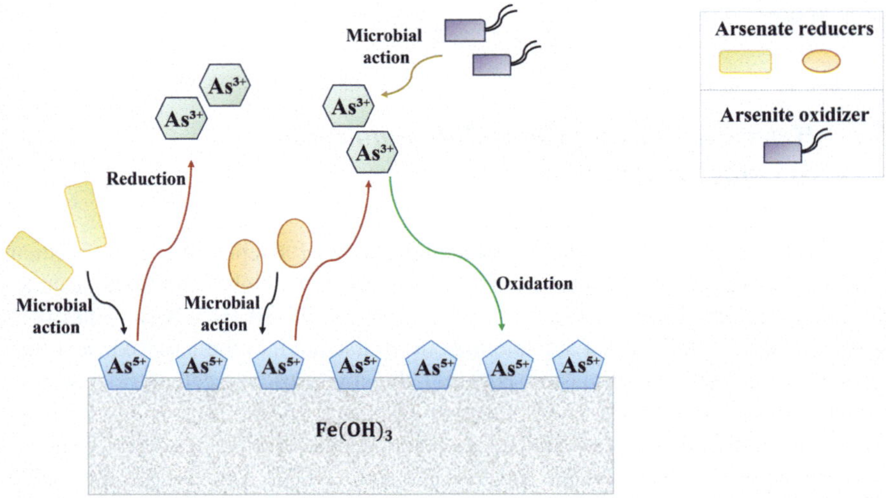

Fig. 1 Oxides of Fe with adsorbed arsenates (As^{5+}) when acted upon by arsenate-reducing microbes trigger the release of arsenites (As^{3+}), which get released into the environment unless acted upon by arsenite oxidizers, which get adsorbed on the surface of the iron hydroxides again

5 Microbial Roles in Arsenic Transformation

Despite the apparent toxicity, microorganisms play an active part in the dissolution of arsenic-bearing minerals, suggesting the presence of arsenic-evading mechanisms in the microbial cells. Dissolving minerals releases metal species from their crystal lattices that are either used as sources of energy (e.g., Fe^{2+} or S^{2-}) or as terminal electron acceptors (e.g., O_2, Fe^{3+}, SO_4^{2-}) (Lukasz et al. 2014; Stolz and Oremland 1999). Microbes are capable of physically inducing mineral dissolution by the secretion of extracellular polymeric substances (EPS), which serve as a matrix for microbial colonization. Since EPS retains water fairly effectively, this also contributes to mineral fracturing, weathering, and metal leaching (Barker et al. 1997; Tourney and Ngwenya 2014; Welch et al. 1999).

Apart from EPS, microbially produced low molecular mass organic acids like acetic, oxalic, and citric acid also aid in the dissolution of minerals that are complexed with arsenic (Ash et al. 2016; Collins 2004). Low molecular, high-affinity compounds like siderophores, previously known to be iron-binding, have also demonstrated affinity for other metals and metalloids like Mn, Mg, Cu, Ni, Zn, Cr(III), Pb, Cd, Sn and, most importantly, As. However, siderophores contribute to arsenite mobility by complexing with arsenates, which are then reduced by dissimilatory arsenate reducers like *Bacillus*, *Shewanella*, *Citrobacter*, *Sulfurospirillum*, and *Desulfomicrobium* species (Cummings et al. 1999; Uhrynowski et al. 2017; Zobrist et al. 2000).

5.1 Heterotrophic and Chemolithoautotrophic Arsenite Oxidizers

Most bacteria are able to evade the damaging effects of arsenic on account of their intrinsic detoxification pathways by oxidizing the highly soluble and toxic arsenite. These bacteria are classified as heterotrophic arsenite oxidizers (HAOs), and they possess metabolic pathways that are able to convert arsenite into the less toxic arsenate via surface interaction (periplasmic enzyme), thereby restricting cell entry of arsenic (Oremland and Stolz 2003). Gram-negative *Comamonads* and *Hydrogenophaga* sp. have been reported in a few studies as heterotrophic arsenite oxidizers (Garcia-Dominguez et al. 2008; vanden Hoven and Santini 2004). Chemolithoautotrophic arsenite oxidizers (CAOs), on the other hand, comprised of either aerobic or anaerobic bacteria, depend only on carbon dioxide as their C-source (carbon source) and reduce arsenic-containing minerals, such as realgar, arsenopyrite, orpiment, and enargite with the help of terminal electron acceptors (organic sources) to produce ATP. Arsenite acts as an electron donor (e^- donor) in cellular respiration while reducing either oxygen or nitrate (Shi et al. 2020). Haloalkaliphilic *Ectothiorhodospira* MLHE-1 isolated from an anoxic bottom water in Lake Mono, California, demonstrated the conversion of supplemented nitrate into nitrite almost entirely complemented by the oxidation of arsenite into arsenate (Oremland et al. 2002). The following equation represents the electron transfer between arsenite and nitrate:

$$H_2AsO_3^- + NO_3^- \rightarrow H_2AsO_4^- + NO_2^- \left(\Delta G_0' = -87.16\,\text{kJ / mol}\right)$$

Dong et al. (2014) presented the facultative chemolithoautotrophic properties of *Sinorhizobium meliloti* KGO-5, which could efficiently oxidize arsenite in both autotrophic and heterotrophic conditions. The periplasmic arsenite oxidase enzyme was purified from the mentioned bacterium NT-26, which belongs to the *Agrobacterium/Rhizobium* branch of Class α-Proteobacteria and partially characterized in a study which revealed the presence of two heterodimeric subunits, AroA (98 kDa) and AroB (14 kDa) and also molybdenum and iron atoms in the subunits (Santini et al. 2000; Santini and vanden Hoven 2004). Another gram-negative bacterium *Ancylobacter* sp. TS-1 originally isolated from a high Andean watershed in northern Chile and analyzed at molecular, kinetic, and phylogenetic levels revealed its chemolithoautotrophic arsenite-oxidizing nature (Anguita et al. 2018). Table 1 lists several bacteria and their respective arsenite oxidase genes as reported in different studies.

5.2 Dissimilatory Arsenate-Reducing Prokaryotes

Similar to the arsenite-oxidizing detoxification mechanism, bacteria are also capable of removing arsenate from their cell. Arsenate, which is a phosphate analogue, is taken up into the bacterial cell via specialized phosphate transporters present on the cell membrane. In the cytoplasm, arsenate gets reduced to arsenite by the arsenate reductase enzyme followed by extrusion via the arsenite efflux pump (Rosen 2002). But certain anaerobes utilize arsenate as a terminal electron acceptor while simultaneously oxidizing inorganic and organic donors (such as lactate, acetate, ethanol, glycerol, pyruvate) to generate energy. Such bacteria are termed as dissimilatory arsenate-reducing prokaryotes (DARPs); they are phylogenetically diverse and also widespread in the natural environment (Kumari and Jagadevan 2016) (Table 2). The respiratory arsenate reductase (Arr) is a heterodimeric, periplasmic enzyme with subunits containing an Fe-S cluster, a characteristic of dimethylsulfoxide (DMSO) reductase family of enzymes (Oremland and Stolz 2003). The bacterium utilizes arsenate as an electron acceptor instead of oxygen, thus making it an anaerobic process, *Chrysiogenes arsenatis* uses acetate as an e^- donor to transform arsenate to arsenite catalyzed by the respiratory arsenate reductase (Arr), which is a heterodimer $\alpha_1\beta_1$ (87 and 29 kDa subunits) with a native molecular mass of 123 kDa (Krafft and Macy 1998). The Arr enzyme was also purified from a Gram-positive, haloalkaliphilic bacterium *Bacillus selenitireducens* strain MLS10, from Mono Lake, USA (Switzer Blum et al. 1998). Unlike the *C. arsenatis*, the respiratory arsenate reductase enzyme from *B. selenitireducens* is well bound to the cytoplasmic

Table 1 List of arsenite oxidizers from different environmental sources

Isolate	Source environment	Arsenite oxidation type	Gene	GenBank accession no.	References
Zoogloea sp. str. ULPAs1	Aquatic	Chemoautotrophic	aoxA and aoxB	AF509588	Muller et al. (2003), Weeger et al. (1999)
Ancylobacter sp. str. S-1	Soil/ sediment	Chemoautotrophic	aroAB	EF015460/ EF015461	Rhine et al. (2007)
Thiobacillus sp. str. S1	Soil/ sediment	Chemoautotrophic	aroAB	EF015459	
Hydrogenophaga sp. str. CL3	Petroleum refinery	Chemoautotrophic	aroAB	EF015462	
Bosea sp. str. WAO	Shale	Chemoautotrophic	aroAB	nd	
Azoarcus sp. str. DAO1	Industrial Soil	Heterotrophic	nd	nd	
Sinorhizobium sp. str. DAO10	Industrial Soil	Chemoautotrophic	aroAB	EF015459	
Sinorhizobium sp. str. M14	Gold mine	Heterotrophic, facultative chemoautotrophic	aox	nd	Drewniak et al. (2008)
Alkalilimnicola ehrlichii str. MLHE-1	Aquatic	Chemoautotrophic	arxA	nd	Zargar et al. (2010)
Polaromonas sp. str. GM1	Sub-arctic gold mine	Heterotrophic	aroA	EU106602	Osborne et al. (2010)
Variovorax sp. str. MM-1	Soil near lead smelter	Heterotrophic	aioA	JQ306331	Bahar et al. (2013)
Azoarcus sp. str. CIB	nd	Chemoheterotrophic	arxA	nd	Durante-Rodríguez et al. (2019)

nd not defined

membrane, and further investigation revealed the presence of metal-binding motifs (i.e., iron–sulfur cluster), which also classifies it as a part of the DMSO reductase family (Afkar et al. 2003).

5.3 Microbial Methylation and Volatilization of Arsenic

Microbes are also able to transform inorganic arsenic compounds into organic arsenic species through methylation. This process was first reported in the fungus *Penicillium brevicaule*, which produced the volatile organoarsenical compound, trimethylarsine (also known as the Gosio gas) when it was allowed to grow in the

Table 2 List of dissimilatory arsenate-reducing prokaryotes

Isolate	Source	Electron donor	Gene name	Location of enzyme	References
Desulfotomaculum auripigmentum str. OREX-4	Sediment	Lactate	nd	nd	Newman et al. (1997a, b)
Sulfurospirillum barnesii str. SES-3	Freshwater marsh	Nitrate	AsR	Membrane-bound	Afkar (2012), Stolz et al. (1999), Zobrist et al. (2000)
Geobacter sp. str. OR-1	Paddy field soil	Acetate	*arrA*	nd	Ohtsuka et al. (2013), Tsuchiya et al. (2019)
Anaeromyxobacter sp. str. PSR-1	Soil form abandoned chemical plant	Acetate	*arrA*	Periplasm	Kudo et al. (2013), Muramatsu et al. (2020)
Aeromonas sp. str. O23A	Gold mine	Lactate	nd	nd	Uhrynowski et al. (2017)
Citrobacter sp. str. JH001	Sediment	Lactate	*arr*	nd	Wang et al. (2017a, b)

nd not defined

presence of arsenious acid, sodium methylarsonate [$CH_3AsO(ONa)_2$] or sodium cacodylate [$(CH_3)_2AsO{\cdot}ONa$] (Challenger and Higginbottom 1935). Production of methylated arsenic species like monomethylarsines [MMAs (III) & MMAs(V)], dimethylarsines [DMAs(III) & DMAs(V)], and trimethylarsines [TMAs(V) & TMAs(III)] take place when arsenate is reduced to arsenite, followed by the addition of methyl groups to arsenite, and continuous reduction and methylation of the products formed (Challenger 1945). However, in subsequent years, several prokaryotes have also been reported to methylate arsenic. The key enzyme in this process, arsenite *S*-adenosylmethionine methyltransferases, is endoded by the *arsM* gene (Chen et al. 2017a). In a study conducted by Qin et al. (2006), the *arsM* gene isolated from *Rhodopseudomonas palustris* was able to confer resistance to arsenite when expressed in a arsenite-sensitive *Escherichia coli*. The volatile end product of this process, trimethyl arsine (TMA) gas, which was erstwhile known to be toxic, was debunked in a report compiled by Cullen (2005) who cited several studies that proved the non-toxic nature of trimethylarsine. However, the intermediate products, MMAs and DMAs, are highly genotoxic (Mass et al. 2001).

Interkingdom horizontal gene transfer events of the *arsM* gene may have significantly contributed to adaptation to toxic arsenic-rich environments. The bacterial *arsM* gene, in particular, has been extensively transferred to the metazoans, fungi, diatoms, and even archaea. Horizontal gene transfer was also seen among eukaryotes, anaerobes and within different bacterial phyla (Chen et al. 2017a; Palmgren et al. 2017; Schönknecht et al. 2013). Functional gene analyses of sediments from a borewell in the Hubei Province of China indicated the presence of diverse novel arsenic-methylating microbes. This study also emphasized anaerobic or microaerobic conditions to be favorable for the biomethylation process. Sediments collected from increased depths also showed a greater concentration of DMAs (Zeng et al. 2018). The widespread use of the organoarsenical roxarsone in poultry feed to control diseases and improve poultry health is another contributor of organic arsenic into the environment. Although the roxarsone is almost entirely excreted out by the chickens, the manure then goes on to act as a contaminant in the soil where it is applied (Li and Chen 2005). Zhai et al. (2017) studied the biomethylation process in two pilot-scale compost piles for 60 days during which real-time analysis of MMA and DMA concentrations was observed to gradually increase over time followed by a gradual decrease, suggesting the sequential conversion of MMAs to DMAs and so on. Gene expression studies of *arsM* also exhibited the increase of the copy number with an increase in methylation, thus positively correlating arsM expression with methylation of arsenic.

There are increasing evidences of the diversity of *arsM* genes in the environment, particularly in As-rich sites like acid mine drainage, contaminated paddy soils, and copper mine (Desoeuvre et al. 2016; Jia et al. 2013; Qiao et al. 2023; Xiao et al. 2016; Zhao et al. 2013). Interestingly, *arsM* genes were also found to be abundant in even low-arsenic environments like estuarine wetlands and low-temperature hot springs (Yang et al. 2024; Zhang et al. 2017). The uptake of methylated arsenic by plant roots and its translocation via phloem channels has been well-documented by Carey et al. (2011). In the plant, the rice grains concentrated most of the organic arsenic compounds as compared to other parts like the roots and leaves. A rice rhizospheric bacterium *Burkholderia gladiolus* str. GSRB05 synthesizes a methylarsenate analog of the amino acid glutamic acid called arsinothricin (AST), which has structural similarities with the herbicide phosphinothricin. AST also has antibiotic action against both Gram-negative and Gram-positive bacteria that effectively inhibits glutamine synthetase (Kuramata et al. 2016; Suzol et al. 2020). This carries significant implications in the already existing problem of arsenic toxicity.

5.4 Role of Arsenic-Transforming Microbes in Biomining Processes

Biomining or bioleaching of ores is a metallurgical process where microbes are employed to solubilize minerals to extract metals such as copper, gold, nickel, zinc and cobalt (Johnson 2014). Precious metals like gold are found locked in minerals like pyrite, arsenopyrite, sphalerite, galena, chalcopyrite etc., and microbes are widely used to convert the insoluble forms of metals into soluble ones that can be easily recovered from solution (Rawlings et al. 2003). When these minerals are exposed to water and atmospheric oxygen, it leads to their oxidation and also production of sulfuric acid. The drop in pH further facilitates the process of leaching leading to the release of heavy metals like iron, arsenic, lead, copper, manganese, and a few others (Evangelou and Zhang 1995). This acidic, metal-rich liquid released from mines is called acid-mine drainage (AMD), and it has drastic effects on the surrounding soil and groundwater. Formation of AMD takes place in a sequential manner starting with the oxidation of pyrite leading to the release of Fe^{2+} and H^+ ions into the environment (Cheng et al. 2009).

$$2FeS_2 + 7O_2 + 2H_2O \rightarrow 2Fe^{2+} + 4SO_4^{2-} + 4H^+$$

The Fe^{2+} ions get oxidized to Fe^{3+} when acted upon by acidophilic bacteria like *Acidithiobacillus ferrooxidans* (Meruane and Vargas 2003).

$$4Fe^{2+} + O_2 + 4H^+ \rightarrow 4Fe^{3+} + 2H_2O$$

This resultant Fe^{3+} reacts in water and precipitates as $Fe(OH)_3$ releasing more H^+ ions in the process.

$$Fe^{3+} + 3H_2O \rightarrow Fe(OH)_3 + 3H^+$$

Similarly, oxidation of sulfide minerals of arsenic, copper, cadmium, zinc, and lead takes place in the same sequential manner. AMD contains either arsenite or arsenate species, the interconversion depending on redox and pH conditions. At high redox levels and at $pH > 10$, arsenate is more abundant while under reduced conditions, and at $pH < 8$, arsenite is the predominant species. In fact, the interplay between these conditions is what determines arsenic mobility and bioavailability as arsenate is more strongly adsorbed than arsenite on hydroxides of iron and aluminum and clay minerals at neutral and acidic pH conditions (Dixit and Hering 2003). Microbes also play significant roles in catalyzing arsenic transformation and mediating redox reactions that directly influence arsenic mobility via mechanisms elaborated in the preceding sections (Wang and Mulligan 2006).

6 Genetics of Arsenic Transformation in Microbes

6.1 Arsenic Detoxification in Microbes

Innovation of advanced molecular approaches in recent times, such as genome sequencing and transcriptomic, proteomic, and metabolomic studies, has made possible the extensive study of the distribution of arsenic resistance genes across the various prokaryotic groups and the factors that influence their expression and transfer. As discussed in the preceding sections, prokaryotes are able to metabolize as well as detoxify arsenic from their system. The arsenic detoxification is carried out by the arsenic resistance operon, commonly termed as the *ars* operon which could be chromosomal- or plasmid-borne in both Gram-negative and Gram-positive bacteria (Carlin et al. 1995; Wu and Rosen 1993). Plasmid-mediated *ars* operon when

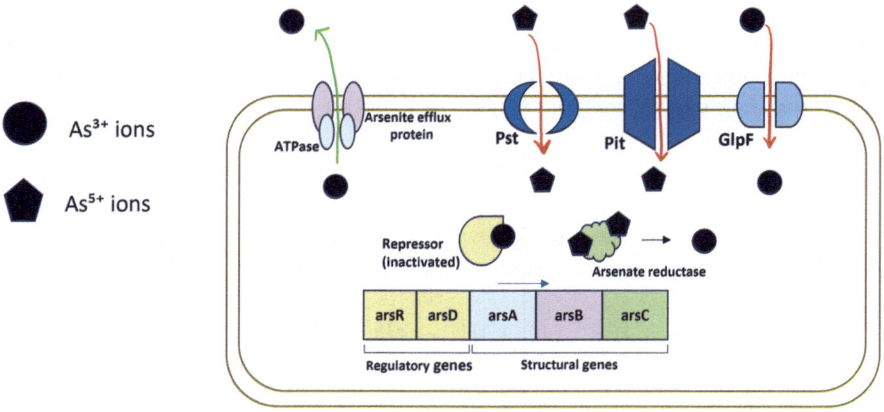

Fig. 2 A general overview of a bacterial cell with the *arsRDABC* operon (plasmid- or chromosome-mediated). Arsenite and arsenate make their way into the cell via phosphate (Pst and Pit) and aquaglyceroporin (GlpF) channels present on the membrane. Arsenite binds to the repressor leading to transcription of the structural genes

expressed in *Staphylococcus aureus* and *Bacillus subtilis* conferred resistance not only to arsenate and arsenite, but also to antimonite by the action of a plasmid-encoded, ATP-powered efflux pump (Kaur and Rosen 1992; Wu and Rosen 1993). As is the case in most inducible operons, ars operon is usually turned off but when the substrate or the inducer or the regulatory molecule (in this case, arsenite, arsenate or antimonite) binds to the repressor, allowing the transcription of the structural genes (Fig. 2). Depending on the bacterial type, the composition of structural genes in the *ars* operon differs. For instance, the *E. coli* genome and Staphylococcus aureus plasmid PI258 contain the *arsRBC* operon whereas the *E. coli* plasmid R773 contain the *arsRDABC* operon (Carlin et al. 1995; Chen et al. 1986; Ji and Silver 1992).

Occasionally, a single bacterial genome may consist of more than one *ars* operon, probably acquired by horizontal gene transfer and duplication (Fekih et al. 2018). The different ars operons may also confer varying degrees of arsenic resistance by encoding for different proteins as was observed in the arsenic-resistant strain *Ochrobactrum tritici* SCII24 which contains two antimony and arsenic resistance operons (*ars1* and *ars2*). After sequence analysis, it was seen that *ars1* operon contains genes encoding for the regulatory proteins ArsR and ArsD, the ATPase ArsA, a cystathionine-β-synthase (CBS-domain) containing protein, which is usually not involved in conferring arsenic resistance and the efflux transporter ArsB; arsA is usually coupled with arsB for functionality. The *ars2* operon, on the other hand, is composed of six genes, $arsR_2C_1Acr3C_2HR_3$, the most striking feature being the presence of two arsR genes in opposite orientations. This operon also encodes for Acr3 protein, a putative arsenite carrier, the arsenate reductase arsC protein and a flavodoxin protein arsH, that catalyzes oxidation of NADPH and generates H_2O_2 as an end product (Branco et al. 2008; Vorontsov et al. 2007; Ye et al. 2007). The arsH protein was subsequently reported to provide resistance to organoarsenicals such as monosodium methylarsenate and roxarsone in the *Pseudomonas putida* KT2440 cells (Chen et al. 2015). *Pantoea* sp. IMH was also reported to contain two ars operons *ars1* ($arsR_1B_1C_1H_1$) and *ars2* ($arsR_2B_2C_2H_2$), the difference being the temperatures at which the two operons show optimum activity (*ars1* at 15 °C and *ars2* at 30 °C). This difference in activity can be attributed to the temperature sensitivity of the $arsR_1$ and $arsR_2$ regulatory proteins (Wang et al. 2016).

6.2 Arsenotrophy in Microbes

Arsenotrophy, or arsenic metabolism, is an ancient trait in microbes, and the genes like *aioA* and *arrA* are abundantly found in the environment. The analysis of organization of arsenite oxidase gene clusters reveals the large subunit *aioA* and small subunit *aioB* to be clustered together with the latter located upstream of *aioA* in aerobic Proteobacteria like *Alcaligenes faecalis, Achromobacter* sp. str. SY8, *Sinorhizobium* sp. str. M14, *Ochrobactrum tritici, Herminiimonas arsenicoxydans* str. ULPAs1 and several others (Osborne and Santini 2012). In addition to these two constant elements, other genes, such as AioS (sensor kinase), AioR (regulator), and c-type cytochrome (electron acceptor), have been observed to play significant roles in transcribing the arsenite oxidase gene operons (Kashyap et al. 2006; Santini et al. 2007; Koechler et al. 2010; Lieutaud et al. 2010). On the other hand, respiratory arsenate reductase *arr* gene encodes for the periplasmic protein that enables anaerobic arsenate respiration while utilizing a variety of electron donors. Arsenate-respiring microbes consist of the *arr* genes in the order *arrAB*, and their expression is negatively regulated by oxygen and nitrate (Saltikov et al. 2005). Interestingly, the *arx* gene translates into a protein that is structurally similar to Arr but functions like an alternate arsenite oxidase (Zargar et al. 2010).

An in-situ study performed using enrichment and mixed cultures from several sources demonstrated arsenite oxidation in denitrifying conditions, and *Azoarcus, Acidovorax, Alicycliphilus*, and *Diaphorobacter* were the dominant bacterial genera responsible for the anoxic arsenite metabolism (Sun et al. 2009). Some thermophilic green non-sulfur bacteria, like *Chloroflexus aurantiacus*, also express arsenite oxidase during their photosynthetic processes, suggesting the wide implications of this enzyme in bioenergetic reactions in the bacterial cell (Lebrun et al. 2003). Metagenomic analysis of the oxygen-deficient zones in the Eastern North Pacific Ocean revealed the abundance and expression of the dissimilatory arsenate reductase *arrAB* and its closely related arsenite chemoautotrophic arsenite oxidase *arxA/aioA* genes in several bacterial groups, suggesting a complete arsenotrophic metabolism by the microbes in the anoxic environment (Saunders et al. 2019). These findings shed light on the possibility of an even higher abundance of arsenotrophic microorganisms in the anoxic Precambrian marine environment, which also had significantly higher arsenic concentrations (Bergman and Kolesov 2012).

In addition to arsenic metabolism, some bacteria, such as *Desulfurispirillum indicum* strain S5, consist of complex multiple respiratory mechanisms that enable them to reduce substrates like selenate, selenite, nitrate, and nitrite (Rauschenbach et al. 2011). Genomic analysis of the arsenic-respiring *Desulfitobacterium hafniense* DCB-2 revealed interesting genetic elements responsible for biofilm formation, metal reduction, nitrogen and carbon fixation, and dehalogenation, among others, making it a desirable bioremediating agent in sites contaminated by multiple pollutants (Kim et al. 2012).

7 Bioremediation Scenarios Using Arsenic-Metabolizing Microbes

Various microbial detoxification and bioenergetic pathways of arsenic explained in the preceding sections highlight the complexities of microbial cellular mechanisms depending on their ecological niches. Arsenite oxidizers and arsenate reducers play crucial roles in the release and immobilization of arsenic depending on other environmental factors, thus meticulously regulating the bioavailability of arsenic. While microbes possess extraordinary ability to resist and respire arsenic, other higher organisms such as humans, plants, and animals face a dire threat due to its carcinogenic properties. Recent bioremediation approaches have seen the use of genetically modified organisms (GMOs) in the bioaccumulation of arsenic; the overexpression of the metalloregulator arsR protein in a genetically engineered *Escherichia coli* led to the accumulation of substantial amounts of arsenic (Kostal et al. 2004).

Apart from bioaccumulation, there are several other microbial mechanisms that are useful tools in bioremediation. In biosorption, bacterial cells interact with the contaminant and either immobilize them in special organelles in the cytoplasm or on the cell surface with the help of cellular entities like lipids, polysaccharides, or specialized proteins (Prasad et al. 2013; Saba et al. 2019; Tariq et al. 2019; Titah et al. 2018). Gao et al. (2018) demonstrated an efficient biosorption and around tenfold biotransformation of trivalent arsenite into pentavalent arsenate by *Acidithiobacillus ferrooxidans* BY3 in the presence of ferrous ions in the reaction solution. Although genetically modified microorganisms (GMOs) seem to be the most efficient bioremediating agents, natural attenuation of contaminated soils is becoming increasingly popular as well due to its low cost and the flexibility of its use in combination with other remediating methods (Mulligan and Yong 2004; Roychowdhury et al. 2018). Another aspect that contributes to its mass appeal is the general trepidation of the public in using organisms that carry genetically modified genes that may or may not be transferable to other non-specific target organisms (Kumar et al. 2013).

In all respects, the use of microbes for bioremediation of arsenic contamination is undoubtedly an added advantage in terms of sustainability and cost-effectiveness when compared with other physical and chemical remediation techniques (Zheng et al. 2022). In situ application of microbes to polluted sites also ensures ease of access as well as reduction of costs in transportation of contaminated material to treatment plants (Yap and Peng 2019).

8 Conclusion

In conclusion, geomicrobes, particularly bacteria, play significant roles in the transformation of arsenic in the environment. In particular, microbial metabolic processes that are involved in redox reactions and methylation of arsenic are pivotal in influencing the speciation and mobility of arsenic and ultimately affecting its bioavailability in soil and water systems. These interconversions of arsenic from toxic to less toxic forms by bacteria shed light on their potential use in bioremediation strategies. This chapter has explained various bacterial mechanisms that are involved in detoxification and respiration of arsenic, which is an essential step in developing effective strategies to mitigate arsenic contamination and ensure safe water and soil resources without much threat to the living organisms.

Competing Interests All the authors declare that they have no competing interests.

References

Abbas G, Murtaza B, Bibi I, Shahid M, Niazi NK, Khan MI et al (2018) Arsenic uptake, toxicity, detoxification, and speciation in plants: physiological, biochemical, and molecular aspects. Int J Environ Res Public Health 15(1):59

Afkar E (2012) Localization of the dissimilatory arsenate reductase in Sulfurospirillum barnesii strain SeS-3. Am J Agric Biol Sci 7:97–105

Afkar E, Lisak J, Saltikov C, Basu P, Oremland RS, Stolz JF (2003) The respiratory arsenate reductase from Bacillus selenitireducens strain MLS10. FEMS Microbiol Lett 226(1):107–112

Al-Abed SR, Jegadeesan G, Purandare J, Allen D (2007) Arsenic release from iron rich mineral processing waste: influence of pH and redox potential. Chemosphere 66(4):775–782

Anawar HM, Akai J, Sakugawa H (2004) Mobilization of arsenic from subsurface sediments by effect of bicarbonate ions in groundwater. Chemosphere 54(6):753–762

Anguita JM, Rojas C, Pastén PA, Vargas IT (2018) A new aerobic chemolithoautotrophic arsenic oxidizing microorganism isolated from a high Andean watershed. Biodegradation 29:59–69

Ascar L, Ahumada I, Richter P (2008) Influence of redox potential (Eh) on the availability of arsenic species in soils and soils amended with biosolid. Chemosphere 72(10):1548–1552

Ash C, Tejnecký V, Borůvka L, Drábek O (2016) Different low-molecular-mass organic acids specifically control leaching of arsenic and lead from contaminated soil. J Contam Hydrol 187:18–30

Bahar MM, Megharaj M, Naidu R (2013) Kinetics of arsenite oxidation by Variovorax sp. MM-1 isolated from a soil and identification of arsenite oxidase gene. J Hazard Mater 262:997–1003

Bari AF, Lamb D, Choppala G, Bolan N, Seshadri B, Rahman MA, Rahman MM (2020) Geochemical fractionation and mineralogy of metal (loid) s in abandoned mine soils: insights into arsenic behaviour and implications to remediation. J Hazard Mater 399:123029

Barker WW, Welch SA, Banfield JF (1997) Biogeochemical weathering of silicate minerals. Rev Mineral 35:391–428
Bergman IA, Kolesov GM (2012) Arsenic, antimony, and bismuth as indicators of the genesis of ore material in Early Precambrian ferrous quartzite formations. Geochem Int 50:816–831
Bissen M, Frimmel FH (2003) Arsenic—a review. Part I: occurrence, toxicity, speciation, mobility. Acta Hydrochim Hydrobiol 31(1):9–18
Blum JS, Bindi AB, Buzzelli J, Stolz JF, Oremland RS (1998) Bacillus arsenicoselenatis, sp. nov., and Bacillus selenitireducens, sp. nov.: two haloalkaliphiles from Mono Lake, California that respire oxyanions of selenium and arsenic. Arch Microbiol 171:19–30
Branco R, Chung AP, Morais PV (2008) Sequencing and expression of two arsenic resistance operons with different functions in the highly arsenic-resistant strain Ochrobactrum tritici SCII24 T. BMC Microbiol 8:1–12
Bucher M (2007) Functional biology of plant phosphate uptake at root and mycorrhiza interfaces. New Phytol 173(1):11–26
Buck WB (1970) Diagnosis of feed-related toxicoses. J Am Vet Med Assoc 156:1434–1443
Burger J, Gaines KF, Boring CS, Stephens WL, Snodgrass J, Dixon C et al (2002) Metal levels in fish from the Savannah River: potential hazards to fish and other receptors. Environ Res 89(1):85–97
Byeon E, Kang HM, Yoon C, Lee JS (2021) Toxicity mechanisms of arsenic compounds in aquatic organisms. Aquat Toxicol 237:105901
Carey AM, Norton GJ, Deacon C, Scheckel KG, Lombi E, Punshon T et al (2011) Phloem transport of arsenic species from flag leaf to grain during grain filling. New Phytol 192(1):87–98
Carlin A, Shi W, Dey S, Rosen BP (1995) The ars operon of Escherichia coli confers arsenical and antimonial resistance. J Bacteriol 177(4):981–986
Challenger F (1945) Biological methylation. Chem Rev 36(3):315–361
Challenger F, Higginbottom C (1935) The production of trimethylarsine by Penicillium brevicaule (Scopulariopsis brevicaulis). Biochem J 29(7):1757
Chen CM, Misra TK, Silver S, Rosen BP (1986) Nucleotide sequence of the structural genes for an anion pump. The plasmid-encoded arsenical resistance operon. J Biol Chem 261(32):15030–15038
Chen J, Bhattacharjee H, Rosen BP (2015) ArsH is an organoarsenical oxidase that confers resistance to trivalent forms of the herbicide monosodium methylarsenate and the poultry growth promoter roxarsone. Mol Microbiol 96(5):1042–1052
Chen SC, Sun GX, Rosen BP, Zhang SY, Deng Y, Zhu BK et al (2017a) Recurrent horizontal transfer of arsenite methyltransferase genes facilitated adaptation of life to arsenic. Sci Rep 7(1):7741
Chen X, Sun Q, Ding S, Chen M, Fan X, Zhang L, Zhang C (2017b) Mobile arsenic distribution and release kinetics in sediment profiles under varying pH conditions. Water Air Soil Pollut 228:1–12
Cheng H, Hu Y, Luo J, Xu B, Zhao J (2009) Geochemical processes controlling fate and transport of arsenic in acid mine drainage (AMD) and natural systems. J Hazard Mater 165(1–3):13–26
Collins RN (2004) Separation of low-molecular mass organic acid–metal complexes by high-performance liquid chromatography. J Chromatogr A 1059(1–2):1–12
Cullen WR (2005) The toxicity of trimethylarsine: an urban myth. J Environ Monit 7(1):11–15
Cummings DE, Caccavo F, Fendorf S, Rosenzweig RF (1999) Arsenic mobilization by the dissimilatory Fe (III)-reducing bacterium Shewanella alga BrY. Environ Sci Technol 33(5):723–729
DasSarma P, Antunes A, Simões MF, DasSarma S (2020) Earth's stratosphere and microbial life. Curr Issues Mol Biol 38(1):197–244
Desoeuvre A, Casiot C, Héry M (2016) Diversity and distribution of arsenic-related genes along a pollution gradient in a river affected by acid mine drainage. Microb Ecol 71:672–685
Dixit S, Hering JG (2003) Comparison of arsenic (V) and arsenic (III) sorption onto iron oxide minerals: implications for arsenic mobility. Environ Sci Technol 37(18):4182–4189

Dong D, Ohtsuka T, Dong DT, Amachi S (2014) Arsenite oxidation by a facultative chemolithoautotrophic Sinorhizobium sp. KGO-5 isolated from arsenic-contaminated soil. Biosci Biotechnol Biochem 78(11):1963–1970
Drahota P, Filippi M (2009) Secondary arsenic minerals in the environment: a review. Environ Int 35(8):1243–1255
Drewniak L, Matlakowska R, Sklodowska A (2008) Arsenite and arsenate metabolism of Sinorhizobium sp. M14 living in the extreme environment of the zloty Stok gold mine. Geomicrobiol J 25(7–8):363–370
Durante-Rodríguez G, Fernández-Llamosas H, Alonso-Fernandes E, Fernández-Muñiz MN, Muñoz-Olivas R, Díaz E, Carmona M (2019) ArxA from Azoarcus sp. CIB, an anaerobic arsenite oxidase from an obligate heterotrophic and mesophilic bacterium. Front Microbiol 10:1699
Environmental Protection Agency (2013) Fed Regist 78(59):18590–18591
Evangelou VP, Zhang YL (1995) A review: pyrite oxidation mechanisms and acid mine drainage prevention. Crit Rev Environ Sci Technol 25(2):141–199
Fekih IB, Zhang C, Li YP, Zhao Y, Alwathnani HA, Saquib Q et al (2018) Distribution of arsenic resistance genes in prokaryotes. Front Microbiol 9:2473
Gao Q, Tang D, Song P, Zhou J, Li H (2018) Bio-adsorption and bio-transformation of arsenic by Acidithiobacillus ferrooxidans BY3. Int Microbiol 21:207–214
Garcia-Dominguez E, Mumford A, Rhine ED, Paschal A, Young LY (2008) Novel autotrophic arsenite-oxidizing bacteria isolated from soil and sediments. FEMS Microbiol Ecol 66(2):401–410
Goldberg S (2002) Competitive adsorption of arsenate and arsenite on oxides and clay minerals. Soil Sci Soc Am J 66(2):413–421
Grimmett RER, Mcintosh IG (1939) Occurrence of arsenic in soils and waters in the Waiotapu Valley, and its relation to stock health. NZJ Sci Technol Sect A 21:137–145
Gulens J, Champ DR, Jackson RE (1973) Chemistry of water supply treatment and distribution. Ann Arbor Science, Ann Arbor, MI
Isayenkov SV, Maathuis FJ (2008) The Arabidopsis thaliana aquaglyceroporin AtNIP7; 1 is a pathway for arsenite uptake. FEBS Lett 582(11):1625–1628
Ji G, Silver S (1992) Regulation and expression of the arsenic resistance operon from Staphylococcus aureus plasmid pI258. J Bacteriol 174(11):3684–3694
Jia Y, Huang H, Zhong M, Wang FH, Zhang LM, Zhu YG (2013) Microbial arsenic methylation in soil and rice rhizosphere. Environ Sci Technol 47(7):3141–3148
Johnson DB (2014) Biomining—biotechnologies for extracting and recovering metals from ores and waste materials. Curr Opin Biotechnol 30:24–31
Kashyap DR, Botero LM, Franck WL, Hassett DJ, McDermott TR (2006) Complex regulation of arsenite oxidation in Agrobacterium tumefaciens. J Bacteriol 188(3):1081–1088
Kaur P, Rosen BP (1992) Plasmid-encoded resistance to arsenic and antimony. Plasmid 27(1):29–40
Kim MJ, Nriagu J, Haack S (2000) Carbonate ions and arsenic dissolution by groundwater. Environ Sci Technol 34(15):3094–3100
Kim SH, Harzman C, Davis JK, Hutcheson R, Broderick JB, Marsh TL et al (2012) Genome sequence of Desulfitobacterium hafniense DCB-2, a gram-positive anaerobe capable of dehalogenation and metal reduction. BMC Microbiol 12:1–20
Koechler S, Cleiss-Arnold J, Proux C, Sismeiro O, Dillies MA, Goulhen-Chollet F et al (2010) Multiple controls affect arsenite oxidase gene expression in Herminiimonas arsenicoxydans. BMC Microbiol 10:1–13
Kostal J, Yang R, Wu CH, Mulchandani A, Chen W (2004) Enhanced arsenic accumulation in engineered bacterial cells expressing ArsR. Appl Environ Microbiol 70(8):4582–4587
Krafft T, Macy JM (1998) Purification and characterization of the respiratory arsenate reductase of Chrysiogenes arsenatis. Eur J Biochem 255(3):647–653
Kudo K, Yamaguchi N, Makino T, Ohtsuka T, Kimura K, Dong DT, Amachi S (2013) Release of arsenic from soil by a novel dissimilatory arsenate-reducing bacterium, Anaeromyxobacter sp. strain PSR-1. Appl Environ Microbiol 79(15):4635–4642

Kumar S, Dagar VK, Khasa YP, Kuhad RC (2013) Genetically modified microorganisms (GMOs) for bioremediation. In: Kuhad R, Singh A (eds) Biotechnology for environmental management and resource recovery. Springer, pp 191–218
Kumar PS, Flores RQ, Sjöstedt C, Önnby L (2016) Arsenic adsorption by iron–aluminium hydroxide coated onto macroporous supports: insights from X-ray absorption spectroscopy and comparison with granular ferric hydroxides. J Hazard Mater 302:166–174
Kumari N, Jagadevan S (2016) Genetic identification of arsenate reductase and arsenite oxidase in redox transformations carried out by arsenic metabolising prokaryotes—a comprehensive review. Chemosphere 163:400–412
Kuramata M, Sakakibara F, Kataoka R, Yamazaki K, Baba K, Ishizaka M et al (2016) Arsinothricin, a novel organoarsenic species produced by a rice rhizosphere bacterium. Environ Chem 13(4):723–731
Lanz H Jr, Wallace PW, Hamilton JG (1950) The metabolism of arsenic in laboratory animals using As74 as a trace. Univ Calif Pub Pharmacol 2:263–282
Lebrun E, Brugna M, Baymann F, Muller D, Lievremont D, Lett MC, Nitschke W (2003) Arsenite oxidase, an ancient bioenergetic enzyme. Mol Biol Evol 20(5):686–693
Li YX, Chen TB (2005) Concentrations of additive arsenic in Beijing pig feeds and the residues in pig manure. Resour Conserv Recycl 45(4):356–367
Lieutaud A, Van Lis R, Duval S, Capowiez L, Muller D, Lebrun R et al (2010) Arsenite oxidase from Ralstonia sp. 22: characterization of the enzyme and its interaction with soluble cytochromes. J Biol Chem 285(27):20433–20441
Lu M, Wang H, Li XF, Lu X, Cullen WR, Arnold LL et al (2004) Evidence of hemoglobin binding to arsenic as a basis for the accumulation of arsenic in rat blood. Chem Res Toxicol 17(12):1733–1742
Lukasz D, Liwia R, Aleksandra M, Aleksandra S (2014) Dissolution of arsenic minerals mediated by dissimilatory arsenate reducing bacteria: estimation of the physiological potential for arsenic mobilization. Biomed Res Int 2014(1):841892
Ma JF, Yamaji N (2006) Silicon uptake and accumulation in higher plants. Trends Plant Sci 11(8):392–397
Ma LQ, Komar KM, Tu C, Zhang W, Cai Y, Kennelley ED (2001) A fern that hyperaccumulates arsenic. Nature 409(6820):579–579
Ma JF, Yamaji N, Mitani N, Tamai K, Konishi S, Fujiwara T et al (2007) An efflux transporter of silicon in rice. Nature 448(7150):209–212
Mandal BK, Suzuki KT (2002) Arsenic round the world: a review. Talanta 58(1):201–235
Manning BA, Goldberg S (1997) Adsorption and stability of arsenic (III) at the clay mineral–water interface. Environ Sci Technol 31(7):2005–2011
Manning BA, Fendorf SE, Bostick B, Suarez DL (2002) Arsenic (III) oxidation and arsenic (V) adsorption reactions on synthetic birnessite. Environ Sci Technol 36(5):976–981
Mass MJ, Tennant A, Roop BC, Cullen WR, Styblo M, Thomas DJ, Kligerman AD (2001) Methylated trivalent arsenic species are genotoxic. Chem Res Toxicol 14(4):355–361
Masscheleyn PH, Delaune RD, Patrick WH Jr (1991) Effect of redox potential and pH on arsenic speciation and solubility in a contaminated soil. Environ Sci Technol 25(8):1414–1419
Meharg AA, Naylor J, Macnair MR (1994) Phosphorus nutrition of arsenate-tolerant and nontolerant phenotypes of velvetgrass. J Environ Qual 23(2):234–238
Meruane G, Vargas T (2003) Bacterial oxidation of ferrous iron by Acidithiobacillus ferrooxidans in the pH range 2.5–7.0. Hydrometallurgy 71(1–2):149–158
Miyazaki K, Watanabe C, Mori K, Yoshida K, Ohtsuka R (2005) The effects of gestational arsenic exposure and dietary selenium deficiency on selenium and selenoenzymes in maternal and fetal tissues in mice. Toxicology 208(3):357–365
Moxham JW, Coup MR (1968) Arsenic poisoning of cattle and other domestic animals. N Z Vet J 16(10–11):161–165
Muller D, Lievremont D, Simeonova DD, Hubert JC, Lett MC (2003) Arsenite oxidase aox genes from a metal-resistant β-proteobacterium. J Bacteriol 185(1):135–141

Mulligan CN, Yong RN (2004) Natural attenuation of contaminated soils. Environ Int 30(4):587–601
Muramatsu F, Tonomura M, Yamada M, Kasahara Y, Yamamura S, Iino T, Amachi S (2020) Possible involvement of a tetrathionate reductase homolog in dissimilatory arsenate reduction by Anaeromyxobacter sp. strain PSR-1. Appl Environ Microbiol 86(23):e0082920
Newman DK, Beveridge TJ, Morel F (1997a) Precipitation of arsenic trisulfide by Desulfotomaculum auripigmentum. Appl Environ Microbiol 63(5):2022–2028
Newman DK, Kennedy EK, Coates JD, Ahmann D, Ellis DJ, Lovley DR, Morel FM (1997b) Dissimilatory arsenate and sulfate reduction in Desulfotomaculum auripigmentum sp. nov. Arch Microbiol 168:380–388
Ohtsuka T, Yamaguchi N, Makino T, Sakurai K, Kimura K, Kudo K et al (2013) Arsenic dissolution from Japanese paddy soil by a dissimilatory arsenate-reducing bacterium Geobacter sp. OR-1. Environ Sci Technol 47(12):6263–6271
Oremland RS, Stolz JF (2003) The ecology of arsenic. Science 300(5621):939–944
Oremland RS, Hoeft SE, Santini JM, Bano N, Hollibaugh RA, Hollibaugh JT (2002) Anaerobic oxidation of arsenite in Mono Lake water and by a facultative, arsenite-oxidizing chemoautotroph, strain MLHE-1. Appl Environ Microbiol 68(10):4795–4802
Osborne TH, Santini JM (2012) Prokaryotic aerobic oxidation of arsenite. In: The metabolism of arsenite. CRC Press, pp 61–72
Osborne TH, Jamieson HE, Hudson-Edwards KA, Nordstrom DK, Walker SR, Ward SA, Santini JM (2010) Microbial oxidation of arsenite in a subarctic environment: diversity of arsenite oxidase genes and identification of a psychrotolerant arsenite oxidiser. BMC Microbiol 10:1–8
Palmgren M, Engström K, Hallström BM, Wahlberg K, Søndergaard DA, Säll T et al (2017) AS3MT-mediated tolerance to arsenic evolved by multiple independent horizontal gene transfers from bacteria to eukaryotes. PLoS One 12(4):e0175422
Panagiotaras D, Nikolopoulos D (2015) Arsenic occurrence and fate in the environment; a geochemical perspective. J Earth Sci Climat Change 6(4):1
Paul NP, Galván AE, Yoshinaga-Sakurai K, Rosen BP, Yoshinaga M (2023) Arsenic in medicine: past, present and future. Biometals 36(2):283–301. https://doi.org/10.1007/s10534-022-00371-y
Prasad KS, Ramanathan AL, Paul J, Subramanian V, Prasad R (2013) Biosorption of arsenite (As+3) and arsenate (As+5) from aqueous solution by Arthrobacter sp. biomass. Environ Technol 34(19):2701–2708
Qiao J, Liu J, Palomo A, Bostick BC, Phan K, Zheng Y, Li F (2023) Prevalence of methylated arsenic and microbial arsenic methylation genes in paddy soils of the Mekong Delta. Environ Sci Technol 57(26):9675–9682
Qin J, Rosen BP, Zhang Y, Wang G, Franke S, Rensing C (2006) Arsenic detoxification and evolution of trimethylarsine gas by a microbial arsenite S-adenosylmethionine methyltransferase. Proc Natl Acad Sci 103(7):2075–2080
Raab A, Feldmann J (2003) Microbial transformation of metals and metalloids. Sci Prog 86(3):179–202
Rauschenbach I, Yee N, Häggblom MM, Bini E (2011) Energy metabolism and multiple respiratory pathways revealed by genome sequencing of Desulfurispirillum indicum strain S5. Environ Microbiol 13(6):1611–1621
Rawlings DE, Dew D, du Plessis C (2003) Biomineralization of metal-containing ores and concentrates. Trends Biotechnol 21(1):38–44
Rhine ED, Chadhain SN, Zylstra GJ, Young LY (2007) The arsenite oxidase genes (aroAB) in novel chemoautotrophic arsenite oxidizers. Biochem Biophys Res Commun 354(3):662–667
Robbins LJ, Konhauser KO (2021) Geobiology and geomicrobiology. In: Encyclopedia of geology. Elsevier, pp 554–568. https://doi.org/10.1016/b978-0-12-409548-9.12532-1
Rodgers J (2009) Human African trypanosomiasis, chemotherapy and CNS disease. J Neuroimmunol 211(1–2):16–22
Rosen BP (2002) Biochemistry of arsenic detoxification. FEBS Lett 529(1):86–92

Roychowdhury R, Roy M, Rakshit A, Sarkar S, Mukherjee P (2018) Arsenic bioremediation by indigenous heavy metal resistant bacteria of fly ash pond. Bull Environ Contam Toxicol 101:527–535

Saalfield SL, Bostick BC (2010) Synergistic effect of calcium and bicarbonate in enhancing arsenate release from ferrihydrite. Geochim Cosmochim Acta 74(18):5171–5186

Saba, Rehman Y, Ahmed M, Sabri AN (2019) Potential role of bacterial extracellular polymeric substances as biosorbent material for arsenic bioremediation. Biorem J 23(2):72–81

Sakurai T (2002) Biological effects of organic arsenic compounds in seafood. Appl Organomet Chem 16(8):401–405

Saltikov CW, Wildman RA Jr, Newman DK (2005) Expression dynamics of arsenic respiration and detoxification in Shewanella sp. strain ANA-3. J Bacteriol 187(21):7390–7396

Santini JM, vanden Hoven RN (2004) Molybdenum-containing arsenite oxidase of the chemolithoautotrophic arsenite oxidizer NT-26. J Bacteriol 186(6):1614–1619

Santini JM, Sly LI, Schnagl RD, Macy JM (2000) A new chemolithoautotrophic arsenite-oxidizing bacterium isolated from a gold mine: phylogenetic, physiological, and preliminary biochemical studies. Appl Environ Microbiol 66(1):92–97

Santini JM, Kappler U, Ward SA, Honeychurch MJ, vanden Hoven RN et al (2007) The NT-26 cytochrome c552 and its role in arsenite oxidation. Biochim Biophys Acta 1767(2):189–196

Saunders JK, Fuchsman CA, McKay C, Rocap G (2019) Complete arsenic-based respiratory cycle in the marine microbial communities of pelagic oxygen-deficient zones. Proc Natl Acad Sci 116(20):9925–9930

Schönknecht G, Chen WH, Ternes CM, Barbier GG, Shrestha RP, Stanke M et al (2013) Gene transfer from bacteria and archaea facilitated evolution of an extremophilic eukaryote. Science 339(6124):1207–1210

Shaji E, Santosh M, Sarath KV, Prakash P, Deepchand V, Divya BV (2021) Arsenic contamination of groundwater: a global synopsis with focus on the Indian Peninsula. Geosci Front 12(3):101079

Shaw D (2006) Mobility of arsenic in saturated, laboratory test sediments under varying pH conditions. Eng Geol 85(1–2):158–164

Shi K, Wang Q, Wang G (2020) Microbial oxidation of arsenite: regulation, chemotaxis, phosphate metabolism and energy generation. Front Microbiol 11:569282

Signes-Pastor A, Burló F, Mitra K, Carbonell-Barrachina AA (2007) Arsenic biogeochemistry as affected by phosphorus fertilizer addition, redox potential and pH in a West Bengal (India) soil. Geoderma 137(3–4):504–510

Spehar RL, Fiandt JT, Anderson RL, DeFoe DL (1980) Comparative toxicity of arsenic compounds and their accumulation in invertebrates and fish. Arch Environ Contam Toxicol 9:53–63

Squibb KS, Fowler BA (1983) The toxicity of arsenic and its compounds. Biol Environ Effects Arsenic 6:233–269

Srivastava M, Santos J, Srivastava P, Ma LQ (2010) Comparison of arsenic accumulation in 18 fern species and four Pteris vittata accessions. Bioresour Technol 101(8):2691–2699

Stachowicz M, Hiemstra T, Van Riemsdijk WH (2007) Arsenic-bicarbonate interaction on goethite particles. Environ Sci Technol 41(16):5620–5625

Stolz JF, Oremland RS (1999) Bacterial respiration of arsenic and selenium. FEMS Microbiol Rev 23(5):615–627

Stolz JF, Ellis DJ, Blum JS, Ahmann D, Lovley DR, Oremland RS (1999) Note: sulfurospirillum barnesii sp. nov. and Sulfurospirillum arsenophilum sp. nov., new members of the Sulfurospirillum clade of the ε-Proteobacteria. Int J Syst Evol Microbiol 49(3):1177–1180

Sun W, Sierra-Alvarez R, Fernandez N, Sanz JL, Amils R, Legatzki A et al (2009) Molecular characterization and in situ quantification of anoxic arsenite-oxidizing denitrifying enrichment cultures. FEMS Microbiol Ecol 68(1):72–85

Suzol SH, Hasan Howlader A, Galván AE, Radhakrishnan M, Wnuk SF, Rosen BP, Yoshinaga M (2020) Semisynthesis of the organoarsenical antibiotic arsinothricin. J Nat Prod 83(9):2809–2813

Tariq A, Ullah U, Asif M, Sadiq I (2019) Biosorption of arsenic through bacteria isolated from Pakistan. Int Microbiol 22:59–68
Thakur N, Singh SP, Zhang C (2022) Microorganisms under extreme environments and their applications. Curr Res Microb Sci 3:100141
Titah HS, Abdullah SRS, Idris M, Anuar N, Basri H, Mukhlisin M et al (2018) Arsenic resistance and biosorption by isolated rhizobacteria from the roots of Ludwigia octovalvis. Int J Microbiol 2018(1):3101498
Tourney J, Ngwenya BT (2014) The role of bacterial extracellular polymeric substances in geomicrobiology. Chem Geol 386:115–132
Tsuchiya T, Ehara A, Kasahara Y, Hamamura N, Amachi S (2019) Expression of genes and proteins involved in arsenic respiration and resistance in dissimilatory arsenate-reducing Geobacter sp. strain OR-1. Appl Environ Microbiol 85(14):e0076319
Uhrynowski W, Debiec K, Sklodowska A, Drewniak L (2017) The role of dissimilatory arsenate reducing bacteria in the biogeochemical cycle of arsenic based on the physiological and functional analysis of Aeromonas sp. O23A. Sci Total Environ 598:680–689
vanden Hoven RN, Santini JM (2004) Arsenite oxidation by the heterotroph Hydrogenophaga sp. str. NT-14: the arsenite oxidase and its physiological electron acceptor. Biochim Biophys Acta 1656(2–3):148–155
Vaughan GT (1993) The environmental chemistry and fate of arsenical pesticides in cattle tick dip sites and banana land plantations. CSIRO Division of Coal Industry, Center for Advanced Analytical Chemistry, NSW, Melboune
Vorontsov II, Minasov G, Brunzelle JS, Shuvalova L, Kiryukhina O, Collart FR, Anderson WF (2007) Crystal structure of an apo form of Shigella flexneri ArsH protein with an NADPH-dependent FMN reductase activity. Protein Sci 16(11):2483–2490
Wang S, Mulligan CN (2006) Natural attenuation processes for remediation of arsenic contaminated soils and groundwater. J Hazard Mater 138(3):459–470
Wang HB, Wong MH, Lan CY, Baker AJM, Qin YR, Shu WS et al (2007) Uptake and accumulation of arsenic by 11 Pteris taxa from southern China. Environ Pollut 145(1):225–233
Wang L, Zhuang X, Zhuang G, Jing C (2016) Arsenic resistance strategy in Pantoea sp. IMH: organization, function and evolution of ars genes. Sci Rep 6(1):39195
Wang J, Zeng XC, Zhu X, Chen X, Zeng X, Mu Y et al (2017a) Sulfate enhances the dissimilatory arsenate-respiring prokaryotes-mediated mobilization, reduction and release of insoluble arsenic and iron from the arsenic-rich sediments into groundwater. J Hazard Mater 339:409–417
Wang N, Xue XM, Juhasz AL, Chang ZZ, Li HB (2017b) Biochar increases arsenic release from an anaerobic paddy soil due to enhanced microbial reduction of iron and arsenic. Environ Pollut 220:514–522
Wang S, Li W, Ding C, Zhang J, Zhang N, Li YC et al (2024) Biochar-supported zero-valent iron enhanced arsenic immobilization in a paddy soil: the role of soil organic matter. Biochar 6(1):26
Weeger W, Lièvremont D, Perret M, Lagarde F, Hubert JC, Leroy M, Lett MC (1999) Oxidation of arsenite to arsenate by a bacterium isolated from an aquatic environment. Biometals 12:141–149
Welch SA, Barker WW, Banfield JF (1999) Microbial extracellular polysaccharides and plagioclase dissolution. Geochim Cosmochim Acta 63(9):1405–1419
White DE, Hem JD, Waring GA (1963) Data of geochemistry: chapter F, chemical composition of subsurface waters. US Government Printing Office
Wu J, Rosen BP (1993) Metalloregulated expression of the ars operon. J Biol Chem 268(1):52–58
Xiao Y, Liu X, Liang Y, Niu J, Zhang X, Ma L et al (2016) Insights into functional genes and taxonomical/phylogenetic diversity of microbial communities in biological heap leaching system and their correlation with functions. Appl Microbiol Biotechnol 100:9745–9756
Xu H, Allard B, Grimvall A (1988) Influence of pH and organic substance on the adsorption of As (V) on geologic materials. Water Air Soil Pollut 40:293–305

Xu W, Dai W, Yan H, Li S, Shen H, Chen Y et al (2015) Arabidopsis NIP3; 1 plays an important role in arsenic uptake and root-to-shoot translocation under arsenite stress conditions. Mol Plant 8(5):722–733
Yang Y, Yu Y, Chen X (2024) Arsenic-methylating microbial community in sediment along the water flow is correlated with the distance to a low-temperature hot spring. Water Supply 24(3):918–930
Yap CK, Peng SHT (2019) Cleaning contaminated soils by using microbial remediation: a review and challenges to the weaknesses. Am J Biomed Sci Res 2(3). https://doi.org/10.34297/AJBSR.2019.02.000589
Ye J, Yang HC, Rosen BP, Bhattacharjee H (2007) Crystal structure of the flavoprotein ArsH from Sinorhizobium meliloti. FEBS Lett 581(21):3996–4000
Zargar K, Hoeft S, Oremland R, Saltikov CW (2010) Identification of a novel arsenite oxidase gene, arxA, in the haloalkaliphilic, arsenite-oxidizing bacterium Alkalilimnicola ehrlichii strain MLHE-1. J Bacteriol 192(14):3755–3762
Zeng XC, Yang Y, Shi W, Peng Z, Chen X, Zhu X, Wang Y (2018) Microbially mediated methylation of arsenic in the arsenic-rich soils and sediments of Jianghan Plain. Front Microbiol 9:1389
Zhai W, Wong MT, Luo F, Hashmi MZ, Liu X, Edwards EA et al (2017) Arsenic methylation and its relationship to abundance and diversity of arsM genes in composting manure. Sci Rep 7(1):42198
Zhang N, Wei C, Yang L (2013) Occurrence of arsenic in two large shallow freshwater lakes in China and a comparison to other lakes around the world. Microchem J 110:169–177
Zhang SY, Su JQ, Sun GX, Yang Y, Zhao Y, Ding J et al (2017) Land scale biogeography of arsenic biotransformation genes in estuarine wetland. Environ Microbiol 19(6):2468–2482
Zhao FJ, Dunham SJ, McGrath SP (2002) Arsenic hyperaccumulation by different fern species. New Phytol 156(1):27–31
Zhao FJ, Harris E, Yan J, Ma J, Wu L, Liu W et al (2013) Arsenic methylation in soils and its relationship with microbial arsM abundance and diversity, and As speciation in rice. Environ Sci Technol 47(13):7147–7154
Zheng W, Cui T, Li H (2022) Combined technologies for the remediation of soils contaminated by organic pollutants. A review. Environ Chem Lett 20(3):2043–2062
Zobrist J, Dowdle PR, Davis JA, Oremland RS (2000) Mobilization of arsenite by dissimilatory reduction of adsorbed arsenate. Environ Sci Technol 34(22):4747–4753.w

Biomineralization of Toxic Chemicals and Radionuclides by Geomicrobes

Abirami Subramanian and Sushmitha Baskar

Contents

Abstract The widespread usage of toxic chemicals and radionuclides for various industrial, agricultural, and nuclear activities poses significant risks to ecosystems and human health. These contaminants remain in the ecosystems, leading to bioaccumulation in food chains and other adverse effects on biodiversity and human populations. Traditional remediation methods, such as physical and chemical treatments, have disadvantages with reference to efficiency, cost, and long-term sustainability. So, to overcome the downside of remediation process, scientific community has harnessed the remarkable metabolic capabilities of geomicrobes to transform various chemical contaminants into stable mineral phases. Furthermore, these biogenic mineral phases exhibit compatibility with

A. Subramanian
Centre for Science of Student Learning, Bengaluru, Karnataka, India

S. Baskar (✉)
Environmental Studies, School of Interdisciplinary and Trans-disciplinary Studies (SOITS), Indira Gandhi National Open University (IGNOU), Maidan Garhi, New Delhi, India
e-mail: sushmithab@ignou.ac.in

S. R. Joshi et al. (eds.), *Mineral Transformation and Bioremediation by Geo-Microbes*, Interdisciplinary Biotechnological Advances,
https://doi.org/10.1007/978-981-96-3033-2_5

synergistic remediation approaches, thereby offering sustainable solutions for mitigating the risks posed by toxic chemicals and radionuclides in contaminated environments. This chapter consolidates insights from the available literature to explore the role of geomicrobes in the biomineralization of toxic chemicals and radionuclides.

Keywords Biomineralization · Geomicrobes · Toxic chemicals · Radionuclides · Bioremediation

Abbreviations

DNB	Denitrifying bacteria
EPS	Extracellular polymeric substances
PSB	Phosphate-solubilizing bacteria
SRB	Sulfate-reducing bacteria

1 Introduction

Microbial biomineralization is the process by which microorganisms produce minerals. But what drives this process, and why is it important for geochemical cycles? To grasp the significance of microbial mineralization, let's consider a hypothetical scenario: What if microbial mineralization did not occur on Earth? In the absence of microbial biomineralization, several critical geochemical processes would be severely disrupted. For instance, the carbon cycle would be significantly affected, as many microorganisms play vital roles in the precipitation of calcium carbonate, a key component of carbon sequestration. Without microbial-mediated precipitation, the removal of CO_2 from the atmosphere would be less efficient, thereby exacerbating global warming and climate change (Dong et al. 2021; Ouyang et al. 2022). Moreover, the absence of biomineralization would impact soil formation and stability. Microorganisms contribute to the formation of soil minerals thus enhancing soil structure and fertility. Without these microbial activities, soil quality would degrade which affects the plant growth and agricultural productivity. Microbial biomineralization plays a vital part in the creation of geological structures, such as stromatolites and cave formations. These formations offer significant knowledge about the Earth's history and the evolution of life. Without microbial involvement and biomineralization, our understanding of these processes would be incomplete.

1.1 Fundamentals of Biomineralization

Biomineralization process can be categorized into biologically induced and biologically controlled processes. Biologically induced mineralization, also called as passive biomineralization, is a result of microbial metabolic by-products that create conditions for mineral formation without direct microbial control over the mineral properties. Sulfate-reducing bacteria (SRB) demonstrate this by producing hydrogen sulfide, which reacts with metal ions to form metal sulfides. Biologically controlled mineralization also called as active biomineralization involves organisms exerting precise control over the mineralization process, leading to well-defined mineral characteristics, such as the magnetite produced by magnetotactic bacteria (Lowenstam 1981; Robbins and Konhauser 2021).

The principles underlying biomineralization include supersaturation, nucleation, and crystal growth. Supersaturation is a prerequisite for mineral formation and is often achieved through microbial metabolic activities that alter local chemical conditions (Bäuerlein 2003). Nucleation involves the initial aggregation of ions to form a stable nucleus, which can occur on microbial cell surfaces or as extracellular polymeric substances (EPS), or metabolic products. Once nucleation is established, crystal growth occurs, as ions continue to deposit onto the growing crystal (Southam and Lovley 2000; Aloisi et al. 2006; Robles-Fernández et al. 2022). Various microbes are involved in biomineralization, including bacteria, fungi, and algae. Ureolytic bacteria like *Sporosarcina pasteurii* precipitate calcium carbonate through urea hydrolysis (Hammes et al. 2003), while sulfate-reducing bacteria such as *Desulfovibrio* species form metal sulphides (Dong et al. 2024). Magnetotactic bacteria, like *Magnetospirillum* sp., synthesize magnetite crystals intracellularly (McCausland et al. 2022). Fungi like *Aspergillus niger* and *Neurospora crassa* can precipitate metals namely carbonates, oxides, or sulfides. Additionally, lignin-degrading fungi like *Phanerochaete chrysosporium* produce extracellular enzymes that modify the microenvironment to promote mineral precipitation (Zhao et al. 2020; He et al. 2022). Algae, such as diatoms and coccolithophores, contribute significantly to the global carbon and silica cycles by producing silica and calcium carbonate structures (Durak et al. 2016).

Biomineralization by geomicrobes has garnered considerable interest as a result of its potential applications in environmental remediation. The pervasive contamination of ecosystems by toxic chemicals and radionuclides are hazardous to human health. Traditional remediation methods, such as physical removal or chemical treatments, are often costly and inefficient and can further harm the environment. Therefore, this review focuses on exploring the role and mechanisms of biomineralization of toxic chemicals and radionuclides by geomicrobes to offer a sustainable alternative for mitigating environmental pollution and enhancing ecosystem health. Table 1 provides some examples of microbes used in biomineralization of toxic chemicals and radionuclides.

Table 1 Examples of geomicrobes involved in biomineralization of heavy metals, toxic chemicals, and radionuclides

Metals	Microbes	Geological source	References
Arsenic	*Shewanella* sp. O23S	Gold mine, Złoty Stok (SW Poland)	Staicu et al. (2022)
Cadmium	*Bacillus paramycoides*- MSR1	Sambhar Lake, Rajasthan, India	Kaur et al. (2023)
	Providencia rettgeri	Soil, Qatar	Al Disi et al. (2022)
	Stenotrophomonas rhizophila	Rhizosphere soil, Zanjan province, Iran	Jalilvand et al. (2020)
	Pseudomonas stutzeri 273	Sediments of East China Sea, China	Wu et al. (2017)
Copper	*Yokenella regensburgei* GXAS49-I	Soil, China	Xu et al. (2024)
Chromium	*Pseudomonas aeruginosa* QZ5	Soil, Qatar	Al Disi et al. (2022)
	Pseudomonas aeruginosa QZ9		
Lead	*Pseudomonas* sp.	Phosphate mine, China	Wang et al. (2019)
	Exiguobacterium sp. JBHLT-3	Surface Lake water, Lake Jibuhulangtu Nuur, China	Bai et al. (2021)
	Lysinibacillus sp.	East lake sediment, China	Zhang et al. (2022)
	Variovorax boronicumulans	Rhizosphere soil, Zanjan province, Iran	Jalilvand et al. (2020)
Cerium	*Bacillus licheniformis*	Lake sediment, China	Cheng et al. (2022)
Nickel	*Acinetobacter* sp. H12	Qujiang artificial lake, China	Wu et al. (2021)
Diuron	*Sphingomonas* sp.	IPU-treated agricultural soil, Denmark and United Kingdom	Sørensen et al. (2013)
Uranium	*Idiomarina loihiensis* MAH1	Surface seawater, Alboran Sea	Morcillo et al. (2014)
	Stenotrophomonas strain Br8	Uranium tailing repositories, Limousin, France	Sánchez-Castro et al. (2020)

2 Heavy Metals Mineralization by Geomicrobes and Key Mechanisms

The study of microbial biomineralization reveals diverse mechanisms through which microorganisms manage toxic metals. Figure 1 illustrates the key mechanisms through which microbes mediate biomineralization. For instance, phosphate solubilization and phosphate-solubilizing bacteria (PSB) play pivotal roles in the biomineralization of toxic metals. Transformation of metal ions into insoluble mineral forms by PSB occurs through several interconnected processes. PSB produce organic acids and enzymes, such as phosphatases, which convert insoluble phosphorus compounds into soluble phosphate ions (PO_4^{3-}), thereby increasing their bioavailability. These ions react with toxic metal ions (e.g., Pb^{2+}, Cd^{2+}, Ni^{2+}) to form

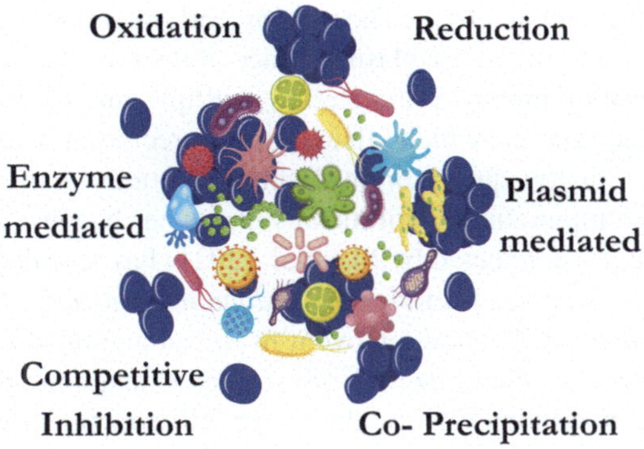

Fig. 1 Key mechanisms through with microbes mediate mineralization

insoluble metal phosphates (e.g., $Pb_3(PO_4)_2$, $Cd_3(PO_4)_2$), precipitating out of solution and reducing metal bioavailability and mobility. Additionally, PSB secrete extracellular polymeric substances (EPS), which chelate metal ions and promote the nucleation and growth of metal phosphate crystals. The surface of PSB cells and their EPS serve as nucleation sites, lowering the energy barrier for crystal formation. PSB enzymes, such as phosphatases, ensure a steady supply of phosphate ions, while the bacteria's ability to modulate pH enhances phosphate solubilization and metal phosphate precipitation (Zhang et al. 2019; Jiang et al. 2022).

Co-precipitation is one of the critical mechanisms in the biomineralization of toxic metals. This process involves the simultaneous precipitation of metals with other minerals, facilitated by microbial activities. For instance, the Fe(II)-oxidizing microbial consortium, like *Acidithiobacillus* spp., *Leptospirillum* spp., *Acidiphilium* spp., and *Acidibacter* spp., oxidizes Fe(II) to Fe(III), leading to an oversaturation of Fe(III) in the medium and subsequent precipitation of Fe(III) minerals. These minerals serve as nucleation sites for heavy metals. As Fe(III) minerals precipitate, heavy metals in the solution are incorporated into the mineral matrix through physical adsorption, chemisorption, and substitution. Specifically, lead (Pb) can react with sulfate ions to form low-solubility Pb sulfates (e.g., anglesite) in low-pH, sulfate-rich environments (Wang et al. 2013; Abramov et al. 2020). Arsenic (As) and chromium (Cr), present as AsO_4^{3-} and CrO_4^{2-}, can be adsorbed by positively charged Fe(III) mineral surfaces or can substitute sulfate ions in minerals like schwertmannite. The low solubility and stable nature of the precipitated Fe(III) minerals ensure long-term immobilization of heavy metals, reducing their bioavailability and mobility (Abramov et al. 2020).

Iron (Fe) oxidation in denitrifying bacteria represents another significant mechanism within the broader context of biomineralization. Fe(II) oxidation in these bacteria like *Acidovorax* sp. is intricately linked with denitrification, a metabolic process where nitrate is reduced to nitrogen gas under anaerobic conditions (Lopez-Adams et al. 2022). In contrast to other bacteria that might oxidize Fe(II) through direct enzymatic pathways, denitrifying bacteria often utilize nitrite as an oxidizing agent. This coupling of Fe(II) oxidation with nitrite reduction not only aids in energy production for the bacteria but also leads to the formation of iron minerals

(Miot et al. 2015). The precipitated iron minerals, often observed at the cell surface or within the periplasmic space, reflect the dual sites of oxidation. This biomineralization process is affected by multiple variables that include hydrophilicity and surface charge, which affect mineral precipitation and stability (Saini and Chan 2013).

In the quest for innovative solutions to mitigate heavy metal pollution, the biomineralization of metal sulfides has become a preferred approach. For instance, study conducted by Ma et al. (2021) has revealed the unique mechanism possessed by deep-sea bacterium *Pseudomonas stutzeri* 273 to enhance its resistance to cadmium (Cd^{2+}) and remove it from contaminated aqueous solutions. This mechanism involves the utilization of L-cysteine, an amino acid that is metabolized into hydrogen sulfide (H_2S) by the enzyme threonine dehydratase. The H_2S then reacts with Cd^{2+} ions to form insoluble CdS nanoparticles. A similar mechanism of forming metal sulphides by producing hydrogen suphide as a byproduct of the enzymatic reaction by *Yokenella regensburgei GXAS49-1* is reported (Xu et al. 2024).

Competitive inhibition plays a critical role in modulating metal uptake and subsequent mineral formation by microorganisms. Competitive inhibition occurs when two or more ions or molecules compete for the same binding sites or transport pathways within a biological system. For instance, the study conducted by Fang et al. (2021) using *Sporosarcina pasteurii* to biomineralize cadmium in the presence of calcium exemplifies this phenomenon. Cd^{2+} and Ca^{2+}, both being divalent cations, can compete for binding sites on the bacterial cell surface and for uptake through similar transport mechanisms. When Ca^{2+} is present, it effectively competes with Cd^{2+} for these binding sites and transport channels, thereby reducing the amount of Cd^{2+} that enters the cell. This competition diminishes the intracellular accumulation of Cd^{2+}, mitigating its toxic effects and allowing the bacterial cells to maintain their metabolic functions, including urease activity.

While much of the research has focused on bacteria, fungi also play a crucial role in carbonate precipitation. Fungi can utilize various microbial pathways, such as ureolysis, to produce carbonates. To support this, in a study, *Aspergillus* sp. UF5 and *Fusarium oxysporum* UF8 isolated from cave speleothems were found to effectively biomineralize calcium carbonate and tolerate high concentrations of heavy metals like lead and strontium (Dhami et al. 2017). This study highlights the importance of fungal biomineralization in environmental remediation.

Proteins significantly contribute to fungal biomineralization. Fungal cells secrete various proteins into the extracellular environment. Proteins, such as enzymes and structural proteins, have the ability to form complexes with metal ions by binding such as copper, zinc, manganese, magnesium, iron, and cobalt due to their affinity for metal cofactors (Aubin-Tam and Hamad-Schifferli 2008). The bound metal ions serve as nucleation sites for the formation of mineral particles. Additionally, proteins act as a scaffold, thereby promoting the nucleation process by stabilizing the initial mineral precursors. This can lead to the formation of metal-protein complexes. Moreover, proteins can prevent the aggregation of minerals by providing steric stabilization that enhances the dispersibility and stability of the biominerals. Electrostatic interactions also play a crucial role, as proteins can attract metal ions through charge complementarity, facilitating the formation of stable metal-protein complexes. This mechanism is clearly studied in *Neurospora crassa* (Liu et al. 2021).

Zhao et al. (2020) conducted a study that highlights a novel mechanism of lead biomineralization mediated by *Phanerochaete chrysosporium*. In this study, when *P. chrysosporium* mycelium was exposed to a lead solution, Pb^{2+} ions were first adsorbed onto the fungal cell surface through chemical bonding with cell surface molecules. Subsequently, intracellular phosphatase enzymes hydrolyzed organic phosphorus compounds to produce inorganic phosphate, which was then secreted into the extracellular environment. This extracellular inorganic phosphate interacted with the adsorbed Pb^{2+} ions, leading to the formation of pyromorphite on the cell surface. Contrastingly, prior research has revealed that metabolites like oxalate, thiol compounds, and glutathione can chelate lead ions and facilitate their precipitation as various lead compounds. For instance, oxalate can form lead oxalate precipitates, while thiol compounds and glutathione can form lead-sulfur complexes (Li et al. 2015; Ye et al. 2018).

2.1 Organic Pollutants Mineralization and Key Mechanisms

Microbial biomineralization mechanisms for organic pollutants are diverse and vary significantly among different microorganisms. These mechanisms are tailored to the specific metabolic capabilities and environmental adaptations of each microbe, thereby enabling the transformation and stabilization of toxic organic compounds into less harmful or inert mineral forms. For instance, the degradation of nitroaromatic compounds by microbial strains typically involves plasmid-mediated pathways where large plasmids may encode the necessary catabolic enzymes (Ma et al. 2007; Bhatt et al. 2021). This has been observed in *Comamonas* sp. (Ma et al. 2007), *Diaphorobacter* sp. (Singh and Ramanathan 2013), *Pseudomonas* sp. (Xu et al. 2019), and various other microbes where specific plasmids carry genes responsible for the degradation of nitrocompounds. For instance, in *Diaphorobacter* sp., the degradation of 3-nitrotoluene proceeds through an oxidative pathway. The initial steps may involve the oxidation of the methyl group or the benzene ring, facilitated by monooxygenases or dioxygenases, leading to intermediates such as 3-nitrobenzaldehyde or 3-nitrobenzyl alcohol. These intermediates can be further processed by catechol 2,3-dioxygenase, resulting in ring cleavage products that enter the Krebs cycle. The ability of certain strains to utilize all isomers of mononitrotoluene and nitrobenzene as carbon and nitrogen sources varies, likely due to differences in their enzymatic systems and plasmid-encoded genes. For example, *Diaphorobacter* sp. can metabolize 2-nitrotoluene, 3-nitrotoluene, 4-nitrotoluene, and nitrobenzene, whereas other strains may exhibit selective utilization patterns. Understanding the genetic and enzymatic basis of these pathways is crucial for elucidating the mechanisms of nitroaromatic compound degradation and can inform bioremediation strategies (Singh and Ramanathan 2013).

Another example is the fungal biomineralization of nanoscale zerovalent iron (nZVI) composites for the treatment of chlorinated pollutants, which was studied using *Neurospora crassa*. The key mechanism observed involved cultivating ureolytic fungi in a medium supplemented with urea and Fe^{2+} ions. Urea was metabolized by the fungi producing carbonate ions, elevating pH, and facilitating iron carbonate precipitates around the fungal hyphae. This process was regulated by extracellular proteins secreted by the fungi, which influenced the nucleation and growth of iron carbonate crystals, eventually forming coral-like structures. Upon heat treatment, the iron carbonate minerals were reduced to form nZVI particles embedded within the carbonized fungal biomass. This configuration prevented aggregation, enhanced stability, and maintained high reactivity. The iron coral composite exhibited high surface area and electron-donating capacity, enabling efficient reductive dechlorination of chlorinated pollutants, such as carbon tetrachloride (CCl_4). The stability of the iron coral in aqueous solutions minimized sedimentation, maintaining high contact with pollutants and resulting in superior degradation efficiency (Li et al. 2020).

Tris(2-chloroisopropyl)phosphate is a chlorinated organophosphate flame retardant commonly found in consumer products, such as furniture, textiles, and electronics. Its persistence and slow degradation result in widespread environmental presence and bioaccumulation in both aquatic and terrestrial organisms, leading to long-term ecological impacts. Huang et al. (2024) present a promising strategy to mitigate these harmful effects through a synergistic bioremediation mechanism. The study highlights how denitrifying bacteria (DNB) can degrade Tris(2-chloroisopropyl)phosphate and stabilize lead ions (Pb^{2+}). The key bacterial strains used in this process are *Achromobacter*, *Pseudomonas*, *Citrobacter*, and *Stenotrophomonas*. Tris(2-chloroisopropyl)phosphate undergoes dechlorination and hydrolysis, producing phosphate (PO_4^{3-}) and chloride ions (Cl^-). The phosphate then reacts with Pb^{2+} to form insoluble lead phosphate compounds, primarily pyromorphite. This biomineralization process is supported by bacterial membrane biosorption, extracellular polymeric substances (EPS) biocomplexation, and efflux mechanisms, which collectively reduce the bioavailability and toxicity of Pb^{2+}. Furthermore, the degradation of Tris(2-chloroisopropyl)phosphate and stabilization of Pb^{2+} are facilitated by TCA cycle-based dissimilation, phosphatase expression, and protein phosphorylation, which help mitigate oxidative stress. In this process, Tris(2-chloroisopropyl)phosphate serves as an electron donor in respiratory redox reactions, with nitrate (NO_3^-) acting as the terminal electron acceptor, thereby providing energy for bacterial metabolism.

3 Biomineralization of Radionuclides by Geomicrobes

Radionuclides emit alpha, beta, or gamma radiation during their decay processes, which can be harmful to living organisms. Their half-life varies widely, from seconds to millions of years, affecting their persistence in the environment. Chemically, radionuclides exhibit diverse properties that influence their behavior in biological and environmental systems. Microbes interact with radionuclides through various mechanisms, leading to either immobilization or mobilization.

3.1 Strontium-90 (Sr-90)

Several studies have demonstrated the potential of different microbial and bacterial strains in precipitating Sr-90 under various environmental conditions. The halotolerant carbonate mineral-producing *Bacillus* sp. strain TK2d, isolated from sediment in Tokyo Bay, has shown a remarkable ability to precipitate Sr-90, especially in highly saline environments. This strain facilitates the remediation process through a series of well-coordinated mechanisms. Initially, Sr^{2+} ions are adsorbed onto the bacterial cell surfaces. This adsorption is mediated by organic acid functional groups and extracellular polymeric substances (EPS) produced by the bacteria. Following adsorption, intracellular ureolysis occurs, releasing bicarbonate ions. These bicarbonate ions react with the adsorbed Sr^{2+}, forming biogenic carbonate minerals on the cell surface. The mineralization process by strain TK2d is distinct from abiotic processes, as the bacterium shows a preference for Sr and Ca over Mg. This selectivity is hypothesized to be due to species-specific ion channels that regulate Ca^{2+}/Mg^{2+} ratios around the microbial cells (Horiike et al. 2017).

Phosphate biomineralization also plays a key part in the immobilization of Sr-90. According to a study by Robinson et al. (2023), remediation involves the dissolution of phosphate amendments into the environment. Phosphate ions react with dissolved Sr-90 to form solid-phase compounds, effectively immobilizing the radionuclide. Microbial activity enhances phosphate availability through the degradation of organic phosphorus compounds, such as citrate and glycerol phosphate, increasing the concentration of free phosphate ions in solution. The microbial community involved in this process includes various bacteria such as *Phalaris aquatica*, *Sphingopyxis macrogoltabida*, *Pseudomonas helmanticensis*, *Ferruginibacter profundus*, *Burkholderia ferrariae*, *Geothrix fermentans*, *Chthoniobacter flavus*, and *Cypionkella psychrotolerans*. The microbial degradation of organic acids elevates the pH, promoting the precipitation of strontium-phosphate minerals. Additionally, the adsorption of phosphate and strontium ions to sediment particles facilitates local mineralization, depending on sediment surface area and microbial community composition.

Research on indigenous ureolytic microbial populations from the sediments of Sellafield, a UK nuclear licensed site by White-Pettigrew et al. (2024), has further highlighted the mechanisms involved in strontium removal and biomineralization. In the absence of urea, Sr removal is primarily achieved through adsorption onto the negatively charged surfaces of mineral particles in the sediment. When urea is introduced, ureolytic bacteria hydrolyze it to produce ammonia, raising the pH and promoting further Sr adsorption and carbonate mineral precipitation. The addition of organic acids like acetate and lactate enhances anaerobic bacterial growth and ureolytic activity, leading to a more rapid increase in pH and improved Sr removal. The presence of Fe(III) provides an electron acceptor for anaerobic bacteria, stimulating the ureolysis process and enhancing Sr removal. Yeast extract acts as an electron donor, further promoting bacterial activity and increasing ureolysis rates, thereby enhancing Sr removal. Elevated pH conditions favor the precipitation of calcium carbonate ($CaCO_3$), with Sr^{2+} ions co-precipitating with $CaCO_3$, effectively removing Sr from solution.

3.2 Uranium-238

Studies have demonstrated how different microorganisms facilitate uranium precipitation through distinct mechanisms. Chandwadkar et al. (2018) investigated uranium biomineralization by *Serratia* sp. strain OT II 7, which promotes uranium phosphate precipitation through a combination of phosphatase activities and cell surface interactions across varying pH conditions. At pH5, the strain exhibits high phosphatase activity, resulting in intracellular deposition of uranyl phosphate and prolonged precipitation kinetics due to decreased enzyme activity and limited nucleation sites. At pH7, rapid uranium precipitation occurs, driven by the sorption of positively charged uranyl species to the negatively charged cell surface, which forms nucleation sites and enhances mineral growth. Conversely, at pH9, extracellular precipitation predominates due to poor biosorption of negatively charged uranyl carbonate complexes, which maintains high cell viability. This study concluded that uranium precipitation effectiveness is influenced by phosphatase activity, availability of nucleation sites, pH-specific enzyme activity, and uranium speciation.

In a similar vein, Song et al. (2019) explored uranium biomineralization using *Bacillus subtilis* ATCC–6633, demonstrating that pH and uranium concentration significantly affect the process. The biosorption capacity of *Bacillus subtilis* increased with pH, peaking at pH5 due to optimal electrostatic interactions between the negatively charged bacterial surface and uranium species. At pH values below 5, the adsorption of positively charged UO_2^{2+} ions were enhanced by attractive forces with the positively charged bacterial surface. However, at pH values above 5, the predominance of negatively charged uranium species led to reduced adsorption due to repulsive interactions. High uranium concentrations exceeding 100 mg/L impaired biosorption efficiency due to toxicity, forming amorphous uranium compounds. X-ray diffraction (XRD) analysis confirmed that at lower concentrations, uranium was primarily mineralized as uramphite $[(NH_4)(UO_2)PO_4{\cdot}3H_2O]$, while higher concentrations led to amorphous compounds, reflecting the detrimental effects of uranium on bacterial metabolic functions.

Additionally, Zhu et al. (2023) investigated the biomineralization of uranium using *Bacillus thuringiensis* X-27, isolated from uranium-contaminated soil. This strain facilitates the precipitation of uranium outside its cells, forming amorphous uranium deposits that gradually transform into crystalline minerals, such as uramphite. The transformation is driven by the release of inorganic phosphate (Pi) into the extracellular environment and a shift in pH to alkaline conditions. The increased Pi concentration promotes uranium phosphate mineral formation, while alkaline pH further supports the crystallization process.

Tu et al. (2019) studied the influence of humic substances on uranium biomineralization induced by *Bacillus* sp. dwc-2. Phosphate ions (PO_4^{3-}), humic acid (HA), and fulvic acid (FA) form complexes with uranyl ions. Although HA and FA can bind with uranyl, their experimental data suggest that they are not as competitive as phosphate ions. The binding strength of uranyl with phosphate is significantly higher than with HA or FA, meaning phosphate ions can more effectively displace uranyl from HA or FA complexes. Consequently, HA and FA do not substantially inhibit the formation of uranium phosphate minerals. Instead, HA and FA influence the morphology of these minerals, causing a transition from small particles to lamellar structures during the biomineralization process. This morphological change occurs because HA and FA affect the second stage of biomineralization, where the uranyl-phosphate complexes nucleate and grow.

4 Biomineralization of Other Radionuclides

The research on biomineralization of other radionuclides is such as cesium, strontium, technetium, plutonium, and neptunium are less common. From the available literature, it can be deduced that microbial interactions with radionuclides, such as cesium, strontium, technetium, plutonium, americium, cerium, curium, and neptunium, involve complex biochemical and physiological mechanisms that enable these microorganisms to tolerate and transform radioactive substances. Microbes can interact with these radionuclides through biosorption, biotransformation, and intracellular accumulation. Biosorption, a metabolism-independent interaction, relies on the binding of radionuclides to cell surface structures. Functional groups such as carboxyl, amine, hydroxyl, phosphate, and sulfhydryl on microbial cell walls interact with radionuclide ions, facilitating their adhesion and concentration on the cell surface (Lloyd and Macaskie 2000; Lopez-Fernandez et al. 2021). Microbes like *Pseudomonas fluorescens*, *Bacillus cereus* (Mao et al. 2011; Kim et al. 2016), *Rhodosporidium fluviale* (Lan et al. 2014), *Actinobacterium nocardiopsis* sp. (Sivaperumal et al. 2018), *Synechococcus* (Yu et al. 2020), *Bacillus pumilus* (Dai et al. 2020), *Aspergillus brasiliensis* (Abass et al. 2024), and marine algae like *Sargassum glaucescens*, *Cystoseira* (Dabbagh et al. 2008), and *Arthrospira platensis* (Zinicovscaia et al. 2020) have demonstrated significant biosorption capacities for radionuclides, including cesium and strontium. Furthermore, biotransformation processes involve the enzymatic reduction or oxidation of radionuclides, altering their valence states and solubility. For example, technetium can be reduced by specific bacterial enzymes to form less soluble species, thereby decreasing its environmental mobility (Chernyh et al. 2007). The process of bioaccumulation is influenced by the radionuclides chemical form, concentration of radionuclides in the environment, and the specific metabolic and physiological characteristics of the microorganism. For instance, certain bacteria and fungi have specialized mechanisms to uptake and accumulate radionuclides efficiently. Bacteria like *Geobacter sulfurreducens*, *Shewanella oneidensis* (Renshaw et al. 2009), *lactobacilli*, *bifidobacteria* (Endo et al. 2024), *Haematococcus pluvialis*, and *Chlorella vulgaris* (Abdelkarim and Imam 2024) have demonstrated significant bioaccumulation of plutonium, cesium, and strontium.

5 Challenges and Future Directions

A significant challenge in microbial biomineralization is the impact of elevated concentrations of chemicals on the biomineralization mechanisms of microbes. Elevated levels of contaminants, whether they are metals, radionuclides, or other chemicals, can disrupt microbial metabolism and influence the efficiency of mineralization processes. High contaminant concentrations often result in adverse consequences that can impede microbial growth and activity. This inhibition can alter or even negate the mineralization process, making it less effective. Consequently, understanding and managing the effects of high contaminant concentrations is crucial for optimizing biomineralization strategies for environmental cleanup and resource recovery.

Another notable impediment emerges from the diversity of metabolic pathways among different microbes. Each microorganism employs unique biochemical pathways to facilitate biomineralization of specific chemicals, metals, or radionuclides. This variability makes it difficult to generalize a single biomineralization mechanism for treating a wide range of contaminants in diverse environments. For example, while certain bacteria may effectively sequester heavy metals through sulfate reduction, others might utilize different metabolic routes to achieve similar results. This heterogeneity of microbial processes complicates the development of universal biomineralization.

To address the challenges associated with high contaminant concentrations, future research should focus on optimizing microbial processes under extreme conditions. Additionally, the development of strategies to generalize biomineralization mechanisms across different types of contaminants will be essential. One promising approach is the engineering of microbial consortia—groups of microorganisms that work together to achieve a broader range of biomineralization processes.

References

Abass MR, Kandeel EM, Abou-Lilah RA, Mohamed MK (2024) Effective biosorption of cesium and strontium ions from aqueous solutions using silica loaded with aspergillus brasiliensis. Water Air Soil Pollut 235(1):61

Abdelkarim MS, Imam N (2024) Radiation hazards and extremophiles bioaccumulation of radionuclides from hypersaline lakes and hot springs. Int J Environ Sci Technol 21(3):3021–3036

Abramov SM, Tejada J, Grimm L, Schädler F, Bulaev A, Tomaszewski EJ et al (2020) Role of biogenic Fe (III) minerals as a sink and carrier of heavy metals in the Rio Tinto, Spain. Sci Total Environ 718:137294

Al Disi Z, Attia E, Ahmad MI, Zouari N (2022) Immobilization of heavy metals by microbially induced carbonate precipitation using hydrocarbon-degrading ureolytic bacteria. Biotechnol Rep 35:e00747

Aloisi G, Gloter A, Kruger M, Wallmann K, Guyot F, Zuddas P (2006) Nucleation of calcium carbonate on bacterial nanoglobules. Geology 34(12):1017–1020

Aubin-Tam ME, Hamad-Schifferli K (2008) Structure and function of nanoparticle–protein conjugates. Biomed Mater 3(3):034001

Bai H, Liu D, Zheng W, Ma L, Yang S, Cao J et al (2021) Microbially-induced calcium carbonate precipitation by a halophilic ureolytic bacterium and its potential for remediation of heavy metal-contaminated saline environments. Int Biodeterior Biodegradation 165:105311

Bäuerlein E (2003) Biomineralization of unicellular organisms: an unusual membrane biochemistry for the production of inorganic nano-and microstructures. Angew Chem Int Ed 42(6):614–641

Bhatt P, Bhandari G, Bhatt K, Maithani D, Mishra S, Gangola S et al (2021) Plasmid-mediated catabolism for the removal of xenobiotics from the environment. J Hazard Mater 420:126618

Chandwadkar P, Misra HS, Acharya C (2018) Uranium biomineralization induced by a metal tolerant Serratia strain under acid, alkaline and irradiated conditions. Metallomics 10(8):1078–1088

Cheng Y, Zhang T, Zhang L, Ke Z, Kovarik L, Dong H (2022) Resource recovery: adsorption and biomineralization of cerium by Bacillus licheniformis. J Hazard Mater 426:127844

Chernyh NA, Gavrilov SN, Sorokin VV, German KE, Sergeant C, Simonoff M et al (2007) Characterization of technetium (vII) reduction by cell suspensions of thermophilic bacteria and archaea. Appl Microbiol Biotechnol 76:467–472

Dabbagh R, Ebrahimi M, Aflaki F, Ghafourian H, Sahafipour MH (2008) Biosorption of stable cesium by chemically modified biomass of sargassum glaucescens and cystoseira indica in a continuous flow system. J Hazard Mater 159(2–3):354–357

Dai Q, Zhang T, Zhao Y, Li Q, Dong F, Jiang C (2020) Potentiality of living Bacillus pumilus SWU7-1 in biosorption of strontium radionuclide. Chemosphere 260:127559

Dhami NK, Quirin MEC, Mukherjee A (2017) Carbonate biomineralization and heavy metal remediation by calcifying fungi isolated from karstic caves. Ecol Eng 103:106–117

Dong X, Liu C, Ma D, Wu Y, Man H, Wu X et al (2021) Organic carbon mineralization and bacterial community of active layer soils response to short-term warming in the Great Hing'an Mountains of Northeast China. Front Microbiol 12:802213

Dong Y, Gao Z, Di J, Wang D, Yang Z, Guo X, Zhu X (2024) Study on the effectiveness of sulfate-reducing bacteria to remove Pb (II) and Zn (II) in tailings and acid mine drainage. Front Microbiol 15:1352430

Durak GM, Taylor AR, Walker CE, Audic S, Schroeder D, Probert I, et al. (2016) A role for diatom-like silicon transporters in calcifying coccolithophores. Nat Commun 7:10543

Endo R, Karasawa S, Aoyagi H (2024) Simultaneous removal of caesium and strontium using different removal mechanisms of probiotic bacteria. Sci Rep 14(1):7630

Fang L, Niu Q, Cheng L, Jiang J, Yu YY, Chu J et al (2021) Ca-mediated alleviation of Cd^{2+} induced toxicity and improved Cd^{2+} biomineralization by Sporosarcina pasteurii. Sci Total Environ 787:147627

Hammes F, Boon N, de Villiers J, Verstraete W, Siciliano SD (2003) Strain-specific ureolytic microbial calcium carbonate precipitation. Appl Environ Microbiol 69(8):4901–4909

He N, Hu L, He Z, Li M, Huang Y (2022) Mineralization of lead by phanerochaete chrysosporium microcapsules loaded with hydroxyapatite. J Hazard Mater 422:126902

Horiike T, Dotsuta Y, Nakano Y, Ochiai A, Utsunomiya S, Ohnuki T, Yamashita M (2017) Removal of soluble strontium via incorporation into biogenic carbonate minerals by halophilic bacterium Bacillus sp. strain TK2d in a highly saline solution. Appl Environ Microbiol 83(20):e00855–e00817

Huang ZS, Tan XQ, Yang HB, Zeng Y, Chen SJ, Wei ZS, Huang YQ (2024) Mechanistic insights into tris (2-chloroisopropyl) phosphate biomineralization coupled with lead (II) biostabilization driven by denitrifying bacteria. Sci Total Environ 189:173927

Jalilvand N, Akhgar A, Alikhani HA, Rahmani HA, Rejali F (2020) Removal of heavy metals zinc, lead, and cadmium by biomineralization of urease-producing bacteria isolated from Iranian mine calcareous soils. J Soil Sci Plant Nutr 20:206–219

Jiang Y, Zhao X, Zhou Y, Ding C (2022) Effect of the phosphate solubilization and mineralization synergistic mechanism of Ochrobactrum sp. on the remediation of lead. Environ Sci Pollut Res 29(38):58037–58052

Kaur M, Sidhu N, Reddy MS (2023) Removal of cadmium and arsenic from water through biomineralization. Environ Monit Assess 195(9):1019

Kim GY, Jang SC, Song YH, Lee CS, Huh YS, Roh C (2016) Screening and identification of a cesium-tolerant strain of bacteria for cesium biosorption. Korean J Environ Biol 34(4):304–313

Lan T, Feng Y, Liao J, Li X, Ding C, Zhang D et al (2014) Biosorption behavior and mechanism of cesium-137 on rhodosporidium fluviale strain UA2 isolated from cesium solution. J Environ Radioact 134:6–13

Li N, Zeng G, Huang D, Huang C, Lai C, Wei Z et al (2015) Response of extracellular carboxylic and thiol ligands (oxalate, thiol compounds) to Pb^{2+} stress in phanerochaete chrysosporium. Environ Sci Pollut Res 22:12655–12663

Li Q, Liu D, Wang T, Chen C, Gadd GM (2020) Iron coral: novel fungal biomineralization of nanoscale zerovalent iron composites for treatment of chlorinated pollutants. Chem Eng J 402:126263

Liu F, Shah DS, Gadd GM (2021) Role of protein in fungal biomineralization of copper carbonate nanoparticles. Curr Biol 31(2):358–368

Lloyd JR, Macaskie LE (2000) Bioremediation of radionuclide-containing wastewaters. In: Environmental microbe-metal interactions. ASM Press, pp 277–327

Lopez-Adams R, Fairclough SM, Lyon IC, Haigh SJ, Zhang J, Zhao FJ et al (2022) Elucidating heterogeneous iron biomineralization patterns in a denitrifying As (III)-oxidizing bacterium: implications for arsenic immobilization. Environ Sci Nano 9(3):1076–1090

Lopez-Fernandez M, Jroundi F, Ruiz-Fresneda MA, Merroun ML (2021) Microbial interaction with and tolerance of radionuclides: underlying mechanisms and biotechnological applications. Microb Biotechnol 14(3):810–828

Lowenstam HA (1981) Minerals formed by organisms. Science 211(4487):1126–1131

Ma YF, Wu JF, Wang SY, Jiang CY, Zhang Y, Qi SW et al (2007) Nucleotide sequence of plasmid pCNB1 from comamonas strain CNB-1 reveals novel genetic organization and evolution for 4-chloronitrobenzene degradation. Appl Environ Microbiol 73(14):4477–4483

Ma N, Cai R, Sun C (2021) Threonine dehydratase enhances bacterial cadmium resistance via driving cysteine desulfuration and biomineralization of cadmium sulfide nanocrystals. J Hazard Mater 417:126102

Mao Y, Hu H, Yan Y (2011) Biosorption of cesium (I) from aqueous solution by a novel exopolymers secreted from pseudomonas fluorescens C-2: equilibrium and kinetic studies. J Environ Sci 23(7):1104–1112

McCausland HC, Wetmore KM, Arkin AP, Komeili A (2022) Global analysis of biomineralization genes in magnetospirillum magneticum AMB-1. Msystems 7(1):e01037–e01021

Miot J, Remusat L, Duprat E, Gonzalez A, Pont S, Poinsot M (2015) Fe biomineralization mirrors individual metabolic activity in a nitrate-dependent Fe (II)-oxidizer. Front Microbiol 6:879

Morcillo F, Gonzalez-Munoz MT, Reitz T, Romero-Gonzalez ME, Arias JM, Merroun ML (2014) Biosorption and biomineralization of U (VI) by the marine bacterium Idiomarina loihiensis MAH1: effect of background electrolyte and pH. PLoS One 9(3):e91305

Ouyang X, Lai DY, Marchand C, Lee SY (2022) Carbon storage and mineralization in coastal wetlands. In: Carbon mineralization in coastal wetlands. Elsevier, pp 295–310

Renshaw JC, Law N, Geissler A, Livens FR, Lloyd JR (2009) Impact of the Fe (III)-reducing bacteria Geobacter sulfurreducens and Shewanella oneidensis on the speciation of plutonium. Biogeochemistry 94:191–196

Robbins LJ, Konhauser KO (2021) Geobiology and geomicrobiology. In: Alderton D, Elias SA (eds) Encyclopedia of geology, 2nd edn. Academic Press, pp 554–568
Robinson C, Shaw S, Lloyd JR, Graham J, Morris K (2023) Phosphate (bio) mineralization remediation of 90Sr-contaminated Groundwaters. ACS ES T Water 3(10):3223–3234
Robles-Fernández A, Areias C, Daffonchio D, Vahrenkamp VC, Sánchez-Román M (2022) The role of microorganisms in the nucleation of carbonates, environmental implications and applications. Fortschr Mineral 12(12):1562
Saini G, Chan CS (2013) Near-neutral surface charge and hydrophilicity prevent mineral encrustation of F e-oxidizing micro-organisms. Geobiology 11(2):191–200
Sánchez-Castro I, Martínez-Rodríguez P, Jroundi F, Solari PL, Descostes M, Merroun ML (2020) High-efficient microbial immobilization of solved U (VI) by the stenotrophomonas strain Br8. Water Res 183:116110
Singh D, Ramanathan G (2013) Biomineralization of 3-nitrotoluene by diaphorobacter species. Biodegradation 24:645–655
Sivaperumal P, Kamala K, Rajaram R (2018) Biosorption of long half-life radionuclide of strontium ion (Sr+) by marine actinobacterium nocardiopsis sp. 13H. Geomicrobiol J 35(4):300–310
Song J, Han B, Song H, Yang J, Zhang L, Ning P, Lin Z (2019) Nonreductive biomineralization of uranium by Bacillus subtilis ATCC–6633 under aerobic conditions. J Environ Radioact 208:106027
Sørensen SR, Juhler RK, Aamand J (2013) Degradation and mineralisation of diuron by Sphingomonas sp. SRS2 and its potential for remediating at a realistic µg L− 1 diuron concentration. Pest Manag Sci 69(11):1239–1244
Southam G, Lovley DR (2000) Environmental microbe-metal interactions. American Society for Microbiology, Washington, DC
Staicu LC, Wójtowicz PJ, Molnár Z, Ruiz-Agudo E, Gallego JLR, Baragaño D, Pósfai M (2022) Interplay between arsenic and selenium biomineralization in *Shewanella* sp. O23S. Environ Pollut 306:119451
Tu H, Lan T, Yuan G, Zhao C, Liu J, Li F et al (2019) The influence of humic substances on uranium biomineralization induced by Bacillus sp. dwc-2. J Environ Radioact 197:23–29
Wang H, Gong L, Cravotta CA III, Yang X, Tuovinen OH, Dong H, Fu X (2013) Inhibition of bacterial oxidation of ferrous iron by lead nitrate in sulfate-rich systems. J Hazard Mater 244:718–725
Wang Y, Xiaoqin N, Wencai C, Faqin D, Yuanyuan Z, Congcong D, et al (2019) A synergistic biosorption and biomineralization strategy for Kocuria sp. to immobilizing U(VI) from aqueous solution. J Mol Liq. 275:215–220, ISSN 0167-7322
White-Pettigrew M, Shaw S, Hughes L, Boothman C, Graham J, Abrahamsen-Mills L et al (2024) Enhanced strontium removal through microbially induced carbonate precipitation by indigenous ureolytic bacteria. ACS Earth Space Chem 8(3):483–498
Wu S, Zheng R, Sha Z, Sun C (2017) Genome sequence of Pseudomonas stutzeri 273 and identification of the exopolysaccharide EPS273 biosynthesis locus. Mar Drugs 15(7):218
Wu Z, Su J, Ali A, Hu X, Wang Z (2021) Study on the simultaneous removal of fluoride, heavy metals and nitrate by calcium precipitating strain Acinetobacter sp. H12. J Hazard Mater 405:124255
Xu W, Zhao Q, Li Z, Lu X, Han S, Ye Z (2019) Biodegradation of dinitrotoluene sulfonates and other nitro-aromatic compounds by Pseudomonas sp. X5 isolated from TNT red water contaminated soil. J Clean Prod 214:782–790
Xu Y, Qi Y, Xie J, Qin Q, Huang G, Tang P et al (2024) A Yokenella regensburgei strain effectively removes Cu (II), Pb (II), and Cd (II) through a combination of absorption and mineralization modes. J Water Process Eng 58:104805
Ye B, Luo Y, He J, Sun L, Long B, Liu Q et al (2018) Investigation of lead bioimmobilization and transformation by penicillium oxalicum SL2. Bioresour Technol 264:206–210
Yu R, Chai H, Yu Z, Wu X, Liu Y, Shen L et al (2020) Behavior and mechanism of cesium biosorption from aqueous solution by living synechococcus PCC7002. Microorganisms 8(4):491

Zhang K, Xue Y, Xu H, Yao Y (2019) Lead removal by phosphate solubilizing bacteria isolated from soil through biomineralization. Chemosphere 224:272–279

Zhang K, Zhang D, Li X, Xue Y (2022) Biomineralization of lead in wastewater: bacterial reutilization and metal recovery. J Hazard Mater 421:126765

Zhao WW, Zhu G, Daugulis AJ, Chen Q, Ma HY, Zheng P et al (2020) Removal and biomineralization of Pb^{2+} in water by fungus Phanerochaete chrysoporium. J Clean Prod 260:120980

Zhu T, Zeng Q, Zhao C, Wen Y, Li S, Li F et al (2023) Extracellular biomineralization of uranium and its toxicity alleviation to Bacillus thuringiensis X-27. J Environ Radioact 261:107126

Zinicovscaia I, Safonov A, Zelenina D, Ershova Y, Boldyrev K (2020) Evaluation of biosorption and bioaccumulation capacity of cyanobacteria Arthrospira (Spirulina) Platensis for radionuclides. Algal Res 51:102075

Microbial Diversity and Mineral Transformation with Reference to Endosymbiotic Niches

Piyali Das and Santa Ram Joshi

Contents

Abstract Microbial diversity, including prokaryotes and eukaryotes and their symbiotic association with each other, can contribute to geomicrobial processes such as metal precipitation, biotransformation, biogeochemical cycling, mineral transformation, decomposition, bioweathering, soil, and sediment formation. Bacteria contribute considerably to mineral weathering by secreting acids and enzymes that break down rocks and minerals, which in turn release vital nutrients like phosphorus

P. Das · S. R. Joshi (✉)
Microbiology Laboratory, Department of Biotechnology and Bioinformatics, North-Eastern Hill University, Shillong, Meghalaya, India

S. R. Joshi et al. (eds.), *Mineral Transformation and Bioremediation by Geo-Microbes*, Interdisciplinary Biotechnological Advances,
https://doi.org/10.1007/978-981-96-3033-2_6

and potassium into the soil. They also play important roles in element cycling, includes nitrogen fixation, which turns nitrogen from the atmosphere into a substance that plants can use, and organic matter breakdown, which releases nutrients back into the soil for plant uptake. Endosymbiotic bacteria likely *Rhizobium* sp. can expedite weathering and increase accessible silica in the rhizosphere. During weathering, polymeric insoluble silica in soils dissolves and releases monosilicic acid, a bioavailable form of silicon that plants absorb. Symbiotic fungi play crucial roles in aerobic terrestrial environments when considering rock surfaces, soil, and plant root-soil interfaces. Mycorrhizal fungi associated with roots play a significant role in mineral transformation and inorganic nutrient redistribution, including essential metals and phosphate, and regulate environmental carbon flow. Lichens are fungal growth form that create a symbiotic relationship with photosynthetic organism. Different microbial groups contribute to mineral transformations and nutrient availability in the environments they inhabit. This chapter includes the significance of endosymbiotic niche in mineral transformation.

Keywords Microbial diversity · Mineral transformation · Endosymbionts · Endosymbiotic niche

Abbreviations

C	Carbon
Ca	Calcium
Co	Cobalt
Cu	Copper
DOI	Digital object identifier
Ed.	Editor
edn.	Edition
Eds.	Editors
ELF	Endolichenic fungi
Fe	Iron
H	Hydrogen
K	Potassium
Mn	Manganese
Mo	Molybdenum
MSM	Mineral-solubilizing microorganism
N	Nitrogen
Na	Sodium
Ni	Nickel
O	Oxygen
P	Phosphorus
Pb	Lead
RNA	Ribonucleic acid

S	Sulphur
Si	Silicon
SSU rRNA	Small subunit ribosomal ribonucleic acid
Zn	Zinc

1 Introduction

The excitement in this new era of exploration is tangible, as molecular tools help uncover the microbial "dark matter" that drives global biogeochemical cycles, supports the health of multicellular organisms, and sustains our ~3.8-billion-year-old biosphere. The shift from simply describing microbial communities to gaining a deeper, quantitative understanding of how they assemble, evolve, and achieve biological and ecological functionality happened swiftly and with remarkable impact (Gibbons and Gilbert 2015).

Microorganisms were the earliest known cellular life forms, appearing more than three billion years ago, long before macro-organisms emerged. In the 1970s, Woese and colleagues developed a universal tree of life that depends on the sequence data of small-subunit ribosomal RNA (SSU rRNA), which led to the discovery of the three domains of life: Eukarya, Archaea, and Bacteria. Before this, organisms were classified as prokaryotes or eukaryotes depending on whether they had a membrane-bound nucleus (Bradley and Martiny 2007).

The domains of Bacteria and Archaea encircle an astonishing diversity of organisms, each uniquely adapted to thrive in Earth's most extreme and varied environments. From searing volcanic hot springs to the crushing depths of the ocean, these microorganisms have evolved a remarkable array of energy sources, carbon and nitrogen pathways, and metabolic strategies. Halophiles flourish in salt-crusted brine ponds, thermophiles thrive on smoldering coal piles, and barophiles endure the immense pressures of the deep sea. Some bacteria form symbiotic relationships with plants, while others inhabit mammalian cells as intracellular parasites or form stable consortia with other microorganisms. This vast microbial diversity offers an unparalleled reservoir of potential for applied microbiology, opening doors to countless innovations (Glazer and Nikaido 2012).

At both global and local levels, fungi play a crucial role in the biogeochemical processes, including inorganic and organic transformations, mineral and rock interactions, nutrient cycling and elemental cycling (Gadd and Raven 2010) significantly impacting living organisms, plant growth, and human health. Their unique ability to transform toxic metals and metalloids offers promising potential for the remediation of contaminated land and environments (Gadd et al. 2007). They help to promote bioweathering, interactions between fungal and clay minerals, mycogenic mineral production, and metal transformations (Gadd and Raven 2010; Deng and Cao 2017). While these activities take place in aquatic as well as terrestrial environments, fungi have the greatest effect on terrestrial habitats, particularly rocks, soil, mineral surfaces, and rhizosphere (Gadd and Raven 2010).

Plants have natural interactions with microbes, both on the surface and within. Some fungi and bacteria from the rhizosphere (root zone) and phyllosphere (leaf surface) can invade the plant, colonizing intercellular gaps and vascular tissues. These microbes can have positive, neutral, or adverse effects on the plant (Govindasamy et al. 2013).

Symbiotic connections in marine ecosystems are ubiquitous in the oceans. Such interactions between animals and microorganisms have a profound impact on biological diversity and environmental equilibrium. These interactions are occasionally disregarded, yet they are essential for preserving ecological health and stability. Sponges are an especially valuable evolutionary model for investigating these microbial relationships. Sponges are one of the oldest existing multicellular organisms (metazoans), which means they have existed for million years. Their historic lineage provides an important point of reference for knowing how life on Earth developed symbiotic connections with microorganisms. Furthermore, sponges are known to reside a wide range of microbes (fungi, bacteria and archaea), which makes them an excellent resource for researching how varied microbial populations influence the behavior of marine organisms across time (Garate et al. 2017).

Microbes within endosymbiotic niches can play significant roles in mineral transformation processes. For example, certain bacteria are capable of catalyzing the oxidation or reduction of minerals as part of their metabolic processes. This can lead to the transformation of minerals and the release of different chemical compounds, which may have implications for nutrient cycling and ecosystem dynamics.

2 Microbial Diversity

New explorers encountered a vast and mysterious world of single-celled organisms, once imperceptible and largely inaccessible. This hidden world, underestimated for so long, has only become accessible through the development of high-throughput sequencing and molecular fingerprinting technologies that enable us to bypass traditional practices (Gibbons and Gilbert 2015).

Microbial diversity encompasses a vast array of organisms, including a wide range of microorganisms such as fungi, bacteria, and viruses, each playing a crucial role in the natural world. These microorganisms serve as the foundation of life on Earth, representing an unparalleled reservoir of chemical and molecular diversity. This diversity forms the backbone of essential ecological processes like biogeochemical cycles and food chains, while sustaining crucial links between individuals and with other organisms, thereby maintaining an intricate balance of ecosystems (Onen et al. 2020).

Despite the difficult conditions, harsh environments such as acidic mine effluent, hot springs, glaciers, and submerged vents maintain a wide variety of microbes from three life domains, i.e., Bacteria, Archaea, and Eukarya. The environmental variation and assorted variety of these microbes in a range of harsh environments have been demonstrated by decades of research, made possible by high-throughput sequencing (Shu and Huang 2022).

2.1 Plant–Microbe Interaction

Plants have an important role in moderating interactions among their associated microbial ecosystems. Plant–microbe interaction research sheds light on various molecular interactions that occur between plants and diseases. Microbes linked with plants can improve resilience to stress by activating a variety of pathways (Islam et al. 2023). Chemical interactions in soil promote microbe-plant relationships (Hassani et al. 2018). Plants influence their interactions with microbes and soil by releasing rhizodeposits, which contain a variety of components including border cells, soluble exudates, root caps, lost carbon, and volatile organic compounds (Farrar et al. 2014).

Plant growth promoting bacteria increase plant resistance to salt and drought and also can aid in the phytoremediation of metal-polluted soils (Schirawski and Perlin 2018; Fig. 1).

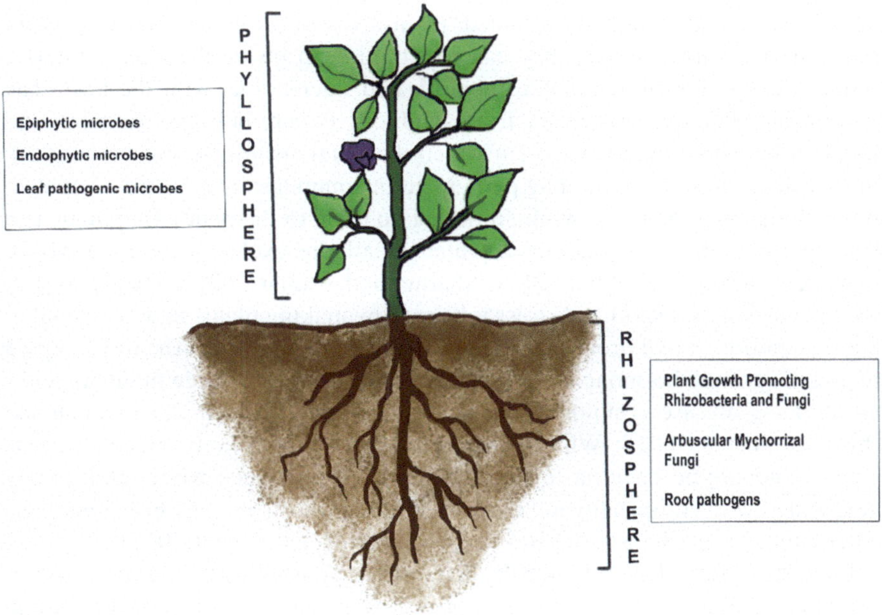

Fig. 1 Overview of plant–microbe interaction

3 Endosymbionts

Endosymbionts are interesting organisms that engage in a symbiotic relationship with another cell or organism, often to benefit both parties. These symbiotic partners can be found in various locations within the host. Some endosymbionts reside within the cells of their host, a relationship known as intracellular endosymbiosis, where they may be integrated into the host's cellular machinery, providing essential functions such as nutrient production or energy generation. Others remain outside the cells, adhering to the surface, known as extracellular endosymbiosis, where they might play critical roles in protecting the host, facilitating nutrient absorption, or engaging in other beneficial interactions (Casem 2016).

Microorganisms that inhabit the tissues of plants, often in a symbiotic association, can include various groups such as fungi, bacteria, and actinobacteria (Pimentel et al. 2011; Singh and Dubey 2015, 2018).

3.1 Endophytes

Endophytes are bacteria and fungi that live within plant tissues without causing harm (Govindasamy et al. 2013). They thrive beneath the cell surface of the epidermal layers and frequently establish themselves in different plant tissues. In these internal surroundings, they can penetrate deep inside the plant's interior tissues and form various kinds of complicated interactions with the host. The relationships that exist may vary from symbiotic (where both the microbes and the plant benefit) to mutualistic (in which the relationships benefit both sides). Furthermore, these bacteria may participate in commensalism, a sort of contact in which the bacteria profit while inflicting no harm to the plant. They may also develop trophobiotic associations, which entail the exchange of minerals or resources (Schulz and Boyle 2006; Sharma and Kumar 2021). They play key roles in nutrient transport and stress tolerance by making plants more resistant to abiotic conditions like drought, salt, and high temperatures, in part by changing the plant's growth hormone metabolism, often occurring in large numbers without inducing disease symptoms (Govindasamy et al. 2013; Suryanarayanan and Thirunavukkarasu 2017; White et al. 2019). This multiplicity of connections enables endophytic bacteria to flourish inside plant tissues while additionally enhancing the overall health and endurance of their hosts in a variety of ecological circumstances (Schulz and Boyle 2006; Sharma and Kumar 2021).

Endophytic fungi have additionally demonstrated the ability to improve phytoremediation. Unlike bacteria, many fungi grow in filamentous form, allowing them to acquire both exploitative and exploratory growth tactics. They can develop linear structures composed of aggregated hyphae, which aid in the protection of the translocation process of the fungus (Deng and Cao 2017). Fungi also help with environmental biotechnology by recovering metals and radionuclides, detoxifying, and degrading contaminants (Gadd and Raven 2010).

3.2 *Endolichenic Fungi: Symbiotic Fungal Partners of Lichens*

Lichens are produced when fungus (mycobiont) form symbiotic relationships with their photosynthetic organisms (photobiont such as single-celled or cyanobacterial cells or filamentous algae) (Suryanarayanan and Thirunavukkarasu 2017). Aside from the photobiont and the mycobiont present in lichen thalli, non-obligate microorganisms reside in the thalli that have lately been discovered. Endosymbionts are microorganisms that dwell asymptomatically in the thalli (Vinayaka 2020). Lichens host these varied population of microorganisms, which involve endolichenic fungi (ELF). These ELF colonize in lichen thalli without generating symptoms associated with disease, and their way of living is similar to those of endophytic fungi. ELF are heterotrophs that rely on the lichen's photosynthetic partner for carbon. As a result, ELF are commonly found in numerous quantities within the algal layer of the lichen thallus (Yang et al. 2021).

Endolichenic microorganisms are sometime known as lichen parasymbionts (Vinayaka 2020; Fig. 2).

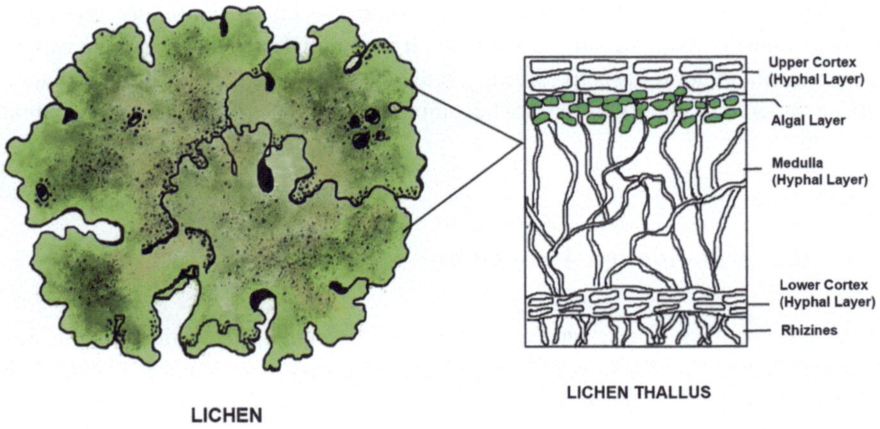

Fig. 2 Lichen: fungal growth form that create a symbiotic relationship with photosynthetic organism. Endolichenic fungi are commonly found in numerous quantities within the algal layer of the lichen thallus

3.3 Symbiotic Association: Exploring Endosymbiotic Relationships in Protists

3.3.1 Dinoflagellates-Coral Reef Symbiotic Relationship

The endosymbiotic association between corals and dinoflagellates is crucial for the shallow water corals to survive, affecting the entire coral reef ecosystem (Teschima et al. 2019). The symbiotic association of coral reefs are extremely efficient, supporting over 33% of the oceanic diversity while occupying just 0.1% of the overall marine environment. The symbiotic algae within the corals harness light energy and carbon dioxide to generate sugars by means of photosynthesis, which are then transmitted to the corals. This association enables corals to generate calcium carbonate, a critical component in reef formation; in exchange, corals provide shelter, carbon dioxide, and nitrogen to their algal symbionts, producing a mutual connection that promotes the survival and expansion of both the organisms (Maor-Landaw et al. 2023).

Corals can survive in nutrient-poor, humid conditions due to their symbiotic association because the community of endosymbionts supplies the majority of the energy needs for the coral host, boosting calcification and promoting coral growth. Numerous factors, including exposure to sunlight, irradiance, host transmission technique, depth, coral species, and symbionts in the water column, affect the relationship between corals and the symbiotic dinoflagellates, or zooxanthellae. To assist corals, in adapting to changes in their environment, a variety of coral-symbiont interactions are essential. Due to their shorter generation cycles than corals, symbionts have a higher propensity to adapt and support coral adaptability rapidly. In addition to helping corals endure bleaching matters, certain symbionts also assist them to adjust to rising ocean temperatures. It is essential to comprehend the variety of symbionts in coral hosts to forecast coral survival in the face of climate change since temperature abnormalities are becoming more frequent and severe (Teschima et al. 2019; Fig. 3).

3.4 Bacteria–Sponge Association

Particular focus is placed on the evidence of potential symbiotic relationships between bacteria and sponges. Sponges are stationary invertebrates that remain attached to solid surfaces and are extremely effective at sifting food from water. They actively pull bacteria from nearby water through sponge-generated currents. The sponge's body captures and retains the large number of bacteria and similar-sized particles (more than 95%) (Imhoff and Stöhr 2003). Sponges exhibit the greatest diversity of associated bacteria among marine invertebrates. Sponges, basic multicellular organisms, often create skeletons consisting of calcium carbonate (calcareous) or silica (siliceous), depending on their classification. In most

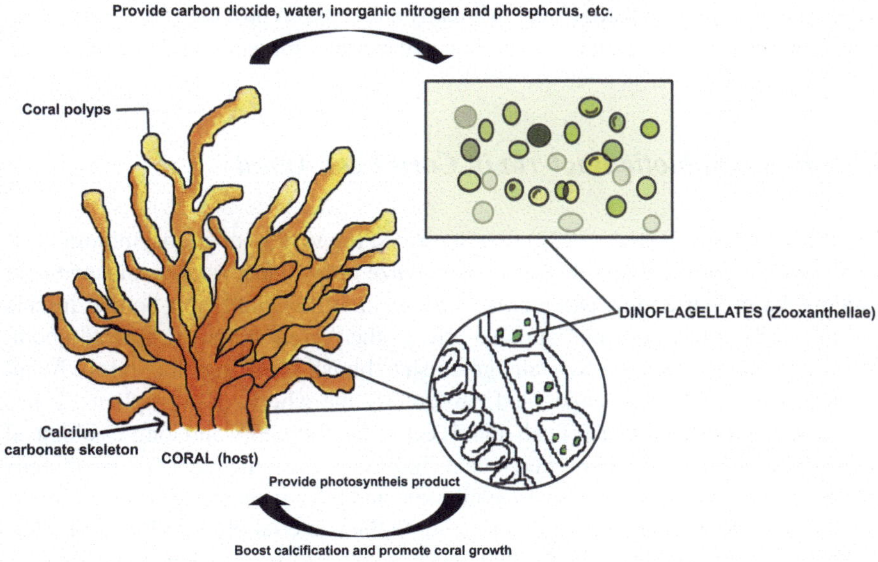

Fig. 3 Coral reef association with Dinoflagellates

circumstances, these two kinds of skeletal structures do not coexist within the same sponge. Sclerosponges, a type of demosponge, are an exception to the norm because they have both siliceous and calcareous skeletons (Uriz et al. 2012).

3.4.1 Bacterial Calcification in Endosymbiosis Niche

Uriz et al. (2012) first reported of symbiotic connections between intracellular calcifying bacteria and marine invertebrates. Using a specific targeted probe, researchers have found an interaction between a sponge species known as *Hemimycale columella* and a particular kind of bacterium that resides within the sponge's cells (Garate et al. 2017). Calcification in sponges is aided by endosymbiotic bacteria known as calcibacteria, which live in sponge cells similar to archeocytes. These bacteria, which do not have cell walls, proliferate within sponge cells before being encased in a calcitic layer and ejected to the sponge's subectosomal (sub-epithelial) zone (Uriz et al. 2012). The probe not just confirmed the identification of these calcified bacteria, but it also allowed scientists to estimate their number inside the tissues of the sponge called *H. columella*. This bacterium has the remarkable ability to manufacture huge amounts of calcareous spherules, which are tiny, spherical structures composed of calcium carbonate, a substance present across numerous biological tissue types. The existence of these calcite-producing bacterium has prompted scientists to investigate if similar kinds of intracellular bacteria may have played an important part in the process of evolution that resulted in skeletal bones in early mammals (Garate et al. 2017). This mutually beneficial relationship most likely originated as a calcium detoxifying mechanism, but it also provided the sponge with other benefits such as skeletal construction and predator protection (Uriz et al. 2012).

3.5 Endosymbiotic Bacteria in Corn Leaf Aphid

The study by Csorba et al. (2022) first described the variety of endosymbiotic bacteria found in corn leaf aphids that invade maize crops under various management systems. Corn leaf aphids (*Rhopalosiphum maidis*) are small, soft-bodied insects that primarily infest corn but can also affect other cereal crops such as sorghum, millet, and barley. They are typically greenish to bluish-black in color and are found in clusters on the leaves, stalks, and particularly the whorls of corn plants. It has been widely reported that nearly all aphids carry the obligatory and long-established endosymbiont *Buchnera*. However, recent studies have revealed that *Geopemphigus* aphids no longer possess *Buchnera* symbionts and have been replaced by an inherited through the mother symbiont belonging to the bacterial phylum *Bacteroidetes*.

Many earlier studies indicate that these symbiotic bacterial groups, such as *Buchnera aphidicola*, *Spiroplasma* spp., *Serratia symbiotica*, etc., play a vital role in characteristics associated with aphid biology, such as resistance to parasitoids, heat stress tolerance, and shifts in host plant range. Heat exposure experiments on *Buchnera* endosymbionts have revealed their impact on the fitness parameters of aphid hosts, such as survival, maturation time, and fertility. Interestingly, these studies showed that different aphid species exhibit different degrees of heat adaptation (Csorba et al. 2022).

The operation role of bacterial endosymbionts can be categorized into two main areas: (1) traits that provide advantages to the host aphid in particular ecological contexts, including safeguarding against high temperature or natural enemies, and (2) roles related to the aphid's metabolism, where the endosymbionts produce and secrete essential nutrients for the host (Csorba et al. 2022).

4 Mineral Transformation

Minerals undergo several transformations due to alterations in their physical and chemical environment (Putnis 1992). Minerals are the primary source of components essential for civilization's growth, including both contaminants and vital plant nutrients. Chemical weathering and human activities like mining and energy extraction release these elements from minerals. Minerals play a crucial role in the biogeochemical cycling of elements by isolating and releasing them. As primary minerals in crustal rocks undergo structural and compositional changes due to physical, chemical, and biological processes, they transform into secondary minerals and soils (Wang et al. 2019). Minerals and rocks contain essential metals such as Mn, Mo, Fe, Co, Ni, Cu, and Zn, as well as phosphorus, which are necessary for the growth of microbes and plants. Elements like C, N, H, O, P, S, Na, K, Ca, Fe, Cu, Zn, Mo, and Si play crucial roles in biological synthesis, information transmission, catalysis, electron transfer, and the formation of solid structures (Gadd et al. 2007).

Certain microorganisms can break down minerals in various ways: using them as terminal electron acceptors, as an energy source, to meet trace element requirements, or to gain a competitive advantage in their environment (Ehrlich 1996).

4.1 *Bioweathering*

Bioweathering is the process by which plants and microorganisms break down minerals by mechanical and chemical processes. This process, while necessary for earth's geological existence of evolution, also has an impact on human life by affecting the quality of water, the formation of soil, and monument preservation (Mapelli et al. 2012).

The main dissolution-re-precipitation processes that lead to mineral substitution reactions in rocks are leaching, cation exchange, diagenesis, and chemical weathering. These reactions involve the substitution of a more stable mineral for one or more of the existing minerals (Gadd 2007).

4.2 *Bioaccumulation*

Bioaccumulation is the growth in contamination levels inside aquatic species as they absorb toxins from their surroundings. Several exposure sources contribute to this process. In bioaccumulation, key toxicokinetic parameters include dissolved absorption rate, efficiency, elimination rate, and food assimilation (Wang 2016).

Thus, bioaccumulation arises from complex interactions between various mechanisms of absorption, excretion, passive release, and metabolism (Streit 1998).

4.3 *Bioremediation*

Bioremediation is the process of employing microorganisms to reduce or remove toxic waste concentrations in polluted areas. This approach uses biological activity to reduce, and eventually eliminate, the negative impacts of environmental contaminants at specified sites. Bioremediation techniques are often divided into two categories: ex situ and in situ (Wu et al. 2014).

5 Role of Different Microbes in Mineral Transformation

Industrial waste products are a serious global concern, affecting the health of countless people. Metallic contaminants from sources that are both natural and artificial affect the environment, polluting food sources such as crop products and seafood via soil, air, and water contamination (Okolo et al. 2016; Sharma and Kumar 2021). The practice of utilizing living things, mainly microbes, that degrade organic materials into non-toxic materials or less harmful compounds under regulated conditions is termed bioremediation. It uses different microbes like fungus or bacteria and plants to detoxify potentially hazardous environmental toxins, making them safer for both human beings and the environment (Kensa 2011).

5.1 Endophytes in Remediation of Heavy Metal Contaminants

Endophytes (that live within plant tissues) aid in heavy metal bioremediation throughout a variety of processes because of their ability to tolerate heavy metals without affecting host plants. These include phytohormone and siderophores synthesis, fixing nitrogen, mineral solubilization, the use of 1-aminocyclopropane-1-carboxylic acid as a source of nitrogen, and nutritional element transformation, which aids in the cleaning of polluted soils and water (Zhan et al. 2011).

"Phytoremediation" is a word that is derived from both Greek and Latin roots. The Greek word "phyto," meaning plant, is paired with the Latin word "remedium," which means to rectify, treat, or remove something unpleasant or undesired. In this sense, "phytoremediation" refers to the use of plants as natural agents to reduce, cleanse, or eliminate environmental contaminants, successfully addressing and repairing environmental imbalances or contamination (Cunningham et al. 1996; Etim 2012). Plants remove contaminants through a variety of methods, including transformation into less toxic forms, absorption and concentration, soil stabilization, and stimulating the growth of microorganisms that break down toxins in the area around the roots (rhizosphere) (Etim 2012). Endophytic fungi have additionally demonstrated the ability to improve phytoremediation (Deng and Cao 2017).

Plants are capable of withstanding organic contaminants, but their development rate is decreased, which makes them unable to heal the damaged soils. This occurs due to the hydrophobicity and phytotoxicity effects of the contaminants, which reduce the plants' ability to absorb nutrients as well as water. Endophytic bacteria or fungi, on the other hand, have been extensively examined for their beneficial effects on plants growing in contaminated soils (Khan et al. 2013; Feng et al. 2017).

Reported by Zahoor et al. (2017), MHR-7, a strain of endophytic fungi found in plants that grew in heavy metal-polluted soil, showed high resistance to various heavy metals, surviving concentrations up to 900 μg/mL without substantial growth loss. As metal concentrations grew, fungal growth slowed. The endophytic fungal strain effectively eliminated 60–87% of heavy metals from a liquid culture containing 300 μg/mL, primarily mediated by bioconversion or metal buildup in its hyphae.

5.2 *Insect-Associated Bacteria That Detoxify Heavy Metals*

Several microbial partners are widespread in insects and can offer functional characteristics to their hosts, enabling them to adapt to new ecological environments (Sudakaran et al. 2017; Lemoine et al. 2020).

Heavy metals are present in soils, either naturally or due to human activities, and can be absorbed by plants, passing through the food chain to insects and other animals (Salt et al. 1998; Gall et al. 2015; Lemoine et al. 2020).

Two microbes associated with bumblebees (*Snodgrasella alvi* and *Lactobacillus bombicola*) showed resistance to selenate in lab conditions and had genes that can detoxify selenate in their genomes (Rothman et al. 2019; Lemoine et al. 2020). Detoxifying bacteria appear to protect the egg masses of aquatic non-biting midges (Diptera, Chironomidae) against heavy metal contamination. This was established by isolating bacteria capable of detoxifying heavy metals in vitro from both eggs and larvae, coupled with bioassays that showed decreased survival in individuals deprived of symbionts following exposure to chromium or lead, compared to those with symbiotic associations (Senderovichla and Halpern 2012; Senderovich and Halpern 2013; Lemoine et al. 2020).

5.3 *Microbes Interact Closely with Clay Minerals*

Clay minerals are naturally abundant and extremely reactive, typically forming due to the weathering of the Earth's most prevalent silicate minerals (Fomina and Skorochod 2020).

Clay minerals and microorganisms can be found almost everywhere on Earth, especially in soils, sediments, and aquatic habitats. Their interactions serve as vital components of many ecological systems. Overall, these interactions have a major impact on Earth's ecosystems, agriculture, industrial processes, and environmental management.

5.4 *Role of Soil Microbes in Mineral Transformation*

Heavy metals exhibit a restricted mobility in soil, causing difficulty for the roots of plants to absorb them directly. However, soil microbes and roots can interact to improve absorption of metals within root system (Sharma and Kumar 2021).

Soil microorganisms play a key role in mineral weathering, nutrient bioavailability, and bioremediation. They break down rocks through processes like biocorrosion, bioerosion, and bioperforation. In the rhizosphere, root exudates enhance the activity of mineral-solubilizing microorganisms (MSM), which release minerals by dissolving them, performing redox reactions, and producing enzymes like acid phosphatases. These microorganisms enhance plant growth by stimulating root development, modulating plant hormones, generating helpful compounds like siderophores, and defending plants against pathogens in favorable and stressful conditions (Mumtaz et al. 2024).

5.5 Microalgae in Remediation of Heavy Metals

Bacteria and algae interact in which algae offer habitat and nutrients to bacteria via natural mechanisms, promoting bacterial growth.

Microalgae are microscopic algae that consist of a broad class of photosynthetic microorganisms, both prokaryotic and eukaryotic that live in various habitats. They can be found both in freshwater and marine environments. Microalgae comprise eukaryotic forms like diatoms, green algae, and prokaryotic cyanobacteria (blue-green algae) (Encarnacao et al. 2023).

Phycoremediation, which involves applying microalgae to clean wastewater and polluted aquatic environments, is gaining popularity as a beneficial, cost-effective, and environmentally conscious alternative to physicochemical processes. Microalgae may absorb a wide range of inorganic and organic pollutants, including heavy metals. Their ability to detoxify heavy metals is attributed to adaptive mechanisms that evolved throughout time and divided into metabolism-independent and dependent routes. The efficacy of heavy metal biosorption varies by microalgae species with some species, such as *S. obliquus* and *C. sorokiniana*, responding differentially to metal pollution from lead, copper, chromium, and cadmium (Encarnacao et al. 2023).

5.6 Role of Lichens in Mineral Transformation

Lichen-forming fungi are essential to biogeochemical cycles because of their mutualistic relationships with cyanobacteria or photosynthetic algae. As early rock colonizers, they aid in the synthesis of biominerals, the cycling of nutrients, and the modification of bedrock mineral composition. Lichens can also accumulate high concentrations of metals and other elements that are important to the environment, like radionuclides (Gadd 2007; Fig. 4).

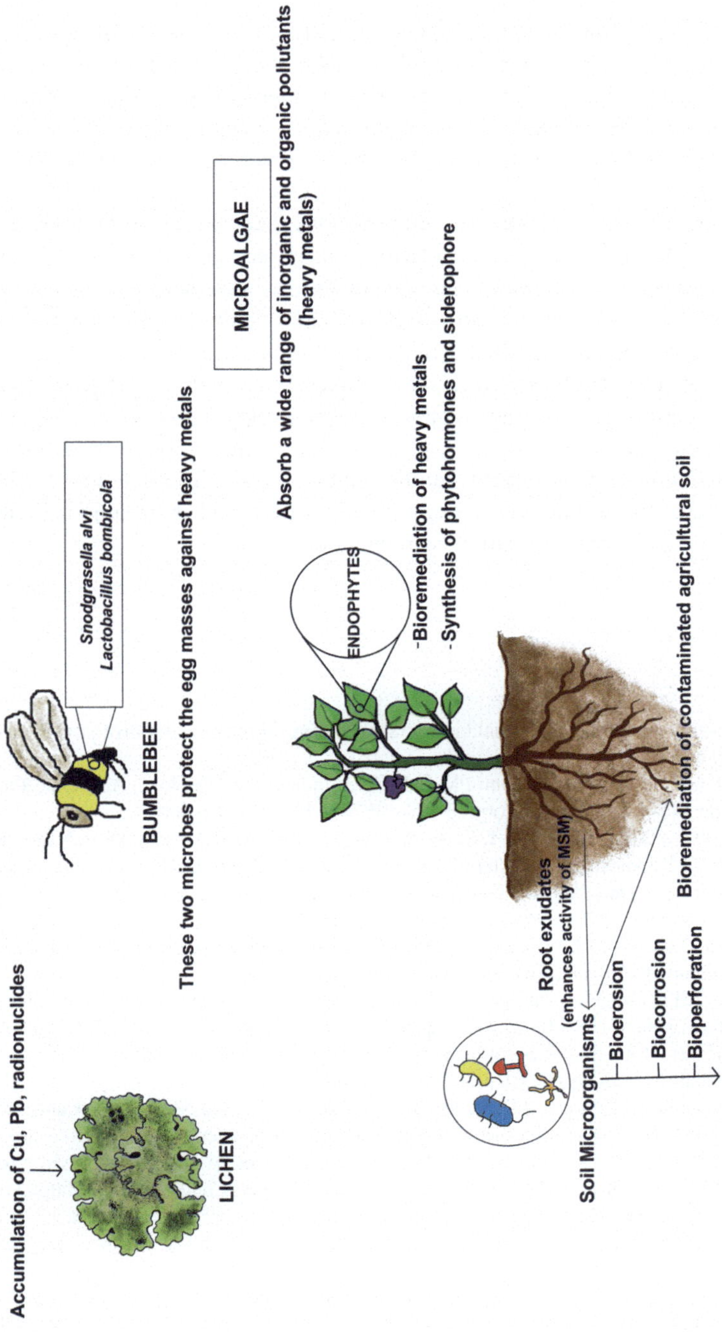

Fig. 4 Role of microbes in mineral transformation

6 Conclusion

Endosymbiotic niches are dynamic, highly specialized conditions where microorganisms and their hosts have symbiotic relationships that significantly impact geochemical cycles. The ability of the host organism to metabolize or convert minerals is often enhanced by the presence of endosymbionts, especially through new enzymatic pathways or biochemical processes that are not observed in free-living microorganisms.

An optimal exchange of nutrients, mineral ions, and energy is facilitated by mutualistic interactions, and this helps with vital processes like the metal cycle (e.g., manganese and iron). These changes are essential for the health of the host and the stability of the environment, particularly in harsh or nutrient-limited environments where conventional microbial activity might not be effective.

Microbial diversity evolves as a result of mineral processing processes, which are significantly influenced by endosymbiotic relationships. Through intricate, precisely calibrated interactions, these connections not only support biochemical cycling but also enhance ecosystem stability and resilience. The potential for biotechnological uses like sustainable agriculture, biomining, and environmental remediation is enhanced by delving into these interactions.

References

Bradley DJ, Martiny JB (2007) Microbial biodiversity. In: Encyclopedia of biodiversity. Elsevier, Amsterdam, pp 1–9

Casem ML (2016) Cell system. In: Conn PM (ed) Case studies in cell biology. Academic Press, New York, pp 345–371. https://doi.org/10.1016/B978-0-12-801394-6.00015

Csorba AB, Fora CG, Bálint J, Felföldi T, Szabó A, Máthé I, Loxdale HD, Kentelky E, Nyaradi II, Balog A (2022) Endosymbiotic bacterial diversity of corn leaf aphid, *Rhopalosiphum maidis* Fitch (Hemiptera: Aphididae) associated with maize management systems. Microorganisms 10(5):939. https://doi.org/10.3390/microorganisms10050939

Cunningham SD, Anderson TA, Schwab AP, Hsu FC (1996) Phytoremediation of soils contaminated with organic pollutants. Adv Agron 56(1):55–114

Deng Z, Cao L (2017) Fungal endophytes and their interactions with plants in phytoremediation: a review. Chemosphere 168:1100–1106. https://doi.org/10.1016/j.chemosphere.2016.10.097

Ehrlich HL (1996) How microbes influence mineral growth and dissolution. Chem Geol 132(1–4):5–9. https://doi.org/10.1016/S0009-2541(96)00035-6

Encarnacao T, Ramos P, Mohammed D, McDonald J, Lizzul M, Nicolau N, da Graca Campos M, Sobral AJ (2023) Bioremediation using microalgae and cyanobacteria and biomass valorisation. In: Encarnacao T, Canelas Pais A (eds) Marine organisms: A solution to environmental pollution? Environmental challenges and solutions. Springer, Cham, pp 5–28. https://doi.org/10.1007/978-3-031-17226-7_2

Etim EE (2012) Phytoremediation and its mechanisms: a review. Int J Environ Bioenergy 2(3):120–136

Farrar K, Bryant D, Cope-Selby N (2014) Understanding and engineering beneficial plant–microbe interactions: plant growth promotion in energy crops. Plant Biotechnol J 12(9):1193–1206. https://doi.org/10.1111/pbi.12279

Feng NX, Yu J, Zhao HM, Cheng YT, Mo CH, Cai QY, Li YW, Li H, Wong MH (2017) Efficient phytoremediation of organic contaminants in soils using plant–endophyte partnerships. Sci Total Environ 583:352–368. https://doi.org/10.1016/j.scitotenv.2017.01.075

Fomina M, Skorochod I (2020) Microbial interaction with clay minerals and its environmental and biotechnological implications. Fortschr Mineral 10(10):861. https://doi.org/10.3390/min10100861

Gadd GM (2007) Geomycology: biogeochemical transformations of rocks, minerals, metals and radionuclides by fungi, bioweathering and bioremediation. Mycol Res 111(1):3–49. https://doi.org/10.1016/j.mycres.2006.12.001

Gadd GM, Raven JA (2010) Geomicrobiology of eukaryotic microorganisms. Geomicrobiol J 27(6–7):491–519. https://doi.org/10.1080/01490451003703006

Gadd GM, Burford EP, Fomina M, Melville K (2007) Mineral transformations and biogeochemical cycles: a geomycological perspective. In: Gadd GM, Watkinson SC, Dyer PS (eds) Fungi in the environment. Cambridge University Press, New York, pp 78–111. https://doi.org/10.1017/CBO9780511541797.006

Gall JE, Boyd RS, Rajakaruna N (2015) Transfer of heavy metals through terrestrial food webs: a review. Environ Monit Assess 187:1–21. https://doi.org/10.1007/s10661-015-4436-3

Garate L, Sureda J, Agell G, Uriz MJ (2017) Endosymbiotic calcifying bacteria across sponge species and oceans. Sci Rep 7(1):43674. https://doi.org/10.1038/srep43674

Gibbons SM, Gilbert JA (2015) Microbial diversity—exploration of natural ecosystems and microbiomes. Curr Opin Genet Dev 35:66–72. https://doi.org/10.1016/j.gde.2015.10.003

Glazer AN, Nikaido H (2012) Microbial diversity. In: Microbial biotechnology fundamental of applied microbiology, 2nd edn. Cambridge University Press, New York, pp 1–44. https://doi.org/10.1017/CBO9780511811227.002

Govindasamy V, Franco CM, Gupta VV (2013) Endophytic actinobacteria: diversity and ecology. In: Verma V, Gange A (eds) Advances in endophytic research. Springer, New Delhi, India, pp 27–59. https://doi.org/10.1007/978-81-322-1575-2_2

Hassani MA, Durán P, Hacquard S (2018) Microbial interactions within the plant holobiont. Microbiome 6:1–17. https://doi.org/10.1186/s40168-018-0445-0

Imhoff JF, Stöhr R (2003) Sponge-associated bacteria: general overview and special aspects of bacteria associated with *Halichondria panicea*. In: Sponges (Porifera). Springer, pp 35–57. https://doi.org/10.1007/978-3-642-55519-0_2

Islam T, Fatema, Hoque MN, Gupta DR, Mahmud NU, Sakif TI, Sharpe AG (2023) Improvement of growth, yield and associated bacteriome of rice by the application of probiotic *Paraburkholderia* and *Delftia*. Front Microbiol 14:1212505. https://doi.org/10.3389/fmicb.2023.1212505

Kensa VM (2011) Bioremediation—an overview. J Ind Pollut Control 27(2):161–168

Khan S, Afzal M, Iqbal S, Khan QM (2013) Plant–bacteria partnerships for the remediation of hydrocarbon contaminated soils. Chemosphere 90(4):1317–1332. https://doi.org/10.1016/j.chemosphere.2012.09.045

Lemoine MM, Engl T, Kaltenpoth M (2020) Microbial symbionts expanding or constraining abiotic niche space in insects. Curr Opin Insect Sci 39:14–20. https://doi.org/10.1016/j.cois.2020.01.003

Maor-Landaw K, Eisenhut M, Tortorelli G, van de Meene A, Kurz S, Segal G, van Oppen MJ, Weber AP, McFadden GI (2023) A candidate transporter allowing symbiotic dinoflagellates to feed their coral hosts. Int Soc Microb Ecol Commun 3(1):7. https://doi.org/10.1038/s43705-023-00218-8

Mapelli F, Marasco R, Balloi A, Rolli E, Cappitelli F, Daffonchio D, Borin S (2012) Mineral–microbe interactions: biotechnological potential of bioweathering. J Biotechnol 157(4):473–481. https://doi.org/10.1016/j.jbiotec.2011.11.013

Mumtaz MZ, Ahmad M, Etesami H, Mustafa A (2024) Mineral solubilizing microorganisms (MSM) and their applications in nutrient bioavailability, bioweathering and bioremediation, volume II. Front Microbiol 14:1345161. https://doi.org/10.3389/fmicb.2023.1345161

Okolo NV, Olowolafe EA, Akawu I, Okoduwa SIR (2016) Effects of industrial effluents on soil resources in Challawa industrial area, Kano, Nigeria. J Global Ecol Environ 5(1):1–10
Onen OI, Aboh AA, Mfam AN, Akor MO, Nweke CN, Osuagwu AN (2020) Microbial diversity: values and roles in ecosystems. Asian J Biol 9(1):10–22. https://doi.org/10.9734/AJOB/2020/v9i130075
Pimentel MR, Molina G, Dionisio AP, Maróstica MR, Pastore GM (2011) Use of endophytes to obtain bioactive compounds and their application in biotransformation process. Biotechnol Res Int 2011(1):576286. https://doi.org/10.4061/2011/576286
Putnis A (1992) Transformation processes in minerals I: exsolution. An introduction to mineral sciences. Cambridge University Press, New York, pp 333–384. https://doi.org/10.1017/CBO9781139170383.013
Rothman JA, Leger L, Graystock P, Russell K, McFrederick QS (2019) The bumble bee microbiome increases survival of bees exposed to selenate toxicity. Environ Microbiol 21(9):3417–3429. https://doi.org/10.1111/1462-2920.14641
Salt DE, Smith RD, Raskin I (1998) Phytoremediation. Annual review. Plant Physiol 49(1):643–668. https://doi.org/10.1146/annurev.arplant.49.1.643
Schirawski J, Perlin MH (2018) Plant–microbe interaction 2017—the good, the bad and the diverse. Int J Mol Sci 19(5):1374. https://doi.org/10.3390/ijms19051374
Schulz B, Boyle C (2006) What are endophytes? In: Schulz BJE, Boyle CJC, Sieber TN (eds) Microbial root endophytes, soil biology, vol 9. Springer, Berlin, pp 1–13
Senderovich Y, Halpern M (2013) The protective role of endogenous bacterial communities in chironomid egg masses and larvae. Int Soc Microb Ecol J 7(11):2147–2158. https://doi.org/10.1038/ismej.2013.100
Senderovichla Y, Halpern M (2012) Bacterial community composition associated with chironomid egg masses. J Insect Sci 12(1):149. https://doi.org/10.1673/031.012.14901
Sharma P, Kumar S (2021) Bioremediation of heavy metals from industrial effluents by endophytes and their metabolic activity: recent advances. Bioresour Technol 339:125589. https://doi.org/10.1016/j.biortech.2021.125589
Shu WS, Huang LN (2022) Microbial diversity in extreme environments. Nat Rev Microbiol 20(4):219–235. https://doi.org/10.1038/s41579-021-00648-y
Singh R, Dubey AK (2015) Endophytic actinomycetes as emerging source for therapeutic compounds. Indo Global J Pharm Sci 5:106–116. https://doi.org/10.1038/ja.2017.20
Singh R, Dubey AK (2018) Diversity and applications of endophytic actinobacteria of plants in special and other ecological niches. Front Microbiol 9:1767. https://doi.org/10.3389/fmicb.2018.01767
Streit B (1998) Bioaccumulation of contaminants in fish. In: Fish ecotoxicology. Birkhäuser, pp 353–387. https://doi.org/10.1007/978-3-0348-8853-0_12
Sudakaran S, Kost C, Kaltenpoth M (2017) Symbiont acquisition and replacement as a source of ecological innovation. Trends Microbiol 25(5):375–390. https://doi.org/10.1016/j.tim.2017.02.014
Suryanarayanan TS, Thirunavukkarasu N (2017) Endolichenic fungi: the lesser known fungal associates of lichens. Mycology 8(3):189–196. https://doi.org/10.1080/21501203.2017.1352048
Teschima MM, Garrido A, Paris A, Nunes FL, Zilberberg C (2019) Biogeography of the endosymbiotic dinoflagellates (symbiodiniaceae) community associated with the brooding coral *Favia Gravida* in the Atlantic Ocean. PLoS One 14(3):e0213519. https://doi.org/10.1371/journal.pone.0213519
Uriz MJ, Agell G, Blanquer A, Turon X, Casamayor EO (2012) Endosymbiotic calcifying bacteria: a new cue to the origin of calcification in metazoa? Evolution 66(10):2993–2999. https://doi.org/10.1111/j.1558-5646.2012.01676.x
Vinayaka KS (2020) Lichen endosymbionts an untapped bioresources for biotechnological products. Acta Sci Microbiol 3(1):95–96. https://doi.org/10.3100/ASMI.2020.03.0463
Wang WX (2016) Bioaccumulation and biomonitoring. In: Marine ecotoxicology. Academic Press, pp 99–119. https://doi.org/10.1016/B978-0-12-803371-5.00004-7

Wang Q, Yang P, Zhu M (2019) Effects of metal cations on coupled birnessite structural transformation and natural organic matter adsorption and oxidation. Geochim Cosmochim Acta 250:292–310. https://doi.org/10.1016/j.gca.2019.01.035

White JF, Kingsley KL, Zhang Q, Verma R, Obi N, Dvinskikh S, Elmore MT, Verma SK, Gond SK, Kowalski KP (2019) Endophytic microbes and their potential applications in crop management. Pest Manag Sci 75(10):2558–2565. https://doi.org/10.1002/ps.5527

Wu Y, Xia L, Yu Z, Shabbir S, Kerr PG (2014) In situ bioremediation of surface waters by periphytons. Bioresour Technol 151:367–372. https://doi.org/10.1016/j.biortech.2013.10.088

Yang JH, Oh SY, Kim W, Woo JJ, Kim H, Hur JS (2021) Effect of isolation conditions on diversity of endolichenic fungal communities from a foliose lichen, *Parmotrema tinctorum*. J Fungi 7(5):335. https://doi.org/10.3390/jof7050335

Zahoor M, Irshad M, Rahman H, Qasim M, Afridi SG, Qadir M, Hussain A (2017) Alleviation of heavy metal toxicity and phytostimulation of *Brassica campestris* L. by endophytic *Mucor* sp. MHR-7. Ecotoxicol Environ Saf 142:139–149. https://doi.org/10.1016/j.ecoenv.2017.04.005

Zhan P, Fu JM, Li SZ (2011) Advances in the application of endophyte on the remediation in heavy metal contaminated soils. In: 2011 International conference on remote sensing, environment and transportation engineering. IEEE, pp 3807–3810. https://doi.org/10.1109/RSETE.2011.5965068

Geo-Microbes in Rhizoremediation: Harnessing Potential of Sulphate-Reducing Bacteria for Mitigation of Xenobiotics Compounds

Ankita Das, Nandita Das, Sandeep Das, and Piyush Pandey

Contents

Abstract Sulphur-reducing bacteria (SRB) play a pivotal role in the bioremediation of xenobiotics through rhizoremediation, leveraging their metabolic pathways to detoxify pollutants. This chapter explores the interlinkages between SRB activity, biogeochemical cycles, and rhizosphere processes, emphasizing their influence on

---

A. Das · N. Das · S. Das · P. Pandey (✉)
Soil and Environment Microbiology Laboratory, Department of Microbiology, Assam University, Silchar, Assam, India

S. R. Joshi et al. (eds.), *Mineral Transformation and Bioremediation by Geo-Microbes*, Interdisciplinary Biotechnological Advances,
https://doi.org/10.1007/978-981-96-3033-2_7

xenobiotic degradation. Advances in molecular techniques, including genomics, metagenomics, proteomics, and metabolomics, have enhanced our understanding of SRB populations and their role in rhizoremediation. Strategies to enhance SRB activity, such as the addition of zero-valent iron, cations, carrier materials, and genetic engineering, offer practical solutions to boost xenobiotic degradation. Furthermore, synthetic bacterial consortia present new opportunities for targeted applications in wastewater treatment, acid mine remediation, and crude oil-contaminated sites. Despite the promising applications of SRB in environmental cleanup, challenges are still there, which necessitates future research into optimizing and scaling these technologies for more real-world applications. Thus, a deep comprehensive report is needed that explores all these components. This chapter provides an inclusive overview of above-mentioned aspects, which highlights SRB's potential in advancing sustainable rhizoremediation.

Keywords SRB · Metabolic pathway · Geo-microbe · Omics · Genetic engineering · Rhizoremediation

Abbreviations

AHL	Acyl-homoserine lactone
AMD	Acid mine drainage
APS	Adenosine 5′-phosphosulfate
As	Arsenic
Cd	Cadmium
c-di-GMP	Cyclic di-guanosine monophosphate
CO_2	Carbon dioxide
COD	Chemical oxygen demand
Cu	Copper
DNA	Deoxyribonucleic acid
EPS	Extracellular polymeric substances
Fe	Iron
FeS	Ferrous sulphide
FISH	Fluorescence in situ hybridization
g/day	Gram per day
GMOs	Genetically modified organisms
GO	Graphene oxide
H_2	Molecular hydrogen
H_2S	Sulphide
Hg	Mercury
HRT	Hydraulic retention time
mM	Millimolar
Mn	Manganese
mW/m^2	Milliwatt per square meter

NADH	Nicotinamide adenine dinucleotide (NAD) + hydrogen (H)
NGS	Next-generation sequencing
Ni	Nickel
PAHs	Polycyclic Aromatic Hydrocarbons/ Polyaromatic Hydrocarbon
Pb	Lead
pH	Potential of hydrogen
POPs	Persistent organic pollutants
qPCR	Real-time quantitative reverse transcription polymerase chain reaction
rGO	Reduced graphene oxide
RNA	Ribonucleic acid
rRNA	Ribosomal ribonucleic acid
S^{2-}	Sulphide
Sb	Antimony
SFC	Sulphide fuel cells
SIP	Stable isotope probing
SO_3^{2-}	Sulphite
SO_4^{2-}	Sulphate
SRB	Sulphur-reducing bacteria
SynComs	Synthetic communities
TCA cycle	Tricarboxylic Acid Cycle
VOCs	Volatile organic compounds
Zn	Zinc
ZnS	Zinc sulphide
ZVI	Zero-valent iron
μm	Micrometre

1 Introduction

In an era marked by rapid industrialization and urbanization, the proliferation of xenobiotic compounds poses a significant threat to environmental health and ecosystem stability. These compounds include a wide array of synthetic chemicals, such as pesticides, hydrocarbons, heavy metals, pharmaceuticals, and industrial solvents, that are often resistant to natural degradation processes and can accumulate in various ecosystems (Bradu et al. 2022). Traditional remediation techniques often fall short in effectively addressing the complex challenges posed by these pollutants, as these are expensive and might have toxic effects in one or another form (Das et al. 2024). In this context, the utilization of biological processes, particularly rhizoremediation, emerges as a promising strategy that leverages the metabolic capabilities of microorganisms to detoxify and transform these harmful substances into less toxic forms (Singha and Pandey 2021).

Among the various microbial agents utilized in bioremediation, sulphate-reducing bacteria (SRB) play a crucial role due to their unique metabolic pathways that facilitate the anaerobic degradation of xenobiotics. These geo-microbes are expert at utilizing sulphate (SO_4^{2-}) as a terminal electron acceptor in their metabolic processes, which not only facilitates the reduction of sulphate to sulphide (S^{2-}) but also promotes the breakdown of various organic pollutants (Novair et al. 2024). A geo-microbe refers to a microorganism that plays a role in geochemical processes, particularly in the transformation of minerals and environmental contaminants (Maduwuba et al. 2021). The interlinkage between biogeochemical cycles and the activities of SRB within the rhizosphere highlights their dual role in nutrient cycling and pollutant degradation (Zak et al. 2021). SRB are likely descendents of ancestral microorganisms that played a key role in the early biogeochemical sulphur cycle from the time life first emerged on Earth. These are both phylogenetically and metabolically diverse, likely serving as some of the earliest respiring microorganisms that influence the biogeochemistry of the various environments these organisms inhabit (Shen and Buick 2004). SRB have adapted remarkably well to nearly all ecosystems on Earth, including extreme environments like oceanic hydrothermal systems and oil fields. In these harsh environments, these microorganisms face severe physicochemical challenges, including changing pH, high temperatures, and pressures (Barton and Fauque 2022). They play an important function in the complete oxidation of biomass or organic matter and contribute to the biogeochemistry of these extreme habitats through sulphide production and metal reduction (Banfield and Nealson 2018). These organisms utilize various metabolic pathways during the degradation of toxic xenobiotic compounds, such as the dissimilatory sulphate reduction pathway, as a primary energy source while simultaneously degrading organic pollutants through anaerobic respiration. The metabolic versatility of SRB helps in their adaptation to diverse environmental conditions and utilization of a range of substrates, which makes them effective agents in bioremediation efforts. SRB can utilize over a hundred compounds as electron donors, including sugars (e.g., glucose, fructose), amino acids (e.g., glycine, serine), monocarboxylic acids (e.g., acetate, propionate), dicarboxylic acids (e.g., fumarate, succinate), alcohols (e.g., methanol, ethanol), and aromatic compounds (e.g., benzoate, phenol) (Barton and Fauque 2009). SRB reduce a wide range of terminal electron acceptors, including inorganic sulphur compounds and other organic and inorganic substances. The genus *Desulfovibrio* spp. has been the focus of most biochemical and physiological studies on SRB because it is easier and faster to grow them than other sulphate reducers (Keller et al. 2011; Vasantharaj et al. 2017).

The presence of SRB in the rhizosphere significantly influences soil microbial communities and nutrient cycling. By increasing the availability of nutrients through the breakdown of complex organic materials, their metabolic activities have the potential to stimulate plant development with positive effects (Kotoky and Pandey 2018). Additionally, SRB can modulate pH levels and redox potential in the rhizosphere, creating favorable conditions for other beneficial microorganisms involved in bioremediation (Novair et al. 2024). Advancements in molecular techniques have revolutionized our understanding of microbial dynamics in the rhizosphere during rhizoremediation. Identifying specific gene targets associated with SRB activity is critical for understanding their role in xenobiotic degradation. Genes involved in sulphate reduction and those encoding key enzymes for xenobiotic metabolism are often the focus of such studies. The integration of omics technologies such as genomics, transcriptomics, proteomics, and metabolomics offers a comprehensive framework for studying SRB interactions within the rhizosphere (Rodríguez et al. 2015). These methods allow researchers to understand intricate microbial interactions and metabolic networks, which is important for effective rhizoremediation treatments. Moreover, SRB do not give maximum results independently, and their interactions with other soil bacteria are essential for improving bioremediation effectiveness. Collaborative interactions among several bacterial species can result in synergistic effects that enhance pollutant degradation rates (Sahinkaya et al. 2015).

This review delves into the intricate mechanisms by which SRB contribute to rhizoremediation, focusing on their metabolic pathways, influence on rhizosphere processes, and interactions with other soil microbes. Further, advanced molecular techniques that enable the monitoring of SRB populations and their metabolic activity in situ have been explored, including various omics approaches that provide insights into the genetic, transcriptomic, proteomic, and metabolomic dimensions of SRB functions. Additionally, the methods for enhancing SRB activity, including the strategic addition of zero-valent iron, inorganic cations, and other additives, as well as the potential of genetic engineering to optimize these processes, have been discussed. Furthermore, the applications of engineered bacterial consortia in the degradation of xenobiotics and the significant role of SRB in wastewater treatment, acid mine bioremediation, and crude oil contamination have also been highlighted. However, there are certain limitations of current methodologies that are explained and future prospects for research were also discussed. Through this comprehensive exploration, we aim to underscore the critical contributions of SRB to environmental sustainability.

2 Interlink Between Biogeochemical Cycles and Rhizoremediation of Xenobiotics by SRB

Biogeochemical cycles, like the sulphur, carbon, nitrogen, oxygen, phosphorus and oxygen cycles, are fundamental to the Earth's ecosystems, regulating the flow of essential elements and sustaining life. Geo-microbes play pivotal roles in these cycles, driving processes like decomposition, nutrient recycling, and the transformation of minerals (Banfield and Nealson 2018). The interlink between biogeochemical cycles and xenobiotic degradation lies in the ability of geo-microbes to use these pollutants as carbon or energy sources and converting those in lesser toxic forms, while contributing to nutrient cycling (Fig. 1) (Gadd 2010). SRB are a key group of geo-microbes that play an important role in both the sulphur cycle and the degradation of xenobiotics. In anoxic conditions, SRB are able to assist the conversion of sulphate to hydrogen sulphide. In most cases, organic molecules or hydrogen are used as electron donors (Fan et al. 2023). This metabolic process allows them to participate in the remediation of xenobiotic contaminants, particularly those

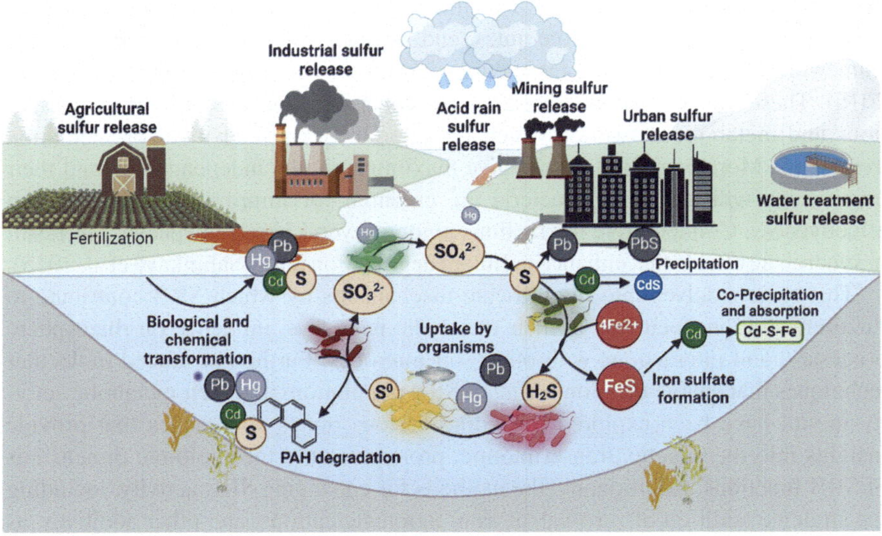

Fig. 1 Source of pollutants and integration of sulphur cycle with xenobiotic degradation by SRB

containing sulphur or heavy metals. SRB can transform and immobilize heavy metals such as cadmium (Cd), lead (Pb), and mercury (Hg) by precipitating them as metal sulphides, reducing their toxicity and mobility in the environment (Xu and Chen 2020). Their involvement in xenobiotic degradation is further enhanced by their ability to mineralize organic pollutants, making them highly relevant in the context of bioremediation strategies (Novair et al. 2024).

The oxidation state of sulphur determines its electron accepting or donating capabilities. Microbial sulphate reduction will take place when sulphate diffuses into anaerobic environments, with sulphate serving as the terminal electron acceptor in anaerobic respiration for microorganisms, such as *Archaeoglobus* spp., *Desulfovibrio* spp., and *Desulfonema* spp. (Birkeland 2005). This process produces sulphide, which can then be used as source of electron by photosynthetic microbes growing in anoxygenic environment and chemolithoautotrophs such as *Chlorobi* spp. and *Thiobacillus* spp., which converts sulphides back into sulphur and sulphate. Some bacteria also reduce elemental sulphur via dissimilatory processes, such as *Desulfuromonas*, thermophilic archaea, and cyanobacteria in hypersaline sediments. Sulphite (SO_3^{2-}) is another key intermediate in the pathway, which is converted to sulphides by microbes like *Clostridium* spp., *Desulfovibrio* spp., *Desulfotomaculum* spp., and *Alteromonas* spp. (Novair et al. 2024). Additionally, microbial activity leads to the production of dimethylsulfide, which is released into the atmosphere and contributes to cloud formation by acting as nuclei for water droplets. Enhanced dimethylsulfide production may help mitigate global climate change by increasing cloud cover, which ultimately helps in reducing the temperature of the Earth's surface (Kloster et al. 2007; Thabet et al. 2017).

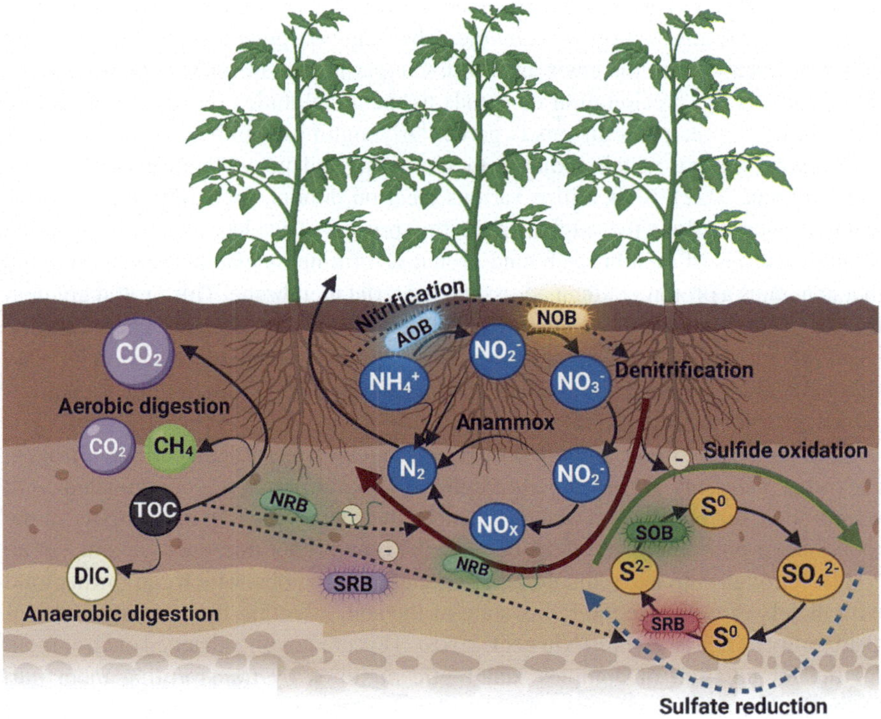

Fig. 2 Interlinked metabolic pathway of SRB with other organisms

The metabolic versatility of SRB enables them to degrade various pollutants in both direct and co-metabolic processes (Fig. 2). The primary metabolic pathway in SRB is the dissimilatory sulphate reduction process, which involves several key biochemical steps and plays a crucial role in bioremediation efforts (Kushkevych et al. 2020b). The process of sulphate reduction begins with the activation of sulphate by the enzyme ATP sulfurylase, which converts sulphate into adenosine 5′-phosphosulfate (APS). This activation step is critical because sulphate is highly stable and requires energy input to be transformed into a more reactive intermediate (Kushkevych et al. 2020a). The conversion of sulphate to APS represents the first step in making sulphate available for further reduction. Following the activation of sulphate, APS is reduced to sulphite by the enzyme APS reductase. This step is pivotal as it transitions the sulphate from a less reactive form to a more easily reducible form, sulphite (De Bont et al. 2022). A significant stage in the process is the reduction of APS to sulphite, which calls for the transfer of electrons, which are often obtained from molecular hydrogen (H_2) or organic molecules like lactate or acetate (Kushkevych et al. 2020b). The selection of the electron donor is contingent upon the particular SRB species as well as the characteristics of the surrounding environment. The final and most crucial step in sulphate reduction is the conversion of sulphite to hydrogen sulphide (H_2S), catalyzed by sulphite reductase (*dsrA*, *dsrB*)

(Han et al. 2022). This process involves the transfer of six electrons to sulphite, resulting in the production of sulphide. The sulphide generated in this step can either be released into the environment or can participate in further chemical reactions, such as the precipitation of metals as metal sulphides (Novair et al. 2024). This ability to precipitate metals is particularly significant in bioremediation, as it can immobilize toxic metals, reducing their mobility and bioavailability (Bryukhanov and Khijniak 2022). SRB utilize various electron donors, including organic acids and hydrogen, to drive the sulphate reduction process. The electrons are transported through a series of membrane-bound proteins, forming an electron transport chain that generates a proton gradient across the bacterial membrane. This proton gradient is used to synthesize ATP via chemiosmosis, providing the necessary energy for the bacterial cell. The ATP generated is not only crucial for sulphate reduction but also supports cellular growth and maintenance, enabling SRB to thrive in their ecological niches (Novair et al. 2024).

SRB participate in the biogeochemical cycling such as sulphur, carbon, nitrogen, and other elements within the rhizosphere. In environments contaminated with xenobiotics, the presence of plants enhances microbial diversity and activity in the root zone, promoting interactions between SRB and other microbes (Arantza et al. 2022). Through sulphate reduction, SRB generate sulphides, which can further bind with metals or influence the redox state of the soil, affecting the mobility and bioavailability of heavy metals (Niu et al. 2018). This creates an enhanced environment for the remediation of complex xenobiotic mixtures by transforming them into forms that are more readily absorbed by plants or other microorganisms (Shaji et al. 2021). Additionally, SRB contribute to the carbon cycle by utilizing organic matter as an energy source and generate carbon dioxide (CO_2) and biomass, which is eventually incorporated into the soil matrix. This interaction highlights the interconnectedness of sulphur and carbon cycling, where SRB, along with other rhizobacteria, perform a crucial task in maintaining ecosystem balance and improving soil health during the remediation process (Gao and Fan 2023). Their ability to function in anoxic or low-oxygen environments makes them especially useful in remediating contaminated soils with limited oxygen diffusion (Liao et al. 2024). Moreover, SRB such as *Desulfovibrio desulfuricans* can also influence the nitrogen cycle by affecting nitrification and denitrification processes (Grubba et al. 2022). Their activity in conjunction with nitrogen-fixing bacteria helps create nutrient-enriched zones, promoting plant growth and enhancing rhizoremediation efficiency (Guo et al. 2024). SRB have a very high affinity for nitrate, with a Km of 0.05 μM compared to 1.8–13.7 μM for denitrifiers. This allows SRB to outcompete denitrifiers for nitrate at low concentrations (Dalsgaard and Bak 1994). In shallow coastal sediments, SRB like *Desulfobacterales* and *Desulfovibrionales* were responsible for most of the heterotrophic nitrogen fixation and denitrification (Liesirova et al. 2023). By working within these biogeochemical cycles, SRB not only degrade xenobiotics but also improve soil structure and fertility, contributing to long-term ecosystem restoration.

3 Influence of SRB on Rhizosphere Processes

SRB play a significant role in influencing rhizosphere processes through several key mechanisms that impact both the chemical environment and the interactions within microbial communities (Santana et al. 2021). Niu et al. (2018) reported that SRB were more abundant in rhizosphere soil compared to non-rhizosphere soil, using the *dsrB* gene as a marker for SRB abundance. This increased presence of SRB in the rhizosphere influences the biological and chemical processes occurring in that environment (Niu et al. 2018). One of the primary ways SRB exert their influence is by modifying the redox conditions in the rhizosphere. By reducing sulphate to hydrogen sulphide under anaerobic conditions, SRB decrease the redox potential of the soil. This shift toward more reducing conditions can alter the speciation of various elements, such as iron (Fe), manganese (Mn), and heavy metals like arsenic (As) and Cd (Qian et al. 2022; Novair et al. 2024). These changes in the oxidation state can either immobilize these elements, reducing their bioavailability, or increase their bioavailability, making them more accessible for degradation or transformation by other microorganisms (Xi et al. 2020; Zhang et al. 2023). For example, Zhang et al. (2023) reported *Desulfovibrio desulfuricans* stabilizes antimony toxicity in microcosm experiments, which was further elevated by the addition of humic acid and sodium sulphate. Another critical function of SRB is the precipitation and immobilization of metals. The H_2S produced by SRB can react with heavy metals present in the soil, leading to the formation of insoluble metal sulphides (e.g., FeS, ZnS) (Niu et al. 2018; Su et al. 2022). This process effectively reduces the mobility and bioavailability of toxic metals, thereby preventing their uptake by plants and minimizing their harmful effects on beneficial rhizosphere microorganisms. By immobilizing these metals, SRB help to maintain a less toxic environment in the rhizosphere, which is essential for the survival and activity of other microbial communities involved in rhizoremediation (Li et al. 2016). The study by Yang et al. (2023) demonstrated that a consortium of SRB, particularly *Desulfovibrio* and *Desulfobulbus*, in the rhizosphere of *Acacia dealbata* and *Pisum sativum* significantly increased the conversion of metals (23.47%) to their residual forms (75.98%) and immobilized heavy metals, such as zinc, copper, arsenic, and lead in the soil. These findings highlighted the potential of SRB-plant associations in immobilization and transformation of metals into less bioavailable forms, thereby enhancing the rhizoremediation of metal-contaminated soils (Yang et al. 2023).

SRB also support microbial consortia in the rhizosphere by creating anaerobic, sulphide-rich microenvironments that favor the growth and activity of specific microorganisms, such as *Desulfobacter* spp., *Desulfovibrio* spp., *Thiobacillus* spp., etc., capable of degrading complex organic pollutants (Vasantharaj et al. 2017; Vernans et al. 2019). The interaction between SRB and these other microbial taxa can result in synergistic effects, enhancing the overall capacity of the rhizosphere to remediate contaminants. For example, *Thiohalophilus*, *Chromatium*, and *Desulfatibacillum* serve to stabilize the rhizosphere of *Sonneratia apetala* and *Kandelia obovate* mangrove shrubs. Furthermore, they augment the population of other microbial communities, such as *Thiohalobacter*, *Chromatium*, and *Desulfatibacillum*, which also play a role in maintaining anoxic conditions in sediments (Li et al. 2021). This interaction is particularly important in the breakdown of recalcitrant compounds, such as hydrocarbons or chlorinated solvents, which require a consortium of microorganisms for complete degradation (Das et al. 2023b). Furthermore, SRB interact directly with plant roots, influencing root health and function. The reduction of sulphate and subsequent production of sulphide can affect nutrient availability in the soil and reduce the uptake of toxic heavy metals by plants (Yang et al. 2023). For instance, adding SRB to the rice rhizosphere lowers the soil's redox potential and raises its pH. This stabilizes Pb toxins by changing to sulphides and prevents uptake of Pb by rice plants (Lin et al. 2010). Additionally, the presence of SRB in the rhizosphere can alter the pattern of root exudation, as plants may modify their exudates in response to the changing redox conditions or to recruit specific microbial communities that interact symbiotically with SRB. They produce enzymes, such as NADH, threonine protein kinase, and ubiquinone reductase, during stress, which further assists in energy metabolism and promotes the adaptability of overall bacterial community (Yang et al. 2023). This interaction between plant roots and SRB can have significant implications for growth promotion in plants and the overall efficacy of rhizoremediation processes.

SRB can enhance phytoremediation by facilitating the uptake of certain contaminants by plants or by reducing the phytotoxicity of heavy metals through immobilization. SRB have been shown to reduce the concentration of toxic methylmercury in rice rhizospheres; however, in the same study, endophytic SRB, such as *Methylocaldum* spp., *Hyphomicrobium* spp., and *Methylocystis* spp., have been reported to enhance the uptake of methylmercury in rice plants (Guo et al. 2024). By mitigating metal stress, SRB can indirectly promote more robust plant root systems, which in turn can exude organic compounds that stimulate further microbial activity in the rhizosphere. This positive feedback loop underscores the importance of SRB in supporting both microbial and plant components of the rhizosphere, particularly in contaminated environments where bioremediation is necessary.

4 Advanced Techniques for Monitoring SRB Populations and Activities in Rhizosphere During Rhizoremediation

The study of SRB in the rhizosphere has evolved significantly with the advent of advanced molecular and imaging techniques. These methodologies allow scientists to explore the multifaceted interactions among SRB, plants, and the surrounding microbial communities, thereby enhancing our understanding of their roles in bioremediation of xenobiotics and nutrient cycling.

4.1 Gene Targets in the Study of SRB Within the Rhizosphere

When studying SRB in the rhizosphere, several genes are commonly targeted to understand their abundance and functional roles. By using qPCR, changes in SRB populations in response to varying environmental conditions, such as heavy metal contamination and plant growth, can be effectively monitored (Kushkevych et al. 2021; Yang et al. 2023). The most frequently studied genes include the 16S rRNA gene, which is often used for general bacterial identification and quantification, although it may not accurately reflect the diversity of SRB in environmental samples due to its presence in all bacterial phyla (Zhang and Fang 2006). The dissimilatory sulphite reductase (*dsr*) genes, specifically *dsrA* and *dsrB*, are crucial for SRB metabolic pathways, making them reliable markers for quantifying SRB populations (Neukirchen et al. 2023). Bahr et al. (2005) targeted the *dsrAB* gene to quantify SRB abundance in a constructed wetland treating mine drainage. They found that SRB numbers were significantly higher in the rhizosphere of wetland plants compared to bulk sediments, highlighting the importance of plant-SRB interactions in metal immobilization (Bahr et al. 2005). Studies have shown that quantifying these genes provides a better assessment of SRB abundance compared to 16S rRNA alone (Ben-Dov et al. 2007). The adenosine-5′-phosphosulfate reductase (*aps*) gene, particularly *apsA*, is also targeted in qPCR studies as it plays a significant role in the energy metabolism of SRB and is considered a reliable indicator of xenobiotic degradation by SRB in various environments, including the rhizosphere (Zhang and Fang 2006; Ben-Dov et al. 2008). Additionally, bisulphate reductase genes are studied for their role in the reduction of sulphate to sulphide, contributing to the understanding of SRB's ecological functions in the rhizosphere (Agrawal and Lal 2009). Research has demonstrated that targeting these specific genes through PCR and qPCR techniques allows for a more accurate quantification of SRB in environmental samples, enhancing our understanding of their ecological roles.

4.2 Omics Approaches

The application of omics technologies, including genomics, transcriptomics, proteomics, and metabolomics, provides a comprehensive understanding of the interactions of SRB within the rhizosphere. By combining multiple omics approaches, researchers can gain a comprehensive understanding of the molecular mechanisms underlying the synergistic relationship between SRB and plants in xenobiotic degradation (Kumar et al. 2021).

4.2.1 Genomics and Metagenomics

Metagenomics allows for the comprehensive analysis of microbial communities without the need for culturing of individual microorganisms. By sequencing the total DNA extracted from soil samples, researchers can identify the diversity of SRB and their functional genes (Fig. 3). Metagenomic studies have revealed the potential metabolic pathways utilized by SRB in the rhizosphere, providing insights into their ecological roles and interactions during rhizoremediation (Xu et al. 2018). For example, recent studies have identified catabolic genes linked to the degradation of persistent organic pollutants (POPs) by SRB, indicating their potential role in detoxifying environments contaminated with such compounds (Štefanac et al. 2021; Miglani et al. 2022). Additionally, metagenomic analyses have shown that the presence of certain SRB correlates with reduced levels of specific xenobiotics in contaminated sites, further supporting their utility in bioremediation (Agrawal and Shahi 2015). The application of genomics and metagenomics has significantly advanced our understanding of SRB and their interactions with plants in the rhizosphere. Recent studies have utilized next-generation sequencing (NGS) to explore the complex microbial communities associated with plant roots, revealing insights into the functional roles of SRB in nutrient cycling and plant health (Paganin et al. 2021). For example, studies have shown that metagenomic and metatranscriptomic analyses can reveal the presence of xenobiotic-degrading bacteria, such as *Pseudomonas* and *Sphingopyxis*, along with the transcription of important genes, which encodes enzymes like peroxidases and oxygenases that are crucial for breaking down pollutants (Russell et al. 2021). Moreover, machine learning techniques have been integrated with metagenomic data to enhance the predictive capabilities regarding microbial interactions, as shown by Statnikov et al. (2013), emphasizing the potential of these methods to analyze complex microbial datasets (Statnikov et al. 2013). These advancements justifies the significance of genomics and metagenomics in elucidating the dynamic interactions of SRB, paving the way for improved agricultural practices and ecosystem management. Fluorescence in situ hybridization (FISH) has also been a valuable tool in studying SRB, particularly in visualizing microbial consortia in biofilms. FISH has been used to identify SRB genera, such as *Desulfosarcina* and *Desulfococcus*, in marine sediments and wastewater biofilms (Ramsing et al. 1993; Manz et al. 1998). Additionally, the combination of FISH with other techniques, such as rRNA slot blot hybridization and qPCR targeting *dsr* genes, has provided insights into the spatial distribution and specific cellular rRNA contents of SRB in various environments (Ravenschlag et al. 2000). Metagenomics has been applied to study the diversity and functional potential of SRB in various habitats, including revegetated acidic mine wasteland, wastewater, contaminated soils etc. (Zhang et al. 2022). For example, metagenomic studies of contaminated soil samples from Nalabana Island revealed the abundance of genes for xenobiotic biodegradation, sulphur metabolisms, carbon metabolisms, nitrogen metabolisms, and oxidative phosphorylation (Mohapatra et al. 2021). Thus, metagenomics provides valuable insights into the functional capabilities of SRB in xenobiotic degradation.

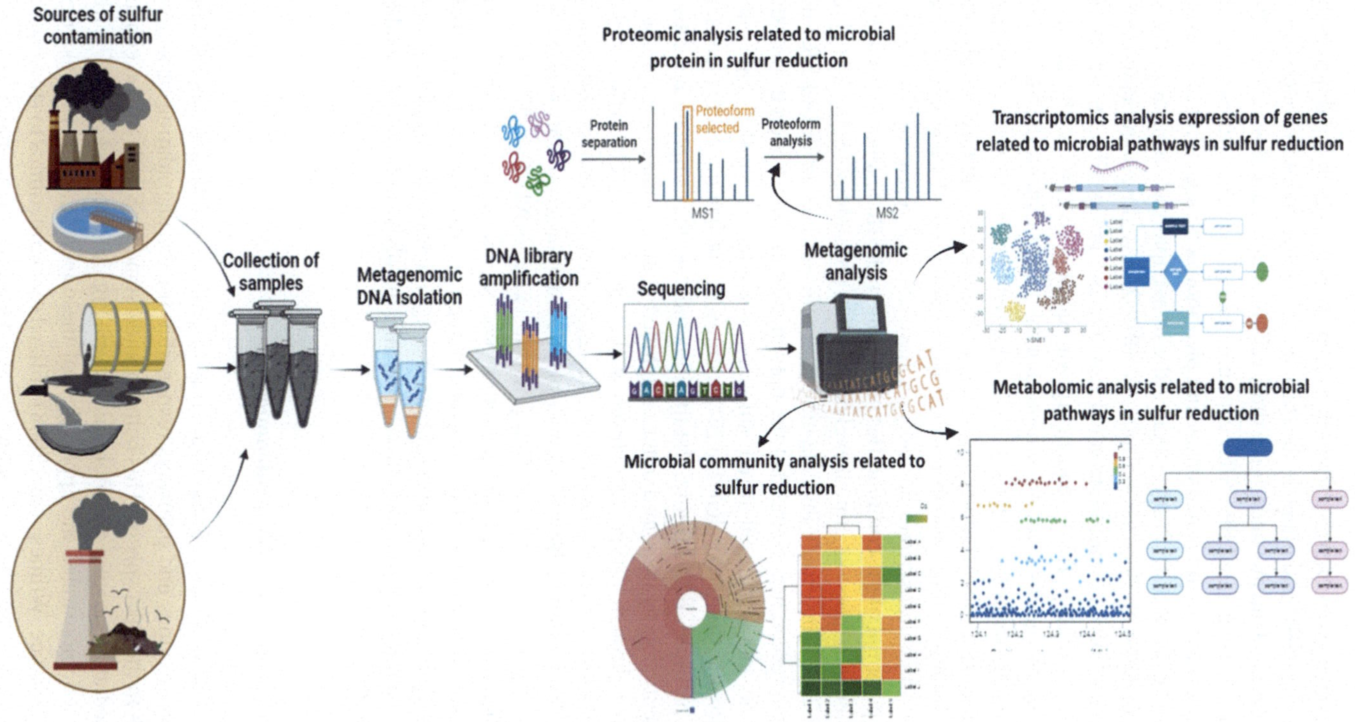

Fig. 3 Cutting-edge molecular techniques for investigating xenobiotic degradation by SRB

4.2.2 Transcriptomics and Metatranscriptomics

Transcriptomics and metatranscriptomics are powerful tools for studying the interactions between SRB and plants in the rhizosphere, which helps in understanding of the molecular mechanisms underlying these relationships (Li and Bao 2021). Recent research has demonstrated that SRB can influence plant gene expression, particularly in the context of stress responses and phytohormone signalling. For example, a study by Park et al. (2017) investigated the transcriptomic responses of soybean plants inoculated with *Bacillus aryabhattai* SRB02, revealing significant upregulation of genes involved in jasmonic acid biosynthesis and signalling pathways (Park et al. 2017). Similarly, RNA sequencing was utilized to explore the interactions between rice and various plant growth-promoting bacteria, including SRB, identifying differentially expressed genes related to nutrient uptake as well as stress tolerance (Nejat et al. 2018; Thomas et al. 2019). The findings highlighted the role of SRB in promoting plant growth through the modulation of metabolic pathways. Furthermore, metatranscriptomic analyses have allowed researchers to assess the functional potential of the communities prevailing in the rhizosphere, revealing how SRB contribute to biogeochemical cycles and plant physiology (Schenk et al. 2012; Dang et al. 2023). In a study investigating the transcriptional profile of SRB during Fe reduction, researchers found that the upregulation of the O-acetylserine sulfhydrylase gene suggested enhanced sulphate assimilation in the soil samples. Additionally, the study revealed that SRB commonly utilize the energy-conserving trithionate pathway. Furthermore, the findings indicate that sulphate not only promotes Fe reduction by its conversion to sulphide but also enhances enzymatic electron transfer through c-type cytochromes (Li and Bao 2021). Transcriptomics studies also used to understand the gene regulation patterns under different stress conditions. For example, the alteration of pH in acid mine treatment system resulted in changes of gene expression of *Desulfovibrio vulgaris* ATCC 7757. The expression of genes associated with the reduction process, including *DVU0499* and *sat*, which encode proteins crucial for sulphate reduction, was reduced at low pH (Yu et al. 2020). These investigations highlight the significance of transcriptomic methodologies in explaining the intricate relationships involved in bacterial metabolism of xenobiotic.

4.2.3 Proteomics and Metaproteomics

Proteomics involves studying the total array of proteins synthesized by SRB and to identify key enzymes and pathways responsible for the degradation of xenobiotic compounds in the rhizosphere. This helps in understanding SRB's role in enhancing the bioremediation process at a molecular level (Fig. 3.) (Dolla et al. 2007). Recent studies have demonstrated the application of these techniques in various environments. For instance, Wilmes et al. (2008) utilized metaproteomics to study activated sludge wastewater treatment, revealing knowledge of the functional roles of microbial communities. The potential SRB *Candidatus Accumulibacter phosphatis* was found to have enzymes important for energy generation, glycogen synthesis, polyhydroxyalkanoate synthesis, glycolysis, TCA cycle, fatty acid synthesis, and phosphate transport (Wilmes et al. 2008). Similarly, a study by Schenk et al. (2012) highlighted the potential of multi-species transcriptomics in unraveling plant-microbe interactions, emphasizing the importance of integrating different omics approaches to gain a comprehensive understanding of these complex relationships (Schenk et al. 2012). *Desulfovibrio vulgaris* Hildenborough is an SRB and widely studied for its role in metal corrosion and heavy-metal bioremediation. A proteomic study for its biofilm-forming ability during heavy metal remediation revealed changes in the abundance of genes and proteins involved in carbon or energy flow and in the maintenance of extracellular structures (Clark et al. 2012). The complete genome, sub-proteomes, and key metabolites of *Desulfobacula toluolica* Tol2 (marine SRB) revealed its ability to metabolize complex aromatic compounds, such as toluene under sulphate-reducing conditions, which highlight its potential for bioremediation in marine environments (Wöhlbrand et al. 2013). In the framework of interaction between SRB and plants, proteomics and metaproteomics can provide valuable understandings of the ongoing molecular mechanisms. By identifying and quantifying proteins expressed by SRB and their plant hosts, researchers can elucidate the functional roles of specific microbial strains, their adaptations regarding alterations in environmental conditions, and their influence on plant growth and stress responses (Tripathi 2022). These findings underscore the potential of proteomics and metaproteomics in the study of the complex interplay prevailing in ecosystem.

4.2.4 Metabolomics

Metabolomics studies in rhizoremediation of xenobiotics by SRB involve the comprehensive analysis of metabolic changes in microbes as they interact to degrade or transform toxic xenobiotic compounds in the soil. It helps in identifying metabolic pathways used by SRB to facilitate detoxification of pollutants (Fig. 3) (Rodríguez et al. 2015). Metabolomics enables comprehensive profiling of primary and secondary metabolites produced by SRB during degradation of any xenobiotic compound (Miglani et al. 2022). It helps in identifying novel metabolites and pathways involved in metabolism that may not be detected by targeted biochemical assays (Qiu et al. 2023). Metabolomics combined with multivariate data analysis and LC-MS has been successfully applied to numerous xenobiotic studies, which thereby helps in reporting of novel metabolites and pathways. Metabolite profiling, flux analysis, fingerprinting, and target analysis are techniques used to identify, analyze, and quantify a wide range of metabolites (Parisi et al. 2009; Antoniewicz 2021; Sharma et al. 2022a). Metabolomics has also been employed to assess variations in the expression of genes, to analyze the physiological responses of organisms in different environmental conditions, and to identify the species responsible for the degradation of xenobiotic pollutants (Patankar et al. 2022). Metagenome predictions indicated a rise in genes linked to amino acid, carbohydrate, lipid, and xenobiotic metabolism, along with increased metabolites like carboxylic acids, fatty acids, and amino acids, after microbial treatment of produced water (Kotu et al. 2024). A study on *Oleidesulfovibrio alaskensis* G20 revealed the *sat* gene as a key link between enriched sulphur metabolism pathways and gene clusters involved in seleno compound metabolism, amino acid metabolism, secondary metabolite synthesis, and cofactor biosynthesis during bioremediation (Saxena et al. 2023). Stable isotope-resolved metabolomics is one of the current approaches to study xenobiotic metabolism in SRB, enabling network-wide metabolic pathway elucidation (Kim et al. 2023). Metabolomics can be applied to study the degradation mechanisms of various xenobiotics like carbaryl, N-methyl carbamates pesticides, heavy metals, hydrocarbons, and other pollutants by SRB (Sharma et al. 2022b; Das et al. 2023b).

Stable isotope probing (SIP) has emerged as a highly effective metabolomic technique for investigating the interactions of SRB. This technique provides valuable insights into the ecological roles and metabolic activities of SRB in the rhizosphere. By incubating environmental samples with stable isotope-labelled substrates (e.g., ^{13}C-labeled plant root exudates), the labelled DNA or RNA of actively growing microorganisms can be separated and analyzed. This approach allows researchers to identify the SRB species that are actively involved in the degradation of specific compounds or the utilization of plant-derived carbon sources in the rhizosphere (Dumont and Murrell 2005; Cupples 2016). Bombach et al. (2009) used the SIP technique with ^{13}C to label toluene, which helps in identification of the SRB capable of toluene degradation in contaminated aquifer sediments. This technique can be adapted to study the interactions between SRB and plants in the rhizosphere by using labelled plant materials or root exudates as substrates (Bombach et al. 2009). In another study, Cupples (2016) utilized SIP to investigate the metabolic pathways of SRB in the presence of plant exudates, demonstrating that SRB preferentially utilized root-derived organic compounds, thereby enhancing their growth and activity in the rhizosphere. Additionally, single-cell SIP (SC-SIP) techniques have been employed to examine the spatial distribution and physiological diversity of SRB within plant-associated microbial communities (Alcolombri et al. 2022). This approach has provided valuable insights into how SRB contribute to nutrient cycling and plant health, particularly under varying environmental conditions. Furthermore, SIP has been instrumental in elucidating the roles of SRB in promoting plant growth by enhancing nutrient availability and modulating plant hormone levels, thereby highlighting the potential of SRB as beneficial microorganisms in sustainable agriculture (Gamalero et al. 2022).

Thus, metabolomics is an essential technique for investigating the molecular effects of xenobiotics on SRB and elucidating the metabolic pathways involved in their degradation. The integration of metabolomics with other omics approaches provides a comprehensive understanding of SRB-mediated xenobiotic degradation processes.

5 SRB's Interactions with Other Soil Microbes

SRB play a crucial role in bioremediation of xenobiotics from contaminated soils by immobilizing toxicants through the formation of insoluble sulphides and are important group of geo-microbes involved in nutrient cycling (Zak et al. 2021). However, their interactions with other soil microbes during this process are complex and not fully understood. SRB inoculation significantly affects soil bacterial abundance, diversity, and community structure compared to non-inoculated controls. The relative abundance of SRB, such as *Desulfobacterota*, increases dramatically in SRB-inoculated treatments during bioremediation of mine soil (Yan et al. 2023). SRB also promote the enrichment of functional microbes involved in sulphur cycling, like *Desulfosporosinus*, *Desulfitobacterium*, *Desulfobulbus*, and *Dethiosulfovibrio* (Lashani et al. 2023; Yan et al. 2023).

Soil harbors a wide variety of bacterial communities that may either compete with or make synergy with SRB for available resources before reaching a stable equilibrium and coexistence. Surface soils demonstrate greater microbial diversity and activity due to higher availability of organic matter, oxygen, and moisture (Xia et al. 2020; Du et al. 2024). In different soil strata, SRB interacts extensively with majorly aerobic and facultative anaerobic microbes, including abundant bacteria, fungi, and protozoa (Hu et al. 2024). Further, anaerobic conditions exist in deeper soil layers which favor the growth and survival of obligate anaerobic bacteria and archaea, for example, *Clostridium* spp., *Geobacter* spp., *Acetobacterium* spp. etc. (Hu et al. 2024). Notably, *Acetobacterium* supports SRB's growth by converting CO_2 and hydrogen into acetate, which SRB further use for sulphate reduction (Hao et al. 2021). Whereas *Geobacter* spp. are competitor of SRB, they compete for electron donors, for example, acetate, lactate, ethanol, and hydrogen, potentially impacting the activities and abundance of SRB (Zhang et al. 2023). However, it was found that with time abundance of *Geobacter* reduced and SRB increased in a uranium degradation study. Further, the addition of metal ions such as Fe^{3+} helps to limit the growth of *Geobacter* sp. (Barlett et al. 2012). Fungal species such as *Nitrospira* converts ammonia to nitrate (nitrification), thereby supporting SRB through their influence on the nitrogen cycle (Peng et al. 2020). Meanwhile, ectomycorrhizal fungi such as *Pisolithus* and *Laccaria* enhance the uptake of phosphorous and also decompose organic matter from soil. Thus, enhance nutrient availability to SRB and increase their metabolism (Zhang et al. 2019; Mundra et al. 2021).

The introduction of plants further increases the bioremediation potential but also adds complexity, as they release various root exudates, which can alter the interaction between metals and ligand. For example, plants like *Zea mays*, *Brassica juncea*, *Medicago sativa*, and *Ricinus communis* release root exudates that alter soil's physiochemical properties, thereby affecting the overall microbial community structures associated with SRB (Garbeva et al. 2004; Dhir 2017). Within the rhizosphere of these plants, the release of root exudates significantly boosts carbon availability, which promotes the growth of both aerobic and facultative anaerobic bacteria. This condition supports complex interactions of SRB in nutrient cycling and the rhizoremediation. However, elevated levels of metal cations in the soil can exert toxic or inhibitory effects on microbial populations (Hu et al. 2024).

SRB can significantly alter the soil microbial community structure and enrich functional microbes, while also being influenced by competition, collaboration, and metal toxicity from other soil microbes. Understanding these intricate interactions is crucial for optimizing SRB-driven bioremediation strategies in xenobiotic contaminated soils.

6 Methods for Enhancement of SRB Activity During Rhizoremediation of Xenobiotics

6.1 Addition of Zero-Valent Iron (ZVI)

The addition of zero-valent iron (ZVI) enhances SRB's activity and increases the removal of xenobiotic compounds, particularly heavy metals from wastewater (Chen et al. 2008a). ZVI serves as an electron donor, thereby reducing competition for electrons between methanogens and SRB in sulphate-rich wastewater (Zhang et al. 2011). Moreover, SRB attached to ZVI can also gain electrons from hydrogen produced during ZVI corrosion, which lowers the chemical oxygen demand (COD) for sulphate reduction (Kong et al. 2014; Liu et al. 2015). The lowered oxidation-reduction potential (ORP) from ZVI creates a more reductive environment, supporting the growth and survival of SRB (Zhao et al. 2021). Hydrogen, along with organic compounds, acts as an electron donor for SRB, and the consumption of hydrogen helps in neutralizing the pH of environment by producing alkaline by-products (Zhang et al. 2011). For example, H_2 and organic compounds act as electron donor and facilitate the reducing capabilities of *Desulfobacterium* sp. and *Desulfovibrio* sp. (Seth and Edyvean 2006; Qian et al. 2019). Further, ZVI can directly react with contaminants like Cr^{6+} and reduce their toxicity (Gheju 2011). Recent research reported that 2.5 times more Cr^{6+} was remediated from groundwater in the SRB-ZVI system as compared to individual systems (Xu et al. 2024). ZVI also helps SRB to adapt to a wider pH and temperature range. This combination forms stable heavy metal precipitates that do not leach under varying pH or redox conditions (Xu and Chen 2020). While ZVI improves sulphate and heavy metal removal, its cost may be a limiting factor in large-scale applications.

6.2 Supply of Additives

Enhancing SRB activities during xenobiotic degradation can be achieved through various additives that improve microbial metabolism and efficiency. Organic acids like acetic and propionic acid stimulate SRB activity by providing readily available carbon sources, which enhances their metabolic processes and increase sulphate reduction rates (Madirisha et al. 2022). For instance, the addition of acetic acid has been shown to significantly boost the growth and activity of SRB in anaerobic environments, facilitating the degradation of xenobiotic compounds such as phenol and toluene (Guo et al. 2015). In another study by Gandy et al. (2023), it was reported that propionic acid enhanced the activity of SRB, resulting 99% effective zinc removal through bacterial sulphate reduction (Gandy et al. 2023). Nutrient supplements, including nitrogen and phosphorus sources, are crucial for the optimal growth of SRB (Santos et al. 2023). Further, the addition of ammonium salts (e.g., ammonium sulphate) can enhance the sulphate reduction process by providing essential nitrogen, which is often a limiting factor in SRB metabolism. The presence of nitrogen sources can contribute to higher rates of sulphate reduction and improved degradation of toxic compounds in contaminated environments (Novair et al. 2024). In a study, the addition of the strain *Ochrobactrum* sp. POC9 to hydrometallurgical leachates resulted in an increase in the generation of ammonia to 984.7 mg/L as well as vitamin B, both of which contribute to an improvement in the efficiency of wastewater treatment (Yang et al. 2021). Electron donors, such as lactate and ethanol, are commonly used to enhance SRB activity. These compounds serve as substrates for SRB, promoting their growth and metabolic activity. For example, lactate has been effectively used in bioremediation processes to enhance the degradation of chlorinated solvents, as it supports the growth of SRB that can reduce sulphate to sulphide while simultaneously degrading the contaminants (Cao et al. 2012). Moreover, lactate and ethanol also enhance the SRB (*Desulfovibrio* sp., *Clostridium* sp., and *Syntrophobacter* sp.) population (Santos et al. 2022). The addition of lactate and ethanol to SRB resulted in a successful removal of antimony (97.6% and 97.8%) and arsenic (27.8% and 26.4%); the addition of Fe^{2+} further significantly enhanced the overall removal efficiency (Liu et al. 2018).

Biosurfactants can improve the bioavailability of hydrophobic xenobiotic compounds, making them more accessible to SRB. For instance, non-ionic surfactants like Tween 80 have been shown to enhance the degradation of PAHs by increasing their solubility and facilitating microbial access (Das et al. 2024). Enzymatic additives, such as quorum quenching enzymes, can disrupt biofilm formation and enhance microbial interactions in mixed cultures. These enzymes can improve the overall metabolic efficiency of SRB by reducing competition and promoting synergistic relationships with other microorganisms involved in xenobiotic degradation. Research has indicated that the use of quorum quenching enzymes can lead to enhanced degradation rates of complex pollutants in anaerobic environments (Lashani et al. 2023). The strategic use of various additives can significantly enhance the activity of SRB during the degradation of xenobiotic compounds. Future research should continue to explore the optimal combinations and concentrations of these additives to maximize their effectiveness in different environmental contexts.

6.3 Addition of Inorganic Cations

The addition of inorganic cations such as calcium, copper, magnesium, iron, etc.to the medium aids in mitigation of the heavy metal's toxicity on SRB. This is facilitated by their binding to anionic sites on bacterial cell surfaces, thereby blocking the heavy metal adsorption in bacterial cell (Xu and Chen 2020). Fe acts as both an electron acceptor and a cofactor for enzymes involved in xenobiotic degradation by SRB (Kumari and Das 2023). The Fe^{2+} ion serves as a trace element essential for the synthesis of certain enzymes that facilitate contaminant removal. Likewise, the precipitation of Fe^{2+} can take place with sulphide ions and also reduce their toxicity. Further, the remaining insoluble sulphides formed with Fe^{2+} exhibit enhanced capacity for adsorbing and removing heavy metals (González-Paz et al. 2022). For example, addition of 200 mgL^{-1} of Fe^{2+} can increase As and antimony (Sb) removal efficiency from 78% and 98.8% to 98.2% and 99.4%, respectively (Liu et al. 2018). Magnesium ion (Mg^{2+}) is a vital cofactor for numerous enzymes and is necessary for the optimal growth and activity of SRB. Supplementing growth media with Mg^{2+} has been shown to enhance anaerobic biodegradation of xenobiotics by SRB (Cao et al. 2009). High concentrations of Mg^{2+} stimulate SRB activity by enhancing sulphate (SO_4^{2-}) diffusion into the cells and improving mass transfer, rather than inhibiting it (Cao et al. 2009). Calcium ion (Ca^{2+}) plays a structural role in maintenance of cell walls and membranes of SRB. Increasing the availability of Ca^{2+} can enhance the tolerance of SRB to xenobiotics and also improve their biodegradation capabilities (Zhao et al. 2019; Eltarahony et al. 2020). Copper (Cu) is a vital micronutrient and also needed as a co-factor for many enzymes, but excessive Cu can cause harmful effects in microbial cells. Cu leaching from pipelines elevates environmental Cu levels, posing health risks, but SRB can thrive in toxic Cu conditions. Comparative transcriptomic analysis between control and 5 μM Cu^{2+} exposure resulted in the downregulation of 61.43% and the upregulation of 38.57% genes in model organism *Desulfovibrio alaskensis* G20. Significant changes were seen in the expression of inorganic ion transporters, translation machinery, DNA transcription, and signal transduction, particularly at high Cu^{2+} concentrations (Tripathi et al. 2022). Thus, careful optimization of these cation concentrations is essential to maximize biodegradation efficiency while minimizing toxicity to SRB.

6.4 Introduction of Carrier Materials

Carrier materials can enhance microbial activity in SRB systems by improving electron transfer (Zhao et al. 2023). Graphene oxide (GO) has been shown to significantly enhance this transfer compared to conventional carbon materials (Zhang et al. 2018). The population of SRB in the effluent containing GO was observed to be 277 times more than in effluents without GO. Additionally, in microbial electrolysis cells using GO, the sulphate reduction rate increased significantly, achieving a 70% improvement over cells lacking GO. These findings indicate that GO enhances SRB growth by facilitating electron transfer (Yan et al. 2018; Hu et al.

2019). GO can also serve as an electron acceptor, improving sulphate removal efficiency when used as a sole electron donor for SRB. Reduced graphene oxide (rGO) exhibits superior electron transfer capabilities and structural properties compared to GO (Wu et al. 2019). It can promote the development of pili in bacterial sulphate-reducing biofilms, accelerating electron transfer due to its positive surface charge, which enhances electrostatic interactions with negatively charged bacterial membranes (Bueno-López 2020). This electron inflow can also reduce toxic heavy metals, mitigating their harmful effects on microbes (Cheng et al. 2018). The addition of rGO also enhances the extracellular polymeric substances (EPS) production (54%), which thereby protects the bacterium in heavy metal-contaminated environment. In the same study, it was found that Cr(VI) was reduced to Cr(III) by 52% and to elemental Cr by 28% with the action of rGO in SRB-contained bioreactor (Yan et al. 2020). Similarly, *Enterococcus avium* and rGO assembly remarkably reduced heavy metal toxicity by 50% (Yan et al. 2018). The microbial conversion of GO to its reduced form, rGO, by SRB offers a highly advantageous approach, as this process eliminates the need for expensive chemical reductants and complex reaction conditions often required in traditional GO reduction methods. Additionally, it leverages natural microbial activity, making it safer and more sustainable. As a green alternative, SRB-driven GO reduction minimizes harmful byproducts, aligns with eco-conscious practices, and supports broader applications in fields like environmental remediation, electronics, and biotechnology, where non-toxic, renewable processes are increasingly in demand.

Several other carrier materials can effectively support SRB in bioremediation processes. Research has demonstrated that SRB immobilized on zeolite carriers significantly enhance the removal efficiency of heavy metals. For instance, an SRB-zeolite carrier achieved over 90% removal of nickel (Ni^{2+}), outperforming natural zeolite by at least 2.5 times (Kim et al. 2015). Sand has been used as a carrier matrix in upflow packed-bed bioreactors for anaerobic sulphate reduction. This approach demonstrates that sand is a low-cost and efficient support material for effective sulphate reduction using mixed cultures of SRB, while plastic ballast rings enhance biofilm formation in fixed-bed biofilm column reactors to facilitate anaerobic sulphate reduction (Brahmacharimayum et al. 2019). Activated carbon materials also has been explored for its ability to immobilize SRB, as it enhances their metabolic activity which thereby enhance effectiveness in degrading xenobiotics (Żur et al. 2016).

7 Genetic Engineering as Prospect to Enhance SRB Activity During Rhizoremediation

Genetic engineering can significantly enhance the activity of SRB in rhizoremediation of xenobiotics and make the process more effective. Overexpression of genes involved in sulphate-reduction is one of the techniques where key genes involved in the sulphate-reducing pathway, such as those encoding sulphate adenylyltransferase (*sat*), APS reductase (*aprAB*), and dissimilatory sulphite reductase (*dsrAB, aprBA*), can be overexpressed in the bacterium (Haveman et al. 2005; Meyer and Kuever 2007). This would accelerate the reduction of sulphate to sulphide, thereby enhancing the detoxification of xenobiotics. Moreover, certain genes responsible for breaking down specific xenobiotics (e.g., hydrocarbons, chlorinated compounds, or other toxic organic pollutants) can be inserted into SRB to enable them to degrade complex pollutants that are usually beyond their native capability (Siguier et al. 2014). Engineering SRB to express multi-functional enzymes, such as oxygenase or reductases, which are readily involved in degradation or transformation of a wide range of xenobiotics, would expand the scope of contaminants that can be treated. For example, genes encoding enzymes, such as cytochrome P450 or monooxygenases, can help detoxify complex organic pollutants (Esteves et al. 2021; Das et al. 2024).

SRB often dwell in harsh conditions, such as low pH, high concentrations of heavy metals, and nutrient deficiencies in contaminated environments. Genetic modification can help to enhance the tolerance of SRB in these conditions by incorporating stress-tolerance genes, such as those conferring resistance to oxidative stress or high metal concentrations (Jawaharraj et al. 2023). By engineering SRB with traits that improve their colonization and interaction with plant roots, such as genes that enhance chemotaxis or biofilm formation, SRB can become more efficient in the rhizosphere, where they facilitate plant-microbe interactions for more effective xenobiotic degradation (Tripathi et al. 2021). Genetic modification can help SRB better interact with plant roots by improving their response to root exudates. This can be achieved by engineering SRB to express genes that respond to plant-derived signalling molecules, such as flavonoids or organic acids, enhancing their ability to locate and colonize the rhizosphere (Morgan et al. 2013; Nabavi et al. 2020). By enhancing SRB's ability to use plant-derived carbon sources (root exudates) as co-metabolic substrates, SRB can be engineered to degrade xenobiotics more efficiently through pathways that utilize these exudates. Genetic modifications can be targeted to reduce the formation of undesirable byproducts during sulphate reduction such as H_2S, which can be toxic at high concentrations (Kushkevych et al. 2019). Engineering SRB with genes for scavenging or detoxifying H_2S could minimize the negative impacts of bioremediation on both plants and the surrounding environment. Incorporating these genetic enhancements into SRB used for rhizoremediation could greatly improve their efficiency in detoxifying environments contaminated with xenobiotics while simultaneously promoting plant health and growth.

Species of the genus *Desulfovibrio* have been extensively studied for genetic engineering applications (Keller et al. 2011). The initial report on molecular analysis of SRB was reported in 1985 by Voordouw and Brenner, who cloned hydrogenase genes into *Escherichia coli* (Voordouw and Brenner 1985). One of the key challenges in molecular studies of SRB is the low and inconsistent plating efficiency, likely due to oxygen contamination, since individual cells are more sensitive to oxygen than biofilms. The production of sulphide and the release of hydrogen sulphide raise the pH to levels inhibitory to growth (Bender et al. 2006). Despite plating difficulties, some stable and persistent vectors were introduced into *Desulfovibrio* via conjugation in 1989 (Powell et al. 1989; Rousset et al. 1998), enabling the overexpression of native genes to produce functional proteins, such as type-I tetraheme cytochrome c and prismane proteins (Bender et al. 2006). Conjugation is effective, but counter-selection of the *E. coli* donor remains problematic due to aminoglycoside resistance under anaerobic conditions (Ojdana et al. 2018). Only one case of transduction in SRB has been reported, involving a defective bacteriophage in *D. desulfuricans* that transferred antibiotic resistance and auxotrophic markers within the species through bacteriophage-like particles. However, this vector had a limited host range (Rapp and Wall 1987). Transposons, like mini-Tn5 derivatives, have been used successfully for random mutagenesis in *D. vulgaris* and *D. desulfuricans*, though efficiencies are typically low. Electroporation has also been explored, with initial success in *D. fructosovorans*, though it remains less effective in other species. Gene interruption via plasmid integration is another method, as demonstrated in *D. desulfuricans* G20 by disrupting the tetraheme cytochrome C_3 gene (Bender et al. 2006). Additionally, peptide tagging, such as the use of Strep-tag®, is being tested for isolating functional protein complexes following genome integration via electroporation (Chhabra et al. 2011).

While genetic engineering offers promising advancements, several challenges also need to be addressed. The introduction of genetically modified organisms (GMOs) into natural ecosystems raises concerns about potential unintended consequences, including horizontal gene transfer between engineered and native microbial populations (Plavec and Berlec 2020). Moreover, the deployment of GMOs in environmental applications is also a subject to stringent regulations due to safety and ecological concerns. In this case, ensuring compliance with biosafety standards is crucial before any field trials and applications (Saxena et al. 2020). Therefore, developing systems for monitoring the performance and ecological impact of engineered SRB is essential in this regard (Sánchez-Paniagua López et al. 2018). Strategies such as programmed cell death mechanisms might be implemented to ensure that GMOs do not persist in the environment after completing their remediation tasks or after certain period of time.

8 Synthetic Bacterial Consortium in Xenobiotic Degradation by SRB

Microbial consortia play a crucial role in the bioremediation of xenobiotic compounds and in enhancing survival in harsh environments, such as wastewater, acid mine tailings, and crude oil-contaminated sites. By adapting to environmental fluctuations, these consortia enable syntrophic degradation of complex substances, allowing complete metabolic reactions through collaboration among different microorganisms (Latorre et al. 2016). Engineered consortia, including synthetic communities (SynComs) designed to mimic natural microbiomes, could be employed in industrial bioreactors to improve the removal of xenobiotics and sulphate through their unique detoxification properties. SynComs, while simplifying the natural microbiome, retain key interactions between bacteria and hosts, providing capabilities beyond what a single bacterium can achieve. The synergistic interactions within SynComs can enhance community stability, and the effective application of such microbial consortia necessitates the presence of highly resilient microorganisms that possess either natural or engineered degradation pathways. These robust organisms are essential for ensuring optimal performance in bioremediation processes, as they can effectively break down a wide range of contaminants under varying environmental condition (Maheshwari et al. 2023).

Brenner et al. (2008) initially proposed synthetic biology for generating microbial consortia, highlighting that cell signalling and communication are crucial initial steps in the establishment of engineered consortia (Brenner et al. 2008). Cell communication enables division of labor among microbial populations, which allows them to undertake difficult degrading steps. Moreover, this division of work also delivers cumulative benefits to the involved species (Hays et al. 2015). In bacteria, communication happens through tiny peptides in gram-positive species and acyl-homoserine lactone (AHL) signalling molecules in gram-negative organisms (Ayangbenro et al. 2018). For example, the production of extracellular polymeric substances (EPS) by the bioleaching organism *Acidithiobacillus ferrooxidans* indicates that communication plays a key role in biofilm formation, which is essential for the leaching mechanism. This organism secretes and responds to compounds from the acyl AHLs family as part of the auto-inducer quorum sensing system, along with the c-di-GMP signalling pathway also involved in quorum sensing (Vu et al. 2009; Ayangbenro et al. 2018). Modifying AHL and c-di-GMP levels through synthetic biology could enhance the organism's attachment to ore particles, thereby improving the leaching process (Brune and Bayer 2012). Additionally, inter-population communication involves the exchange of compounds related to metabolism and growth. A SynCom of five bacteria (*Acidithiobacillus ferrooxidans*, *Acidithiobacillus thiooxidans*, *Acidiphilium multivorum*, *Leptospirillum ferriphilum*, and *Sulfobacillus thermosulfidooxidans*) was evaluated for its copper bioleaching potential based on metal tolerance and oxidation activity. *L. ferriphilum* and *A. thiooxidans* showed higher copper recovery efficiency with the ability to oxidize iron and sulphur compounds. This consortium demonstrates a unique capability to leach copper sulphide ores and heavy metals like arsenic and copper, which was further confirmed by analysis of their whole genomes (Latorre et al. 2016). The use of these organisms for addressing persistent contaminants in the environment appears promising. However, despite developments in synthetic biology for the development of designed consortia for xenobiotic remediation, stresses persist concerning their possible unknown implications on ecosystems.

9 Application of SRB for Addressing Environmental Issues

9.1 Wastewater Treatment

SRB are crucial in wastewater treatment, providing numerous advantages that improve both the efficiency and effectiveness of the process. A key benefit of using SRB is their ability to produce minimal sludge compared to conventional aerobic treatment methods. This reduction in sludge not only decreases the overall volume of waste generated but also lowers the costs associated with sludge disposal (Van Den Brand et al. 2015). Additionally, SRB contribute to pathogen reduction in wastewater, which is crucial for improving water quality and ensuring public health. A study reported that a sulphate-enriched bioreactor achieved a higher removal of coliforms (2.2 log) compared to conventional activated sludge systems (1.5–1.7 log). The reduction in coliforms was primarily attributed to contact with sulphide, rather than factors such as pH or contact time. This suggests that active sulphide production by SRB in wastewater treatment systems can effectively reduce fecal coliforms and other pathogens in the effluent of wastewater treatment plants (Abdeen et al. 2010). They are also effective at removing heavy metals from wastewater, further enhancing the treatment process. Under anaerobic conditions with organic substrates, SRB reduce sulphate and facilitate the removal of metals like Cu, Pb, Zn, Cd, Ni, and Fe through metal sulphide precipitation (Table 1). Biological sulphide precipitation offers advantages over chemical methods, including lower solubility of sulphides at low pH, reduced sludge volumes, cost-effectiveness, and efficient removal even at low metal concentrations (Kiran et al. 2017). SRB can be used as a pre-treatment before anaerobic digestion to remove sulphate from wastewater, which improves biogas quality by reducing hydrogen sulphide and increasing methane production. By depleting sulphate and using a small amount of COD, SRB help enhance methanogenic activity, resulting in cleaner and more efficient energy recovery during anaerobic sludge digestion (Chen et al. 2008b; Das et al. 2023a).

The application of SRB in wastewater treatment is optimized under specific conditions. Domestic wastewater typically provides an environment conducive to SRB activity, characterized by favorable levels of pH, organic substrates, sulphate, salt, temperature, and limited oxygen presence. Key parameters such as the chemical oxygen demand (COD) to sulphate ratio and hydraulic retention time (HRT) must be carefully managed to maximize SRB performance (Van Den Brand et al. 2015). For instance, maintaining an appropriate COD/sulphate ratio can significantly influence the efficiency of sulphate reduction and overall treatment effectiveness. An innovative approach to integrating SRB into wastewater treatment systems is through the development of sulphate-reducing bacteria-based wastewater treatment systems, which can also incorporate sulphide fuel cells (SFC). This integration allows for simultaneous wastewater treatment and electricity generation (Rodriguez-Freire et al. 2016). For example, a study has shown that at a COD/sulphate ratio of 0.67 and an HRT of 4 h, COD and sulphate removal rates can reach impressive levels of 83–84%. Furthermore, at an eight-hour HRT, these systems can generate a voltage of 0.02 V and a power density of 130 mW/m^2 while achieving nearly complete sulphide removal (Yun et al. 2019). The diversity of SRB communities present in anaerobic digesters is another fascinating aspect of their role in wastewater treatment. These communities include various genera such as *Desulfomonile*, *Desulfomicrobium*, *Desulfococcus*, *Desulfobulbus*, and *Desulforhopalus*. Notably, species from the *Desulfobulbus* and *Desulfovibrio* genera often dominate these communities due to their ability to tolerate oxygen, which enhances their survival and activity in fluctuating environmental conditions (Table 1) (El Houari et al. 2020).

Table 1 Recent studies on application of SRB in degradation of different xenobiotic compounds

Waste source	Supplied carbon source	Microorganism and its source	Contaminant removed	Sulphate removed	Duration	Temperature and pH	Reference
Wastewater	Ethanol, nitrate, sulphate, and iron	*Clostridium* spp., *Alkaliphilus* spp., and *Bacillus* spp.	Linear alkylbenzene sulfonate (53 ± 4% and 75 ± 3%)		49 days		Costa et al. (2024)
Synthetic wastewater	Humic acid or sodium sulphate	*Desulfovibrio desulfuricans*	Sb (58.40%)				Zhang et al. (2023)
Synthetic-enriched wastewater		*Paracoccus*, *Thiobacillus*, *Desulfomicrobium*, and *Sulfurospirillum*	Sulfamethoxazole (20–60%)		More than 60 days		Su et al. (2023)
Acidic coal mine drainage		Acidophilic SRB *Desulfosporosinus meridiei*	Cr (100%), Fe (73%), Al (79%), Zn (98%), co (95%), Ni (99%) and Mn (21%)	45%	218 days	Initial pH: 2.2 Final pH: 5.1–6.8	Singh and Chakraborty (2022)
Pharmaceutical laden sludge		*Paraclostridium* sp.	Ciprofloxacin		24 h		Fang et al. (2021)
Wastewater		Fermentative acidogenic bacteria and SRB	Carbamazepine (46%)	98%			Tahir et al. (2021)
Acid mine drainage wastewater		Strictly anaerobic SRB and facultative *Firmicutes* sp., *Bacillus* sp.	Cd (98%), Al (97%), Mn (95%), Pb (94%), Sr (94%), and cu (91%)	85%	21 days	pH: 5.6–6.5	Akinpelu et al. (2021)
Activated sludge	Tryptophan	Indigenous bacteria	Sulfadiazine (10.20–16.00%) and sulfamethoxazole (65.44–86.26%)		200 days		Zhao et al. (2022)

Waste source	Supplied carbon source	Microorganism and its source	Contaminant removed	Sulphate removed	Duration	Temperature and pH	Reference
Industrial DDX-polluted soil	Sugarcane bagasse, glucose, ethanol, and lactic acid	Firmicutes, Proteobacteria, and Actinobacteria	DDT (96.1%)		20 days	pH: 7	Velasco et al. (2020)
Synthetic wastewater solution	Lactate and Ethanol	*Escherichia coli*, *Clostridium* sp. (SRB), and Ruminococcaceae	Sb (97.6–99.4%) As (27.8–98.4%)		12 days	Temp: 30 pH: 7	Liu et al. (2018)
Urban pond sediments	Lactate Glycerol protein biomass	*Pantoea agglomerans*, *Enterobacter* sp., *Citrobacter* sp., *Cupriavidus metallidurans*, *Ralstonia* sp., and *Burkholderia cepacian*	As (74.8%)	80%	100 days	pH: 5	de Matos et al. (2018)
Alluvial aquifers of arid and semi-arid zones in Mexico	Lactate, sulphate, and arsenate	Indigenous microorganisms	As (52–90.9%)		30 days	Temp: 30 °C pH: 6.8	Rios-Valenciana et al. (2017)
Synthetic mine waste	Ethanol	*Desulfovibrio desulfuricans* and *Desulfomicrobium baculatum*	As (98–100%), Cu, Fe, Ni, and Zn (99%)	90%	500 days	Initial pH: 3–4 Final pH: 6–8	Sahinkaya et al. (2015)
Acid mine drainage		*Desulfovibrio desulfiricans* and *Clostridium* Spp.	Fe, Cu, Zn, and Cd (99.9%) Mn (42.1–99.3%)	80.2%	152 h	Initial pH: 2.8 Final pH: 7.8–8.3	Zhang and Wang (2016)
Contaminated groundwater	Ethanol	SRB	As (91.2%)		250 days	Temp: 30 °C pH: 6.25–6.50	Rodriguez-Freire et al. (2016)

(continued)

Table 1 (continued)

Waste source	Supplied carbon source	Microorganism and its source	Contaminant removed	Sulphate removed	Duration	Temperature and pH	Reference
Acid mine drainage	Chicken manure, dairy manure and sawdust	SRB	Cd, Cu, Fe, Mn, Ni, Zn	79.04% for chicken manure, 64.78% for dairy manure, and 50.27% for sawdust	35 days	pH: 3–3.5	Zhang and Wang (2014)
Sludge from a sulphate-reducing anaerobic baffled reactor	Ethanol	*Desulfomicrobium baculatum*, *Desulfovibrio desulfiricans*, *Desulfovibrio africanus*, *Desulfovibrio magneticus*	As (63–96%)		245 days	Temp: 30–32 °C pH: 7.4	Altun et al. (2014)
Synthetic wastewater and acid mine drainage from copper mine	Lactate and Fe^0	SRB	Cu (99%), Fe (86%), Mn	61%	56 days	Temp: 25 °C Initial pH:2.75 Final pH: 6.2	Bai et al. (2013)
Synthetic wastewater		Consortium of SRB	Cu, Cr, Ni, Zn (94–100%)	67%	12 weeks	Initial pH: 6.0	Kieu et al. (2011)
Contaminated soil	Lactate	Mixture of SRB	Crude oil (84.4%)		54 days		Al-Asadi (2011)

However, SRB activity also presents challenges in wastewater treatment. A major issue is the production of hydrogen sulphide, which is a toxic and malodorous gas. The presence of hydrogen sulphide can lead to infrastructure corrosion, especially in sewer systems and treatment plant equipment, posing significant maintenance and health risks (Barton et al. 2014). Additionally, in anaerobic digesters, SRB often compete with methanogens for substrates like acetate and hydrogen, which can lead to a decrease in methane production, a valuable byproduct of wastewater treatment (Paulo et al. 2015). Thus, SRB represent a promising green technology for wastewater treatment. Their ability to reduce sludge production, remove pathogens and heavy metals, and even generate electricity when integrated with fuel cells makes them a valuable asset in modern wastewater management strategies.

9.2 Bioremediation of Acid Mine Drainage

Acid mine drainage (AMD) is a severe environmental hazard produced by the oxidation of sulphide minerals, particularly pyrite, during mining activities (Pierre Louis et al. 2015). The exposure of these minerals to air and water leads to the formation of sulfuric acid, which leaches heavy metals into neighboring water bodies, generating highly acidic and hazardous conditions. This poses significant risks to aquatic ecosystems, soil health, and human health (García-Gómez et al. 2014). Conventional treatments for acid mine tailings waste are costly, often involve polluting chemicals, and aren't always sustainable. Using plants and microorganisms (rhizoremediation strategy) offers a more cost-effective and eco-friendlier alternative (Panda et al. 2016).

SRB-based bioreactors offer a promising solution by simultaneously removing sulphate and metals, with the added advantage of reusability. However, their efficiency depends on factors like microbial diversity, carbon source, presence of micronutrients, and reactor design (Neculita et al. 2007; Martins et al. 2011). Overall, bioremediation using SRB is a environmental friendly and sustainable approach, effectively neutralizing acidity, buffering mine tailings, and removing heavy metals at a lower operating cost. Acid mine drainage is often controlled by adding alkaline minerals to sulphide-rich mine waste. Microbial-driven alkalinity generation primarily involves reduction processes, such as sulphate reduction, iron reduction, manganese reduction, methanogenesis, ammonification, and denitrification (Singh and Chakraborty 2022). However, acid mine tailings generally contain low levels of organic carbon, which can limit microbial activity. To effectively stimulate this activity, it is often necessary to add external electron donors, such as molecular hydrogen or various organic compounds. This supplementation provides the essential nutrients that microbes need to thrive and promotes more efficient bioremediation processes in these challenging environments (Novair et al. 2024). The isolation of acid-tolerant SRB, which are capable of reducing sulphate to sulphide and generating alkalinity, has enabled the development of innovative treatment methods (Table 1). The process of sulphate reduction serves as a rate-limiting step in the selective removal of toxic metals from acid mine tailings. This is because many of these metals tend to form highly insoluble sulphides, which complicates their extraction and necessitates effective bioremediation strategies (Hallberg 2010). Additionally, metal recovery is achieved by controlling sulphate concentrations through pH regulation in bioreactors. Various microorganisms had been reported to be used in microbial leaching of metals from ores and waste, including *Acidithiobacillus ferrooxidans*, *At. thiooxidans*, *Clostridium*, *Cronobacter*, *Thiobacillus denitrificans*, *Sulfolobus* sp., *Gallionella ferruginea*, *Sulfobacillus thermosulfidooxidans*, *Sulfobacillus acidophilus*, *Leptospirillum ferrooxidans*, *Acidiphilium cryptum*, *Acidianus brierleyi*, *Ferribacterium limneticum*, etc. (Table 1) (Torrentó et al. 2012; Ayangbenro et al. 2018; Tavakoli et al. 2021). The effectiveness of remediation using SRB largely depends on optimizing the supply of salts and organic substrates, such as manure, compost, sugarcane wastes, sawdust, yeast extract, molasses, etc. These materials play a crucial role in supporting bacterial metabolism, thereby enhancing the overall efficiency of the bioremediation process (Zhang and Wang 2014). For example, a study evaluated sulphate removal from acid mine drainage containing As, Cu, Fe, Ni, and Zn in an anaerobic system, using ethanol as a carbon and electron source. After incubation of 500 days, the sulphate removal efficiency reached 99%, with metal removal at 98–100% for As and 99% for Cu, Fe, Ni, and Zn. The identified SRB species in the reactor were *Desulfovibrio desulfuricans* and *Desulfomicrobium baculatum* (Sahinkaya et al. 2015). Furthermore, SRB have been effectively employed on an industrial scale for the remediation of acid mine wastewaters. Notably, two patented bioreactor systems, BioSulphide® and Thiopaq®, have been developed specifically for this purpose, demonstrating the practical applications of SRB in treating contaminated water from mines (Ayangbenro et al. 2018).

9.3 Treatment of Crude Oil Contaminated Sites Using SRB

The world economy and advancing civilization are heavily reliant on the petroleum industry for energy production and for source of raw materials for products, such as fertilizers and chemicals (Das et al. 2024). Crude oil contains a complex mixture of hydrocarbons, including volatile organic compounds (VOCs) and PAHs, which present considerable risks to both the environment and human health. When crude oil spills into aquatic ecosystems, it disrupts marine life by smothering organisms, reducing oxygen levels, and also contaminate food chains (Kotoky et al. 2018). SRB are gaining attention for their ability to metabolize hydrocarbons, positioning them as promising agents for bioremediation in anaerobic and oil-contaminated areas. Alkanes are generally more easily degraded by microbes, while resins and asphaltenes are very hard to metabolize (Das et al. 2023c). Aromatic compounds like PAHs are very toxic and can build up in animals and plants at higher trophic levels. While they can be partially broken down, they tend to stay in the environment for a long time (Nikitha et al. 2017; Zhou et al. 2024). Two mesophilic strains,

Hxd3 and Pnd3, have the ability to use *n*-alkanes while simultaneously reducing sulphates to sulphides. Specifically, Hxd3 thrives in the presence of alkanes with chain lengths of C_{15}–C_{18}, whereas Pnd3 prefers those with chain lengths of C_8–C_{12}. Additionally, a thermophilic community of SRB was discovered in water near crude oil reservoirs in Vietnam, which can utilize *n*-alkanes ranging from C_{12} to C_{16} at a temperature of 90 °C. *Desulfobacula toluolica* and *Desulfobacterium cetonicum* were isolated from anaerobic marine sediments and were able to utilize toluene as their only carbon source, resulting in the production of 20 mM of sulphides over a period of 25 days. Similarly, they were capable of metabolizing 0.3 mM of toluene over the course of 12 days. During this period, it produced around 1.75 mM of sulphide, indicating its potential utility in bioremediation processes in toluene-polluted environments (Wolicka and Borkowski 2007). More recently, in a wastewater remediation study by SRB, it was reported that toluene, benzene, and methyl-*t*-butyl ether were degraded up to 92, 65, and 45%, respectively, and toluene was more susceptible to the bacterial action among them (Huang and Kao 2016). The performance was further enhanced with the addition of sulphate reduction materials starch, magnesium sulphate, and polylactic acid (Hsia et al. 2021). Many SRB species can utilize aromatic compounds like phenol, catechol, and indole. *Desulfobacterium phenolicum*, an isolate of marine sediments, metabolized 2 mM of phenol in 4 days, producing 7.2 mM HS^-. *Desulfobacterium indolicum* used 1 mM indole over 150 h and produced 5 mM HS^-. *Desulfobacterium cetonicum* oxidized m-cresol at 0.24 g/day. The mesophilic *Desulfotomaculum gibsoniae* from freshwater sediments metabolized phenol, catechol, and cresols, using sulphate, sulphite, and thiosulfates as electron acceptors (Wolicka and Borkowski 2007). Thus, the treatment of crude oil-contaminated sites using SRB offers a promising bioremediation approach due to their ability to degrade hydrocarbons anaerobically. SRB-driven processes can efficiently reduce toxic compounds while enhancing environmental recovery through natural attenuation mechanisms.

10 Limitations and Future Prospects

Rhizoremediation utilizing SRB is a cost effective and ecofriendly approach for control of xenobiotic pollution. One significant limitation is their relatively slow growth rate compared to other microorganisms, which can hinder rapid bioremediation processes (Zhang et al. 2022). Additionally, SRB are highly vulnarable to environmental factors, like pH, temperature, and oxygen levels, which make them challenging to sustain ideal growth conditions in field applications (Tran et al. 2021). The accumulation of sulphide, a byproduct of their metabolic processes, can also inhibit their growth and activity, further reducing their effectiveness (Sheng et al. 2011). Moreover, competing microbial processes like methanogenesis can limit the activity of SRB by utilizing available electron donors (Spanjers et al. 2002). While SRB are capable of removing heavy metals through precipitation, they may not achieve complete removal of all contaminants. Mercury and its compounds, particularly methylmercury, are highly toxic to humans and ecosystems. SRB, such as *Desulfovibrio*, *Desulfococcus*, and *Desulfobacterium*, are crucial for the methylation of Hg in anaerobic environments like aquatic sediments and soils (Li et al. 2018).

Environment has diverse groups of toxicants that can inhibit the growth and activity of even highly tolerant microorganisms like SRB (Guo et al. 2017). Additionally, the presence of sulphate compounds in wastewater can form complexes with other toxicants, making it challenging for SRB to effectively remove sulphate from the wastewater matrix (Novair et al. 2024). Achieving optimal plant-microbe synergy is another hurdle, as the compatibility between plants and SRB can vary depending on soil type and pollutant levels (Asmarlaili et al. 2018). Moreover, scaling up these processes from lab settings to large contaminated sites remains a logistical and economic challenge.

Despite these limitations, the future prospects for SRB and rhizoremediation are promising. Research aimed at optimizing growth conditions could significantly improve their rhizoremediation potential. Genetic engineering holds promise for developing more efficient and robust SRB strains tailored for specific contaminants (Keller et al. 2011). Engineering plant-microbe consortia with complementary degradation pathways offers another promising approach to improving efficiency (Raturi et al. 2023). Nanotechnology could also be integrated to enhance contaminant bioavailability and to minimize the toxicity of by-products (Assunção et al. 2016). Bioreactor systems offer potential for overcoming environmental constraints by creating controlled conditions for SRB and plant growth. Finally, large-scale field trials and government support will be crucial in validating these approaches and facilitating their implementation in real-world scenarios. With continued research and innovation, the combination of SRB in rhizoremediation approach could become a powerful tool for sustainable bioremediation of xenobiotics.

11 Conclusion

The role of SRB in the rhizoremediation of xenobiotics presents a promising frontier in environmental biotechnology. By leveraging the unique metabolic pathways of SRB, we can significantly influence the detoxification of harmful contaminants within the rhizosphere. The ability of SRB to interact with other soil microbes, modify rhizosphere dynamics, and enhance nutrient cycling further underscores their role as a vital component in bioremediation strategies. With the advancement of molecular techniques such as genomics, metagenomics, proteomics, and metabolomics, our understanding of SRB populations and their activity within the rhizosphere has deepened greatly, which offers insights into how their activity can be tracked and optimized during rhizoremediation efforts. There are plenty of approaches that enhance SRB activity; these include the use of zero-valent iron, additives, inorganic cations, and carrier materials, which present practical solutions for improving the efficacy of SRB-mediated xenobiotic degradation. Genetic engineering further extends the possibilities, allowing for the construction of recombinant bacteria and bacterial consortia designed for targeted xenobiotic degradation. While the potential applications of SRB in wastewater treatment, acid mine bioremediation, and crude oil contamination remediation are vast, some challenges still exist in scaling these technologies for large-scale implementation. Future research must focus on overcoming the limitations associated with the stability, efficiency, and ecological impacts of SRB-based technologies. By integrating omics approaches, innovative biotechnological advancements, and engineered microbial consortium, SRB-mediated rhizoremediation can emerge as an important innovation in sustainable environmental cleanup strategies.

Acknowledgments The authors acknowledge Assam University, Silchar, for providing necessary facilities. A.D. acknowledges the Department of Science and Technology (DST), Government of India, for the DST-INSPIRE fellowship (ID: IF190463); N.D, S.D., and P.P. are grateful to Department of Biotechnology (DBT), Government of India (BT/PR/30261/BCE/08/1495/2018; BT/560/NE/U-Excel/2016), for financial support.

Competing Interests All the authors declare that they have no competing interests.

References

Abdeen S, Di W, Hui L, Chen G-H, van Loosdrecht MC (2010) Fecal coliform removal in a sulfate reduction, autotrophic denitrification and nitrification integrated (SANI) process for saline sewage treatment. Water Sci Technol 62:2564–2570

Agrawal A, Lal B (2009) Rapid detection and quantification of bisulfite reductase genes in oil field samples using real-time PCR. FEMS Microbiol Ecol 69:301–312

Agrawal N, Shahi SK (2015) An environmental cleanup strategy-microbial transformation of xenobiotic compounds. Int J Curr Microbiol App Sci 4:429–461

Akinpelu EA, Ntwampe SK, Fosso-Kankeu E, Nchu F, Angadam JO (2021) Performance of microbial community dominated by bacillus spp. in acid mine drainage remediation systems: a focus on the high removal efficiency of SO42-, Al3+, CD2+, Cu2+, Mn2+, Pb2+, and Sr2+. Heliyon 7:e07241

Al-Asadi AAB (2011) Anaerobic degradation of crude oil by sulphate reducing bacteria isolated from soils contaminated with petroleum hydrocarbons. Univ Thi-Qar J Sci 2:3–12

Alcolombri U, Pioli R, Stocker R, Berry D (2022) Single-cell stable isotope probing in microbial ecology. ISME Comm 2:55

Altun M, Sahinkaya E, Durukan I, Bektas S, Komnitsas K (2014) Arsenic removal in a sulfidogenic fixed-bed column bioreactor. J Hazard Mater 269:31–37

Antoniewicz MR (2021) A guide to metabolic flux analysis in metabolic engineering: methods, tools and applications. Metab Eng 63:2–12

Arantza S-J, Hiram M-R, Erika K, Chávez-Avilés MN, Valiente-Banuet JI, Fierros-Romero G (2022) Bio-and phytoremediation: plants and microbes to the rescue of heavy metal polluted soils. SN Appl Sci 4:59

Asmarlaili S, Rauf A, Hanafiah D, Sudarno Y, Abdi P (2018) The soil sulphate effect and maize plant (Zea mays L.) growth of sulphate reducing bacteria (SRB) inoculation in acid sulfate soils with the different soil water condition. In: IOP conference series: earth and environmental science. IOP Publishing, p 012037

Assunção A, Vieira B, Lourenço JP, Costa MC (2016) Recovery of gold (0) nanoparticles from aqueous solutions using effluents from a bioremediation process. RSC Adv 6:112784–112794
Ayangbenro AS, Olanrewaju OS, Babalola OO (2018) Sulfate-reducing bacteria as an effective tool for sustainable acid mine bioremediation. Front Microbiol 9:1986
Bahr M, Crump BC, Klepac-Ceraj V, Teske A, Sogin ML, Hobbie JE (2005) Molecular characterization of sulfate-reducing bacteria in a New England salt marsh. Environ Microbiol 7:1175–1185
Bai H, Kang Y, Quan H, Han Y, Sun J, Feng Y (2013) Treatment of acid mine drainage by sulfate reducing bacteria with iron in bench scale runs. Bioresour Technol 128:818–822
Banfield JF, Nealson KH (2018) Geomicrobiology: interactions between microbes and minerals. Walter de Gruyter GmbH & Co KG, Berlin
Barlett M, Zhuang K, Mahadevan R, Lovley D (2012) Integrative analysis of Geobacter spp. and sulfate-reducing bacteria during uranium bioremediation. Biogeosciences 9:1033–1040
Barton LL, Fauque GD (2009) Biochemistry, physiology and biotechnology of sulfate-reducing bacteria. Adv Appl Microbiol 68:41–98
Barton LL, Fauque GD (2022) Ecology of dissimilatory sulfate reducers: life in extreme conditions and activities of SRB. In: Sulfate-reducing bacteria and archaea. Springer, Cham, pp 463–527
Barton LL, Fardeau M-L, Fauque GD (2014) Hydrogen sulfide: a toxic gas produced by dissimilatory sulfate and sulfur reduction and consumed by microbial oxidation. In: The metal-driven biogeochemistry of gaseous compounds in the environment. Springer, Dordrecht, pp 237–277
Bender KS, Cheyen H, Wall JD (2006) Analysing the metabolic capabilities of Desulfovibrio species through genetic manipulation. Biotechnol Genet Eng Rev 23:157–174
Ben-Dov E, Brenner A, Kushmaro A (2007) Quantification of sulfate-reducing bacteria in industrial wastewater, by real-time polymerase chain reaction (PCR) using dsrA and apsA genes. Microb Ecol 54:439–451
Ben-Dov E, Shapiro OH, Gruber R, Brenner A, Kushmaro A (2008) Changes in microbial diversity in industrial wastewater evaporation ponds following artificial salination. FEMS Microbiol Ecol 66:437–446
Birkeland NK (2005) Sulfate-reducing bacteria and archaea. In: Petroleum microbiology. ASM Press, pp 35–54
Bombach P, Chatzinotas A, Neu TR, Kästner M, Lueders T, Vogt C (2009) Enrichment and characterization of a sulfate-reducing toluene-degrading microbial consortium by combining in situ microcosms and stable isotope probing techniques. FEMS Microbiol Ecol 71:237–246
Bradu P, Biswas A, Nair C, Sreevalsakumar S, Patil M, Kannampuzha S, Mukherjee AG, Wanjari UR, Renu K, Vellingiri B (2022) Recent advances in green technology and industrial revolution 4.0 for a sustainable future. Environ Sci Pollut Res Int 1:124488
Brahmacharimayum B, Mohanty M, Ghosh P (2019) Theoretical and practical aspects of biological sulfate reduction: a review. Global NEST J 21:222–244
Brenner K, You L, Arnold FH (2008) Engineering microbial consortia: a new frontier in synthetic biology. Trends Biotechnol 26:483–489
Brune KD, Bayer TS (2012) Engineering microbial consortia to enhance biomining and bioremediation. Front Microbiol 3:203
Bryukhanov A, Khijniak T (2022) The application of sulfate-reducing bacteria in the bioremediation of heavy metals and metalloids. Appl Biochem Microbiol 58:S1–S15
Bueno-López JI (2020) Elucidating the role of graphene oxide on the methanogenic activity of anaerobic consortia
Cao J-Y, Zhang G-J, Mao Z-S, Fang Z-H, Yang C, Han B-L (2009) Influence of Mg2+ on the growth and activity of sulfate reducing bacteria. Hydrometallurgy 95:127–134
Cao J, Zhang G, Mao Z-S, Li Y, Fang Z, Yang C (2012) Influence of electron donors on the growth and activity of sulfate-reducing bacteria. Int J Miner Process 106:58–64
Chen S, Chen W, Shih C (2008a) Heavy metal removal from wastewater using zero-valent iron nanoparticles. Water Sci Technol 58:1947–1954

Chen Y, Cheng JJ, Creamer KS (2008b) Inhibition of anaerobic digestion process: a review. Bioresour Technol 99:4044–4064
Cheng Y, Mallavarapu M, Naidu R, Chen Z (2018) In situ fabrication of green reduced graphene-based biocompatible anode for efficient energy recycle. Chemosphere 193:618–624
Chhabra SR, Joachimiak MP, Petzold CJ, Zane GM, Price MN, Reveco SA, Fok V, Johanson AR, Batth TS, Singer M (2011) Towards a rigorous network of protein-protein interactions of the model sulfate reducer Desulfovibrio vulgaris Hildenborough. PLoS One 6:e21470
Clark ME, He Z, Redding AM, Joachimiak MP, Keasling JD, Zhou JZ, Arkin AP, Mukhopadhyay A, Fields MW (2012) Transcriptomic and proteomic analyses of Desulfovibrio vulgaris biofilms: carbon and energy flow contribute to the distinct biofilm growth state. BMC Genomics 13:1–17
Costa JL, Silva LG, Veras ST, Gavazza S, Florencio L, Motteran F, Kato MT (2024) Use of nitrate, sulphate, and iron (III) as electron acceptors to improve the anaerobic degradation of linear alkylbenzene sulfonate: effects on removal potential and microbiota diversification. Environ Sci Pollut Res:1–16
Cupples AM (2016) Contaminant-degrading microorganisms identified using stable isotope probing. Chem Eng Technol 39:1593–1603
Dalsgaard T, Bak F (1994) Nitrate reduction in a sulfate-reducing bacterium, Desulfovibrio desulfuricans, isolated from rice paddy soil: sulfide inhibition, kinetics, and regulation. Appl Environ Microbiol 60:291–297
Dang H, Ewald JM, Mattes TE (2023) Genome-resolved metagenomics and Metatranscriptomics reveal insights into the ecology and metabolism of anaerobic microbial communities in PCB-contaminated sediments. Environ Sci Technol 57:16386–16398
Das A, Das S, Das N, Pandey P, Ingti B, Panchenko V, Bolshev V, Kovalev A, Pandey P (2023a) Advancements and innovations in harnessing microbial processes for enhanced biogas production from waste materials. Agriculture 13:1689
Das N, Das A, Das S, Bhatawadekar V, Pandey P, Choure K, Damare S, Pandey P (2023b) Petroleum hydrocarbon catabolic pathways as targets for metabolic engineering strategies for enhanced bioremediation of crude-oil-contaminated environments. Fermentation 9:196
Das S, Das N, Choure K, Pandey P (2023c) Biodegradation of asphaltene by lipopeptide-biosurfactant producing hydrocarbonoclastic, crude oil degrading bacillus spp. Bioresour Technol 382:129198
Das A, Das N, Rajkumari J, Pandey P, Pandey P (2024) Exploring the bioremediation potential of Bacillus spp. for sustainable mitigation of hydrocarbon contaminants. Environ Sustain 7:1–22
De Bont L, Donnay N, Couturier J, Rouhier N (2022) Redox regulation of enzymes involved in sulfate assimilation and in the synthesis of sulfur-containing amino acids and glutathione in plants. Front Plant Sci 13:958490
de Matos LP, Costa PF, Moreira M, Gomes PCS, de Queiroz Silva S, Gurgel LVA, Teixeira MC (2018) Simultaneous removal of sulfate and arsenic using immobilized non-traditional SRB mixed culture and alternative low-cost carbon sources. Chem Eng J 334:1630–1641
Dhir B (2017) Bioremediation technologies for the removal of pollutants. In: Advances in environmental biotechnology. Springer, Singapore, pp 69–91
Dolla A, Kurtz J, Teixeira M, Voordouw G (2007) Biochemical, proteomic and genetic characterization of oxygen survival mechanisms in sulphatereducing bacteria of the genus *Desulfovibrio*. In: Sulphate-reducing bacteria: Environmental and engineered systems. LL Barton & WA Hamilton, Cambridge, pp 185–214
Du C, Bai X, Li Y, Tan Q, Zhao C, Luo G, Wu L, Chen F, Li C, Ran C (2024) Storage, form, and influencing factors of karst inorganic carbon in a carbonate area in China. Sci China Earth Sci 67:725–739
Dumont MG, Murrell JC (2005) Stable isotope probing—linking microbial identity to function. Nat Rev Microbiol 3:499–504
El Houari A, Ranchou-Peyruse M, Ranchou-Peyruse A, Bennisse R, Bouterfas R, Goni Urriza MS, Qatibi A-I, Guyoneaud R (2020) Microbial communities and sulfate-reducing microorganisms

abundance and diversity in municipal anaerobic sewage sludge digesters from a wastewater treatment plant (Marrakech, Morocco). PRO 8:1284
Eltarahony M, Zaki S, Abd-El-Haleem D (2020) Aerobic and anaerobic removal of lead and mercury via calcium carbonate precipitation mediated by statistically optimized nitrate reductases. Sci Rep 10:4029
Esteves F, Rueff J, Kranendonk M (2021) The central role of cytochrome P450 in xenobiotic metabolism—a brief review on a fascinating enzyme family. J Xenobio 11:94–114
Fan K, Wang W, Xu X, Yuan Y, Ren N, Lee D-J, Chen C (2023) Recent advances in biotechnologies for the treatment of environmental pollutants based on reactive Sulfur species. Antioxidants 12:767
Fang H, Oberoi AS, He Z, Khanal SK, Lu H (2021) Ciprofloxacin-degrading Paraclostridium sp. isolated from sulfate-reducing bacteria-enriched sludge: optimization and mechanism. Water Res 191:116808
Gadd GM (2010) Metals, minerals and microbes: geomicrobiology and bioremediation. Microbiology 156:609–643
Gamalero E, Bona E, Glick BR (2022) Current techniques to study beneficial plant-microbe interactions. Microorganisms 10:1380
Gandy CJ, Gray ND, Mejeha OK, Sherry A, Jarvis AP (2023) Use of propionic acid additions to enhance zinc removal from mine drainage in short residence time, flow-through sulfate-reducing bioreactors. J Environ Manag 327:116862
Gao P, Fan K (2023) Sulfur-oxidizing bacteria (SOB) and sulfate-reducing bacteria (SRB) in oil reservoir and biological control of SRB: a review. Arch Microbiol 205:162
Garbeva PV, Van Veen JA, Van Elsas JD (2004) Microbial diversity in soil: selection of microbial populations by plant and soil type and implications for disease suppressiveness. Annu Rev Phytopathol 42:243–270
García-Gómez C, Sánchez-Pardo B, Esteban E, Peñalosa JM, Fernández MD (2014) Risk assessment of an abandoned pyrite mine in Spain based on direct toxicity assays. Sci Total Environ 470:390–399
Gheju M (2011) Hexavalent chromium reduction with zero-valent iron (ZVI) in aquatic systems. Water Air Soil Pollut 222:103–148
González-Paz JR, del Carmen Monterrubio–Badillo M, Ordaz A, García-Peña EI, Guerrero-Barajas C (2022) Influence of Fe2+ and Fe3+ on the performance and microbial community composition of a MFC inoculated with sulfate-reducing sludge and acetate as electron donor. J Chem 2022:5685178
Grubba D, Yin Z, Majtacz J, Al-Hazmi HE, Mąkinia J (2022) Incorporation of the sulfur cycle in sustainable nitrogen removal systems-a review. J Clean Prod 372:133495
Guo X, Lu Z, Wang P, Li H, Huang Z, Lin K, Liu Y (2015) Diversity and degradation mechanism of an anaerobic bacterial community treating phenolic wastewater with sulfate as an electron acceptor. Environ Sci Pollut Res 22:16121–16132
Guo J, Kang Y, Feng Y (2017) Bioassessment of heavy metal toxicity and enhancement of heavy metal removal by sulfate-reducing bacteria in the presence of zero valent iron. J Environ Manag 203:278–285
Guo P, Du H, Mao Q, Deng Y, Wang X, Li J, Xiong B, Fan X, Wang D, Agathokleous E (2024) The duality of sulfate-reducing bacteria: reducing methylmercury production in rhizosphere but enhancing accumulation in rice plants. J Hazard Mater 476:135049
Hallberg K (2010) New perspectives in acid mine drainage microbiology. Hydrometallurgy 104:448–453
Han S, Li Y, Gao H (2022) Generation and physiology of hydrogen sulfide and reactive sulfur species in bacteria. Antioxidants 11:2487
Hao J, Chai YN, Lopes LD, Ordóñez RA, Wright EE, Archontoulis S, Schachtman DP (2021) The effects of soil depth on the structure of microbial communities in agricultural soils in Iowa (United States). Appl Environ Microbiol 87:e02673–e02620

Haveman SA, Greene EA, Voordouw G (2005) Gene expression analysis of the mechanism of inhibition of Desulfovibrio vulgaris Hildenborough by nitrate-reducing, sulfide-oxidizing bacteria. Environ Microbiol 7:1461–1465

Hays SG, Patrick WG, Ziesack M, Oxman N, Silver PA (2015) Better together: engineering and application of microbial symbioses. Curr Opin Biotechnol 36:40–49

Hsia K, Chen C, Ou J, Lo K, Sheu Y, Kao C (2021) Treatment of petroleum hydrocarbon-polluted groundwater with innovative in situ sulfate-releasing biobarrier. J Clean Prod 295:126424

Hu J, Zeng C, Liu G, Lu Y, Zhang R, Luo H (2019) Enhanced sulfate reduction accompanied with electrically-conductive pili production in graphene oxide modified biocathodes. Bioresour Technol 282:425–432

Hu C, Yang Z, Chen Y, Tang J, Zeng L, Peng C, Chen L, Wang J (2024) Unlocking soil revival: the role of sulfate-reducing bacteria in mitigating heavy metal contamination. Environ Geochem Health 46:417

Huang W, Kao C (2016) Bioremediation of petroleum-hydrocarbon contaminated groundwater under sulfate-reducing conditions: effectiveness and mechanism study. J Environ Eng 142:04015089

Jawaharraj K, Peta V, Dhiman SS, Gnimpieba EZ, Gadhamshetty V (2023) Transcriptome-wide marker gene expression analysis of stress-responsive sulfate-reducing bacteria. Sci Rep 13:16181

Keller KL, Wall JD, Chhabra S (2011) Methods for engineering sulfate reducing bacteria of the genus Desulfovibrio. Method Enzymol 497:503–517

Kieu HT, Müller E, Horn H (2011) Heavy metal removal in anaerobic semi-continuous stirred tank reactors by a consortium of sulfate-reducing bacteria. Water Res 45:3863–3870

Kim IH, Choi J-H, Joo JO, Kim Y-K, Choi J-W, Oh B-K (2015) Development of a microbe-zeolite carrier for the effective elimination of heavy metals from seawater. J Microbiol Biotechnol 25:1542–1546

Kim J, Hwangbo M, Shih C-H, Chu K-H (2023) Advances and perspectives of using stable isotope probing (SIP)-based technologies in contaminant biodegradation. Water Res X 20:100187

Kiran MG, Pakshirajan K, Das G (2017) An overview of sulfidogenic biological reactors for the simultaneous treatment of sulfate and heavy metal rich wastewater. Chem Eng Sci 158:606–620

Kloster S, Six K, Feichter J, Maier-Reimer E, Roeckner E, Wetzel P, Stier P, Esch M (2007) Response of dimethylsulfide (DMS) in the ocean and atmosphere to global warming. J Geophys Res Biogeosci 112

Kong Q, Ngo HH, Shu L, Fu R-S, Jiang C-H, Miao M-S (2014) Enhancement of aerobic granulation by zero-valent iron in sequencing batch airlift reactor. J Hazard Mater 279:511–517

Kotoky R, Pandey P (2018) Plant-microbe symbiosis as an instrument for the mobilization and removal of heavy metals from contaminated soils-a realistic approach. Curr Biotechnol 7:71–79

Kotoky R, Rajkumari J, Pandey P (2018) The rhizosphere microbiome: significance in rhizoremediation of polyaromatic hydrocarbon contaminated soil. J Environ Manag 217:858–870

Kotu SP, Yang F, Klemashevich C, Mannan MS, Jayaraman A (2024) Metagenomic and metabolomic analysis of microbiologically influenced corrosion of carbon steel in produced water. In: Petroleum microbiology. CRC Press, pp 96–119

Kumar V, Singh K, Shah MP, Singh AK, Kumar A, Kumar Y (2021) Application of omics technologies for microbial community structure and function analysis in contaminated environment. In: Wastewater treatment. Elsevier, pp 1–40

Kumari S, Das S (2023) Bacterial enzymatic degradation of recalcitrant organic pollutants: catabolic pathways and genetic regulations. Environ Sci Pollut Res 30:79676–79705

Kushkevych I, Dordević D, Vítězová M (2019) Toxicity of hydrogen sulfide toward sulfate-reducing bacteria Desulfovibrio piger Vib-7. Arch Microbiol 201:389–397

Kushkevych I, Abdulina D, Kováč J, Dordević D, Vítězová M, Iutynska G, Rittmann SK-M (2020a) Adenosine-5′-phosphosulfate-and sulfite reductases activities of sulfate-reducing bacteria from various environments. Biomol Ther 10:921

Kushkevych I, Cejnar J, Treml J, Dordević D, Kollar P, Vítězová M (2020b) Recent advances in metabolic pathways of sulfate reduction in intestinal bacteria. Cells 9:698

Kushkevych I, Hýžová B, Vítězová M, Rittmann SK-M (2021) Microscopic methods for identification of sulfate-reducing bacteria from various habitats. Int J Mol Sci 22:4007

Lashani E, Amoozegar MA, Turner RJ, Moghimi H (2023) Use of microbial consortia in bioremediation of metalloid polluted environments. Microorganisms 11:891

Latorre M, Cortés MP, Travisany D, Di Genova A, Budinich M, Reyes-Jara A, Hödar C, González M, Parada P, Bobadilla-Fazzini RA (2016) The bioleaching potential of a bacterial consortium. Bioresour Technol 218:659–666

Li G-X, Bao P (2021) Transcriptomics analysis of the metabolic mechanisms of iron reduction induced by sulfate reduction mediated by sulfate-reducing bacteria. FEMS Microbiol Ecol 97:fiab005

Li X, Wu Y, Zhang C, Liu Y, Zeng G, Tang X, Dai L, Lan S (2016) Immobilizing of heavy metals in sediments contaminated by nonferrous metals smelting plant sewage with sulfate reducing bacteria and micro zero valent iron. Chem Eng J 306:393–400

Li X, Lan S-M, Zhu Z-P, Zhang C, Zeng G-M, Liu Y-G, Cao W-C, Song B, Yang H, Wang S-F (2018) The bioenergetics mechanisms and applications of sulfate-reducing bacteria in remediation of pollutants in drainage: a review. Ecotoxicol Environ Saf 158:162–170

Li M, Fang A, Yu X, Zhang K, He Z, Wang C, Peng Y, Xiao F, Yang T, Zhang W (2021) Microbially-driven sulfur cycling microbial communities in different mangrove sediments. Chemosphere 273:128597

Liao L, Li Q, Yang Y, Xu R, Zhang Y (2024) Immobilization behavior and mechanism of Cd2+ by sulfate-reducing bacteria in anoxic environments. Water 16:1086

Liesirova T, Aarenstrup-Launbjerg T, Hallstrøm S, Bittner MJ, Riemann L, Voss M (2023) Nitrogen-fixing sulfate reducing bacteria in shallow coastal sediments under simulated resuspension. Estuar Coast Shelf Sci 280:108165

Lin H, Shi J, Chen X, Yang J, Chen Y, Zhao Y, Hu T (2010) Effects of lead upon the actions of sulfate-reducing bacteria in the rice rhizosphere. Soil Biol Biochem 42:1038–1044

Liu Y, Zhang Y, Ni B-J (2015) Zero valent iron simultaneously enhances methane production and sulfate reduction in anaerobic granular sludge reactors. Water Res 75:292–300

Liu F, Zhang G, Liu S, Fu Z, Chen J, Ma C (2018) Bioremoval of arsenic and antimony from wastewater by a mixed culture of sulfate-reducing bacteria using lactate and ethanol as carbon sources. Int Biodeterior Biodegradation 126:152–159

Madirisha M, Hack R, Van der Meer F (2022) The role of organic acid metabolites in geo-energy pipeline corrosion in a sulfate reducing bacteria environment. Heliyon 8:e09420

Maduwuba MC, Okpokwasili GC, Ibiene AA (2021) Bacteriological and heavy metal evaluation of abandoned crude oil–contaminated sites in Gio community, Ogoniland, Nigeria. GSC Biol Pharm Sci 16:267–273

Maheshwari DK, Das A, Dheeman S, Pandey P (2023) An overall insight into the attributes, interactions, and future applications of "microbial consortium" for plant growth promotion with contemporary approaches. In: Sustainable agrobiology: Design and development of microbial consortia. Springer, Singapore, pp 3–22

Manz W, Eisenbrecher M, Neu TR, Szewzyk U (1998) Abundance and spatial organization of Gram-negative sulfate-reducing bacteria in activated sludge investigated by in situ probing with specific 16S rRNA targeted oligonucleotides. FEMS Microbiol Ecol 25:43–61

Martins M, Santos ES, Faleiro ML, Chaves S, Tenreiro R, Barros RJ, Barreiros A, Costa MC (2011) Performance and bacterial community shifts during bioremediation of acid mine drainage from two Portuguese mines. Int Biodeterior Biodegradation 65:972–981

Meyer B, Kuever J (2007) Molecular analysis of the distribution and phylogeny of dissimilatory adenosine-5′-phosphosulfate reductase-encoding genes (aprBA) among sulfur-oxidizing prokaryotes. Microbiology 153:3478–3498

Miglani R, Parveen N, Kumar A, Ansari MA, Khanna S, Rawat G, Panda AK, Bisht SS, Upadhyay J, Ansari MN (2022) Degradation of xenobiotic pollutants: an environmentally sustainable approach. Meta 12:818

Mohapatra M, Yadav R, Rajput V, Dharne MS, Rastogi G (2021) Metagenomic analysis reveals genetic insights on biogeochemical cycling, xenobiotic degradation, and stress resistance in mudflat microbiome. J Environ Manag 292:112738

Morgan MJ, Osorio S, Gehl B, Baxter CJ, Kruger NJ, Ratcliffe RG, Fernie AR, Sweetlove LJ (2013) Metabolic engineering of tomato fruit organic acid content guided by biochemical analysis of an introgression line. Plant Physiol 161:397–407

Mundra S, Kjønaas OJ, Morgado LN, Krabberød AK, Ransedokken Y, Kauserud H (2021) Soil depth matters: shift in composition and inter-kingdom co-occurrence patterns of microorganisms in forest soils. FEMS Microbiol Ecol 97:fiab022

Nabavi SM, Šamec D, Tomczyk M, Milella L, Russo D, Habtemariam S, Suntar I, Rastrelli L, Daglia M, Xiao J (2020) Flavonoid biosynthetic pathways in plants: versatile targets for metabolic engineering. Biotechnol Adv 38:107316

Neculita CM, Zagury GJ, Bussière B (2007) Passive treatment of acid mine drainage in bioreactors using sulfate-reducing bacteria: critical review and research needs. J Environ Qual 36:1–16

Nejat N, Ramalingam A, Mantri N (2018) Advances in transcriptomics of plants. In: Plant genetics and molecular biology. Springer, Cham, pp 161–185

Neukirchen S, Pereira IA, Sousa FL (2023) Stepwise pathway for early evolutionary assembly of dissimilatory sulfite and sulfate reduction. ISME J 17:1680–1692

Nikitha T, Satyaprakash M, Vani SS, Sadhana B, Padal S (2017) A review on polycyclic aromatic hydrocarbons: their transport, fate and biodegradation in the environment. Int J Curr Microbiol Appl Sci 6:1627–1639

Niu Z-S, Pan H, Guo X-P, Lu D-P, Feng J-N, Chen Y-R, Tou F-Y, Liu M, Yang Y (2018) Sulphate-reducing bacteria (SRB) in the Yangtze estuary sediments: abundance, distribution and implications for the bioavailibility of metals. Sci Total Environ 634:296–304

Novair SB, Atigh ZBQ, Lajayer BA, Shu W, Price G (2024) The role of sulphate-reducing bacteria (SRB) in bioremediation of sulphate-rich wastewater: focus on the source of electron donors. Process Safe Environ Protect 184:190–207

Ojdana D, Sieńko A, Sacha P, Majewski P, Wieczorek P, Wieczorek A, Tryniszewska E (2018) Genetic basis of enzymatic resistance of E. coli to aminoglycosides. Adv Med Sci 63:9–13

Paganin P, Alisi C, Dore E, Fancello D, Marras PA, Medas D, Montereali MR, Naitza S, Rigonat N, Sprocati AR (2021) Microbial diversity of bacteria involved in biomineralization processes in mine-impacted freshwaters. Front Microbiol 12:778199

Panda S, Mishra S, Akcil A (2016) Bioremediation of acidic mine effluents and the role of sulfidogenic biosystems: a mini-review. Euro-Mediterr J Environ Integr 1:1–9

Parisi VA, Brubaker GR, Zenker MJ, Prince RC, Gieg LM, Da Silva ML, Alvarez PJ, Suflita JM (2009) Field metabolomics and laboratory assessments of anaerobic intrinsic bioremediation of hydrocarbons at a petroleum-contaminated site. Microb Biotechnol 2:202–212

Park Y-G, Mun B-G, Kang S-M, Hussain A, Shahzad R, Seo C-W, Kim A-Y, Lee S-U, Oh KY, Lee DY (2017) Bacillus aryabhattai SRB02 tolerates oxidative and nitrosative stress and promotes the growth of soybean by modulating the production of phytohormones. PLoS One 12:e0173203

Patankar RS, Reshi N, Kutty R (2022) Bioremediation: gaining insights through metabolomics. In: Omics Insights in Environmental Bioremediation. Springer, Singapore, pp 37–46

Paulo LM, Stams AJ, Sousa DZ (2015) Methanogens, sulphate and heavy metals: a complex system. Rev Environ Sci Biotechnol 14:537–553

Peng W, Li X, Lin M, Fan W (2020) Microbiological analysis of cadmium-contaminated sediments during biostabilization with indigenous sulfate-reducing bacteria. J Soils Sediments 20:584–593

Pierre Louis A-M, Yu H, Shumlas SL, Van Aken B, Schoonen MA, Strongin DR (2015) Effect of phospholipid on pyrite oxidation and microbial communities under simulated acid mine drainage (AMD) conditions. Environ Sci Technol 49:7701–7708

Plavec TV, Berlec A (2020) Safety aspects of genetically modified lactic acid bacteria. Microorganisms 8:297

Powell B, Mergeay M, Christofi N (1989) Transfer of broad host-range plasmids to sulphate-reducing bacteria. FEMS Microbiol Lett 59:269–273

Qian Z, Tianwei H, Mackey HR, van Loosdrecht MC, Guanghao C (2019) Recent advances in dissimilatory sulfate reduction: from metabolic study to application. Water Res 150:162–181

Qian D, Liu H, Hu F, Song S, Chen Y (2022) Extracellular electron transfer-dependent Cr (VI)/sulfate reduction mediated by iron sulfide nanoparticles. J Biosci Bioeng 134:153–161

Qiu S, Cai Y, Yao H, Lin C, Xie Y, Tang S, Zhang A (2023) Small molecule metabolites: discovery of biomarkers and therapeutic targets. Signal Transduct Target Ther 8:132

Ramsing NB, Kühl M, Jørgensen B (1993) Distribution of sulfate-reducing bacteria, O_2, and H_2S in photosynthetic biofilms determined by oligonucleotide probes and microelectrodes. Appl Environ Microbiol 59:3840–3849

Rapp BJ, Wall JD (1987) Genetic transfer in Desulfovibrio desulfuricans. Proc Natl Acad Sci 84:9128–9130

Raturi G, Chaudhary A, Rana V, Mandlik R, Sharma Y, Barvkar V, Salvi P, Tripathi DK, Kaur J, Deshmukh R (2023) Microbial remediation and plant-microbe interaction under arsenic pollution. Sci Total Environ 864:160972

Ravenschlag K, Sahm K, Knoblauch C, Jørgensen BB, Amann R (2000) Community structure, cellular rRNA content, and activity of sulfate-reducing bacteria in marine Arctic sediments. Appl Environ Microbiol 66:3592–3602

Rios-Valenciana EE, Briones-Gallardo R, Cházaro-Ruiz LF, Martínez-Villegas N, Celis LB (2017) Role of indigenous microbiota from heavily contaminated sediments in the bioprecipitation of arsenic. J Hazard Mater 339:114–121

Rodríguez E, García-Encina PA, Stams AJ, Maphosa F, Sousa DZ (2015) Meta-omics approaches to understand and improve wastewater treatment systems. Rev Environ Sci Biotechnol 14:385–406

Rodriguez-Freire L, Moore SE, Sierra-Alvarez R, Root RA, Chorover J, Field JA (2016) Arsenic remediation by formation of arsenic sulfide minerals in a continuous anaerobic bioreactor. Biotechnol Bioeng 113:522–530

Rousset M, Casalot L, Rapp-Giles BJ, Dermoun Z, de Philip P, Bélaich J-P, Wall JD (1998) New shuttle vectors for the introduction of cloned DNA in Desulfovibrio. Plasmid 39:114–122

Russell JN, Perry BJ, Bergsveinson J, Freeman CN, Sheedy C, Nilsson D, Braul L, Yost CK (2021) Metagenomic and metatranscriptomic analysis reveals enrichment for xenobiotic-degrading bacterial specialists and xenobiotic-degrading genes in a Canadian Prairie two-cell biobed system. Environ Microbiol Rep 13:720–727

Sahinkaya E, Yurtsever A, Toker Y, Elcik H, Cakmaci M, Kaksonen AH (2015) Biotreatment of As-containing simulated acid mine drainage using laboratory scale sulfate reducing upflow anaerobic sludge blanket reactor. Miner Eng 75:133–139

Sánchez-Paniagua López M, Manzanares-Palenzuela CL, López-Ruiz B (2018) Biosensors for GMO testing: nearly 25 years of research. Crit Rev Anal Chem 48:391–405

Santana MM, Dias T, Gonzalez JM, Cruz C (2021) Transformation of organic and inorganic sulfur–adding perspectives to new players in soil and rhizosphere. Soil Biol Biochem 160:108306

Santos AMd, Costa JM, Braga JK, Flynn TM, Brucha G, Sancinetti GP, Rodriguez RP (2022) Lactate as an effective electron donor in the sulfate reduction: impacts on the microbial diversity. Environ Technol 43:3149–3160

Santos AMd, Costa JM, Sancinetti GP, Rodriguez RP (2023) Impacts of phosphorus and nitrogen absence on microbial diversity and sulfate removal in anaerobic batch reactors. J Environ Sci Health A 58:563–569

Saxena G, Kishor R, Saratale GD, Bharagava RN (2020) Genetically modified organisms (GMOs) and their potential in environmental management: constraints, prospects, and challenges. In: Bioremediation of industrial waste for environmental safety: Volume II: Biological agents and methods for industrial waste management. Springer, Singapore, pp 1–19

Saxena P, Rauniyar S, Thakur P, Singh RN, Bomgni A, Alaba MO, Tripathi AK, Gnimpieba EZ, Lushbough C, Sani RK (2023) Integration of text mining and biological network analysis: identification of essential genes in sulfate-reducing bacteria. Front Microbiol 14:1086021

Schenk PM, Carvalhais LC, Kazan K (2012) Unraveling plant–microbe interactions: can multi-species transcriptomics help? Trends Biotechnol 30:177–184

Seth A, Edyvean R (2006) The function of sulfate-reducing bacteria in corrosion of potable water mains. Int Biodeterior Biodegradation 58:108–111

Shaji P, Vincent SG, Palanisami T (2021) Bioremediation of wastewater by Sulphate reducing bacteria. In: Bioremediation science. CRC Press, pp 337–343

Sharma P, Bano A, Singh SP, Dubey NK, Chandra R, Iqbal HM (2022a) Microbial fingerprinting techniques and their role in the remediation of environmental pollution. Clean Chem Eng 2:100026

Sharma P, Singh SP, Iqbal HM, Tong YW (2022b) Omics approaches in bioremediation of environmental contaminants: An integrated approach for environmental safety and sustainability. Environ Res 211:113102

Shen Y, Buick R (2004) The antiquity of microbial sulfate reduction. Earth Sci Rev 64:243–272

Sheng Y, Cao H, Li Y, Zhang Y (2011) Effects of sulfide on sulfate reducing bacteria in response to Cu (II), Hg (II) and Cr (VI) toxicity. Chin Sci Bull 56:862–868

Siguier P, Gourbeyre E, Chandler M (2014) Bacterial insertion sequences: their genomic impact and diversity. FEMS Microbiol Rev 38:865–891

Singh S, Chakraborty S (2022) Biochemical treatment of coal mine drainage in constructed wetlands: influence of electron donor, biotic–abiotic pathways and microbial diversity. Chem Eng J 440:135986

Singha LP, Pandey P (2021) Rhizosphere assisted bioengineering approaches for the mitigation of petroleum hydrocarbons contamination in soil. Crit Rev Biotechnol 41:749–766

Spanjers H, Weijma J, Abusam A (2002) Modelling the competition between sulphate reducers and methanogens in a thermophilic methanol-fed bioreactor. Water Sci Technol 45:93–98

Statnikov A, Henaff M, Narendra V, Konganti K, Li Z, Yang L, Pei Z, Blaser MJ, Aliferis CF, Alekseyenko AV (2013) A comprehensive evaluation of multicategory classification methods for microbiomic data. Microbiome 1:1–12

Štefanac T, Grgas D, Landeka Dragičević T (2021) Xenobiotics—division and methods of detection: a review. J Xenobio 11:130–141

Su Z, Li X, Xi Y, Xie T, Liu Y, Liu B, Liu H, Xu W, Zhang C (2022) Microbe-mediated transformation of metal sulfides: mechanisms and environmental significance. Sci Total Environ 825:153767

Su X, Sun F, Zhang J, Xing D, Li X, Song Z, Feng L, Huang Z, Li A (2023) Characterization and shifting of microbial community to denitrification for anaerobic sulfamethoxazole biodegradation with different electron acceptors. J Clean Prod 387:135870

Tahir K, Miran W, Jang J, Shahzad A, Moztahida M, Kim B, Lim SR, Lee DS (2021) Carbamazepine biodegradation and volatile fatty acids production by selectively enriched sulfate-reducing bacteria and fermentative acidogenic bacteria. J Chem Technol Biotechnol 96:592–602

Tavakoli HZ, Bahrami-Bavani M, Miyanmahaleh Y, Tajer-Mohammad-Ghazvini P (2021) Identification and characterization of a metal-resistant Acidithiobacillus ferrooxidans as important potential application for bioleaching. Biologia 76:1327–1337

Thabet OBD, Gtari M, Sghaier H (2017) Microbial diversity in phosphate rock and phosphogypsum. Waste Biomass Valor 8:2473–2483

Thomas J, Kim HR, Rahmatallah Y, Wiggins G, Yang Q, Singh R, Glazko G, Mukherjee A (2019) RNA-seq reveals differentially expressed genes in rice (Oryza sativa) roots during interactions with plant-growth promoting bacteria, *Azospirillum brasilense*. PloS one 14:e0217309

Torrentó C, Urmeneta J, Edwards KJ, Cama J (2012) Characterization of attachment and growth of Thiobacillus denitrificans on pyrite surfaces. Geomicrobiol J 29:379–388

Tran TTT, Kannoorpatti K, Padovan A, Thennadil S (2021) Sulphate-reducing bacteria's response to extreme pH environments and the effect of their activities on microbial corrosion. Appl Sci 11:2201

Tripathi AK (2022) Applications of omics technology in deciphering microbial mechanisms in environmental waste management. South Dakota School of Mines and Technology, Rapid City

Tripathi AK, Thakur P, Saxena P, Rauniyar S, Gopalakrishnan V, Singh RN, Gadhamshetty V, Gnimpieba EZ, Jasthi BK, Sani RK (2021) Gene sets and mechanisms of sulfate-reducing bacteria biofilm formation and quorum sensing with impact on corrosion. Front Microbiol 12:754140

Tripathi AK, Saxena P, Thakur P, Rauniyar S, Samanta D, Gopalakrishnan V, Singh RN, Sani RK (2022) Transcriptomics and functional analysis of copper stress response in the sulfate-reducing bacterium Desulfovibrio alaskensis G20. Int J Mol Sci 23:1396

Van Den Brand TP, Roest K, Chen G, Brdjanovic D, Van Loosdrecht M (2015) Potential for beneficial application of sulfate reducing bacteria in sulfate containing domestic wastewater treatment. World J Microbiol Biotechnol 31:1675–1681

Vasantharaj K, Jerold M, Vigneshwaran C, Velan M, Sivasubramanian V (2017) Bioremediation of sulfate-rich wastewater using lactate-fed sulfidogenic enrichment culture predominantly Desulfovibrio sp.: box-Behnken design optimization. Desalin Water Treat 83:30–39

Velasco A, Aburto-Medina A, Ortíz I (2020) Enhancement of the DDT reductive dehalogenation by different cosubstrates: role of sulfidogenic and biogeochemical processes in soil. Appl Geochem 117:104604

Vernans B, Iswanto B, Rinanti A (2019) Bioremediation of soil polluted with copper (Cu2+) by mixed culture bacteria Thiobacillus sp. and clostridium sp. Int J Sci Technol Res 8:3915–3919

Voordouw G, Brenner S (1985) Nucleotide sequence of the gene encoding the hydrogenase from Desulfovibrio vulgaris (Hildenborough). Eur J Biochem 148:515–520

Vu B, Chen M, Crawford RJ, Ivanova EP (2009) Bacterial extracellular polysaccharides involved in biofilm formation. Molecules 14:2535–2554

Wilmes P, Wexler M, Bond PL (2008) Metaproteomics provides functional insight into activated sludge wastewater treatment. PLoS One 3:e1778

Wöhlbrand L, Jacob JH, Kube M, Mussmann M, Jarling R, Beck A, Amann R, Wilkes H, Reinhardt R, Rabus R (2013) Complete genome, catabolic sub-proteomes and key-metabolites of D esulfobacula toluolica Tol2, a marine, aromatic compound-degrading, sulfate-reducing bacterium. Environ Microbiol 15:1334–1355

Wolicka D, Borkowski A (2007) The geomicrobiological role of sulphate-reducing bacteria in environments contaminated by petroleum products. Geomicrobiol J 24:599–607

Wu Y, Wang L, Jin M, Kong F, Qi H, Nan J (2019) Reduced graphene oxide and biofilms as cathode catalysts to enhance energy and metal recovery in microbial fuel cell. Bioresour Technol 283:129–137

Xi Y, Lan S, Li X, Wu Y, Yuan X, Zhang C, Yunguo L, Huang Y, Quan B, Wu S (2020) Bioremediation of antimony from wastewater by sulfate-reducing bacteria: effect of the coexisting ferrous ion. Int Biodeterior Biodegradation 148:104912

Xia Q, Rufty T, Shi W (2020) Soil microbial diversity and composition: links to soil texture and associated properties. Soil Biol Biochem 149:107953

Xu Y-N, Chen Y (2020) Advances in heavy metal removal by sulfate-reducing bacteria. Water Sci Technol 81:1797–1827

Xu W, Zhang Y, Cao H, Sheng Y, Li H, Li Y, Zhao H, Gui X (2018) Metagenomic insights into the microbiota profiles and bioaugmentation mechanism of organics removal in coal gasification wastewater in an anaerobic/anoxic/oxic system by methanol. Bioresour Technol 264:106–115

Xu H, Qin C, Zhang H, Zhao Y (2024) New insights into long-lasting Cr (VI) removal from groundwater using in situ biosulfidated zero-valent iron with sulfate-reducing bacteria. J Environ Manag 355:120488

Yan J, Ye W, Jian Z, Xie J, Zhong K, Wang S, Hu H, Chen Z, Wen H, Zhang H (2018) Enhanced sulfate and metal removal by reduced graphene oxide self-assembled enterococcus avium sulfate-reducing bacteria particles. Bioresour Technol 266:447–453

Yan J, Ye W, Liang X, Wang S, Xie J, Zhong K, Bao M, Yang J, Wen H, Li S (2020) Enhanced reduction of sulfate and chromium under sulfate-reducing condition by synergism between extracellular polymeric substances and graphene oxide. Environ Res 183:109157

Yan X, Gao B, Wang J, Zhu X, Zhang M (2023) Insights into remediation effects and bacterial diversity of different remediation measures in rare earth mine soil with SO42− and heavy metals. Front Microbiol 14:1050635

Yang Z, Liu Z, Dabrowska M, Debiec-Andrzejewska K, Stasiuk R, Yin H, Drewniak L (2021) Biostimulation of sulfate-reducing bacteria used for treatment of hydrometallurgical waste by secondary metabolites of urea decomposition by Ochrobactrum sp. POC9: from genome to microbiome analysis. Chemosphere 282:131064

Yang Z, Wu Q, Liu Z, Qi X, Zhang Z, He M, Peng C, Zeng L, Wang J, Li F (2023) Harnessing sulfate-reducing bacteria with plants growing to revitalize metal-tainted coal mine soils in Midwest China: metal sequestration performance, ecological networking interaction, and functional enzymatic prediction. Front Microbiol 14:1306573

Yu H, Jiang Z, Lu Y, Yao X, Han C, Ouyang Y, Wang H, Guo C, Ling F, Dang Z (2020) Transcriptome analysis of the acid stress response of Desulfovibrio vulgaris ATCC 7757. Curr Microbiol 77:2702–2712

Yun Y-M, Lee E, Kim K, Han J-I (2019) Sulfate reducing bacteria-based wastewater treatment system integrated with sulfide fuel cell for simultaneous wastewater treatment and electricity generation. Chemosphere 233:570–578

Zak D, Hupfer M, Cabezas A, Jurasinski G, Audet J, Kleeberg A, McInnes R, Kristiansen SM, Petersen RJ, Liu H (2021) Sulphate in freshwater ecosystems: a review of sources, biogeochemical cycles, ecotoxicological effects and bioremediation. Earth Sci Rev 212:103446

Zhang T, Fang HH (2006) Applications of real-time polymerase chain reaction for quantification of microorganisms in environmental samples. Appl Microbiol Biotechnol 70:281–289

Zhang M, Wang H (2014) Organic wastes as carbon sources to promote sulfate reducing bacterial activity for biological remediation of acid mine drainage. Miner Eng 69:81–90

Zhang M, Wang H (2016) Preparation of immobilized sulfate reducing bacteria (SRB) granules for effective bioremediation of acid mine drainage and bacterial community analysis. Miner Eng 92:63–71

Zhang J, Zhang Y, Quan X, Liu Y, An X, Chen S, Zhao H (2011) Bioaugmentation and functional partitioning in a zero valent iron-anaerobic reactor for sulfate-containing wastewater treatment. Chem Eng J 174:159–165

Zhang P, Li Z, Zhang S, Shao G (2018) Recent advances in effective reduction of graphene oxide for highly improved performance toward electrochemical energy storage. Energy Environ Mater 1:5–12

Zhang H, Song S, Jia Y, Wu D, Lu H (2019) Stress-responses of activated sludge and anaerobic sulfate-reducing bacteria sludge under long-term ciprofloxacin exposure. Water Res 164:114964

Zhang Z, Zhang C, Yang Y, Zhang Z, Tang Y, Su P, Lin Z (2022) A review of sulfate-reducing bacteria: metabolism, influencing factors and application in wastewater treatment. J Clean Prod 376:134109

Zhang M, Xiong J, Zhou L, Li J, Fan J, Li X, Zhang T, Yin Z, Yin H, Liu X (2023) Community ecological study on the reduction of soil antimony bioavailability by SRB-based remediation technologies. J Hazard Mater 459:132256

Zhao Y, Yan H, Zhou J, Tucker ME, Han M, Zhao H, Mao G, Zhao Y, Han Z (2019) Bio-precipitation of calcium and magnesium ions through extracellular and intracellular process induced by bacillus licheniformis SRB2. Fortschr Mineral 9:526

Zhao Q, Li X, Xiao S, Peng W, Fan W (2021) Integrated remediation of sulfate reducing bacteria and nano zero valent iron on cadmium contaminated sediments. J Hazard Mater 406:124680

Zhao Q, Guo W, Luo H, Xing C, Wang H, Liu B, Si Q, Li D, Sun L, Ren N (2022) Insights into removal of sulfonamides in anaerobic activated sludge system: mechanisms, degradation pathways and stress responses. J Hazard Mater 423:127248

Zhao C, Chen N, Liu T, Feng C (2023) Effects of adding different carbon sources on the microbial behavior of sulfate-reducing bacteria in sulfate-containing wastewater. J Clean Prod 392:136332

Zhou J, Lin W-H, Yu Y-L, Dong C-D, Zhang H, Hu Z, Kao C-M (2024) Transitioning weathered oil fields towards new energy: a review on utilizing hydrogenotrophic methanogens for petroleum hydrocarbons remediation. J Hazard Mater 477:135279

Żur J, Wojcieszyńska D, Guzik U (2016) Metabolic responses of bacterial cells to immobilization. Molecules 21:958

Geomycology: Exploring Fungal Roles in Elemental Cycling

Sheetal Joshi and Santa Ram Joshi

Contents

Abstract The ubiquitous presence of microbes in the environment inevitably leads to their interaction with metals and minerals present in their surroundings. Microbes exhibit the remarkable capability to modify the physical and chemical properties of minerals, thereby utilising them for growth and survival in harsh conditions. Geomycology delves into exploring these interactions, particularly focusing on fungal microbes and their relationship with geological processes. Fungi, often overlooked in traditional ecological frameworks, play crucial roles in biogeochemical

S. Joshi · S. R. Joshi (✉)
Microbiology Laboratory, Department of Biotechnology and Bioinformatics, North Eastern Hill University, Shillong, Meghalaya, India

S. R. Joshi et al. (eds.), *Mineral Transformation and Bioremediation by Geo-Microbes*, Interdisciplinary Biotechnological Advances,
https://doi.org/10.1007/978-981-96-3033-2_8

cycles by mediating the transformation and redistribution of essential elements such as carbon, nitrogen, and phosphorus. This chapter aims to highlight the intricate mechanisms by which fungi modulate nutrient dynamics and elemental cycles in soil, sediments and other environments. From the breakdown of organic matter through saprotrophic decomposition to the establishment of symbiotic relationships mediated by mycorrhiza and lichens, fungi demonstrate remarkable adaptability in shaping and contributing to geochemical cycles. Various biological processes such as biomineralisation, biological element cycle, rock and mineral transformation, carbon sequestration, bioweathering and metal-fungi interaction influence the dynamics of microbial communities by contributing to mineral weathering and nutrient availability. Moreover, certain fungal transformations hold practical significance in environmental biotechnology, facilitating metal leaching, recovery and detoxification alongside the degradation of xenobiotic and organic pollutants. However, these mechanisms can also lead to adverse effects when involved in the degradation of natural products and building materials. This chapter draws attention to a deeper understanding of fungal contributions to ecological dynamics and the potential for the utilisation of fungal resources for applications in diverse biotechnological fields.

Keywords Geomycology · Biomineralisation · Macrofungi · Metals · Lichen · Bioweathering

Abbreviations

N	Nitrogen
H	Hydrogen
C	Carbon
K	Potassium
Cd	Cadmium
Au	Gold
Co	Cobalt
S	Sulphur
Cr	Chromium
Th	Thorium
Fe	Iron
O	Oxygen
Cu	Copper
Se	Selenium
Hg	Mercury
P	Phosphorous
Mn	Manganese
Ag	Silver

Rb	Rubidium
Ca	Calcium
Pb	Lead
Ti	Titanium
As	Arsenic
U	Uranium
Na	Sodium
Zn	Zinc
Te	Telerium
Pg	Petagram
CO_2	Carbon dioxide
CH_4	Methane
NH_4	Ammonium
NO_3	Nitrate
NO_2	Nitrite
N_2O	Nitrous oxide
nos	Nitrous oxide reductase
nor	Nitric oxide reductase
nirK	Nitrite reductase
AMF	Arbuscular mycorrhizal fungi
EcM	Ectomycorrhizal fungi
ErM	Ericoid mycorrhiza
Gt	Gigatonne
NADP	Nicotinamide adenine dinucleotide phosphate
$CaCO_3$	Calcium carbonate
HNO_3	Nitric acid
H_2SO_4	Sulphuric acid
UV	Ultra violet
OH^-	Hydroxyl
PO_4^{2-}	Phosphate
SO_4^{2-}	sulphate
COO^-	Carboxyl
NH_2^-	Amine
SOM	Soil organic matter
DOM	Dissolved organic matter

1 Introduction

The biosphere is an amalgamation of the lithosphere, atmosphere and hydrosphere, encompassing all ecosystems on Earth and binding them into a cohesive whole (Goltsov and Goltsova 2014). Within this interconnected web of life, abiotic and biotic components interact to drive essential life cycles and facilitate the movement of vital inorganic minerals. The process of geological minerals interacting with the biotic components of the ecosystem, traversing biochemical pathways within living organisms and their environment, is the driving force of biogeochemical cycles. These complex processes recycle and redistribute minerals essential for growth and metabolism, ensuring the ecosystems' productivity (Huang et al. 2024). Geomicrobiology centres on the roles of microscopic organisms in these intricate processes and examines the microbial-mediated interactions with minerals in the environment. The interaction of the abiotic metal-mineral complexes with the biotic microbial fractions causes alterations in the physical and chemical properties of the metals, which in turn affect the growth and survival of the microbes (Gadd 2007). The sub-discipline of geomycology investigates the intricate relationship between fungi and geological processes. Geomycology encompasses nutrient cycling, soil formation, mineral weathering and biogeochemical processes (Gadd 2013). The process of mineralisation and biogenic minerals is a testament to microbes' diversified and vital contributions to the geological processes.

The emerging field of geomycology highlights the historical oversight of fungi's role in driving geological cycles. Traditionally viewed merely as agents of decomposition and carbon turnover, fungi are now recognised as integral to aerobic terrestrial ecosystems and extreme environments subjected to severe stressors such as pH, salinity and temperature fluctuations. Fungi have been estimated to constitute a significant portion of microbial biomass in ecosystems, ranging from an estimated 55–89%. Thus, it becomes crucial to acknowledge the impact of fungi on global biogeochemical cycles (Coleine et al. 2022).

2 Importance of Studying Fungi from a Geological Prospect

In the pre-cambium era, when life on Earth was thought to have originated, the microorganisms utilised carbon (C) and affected its cycling. Since a cell is also made of nitrogen (N), sulphur (S) phosphorus (P), oxygen (O) and hydrogen (H), the microbes also affect the distribution of these elements. When the nutrient sources started diminishing, other available inorganic elements occurring in more than one oxidation state were utilised to generate energy by electron flow. These elements included H, N, iron (Fe), S, manganese (Mn) and others, which gave rise to chemolithotrophy. Thus, in the present times, these elements are closely related to biological life, and their cycling is dependent on microbial action (Ehrlich 1998). Fungi are heterotrophic eukaryotes that rely on the chemoorganotrophic mode of nutrition. They represent a diverse group of organisms that share close evolutionary ties with the animal kingdom. They consist of unicellular yeast and multicellular filamentous moulds, which reproduce via spore production through sexual and asexual means. They grow through solid substrates by dissolving them using extracellular metabolites while also causing physical damage induced by the growth of mycelium (Naranjo-Ortiz and Gabaldón 2019). The kingdom fungi are divided into seven phyla based on their modes of reproduction, life cycles, cell wall and septum structure (Hibbett et al. 2007).

Fungi inhabit diverse environments and are conventionally regarded as terrestrial aerobes; they thrive in soil and decompose dead organic matter, thus contributing to the redistribution of organic elements (C, N, P, S, O and other trace elements). However, they have also been isolated from harsher habitats like the Artic Poles, hot deserts, acidic environments, metal-polluted aquatic and terrestrial zones and hypersaline regions. Fungi also inhabit deep-sea sediments, hydrothermal vents, methane cold-seeps and the igneous oceanic crust (Gadd 2017a). Fungi can maintain soil structure by producing exopolymers and metabolites by their filamentous growth (Bohu et al. 2022). The soil fungi *Penicillium glandicola*, isolated from caves and mines, could tolerate temperatures ranging from 5 to 30°C with a water potential of −100 bars (Cecchi et al. 2018).

Additionally, fungi are notorious pathogens capable of causing plant and animal diseases. However, beyond these recognised roles, fungi are increasingly acknowledged as significant contributors to biogeochemical cycles in both aquatic and terrestrial ecosystems. They function as geoactive agents shaping various biogeochemical processes like element cycling, formation and dissolution of minerals and transformations of metals, among others (Gadd 2017b). Due to their ubiquitous presence throughout the various soil and regolith layers, they can mediate biogeochemical processes like rock weathering and mineral cycling. Their presence in diverse niches enables them to obtain a high genetic diversity, estimated to be 2.2–3.8 million species (Debeljak and Baltar 2023). Fungi have evolved to tolerate and change metal mobility by interacting with minerals and rocks, driving critical geological processes like biomineralisation, bioweathering, bioleaching and bioremediation (Bohu et al. 2022). However, out of the estimated diversity of fungal species, a mere 1,20,000 fungal species have been described, thus limiting the understanding of their functional roles (Bahram and Netherway 2022). Geomycology is critical for various environmental and industrial applications, including the reclamation and restoration of contaminated sites, bioremediation, the decommissioning of nuclear facilities and radionuclide containment, the bio-recovery of essential elements and the development of novel biomaterials (Gadd 2017a).

2.1 *Fungi and Organic Matter Decomposition*

Organic matter consists of a complex mixture of non-living biological compounds, primarily composed of carbon, hydrogen, nitrogen, oxygen, phosphorous and sulphur. It can be found in various forms, including particulates in the atmosphere, dissolved organics in water and humus in soils. Soil organic matter (SOM), which supports plant life, results from an amalgamation of plant, microbial and rhizospheric compounds released at different degradation stages. SOM is notable for containing three times more carbon than all life on Earth combined, while aquatic systems are estimated to hold twice the amount of organic matter found in living organisms. Within SOM, analytical techniques such as NMR reveal that the humic fraction is a complex mixture of microbial and plant biopolymers, making it difficult to classify into a specific chemical category. The polar and hydrophobic elements of SOM and dissolved organic matter (DOM) in terrestrial and aquatic systems are crucial for transporting heavy metals and hydrophobic anthropogenic chemicals. A major component of DOM identified through NMR and FTIR analysis is carboxyl-rich acyclic molecules (CRAM) (Simpson et al. 2011).

Since fungi have been known to decompose organic matter, their role in the global C-cycle is well studied. However, this has drawn focus away from the contributions of fungi in other geochemical cycles. Fungi not only degrade complex natural molecules like lignin, chitin and starch, but they also decompose anthropogenic compounds like pesticides, hydrocarbons and organometals. Plant, animal and microbial biomass decay contributes to organic matter that contains various elements essential for life. The major and minor elements present in the organic matter, along with trace elements found in low concentrations within living systems, are inherently linked to biomass. Consequently, their release should be connected to fungal-mediated degradation, which plays a significant role in geomycological cycles (Gadd 2017a, b).

3 Elemental Cycling

Biogeochemical cycles are processes through which elements cycle between the biotic and abiotic spheres, changing their forms while maintaining their balances in the Earth's reservoirs. The movement and transformation of elements within ecosystems and the Earth's environment are governed by a range of geological, biological and chemical processes (Huang et al. 2024). Biogeochemical cycles have co-evolved with specific biological metabolic pathways; hence, they serve as a selective force in the metabolic diversity of microbes. Biogeochemical cycles over geological epochs significantly alter the chemical constitution of the oceans, atmosphere and terrestrial ecosystems (Falkowski 2001).

The contribution of fungi towards the elemental cycle majorly occurs through the breakdown of organic matter in soil and aquatic bodies. They are thus instrumental in the cycling of C, N and P as they are the components of organic litter; however, they are also involved in the elemental cycling and mineralisation of other elements like Mn, Fe, S, sodium (Na), potassium (K), copper (Cu), Zinc (Zn), cadmium (Cd), silver (Ag), selenium (Se), uranium (U), tellurium (Te), nickel (Ni), gold (Au) and some metalloids. Some fungi that have been reported to partake in the biogeochemical cycling of different elements have been demonstrated in a tabular form in Table 1. For instance, *Trichoderma harzianum* was reported to be involved in the bioaccumulation and biosorption of metals such as Cd, Cu, U, Zn, Ni and Ag (Cecchi et al. 2018). The movement of these elements is dependent and influenced by biological and physical factors like metabolic and morphological adaptation of the fungal species, symbiotic associations, pH, temperature, availability of water and the concentration and toxicity potential of the metal. The diversity of the fungi depends on environmental factors such as temperature, salinity, dissolved oxygen,

Table 1 Shows the involvement of fungi in the cycling of different elements

Element	Examples of fungi involved	Reference
Carbon (C)	Basidiomycetes and ascomycetes	Zeiner et al. (2017)
Nitrogen (N)	*Bipolar sorokiniana*, *Fusarium oxysporium* and *Cylindrocarpon tonkinense*	Singh (2021)
Phosphorous (P)	*Aspergillus niger*	Geng et al. (2022)
Manganese (Mn)	*Aspergillus niger*, *Serpula himantioides*	Wei et al. (2012)
Silver (Ag)	*Trichoderma harzianum*	Cecchi et al. (2018)
Potassium (K)	*Fusarium*	Richardson et al. (2023)
Copper (Cu)	*Trichoderma harzianum*	Cecchi et al. (2018)
Sulphur (S)	*Trichoderma*, *Aspergillus*, *Mucor*, *Fusarium*, *Penicillium*	Grayston and Wainwright (1988), Cecchi et al. (2018)
Iron (Fe)	*Trichoderma guizhouense*	Yu et al. (2019)
Zinc (Zn)	*Trichoderma harzianum*	Wei et al. (2013), Cecchi et al. (2018)
Cadmium (Cd)	*Trichoderma harzianum*	Cecchi et al. (2018)
Uranium (U)	*Trichoderma harzianum*	Akhtar et al. (2007)
Nickle (Ni)	*Trichoderma harzianum*	Cecchi et al. (2018)
Gold (au)	*Fusarium oxysporum*, *Colletotrichum* sp.	Bohu et al. (2022)
Selenium (se)	*Aspergillus* sp.	Ruiz-Fresneda et al. (2024)
Tellurium (Te)	*Aspergillus* sp.	Ruiz-Fresneda et al. (2024)

pH, total organic carbon, nitrogen and phosphorous and pollution. Thus, fungal diversity can be a bioindicator since a region's taxonomic composition could indicate environmental changes (Wu et al. 2024).

Elemental cycling is essential, especially in ecosystems without light input, such as cave ecosystems. *Penicillium*, *Aspergillus* and *Mortierella* genera of fungi have been reported from air, rock, sediment and water samples of low-light environments of caves (Zhu et al. 2022). Like their bacterial counterparts, fungi are known to attach to rock surfaces and produce biofilms, which become a site of mineral precipitation. Soil fungi, *Trichoderma*, *Aspergillus*, *Mucor*, *Fusarium* and *Penicillium*, have been observed to release sulphate to the environment by the oxidation of sulfur, dimethylsulfide, dimethylsulfoxide and thiosulfate (Cecchi et al. 2018).

3.1 Carbon Cycling

Carbon is the basis of all life, making up about 45–50% of dry biomass. The carbon circulates between living beings, forming the long and short carbon cycle. The long carbon cycle results from sedimentation and burial, where fossilisation occurs. Coal, oil and natural gas are organic carboniferous reservoirs millions of years old. Photosynthesis, respiration and decomposition contribute to the short carbon cycle, where carbon is stored in biomass and soil (Hannula and Morriën 2022). Soil is the second largest carbon reservoir on earth, estimated to constitute approximately 2000 Pg C in soil organic carbon (Wu et al. 2024). Degradation of the organic matter in the soil produces humus, the reservoir for carbon in the soil. Humus increases water-holding capacity and promotes cation exchange and nitrogen binding, enhancing the soil structure (Boddy 2016). The share of carbon stored in the atmosphere and plant biomass is less than 25%; approximately 75% of the terrestrial carbon is contained underground in the soil (Hawkins et al. 2023).

The C stored in biomass and cellular systems of living organisms in the form of root exudates, dead plant material, fertilisers and microbial necrotic mass is introduced to the soil, and root exudates constitute the most significant carbon pool in the soil. These exudates influence microbial diversity around plant roots since they are rich in sugars and organic acids (Hannula and Morriën 2022). Plants and other autotrophs utilise atmospheric carbon to fix organic carbon molecules for their growth. The carbon fixed by the autotrophs is transformed to the gaseous form and liberated back into the atmosphere by respiration and, ultimately, by decomposing autotrophs and heterotrophs with the help of soil microbes. The redox cycling of manganese, a continuous process occurring in soil, is also thought to influence terrestrial decomposition rates (Zeiner et al. 2021). The form in which carbon is released back into the atmosphere depends on the oxygen levels in which the decomposition occurs. In aerobic conditions, carbon is released as carbon dioxide (CO_2), while in anoxic regions, methane (CH_4) is released through anaerobic respiration and fermentation processes (Qiu et al. 2008).

Fungi efficiently sense their environment and can colonise regions containing diverse nutrient sources. Carbon in biological systems is locked in complex molecules like cellulose, hemicellulose, pectin and lignin. These complex carbon-rich plant materials are degraded and recycled by releasing extracellular enzymes and metabolites. Fungi secrete cellulase, hemicelluloses, pectinases and lignin-degrading enzymes into the environment to degrade these recalcitrant biomolecules to simpler sugars for their use in cellular activity (Sabuda et al. 2022). The gene expression patterns and enzyme production in fungal cells change to effectively switch up the type of enzymes and metabolites needed for the most efficient utilisation of the available nutrients. The white rot fungi *Ganoderma lucidum* produces the enzyme manganese peroxidase, which is crucial in lignin degradation (Zhang et al. 2019). The same C source triggers different secretosomes to effectively utilise the available carbon, as seen in *Alternaria alternata*, *Paraphaeosphaeria sporulosa* and *Pyrenochaeta* sp. (Zeiner et al. 2017). Fungi, like other eukaryotes, employ glycolysis, glycerol metabolism, fermentation and tricarboxylic acid cycle for the metabolisation of the carbon (Sabuda et al. 2022). The arsenal of carbohydrate-active enzymes enables fungi to exploit a wide range of nutrient sources, a diversity that has been likened to that of angiosperms and gymnosperms (Bahram and Netherway 2022).

Fungi are involved in carbon cycling and are crucial in mitigating climate change. The enhancement of soil aggregation by fungi creates pockets where soil carbon can be stored, inaccessible to microbial breakdown and plant root uptake, thereby preventing its addition to the global carbon chain. The aggregation process depends on the organic acids and sugars from fungal metabolism, which also prints in root exudates (Husain et al. 2024). Mycorrhizal fungi release iron-rich glycoprotein, glomalin, which attracts small soil particles and provides stability, aiding the aggregation process (Wilkes 2021). All of these compounds bind soil particles, forming stable clusters that promote the absorption and binding of carbon onto clay and silt, thereby forming larger aggregates (Husain et al. 2024).

Carbon cycling is not limited to terrestrial environments; oceans are responsible for nearly half of the global net primary productivity. Diatom-derived polysaccharides are significant contributors to the marine carbon cycle. Fungi are prominent drivers of aquatic carbon turnover, as shown in a study where the diatomic carbon was labelled with ^{13}C and was seen to be assimilated by fungi related to the yeast *Malassezia*. However, fungi from the sea-floor sediments, which included Chytridiomycota and Mucormycota fungi, assimilated carbon from chemosynthetic bacterial necromass with the action of cellulolytic and proteolytic enzymes in anaerobic conditions (Orsi et al. 2022).

3.2 Nitrogen Cycling

The cyclic journey of nitrogen (N) involves transformative processes such as mineralisation, oxidations, reductions, immobilisation and nitrogen fixation. Mineralisation is the conversion of soil organic N into inorganic or mineral N. This is also called ammonification because of the large quantities of ammonia (NH_4) released. Fungi decompose the organic N present in the form of amino acids, nucleic acids and hexoamines and take up a large portion of the released N for themselves. Thus, fungal contributions to N mineralisation are offset by their greater assimilation capacities compared to bacteria. The assimilation of inorganic N by microorganisms is also known as the N-immobilisation process. Fungi such as *Alternaria*, *Fusarium*, *Cephalosporium*, *Monailia*, *Trichoderma* and *Mucor* have strong ammonification ability, with reports suggesting fungi contribute 65–80% of ammonification in forest soils.

The bioavailable forms of nitrogen include ammonium (NH_4^+), nitrate (NO_3^-) and nitrite (NO_2^-) ions, which play a crucial role throughout this cycle. Fungi are involved in the denitrification process, in which NO_3 or NO_2 is reduced to nitric oxide (NO), nitrous oxide (N_2O) and ultimately to nitrogen (N_2), which is released back into the atmosphere. Ascomycota fungi, *Cylindrocarpon tonkinense*, *Gibberella fujikuroii* and the basidiomycota *Trichosporon cutaneum* denitrify nitrates in the soil. Soil fungi *Fusarium oxysporum* and *Fusarium solani*, at low O_2 concentrations, can reduce NO_2 to N_2O (Singh 2021).

The enzymes nitrous oxide reductase (nos), nitric oxide reductase (nor) and nitrite reductase (nirK) are involved in the denitrification process. However, the enzyme responsible for the conversion of $N_2O \rightarrow N_2$ has not been identified in fungi, and thus, the end product of NO_3 and NO_2 reduction is mostly N_2O. Thus, nor and nirK are the key enzymes for the fungal denitrification process (Aldossari and Ishii 2021).

The chemical reactions involved in the denitrification process are given under,

$NO_3^- \rightarrow NO_2^-$
$NO_2 \rightarrow NO + N_2O$
$2NO_3 + 10\ e^- + 12H^+ \rightarrow N_2 + 6H_2O$ (Singh 2021)

Fungi can also co-denitrify the nitrate by which hybrid N_2O or N_2 compounds are formed by associating two nitrogen sources. The form of organic nitrogen used determined the end product of the reaction. Thus, imines and azide N-containing compounds generate N_2O, whereas amines lead to the formation of N_2. This process is known as bionitrosation since part of the reaction is speculated to occur via abiotic processes. The NO_2 or NO produced by fungal denitrification can abiotically react with other N sources to form N_2O or N_2. *Bipolar sorokiniana* decomposes organic matter, producing N_2O or N_2, which is released into the atmosphere (Singh 2021).

The process involved is demonstrated under,

$NO_2^- + RNH_2 \rightarrow N_2$
$NO_2^- + RNHOH \rightarrow N_2O$

Fungi play a more substantial role in the emissions of N_2O compared to bacteria and are influenced by conditions such as soil moisture, organic matter, temperature and pH (Aldossari and Ishii 2021).

In terrestrial systems, N is bound in recalcitrant forms within organic matter, often associated with clay particles and protected from microbial degradation. Since there is competition between decomposers and nitrifiers, plants explore alternative strategies for nitrogen acquisition to ensure survival. Ectomycorrhizal fungi play a crucial role in this process by mobilising nitrogen from organic matter and using it as a primary growth substrate through the extensive production of extracellular hydrolytic enzymes, such as proteinase, chitinase and ribonuclease. In return for carbohydrates from their host plant, these fungi assimilate and transfer N facilitated by specialised carrier transport proteins (Chalot and Brun 1998). Arbuscular mycorrhizal fungi (AMF) are unable to fix atmospheric nitrogen for the plants; however, their symbiotic associations are beneficial for plants and the nodulating rhizobacteria *Rhizobium*. The process of denitrification releases nitrogen, which is assimilated by the processes of nitrification and ammonification (Salazar et al. 2020). The elaborate root systems of mycorrhizal fungi transport 20% of the required plant nitrogen (Hawkins et al. 2023). Nitrogen, an essential plant nutrient, is sensitive to temperature, and its movements between the various nitrogen pools in the environment are subject to change with the increasing trends of global basal temperature. This change can alter the biogeochemistry of soil nitrogen in particular ecosystems by releasing or intensifying nitrogen limitation for biological species (Salazar et al. 2020).

In aquatic environments, fungi such as hyphomycetes attach to leaf litter and wood debris in flowing freshwater. Other fungi include zoosporic chytridomycetes and oomycetes found in lakes, while Trichomycetes inhabit the guts of different aquatic insects. The concentration of dissolved nitrogen in water is low and depends on the bedrock, soil composition, land use and biotic activity in the water body. NO_3 is at higher concentrations in oxygenated water, while NH_4 and NO_2 levels are higher in anoxic regions. Aquatic hyphomycetes can utilise organic and inorganic nitrogen sources, while chytridomycetes and oomycetes cannot reduce NO_3. Fungi are thought to contribute approximately 95% to the total microbial biomass of the leaf-decomposing microbes and are significant contributors to nitrogen immobilisation. Fungi have not been elaborated for their role in denitrification in aquatic bodies; however, since *Fusarium*, *Cylindrocarpon* and *Nectria* are commonly found in aquatic fungal genera, their association with terrestrial nitrogen metabolism is indicative of their potential role in aquatic systems (Gulis et al. 2006).

3.3 Phosphorous Cycling

Phosphorous (P) is a significant constituent of biomolecules such as nucleic acid, lipids, proteins and coenzymes and contributes to cell respiration and photosynthesis. Unlike other elements in the biogeochemical cycles, the atmosphere is not the reservoir of P. It is compounded and stored in the lithosphere in rock and sedimentary deposits, available to organisms by weathering these inorganic phosphorus minerals, mainly apatite. These minerals contain phosphate oxyanions, which are attracted by the Ca^{2+} cations, forming a hexagonal structure (Filippelli 2002). However, the bioavailability of phosphorous is reduced by its retention in soil particles. It is estimated that 43% of the lands are phosphorous deficient, and P levels restrict the carbon sequestration capacity of forests (Du et al. 2020). Phosphorous gets washed away by rain and other factors to the aquatic bodies where the cycle continues. It is an essential macronutrient for phytoplanktons, algae and sea plants, and its availability determines the primary productivity of aquatic systems. The average concentration of inorganic P in oceans is 2.3 micromolars, which accounts for 50–100% of the dissolved P. In marine systems, the autotrophs have been estimated to incorporate 3–4 times higher P than the terrestrial levels. In the geological time scale, P is removed from oceans via long-term burial in marine sediments (Canfield et al. 2005). In the soil, phosphate anions are adsorbed onto Al and Fe oxides and precipitate with them in acidic pH; the same phenomenon is observed in neutral soil where phosphate anions precipitate with Ca^{2+}. Phosphorous-rich minerals are used as fertilisers for plants by the chemical acidification of rock-phosphates (calcium phosphate). Since fungi also produce organic acids, the phosphates fixed onto soil particles are accessible to fungal mycelium. The organic acids replace P on the surface of metallic hydroxides by ligand exchange. The acids dissolve hydroxides onto which P is adsorbed and complex themselves with the metals, preventing

P from precipitating. *Aspergillus niger*, *Penicillium islandicum* and *Penicillium purpurogenum* have been reported to have potent P solubilisation ability. Various organic acids, such as oxalic acid, citric acid and gluconic acid, produced by fungi show different levels of P-solubilisation (de Oliveira Mendes et al. 2014). Plants increase their P uptake with the help of mycorrhizal fungi, which extend the root surface area of plants (Filippelli 2002). About 80% of the P requirement of plants is fulfilled by the nutrient supply by the mycorrhizal fungi (Hawkins et al. 2023). Different types of mycorrhizal fungi colonise the roots of plants. Plant roots lack any mycorrhizal associations in conditions where the soil is severely phosphorous deficient. Ectomycorrhizal fungi have been documented to transport the bioavailable form of phosphorous, orthophosphate, as $H_2PO_4^-$ ions, to distances up to 40 cm away from the root system. Fungal P-transport occurs via transporters on the hyphae, while the acquired P is stored in the vacuoles. The endophyte *Glomus mosseae* shows varied expression levels of the P transporter, *Gmos*PT, depending on the different concentrations of P in the soil. Another transporter protein, RiPT7, found in *Rhizophagus irregularis*, when inhibited, was observed to upregulate the activity of genes involved in the synthesis of polyphosphate. Polyphosphates are polymers of P used by the fungus to mobilise and sustain P levels in case of deficiency. The endophytic fungi *Colletotrichum tofieldiae* transfers P to plants under low P stress, enhancing the expression of P-transporters PHT1.2 and PHT1.3 (Luo et al. 2024). In the yeast *Saccharomyces cerevisiae*, it was discovered that cell-cycle activities regulate P homeostasis. Pho4 and Pho2 transcription factors activate the set of P-transport genes when P is limited. These transcription factors activate genes for the accumulation, mobilisation and attainment of P (Bhalla et al. 2022).

4 Macrofungi and Metal Interaction

Mushrooms are saprophytic basidiomycetes or ascomycetes that also establish ectomycorrhizal relations with the roots of vascular trees. They facilitate the transfer of elements from the soil into organic systems within terrestrial habitats and contribute to nutrient and carbon cycling (Cairney 2005; Andronikov et al. 2023). A significant part of the fixed C in terrestrial ecosystems is held in C-rich lignocellulosic biomass, which is low in N concentrations. The white-rot basidiomycota, such as *Armillaria*, *Ganoderma* and *Phlebia*, degrade lignin. Saprophytic soft-rot fungi belonging to Ascomycota, such as *Xylaria* and *Anthostoma*, are found to utilise cellulose and hemicellulose constituents in dead logs more preferentially than other carbon sources. In boreal forests where N is a limited resource, primary colonisers of deadwood, such as *Fomitopsis*, are accustomed to N-limited conditions, with their mycelia containing only about 0.8% N. Conversely, late colonisers like *Lycoperdon* have higher N requirements, up to 6.4%. Bacterial nitrogen fixation helps alleviate N limitation and facilitates late decomposers' complete utilisation of organic matter (TlÁskal et al. 2021). In grasslands, mushrooms growing in moss and leaf litter, such as *Galerina* and *Mycena*, are involved in organic matter decomposition and lignin degradation. Since herbivores consume a significant part of the above-ground plant material, the partially digested plant litter in the form of dung is available as a rich nutrient source. Basidiomycetes specialised in decomposing animal dung have evolved pigmented spores to withstand the digestive processes of herbivores, ensuring their presence in dung. Various mushroom species, such as *Conocybe*, *Panaeolus*, *Psathyrella*, *Psilocybe* and *Stropharia*, are known to decompose dung and form basidiocarps, enriching the nutrient content of the soil. *Coprinus* fungi produce inhibitory molecules to inhibit the hyphal growth of other competing species (Griffith and Roderick 2008). The changing environmental patterns due to the increasing use of fertilisers and excess nitrogen runoffs from agricultural practices have been shown to influence the growth of macrofungi. Some macrofungi exhibiting ectomycorrhizal associations respond differently to nitrogen levels, with some species classified as nitrophilic and others as nitrophobic. For instance, *Amphinema byssoides* is nitrophobic, and studies have reported a decline in its frequency in forest plots subjected to nitrogen fertilisation. Similarly, waxy caps such as *Hygrocybe*, which thrive in nutrient-poor grasslands, have experienced a decrease in abundance in intensively managed grasslands due to nitrogen fertiliser input, reflecting their strong nitrophobic tendencies. Conversely, species of *Russula* are favoured by increased nitrogen availability and have shown an increase in frequency in forests, where they form ectomycorrhizal associations with various host trees (O'Hanlon 2017).

Besides their contribution to the decomposition and release of C and N compounds back into the soil, macrofungi accumulate various metals, including toxic heavy metals. Metals from the soil are absorbed by plants and fungi as nutrients and are translocated throughout their system. It is hypothesised that metals tend to aggregate and concentrate within organisms through the effective vascular uptake system (Clemens et al. 2002). However, compared to plants, mushrooms can accumulate significantly higher levels of metals, even in areas with low soil metal concentrations. The fruiting body of macrofungi *Xerocomellus chrysenteron* was found to accumulate metals, such as Ag, Cd, K, P, Rb, Se and Zn, which did not reflect the geochemistry of the bedrock soil on which the mushroom was found growing (Andronikov et al. 2023). The analysis of the mushroom's soil revealed that the metals Ag, Cd, K and Rb were depleted up to 14–33% compared to soil levels from regions of no mushroom growth. The ability of the mushrooms to hyperaccumulate heavy metals may depend on various factors, including the total metal content in the soil, the age of the mycelium and the genetic predisposition of the macrofungi to metal accumulation. *Amanita strobiliformis* and *Amanita solitaria* accumulate Ag in their tissues up to 2500 times higher than concentrations in the neighbouring soil (Borovička et al. 2007). A small cystine-rich protein, metallothionein, binds heavy metals and is involved in the metal-protein complex formation in these hyperaccumulators (Byrne and Tušek-Žnidarič 1990). The pileus may accumulate higher levels of metals in their cells than the stipe, possibly due to the movement or transport of minerals and nutrients from the underground mycelium travelling upwards (Liu et al. 2022).

5 Aquatic Fungi and Cycling

Fungi have been estimated to have about 5 million species globally; however, less than 10% have likely been identified. Aquatic fungi are a minor proportion of the identified final species, with an estimate of a modest 1900 species collected exclusively from marine niches (Amend et al. 2019; Breyer and Baltar 2023). The most abundant fungi found in aquatic bodies belong to the phyla Ascomycota (especially the classes Eurotiomycetes, Dothideomycetes, Sordariomycetes, Leotiomycetes, Saccharomycetes and Pezizomycetes) followed by Basidiomycota (e.g., Agaricomycetes, Exobasidiomycetes, Malasseziomycetes, Microbotryomycetes). In coastal regions, Chytridiomycota populations fluctuate with phytoplankton blooms since fungi belonging to this division are parasitic to the phytoplanktons. Similarly, an abundance of Rozellomycota (i.e., Cryptomycota), Muccoromycota and Zoopagomycota contributes to mycoplankton diversity in aquatic systems (Breyer and Baltar 2023).

Fungi have been identified from a wide variety of marine environments, such as subseafloor basalt, deep continental crust and anoxic deep-sea regions. The significant genera identified from these environments include *Aspergillus*, *Aureobasidium*, *Candida*, *Cladosporium*, *Cryptococcus*, *Malassezia*, *Penicillium* and *Rhodotorula* (Amend et al. 2019). The deep biosphere has anoxic conditions and fungi belonging to Ascomycota, Basidiomycota and Chyritidomycota, found in these regions provide the limiting nutrient hydrogen to the methanogenic bacteria. Thus, fungi play essential roles in energy cycling in regions that predominantly harbour only prokaryotes. Aquatic fungi are considered the biological carbon pump because they are detritus feeders of leaf litter in water systems. They have been known to form associations with diatoms. They can enhance the nutritional content of leaf litter by optimising the C:N:P ratios in the decomposed organic matter, with their mycelia serving as a source of nutrition for aquatic invertebrates (Grossart et al. 2019). These fundamental roles position fungi as key regulators of food webs and biogeochemical cycles. By facilitating the availability of resources to higher-order consumers, fungi enhance the efficiency of energy transfer in the aquatic system (Bahram and Netherway 2022). In oxygen-limited zones like the deep biosphere, fungi are known to supplement hydrogen to the methanogenic bacteria for the generation of methane gas (Ivarsson et al. 2016). Mycoplanktons float along the water column and participate in C-cycling by decomposing phytoplanktons; for instance, *Cladosporium* secretes enzyme glutan 1,3-β-glucosidase for the assimilation of phytoplankton organic carbon. However, limited data on the fungi-dependent biogeochemical flux of elements in oceans and other aquatic systems highlights a need for detailed study (Breyer and Baltar 2023).

6 Fungal Symbiosis

Fungi establish strong associations with various organisms often blurring the boundaries between species. This is particularly evident in lichens, mycorrhiza, endophytes and mixed associations within biofilms (Bahram and Netherway 2022). Geological transformations can also result from these collaborative relationships. In these partnerships, fungi receive photosynthetically derived carbon from their autotrophic partners in exchange for providing inorganic nutrients. These associations include mycorrhizal partnerships with plants and symbiotic relationships with cyanobacteria in lichens (Gadd 2017a).

6.1 Mycorrhizae

Mycorrhizae are soil fungi that can associate with plant roots and engage in nutrient exchange. They fall into four categories based on their evolutionary ecology, host plant associations and anatomical structures: arbuscular mycorrhiza, ectomycorrhiza, ericoid mycorrhiza and orchid mycorrhiza. Fungi belonging to the phyla Glomeromycota, Basidiomycota and Ascomycota can establish these mycorrhizal connections (Mohan and Joshi 2024). They colonise the inner root volume, extending to the root surface and surrounding soil, increasing the soil contact of the plants (Gadd 2010). In the form of Rhynie chert, fossilised evidence suggests that mycorrhiza-like fungi formed symbiotic associations with early land plants and supported their establishment in the terrestrial environment (Hawkins et al. 2023). The extensive association of mycorrhizal fungi with over 90% of plant families support their evolutionary significance in plant development and diversification (Bahram and Netherway 2022). The functioning of mycorrhiza has implications for the geochemical cycling of elements. Arbuscular mycorrhizal fungi (AMF) are symbiotically associated with almost 70% herbaceous and woody terrestrial plants. The chemical signalling between the plant and the mycorrhiza allows the fungi to grow arbuscules, which are the sites for the trade-off of fatty acids and lipids in return for nitrogen and phosphorous from the soil (Keymer et al. 2017). A study found that AMF associated with the herbaceous plant *Lindenbergia philippensis* growing in Zn-contaminated soil enhanced Zn accumulation in the rhizosphere compared to plants growing without AMF (Kangwankraiphaisan et al. 2013). Ectomycorrhizal fungi (EcM), which develop a mycelial mantle around plant roots, have evolved from saprotrophic fungal lineages to form symbiotic relationships with plants. These fungi do not penetrate plant root cells but instead secrete exoenzymes to break down complex organic materials, which are then supplied to the plant. Approximately 2% of plants, including certain gymnosperms and some angiosperm trees, are associated with EcM fungi (Hawkins et al. 2023). EcM fungi release oxalic acid and siderophores, which create narrow pores in minerals and release elements such as Ca, K, Mn, Pb and Ti from the dissolution of apatite, wood ash and other minerals (Gadd 2017a). In contrast, ericoid mycorrhiza (ErM) forms coils inside the roots of the Ericaceae family of plants found in acidic and infertile soils. They produce exoenzymes to dissolve minerals and other organic material to access the nutrients required for plant exchange (Hawkins et al. 2023).

Oidiodendron maius, which is an ErM fungi, can solubilise Zn oxides and phosphates and promote their release (Gadd 2017a).

For the mitigation of climatic factors, mycorrhizal fungi have been shown to alter the emissions of the greenhouse gas N_2O into the atmosphere by down-regulating the nirK gene and upregulation of nos genes, which are associated with the utilisation of N_2O (Singh 2021). They also serve as a terrestrial carbon sink by trapping C in their mycelium, which temporarily prevents its release as CO_2. In a report based on the analysis of 200 datasets, the global plant communities were estimated to allocate 3.93Gt CO_2/year to AMF, 9.07Gt CO_2/year to EcM fungi and 0.12Gt CO_2/year to ErM fungi. Thus, mycorrhizal fungi hold 36% of annual CO_2 emissions from fossil fuels (Hawkins et al. 2023).

6.2 Lichens

Lichens are the symbiotic associations of fungi with green algae or cyanobacteria known for their slow growth, stress tolerance and longevity. They are found in nearly all terrestrial ecosystems, covering 8% of Earth's land surface and contributing significantly to the biomass of tundra and grassland ecosystems. With over 18,000 species worldwide, lichens outnumber higher plants in many higher latitude regions. These organisms combine the functions of cyanobacteria, which fix atmospheric N and fungi, which contribute to the weathering of minerals. Around 10% of all lichens contain cyanobacteria enabling them to fix atmospheric N (Purvis and Pawlik-Skowrońska 2008). In a study, it was observed that the lichen *Lobar oregana* contributed to almost 50% of the total N in the *Pseudotsuga menzeisii* forest of Oregon (Asplund and Wardle 2017).

The weathering of rocks by lichens occurs as a physical and chemical process. Crustose lichens that tightly adhere to their substrates and produce very little biomass primarily induce rock weathering through hyphal penetration and the chemical dissolution caused by organic acids (Asplund and Wardle 2017). These lichens are abundant in metal-contaminated sites, and some metallophyte lichens are found to colonise minerals such as Ur, Cu, Pb and As-phosphates directly. The mycobiont can produce metabolites that corrode the minerals and leach out the metal; these include the respiratory CO_2, secretion of oxalic acid and the production of metabolites that can complex with metals (Chen et al. 2000). The lichen *Diploschistes muscorum* found growing in calamine-rich soil can dissolve Zn and complex it with diploschistesic and lecanoric acid. *Aureobasidium pullulans* produces high levels of melanin, which can bind Cu^{2+} and Fe^{2+}. Lichens can absorb nutrients from dew and fog by their large surface-to-biomass ratio. For instance, *Ramalina menziesii*, a chlorolichen, has been observed to capture 2.85 kg of N and 0.15 kg of P per hectare yearly. In the oak woodlands where this lichen thrives, it absorbs 13% of the annual N turnover, 4% of P, 7% of K, 1% of Ca, 3% of Mg and 8% of Na. In areas undergoing early succession following a destructive event, lichens are often the first colonisers in regions limited by N-availability. For instance, the fruticose lichen, *Stereocaulon* spp., which fixes N, frequently dominates landscapes affected by lava flow. Additionally, lichens contribute to soil formation through rock mineralisation; secondary clay particles that develop beneath areas of lichen growth reflect the role of organic acids and mucopolysaccharides excretion by the lichen hyphae in dissolving minerals (Purvis and Pawlik-Skowrońska 2008; Asplund and Wardle 2017).

7 Geomycology and Human Development

Fungi and bacteria are studied as different disciplines in science; however, in the natural environment, they work in synergy. Various fungal processes that aid in the geochemical cycling serve applicative purposes for humans. Some of the applications of fungi are elaborated as follows.

7.1 Biomineralisation

Biominerals are inorganic solids produced by biological organisms to harden and stiffen tissues. These mineralised tissues play a crucial role in the protection, support and acquisition of resources for microorganisms (Gadd 2010). Mineralisation occurs as calcification, silicification and iron mineralisation in extracellular and intercellular spaces of microbes (Qin et al. 2020). Fungal biomineralisation occurs as fungi alter their external environment, leading to mineral precipitation through changes in pH, O_2 levels, redox potential, metals redox transformations and the secretion of metabolites, such as CO_2, H^+ or organic acids. Fungal cell walls and extracellular minerals have a high affinity for metal binding, which promotes nucleation and mineral deposition. The type of metabolite dictates the interaction with the mineral, resulting in either mineral dissolution or mineral diagenesis through redox reactions or acid-base attacks (Gadd 2021). Fungi require high concentrations of Ca^{2+} ions for their apical growth. Chitin and glycoproteins secreted by the fungi can adsorb Ca^{2+}, which initiates the biomineralisation process. The study of the biomineralisation process could lead to the discovery of novel biomaterials with unique properties, such as catalytic abilities. These materials can be utilised as colloids, nanoparticles or crystals and can have various applications, including biomaterials for structural, technological and environmental purposes or even antimicrobial drugs (Gadd 2007). Fungi are recognised as one of the significant biological forms that can carry out mineralisation and form secondary minerals and common amorphous complexes. The type of metal and the species of fungi involved determines the mineralisation function in most fungi (Bohu et al. 2022). Secondary mineral formation can occur due to the redox reactions of metals and metabolic activity of the microbe where metabolites such as oxalates, CO_2 and sulphide precipitate metals in the cells (Gadd 2010). Fungi can form minerals at moderate temperatures and pressures, which was thought to occur at only high temperatures and pressures (Ehrlich 1998).

The most common minerals of fungal origin include oxides, carbonates, phosphates, sulfates, sulfides and oxalates. The formation and weathering of carbonates, particularly calcium carbonate ($CaCO_3$), are closely linked to the global carbon and calcium cycles. Fungi play a significant role in precipitating carbonates, which are found in fungal biofilms and crusts that develop on limestone and marble. Notably, biological processes, including those involving fungi, contribute more substantially to the global $CaCO_3$ pool than physiological processes (Bindschedler et al. 2016). Cave environments lack light and thus depend on energy transfer from inorganic compounds to sustain life. Fungi are essential to the mineralisation process and have been reported to precipitate calcium carbonates in such environments. In a study, *Aspergillus versicolor* was isolated from the cave ecosystem and was shown to mineralise Ca in the form of calcium carbonate ($CaCO_3$) precipitates evident from its ureolytic activity (Akoijam et al. 2021).

Fungi secrete copious amounts of oxalic acid, which can be complex with divalent metals and precipitate them. Calcium oxalate layers the lichens growing on rock minerals, and manganese oxalate is formed due to the fungal transformation of manganese oxides. Copper oxalates are found in lichens growing on copper-rich rocks. In arid and semi-arid regions, fungal cells growing on calcareous soils often exhibit biomineralization of calcite and calcium oxalate. Calcium oxalate typically does not accumulate in the environment primarily because oxalotrophic microbes convert it into carbonate, facilitating calcite precipitation. Other metal oxalates such as manganese and copper are formed depending on the substrate. Copper oxalate commonly forms when wood-rotting fungi grow on wood treated with copper-containing preservatives, like copper-chrome arsenate, potentially enhancing copper tolerance. It can also arise from fungal biodeterioration of copper metal. Its durability may create a protective crust on copper, making it valuable for preserving historical artefacts. (Gadd 2021).

Fungi are involved in the oxide formation, some of which can transform Mn(II) to Mn(III) and Mn(IV), which are insoluble. These oxides have a strong metal sorption capacity for binding Ni, Zn, Co, Cu, Mn, Pb and Cd; they also contain copious amounts of organic C that supports biomineralisation. Phosphates are released from minerals when they react with metals such as lead and uranium, and due to the insolubility of some metal-phosphates, can immobilise toxic heavy metals. Pyromorphite, a lead-phosphate mineral formed by fungal action, immobilises Pb from contaminated soil (Gadd 2021).

Enzymes intrinsic to metabolic processes, like NADP-dependent reductases, were reported to be involved in the biomineralisation pathway for the reduction of gold ions. Other metabolites formed during respiratory and other processes, such as very reactive superoxides, can dissolve colloidal gold, forming soluble gold-containing molecules (Bohu et al. 2022).

Precipitation of soluble metals into insoluble compounds limits their mobility and toxicity by rendering them unavailable to react in the environment. Fungi play an essential role in bioprecipitation, as demonstrated by *Aspergillus niger* and *Paecilomyces javanicus*, when grown in a lead-containing medium, were able to remove lead from the solution thoroughly, most of which were precipitated as lead-containing minerals around the biomass. In the presence of phosphate, fungi have been found to precipitate heavy metals as metal phosphates. For instance, the fungal mould *Penicillium ochrochloron* was found to precipitate cupric oxalate and cupric phosphate in the presence of phosphate. When fungi are exposed to highly toxic heavy metals in the environment, they defend themselves by the production of specific proteins; for instance, *Xerocomus chrysenteron* was found to produce acid phosphatase in the presence of Cd and Cu, which were involved in the precipitation of the metal and providing tolerance to the fungi. Other ectomycorrhizal basidiomycetes, such as *Laccaria lacata* and *Suillus bovinus*, have also been shown to enhance the production of acid phosphates during elevated levels of the heavy metal Cr (Liang et al. 2016).

7.2 Bioweathering

Bioweathering by fungi is a well-established phenomenon that involves the transformation of rocks and minerals through chemical and biomechanical processes. Fungi are well-adapted to harsh environments, such as extreme UV radiation, metal toxicity, and desiccation, which often characterise rocky surfaces. Fungi scavenge for additional nutrient substrate within rocks by spreading their hyphal network in rock crevices and microfractures by penetration, boring and burrowing. This exploration exploits the weak structures in the crystal plane of the mineral; it enhances the weathering process by increasing intracellular turgor pressure, aiding in the mobility of metals, formation of secondary minerals and release of essential nutrients. The release of organic acids facilitates the breakdown of rock materials, forming patinas, films, varnishes and crusts. By contributing to these processes, fungi significantly impact geological transformations and nutrient cycling within rocky environments. Protonolysis, carbonic acid formation, iron-binding siderophores and methylation contribute to metal mobility (Gadd 2017a).

Despite the challenging conditions—nutrient deficiencies, low moisture, high salinity and extreme pH—rocks can become niches for fungi due to their resilience and adaptability. Igneous, sedimentary rocks, siliceous rocks containing silica, silicates, aluminosilicates and calcareous carbonate rocks are susceptible to weathering; two primary ways by which weathering can occur are weathering by metabolic products and weathering as a result of oxidative or reductive reactions. Fungi produce metabolites which include acids such as HNO_3, H_2SO_4, citric acid, oxalic acid, gluconic acid, lobaric and physolic acid from lichens that corrode and dissolve minerals. Environmental factors such as pH, sources of C, P and N and certain metals influence the nature and quantity of metabolites produced by fungi. *Penicillium brevicompactum* secretes organic acids in higher concentrations, which leads to metal corrosion and bioleaching. The fungi also uptake high amounts of metals such as Cu^{2+}, Cd^{2+}, Co^{2+} and Zn^{2+} (Cecchi et al. 2018).

Fungal transformation of minerals includes major minerals carbonates, phosphates, oxalates, oxides, silicates, mycogenic minerals, redox of metals and metalloids, and halide transformation.

7.2.1 Carbonates

Bioweathering of carbonate rocks promotes inorganic weathering, which increases the surface area available to microbes for biological abrasion, speeding the process of deterioration. Fungi such as *Ostpacoblabe* and *Lithophythium* can bore into the rocks with mycelia and corrode rock minerals (Bin et al. 2008). Bioweathering by fungi is also involved in the deterioration of human-made structures rich in carbonates, such as natural building stone, concrete and stone sculptures, mainly made of limestone and sandstone (Ehrlich 1998). A study showed that the fungi *Serpula himantioides* could crystallize calcite and calcium oxalate from limestones (Gadd 2017a). These interactions hint at the potential for significant damage over time, impacting the integrity and longevity of such materials (Burford et al. 2006).

7.2.2 Oxalates

Oxalic acid is a strong organic acid and is produced ubiquitously by fungi. Oxalic acid erodes minerals and can also precipitate metals. Fungi degrade wood, plaster boards and even paper with the production of oxalic acid, which causes the deposition of calcium oxalates in the form of whewellite and weddellite (Gadd 2021). In a study, *Aspergillus niger* could breakdown mimetite, a lead-apatite mineral, and result in the precipitation of lead oxalate as a myogenic biomineral. *A. niger* also bioleaching other lead-containing minerals, such as pyromorphite and vanadinite (Ceci et al. 2015).

7.2.3 Silicates

Silicates comprise 30% of all minerals, and fungi are known to dissolve these minerals, resulting in clay, soil and sediment formation. Fungi are essential for the mobilization of silica and silicates due to their production of organic acids and chelators (Gadd 2011). The action of organic acid on the silicate rock surfaces also releases other limiting nutrients, such as P and Fe (Gadd 2017b). The phyllosphere mineral asbestos, which is considered hazardous to human health, is highly resistant to weathering but has been reported to be degraded by fungi, such as *Aspergillus fumigatus*, *Fusarium oxysporium*, *Verticellum leptobactrum*, *A. tubingenesis*,

Coemansia reversa and *Paecilomyces lilacinus* (Bhattacharya et al. 2016). Silicate minerals such as mica and chlorite undergo a more significant transformation into clay minerals in ectomycorrhizosphere soil likely due to the production of organic acids and the direct uptake of K and Mg by fungal hyphae. The interactions between fungi and clay minerals are essential for soil development, aggregation and stabilization. They change the sorptive properties of both the clay minerals and the fungal hyphae, which in turn affects the size, shape and structure of mycelial pellets (Gadd 2017b). A study on the fungus *Knufia petricola* revealed that fungi rely on melanin and exopolysaccharides to thrive in harsh conditions. In situations where these fungi are mutants lacking these essential metabolites, their ability to attach to and dissolve the mineral olivine, an orthosilicate, is significantly compromised (Gerrits et al. 2020).

7.2.4 Phosphates

Fungi deteriorate phosphate minerals, such as hydroxyapatite and fluorapatite, which influences the bioavailability of P when used as fertilisers (Gadd 2021). Free-living and symbiotic fungi influence P-availability in the rhizosphere by the solubilisation of phosphate minerals. Secondary minerals, such as Ca and Mn-oxalates, are formed as a result of P solubilisation. In a study, *Aspergillus niger* was found to secrete oxalic acid, which resulted in the transformation of apatite rock phosphate, resulting in the formation of the myogenic mineral whewellite, with the release of soluble phosphate in the medium. Other fungi, such as *Serpula himantioides* and *Trametes versicolor*, also produced secondary calcium oxalate minerals upon degradation of rock phosphate (Mendes et al. 2020).

7.3 Bioleaching

The extraction of metals from ores and secondary waste generated as a result of industrial or domestic processes is called bioleaching. Fungi are favoured for their capacity to extract metals due to the production of metabolites that can operate in moderately acidic conditions (Pathak et al. 2021). Fungi convert the carbon available for their growth to organic acids, for instance, the conversion of glucose to citric acid. Soil fungi such as *Aspergillus niger*, *A. flavus*, *Penicillium chrysogenum* and *P. simplicissimum* are effective in bioleaching under mildly acidic conditions, while macrofungi such as *Marasmius oreades*, *Clitocybe* sp. and *Polysporus* sp. thrive at an alkaline pH. *Aspergillus* fungi were also found to efficiently perform bioleaching from Zn-Mn and Ni-Cd batteries with maltose or sucrose as the substrate for organic acid production (Srichandan et al. 2019). These fungi can tolerate a broad range of process conditions, exhibit high metal tolerance and require inexpensive organic substrates for growth (Pathak et al. 2021).

The metals, present as sulphides and oxides, are dissolved and leached out as metal cations. The solubilisation of metals by the action of fungal metabolites follows chemical reactions, such as acidolysis, redoxolysis, bioaccumulation, biosorption and alkalolysis. In acidolysis, the organic acids produced by the fungi reduce the pH of the surrounding environment, which facilitates the reaction with metal oxides. This action displaces the oxygen to liberate the metal and result in the formation of water. Redox lysis causes the oxidation of metals to increase their mobility due to a change in their oxidation state, for easy uptake by the fungi. The mycelium contains anions such as hydroxyl (OH^-), phosphate (PO_4^{2-}), sulphate (SO_4^{2-}), carboxyl (COO^-) and amine (NH_2^-). These functional groups react with metal cations, and the fungi take up these metals via the mycelium as illustrated in Fig. 1. Alkaloysis occurs when fungi use hydrolysis of urea or deamination of amino acids as an energy source. This process effectively mobilises metals in silicate or amino silicate compounds at alkaline pH. Fungal metal-transformation processes have many applications, including their use in mining expeditions to recover minerals and the reclamation of polluted soil. For instance, metals originating in municipal waste, including Cd, Cu, Ni and Zn, can be solubilised by the action of fungi

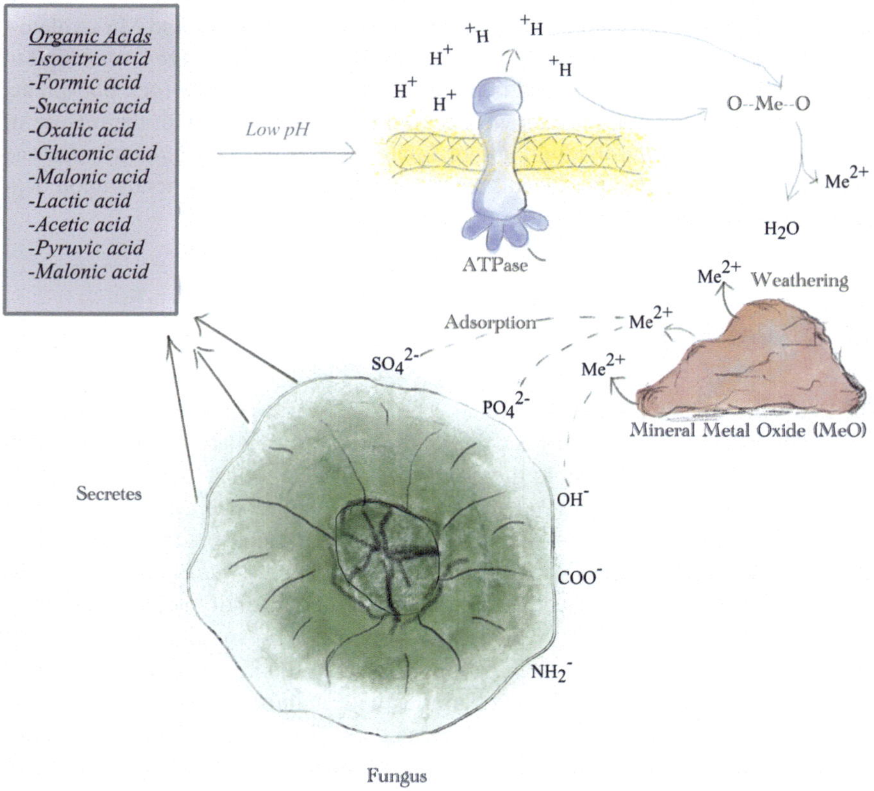

Fig. 1 Action of fungi on minerals, facilitating metal leaching

(Dusengemungu et al. 2021). *Aspergillus* fungi were also found to efficiently bioleach from Zn-Mn and Ni-Cd batteries with maltose or sucrose as the substrate for organic acid production (Gadd 2011).

7.3.1 Bioremediation

Bioremediation uses biological organisms to remove organic and inorganic pollutants from the environment (Singh et al. 2020). Bacterial and fungal systems play important roles in bioremediation, with the fungi being particularly effective at detoxifying, immobilising or reclamation of metallic pollutants. The degradation of fungi is attributed to the production of oxidative enzymes which enables them to utilise a wide range of compounds as nutrient sources. Some fungi, such as *Phanerochaete chrysosporium*, have been observed to contain almost 150 genes related to cytochrome P450, which carry out intracellular oxidation reactions (Ning et al. 2010). Various fungi, including white rot, brown rot, filamentous fungi and yeasts, have been shown to effectively degrade pesticides, pharmaceutical compounds, plastic polymers, and heavy metals. Genera such as *Trametes*, *Ganoderma*, *Aspergillus*, *Fusarium*, *Pleurotus*, *Cladosporium*, *Rhizopus*, *Trichoderma*, *Penicillium*, *Phlebia* and *Mortierella* have been extensively studied and established as capable of breaking down a wide range of diverse organic pollutants (Vaksmaa et al. 2023).

The transformation of metalloids, especially selenium, utilising various processes, including chelation, compartmentalisation, exclusion and immobilisation by fungi in synergy with bacteria, has been shown to bioremediate contaminated land and water. Plants also contribute to the bioremediation process by association with mycorrhizae. These associations improve phytoextraction by increasing plant biomass. The factors that affect efficient transformation depend on the tolerance of fungi toward the metal and the nutrients available at the contaminated site for use by the fungi. These fungi form barriers that prevent the accumulation of metals in plants; this was observed in the study where ectomycorrhiza could fix and reduce the movement of Cd(II) and Pb(II) in birch tree tissues. The fungi can tolerate metals due to the activation of mechanisms such as redox transformations, production of metal-binding peptides and proteins(metallothioneins, phytochelatins), organic and inorganic precipitation, active transport, efflux and intracellular compartmentalisation (Gadd 2010).

Fungi work in synergy with prokaryotes and other microbes for the complete degradation of high molecular weight hydrocarbons and generas, such as *Cladosporium*, *Aspergillus*, *Cunninghamella*, *Penicillium*, *Fusarium* and *Mucor*, are known to partake in the degradation of aliphatic and aromatic hydrocarbons. Fungi have been shown to populate in abundance the areas under oil spills in a study conducted from coastal sediments in the Gulf of Mexico. Fungi are considered the primary hydrocarbon degraders of oil spills due to the extracellular secretion of enzymes, as they have a higher tolerance to hydrocarbons (Amend et al. 2019). Production of biosurfactants enhances fungal adhesion to form biofilms and facilitates the degradation of these recalcitrant molecules. Fungal genera such as *Candida*, *Pseudozyma*, *Rhodotorula*, *Cunninghamella*, *Penicillium*, *Fusarium*, *Phoma*, *Exophiala*, *Aspergillus* and *Cladophialophora* are known to produce biosurfactants (Burgaud et al. 2022).

7.3.2 Biosorption and Bioaccumulation

Biosorption is a process that involves the adsorption of metals, radionuclides and organic pollutants by living or dead biomass, facilitating their removal, recovery and deposition. This spontaneous process occurs through the interaction of fungi's organic components and cellular structures with the metal and radioactive nuclei, resulting in the formation of complexes (Kapoor and Viraraghavan 1995). Fungi interact with metals through various mechanisms, including mobilisation, biosorption to cell walls, metal uptake, intracellular metal immobilisation and metal transformations (Siddiquee et al. 2015). The essential variables that influence the adsorption of metals by fungi are pH, reaction temperature, concentration of metal and contact time (Tamjidi et al. 2023).

Fungi can produce metal-binding peptides in response to metal stress. Phytochelatin, which are cysteine-rich peptides, are synthesised from glutathione by an enzymatic process involving phytochelatin synthase and are involved in the biosorption process. The fungal biomass is an excellent biosorbent for Ni, Zn, Ag, Cu, Cd, Pb, U and Th. This remarkable ability is thought to be enhanced due to the presence of chitin and melanin pigments in the cellular framework. Fungal cells and hyphal surfaces provide active sites for biosorption, and thus, they can enhance the formation of metal precipitates (Gadd 2017a). Some fungi belonging to genera, such as *Rhizopus*, *Trametes*, *Aspergillus*, *Phanerochaete*, *Penicillium*, *Trichoderma* and *Phlebia*, have been extensively studied for the bioabsorption potential using their metal removal capabilities from aqueous solutions of heavy metals (Tamjidi et al. 2023).

Bioaccumulation is a metabolically active process that entails the uptake and internalisation of heavy metals into the fungal cells. These metals after being bound to cell walls get transported into the cell and are sequestered in vesicles or bound to intercellular proteins and non-protein thiols. Fungi employ intracellular strategies to minimise metal toxicity, which include enhanced transporters in cell walls, antioxidant processes and biochemical transformation into less toxic forms. Fungi have been found to effectively remove metal contamination from both terrestrial and aquatic systems. The fungi *Gloeophyllum sepiarium* was found to remove 94% of Cr in 6 months from contaminated soils, while *Asperguillus fumigatus* and *A. niger* when used as biosorbents coated on magnetic nanoparticles could remove Cr from aquatic environments. *Corollospora lacera* and *Monodictys pelagica*, two marine fungi, have also been found to effectively sequester Cd and Pb (Vaksmaa et al. 2023).

8 Geomycology as an Emerging Science

The multidisciplinary field of geomycology investigates the interactions between fungi and geological processes, focusing on fungal roles in mineral transformations. While prokaryotes have been extensively studied for their influence on geological processes, attention is increasingly turning to fungi. Their contributions to rock weathering, metal accumulation, nutrient cycling, soil formation and their effects on other microbial communities in mineral substrates are central to geomycological research (Gadd 2011).

Recent advances in molecular techniques and the integration of omics technologies can enhance our understanding of fungal diversity and function, revealing hidden genetic and metabolic pathways. Fungi previously believed unable to survive in harsh environments, dominated solely by prokaryotes, have now been isolated from acidic mines with very low pH, such as *Penidiella* sp. and *Coniochaeta fodinicola.* Fungal biomineralisation often results in the formation of nanomaterials, with significant chemical reactivity, warranting future research into the physical and chemical factors that optimize the production and size of mycomineral particles for specific industrial applications (Gadd 2017b).

Rock-inhabiting fungi create biodeteriorating biofilms on exposed desert rocks and anthropogenic surfaces, such as solar panels, highlighting their relevance in material sciences and geomycology. Their interactions can provide insights into ecological processes and the durability of materials in various environments (Gerrits et al. 2020). Although bacteria and fungi are typically studied separately, examining their coexistence and interactions could yield significant insights into geochemistry and applications in bioremediation, bio-recovery and biomineral production (Gadd 2017b). Different roles that fungi play in geomycological processes have been depicted in Fig. 2. Microbial carbonate precipitation has garnered interest for its potential soil stabilization, sand cementation, concrete repair, self-healing and the immobilization of toxic metals (Gadd 2021). Furthermore, fungi pose a significant threat to preserving historical art and buildings through their role in weathering. Epilithic fungi on rock surfaces and endolithic fungi within stone crevices contribute to deterioration. It is crucial to understand their impact on effective preservation strategies to protect these timeless relics (Gadd 2017b). The deepened understanding of Earth's ecosystems is unveiled through the study of the intricate relationships that sustain life; thus, geomycology heralds a new frontier in environmental science.

9 Conclusion

Geomycology builds on the intricate interplay between fungi and geological processes, and the profound impact these organisms has on ecological dynamics and biogeochemical cycles. The adaptability and resilience of fungi to change minerals and elements highlight their critical role as decomposers and contributors to soil

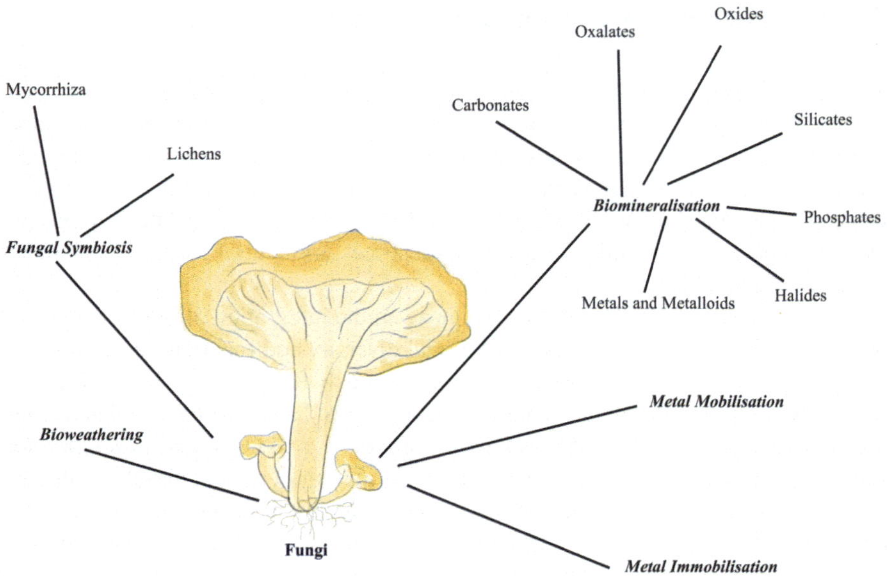

Fig. 2 Fungal roles in geomycology

formation, nutrient cycling and environmental biotechnology. For instance, fungal-mediated mineralisation can be utilised to develop nanoparticles since these are relatively less toxic when compared to the physical or chemically synthesised methods. These fungal actions likely solve some of the most pressing environmental challenges. Thus, a deeper appreciation and understanding of geomycology would only enrich our knowledge of the natural world and unveil potential pathways for enhancing ecological resilience and human well-being.

References

Akhtar K, Waheed Akhtar M, Khalid AM (2007) Removal and recovery of uranium from aqueous solutions by *Trichoderma harzianum*. Water Res 41:1366–1378. https://doi.org/10.1016/j.watres.2006.12.009

Akoijam N, Dutta S, Joshi SR (2021) Biomineralization potential of a Ureolytic fungus isolated from Mawsmai Cave in Meghalaya. NEHU J 12:31–48

Aldossari N, Ishii S (2021) Fungal denitrification revisited—recent advancements and future opportunities. Soil Biol Biochem 157:108250. https://doi.org/10.1016/j.soilbio.2021.108250

Amend A, Burgaud G, Cunliffe M, Edgcomb VP, Ettinger CL, Gutiérrez MH et al (2019) Fungi in the marine environment: open questions and unsolved problems. MBio 10(2):10–1128. https://doi.org/10.1128/mbio.01189-18

Andronikov AV, Andronikova IE, Sebek O, Martinkova E, Stepanova M (2023) Accumulation and within-mushroom distribution of elements in red cracking bolete (*Xerocomellus chrysenteron*) collected over the extended period from compositionally contrasting substrates. Environ Monit Assess 195:1157. https://doi.org/10.1007/s10661-023-11786-6

Asplund J, Wardle DA (2017) How lichens impact on terrestrial community and ecosystem properties. Biol Rev 92:1720–1738. https://doi.org/10.1111/brv.12305

Bahram M, Netherway T (2022) Fungi as mediators linking organisms and ecosystems. FEMS Microbiol Rev 46:58. https://doi.org/10.1093/femsre/fuab058

Bhalla K, Qu X, Kretschmer M, Kronstad JW (2022) The phosphate language of fungi. Trends Microbiol 30:338–349. https://doi.org/10.1016/j.tim.2021.08.002

Bhattacharya S, John PJ, Ledwani L (2016) Fungal weathering of asbestos in semi arid regions of India. Ecotoxicol Environ Saf 124:186–192. https://doi.org/10.1016/j.ecoenv.2015.10.022

Bin L, Ye C, Lijun Z, Ruidong Y (2008) Effect of microbial weathering on carbonate rocks. Earth Sci Front 15(6):90–99. https://doi.org/10.1016/S1872-5791(09)60009-9

Bindschedler S, Cailleau G,Verrecchia E (2016) Role of Fungi in the Biomineralization of Calcite. Minerals 6(2):41. https://doi.org/10.3390/min6020041

Boddy L (2016) Fungi, ecosystems, and global change. In: The Fungi. Elsevier, pp 361–400. https://doi.org/10.1016/B978-0-12-382034-1.00011-6

Bohu T, Anand RR, Kaksonen AH, González-Álvarez I, Pages A, Noble RRP et al (2022) The role of fungi in the biogeochemical cycling of supergene gold and satellite transition metals: a potential new exploration tool. Ore Geol Rev 140:104595. https://doi.org/10.1016/j.oregeorev.2021.104595

Borovička J, Řanda Z, Jelínek E, Kotrba P, Dunn CE (2007) Hyperaccumulation of silver by *Amanita strobiliformis* and related species of the section Lepidella. Mycol Res 111:1339–1344. https://doi.org/10.1016/j.mycres.2007.08.015

Breyer E, Baltar F (2023) The largely neglected ecological role of oceanic pelagic fungi. Trend Ecol Evol 38:870–888. https://doi.org/10.1016/j.tree.2023.05.002

Burford EP, Hillier S, Gadd GM (2006) Biomineralization of Fungal Hyphae with Calcite (CaCO3) and Calcium Oxalate Mono- and Dihydrate in Carboniferous Limestone Microcosms. Geomicrobiol J 23(8): 599–611. https://doi.org/10.1080/01490450600964375

Burgaud G, Edgcomb V, Hassett BT, Kumar A, Li W, Mara P et al (2022) Marine fungi. In: Stal LJ, Cretoiu MS (eds) The marine microbiome. The microbiomes of humans, animals, plants, and the environment, vol 3. Springer, pp 243–295. https://doi.org/10.1007/978-3-030-90383-1_5

Byrne AR, Tušek-Žnidarič M (1990) Studies of the uptake and binding of trace metals in fungi. Part I: accumulation and characterization of mercury and silver in the cultivated mushroom, *Agaricus bisporus*. Appl Organometal Chem 4:43–48. https://doi.org/10.1002/aoc.590040108

Cairney JWG (2005) Basidiomycete mycelia in forest soils: dimensions, dynamics and roles in nutrient distribution. Mycol Res 109:7–20. https://doi.org/10.1017/S0953756204001753

Canfield DE, Kristensen E, Thamdrup B (2005) The phosphorus cycle. Adv Marine Biol 48:419–440. https://doi.org/10.1016/S0065-2881(05)48011-6

Cecchi G, Ceci A, Marescotti P, Persiani AM, Di Piazza S, Ballirano P et al (2018) The geological roles played by microfungi in interaction with sulfide minerals from libiola mine, Liguria, Italy. Geomicrobiol J 35:564–569. https://doi.org/10.1080/01490451.2018.1429507

Ceci A, Kierans M, Hillier S, Persiani AM, Gadd GM (2015) Fungal bioweathering of mimetite and a general geomycological model for lead apatite mineral biotransformations. Appl Environ Microbiol 81(15):4955–4964. https://doi.org/10.1128/AEM.00726-15

Chalot M, Brun A (1998) Physiology of organic nitrogen acquisition by ectomycorrhizal fungi and ectomycorrhizas. FEMS Microbiol Rev 22(1):21–44. https://doi.org/10.1111/j.1574-6976.1998.tb00359.x

Chen J, Blume H-P, Beyer L (2000) Weathering of rocks induced by lichen colonization—a review. Catena 39(2):121–146. https://doi.org/10.1016/S0341-8162(99)00085-5

Clemens S, Palmgren MG, Krämer U (2002) A long way ahead: understanding and engineering plant metal accumulation. Trends Plant Sci 7:309–315. https://doi.org/10.1016/S1360-1385(02)02295-1

Coleine C, Stajich JE, Selbmann L (2022) Fungi are key players in extreme ecosystems. Trend Ecol Evol 37:517–528 de Oliveira Mendes G, Moreira de Freitas AL, Liparini Pereira O, Ribeiro da Silva I, Bojkov Vassilev N, Dutra Costa M (2014) Mechanisms of phosphate solubilization by fungal isolates when exposed to different P sources. Ann Microbiol 64:239–249. https://doi.org/10.1007/s13213-013-0656-3

Debeljak P, Baltar F (2023) Fungal diversity and community composition across ecosystems. J Fungi 9:510. https://doi.org/10.3390/jof9050510

Du E, Terrer C, Pellegrini AFA, Ahlström A, van Lissa CJ, Zhao X et al (2020) Global patterns of terrestrial nitrogen and phosphorus limitation. Nat Geosci 13:221–226. https://doi.org/10.1038/s41561-019-0530-4

Dusengemungu L, Kasali G, Gwanama C, Mubemba B (2021) Overview of fungal bioleaching of metals. Environ Adv 5:100083. https://doi.org/10.1016/j.envadv.2021.100083

Ehrlich H (1998) Geomicrobiology: its significance for geology. Earth Sci Rev 45:45–60. https://doi.org/10.1016/S0012-8252(98)00034-8

Falkowski PG (2001) Biogeochemical cycles. In: Encyclopedia of biodiversity. Elsevier, pp 552–564. https://doi.org/10.1016/B978-0-12-384719-5.00013-7

Filippelli GM (2002) The global phosphorus cycle. Rev Mineral Geochem 48(1):391–425. https://doi.org/10.2138/rmg.2002.48.10

Gadd GM (2007) Geomycology: biogeochemical transformations of rocks, minerals, metals and radionuclides by fungi, bioweathering and bioremediation. Mycol Res 111:3–49. https://doi.org/10.1016/j.mycres.2006.12.001

Gadd GM (2010) Metals, minerals and microbes: geomicrobiology and bioremediation. Microbiology 156:609–643. https://doi.org/10.1099/mic.0.037143-0

Gadd GM, Rhee YJ, Stephenson K, Wei, Z (2011) Geomycology: metals, actinides and biominerals. Environ Microbiol Rep 4(3): 270–296. https://doi.org/10.1111/j.1758-2229.2011.00283.x

Gadd GM (2013) Geomycology: fungi as agents of biogeochemical change. Biol Environ Proceed Royal Irish Acad 113B:139–153. https://doi.org/10.1353/bae.2013.0006

Gadd GM (2017a) The geomycology of elemental cycling and transformations in the environment. Microbiol Spect 5:1–16. https://doi.org/10.1128/microbiolspec.FUNK-0010-2016

Gadd GM (2017b) New horizons in geomycology. Environ Microbiol Rep 9:4–7. https://doi.org/10.1111/1758-2229.12480

Gadd GM (2021) Fungal biomineralization. Curr Biol 31:R1557–R1563. https://doi.org/10.1016/j.cub.2021.10.041

Geng Y, Pan S, Zhang L, Qiu J, He K, Gao H, Li Z, Tian D (2022) Phosphorus biogeochemistry regulated by carbonates in soil. Environ Res 214:113894. https://doi.org/10.1016/j.envres.2022.113894

Gerrits R, Pokharel R, Breitenbach R, Radnik J, Feldmann I, Schuessler JA (2020) How the rock-inhabiting fungus *K. petricola* A95 enhances olivine dissolution through attachment. GCA 282: 76–97.

Goltsov VA, Goltsova LF (2014) Biosphere synergism and the humankind virtual path to the hydrogen civilization era. Int J Hydrog Energy 39:9931–9942

Grayston SJ, Wainwright M (1988) Sulphur oxidation by soil fungi including some species of mycorrhizae and wood-rotting basidiomycetes. FEMS Microbiol Lett 53:1–8. https://doi.org/10.1111/j.1574-6968.1988.tb02641.x

Griffith GW, Roderick K (2008) Chapter 15: Saprotrophic basidiomycetes in grasslands: distribution and function. In: British mycological society symposia series, vol 28. Elsevier, pp 277–299. https://doi.org/10.1016/S0275-0287(08)80017-3

Grossart HP, Van den Wyngaert S, Kagami M, Wurzbacher C, Cunliffe M, Rojas-Jimenez K (2019) Fungi in aquatic ecosystems. Nat Rev Microbiol 17:339–354. https://doi.org/10.1038/s41579-019-0175-8

Gulis V, Kuehn K, Suberkropp K (2006) The role of fungi in carbon and nitrogen cycles in freshwater ecosystems. In: Fungi in biogeochemical cycles. Cambridge University Press, Cambridge, pp 404–435. https://doi.org/10.1017/CBO9780511550522.018

Hannula SE, Morriën E (2022) Will fungi solve the carbon dilemma? Geoderma 413:115767. https://doi.org/10.1016/j.geoderma.2022.115767

Hawkins HJ, Cargill RIM, Van Nuland ME, Hagen SC, Field KJ, Sheldrake M et al (2023) Mycorrhizal mycelium as a global carbon pool. Curr Biol 33:R560–R573. https://doi.org/10.1016/j.cub.2023.02.027

Hibbett DS, Binder M, Bischoff JF, Blackwell M, Cannon PF, Eriksson OE et al (2007) A higher-level phylogenetic classification of the fungi. Mycol Res 111:509–547. https://doi.org/10.1016/j.mycres.2007.03.004

Huang T, Hu Q, Shen Y, Anglés A, Fernández-Remolar DC (2024) Biogeochemical cycles. In: Encyclopedia of biodiversity. Elsevier, pp 393–407. https://doi.org/10.1016/B978-0-12-822562-2.00347-9

Husain H, Keitel C, Dijkstra FA (2024) Fungi are more important than bacteria for soil carbon loss through priming effects and carbon protection through aggregation. Appl Soil Ecol 195:105245. https://doi.org/10.1016/j.apsoil.2023.105245

Ivarsson M, Schnürer A, Bengtson S, Neubeck A (2016) Anaerobic fungi: a potential source of biological H_2 in the oceanic crust. Front Microbiol 7. https://doi.org/10.3389/fmicb.2016.00674

Kangwankraiphaisan T, Suntornvongsagul K, Sihanonth P, Klysubun W, Gadd GM (2013) Influence of arbuscular mycorrhizal fungi (AMF) on zinc biogeochemistry in the rhizosphere of *Lindenbergia philippensis* growing in zinc-contaminated sediment. Biometals 26:489–505. https://doi.org/10.1007/s10534-013-9634-2

Kapoor A, Viraraghavan T (1995) Fungal biosorption—an alternative treatment option for heavy metal bearing wastewaters: a review. Bioresour technol 53(3): 195–206.

Keymer A, Pimprikar P, Wewer V, Huber C, Brands M, Bucerius SL et al (2017) Lipid transfer from plants to arbuscular mycorrhiza fungi. elife 6. https://doi.org/10.7554/eLife.29107

Liang X, Kierans M, Ceci A, Hillier S, Gadd GM (2016) Phosphatase-mediated bioprecipitation of lead by soil fungi. Environ Microbiol 18:219–231. https://doi.org/10.1111/1462-2920.13003

Liu S, Liu H, Li J, Wang Y (2022) Research Progress on elements of wild edible mushrooms. J Fungi 8:964. https://doi.org/10.3390/jof8090964

Luo Y, Ma L, Feng Q, Luo H, Chen C, Wang S et al (2024) Influence and role of fungi, bacteria, and mixed microbial populations on phosphorus Acquisition in Plants. Agriculture 14:358. https://doi.org/10.3390/agriculture14030358

Mendes, Gde O, Bahri-Esfahani J, Csetenyi, L, Hillier, S, George, TS, Gadd, GM (2020) Chemical and Physical Mechanisms of Fungal Bioweathering of Rock Phosphate. Geomicrobiol J 38(5): 384–394. https://doi.org/10.1080/01490451.2020.1863525

Mohan VK, Joshi SR (2024) Mycorrhizal fungi as ecofriendly interventions for crop productivity. In: Industrial microbiology and biotechnology. Springer Nature, Singapore, pp 663–677. https://doi.org/10.1007/978-981-97-1912-9_26

Naranjo-Ortiz MA, Gabaldón T (2019) Fungal evolution: major ecological adaptations and evolutionary transitions. Biol Rev 94:1443–1476. https://doi.org/10.1111/brv.12510

Ning D, Wang H, Zhuang Y (2010) Induction of functional cytochrome P450 and its involvement in degradation of benzoic acid by phanerochaete chrysosporium. Biodegradation 21:297–308. https://doi.org/10.1007/s10532-009-9301-z

O'Hanlon R (2017) Fungi in the environment. In: Fungi: Biology and applications. Wiley, Chichester, pp 333–353. https://doi.org/10.1002/9781119374312.ch13

Orsi WD, Vuillemin A, Coskun ÖK, Rodriguez P, Oertel Y, Niggemann J et al (2022) Carbon assimilating fungi from surface ocean to subseafloor revealed by coupled phylogenetic and stable isotope analysis. ISME J 16:1245–1261. https://doi.org/10.1038/s41396-021-01169-5

Pathak A, Kothari R, Vinoba M, Habibi N, Tyagi VV (2021) Fungal bioleaching of metals from refinery spent catalysts: a critical review of current research, challenges, and future directions. J Environ Manag 280:111789. https://doi.org/10.1016/j.jenvman.2020.111789

Purvis OW, Pawlik-Skowrońska B (2008) Chapter 12: Lichens and metals. In: British Mycological Society Symposia Series, vol 27. Elsevier, pp 175–200. https://doi.org/10.1016/S0275-0287(08)80054-9

Qin W, Wang C, Ma Y, Shen M, Li J, Jiao K et al (2020) Microbe-mediated extracellular and intracellular mineralization: environmental, industrial, and biotechnological applications. Adv Mater 32:e1907833. https://doi.org/10.1002/adma.201907833

Qiu Q, Noll M, Abraham WR, Lu Y, Conrad R (2008) Applying stable isotope probing of phospholipid fatty acids and rRNA in a Chinese rice field to study activity and composition of the methanotrophic bacterial communities in situ. ISME J 2:602–614. https://doi.org/10.1038/ismej.2008.34

Richardson JA, Anderton CR, Bhattacharjee A (2023) Saprotrophic fungus induces microscale mineral weathering to source potassium in a carbon-limited environment. Fortschr Mineral 13:641. https://doi.org/10.3390/min13050641

Ruiz-Fresneda MA, Morales-Hidalgo M, Povedano-Priego C, Jroundi F, Hidalgo-Iruela J, Cano-Cano M et al (2024) Unlocking the key role of bentonite fungal isolates in tellurium and selenium bioremediation and biorecovery: implications in the safety of radioactive waste disposal. Sci Total Environ 912:169242. https://doi.org/10.1016/j.scitotenv.2023.169242

Sabuda MC, Mejia J, Wedal M, Kuester B, Xu T, Santelli CM (2022) The effect of organic carbon form and concentration on fungal selenite reduction. Appl Geochem 136:105163. https://doi.org/10.1016/j.apgeochem.2021.105163

Salazar A, Rousk K, Jónsdóttir IS, Bellenger J, Andrésson ÓS (2020) Faster nitrogen cycling and more fungal and root biomass in cold ecosystems under experimental warming: a meta-analysis. Ecology 101:e02938. https://doi.org/10.1002/ecy.2938

Siddiquee S, Rovina K, Azad SA, Naher L, Suryani S, Chaikaew P (2015) Heavy metal contaminants removal from wastewater using the potential filamentous fungi biomass: a review. J Microb Biochem Technol 7:384–393. https://doi.org/10.4172/1948-5948.1000243

Simpson AJ, McNally DJ, Simpson MJ (2011) NMR spectroscopy in environmental research: From molecular interactions to global processes. Prog Nucl Magn Reson Spectrosc 58(3–4): 97–175.

Singh M (2021) Fungi and nitrogen cycle: symbiotic relationship, mechanism and significance. In: Soil nitrogen ecology. Soil biology. Springer, Cham, pp 391–406. https://doi.org/10.1007/978-3-030-71206-8_20

Singh YD, Singh MC, Panda MK (2020) Biotechnological aspects of mangrove microorganisms. In: Biotechnological utilization of mangrove resources. Academic Press, pp 381–398. https://doi.org/10.1016/B978-0-12-819532-1.00018-4

Srichandan H, Mohapatra RK, Parhi PK, Mishra S (2019) Bioleaching approach for extraction of metal values from secondary solid wastes: a critical review. Hydrometallurgy 189:105122. https://doi.org/10.1016/j.hydromet.2019.105122

Tamjidi S, Ameri A, Esmaeili H (2023) A review of the application of fungi as an effective and attractive bio-adsorbent for biosorption of heavy metals from wastewater. Environ Monitor Assess 195:91. https://doi.org/10.1007/s10661-022-10687-4

TlÁskal V, BrabcovÁ V, Větrovský T, Jomura M, López-Mondéjar R, Oliveira Monteiro LM et al (2021) Complementary roles of wood-inhabiting fungi and bacteria facilitate deadwood decomposition. MSystems 6:10–1128. https://doi.org/10.1128/mSystems.01078-20

Vaksmaa A, Guerrero-Cruz S, Ghosh P, Zeghal E, Hernando-Morales V, Niemann H (2023) Role of fungi in bioremediation of emerging pollutants. Front Mar Sci 10:1070905. https://doi.org/10.3389/fmars.2023.1070905

Wei Z, Hillier S, Gadd GM (2012) Biotransformation of manganese oxides by fungi: solubilization and production of manganese oxalate biominerals. Environ Microbiol 14:1744–1753. https://doi.org/10.1111/j.1462-2920.2012.02776.x

Wei Z, Liang X, Pendlowski H, Hillier S, Suntornvongsagul K, Sihanonth P et al (2013) Fungal biotransformation of zinc silicate and sulfide mineral ores. Environ Microbiol 15:2173–2186. https://doi.org/10.1111/1462-2920.12089

Wilkes TI (2021) Arbuscular mycorrhizal fungi in agriculture. Encyclopedia 1:1132–1154. https://doi.org/10.3390/encyclopedia1040085

Wu H, Cui H, Fu C, Li R, Qi F, Liu Z et al (2024) Unveiling the crucial role of soil microorganisms in carbon cycling: a review. Sci Total Environ 909:168627. https://doi.org/10.1016/j.scitotenv.2023.168627

Yu G-H, Chi Z-L, Teng HH, Dong H-L, Kappler A, Gillings MR et al (2019) Fungus-initiated catalytic reactions at hyphal-mineral interfaces drive iron redox cycling and biomineralization. Geochim Cosmochim Acta 260:192–203. https://doi.org/10.1016/j.gca.2019.06.029

Zeiner CA, Purvine SO, Zink EM, Paša-Tolić L, Chaput DL, Wu S et al (2017) Quantitative iTRAQ-based secretome analysis reveals species-specific and temporal shifts in carbon utilization strategies among manganese(II)-oxidizing Ascomycete fungi. Fungal Genet Biol 106:61–75. https://doi.org/10.1016/j.fgb.2017.06.004

Zeiner CA, Purvine SO, Zink E, Wu S, Paša-Tolić L, Chaput DL et al (2021) Mechanisms of manganese(II) oxidation by filamentous ascomycete fungi vary with species and time as a function of secretome composition. Front Microbiol 12:610497. https://doi.org/10.3389/fmicb.2021.610497

Zhang B, Zhou J, Li Q, Gan B, Peng W, Zhang X et al (2019) Manganese affects the growth and metabolism of *Ganoderma lucidum* based on LC-MS analysis. PeerJ 7:e6846–e6846. https://doi.org/10.7717/peerj.6846

Zhu H-Z, Jiang C-Y, Liu S-J (2022) Microbial roles in cave biogeochemical cycling. Front Microbiol 13. https://doi.org/10.3389/fmicb.2022.950005

Ecological Restoration of Salt Marshes by Geomicrobes

Prasann Kumar and Komal Agrawal

Contents

P. Kumar (✉)
Department of Agronomy, School of Agriculture, Lovely Professional University, Phagwara, Punjab, India

K. Agrawal
Department of Microbiology, School of BioEngineering and Biosciences, Lovely Professional University, Phagwara, Punjab, India

S. R. Joshi et al. (eds.), *Mineral Transformation and Bioremediation by Geo-Microbes*, Interdisciplinary Biotechnological Advances,
https://doi.org/10.1007/978-981-96-3033-2_9

Abstract Salt marshes are vital coastal ecosystems that offer crucial services, including carbon sequestration, shoreline stabilisation, and habitat provision for diverse species. Nevertheless, these habitats are progressively endangered by human activities and the effects of climate change. Conventional restoration techniques commonly prioritise physical and chemical approaches, which may not comprehensively tackle the fundamental ecological processes. The current chapter examines the novel use of geomicrobes, microorganisms that impact geological and chemical processes, to restore salt marshes in the environment. Geomicrobes can improve the process of nutrient cycling, stimulate the growth of plants, and stabilise sediments. This makes them a valuable and comprehensive method for restoration that is environmentally friendly and can be maintained over time. We thoroughly examine the latest scientific studies on microorganisms in salt marshes, elucidate their potential contributions to ecological restoration, and suggest a comprehensive approach for incorporating these microorganisms in restoration projects. This chapter aims to integrate microbial ecology into restoration practices by focusing on the environmental functions of geomicrobes and their potential to enhance salt marsh restoration. It provides a comprehensive and innovative framework for this purpose. This approach holds the potential to deliver enduring and robust solutions to the challenges salt marsh ecosystems face.

Keywords Salt marshes · Restoration · Ecosystem · Geomicrobes · Ecology

Abbreviations

N_2	Atmospheric nitrogen
NH_3	Ammonia
NO_3^+	Nitrates
N_2	Nitrogen gas
SO_4^{2+}	Sulfate
H_2S	Hydrogen sulfide

1 Introduction

Salinity marsh ecosystems are highly productive and ecologically important habitats on Earth. Originating from the intertidal zones between land and saltwater, these ecosystems function as inherent barriers against coastal erosion, offer habitats for diverse fauna, and significantly contribute to global nutrient cycles. Salt marshes are distinguished by halophytic (salt-tolerant) flora, including grasses and shrubs, which have evolved to thrive in the high-salinity environments in coastal regions. These plants are essential for soil stabilisation, erosion mitigation, and the sustenance of diverse organisms, including avifauna, aquatic species, invertebrates, and microorganisms. The significance of salt marshes in ecology cannot be exaggerated. These coastal ecosystems offer vital services that benefit the environment and human society (Devi et al. 2024e; Kumar et al. 2023a, 2024j; Sharma et al. 2024). An essential role of salt marshes is their capacity to sustain biodiversity. These habitats' dense vegetation and intricate structures offer sustenance, refuge, and reproductive sites for many species, including economically important fish and shellfish. Salt marshes serve as a crucial food source and rest for migratory birds during their extensive migrations. Simultaneously, the sediments' abundant nutrients provide essential support for a diverse range of invertebrates, which serve as food for larger predators. In addition to promoting biodiversity, salt marshes profoundly impact carbon sequestration. Salt marshes, known as "blue carbon" ecosystems, specialise in efficiently capturing and storing atmospheric carbon dioxide. The abundant plant life sequesters carbon in the soil, which can persist for an extended period, rendering these ecosystems crucial in mitigating climate change. Evidence has demonstrated that salt marshes can capture carbon at rates equal to or higher than terrestrial forests. Apart from carbon sequestration, salt marshes also play a role in nutrient cycling, namely nitrogen, using microbial processes that eliminate surplus nitrogen from the water, mitigating the potential for eutrophication in coastal waters. Aquatic filtration is another vital ecological role performed by salt marshes. These ecosystems function as inherent screening mechanisms, capturing contaminants, sediments, and nutrients before entering the open ocean. Salt marshes are crucial in preserving water quality in coastal areas and mitigating detrimental algal blooms by sequestering surplus nutrients, including nitrogen and phosphorus, from agricultural runoff. The usefulness of this water filtration capacity to human populations lies in its ability to enhance the quality of drinking water and sustain fisheries and tourism sectors that rely on pristine and thriving oceans. Despite their ecological significance, salt marshes face substantial threats from various factors. Urbanisation, agriculture, and industrial pollution are examples of anthropogenic pressures that have caused extensive deterioration of these ecosystems. Coastal development frequently destroys salt marshes as land is reclaimed for infrastructure or agriculture. Construction of seawalls, dikes, and other barriers can impede the natural movement of water and sediment, resulting in erosion and the loss of habitats. Moreover, industrial operations can potentially introduce contaminants, such as heavy metals and hydrocarbons, into salt marshes, adversely impacting plant and microbial populations. A major contributor to the deterioration of salt marsh ecosystems is agricultural runoff. Intensive application of fertilisers in agriculture results in nutrient contamination, namely nitrogen and phosphorus, which can induce eutrophication in coastal seas. Eutrophication leads to algae proliferation, drastically reducing oxygen levels in the water and forming oxygen-depleted areas where aquatic organisms cannot survive. This phenomenon impacts the biodiversity of salt marshes and diminishes their ability to offer vital functions, such as water purification and carbon storage. Natural phenomena, especially those associated with climate change, also present substantial risks to salt marshes. The increasingly elevated sea levels, caused by the warming of the Earth's climate, are a highly urgent and critical issue confronting coastal ecosystems. The rise in sea levels can result in salt marshes'flooding, causing vegetation depletion and alterations in the ecological makeup of plant and animal populations. Lacking adequate sediment accumulation to match the increasing water levels, salt marshes are at risk of submerging, resulting in habitat loss and biodiversity (Devi et al. 2024f; Kumar and Singh 2024a, 2024b, 2024c, 2024d). Climate change is projected to cause an increase in both the frequency and intensity of storm surges, which also pose a threat to salt marshes. Storm surges during severe weather phenomena, such as hurricanes or typhoons, can erode the soil, displace vegetation, and introduce substantial quantities of saltwater into freshwater systems, disrupting the intricate equilibrium of salinity that salt marshes rely on. Under certain circumstances, these occurrences can result in enduring alterations in the physical environment, diminishing salt marsh ecosystems' ability to withstand adverse impacts. Salt marshes are significantly impacted by salinity imbalances, which pose a paramount concern for their health. Despite the inherent adaptability of these ecosystems to varying salinity levels, substantial alterations in salinity caused by human activities, such as water diversion or heightened freshwater input, can disturb the fragile equilibrium of these systems. Elevated salinity levels can impede the growth of plants, diminish microbial activity, and have a detrimental impact on the net productivity of salt marshes. In contrast, a decrease in salinity caused by the arrival of fresh water can result in the invasion of non-indigenous plant species, which can surpass native vegetation and modify the composition and operation of the ecosystem. Given these difficulties, the pursuit of ecological restoration has grown progressively essential for preserving and rehabilitating salt marshes. An up-and-coming method for restoring salt marshes is the utilisation of geomicrobes, which are microorganisms that have a crucial function in the biogeochemical cycling of nutrients and the general well-being of ecosystems. Geomicrobes refer to microorganisms that exert substantial influence on the geological and chemical processes of the environment, namely in soil and sediment. In salt marshes, these microorganisms play a crucial role in sustaining nutrient cycles, promoting plant development, and strengthening the capacity of the ecosystem to withstand environmental pressures. The microbial ecology in salt marshes is intricate, encompassing many bacteria, archaea, fungi, and other microecological species interacting with plants, sediments, and one another (Devi and Kumar 2024a, 2024b; Kumar et al. 2024a, 2024d, 2024e, 2024f, 2024g; Kumar and Choudhury 2024a). These microorganisms carry out vital chemical reactions such as nitrogen fixation, sulfate reduction, and methane oxidation. As an example, nitrogen-fixing bacteria transform atmospheric nitrogen into forms that plants can utilise while sulfate-reducing bacteria are essential in sulfur cycling, which affects the composition of sediments and promotes plant metabolism. Methanotrophic bacteria, by metabolising methane generated by anaerobic processes in the soil, play a crucial role in controlling greenhouse gas emissions from salt marshes, contributing to their efficacy as carbon sinks. It is impossible to overstate the significance of geomicrobes in biogeochemical cycling and ecosystem resilience. Microorganisms play a crucial role in nutrient cycling by regulating the availability of vital nutrients, including nitrogen, phosphorus, and sulfur, necessary for plant growth and overall ecological productivity. Geomicrobes enhance the long-term stability and biodiversity of salt marsh ecosystems by promoting the decomposition of organic matter and the recycling of nutrients. Moreover, geomicrobes bolster the ability of salt marshes to withstand environmental pressures, such as changes in salinity and pollution, by facilitating interactions between plants and microbes that promote plant life and tolerance to stress. Marine salt marsh ecosystems are crucial in preserving biodiversity, storing carbon, and purifying water in coastal areas. Nevertheless, these ecosystems face many hazards from human activities and climate change. Restoring salt marshes, specifically using geomicrobes, presents a promising approach to improving these vital habitats' resilience and ecological function. By comprehending and utilising the potential of geomicrobes, we can formulate more efficient approaches to the restoration and conservation of salt marsh ecosystems for future generations.

2 Salt Marsh Degradation and the Need for Restoration

2.1 Key Drivers of Salt Marsh Degradation

Variations significantly influence the health and functionality of salt marsh ecosystems in soil and water salinity. Salt marshes have inherent adaptations to regulate salinity levels in response to tidal flows and seasonal variations. However, excessive fluctuations, frequently caused by human activities, can disrupt these ecosystems. Elevated salinity, usually resulting from upstream water diversions, agricultural runoff, or decreased freshwater input, can impede the development of salt-tolerant plants, decrease microbial activity, and modify the makeup of plant and animal populations. Conversely, a substantial decrease in salinity, typically caused by the excessive contribution of freshwater from urban drainage or storms, can facilitate the invasion of non-native species that surpass native salt-tolerant plants in competition, further disturbing the marsh's ecological equilibrium. Eutrophication and nutrient overload pose a significant risk to salt marshes. Eutrophication is the introduction of excessive nutrients, namely nitrogen and phosphorus, carried by agricultural runoff into coastal waters. The presence of these nutrients stimulates the proliferation of algae, resulting in detrimental algal blooms. When the algae perish and undergo decomposition, the oxygen levels in the water decrease significantly, resulting in hypoxic or anoxic conditions that are fatal to aquatic organisms. This process disturbs the ecological balance of the food chain, damages the fish populations, and diminishes the levels of biodiversity in the marsh. Moreover, the surplus nutrients can disrupt the equilibrium of nitrogen cycling in the ecosystem, impeding the activity of advantageous microorganisms and diminishing the marsh's ability to filter and purify water (Choudhury et al. 2024; Dey et al. 2024; Kumar et al. 2024h, 2024i, 2024k; Pandey et al. 2024). Changes in sedimentation and erosion, worsened by human activities, intensify the difficulties encountered by salt marshes. These ecosystems' structural integrity and plant growth support depend on a precise equilibrium between sediment deposition and erosion. Nevertheless, anthropogenic actions such as dredging, dam construction, and land reclamation can disturb this equilibrium, resulting in excessive erosion and soil loss. With the removal of sediments or the reduction of their natural deposition, the capacity of marsh vegetation to stabilise the soil diminishes, resulting in the collapse of the ecosystem. Erosive processes also deplete the uppermost layer of soil, which is abundant in nutrients and microorganisms, impeding the marsh's ability to regenerate. Furthermore, without adequate sediment accumulation, salt marshes cannot match the global sea level rise rate, exacerbating the vulnerability to habitat degradation. The impacts of climate change intensify all of these elements, as increasing temperatures, modified precipitation patterns, and more frequent severe weather events present novel obstacles to the resilience of salt marshes. Elevated temperatures can facilitate the acceleration of evaporation rates, exacerbating the variations in salinity within marshes. Alterations in precipitation patterns, such as increased occurrence of droughts or excessive rainfall, can disturb the equilibrium of water, resulting in drought-induced strain or inundation across marshes. Moreover, climate change is amplifying the occurrence and strength of storms and storm surges, expediting erosion, disturbing plant communities, and imperilling the natural infrastructure of salt marshes. Consequently, the collective impact of these climate-related elements poses a greater risk to the long-term sustainability of salt marsh ecosystems, rendering the need for restoration and management initiatives more pressing.

2.2 Traditional Restoration Methods

The foundation of salt marsh restoration is the implementation of hydrological restoration methods, including dredging and channel modification. Implementing these techniques, which target the restoration of the inherent movement of water and sediment within the ecosystem, is crucial for preserving the fragile equilibrium of salt marsh habitats. Dredging is extracting accumulated sediments from waterways to restore the natural tidal flows, mitigate flooding risks, and improve the marsh's ability to aggregate sediment. This is especially vital in regions where the sedimentation process has been disturbed by upstream engineering, urban development, or other human activities. Channel modification refers to changing the water flow's form, dimensions, or direction in natural or artificial water channels to enhance water circulation and sediment transport. These interventions are crucial in restoring the hydrological conditions required for salt marsh plants' development and long-term viability. Nevertheless, hydrological restoration methods are frequently inadequate in completely rehabilitating a deteriorated salt marsh. Both vegetation planting and soil stabilisation are integral aspects of the restoration process. The restoration of hydrological flow is followed by planting native salt-tolerant vegetation, stabilising the soil, mitigating erosion, and facilitating the restoration of the ecosystem's biodiversity. Spartina species, native marsh plants found in temperate geographical areas, play a crucial role in sediment retention, wildlife habitat provision, and nutrient cycling. Furthermore, these plants have a crucial function in carbon sequestration by efficiently capturing and storing carbon dioxide from the atmosphere. Aside from direct planting, soil stabilisation efforts, such as applying biodegradable erosion control mats or sediment fencing, are supplementary measures and essential strategies that can safeguard young plants from being carried away by tides or storm surges. Implementing these methods gives us confidence that we are progressing towards restoring the inherent mechanisms that sustain the long-term well-being of the marsh. This includes the accumulation of organic matter that can facilitate future plant development and enhance the ability of the ecosystem to withstand adverse events (Devi et al. 2024a, 2024c, 2024d; Kumar and Choudhury 2024b; Kumar and Raut 2024; Saurabh et al. 2024; Sharma and Kumar 2024). Notwithstanding the significance of hydrological and vegetative restoration endeavours, conventional restoration methods encounter various constraints. An inherent obstacle is the element of cost. Implementing dredging, channel modification, and large-scale planting projects often incurs substantial costs, necessitating substantial financial commitments. The exorbitant expenses associated with restoration can restrict the feasibility of implementing these methods on a significant magnitude, especially in areas with restricted financial means or conflicting funding priorities. Moreover, the continuous expenses for maintenance are often essential to guarantee the sustained effectiveness of the restoration programs, particularly in regions prone to recurrent flooding, erosion, or pollution. Developing more cost-effective and sustainable restoration methods to address these constraints is crucial. One further notable constraint of conventional restoration techniques is their lack of scalability. Although hydrological restoration and vegetation planting can be successful locally, implementing these methods in larger, more extensive regions can provide difficulties. Implementing restoration measures in a limited marsh area can result in specific enhancements. However, suppose the larger landscape-scale constraints, such as watershed management or regional sediment supply, are not addressed. In that case, the advantages of these efforts may be limited in duration or limited to specific locations. The successful implementation of large-scale restoration projects necessitates the synchronised efforts of several jurisdictions, stakeholders, and ecosystems, which can present practical challenges. Furthermore, apart from their high cost and limited scalability, conventional restoration methods cannot often adjust to evolving environmental conditions, especially in response to climate change. Restoration interventions are usually planned according to the existing environmental conditions. However, the effectiveness of these methods may be undermined by rising sea levels, rising temperatures, and shifting precipitation patterns. Specifically, the selected plant species for revegetation may not withstand the elevated salinity levels or heightened flooding linked to future climatic conditions. Furthermore, the current hydrological adjustments may not adequately control the anticipated increase in the frequency and severity of storm surges. The lack of adaptability of conventional methods to these evolving circumstances can diminish the long-term efficacy of restoration programs. Ongoing modifications or interventions may be required as environmental conditions change (Devi et al. 2024b; Jagadesh et al. 2024; Kumar 2023, 2024a, 2024b, 2024c; Kumar et al. 2024c; Kumar and Choudhury

2024c; Raj et al. 2024). Furthermore, conventional restoration initiatives frequently fail to consider the possible contributions of geomicrobes and other natural processes that can strengthen the ability of salt marsh ecosystems to withstand challenges. By prioritising physical modifications, such as dredging or planting, these methods may overlook chances to utilise biological processes, such as microbial-mediated nutrient cycling, which can enhance the long-term water quality of the marsh. Incorporating microbial ecology into restoration strategies can enhance the adaptability and resilience of restored wetlands, enabling them to endure forthcoming environmental changes better. Although hydrological restoration mechanisms and vegetation planting are crucial elements of salt marsh restoration, they encounter substantial cost, scalability, and adaptability constraints. Given the ongoing shifts in environmental conditions caused by climate change and other variables, it is evident that there is a pressing requirement for improved and adaptable strategies. Integrating biological processes, such as using geomicrobes, can augment the resilience of ecosystems and facilitate sustainable remediation. This mandate for innovation should motivate us to investigate novel paths and adjust our approaches to changing environmental issues.

3 Geomicrobes: Agents of Ecological Restoration

3.1 Overview of Geomicrobes

The geomicrobes, comprising bacteria, archaea, and fungi, are essential for preserving the ecological equilibrium of salt marsh ecosystems. The various categories of microorganisms mentioned above make unique yet interrelated contributions to nutrient cycling and the overall functioning of the ecosystem. The microorganisms that exhibit the highest diversity and abundance in salt marshes are bacteria, which frequently serve as the main catalysts for nutrient-cycling processes. Although less common, archaea are highly specialised and fulfil crucial functions in methane generation and nitrogen cycling processes. Fungi, while often linked to land-based ecosystems, also play significant roles in salt marshes, namely in the breakdown of organic biomass and the enhancement of nutrient accessibility. Nutrient cycling is crucial in salt marshes as it vitally supports the productivity and health of the ecosystem. Bacteria have a vital function within the carbon cycle by decomposing organic matter, including plant material, into more basic compounds. The decomposition process results in carbon dioxide emission into the atmosphere and facilitates the recycling of nutrients into the soil. Certain bacteria, called methanotrophs, oxidise methane, a powerful greenhouse gas, decreasing methane release from salt marsh sediments. Methanogenic archaea generate methane in anaerobic environments, contributing to the carbon cycle within these environmental systems (Devi et al. 2023; Devi and Kumar 2023a; Dey et al. 2023b; Karam et al. 2023; Kumar and Choudhury 2024d; Kumar and Kumar 2023; Sharma and Kumar 2023; Srivastav et al. 2024). The process of nitrogen cycling, which is undertaken by geomicrobes in salt marshes, is of paramount significance. The availability of nitrogen, a vital element for the development of plants, is frequently restricted in systems. Nitrogen-fixing bacteria and archaea, including specific species of *Rhizobium* and *Frankia*, play a crucial role in transforming atmospheric nitrogen (N_2) into plant-utilisable formats, such as ammonia (NH_3). On the other hand, denitrifying bacteria, such as *Pseudomonas* and *Bacillus*, catalyse denitrification by transforming nitrates (NO_3^+) into nitrogen gas (N_2), which is subsequently emitted into the atmosphere. This process controls the nitrogen content in the soil and inhibits the buildup of surplus nitrates, averting the potential eutrophication of coastal waters. Within salt marsh ecosystems, sulfur cycling is a vital biochemical process in which sulfate-reducing bacteria have a pivotal function. In anaerobic environments, namely in saturated soils, bacteria like *Desulfovibrio* and *Desulfobacter* catalyse the reduction of sulfate (SO_4^{2+}) to hydrogen sulfide (H_2S). Hydrogen sulfide is crucial in the sulfur cycle despite its toxicity to many organisms. It can undergo additional metabolism by sulfur-oxidising bacteria, such as *Beggiatoa*, which enzymatically transform hydrogen sulfide back into sulfate. The process of sulfur cycling plays a crucial role in preserving the chemical equilibrium of salt marsh sediments and facilitating the development of halophytic plants. Geomicroorganisms also regulate the process of phosphorus cycling in salt marshes. In soils, phosphorus, an indispensable plant nutrient, is frequently restricted. Phosphate-solubilising organisms, defined as specific bacteria and fungi, can transform insoluble phosphorus compounds in the soil into forms that plants can absorb. To enhance the accessibility of phosphorus compounds to plants, these microorganisms secrete organic acids or enzymes that facilitate the breakdown of these compounds. This process is essential in nutrient-poor salt marsh soils, where phosphorus content may otherwise constrain plant growth. An outstanding characteristic of geomicrobes is their capacity to acclimate to the extremely salty conditions in salt marshes. Extreme salinity poses a significant obstacle for most living species, but geomicrobes have developed a range of mechanisms to deal with elevated salinity levels. Numerous bacteria, archaea, and fungi found in salt marshes are halophiles, possessing specific adaptations to flourish in environments with high salt concentrations. These organisms have evolved intricate systems to control the internal salt concentration and preserve osmotic equilibrium, a crucial element for survival in extremely harsh conditions. Halophilic bacteria and archaea frequently synthesise compatible solutes, such as glycine betaine and proline, to enhance the stability of their cellular proteins and enzymes in environments with elevated salt concentrations. Inhibiting water loss from the cell, these solutes shield cellular structures from the detrimental consequences of osmotic stress. Some halophilic archaea, including those of the genus *Halobacterium*, possess specialised membrane proteins that enable them to expel surplus salt from their cells, providing additional protection against the osmotic pressures of saline environments. Despite being less extensively researched in salt marshes than bacteria and archaea, fungi also demonstrate adaptations to saline environments. Halotolerant fungi, including species of *Aspergillus* and *Penicillium*, can thrive in both saline and non-saline conditions, ensuring their remarkable adaptability to the varying salinity levels found in salt marshes. These fungi facilitate the process of organic matter decomposition by enzymatically breaking down plant material and liberating nutrients that other organisms can utilise. Their capacity to flourish in such difficult circumstances is evidence of the remarkable adaptability of microbial communities in salt marsh ecosystems. Bacteria, archaea, and fungi play a crucial role in the nutrient cycles of salt marshes by enabling the metabolic breakdown and recycling of carbon, nitrogen, sulfur, and phosphorus. Their specialised adaptations to high salinity levels enable them to flourish in these environments and carry out vital functions that sustain salt marsh ecosystems' overall health and productivity. Enhanced comprehension of the functions and adaptations of geomicrobes allows for a deeper recognition of their significance in preserving the resilience of salt marshes and their potential in ecological restoration endeavours (Devi and Kumar 2023b; Kumar et al. 2024b; Kumar and Choudhury 2024e; Kumar and Dey 2024; Kumar and Singh 2024e; Kumari et al. 2023; Sharma et al. 2023a).

3.2 Functions of Geomicrobes in Salt Marshes

Microorganisms have a crucial function in the carbon cycling mechanism in salt marsh ecosystems, exerting influence on both the breakdown and storage of carbon. Within these ecological systems, deceased plant material and organic substances undergo decomposition by various bacteria and fungi. Microbial decomposition involves the conversion of intricate organic compounds into more basic forms, resulting in the release of carbon dioxide (CO_2) as a secondary chemical product. Microbial activity facilitates the carbon cycle by reintroducing carbon stored in organic biomass into the atmosphere or soil. Despite the emission of CO_2, salt marshes are renowned for their capacity to sequester carbon, retaining it in sediments for extended durations. The sequestration process is facilitated by microorganisms that actively participate in the creation and stabilisation of soil organic matter, so carbon within the sediments of the marsh is effectively captured. Furthermore, certain microorganisms, namely methanogens, participate in methanogenesis, a metabolic process where organic substances are decomposed without oxygen to generate methane (CH_4), another greenhouse gas. Methanotrophic bacteria metabolise methane, transforming it into CO_2, mitigating methane release from marsh sediments into the atmosphere. The interplay between methanogenesis and methanotrophy is crucial in controlling greenhouse gas emissions in salt marsh ecosystems. Microbiological communities play a crucial role in driving the nitrogen cycling process in salt marshes. Microorganisms, such as bacteria and archaea, engage in nitrification, denitrification, and nitrogen fixation, three essential stages of the nitrogen cycle metabolism. The process of nitrification involves the transformation of ammonia (NH_3) into nitrites (NO_3^+) and, subsequently, nitrates (NO_3^+), which is facilitated by ammonia-oxidising bacteria (AOB) and ammonia-oxidising archaea (AOA). These microorganisms are essential in facilitating nitrogen availability to plants in readily absorbable forms. Through denitrification, carried out by bacteria like *Pseudomonas*, nitrates are transformed into nitrogen gas (N_2) and subsequently emitted into the atmosphere, mitigating the risk of excessive nutrients in the marsh. Nitrogen fixation is the process by which diazotrophic bacteria and archaea transform atmospheric nitrogen (N_2) into ammonia, a chemically usable form of nitrogen for plants. The combined mechanisms of these processes effectively control nitrogen availability in salt marsh ecosystems, ensuring efficient cycling of nitrogen among the atmosphere, soil, and plants. Sulfur cycling is an essential process facilitated by salt marsh microorganisms, namely bacteria, that reduce sulfate and conduct sulfur oxidation (Devi et al. 2023a; Dey et al. 2023a, 2023c; Hidangmayum et al. 2023; Kumar et al. 2023c; Saini et al. 2023; Sharma et al. 2023; Sinam et al. 2023). Under anaerobic conditions, sulfate-reducing bacteria, including those belonging to the *Desulfovibrio* genus, transform sulfate (SO_4^{2+}) into hydrogen sulfide (H_2S). This hazardous gas has the potential to build up in river sediments. It is important to note that this process substantially affects sediment chemistry and plant health since hydrogen sulfide can impede the growth of plant roots and cause soil hypoxia. Nevertheless, sulfur-oxidising bacteria, such as *Beggiatoa* and *Thiobacillus*, convert hydrogen sulfide back into sulfate, decreasing its toxicity and preserving equilibrium in the sulfur cycle. These microorganisms function crucially in reducing the detrimental impacts of hydrogen sulfide accumulation, promoting plant well-being and soil integrity. Phosphorus cycling, while less conspicuous than carbon, nitrogen, or sulfur cycles, is equally essential for the productivity of salt marshes. Microbial organisms are crucial in facilitating the availability of phosphorus, a frequently scarce nutrient, to plants. Many microorganisms can solubilise phosphorus, transforming it from insoluble substances in the soil into soluble forms that plants can assimilate. Phosphate-solubilising bacteria (PSB) and fungi efflux organic acids or enzymes that degrade phosphorus compounds, thus increasing the accessibility of phosphorus in salt marsh soils with low nutrient content. Microbiological activity is vital in sustaining plant growth and productivity in ecosystems where phosphorus can be constrained. Microbial mechanisms enable the efficient release and cycling of phosphorus, assuring this vital nutrient's constant availability to sustain plant and microbial populations. In summary, the intricate relationships between microbial communities and biogeochemical cycles in salt marshes highlight these organisms' significance in maintaining the ecosystem's health. Microorganisms facilitate carbon, nitrogen, sulfur, and phosphorus cycling, which is crucial for salt marsh ecosystems' productivity, stability, and resilience. Their capacity to facilitate crucial changes in these nutrient cycles renders them essential for preserving salt marshes'ecological equilibrium and functionality, especially in response to environmental pressures and human influences (Devi et al. 2023b; Kumar et al. 2018a, 2018d, 2023b; Kumar and Dwivedi 2018; Naik and Kumar 2020; Pankaj et al. 2012a; Sharma et al. 2023b, 2023c).

3.3 Microbial Interactions with Plants and Sediments

Geomicroorganisms and marsh plants have complex symbiotic interactions, especially in the rhizosphere, the concentrated soil area affected by plant roots. This interaction, characterised by a fragile equilibrium of mutual exchange, is essential for the process of nutrient cycling and the well-being of plants in salt marsh ecosystems. The microorganisms in the rhizosphere, such as bacteria, fungi, and archaea, derive advantages from the organic substances released by plant roots, which supply them with vital carbon sources for their metabolic processes. Furthermore, these microorganisms aid plants in obtaining nitrogen, phosphorus, and other micronutrients often scarce in salt marsh habitats. Nitrogen-fixing bacteria, predominantly belonging to the genus *Rhizobium*, transform atmospheric nitrogen into readily available forms to plants, promoting plant growth in soils with low nutrient content. Symbiotic relationships between mycorrhizal fungus and plant roots extend the plant's root system and enhance phosphorus absorption. These relationships, which demonstrate the delicate equilibrium in nature, are crucial for maintaining the productivity and resilience of salt marsh vegetation, stabilising sediments, and preventing erosion. In addition to their function in nutrient cycling, geomicrobes substantially influence sediment structure and stability. The development of biofilms, which are strata of microorganisms that adhere to sediment particles, is facilitated by bacteria and fungi. The biofilms function as adhesive agents, tightly binding particles together and decreasing the probability of erosion. By synthesising extracellular polymeric substances (EPS), geomicrobes augment the cohesiveness of sediment, thus increasing its resistance to tidal forces and storm surges. Efficient stabilisation of sediments by microorganisms is crucial in salt marshes, where loose soils are susceptible to erosion. Furthermore, geomicrobes make a valuable contribution to the accumulation of organic matter in sediments by facilitating the trapping and retention of fine particles. This, in turn, enhances soil structure and promotes long-term ecological stability. Geomicrobes play a crucial role in ensuring the resilience of salt marsh ecosystems (Dwivedi and Kumar 2011; Kotia et al. 2021; Kumar et al. 2011a, 2011b, 2017, 2022a). The process of bioturbation, which involves the manipulation of soils by organisms such as burrowing animals and microbial activity, has a substantial impact on soil aeration and the availability of nutrients in salt marshes. Microorganisms can participate in bioturbation by locomotion within the soil, but larger organisms such as crabs, worms, and molluscs primarily occupy this role. To facilitate aerobic microbial processes like nitrification, these animals construct caves and tunnels that enable oxygen to permeate further into the sediment. Through the redistribution of organic matter and the fragmentation of compacted soils, bioturbation enhances the availability of nutrients by allowing the microbes to access previously buried or inaccessible nutrients. Promoting nutrient cycling and establishing favourable conditions for microbial and plant communities to flourish contributes to the salt marsh's general health and productivity. Bioturbation contributes to salt marsh ecosystems' dynamic and productive characteristics by influencing soil aeration and nutrient availability.

4 Applications of Geomicrobes in Salt Marsh Restoration

4.1 Bioremediation of Pollutants

Geomicroorganisms and marsh plants have complex symbiotic interactions, especially in the rhizosphere, the concentrated soil area affected by plant roots. This interaction, characterised by a fragile equilibrium of mutual exchange, is essential for the process of nutrient cycling and the well-being of plants in salt marsh ecosystems. The microorganisms in the rhizosphere, such as bacteria, fungi, and archaea, derive advantages from the organic substances released by plant roots, which supply them with vital carbon sources for their metabolic processes. Furthermore, these microorganisms aid plants in obtaining nitrogen, phosphorus, and other micronutrients often scarce in salt marsh habitats. Nitrogen-fixing bacteria, predominantly belonging to the genus *Rhizobium*, transform atmospheric nitrogen into readily available forms to plants, promoting plant growth in soils with low nutrient content. Symbiotic relationships between mycorrhizal fungus and plant roots extend the plant's root system and enhance phosphorus absorption. These relationships, which demonstrate the delicate equilibrium in nature, are crucial for maintaining the productivity and resilience of salt marsh vegetation, stabilising sediments, and preventing erosion. In addition to their function in nutrient cycling, geomicrobes substantially influence sediment structure and stability. The development of biofilms, which are strata of microorganisms that adhere to sediment particles, is facilitated by bacteria and fungi. The biofilms function as adhesive agents, tightly binding particles together and decreasing the probability of erosion. By synthesising extracellular polymeric substances (EPS), geomicrobes augment the cohesiveness of sediment, thus increasing its resistance to tidal forces and storm surges. Efficient stabilisation of sediments by microorganisms is crucial in salt marshes, where loose soils are susceptible to erosion. Furthermore, geomicrobes make a valuable contribution to the accumulation of organic matter in sediments by facilitating the trapping and retention of fine particles. This, in turn, enhances soil structure and promotes long-term ecological stability. Geomicrobes play a crucial role in ensuring the resilience of salt marsh ecosystems (Kumar et al. 2013, 2018c, 2019, 2022c; Kumar and Naik 2020; Kumar and Pathak 2019; Pankaj et al. 2012b; Pathak et al. 2017). The process of bioturbation, which involves the manipulation of soils by organisms such as burrowing animals and microbial activity, has a substantial impact on soil aeration and the availability of nutrients in salt marshes. Microorganisms can participate in bioturbation by locomotion within the soil, but larger organisms such as crabs, worms, and molluscs primarily occupy this role. To facilitate aerobic microbial processes like nitrification, these animals construct caves and tunnels that enable oxygen to permeate further into the sediment. Through the redistribution of organic matter and the fragmentation of compacted soils, bioturbation enhances the availability of nutrients by allowing the microbes to access previously buried or inaccessible nutrients. Promoting nutrient cycling and establishing favourable conditions for microbial and plant communities to flourish contributes to the salt marsh's general health and productivity. Bioturbation contributes to salt marsh ecosystems' dynamic and productive characteristics by influencing soil aeration and nutrient availability.

4.2 *Microbially Mediated Salt Tolerance in Plants*

Utilising salt-tolerant microorganisms to improve plant resistance to salinity stress shows excellent potential for restoring and maintaining salt marsh health. The salt-tolerant or halotolerant microbes, namely bacteria and fungi, are small yet potent organisms capable of mitigating the negative impacts of salinity. Their interaction with plant roots greatly enhances their capacity to tolerate elevated salt concentrations. The presence of these microorganisms is essential for plants to effectively regulate osmotic stress, sustain nutrient absorption, and sustain critical physiological functions even in highly salty environments. By introducing these advantageous microorganisms, the vegetation in salt marshes becomes more resilient to variations in salinity, enhancing the overall stability and productivity of the ecosystem. The processes by which microbes facilitate salt tolerance in plants encompass several crucial mechanisms, such as the synthesis of osmoprotectants and the control of ion transport. Osmoprotectants are small organic molecules, including proline, glycine betaine, and trehalose, that assist cells in preserving osmotic equilibrium in the presence of high salt concentrations. Microbial species that can tolerate high salt concentrations synthesise these chemicals and, in certain instances, stimulate plants to produce them. This phenomenon mitigates the osmotic stress experienced by plant cells, inhibiting water loss and facilitating tissue moisture retention. Furthermore, besides osmoprotectants, these microorganisms can control ion transport in plant cells (Aley et al. 2022; Chakraborty et al. 2021; Das et al. 2022; Goud et al. 2022; Kumar et al. 2018b, 2021, 2021b, 2022b; Kumar and Dwivedi 2020; Kumar and Mistri 2020; Upadhyay et al. 2023). Hyperaccumulation of sodium (Na^+) in plant tissues under high salinity conditions can have toxic effects. By activating or enhancing the expression of ion pumps and channels in plant roots, salt-tolerant microbes can regulate ion transport by sequestering sodium into vacuoles or expelling it from the cells. This facilitates the maintenance of a desirable equilibrium of ions, namely potassium (K^+), crucial for stimulating enzyme activity and promoting general plant well-being. Beneficial microbial inoculations encompass the utilisation of halotolerant bacteria and mycorrhizal fungus to enhance plant resilience in salt marshes and other saline ecologies. Studies have demonstrated that halotolerant bacteria, including *Halomonas* and *Bacillus* species, can improve plant growth by synthesising osmoprotectants and promoting more efficient nutrient absorption in saline environments. Mycorrhizal fungi, namely arbuscular mycorrhizal fungi (AMF), play a vital role in salt tolerance. Fungal species establish symbiotic relationships with plant roots, elongating their root systems and augmenting the uptake of water and nutrients. Under saline conditions, arbuscular mycorrhizal fungi (AMF) can enhance the effectiveness of ion control and assist plants in handling stress caused by salt. By introducing these advantageous microorganisms into plants, restoration endeavours can improve the ability of salt marsh vegetation to withstand and thrive in the face of salinity stresses (Table 1).

Table 1 Microbial species and its mechanism of salt tolerance

No.	Microbial species	Plant species	Mechanism of salt tolerance	Key outcome/benefit
1	*Azospirillum brasilense*	Wheat (*Triticum aestivum*)	Enhances root development, improves nutrient uptake	Increased biomass and yield under saline conditions
2	*Pseudomonas putida*	Barley (*Hordeum vulgare*)	Induces antioxidant enzyme activity	Improved plant growth and stress tolerance
3	*Bacillus subtilis*	Tomato (*Solanum lycopersicum*)	Produces exopolysaccharides that sequester salt ions	Increased plant survival and reduced Na^+ toxicity
4	*Rhizobium meliloti*	Alfalfa (*Medicago sativa*)	Fixes nitrogen, reduces oxidative stress	Enhanced root nodulation and nitrogen fixation, leading to better growth under salt stress
5	*Pseudomonas fluorescens*	Rice (*Oryza sativa*)	Produces phytohormones (auxins)	Improved root architecture and salt tolerance
6	*Arthrobacter nitroguajacolicus*	Cotton (*Gossypium hirsutum*)	Enhances osmoprotectant synthesis (proline)	Better growth and stress resilience in saline soils
7	*Klebsiella variicola*	Maize (*Zea mays*)	Increases potassium uptake, limits sodium absorption	Improved ion homeostasis, reduced salt stress
8	*Bacillus megaterium*	Soybean (*Glycine max*)	Phosphorus solubilisation, enhances nutrient availability	Increased pod formation and overall plant health under saline stress
9	*Glomus intraradices*	Wheat (*Triticum aestivum*)	Mycorrhizal association enhances water uptake	Improved drought and salinity tolerance
10	*Bacillus pumilus*	Sunflower (*Helianthus annuus*)	Produces ACC deaminase, reduces ethylene levels	Improved root growth and plant survival in saline conditions
11	*Azospirillum lipoferum*	Corn (*Zea mays*)	Stimulates nitrogen fixation, enhances stress enzyme activities	Increased plant biomass and yield under salinity
12	*Halomonas elongata*	Rice (*Oryza sativa*)	Produces osmolytes such as ectoine	Improved salt tolerance through osmotic adjustment
13	*Trichoderma harzianum*	Tomato (*Solanum lycopersicum*)	Enhances nutrient uptake and root growth	Reduced salt stress, increased yield

(continued)

Table 1 (continued)

No.	Microbial species	Plant species	Mechanism of salt tolerance	Key outcome/benefit
14	*Enterobacter cloacae*	Cotton (*Gossypium hirsutum*)	Promotes antioxidant defence mechanisms	Improved salt tolerance, enhanced plant vigour
15	*Piriformospora indica*	Rice (*Oryza sativa*)	Facilitates potassium uptake, reduces Na^{+} absorption	Increased plant growth under high salinity
16	*Mesorhizobium ciceri*	Chickpea (*Cicer arietinum*)	Enhances nitrogen fixation and stress enzyme activity	Improved growth and higher yield under salt stress
17	*Bacillus amyloliquefaciens*	Soybean (*Glycine max*)	Secretes auxins, enhances root elongation	Increased salt tolerance and better nutrient acquisition
18	*Pseudomonas stutzeri*	Wheat (*Triticum aestivum*)	Improves water use efficiency, reduces oxidative damage	Enhanced plant height, biomass, and tolerance under saline conditions
19	*Rhizobium leguminosarum*	Pea (*Pisum sativum*)	Nitrogen fixation and oxidative stress reduction	Increased nitrogen availability, better stress response
20	*Frankia* spp.	Casuarina (*Casuarina equisetifolia*)	Nitrogen-fixing bacteria aiding in salt stress resilience	Improved nitrogen availability, enhanced salt tolerance in roots
21	*Mycorrhizae (AMF)*	Pepper (*Capsicum annuum*)	Improves nutrient uptake and water absorption	Reduced salt accumulation in plant tissues, better growth
22	*Azotobacter chroococcum*	Rice (*Oryza sativa*)	Promotes nitrogen fixation and stress defence mechanisms	Enhanced growth and biomass under saline conditions
23	*Paenibacillus polymyxa*	Wheat (*Triticum aestivum*)	Produces antibiotics, enhances stress tolerance	Improved root growth and nutrient uptake under salt stress
24	*Bacillus licheniformis*	Barley (*Hordeum vulgare*)	Produces enzymes that detoxify salt stress-induced ROS	Reduced oxidative damage, improved growth under saline conditions
25	*Sinorhizobium meliloti*	Alfalfa (*Medicago sativa*)	Nitrogen fixation and production of osmoprotectants	Enhanced salt stress tolerance, increased biomass
26	*Pseudomonas aeruginosa*	Tomato (*Solanum lycopersicum*)	Produces siderophores and phytohormones	Increased nutrient uptake and stress tolerance
27	*Bacillus cereus*	Rice (*Oryza sativa*)	Enhances proline accumulation and antioxidant activity	Improved salt tolerance and yield
28	*Halomonas maura*	Tomato (*Solanum lycopersicum*)	Production of extracellular polysaccharides	Enhanced soil aggregation, reduced salt uptake

(continued)

Table 1 (continued)

No.	Microbial species	Plant species	Mechanism of salt tolerance	Key outcome/benefit
29	*Bacillus velezensis*	Cucumber (*Cucumis sativus*)	Produces phytohormones and improves stress response	Increased growth and photosynthesis under saline conditions
30	*Phyllobacterium brassicacearum*	Arabidopsis (*Arabidopsis thaliana*)	Induces systemic tolerance mechanisms	Enhanced root growth, reduced Na^+ uptake
31	*Alcaligenes faecalis*	Rice (*Oryza sativa*)	Enhances iron uptake and root development	Better nutrient acquisition and improved tolerance to salinity
32	*Pseudomonas koreensis*	Wheat (*Triticum aestivum*)	Enhances production of stress-related enzymes	Reduced oxidative damage and better plant growth under salt stress
33	*Bacillus thuringiensis*	Maize (*Zea mays*)	Enhances antioxidant enzymes and reduces oxidative stress	Increased biomass and better stress tolerance
34	*Serratia marcescens*	Sugarcane (*Saccharum officinarum*)	Produces plant hormones, enhances root growth	Better tolerance to salt, improved sugar yield
35	*Phanerochaete chrysosporium*	Wheat (*Triticum aestivum*)	Fungal-mediated detoxification of salt stress-related ROS	Improved growth and photosynthetic efficiency under saline conditions
36	*Paenibacillus mucilaginosus*	Maize (*Zea mays*)	Potassium solubilisation, enhances nutrient availability	Increased salt tolerance and nutrient uptake
37	*Arbuscular mycorrhizae (AMF)*	Soybean (*Glycine max*)	Enhances phosphorus uptake and root development	Better drought and salinity tolerance
38	*Azospirillum halopraeferens*	Rice (*Oryza sativa*)	Nitrogen fixation and osmoregulatory mechanisms	Increased biomass and root length under salt stress
39	*Paenibacillus sabinae*	Tomato (*Solanum lycopersicum*)	Produces exopolysaccharides and promotes osmotic adjustment	Improved tolerance to salinity, enhanced root biomass
40	*Endophytic fungi* (*Fusarium* sp.)	Tomato (*Solanum lycopersicum*)	Induces systemic stress tolerance pathways	Enhanced plant growth and reduced salt stress
41	*Streptomyces griseus*	Barley (*Hordeum vulgare*)	Produces antifungal compounds, enhances stress response	Improved salinity tolerance and pathogen resistance

4.3 Enhancing Biogeochemical Cycles

Microbiological interventions are essential for facilitating nitrogen and carbon sequestration in salt marsh ecosystems, improving nutrient cycling and the general resilience of these ecosystems. Microbial activity is crucial to cycling nitrogen and carbon, vital for plant development and ecological productivity. Microbial activities such as nitrogen fixation and carbon decomposition are critical in providing essential nutrients to plants and facilitating carbon sequestration in the soil, contributing significantly to climate change mitigation. Certain bacteria, known as diazotrophs, convert atmospheric nitrogen (N_2) into ammonia (NH_3), enabling plants to access nitrogen. This microbial nitrogen fixation decreases the requirement for external nitrogen inputs, enhancing nutrient cycle sustainability in marshes. Furthermore, nitrogen-fixing bacteria are crucial in preserving soil fertility, especially in settings where nitrogen is a significant constraint on plant development (Kumar et al. 2018, 2020, 2021, 2021a, 2022b; Kumar and Dwivedi 2020; Kumari et al. 2022; Reddy et al. 2022). Microbial activity facilitates the process of organic matter breakdown by decomposition, so converting it into stable carbon forms that can be stored in the soil contributes to carbon sequestration. Sequestration of carbon as organic matter occurs in salt marshes, where it gradually accumulates in the sediments. Specific microorganisms, particularly those engaged in anaerobic decomposition, inhibit the complete mineralisation of carbon into CO_2, thus facilitating its long-term storage. By augmenting microbial communities that specialise in carbon sequestration, restoration initiatives can enhance the ability of salt marshes to store carbon, therefore contributing to the mitigation of global carbon emissions and the stabilisation of the climate. Sulfate-reducing bacteria (SRBs) are crucial in maintaining soil health and stabilising marshland ecosystems. In the sulfur cycle, these bacteria, which flourish in the anaerobic environments commonly found in marsh sediments, have a crucial function by converting sulfate (SO_4^{2+}) into hydrogen sulfide (H_2S). Although high concentrations of H_2S can be poisonous, the beneficial effects of SRBs on soil health are substantial. Subsurface rhizobacteria (SRBs) promote the production of iron sulfides, which aid in aggregating soil particles and enhancing sediment structure, increasing soil stability and erosion resistance. Moreover, SRBs enhance nutrient cycling by the mechanism of organic matter breakdown, facilitating the availability of sulfur and other essential nutrients to plants. Potential restoration methods may involve modifying hydrological conditions to create anaerobic environments that support the activity of sulfate-reducing bacteria, thus improving the stability of the marsh. Moreover, incorporating organic substrates such as plant litter or compost can supply the essential carbon sources for the robust growth of SRBs. Through the promotion of SRBs, restoration projects have the potential to enhance soil structure, augment nutrient availability, and bolster the overall resilience of salt marsh ecosystems against erosion and environmental stressors. Microbiological-based interventions are essential for preserving salt marshes'health and stability in response to climate change and other human-induced stresses (Devi et al. 2023b; Dey et al. 2023a, 2023c; Kumar et al. 2018a, 2018d, 2023b; Kumar and Dwivedi 2018; Sharma et al. 2023b, 2023c; Sinam et al. 2023).

4.4 Bioaugmentation and Biofertilisation

Introducing beneficial microbes into salt marsh ecosystems to accelerate restoration processes has become an increasingly recognised approach in ecological restoration. Microbial communities play crucial roles in maintaining the health and resilience of salt marshes through their involvement in nutrient cycling, sediment stabilisation, and plant-microbe interactions. By harnessing the power of these naturally occurring microorganisms, restoration efforts can be significantly enhanced, particularly in environments where human activities or natural disturbances have degraded. Bioaugmentation—the deliberate introduction of beneficial microbes to accelerate ecological recovery—has shown promise in improving plant resilience to stressors such as salinity, promoting nutrient availability, and stabilising sediment structures, all of which are critical for restoring the functionality of salt marsh ecosystems. Introducing beneficial microbes involves the injection of salt marsh soils or plants with specific microbial strains, including bacteria, archaea, and fungi, known to enhance ecological processes. These microbes can help plants better tolerate stressful environmental conditions, such as high salinity and nutrient-poor soils, by aiding in nutrient acquisition, producing osmoprotectants, or regulating ion transport in plant tissues. For example, halotolerant bacteria and mycorrhizal fungi are often introduced into marshes to improve plant growth and survival under saline conditions. These microbes form symbiotic relationships with plant roots, enhancing nutrient uptake and water retention while protecting plants from salt-induced stress. By leveraging these microbial processes, restoration practitioners can accelerate vegetation recovery, improve soil health, and promote the long-term sustainability of restored salt marsh ecosystems. Several case studies demonstrate the success of bioaugmentation in salt marsh restoration projects. For instance, in one study, introducing halotolerant bacteria such as *Halomonas* and *Bacillus* species into degraded salt marshes improved the growth and survival of marsh vegetation, even in areas with high salinity levels. These bacteria enhance plant resilience by producing osmoprotectants, such as proline and glycine betaine, which help plants maintain osmotic balance under stressful conditions. In another example, mycorrhizal fungi were inoculated into salt marsh soils to promote the establishment of marsh grasses like *Spartina alterniflora*, which play a critical role in sediment stabilisation. The fungi formed symbiotic associations with plant roots, increasing phosphorus uptake and improving overall plant health, leading to faster vegetation recovery and improved sediment stability. Despite the promising results of microbial inoculation techniques, several prospects and challenges are still associated with their widespread implementation in salt marsh restoration. One of the critical prospects is the potential for microbial inoculation to enhance the resilience of salt marshes to climate change and other environmental stressors (Devi et al. 2024c, 2024d; Kumar 2023, 2024a, 2024b; Raj et al. 2024; Saurabh et al. 2024). By introducing microbes that help plants tolerate salinity, drought, and nutrient limitations, restoration practitioners can increase the likelihood of success in restoring degraded marshes and ensure the long-term survival of these critical ecosystems. Additionally, microbial inoculation offers a cost-effective and scalable approach to restoration, as many beneficial microbes can be cultivated and applied in large quantities at relatively low cost. However, challenges remain in optimising microbial inoculation techniques. One of the main challenges is the difficulty in predicting how introduced microbes will interact with the existing microbial communities and environmental conditions in the salt marsh. In some cases, introduced microbes may not establish themselves effectively or may be outcompeted by native microorganisms, limiting their impact on restoration. Additionally, there is a risk that introduced microbes could have unintended consequences, such as altering nutrient cycling in ways that could negatively affect the ecosystem or outcompeting native species that play critical ecological roles. Another challenge is the variability in microbial inoculation success across different environmental conditions and salt marsh ecosystems. Factors such as salinity, hydrology, and soil composition can influence the effectiveness of microbial injections, making it difficult to generalise results from one restoration site to another. Further research is needed to understand better the interactions between introduced microbes and the specific environmental conditions of each restoration site and to identify the most effective microbial strains for different restoration goals. Introducing beneficial microbes holds excellent potential for accelerating salt marsh restoration processes, but further research and careful management are necessary to ensure its success. By addressing the challenges associated with microbial inoculation, restoration practitioners can harness the power of microbial communities to promote the recovery and long-term sustainability of salt marsh ecosystems (Table 2).

5 Geomicrobe-Driven Restoration Projects

5.1 North American Salt Marsh Restoration Projects

Using microbial applications in coastal restoration projects is a practical approach and a cutting-edge field within ecological engineering. This novel methodology capitalises on the capabilities of advantageous microorganisms to augment the restoration and adaptability of deteriorated coastal ecosystems. In natural settings such as salt marshes, mangroves, and seagrass meadows, these applications enhance soil health, stimulate plant growth, and facilitate the restoration of vital ecological processes. Through microbial processes, restoration strategies can effectively tackle significant issues related to coastal degradation, such as nutrient imbalances, erosion, and habitat loss. Microbial organisms have multiple essential functions in the process of coastal restoration. A critical application is in the field of soil and sediment health. Microbial microorganisms, such as bacteria and fungi, play a crucial role in nutrient cycling by converting organic substances and minerals into forms that plants can effectively use. For instance, nitrogen-fixing bacteria can transform atmospheric nitrogen into ammonia, a compound that plants can assimilate and

Table 2 Microbial species and their mechanism of action

No.	Microbial species	Target plant/ ecosystem	Mechanism of action	Key outcome/benefit
1	*Azotobacter vinelandii*	Maize (*Zea mays*)	Nitrogen fixation	Enhanced nitrogen availability, increased yield and biomass
2	*Bacillus subtilis*	Tomato (*Solanum lycopersicum*)	Produces phytohormones and suppresses pathogens	Improved plant growth and reduced disease incidence
3	*Pseudomonas fluorescens*	Wheat (*Triticum aestivum*)	Phosphate solubilisation and siderophore production	Increased nutrient uptake and disease resistance
4	*Rhizobium leguminosarum*	Pea (*Pisum sativum*)	Symbiotic nitrogen fixation	Higher nodulation, better nitrogen availability, and enhanced growth
5	*Trichoderma harzianum*	Cucumber (*Cucumis sativus*)	Enhances root growth and nutrient uptake	Increased plant vigour and improved tolerance to abiotic stress
6	*Bacillus amyloliquefaciens*	Lettuce (*Lactuca sativa*)	Produces antibiotics, promotes root elongation	Improved growth, reduced pathogen attack
7	*Glomus mosseae*	Soybean (*Glycine max*)	Mycorrhizal association for better nutrient absorption	Increased phosphorus uptake and drought tolerance
8	*Bradyrhizobium japonicum*	Soybean (*Glycine max*)	Symbiotic nitrogen fixation	Enhanced nitrogen availability and better crop yield
9	*Paenibacillus polymyxa*	Wheat (*Triticum aestivum*)	Potassium solubilisation and hormone production	Improved potassium uptake, enhanced plant health
10	*Rhizobium tropici*	Common bean (*Phaseolus vulgaris*)	Nitrogen fixation in root nodules	Improved nitrogen uptake, better pod development
11	*Pseudomonas putida*	Rice (*Oryza sativa*)	ACC deaminase activity, reduces ethylene levels	Better root growth, improved tolerance to salinity
12	*Bacillus megaterium*	Barley (*Hordeum vulgare*)	Phosphate solubilisation	Enhanced phosphorus availability, improved yield
13	*Arbuscular mycorrhizae (AMF)*	Tomato (*Solanum lycopersicum*)	Enhances root colonisation and nutrient acquisition	Increased plant growth and stress resilience

(continued)

Table 2 (continued)

No.	Microbial species	Target plant/ ecosystem	Mechanism of action	Key outcome/benefit
14	*Azospirillum brasilense*	Maize (*Zea mays*)	Nitrogen fixation and phytohormone production	Higher root growth, better nitrogen uptake, and yield increase
15	*Frankia* spp.	Alder (*Alnus* spp.)	Nitrogen-fixing symbiosis in roots	Enhanced nitrogen availability and improved soil health
16	*Streptomyces* sp.	Tomato (*Solanum lycopersicum*)	Produces antifungal compounds, enhances plant immunity	Reduced fungal infection and improved growth
17	*Bacillus pumilus*	Rice (*Oryza sativa*)	Phosphate solubilisation and exopolysaccharide production	Improved nutrient availability and plant vigour under stress
18	*Pseudomonas aeruginosa*	Wheat (*Triticum aestivum*)	Siderophore production and phytohormone secretion	Improved nutrient uptake and root development
19	*Rhizobium etli*	Bean (*Phaseolus vulgaris*)	Nitrogen fixation and phosphate solubilisation	Improved nitrogen uptake and overall plant health
20	*Azotobacter chroococcum*	Wheat (*Triticum aestivum*)	Nitrogen fixation and promotes plant hormone production	Increased biomass and better nitrogen availability
21	*Paenibacillus mucilaginosus*	Sugarcane (*Saccharum officinarum*)	Potassium solubilisation and organic acid production	Improved potassium availability, better root growth
22	*Bacillus cereus*	Pepper (*Capsicum annuum*)	Produces antibiotics and stimulates root growth	Enhanced disease resistance and improved growth
23	*Pseudomonas koreensis*	Tomato (*Solanum lycopersicum*)	Produces ACC deaminase and siderophores	Reduced ethylene levels, improved nutrient uptake
24	*Bacillus thuringiensis*	Maize (*Zea mays*)	Produces bioinsecticides, enhances root growth	Improved pest resistance and higher yield
25	*Sinorhizobium meliloti*	Alfalfa (*Medicago sativa*)	Nitrogen fixation and growth promotion	Higher nitrogen availability and increased forage yield
26	*Phosphobacteria* sp.	Rice (*Oryza sativa*)	Solubilises phosphorus and enhances root development	Better phosphorus uptake and improved plant vigour
27	*Rhizophagus irregularis*	Corn (*Zea mays*)	Mycorrhizal association enhances nutrient absorption	Improved drought and stress tolerance, increased biomass

(continued)

Table 2 (continued)

No.	Microbial species	Target plant/ ecosystem	Mechanism of action	Key outcome/benefit
28	*Enterobacter cloacae*	Wheat (*Triticum aestivum*)	Produces indole-3-acetic acid (IAA) and siderophores	Enhanced root elongation and nutrient uptake
29	*Azospirillum lipoferum*	Rice (*Oryza sativa*)	Nitrogen fixation and phytohormone production	Increased biomass and stress tolerance
30	*Serratia marcescens*	Tomato (*Solanum lycopersicum*)	Produces plant hormones, improves root growth	Enhanced growth and salt tolerance
31	*Halomonas elongata*	Barley (*Hordeum vulgare*)	Produces osmoprotectants like ectoine	Improved salinity tolerance, better growth
32	*Rhizobium phaseoli*	Common bean (*Phaseolus vulgaris*)	Nitrogen fixation in legume nodulesz	Increased nitrogen uptake, enhanced pod and grain production
33	*Bacillus licheniformis*	Lettuce (*Lactuca sativa*)	Phosphate solubilisation, growth promotion	Improved growth, nutrient acquisition
34	*Trichoderma viride*	Soybean (*Glycine max*)	Produces enzymes that degrade pathogenic fungi	Reduced fungal infection and increased growth
35	*Mesorhizobium ciceri*	Chickpea (*Cicer arietinum*)	Symbiotic nitrogen fixation	Enhanced nitrogen uptake, increased yield
36	*Phanerochaete chrysosporium*	Tomato (*Solanum lycopersicum*)	Fungal-mediated organic matter decomposition	Enhanced nutrient availability and plant vigour
37	*Bacillus velezensis*	Rice (*Oryza sativa*)	Produces antibiotics, suppresses pathogens	Increased disease resistance, better yield
38	*Azospirillum amazonense*	Corn (*Zea mays*)	Nitrogen fixation, promotes plant hormones	Improved nitrogen uptake and higher biomass
39	*Rhizobium leguminosarum bv. trifolii*	Clover (*Trifolium* spp.)	Nitrogen fixation, improves soil fertility	Increased nitrogen availability, enhanced growth and soil health
40	*Streptomyces griseus*	Wheat (*Triticum aestivum*)	Produces antifungal and antibacterial compounds	Reduced pathogen load, improved root health
41	*Bacillus firmus*	Tomato (*Solanum lycopersicum*)	Produces lipopeptides, enhances stress tolerance	Increased root biomass and improved stress resilience

utilise for their growth. Mycorrhizal fungus and phosphate-solubilising bacteria enhance the availability of phosphorus in the soil. Microbiological injections can improve soil fertility and facilitate vegetation growth in salt marshes characterised by nutrient-deficient soils or salinity-induced stress. Another noteworthy application is in the field of sediment stabilisation. Shoreline regions are susceptible to erosion, especially when there is a loss of vegetation cover. Microbial activity enhances sediment stabilisation by forming biofilms, effectively binding soil particles together and mitigating erosion. Beneficial microorganisms in salt marshes and mangrove forests contribute to creating a more stable sediment matrix by improving the cohesiveness of soil sediment particles. This is especially crucial in regions affected by human activities or natural disruptions, where microbial interventions can assist in reconstructing soil structure and averting additional erosion. Microbial organisms, often overlooked but essential contributors to coastal restoration, also have a vital function in reducing the impact of pollution and reinstating ecological equilibrium. For example, specific bacteria can break down contaminants like oil and heavy metals, enhancing soil and water quality. This bioremediation technique is especially beneficial in coastal regions impacted by industrial spills or agricultural runoff. By introducing microorganisms possessing specialised abilities to degrade pollutants, restoration practitioners can expedite the process of cleaning and restoring the health of coastal environments, providing a promising prospect for the future. Microbial populations enhance soil health and pollution control and bolster plant resilience, especially under adverse circumstances like high salinity or drought. Within saline habitats such as salt marshes and mangroves, halotolerant microorganisms can assist plants in managing osmotic stress by synthesising osmoprotectants or controlling ion transport (Choudhury et al. 2024; Devi et al. 2024e, 2024f; Kumar et al. 2024a, 2024f, 2024g; Kumar and Choudhury 2024a; Kumar and Singh 2024c). For instance, halotolerant bacteria and mycorrhizal fungus can augment plant growth by enhancing nutrient absorption and inducing stress tolerance. This is essential for establishing and preserving vegetation in coastal regions where salinity can significantly restrict plant survival. Multiple highly successful case studies demonstrate the efficacy of microbial applications in coastal restoration. The introduction of mycorrhizal fungus into salt marsh soils in the Gulf of Mexico has proven to enhance the growth and survival of marsh grasses such as *Spartina alterniflora*. These fungi augment the absorption of nutrients and promote stress tolerance, improving plant establishment and sediment stabilising. Furthermore, the use of halotolerant bacteria in mangrove reforestation initiatives in Florida has effectively enhanced the ability of plants to withstand saline conditions, resulting in the successful rehabilitation of mangrove vegetation. The successes notwithstanding, obstacles are linked to microbial applications in coastal restoration. A primary obstacle is carefully selecting suitable microbial strains for particular environmental conditions. Not all microorganisms will flourish in every environment, so comprehending the indigenous microbial population and environmental conditions is crucial. The acquisition of this knowledge is critical for the successful implementation. Furthermore, continuous surveillance and flexible control are essential to assess the efficacy of microbial treatments and implement required modifications. Microbiological applications in coastal restoration provide novel solutions that have great potential to improve the recovery and resilience of ecosystems. Microbes make a crucial contribution to the restoration of degraded coastal environments by enhancing soil health, stabilising sediments, reducing pollution, and increasing plant resilience. Sustained investigation and development in this domain provide optimism and the capacity to significantly improve restoration methods and attain more sustainable and efficient results for coastal ecosystems.

5.2 European Salt Marshes

Microbiological communities at the local level are crucial for restoring ecosystems, particularly in disrupted or deteriorated settings like coastal wetlands, salt marshes, and mangroves. These communities, comprising a variety of bacteria, archaea, and fungi, play a crucial role in the operation of these ecosystems by participating in nutrient cycling, soil stabilisation, and plant-microbe interactions. Their existence and exertion of influence can greatly impact the course of recovery endeavours, determining the pace and effectiveness of ecological restoration interventions. Nutrient cycling is a key mechanism by which local microbial communities contribute to the restoration of ecosystems. The conversion of inorganic and organic compounds into usable forms for plants and other organisms is a critical function of microbial-mediated processes in coastal ecosystems, such as salt marshes. For example, nitrogen-fixing bacteria convert atmospheric nitrogen into ammonia, which plants can assimilate and utilise for their metabolism. Without this microbial activity, vegetation would have limited access to nitrogen, a vital nutritional element for plant growth. In coastal soils, phosphorus is frequently bound in forms inaccessible to plants. However, phosphate-solubilising bacteria disrupt these complexes, thus increasing the nutrient availability. Through these mechanisms, indigenous microbiological populations guarantee the availability of vital nutrients, facilitating the restoration of plant life, a crucial aspect of ecological restoration. Microbiological communities play a crucial role in facilitating carbon cycling. In coastal ecosystems, microbes facilitate the decomposition of organic matter by breaking down deceased plants and animals into more basic biochemical compounds. This process facilitates the recycling of nutrients and effectively sequesters carbon within the soil, contributing to climate change mitigation by capturing carbon that would otherwise be emitted into the atmosphere as carbon dioxide (CO_2). Native microbiological populations, especially anaerobic bacteria, play a vital role in this process. In environments characterised by low oxygen levels in the sediments, such as salt marshes, anaerobic bacteria decompose organic material in the absence of oxygen, so facilitating the long-term storage of carbon. Effective carbon sequestration is crucial for maintaining ecological health and regulating the global carbon equilibrium. Bioturbation and sediment binding are mechanisms by which local microbial communities impact ecosystems' physical structure and stability. Microbial organisms, especially those that create biofilms, can stabilise sediments by aggregating soil particles, thus mitigating erosion in areas susceptible to tidal forces or storms. In coastal restoration projects, sediment stabilisation is paramount, as erosion control is essential for establishing and maintaining vegetation. Through the development of cohesive soil structures, these microbial communities establish a stable foundation for plant roots to rest on, halting any additional ecosystem deterioration. Furthermore, apart from their direct involvement in nutrient cycling and sediment stabilisation, local microbial communities establish symbiotic associations with plants essential for ecosystem restoration. For instance, mycorrhizal fungi, which form symbiotic relationships with plant roots, elongate the root systems of plants, so augmenting their capacity to uptake water and nutrients from the soil. This relationship is significant in degraded ecological settings with limited soil nutrients or severe conditions. Plant species that form symbiotic relationships with this fungus often possess superior abilities to establish themselves, increase, and endure in rehabilitated ecosystems. The symbiotic relationship between plants and microbes is an essential element of ecological persistence since it facilitates the rapid recovery of vegetation and its ability to endure environmental pressures like salinity or drought. Furthermore, the indigenous microbial populations have evolved to thrive in the unique conditions of their surroundings, rendering them especially well-suited for interventions aimed at restoration. The microorganisms have adapted to thrive in the distinctive maritime environments characterised by varying salinity, water levels, and temperature. The ability to adapt to local stressors confers an advantage upon them compared to introduced species, which may encounter difficulties in survival or carry out their ecological functions with great efficiency. Incorporating indigenous microbial communities into restoration efforts allows practitioners to guarantee that the rehabilitated ecosystem is sustained by organisms already highly adapted to the environmental conditions. Community microbial populations at the local level are essential for restoring ecosystems and maintaining long-term health. By their functions in nutrient cycling, carbon sequestration, sediment stabilisation, and plant symbiosis, these communities establish the conditions essential for successful restoration. Their distinctive adaptations to specific local conditions render them extremely valuable for restoration endeavours, guaranteeing a more rapid and sustainable recovery of ecosystems. By comprehending and harnessing these microbial communities' potential, restoration professionals can enhance the ability of deteriorated ecosystems to repair and function.

5.3 *Asian Salt Marsh Restoration*

Geomicrobes, an assemblage of bacteria, archaea, and fungi, exhibit extraordinary adaptations that enable them to flourish in extremely salty and tropical habitats. The severe conditions present considerable obstacles for microbiological life, including elevated osmotic pressure, extreme temperatures, limited nutrient availability, and changing environmental conditions. Nevertheless, geomicrobes have developed physiological, metabolic, and genetic adaptations that enable them to survive, operate, and even fulfil crucial functions in these ecosystems, including salt marshes, mangroves, and tropical coastal systems. Microbial organisms in extremely salty environments experience osmotic stress, a phenomenon in which the salt concentration in the surrounding environment can induce cellular water loss, resulting in desiccation and cellular injury. In response, halophilic (salt-loving) and halotolerant (salt-tolerant) microorganisms have evolved many mechanisms to preserve their biochemical integrity and metabolic processes. An indispensable adaptation is the synthesis of osmoprotectants, which are alternatively referred to as compatible solutes. Microbiological cells utilise small organic molecules, including glycine betaine, proline, and trehalose, to maintain osmotic equilibrium without disrupting cellular processes. To preserve cell structure and enzymatic activity, geomicrobes can effectively control water loss and sustain turgor pressure by augmenting the internal concentration of these solutes. An alternative approach involves ion transporters to control the intracellular levels of sodium (Na^+) and potassium (K^+) ions. Several halophilic microorganisms actively excrete sodium from their cells while importing potassium, which is less toxic at elevated levels. Specialised proteins and enzymes in these microorganisms are designed to operate in environments with high salt concentrations, enabling metabolic processes to persist even in extremely high salinity. These proteins frequently possess more acidic residues on their surfaces, contributing to their stability and functionality in elevated salt concentrations. Furthermore, specific microorganisms, namely halophilic archaea, have modified their cell membranes to enhance their resistance to the impacts of salt. Frequently, these microorganisms possess membrane lipids that are extensively saturated, so inhibiting excessive fluidity or permeability of the membranes under saline environments. In certain instances, archaea synthesise distinctive lipid structures called ether lipids, which exhibit greater stability in harsh conditions than the ester-linked lipids in bacteria and eukaryotes. Tropical ecosystems, such as mangrove forests, coral reefs, and tropical coastal wetlands, provide geomicrobes with extra obstacles, such as elevated temperatures, varying moisture levels, and strong solar ultraviolet radiation. In these regions, geomicrobes have evolved and made several adaptations to effectively deal with the local conditions. In tropical environments, thermotolerance is a significant adaptation that enables organisms to survive and operate effectively at elevated temperatures. Numerous tropical microorganisms synthesise heat-shock proteins (HSPs), which inhibit protein denaturation and facilitate the refolding of impaired proteins in response to stress. Furthermore, tropical geomicrobes frequently possess thermostable enzymes that maintain functionality even at high temperatures. This enables them to perform vital metabolic activities, such as the cycling of nutrients, even in the high temperatures prevailing in tropical climates. Tropical coastal regions are also susceptible to varying water availability, especially in mangrove ecosystems where tidal activity leads to episodic inundation and aridity. Geomicrobes in these habitats frequently acclimate to aerobic and anaerobic conditions, modifying their metabolic pathways in response to changes in oxygen availability. For instance, several geomicrobes can engage in aerobic respiration in the presence of oxygen and transition to anaerobic processes such as fermentation or sulfate reduction when oxygen levels decrease, even during high tide when sediments become saturated. A further adaptation noted in tropical environments is the ability to withstand UV radiation. Tropical coastal ecosystems, particularly those adjacent to the equator, are subjected to high solar radiation levels, which can result in DNA damage in microorganisms. Numerous tropical geomicrobes have evolved highly effective DNA repair processes, such as photoreactivation, which harnesses sunlight to counteract the damage caused by ultraviolet radiation. Furthermore, certain microorganisms synthesise pigments, such as carotenoids and melanin, that function as sunscreens by absorbing detrimental UV rays and safeguarding cellular constituents from harm. In highly saline and tropical environments, the adaptations of geomicrobes serve as survival mechanisms and facilitate their vital contributions to ecosystem metabolism (Chakraborty et al. 2021; Das et al. 2022; Goud et al. 2022; Kumar et al. 2018b, 2021; Kumar and Mistri 2020; Pathak et al. 2017; Upadhyay et al. 2023). One illustration of this is the role of halophilic microorganisms in salt marshes and saline wetlands in biogeochemical cycling through their facilitation of nitrogen fixation, sulfur reduction, and carbon sequestration. Geomicrobes in tropical mangrove forests decompose organic material, stabilise sediments, and enhance plant health via symbiotic interactions, such as those between mycorrhizal fungus and mangrove roots. The preservation of the health and resilience of saline and tropical ecosystems, which are at risk due to climate change, sea-level rise, and human activity, relies heavily on these crucial microbial adaptations. Through comprehending the processes that enable geomicrobes to flourish in these highly challenging conditions, scientists can devise tactics to improve restoration endeavours and alleviate the consequences of environmental pressures on these vital natural systems. Geomicrobes possess a variety of highly selective adaptations that allow them to survive and operate effectively in the demanding conditions of extremely salty and tropical habitats. By producing osmoprotectants, regulating ions, exhibiting thermotolerance, and employing other mechanisms, these microorganisms endure and perform crucial functions in ecological processes, including nutrient cycling, carbon sequestration, and sediment stabilisation. Their ability to recover and adapt makes them crucial contributors to these ecosystems' health and ecological restoration.

6 Challenges and Limitations

The study of microbial ecology, namely in salt marsh ecosystems, is still evolving in comprehending microbial diversity and the intricate roles the microorganisms carry out. Considerable gaps in knowledge persist despite the acknowledgement of geomicrobes as essential contributors to nutrient cycling, carbon sequestration, and plant resilience. Insufficient knowledge of microbial diversity in these environments is a major source of uncertainty. Salt marshes are dynamic ecosystems marked by varying salinity, water levels, and nutrient availability, providing various ecological niches for microbiologically diverse communities. Nevertheless, the actual magnitude of microbial variety in salt marshes remains largely unexplored, mostly because of the constraints imposed by conventional culturing methods. Most microorganisms in these ecosystems are either non-cultivable or present specific growth needs that are challenging to reproduce in controlled laboratory settings. Consequently, a lack of comprehensive understanding of the microbial environment results in identifying numerous potentially significant microbial species and their respective functions. Moreover, even among the established microorganisms, a considerable amount of knowledge remains to be acquired regarding their particular functions in the functioning of ecosystems. Although sulfate-reducing or nitrogen-fixing bacteria are well recognised for their crucial contributions to biogeochemical cycling, several other microbial groups' functions remain incompletely elucidated. For example, there is an increasing acknowledgement that microbial networks play a role in intricate plant interactions, impacting the absorption of nutrients and the ability to withstand stress. Yet, the precise mechanisms by which these interactions take place remain inadequately understood. Moreover, functional redundancy in microbial communities, where several microbial species carry out similar ecological functions, adds complexity to our predictions of how microbial population changes might impact the ecosystem's overall functioning. To address these gaps in knowledge, there is an urgent requirement for sophisticated molecular methods in microbial research, such as metagenomics and transcriptomics. Without culturing, metagenomics enables the sequencing of whole microbial communities, providing a more comprehensive perspective on microbial diversity. This methodology can detect the species present within a microbial community and the genes linked to important ecological processes, such as carbon and nitrogen cycling. In contrast, transcriptomics offers a valuable understanding of the gene expression patterns within microbial communities, exposing the specific genes that are being actively transcribed in reaction to local environmental factors. Integrating molecular techniques with conventional ecological methods enables researchers to comprehensively comprehend microbial functions and their reactions to ecological changes, augmenting our capacity to effectively manage and restore salt marsh ecosystems. With the advancement of scientific understanding in microbial ecology, expanding microbial restoration initiatives from small-scale experimental plots to extensive environmental systems becomes more complex. Although small-scale studies and pilot projects have shown the capacity of microbial interventions to improve ecosystem restoration, operationalising these achievements on a larger scale is challenging. An inherent obstacle is the logistical complexity of administering microbial inoculants across extensive and diverse landscapes, such as salt marshes, which cover huge areas and exhibit highly fluctuating environmental conditions. The diversity in soil composition poses challenges in devising a universal strategy for microbial restoration since microbial communities that flourish in one section of a marsh may not exhibit the same level of performance in another section, characterised by varying salinity levels, water availability, or nutrient conditions. Moreover, the economic costs of extensive microbial restoration initiatives can be excessive. The large-scale production, transportation, and application of microbial inoculants necessitate substantial financial resources, and the long-term efficacy of these treatments remains highly uncertain. Microbiological restoration has demonstrated potential in enhancing plant growth, stabilising sediments, and improving nutrient cycling in small-scale experimental environments. The long-term persistence and proportionality of these benefits to larger ecosystems remain uncertain. In addition, potential regulatory and permitting issues can provide further barriers, as introducing non-native microbial species may give rise to apprehensions regarding unintended ecological repercussions, such as disturbances to indigenous microbial populations or the dissemination of diseases. Tackling these issues necessitates novel strategies for expanding microbial restoration endeavours. The utilisation of locally sourced microbial communities, which have already acclimated to the particular conditions of the restoration site, presents a plausible solution. Restoration practitioners can enhance the probability of successful establishment and long-term persistence by choosing appropriate microbial inoculants indigenous to the region. Furthermore, progress in microbial production technologies, such as developing economically efficient bioreactors, can potentially decrease the financial obstacles to widespread implementation. Effective collaboration among scientists, policymakers, and restoration practitioners will be crucial in formulating approaches that harmonise the ecological advantages of microbial restoration with practical and economic viability. Beyond the difficulties of scaling, long-term monitoring and adaptive management are crucial for ensuring the sustainability of microbial restoration efforts. Microbiological communities exhibit a remarkable degree of sensitivity to variations in environmental conditions, leading to temporal changes in their composition and function in reaction to parameters such as temperature, salinity, and nutrient availability. Consequently, the limited duration of monitoring may not offer a comprehensive assessment of the sustained effectiveness of microbial restoration approaches. The continuous and long-term monitoring of microbial communities and ecosystem functions is crucial to evaluate the evolution of these factors and ascertain the sustainability of the initial benefits of microbial contamination. Moreover, it is important that long-term monitoring does not only concentrate on microbial communities in isolation. Instead, it should incorporate microbial techniques with other restoration methods, such as hydrological and vegetative interventions. The interconnections within coastal ecosystems, such as salt marshes, are such that alterations in one component of the system can quickly impact others. For example, hydrological restoration methods, such as dredging or channel modification, can potentially modify water flow and salinity patterns, impacting microbial communities and their activities. Furthermore, introducing vegetation in rehabilitated regions can impact microbial populations by providing organic material and modifying soil composition. By systematically observing the interplay among microorganisms, hydrological processes, and plant life, restoration practitioners can formulate comprehensive approaches that augment the overall resilience of the ecosystem. However, to fully realise microbial ecology's potential in improving coastal restoration efforts, it is necessary to address several knowledge gaps and practical challenges. Improving our capacity to monitor and control microbial communities in salt marshes requires a more comprehensive knowledge of microbial diversity and function and incorporating sophisticated molecular techniques. Increasing the scale of microbial restoration initiatives to extensive ecosystems poses logistical and economic obstacles, but inventive strategies and cooperative endeavours can surmount these difficulties. Finally, long-term monitoring and adaptive management are essential for ensuring the sustainability of microbial interventions and combining microbial approaches with other restoration methods to optimise ecosystem recovery and resilience.

7 Future Directions and Research Needs

The development of microbial-based restoration technologies is a state-of-the-art field in environmental restoration, which aims to use the capabilities of microorganisms to rebuild deteriorated ecosystems. This approach is promising in coastal and marine settings, where conventional restoration techniques often prove inadequate due to the intricacy and magnitude of deterioration. Developing new microbial inoculants customised to particular environmental conditions is crucial to microbial-based restoration. These inoculants comprise meticulously chosen or scientifically modified microorganisms capable of flourishing in difficult conditions, such as contaminated waterways, deteriorated soils, or impaired coral reefs. The selection procedure is rigorous, considering the distinct microbial communities found in each ecosystem and determining the species most suitable for improving restoration. Introducing these customised inoculants into deteriorated ecosystems enables researchers to expedite natural processes such as nutrient cycling, carbon sequestration, and sediment stabilisation, facilitating the restoration of ecological functions and biodiversity. Scientific progress in genetic engineering and synthetic biology has significantly transformed the field of microbial-based restoration in recent years. By manipulating the genetic composition of geomicrobes, which are microorganisms that have vital functions in geochemical processes, scientists can augment their capacity to restore ecosystems. Researchers can, for instance, manipulate microorganisms to enhance their ability to degrade pollutants, withstand severe environmental conditions, or even synthesise substances that facilitate the proliferation of indigenous species. Technological tools in synthetic biology facilitate the creation of microbial consortia, in which several species collaborate synergistically to accomplish restoration objectives. This methodology improves the efficiency of restoration achieved by microorganisms and introduces novel opportunities for tackling previously deemed unattainable environmental issues. For example, genetically modified microorganisms could be strategically placed in regions impacted by oil spills, heavy metal pollution, or ocean acidification, offering a potent means of alleviating the consequences of human-induced environmental changes. Although microbial-based restoration technologies show substantial potential, combining them with ecosystem-based approaches is crucial to attain sustainable and long-term restoration results. Ecosystem-based approaches prioritise the restoration of whole habitats rather than exclusively concentrating on individual species or ecological processes. Integrating microbial interventions with habitat restoration techniques, such as rehabilitating oyster reefs, seagrass beds, or mangrove forests, establishes a comprehensive method for ecological recovery. Oyster reefs, for instance, offer essential habitat for a broad range of marine species and play a role in water purification and shoreline stabilisation. By introducing microbial inoculants that augment the growth and resilience of oysters, scientists can expedite the restoration of these ecosystems and enhance their ability to deliver ecological services. Likewise, within seagrass beds, the use of microbiological interventions can facilitate the process of nutrient cycling and improve the health of roots, augmenting the habitat's resilience and productivity. The successful integration of microbial techniques with ecosystem-based approaches necessitates a meticulous partnership among microbiologists, ecologists, and restoration practitioners and a profound comprehension of the interplay between microorganisms and their host ecosystems. Effective microbial-based restoration relies on multidisciplinary collaborations, integrating knowledge from different disciplines to create novel and thorough restoration approaches. Microbiologists offer valuable understanding of the behaviour and functions of microorganisms in various biological settings, while ecologists provide information on the dynamics of ecosystems and the interrelationships among species. Engineers purposefully develop systems to transport microbial inoculants to specific locations, such as bioreactors or dispersal mechanisms. Furthermore, policy experts and social scientists are partners in guiding the regulatory and societal dimensions of restoration projects. Engaging with local communities and stakeholders is crucial to ensure restoration efforts align with local needs and values and establish the required support for large-scale interventions. In coastal areas where livelihoods greatly depend on the environment's well-being, such as in fishing or tourism, microbiological restoration methods can produce substantial social and economic advantages by reinstating ecosystem services that people depend on. The policy and management consequences of restoration driven by microbes are extensive, with the potential to fundamentally change conservation programmes and practices at various levels. The scientific community's ongoing demonstration of the efficacy of microbial-based restoration has generated increasing interest in incorporating these methods into national and international coastal management frameworks. Nevertheless, this necessitates a change in how restoration is perceived and executed, departing from conventional techniques that, mostly, emphasise physical restoration (such as reconstructing habitats) towards more biologically orientated strategies that harness the potential of microorganisms. An essential obstacle in this context is the requirement for effective legislation and guidelines that control the utilisation of genetically modified organisms (GMOs) in restoration initiatives. Although genetically modified microorganisms confer notable benefits regarding restoration potential, their application gives rise to apprehensions regarding unintended ecological consequences, such as disseminating modified characteristics to non-target species or disturbing established microbial communities. Hence, policymakers must collaborate closely with scientists to establish a regulatory structure that effectively harmonises the advantages of microbial-based restoration with the imperative of safeguarding the integrity of the ecosystem. Implementing microbial strategies in coastal management plans at the national level can strengthen the ability of coastal ecosystems to withstand the impacts of climate change and other environmental pressures. In the context of rising sea levels and increasing storm frequency, implementing microbial-based restoration can stabilise shorelines, mitigate erosion, and facilitate the development of vegetation that serves as a protective barrier for coastal communities against storm surges. Integrating microbial approaches into current management strategies enables governments to enhance the efficiency of their conservation endeavours and mitigate the expenses linked to conventional engineering solutions, such as seawalls or artificial reefs. Furthermore, microbial-based restoration is a nature-based solution that prioritises collaborating with nature to tackle environmental complexities. This paradigm confers advantages to the environment and promotes sustainable development by offering economically viable and ecologically sound long-term solutions. Globally, microbial-based restoration can significantly contribute to attaining worldwide conservation objectives, as specified in the United Nations Sustainable Development Goals (SDGs) or the Convention on Biological Diversity (CBD). One potential application of microbial techniques is their integration into the restoration of marine and coastal ecosystems as a component of the Decade of Ocean Science for Sustainable Development (2021–2030), a worldwide endeavour to reverse the deterioration of ocean health. Effective international collaboration is crucial for tackling the transboundary characteristics of several environmental issues, including plastic pollution, ocean acidification, and the deterioration of migratory species' habitats. Through exchanging knowledge, technology, and resources, nations can collaborate to create and execute microbial-based restoration approaches that affect a worldwide scale. Recent advancements in microbial-based restoration technologies signify a fundamental change in environmental restoration, providing novel approaches to tackle ecosystems' intricate and extensive deterioration. By developing innovative microbial inoculants customised for particular ecological conditions, employing genetic engineering and synthetic biology, and combining microbial techniques with ecosystem-based approaches, scientists can improve the efficiency of restoration endeavours and facilitate the restoration of ecological functions and biodiversity. Establishing multidisciplinary partnerships and integrating policies to expand these technologies and guarantee their effectiveness in various environmental settings is crucial. The ongoing development of microbial-based restoration holds promise for revolutionising conservation methods and significantly contributing to the sustainable management of coastal and marine ecosystems globally.

8 Conclusion

In summary, geomicrobes are crucial in rehabilitating salt marshes, providing distinctive and effective remedies to some of the most urgent environmental issues confronting coastal ecosystems. Salt marshes are essential ecosystems that produce diverse ecological services, including coastal protection, carbon sequestration, and support for biodiversity. Nevertheless, these ecosystems face growing risks from climate change, pollution, and the loss of their ecological habitats. Geomicroorganisms play a crucial role in the persistence and restoration of these delicate ecosystems through their inherent activities, which include nutrient cycling, pollutant breakdown, and sediment stabilisation. Applying microbial-based restoration technologies can augment the intrinsic capabilities of salt marshes, so facilitating their capacity to endure and recuperate from environmental stressors. Targeted application of specialised microbial inoculants adapted to the conditions of degraded salt marshes can expedite the restoration of these ecosystems, promoting plant growth, enhancing soil health, and boosting overall ecological stability. The future outlook for microbial-based restoration technologies appears promising. The ongoing progress in genetic engineering and synthetic biology will enable the development of increasingly complex microbial consortia capable of concurrently carrying out several restoration functions. One potential application of genetically modified geomicrobes is their ability to withstand elevated pollution levels while stimulating plant growth and improving sediment stability. This makes them precious assets in the restoration of salt marshes. Moreover, as we enhance our comprehension of the dynamics of microbial communities and their interactions with host ecosystems, the possible uses of geomicrobes in environmental restoration endeavours will continue to grow. In addition to salt marshes, these technologies have the potential to be implemented in other coastal ecosystems, including mangroves, seagrass beds, and coral reefs, so providing scalable remedies for the worldwide deterioration of marine biodiversity. Furthermore, the use of microorganisms in the restoration process is based on the concepts of nature-based solutions. These solutions emphasise utilising natural processes to tackle environmental issues, so establishing a sustainable and ecologically appropriate method for the recovery of ecosystems. Nevertheless, fully harnessing the capabilities of microbial-based restoration technologies necessitates prompt and synchronised efforts. To ensure the recognition and incorporation of geomicrobes into national and international coastal management frameworks, policymakers, scientists, and conservation practitioners must collaborate and integrate these innovative tools into ecological restoration strategies. This will pertain to the formulation of novel regulations and principles that control the utilisation of microbial inoculants, specifically those containing genetically modified organisms (GMOs), to guarantee their safe, efficient, and environmentally sustainable implementation. Furthermore, synergistic partnerships among microbiologists, ecologists, engineers, and social scientists are crucial for the development and execution of comprehensive restoration initiatives that utilise microbial interventions in conjunction with habitat restoration endeavours. It is imperative to issue a call for action to prioritise the engagement of geomicrobes in sustainable coastal management. Microbiological-based restoration offers many advantages, including improved ecological resilience and cost-effective restoration solutions that deliver long-term environmental and social benefits. In light of the ongoing impact of climate change and human activities on coastal ecosystems, there is an urgent requirement for creative and efficient restoration approaches. Incorporating geomicrobes into coastal management strategies guarantees that restoration initiatives are based on the most recent scientific progress and customised to the unique requirements of each ecosystem, resulting in improved and more robust coastal habitats for future generations. It is imperative to take immediate action; microbial-based restoration technologies should be the primary focus of endeavours to restore and safeguard the crucial coastal ecosystems of our planet.

References

Aley P, Singh J, Kumar P (2022) Chapter 23: Adapting the changing environment: microbial way of life. In: Kumar A, Singh J, Ferreira LFRBT-MUCC (eds) Microbiome under changing climate: implications and solutions. Woodhead Publishing, pp 507–525. https://doi.org/10.1016/B978-0-323-90571-8.00023-7

Chakraborty S, Kumar P, Sanyal R, Mane AB, Arvind Prasanth D, Patil M, Dey A (2021) Unravelling the regulatory role of miRNAs in secondary metabolite production in medicinal crops. Plant Gene 27:100303. https://doi.org/10.1016/j.plgene.2021.100303

Choudhury D, Kumar P, Zhimo VY, Sahoo J (2024) Crop rotation patterns and soil health management. In: Bioremediation of emerging contaminants from soils: soil health conservation for improved ecology and food security. Elsevier, Phagwara, India, pp 565–589. https://doi.org/10.1016/B978-0-443-13993-2.00025-6

Das T, Saha SC, Sunita K, Majumder M, Ghorai M, Mane AB, Prasanth DA, Kumar P, Pandey DK, Al-Tawaha AR, Batiha GE-S, Shekhawat MS, Ghosh A, Sharifi-Rad J, Dey A (2022) Promising botanical-derived monoamine oxidase (MAO) inhibitors: pharmacological aspects and structure-activity studies. South Afr J Bot 146:127–145. https://doi.org/10.1016/j.sajb.2021.09.019

Devi P, Kumar P (2023a) Role of oxidative enzymes in the alleviation of oxidative stress in plant system. In: Microbial oxidative enzymes: biotechnological applications. De Gruyter, Phagwara, India, pp 109–128. https://doi.org/10.1515/9783111062235-006

Devi P, Kumar P (2023b) Effect of microbial oxidative enzymes on soil health. In: Microbial oxidative enzymes: biotechnological applications. De Gruyter, Phagwara, India, pp 37–61. https://doi.org/10.1515/9783111062235-003

Devi P, Kumar P (2024a) Challenges in the storage of herbal medicine products and strategies for sustainable management. In: Reference series in phytochemistry. Springer, Phagwara, India, pp 1739–1767. https://doi.org/10.1007/978-3-031-43199-9_37

Devi P, Kumar P (2024b) Herbal medicine and pregnancy. In: Reference series in phytochemistry. Springer, Phagwara, India, pp 693–722. https://doi.org/10.1007/978-3-031-43199-9_25

Devi P, Kumar P, Sharma K, Singh J (2023) Chapter 10: Is there a relationship between myconanotechnology and sustainable development? In: Kuddus M, Ahmad IZ, Hussain CMBT-M et al (eds) Progress in biochemistry and biotechnology. Academic Press, pp 175–207. https://doi.org/10.1016/B978-0-443-15262-7.00009-7

Devi P, Dey SR, Kumar P (2023a) Integration of water resources management in rural areas. In: Water resources management for rural development: challenges and mitigation. Elsevier, Phagwara, India, pp 139–152. https://doi.org/10.1016/B978-0-443-18778-0.00013-1

Devi P, Dey SR, Saini L, Kumar P, Panigrahi S, Dwivedi P (2023b) Toward sustainable agriculture: strategies involving phytoprotectants against reactive oxygen species. In: Reactive oxygen species: prospects in plant metabolism. Springer Nature, Phagwara, India, pp 229–247. https://doi.org/10.1007/978-981-19-9794-5_13

Devi P, Dey SR, Sharma K, Kumar P, Singh J (2024a) Waste management as an opportunity for the inclusion of vulnerable groups. In: Waste management in the circular economy. Springer, Phagwara, India, pp 207–230. https://doi.org/10.1007/978-3-031-42426-7_11

Devi P, Dogra P, Kumar P, Singh J (2024b) Global status of biomass energy programmes-challenges and roadmap. In: Biomass energy for sustainable development. CRC Press, Phagwara, India, pp 394–414. https://doi.org/10.1201/9781003406501-20
Devi P, Kapoor P, Kumar P (2024c) Biomarkers for toxicity detection. In: Biomarkers in environmental and human health biomonitoring: an integrated perspective. Elsevier, Phagwara, India, pp 313–329. https://doi.org/10.1016/B978-0-443-13860-7.00004-5
Devi P, Kaur A, Kumar P (2024d) Phytoremediation of metal contaminated soil using energy crops: soil health maintenance along with biofuel production. In: Bioremediation of emerging contaminants from soils: soil health conservation for improved ecology and food security. Elsevier, Punjab, India, pp 307–333. https://doi.org/10.1016/B978-0-443-13993-2.00015-3
Devi P, Kumar P, Singh J (2024e) Biotechnological approaches as clean technology for the development of a sustainable environment. In: Development in waste water treatment research and processes: role of environmental microbiology in industrial wastewater research. Elsevier, Phagwara, India, pp 381–403. https://doi.org/10.1016/B978-0-443-13609-2.00016-1
Devi P, Kumar P, Singh J (2024f) Application of potential microbes in bioremediation. In: Development in waste water treatment research and processes: role of environmental microbiology in industrial wastewater research. Elsevier, Phagwara, India, pp 201–225. https://doi.org/10.1016/B978-0-443-13609-2.00021-5
Dey SR, Devi P, Kumar P (2023a) Organic contaminants in aquatic environments: sources and impact assessment. In: Metal organic frameworks for wastewater contaminant removal. Wiley, Phagwara, India, pp 299–317
Dey SR, Kumar P, Singh J (2023b) Growth of mustard under open field conditions with salicylic acid and antimicrobial thiourea. Biopestic Int 19:183–192. https://doi.org/10.59467/BI.2023.19.183
Dey SR, Sharma M, Kumar P (2023c) Effects and responses of chromium on plants. In: Environmental science and engineering. Springer, Phagwara, India, pp 385–427. https://doi.org/10.1007/978-3-031-44029-8_14
Dey SR, Sharma M, Kumar P (2024) Phytoremediation of phenolic compounds from soil. In: Bioremediation of emerging contaminants from soils: soil health conservation for improved ecology and food security. Elsevier, Phagwara, India, pp 335–358. https://doi.org/10.1016/B978-0-443-13993-2.00016-5
Dwivedi P, Kumar P (2011) Anti-diabetic medicinal plants and their conservation: waging green war on diabetes. Med Plants 3:181–189. https://doi.org/10.5958/j.0975-4261.3.3.031
Goud EL, Singh J, Kumar P (2022) Climate change and their impact on global food production. In: Microbiome under changing climate. Elsevier, pp 415–436
Hidangmayum A, Dwivedi P, Kumar P, Upadhyay SK (2023) Seed priming and foliar application of chitosan ameliorate drought stress responses in Mungbean genotypes through modulation of morpho-physiological attributes and increased Antioxidative defense mechanism. J Plant Growth Regul 42:6137–6154. https://doi.org/10.1007/s00344-022-10792-1
Jagadesh M, Dash M, Kumari A, Singh SK, Verma KK, Kumar P, Bhatt R, Sharma SK (2024) Revealing the hidden world of soil microbes: metagenomic insights into plant, bacteria, and fungi interactions for sustainable agriculture and ecosystem restoration. Microbiol Res 285:127764. https://doi.org/10.1016/j.micres.2024.127764
Karam N, Kumar P, Anand YR, Choudhury D (2023) Management of rice blast-a review of the strategies available. Biopestic Int 19:107–116. https://doi.org/10.59467/BI.2023.19.107
Kotia A, Rutu P, Singh V, Kumar A, Dhoke S, Kumar P, Singh DK (2021) Rheological analysis of rice husk-starch suspended in water for sustainable agriculture application. In: 2nd international conference on functional materials, manufacturing and performances, ICFMMP 2021. Elsevier Ltd, Phagwara, India, pp 1962–1966. https://doi.org/10.1016/j.matpr.2021.09.325
Kumar P (2023) Degradation of xenobiotics by bacteria and fungi: an overview. Biopestic Int 19:117–132. https://doi.org/10.59467/BI.2023.19.117

Kumar P (2024a) Biochar production from marine algae and its application in the treatment of industrial effluent. In: Marine biomass: biorefinery, bioproducts and environmental bioremediation. De Gruyter, Phagwara, India, pp 57–93. https://doi.org/10.1515/9783111353951-004
Kumar P (2024b) Application of marine biomass for carbon neutrality, sustainable environment, and economy. In: Marine biomass: biorefinery, bioproducts and environmental bioremediation. De Gruyter, Phagwara, India, pp 95–128. https://doi.org/10.1515/9783111353951-005
Kumar P (2024c) Soil conservation for global sustainability. In: World sustainability series. Springer, Phagwara, India, pp 103–128. https://doi.org/10.1007/978-3-031-57456-6_6
Kumar P, Choudhury D (2024a) Optimized phytoremediation process for the sustainable management radionuclides. In: Bioremediation of emerging contaminants from soils: soil health conservation for improved ecology and food security. Elsevier, Phagwara, India, pp 443–464. https://doi.org/10.1016/B978-0-443-13993-2.00020-7
Kumar P, Choudhury D (2024b) Role of indigenous knowledge in agricultural soil reclamation without disturbing other ecosystems. In: Bioremediation of emerging contaminants from soils: soil health conservation for improved ecology and food security. Elsevier, Phagwara, India, pp 465–488. https://doi.org/10.1016/B978-0-443-13993-2.00021-9
Kumar P, Choudhury D (2024c) Microbial native soil bacteria against cadmium toxicity. In: Springer water. Springer Nature, Phagwara, India, pp 193–226. https://doi.org/10.1007/978-3-031-54005-9_9
Kumar P, Choudhury D (2024d) Toxicity of rhizospheric cadmium contaminated soil and its phytoremediation. In: Springer water. Springer Nature, Phagwara, India, pp 227–264. https://doi.org/10.1007/978-3-031-54005-9_10
Kumar P, Choudhury D (2024e) Hydroponic removal of organic contaminants from water. In: Springer water. Springer Nature, Phagwara, India, pp 143–164. https://doi.org/10.1007/978-3-031-53258-0_6
Kumar P, Dey SR (2024) Hydroponics phytoremediation: an overview. In: Springer water. Springer Nature, Phagwara, India, pp 361–396. https://doi.org/10.1007/978-3-031-53258-0_13
Kumar P, Dwivedi P (2018) Putrescine and glomus mycorrhiza moderate cadmium actuated stress reactions in zea mays l. by means of extraordinary reference to sugar and protein. Vegetos 31:74–77. https://doi.org/10.5958/2229-4473.2018.00076.9
Kumar P, Dwivedi P (2020) Lignin estimation in sorghum leaves grown under hazardous waste site. Plant Arch 20:2558–2561
Kumar VA, Kumar P (2023) Revolutionizing agriculture with satellite technology for farmers: a review. Biopestic Int 19:97–106. https://doi.org/10.59467/BI.2023.19.97
Kumar P, Mistri TK (2020) Transcription factors in SOX family: potent regulators for cancer initiation and development in the human body. Semin Cancer Biol 67:105–113. https://doi.org/10.1016/j.semcancer.2019.06.016
Kumar P, Naik M (2020) Biotic symbiosis and plant growth regulators as a strategy against cadmium and lead stress in chickpea. Plant Arch 20:2495–2500
Kumar P, Pathak S (2019) Responsiveness index of sorghum (Sorghum bicolor (L.) Moench) grown under cadmium contaminated soil treated with putrescine and mycorrhiza. Bangladesh J Bot 48:139–143
Kumar P, Raut AM (2024) Microbes-assisted bioaugmentation process in the reduction of emerging industrial pollutants from soil. In: Bioremediation of emerging contaminants from soils: soil health conservation for improved ecology and food security. Elsevier, Phagwara, India, pp 519–540. https://doi.org/10.1016/B978-0-443-13993-2.00023-2
Kumar P, Singh J (2024a) Exploring enzyme biotechnology's role in green chemistry and advancements in pharmaceutical technologies. In: Enzyme biotechnology for environmental sustainability. Elsevier, Phagwara, India, pp 465–495. https://doi.org/10.1016/B978-0-443-22072-2.00015-2
Kumar P, Singh J (2024b) Environmental and health risk assessment of chemical pollutants in drinking water and wastewaters. In: Trends in biological processes in industrial wastewa-

ter treatment. Institute of Physics Publishing, Phagwara, India, pp 5.1–5.30. https://doi.org/10.1088/978-0-7503-5678-7ch5
Kumar P, Singh J (2024c) Bioprocesses in industrial wastewater treatment: trends and prospects. In: Trends in biological processes in industrial wastewater treatment. Institute of Physics Publishing, Department of Agronomy, Phagwara, India, pp 15.1–15.50. https://doi.org/10.1088/978-0-7503-5678-7ch15
Kumar P, Singh J (2024d) Emerging contaminants and ways to reduce them. In: Trends in biological processes in industrial wastewater treatment. Institute of Physics Publishing, Phagwara, India, pp 4.1–4.34. https://doi.org/10.1088/978-0-7503-5678-7ch4
Kumar P, Singh J (2024e) Hydroponic: an eco-friendly future. In: Springer water. Springer Nature, Phagwara, India, pp 223–267. https://doi.org/10.1007/978-3-031-53258-0_9
Kumar P, Mandal B, Dwivedi P (2011a) Heavy metals scavenging of soils and sludges by ornamental plants. J Appl Hortic 13:144–146
Kumar P, Mandal B, Dwivedi P (2011b) Heavy metal scavenging capacity of Mentha spicata and Allium cepa. Med Plants 3:315–318. https://doi.org/10.5958/j.0975-4261.3.4.053
Kumar P, Mandal B, Dwivedi P (2013) Combating heavy metal toxicity from hazardous waste sites by harnessing scavenging activity of some vegetable plants. Vegetos 26:416–425. https://doi.org/10.5958/j.2229-4473.26.2.106
Kumar P, Singh BN, Dwivedi P (2017) Plant growth regulators, plant adaptability and plant productivity: a review on abscisic acid (ABA) signaling in plants under emerging environmental stresses. In: Sustaining future food security in changing environments. Nova Science Publishers, Inc., Varanasi, India, pp 81–97
Kumar P, Pathak S, Kumar M, Dwivedi P (2018) Role of secondary metabolites for the mitigation of cadmium toxicity in sorghum grown under mycorrhizal inoculated hazardous waste site. In: Biotechnological approaches for medicinal and aromatic plants. Springer, Singapore, pp 199–212
Kumar P, Harshavardhan M, Kumar PS, Yumnam J, Jyoti N, Naik M, Misao L, Purnima K (2018a) Effect on chlorophyll a/b ratio in cadmium contaminated maize leaves treated with putrescine and mycorrhiza. Ann Biol 34:281–283
Kumar P, Kumar S, Harshavardhan M, Naik M, Yumnam J, Kumar PS, Jyoti N, Misao L et al (2018b) Evaluation of plant height and leaf length of sorghum grown under different sources of nutrition. Ann Biol 34:284–286
Kumar P, Pathak S, Amarnath KS, Veerendra Brahma Teja P, Dileep B, Kumar K, Singh M, Siddique A (2018c) Effect of growth regulator on morpho-physiological attributes of chilli: a case study. Plant Arch 18:1771–1776
Kumar P, Yumnam J, Kumar PS, Misao L, Jyoti N, Naik M, Purnima, Kumar S, Harshavardhan M (2018d) Cadmium induced changes in germination of maize seed treated with mycorrhiza. Ann Agri Bio Res 23:169–170
Kumar D, Rameshwar SD, Kumar P (2019) Effect of intergated application of inorganic and organic fertilizers on the roots of chickpea. Plant Arch 19:857–860
Kumar P, Kumar T, Singh S, Tuteja N, Prasad R, Singh J (2020) Potassium: a key modulator for cell homeostasis. J Biotechnol 324:198–210
Kumar V, Dwivedi P, Kumar P, Singh BN, Pandey DK, Kumar V, Bose B (2021) Mitigation of heat stress responses in crops using nitrate primed seeds. South African J Bot 140:25–36. https://doi.org/10.1016/j.sajb.2021.03.024
Kumar P, Devi P, Dey SR (2021a) Fungal volatile compounds: a source of novel in plant protection agents. In: Volatiles and metabolites of microbes. Elsevier, pp 83–104
Kumar P, Sharma K, Saini L, Dey SR (2021b) Chapter 8: Role and behavior of microbial volatile organic compounds in mitigating stress. In: Kumar A, Singh J et al (eds) Volatiles and metabolites of microbes. Academic Press, pp 143–161. https://doi.org/10.1016/B978-0-12-824523-1.00010-9

Kumar P, Goud EL, Devi P, Dey SR, Dwivedi P (2022a) Heavy metals: transport in plants and their physiological and toxicological effects. In: Plant metal and metalloid transporters. Springer Nature, Jalandhar, India, pp 23–54. https://doi.org/10.1007/978-981-19-6103-8_2

Kumar P, Goud EL, Devi P, Koul B (2022b) Metal pollutants in the environment. In: Environmental microbiology: emerging technologies. De Gruyter, Jalandhar, India, pp 291–323. https://doi.org/10.1515/9783110727227-012

Kumar P, Koul B, Sharma M (2022c) Phytoremediation of heavy metals. In: Heavy metals in plants: physiological to molecular approach. CRC Press, Jalandhar, India, pp 369–388. https://doi.org/10.1201/9781003110576-17

Kumar P, Choudhury D, Sahoo J, Devi MI (2023a) Exploring the potential for extended investigation: a preliminary examination of Liriomyza trifolii and its bibliometric foundation for analytical pursuits. J Food Chem Nanotechnol 9:S359–S366. https://doi.org/10.17756/jfcn.2023-s1-045

Kumar P, Latha DH, Singh J (2023b) Rhizoremediation of organic emerging soil contaminants: green technology. In: Clean technologies toward a sustainable future: physicochemical, biochemical and biotechnological approaches. IWA Publishing, Phagwara, India, pp 211–229. https://doi.org/10.2166/9781789063783_0211

Kumar P, Saini L, Sharma M (2023c) Assessment of the phytochemical constituents and metabolites in the medicinal plants and herbal medicine used in the treatment and management of respiratory diseases. In: Reference series in phytochemistry. Springer, Phagwara, India. https://doi.org/10.1007/978-3-031-21973-3_19-1

Kumar P, Choudhury D, Dey SR, Sharma M (2024a) Application of genetically modified crops against metallic contaminations. In: Bioremediation of emerging contaminants from soils: soil health conservation for improved ecology and food security. Elsevier, Phagwara, India, pp 541–564. https://doi.org/10.1016/B978-0-443-13993-2.00024-4

Kumar P, Dey SR, Choudhury D (2024b) Effectiveness of cadmium on biochemical shift of pea plant treated with mycorrhiza and putrescine. Nat Environ Pollut Technol 23:275–285. https://doi.org/10.46488/NEPT.2024.v23i01.023

Kumar P, Dwivedi P, Upadhyay SK (2024c) Optimization of polyamine and mycorrhiza in sorghum plant for removal of hazardous cadmium. Plant Physiol Biochem 214:108846. https://doi.org/10.1016/j.plaphy.2024.108846

Kumar P, Karam N, Choudhury D (2024d) Assessment of the phytochemical constituents and metabolites in the medicinal plants and herbal medicine used in the treatment and management of skin diseases. In: Reference series in phytochemistry. Springer, Phagwara, India, pp 533–573. https://doi.org/10.1007/978-3-031-43199-9_21

Kumar P, Khokhar A, Dey SR (2024e) Evidence from the use of herbal medicines in the management and prevention of common eye diseases. In: Reference series in phytochemistry. Springer, Phagwara, India, pp 763–795. https://doi.org/10.1007/978-3-031-43199-9_27

Kumar P, Saini L, Sharma M (2024f) Assessment of the phytochemical constituents and metabolites in the medicinal plants and herbal medicine used in the treatment and management of respiratory diseases. In: Reference series in phytochemistry. Springer, Phagwara, India, pp 481–510. https://doi.org/10.1007/978-3-031-43199-9_19

Kumar P, Saurabh S, Sharma K (2024g) Assessment of the phytochemical constituents and metabolites in medicinal plants and herbal remedies used in the treatment and management of reproductive diseases: polycystic ovary syndrome. In: Reference series in phytochemistry. Springer, Phagwara, India, pp 511–532. https://doi.org/10.1007/978-3-031-43199-9_20

Kumar P, Sharma K, Singh J (2024h) Analysis of complex microbial communities in soil and wastewater treatment processes. In: Microbial metagenomics in effluent treatment plant. Elsevier, Phagwara, India, pp 25–41. https://doi.org/10.1016/B978-0-443-13531-6.00002-1

Kumar P, Sharma K, Singh J (2024i) Response of microbial community to environment changes. In: Microbial metagenomics in effluent treatment plant. Elsevier, Phagwara, India, pp 43–61. https://doi.org/10.1016/B978-0-443-13531-6.00006-9

Kumar P, Singh J, Singh G (2024j) Renewable hydrogen opportunities and challenges. In: Renewable hydrogen: opportunities and challenges in commercial success. Elsevier, Phagwara, India, pp 193–220. https://doi.org/10.1016/B978-0-323-95379-5.00010-9

Kumar P, Srivastav AL, Chaudhary V, van Hullebusch ED, Busquets R (2024k) Bioremediation of emerging contaminants from soils: soil health conservation for improved ecology and food security, bioremediation of emerging contaminants from soils: soil health conservation for improved ecology and food security. Elsevier, Phagwara, India. https://doi.org/10.1016/C2022-0-02758-6

Kumari P, Singh J, Kumar P (2022) Impact of bioenergy for the diminution of an ascending global variability and change in the climate. In: Microbiome under changing climate. Elsevier, pp 469–487

Kumari P, Kumar P, Singh J (2023) Toxic contamination of aquatic habitat and its remediation. In: Spatial modeling of environmental pollution and ecological risk. Elsevier, Phagwara, India, pp 209–220. https://doi.org/10.1016/B978-0-323-95282-8.00021-3

Naik M, Kumar P (2020) Role of growth regulators and microbes for metal detoxification in plants and soil. Plant Arch 20:2820–2824

Pandey SK, Kumar P, Dey SR (2024) Xenobiotic stress management by microbial endophytes. In: Bioremediation of emerging contaminants from soils: soil health conservation for improved ecology and food security. Elsevier, Jabalpur, India, pp 241–259. https://doi.org/10.1016/B978-0-443-13993-2.00012-8

Pankaj PK, Kumar P, Nigam RC, Mishra PK (2012a) Monitoring and surveillance of synthetic pyrethroids and organophosphate in different brands of soft drinks. J Chem Pharm Res 4:3939–3943

Pankaj PK, Kumar P, Nigam RC, Mishra PK (2012b) To studies organochlorine insecticides in different brands of cold drinks. J Chem Pharm Res 4:3934–3938

Pathak S, Kumar P, Mishra PK, Kumar M (2017) Mycorrhiza assisted approach for bioremediation with special reference to biosorption. Pollut Res 36:329–332

Raj A, Goutam E, Kumar P, Singh J (2024) Polyphenol oxidase: an enzyme of plants. In: Polyphenol oxidases: function, wastewater remediation, and biosensors. De Gruyter, Phagwara, India, pp 105–128. https://doi.org/10.1515/9783111033525-004

Reddy SS, Kumar P, Dwivedi P (2022) Heavy metal transporters, phytoremediation potential, and biofortification. In: Plant metal and metalloid transporters. Springer, pp 387–405

Saini L, Sharma K, Kumar P (2023) Water efficiency in agriculture as a vital approach toward water management in rural areas. In: Water resources management for rural development: challenges and mitigation. Elsevier, Phagwara, India, pp 127–138. https://doi.org/10.1016/B978-0-443-18778-0.00020-9

Saurabh K, Meena RK, Kumar P, Upadhyaya A, Das A (2024) Role of hyperaccumulators in the reduction of emerging industrial pollutants from soil. In: Bioremediation of emerging contaminants from soils: soil health conservation for improved ecology and food security. Elsevier, Bihar, India, pp 489–517. https://doi.org/10.1016/B978-0-443-13993-2.00022-0

Sharma K, Kumar P (2023) Microbe-mediated impact on cell membrane stability and leaf pigmentation in pearl millet under toxic mercury stress in soil. Biopestic Int 19:217–228. https://doi.org/10.59467/BI.2023.19.217

Sharma K, Kumar P (2024) Environmental threats posed by xenobiotics. In: Bioremediation of emerging contaminants from soils: soil health conservation for improved ecology and food security. Elsevier, Phagwara, India, pp 183–201. https://doi.org/10.1016/B978-0-443-13993-2.00009-8

Sharma M, Dey SR, Kumar P (2023) Antioxidant defence: a key mechanism of chromium tolerance. In: Environmental science and engineering. Springer, Phagwara, India, pp 91–116. https://doi.org/10.1007/978-3-031-44029-8_5

Sharma K, Devi P, Dey SR, Kumar P (2023a) Mercury phytovolatilization: an overview of the mechanism and mitigation. In: Role of green chemistry in ecosystem restoration to achieve environmental sustainability. Elsevier, Phagwara, India, pp 325–331. https://doi.org/10.1016/B978-0-443-15291-7.00031-6

Sharma K, Devi P, Kumar P (2023b) Metal-organic frameworks for wastewater contaminants removal. In: Metal organic frameworks for wastewater contaminant removal. Wiley, Phagwara, India, pp 95–117

Sharma K, Devi P, Kumar P, Dey A, Dwivedi P (2023c) Hazardous phytotoxic nature of reactive oxygen species in agriculture. In: Reactive oxygen species: prospects in plant metabolism. Springer Nature, Phagwara, India, pp 135–146. https://doi.org/10.1007/978-981-19-9794-5_8

Sharma K, Kumar P, Singh J (2024) Ecology and diversity of microbial communities involved in the removal of priority contaminants and micropollutants in waste. In: Development in waste water treatment research and processes: role of environmental microbiology in industrial wastewater research. Elsevier, Phagwara, India, pp 181–199. https://doi.org/10.1016/B978-0-443-13609-2.00014-8

Sinam V, Kumar P, Singh J (2023) Bioremediation and ecorestoration strategies of aquatic environment. In: Spatial modeling of environmental pollution and ecological risk. Elsevier, Phagwara, India, pp 483–499. https://doi.org/10.1016/B978-0-323-95282-8.00025-0

Srivastav AL, Patel N, Rani L, Kumar P, Dutt I, Maddodi BS, Chaudhary VK (2024) Sustainable options for fertilizer management in agriculture to prevent water contamination: a review. Environ Dev Sustain 26:8303–8327. https://doi.org/10.1007/s10668-023-03117-z

Upadhyay SK, Devi P, Kumar V, Pathak HK, Kumar P, Rajput VD, Dwivedi P (2023) Efficient removal of total arsenic (As3+/5+) from contaminated water by novel strategies mediated iron and plant extract activated waste flowers of marigold. Chemosphere 313:137551. https://doi.org/10.1016/j.chemosphere.2022.137551

Bioconstructions in Hypogean Paleoprecipitation Archives: Possible Models of Extraterrestrial Life

Subhro Banerjee and Santa Ram Joshi

Contents

Abstract A fascinating quest for extraterrestrial microbial life in extreme environments is dealt with in the present chapter. Subsurface environments on extraterrestrial bodies tender shelter to microorganisms from aggressive exterior parameters. Microbial life is profuse and highly prolific in the subsurface of Earth. Caves are typical ecosystems where both microbial growth and secondary mineral precipita-

S. Banerjee
Department of Zoology, SRM University Sikkim, Gangtok, Sikkim, India
e-mail: subhro.b@srmus.edu.in

S. R. Joshi (✉)
Microbiology Laboratory, Department of Biotechnology & Bioinformatics, North-Eastern Hill University, Shillong, Meghalaya, India

S. R. Joshi et al. (eds.), *Mineral Transformation and Bioremediation by Geo-Microbes*, Interdisciplinary Biotechnological Advances,
https://doi.org/10.1007/978-981-96-3033-2_10

tion are pronounced by steady physicochemical parameters. Lava tube caves are prevalent in volcanic territories on Earth, providing entrance to subsurface microbial environments and an abode to widespread microbial communities. Caves are also the sites for the development of exploration strategies and new technologies for subsequent voyages to Mars. Lava tubes and cave-like characteristics have also been spotted on Mars based on orbital imagery and remote sensing records. By resemblance with terrestrial caves, Martian caves might behold secondary mineralization that would enlighten us on aqueous activity long ago, apart from representing the ideal spots to look out for biosignatures.

Keywords Caves · Microbial life · Astrobiology · Mars · Lava tubes · Paleoenvironmental

Abbreviations

NASA	National Aeronautics and Space Administration
CHNOPS	Carbon Hydrogen Nitrogen Oxygen Phosphorus Sulfur
BC	Before Christ
EPS	Exopolysaccharide
km	Kilometer
CO_2	Carbon dioxide
H_2O	Water
MRO	Mars Reconnaissance Orbiter
°C	Degree Celsius
^{13}C	Carbon-13 isotope
^{18}O	Oxygen-18 isotope

1 Background

In the first millennium BC, Babylonians were the earliest people who made an effort to foresee the arrangements of the celestial bodies (Venus, Mercury, Jupiter, Saturn, and Mars) (Leverington 2007). The Greeks also had curiosity about terrestrial movements and attempted to know the reason behind these movements. Interestingly, our acceptance of the universe and the solar system is very much dissimilar from that of the Greek theorists. While their focus was on planetary movements, we looked for microbial existence (Kommedal 2017).

1.1 The Range of Microscopic Life

Earth was shaped 4.5 billion years earlier. Having witnessed several alterations, it ultimately led to circumstances favorable for life's foundation. Nevertheless, it is observed that circumstances that permit initiation of life are not similar situations which maintain life after it has been instigated (Westall 2012; Westall et al. 2015). The Earth carried on evolving and changing as microorganisms surfaced. Life was tailored to garner benefit from diverse ecological circumstances. Consequently, microbes garnered various physiologies and metabolisms to cope with and colonize a wide range of habitats. These microbial actions transformed to generate a significant part in the subsequent evolution of Earth to such an extent that the present biosphere is the outcome of the evolution of microbes ever since its starting point (Madigan et al. 2010).

Astrobiology Institute of NASA has demarcated astrobiology as the learning of the source, distribution, impending of life, and evolution in the cosmos. NASA pursues response to three basic queries: In what way does life evolve and begin? Is there any probability of life outside Earth, and if there is life, by which means can we identify it? What is the prospect of life in the space and on Earth? The foremost action is to search for life which is analogous to that on Earth because it is the sole system for existence identified (Kommedal 2017).

1.2 Prerequisites for Existence

Cockell et al. (2016) reviewed the following five prerequisites that should be met so that life can flourish: (i) suitable physicochemical parameters, (ii) biochemical reactions in the presence of a solvent, (iii) availability of energy, (iv) trace elements, and (v) major elements necessary for existence of life. Water is the mandatory solvent for existence of Earth's life. It is capable of surviving in its solvent form provided fitting ecological circumstances comprising of pressure, chemical composition, and temperature. Microorganisms require energy for continuance, expansion, and duplication. Based on energy requirements, microorganisms are grouped as chemotrophs or phototrophs. Phototrophs make use of sunlight to gain energy, and chemotrophs extract energy from redox reactions where they exploit organic (chemoheterotrophy) or inorganic (chemolithotrophy) electron donors (Cockell et al. 2016).

Even though Earth's life exhibits a complex variety, it is composed of similar molecules and atoms. C, N, H, P, S, and O are the vital elements necessary for Earth's life. To sum up, these six atoms add up to 97% of a typical cell of *E. coli* (Wackett et al. 2004). These elements constitute the basic backbone of carbohydrates, proteins, nucleic acids, lipids, and fatty acids. These comprise the core macromolecules required for life (McKay 2004). Supplementary to these elements, microbes might also rely on further elements which include magnesium and iron, among others (Wackett et al. 2004). Irrespective of anywhere life instigated on Earth, the environments necessitated to put up with liquid water with temperatures that permitted the presence of organic macromolecules (Kommedal 2017). Apart from these, crucial trace and major elements and bioavailable energy need to exist for the foundation of evolution of life (Westall 2012).

2 Introduction

A main portion of biomass on Earth is microorganisms which signify major pools of life's vital six elements (Kommedal 2017). Actually, phosphorus and nitrogen established in microorganisms are approximately ten times greater when matched to the quantities in the biomass of plants (Whitman et al. 1998). Our insights regarding the environment have originated from learning life which is involved with surface. It depends indirectly on exploiting oxygen and biomass generated by photosynthesis or directly on photosynthesis (Edwards et al. 2012). Our overall knowledge regarding life till the 1980s was reliant on photosynthesis and solar energy (Gold 1992). Nevertheless, Corliss et al. (1979) documented microbial populations in hydrothermal vents free of energy derived from the surface. Russell et al. (2014) have also hypothesized that life might have initiated nearby hydrothermal vents.

Caves have been looked upon as the linkage between subsurface and surface locations (Pedersen 2000). Research on subsurface microorganisms have divulged that most of them are heterotrophic. They rely on organic matter (Kommedal 2017). This entry of organic matter has settled on the subsurface which benefits microorganisms in the terrestrial systems (Edwards et al. 2012). Subsurface lithoautotrophic microbial ecosystems (SLiMEs), nevertheless, consume minute quantities of hydrogen and carbon dioxide. This phenomenon is more pronounced when organic matter derived from the surface is controlled from entering, thereby indicative of the existence of life in the dearth of energy bases (Stevens and McKinley 2000; Chapelle et al. 2002). The permeability of rock, ice, and sediment offers the required room for microbial life to thrive (Parnell and McMahon 2016). Till which distance below the surface life exists is unidentified till date. It is thought to be influenced as a result of an upper temperature edge and accessibility of water (Wilkins et al. 2014). Several cave environments on our planet are reported to harbor heterotrophic bacteria which are hypothesized to serve as the food base (Azúa-Bustos et al. 2009).

An example can be cited of the Atacama which is the lone driest and almost certainly the oldest existing desert on Earth, having a hyperaridity environment perhaps for 150 million years (Hartley et al. 2005). To eke out a living in these hyperarid situations, life forms have tailored themselves to utmost low levels of air humidity, virtually total omission of rainfall, exceedingly saline soils, and elevated solar heat (McKay et al. 2003). These ruthless environmental parameters may elaborate the reason for the almost complete absence of microorganisms in the Atacama Desert (Navarro-González et al. 2003). Fascinatingly, the Atacama Desert has been scientifically recognized as a key analog replica of the planet Mars, and quite a handful of investigations have been carried out on a range of subjects of astrobiological significance (Azúa-Bustos et al. 2009). Azúa-Bustos et al. (2009) have reported the occurrence of a member of the earliest eukaryote red algae *Cyanidium* group in a coastal cave of the hyperarid Atacama Desert. This microorganism formed a monospecific biofilm developing under exceptionally low levels of photon flux. Their investigations suggested that this species of *Cyanidium* is a different associate of a newly projected unique mesophilic *Cyanidium* sp., discrete from the residual three other descents which are all thermo-acidophilic. The cave thus elucidated by Azúa-Bustos et al. (2009) might work as a budding islet of life in the midst of unhospitable environment.

3 Microbial Role in Speleothem Growth: Paleoenvironmental Propositions

The development of speleothems which includes stalagmites and stalactites by the precipitation of calcite is normally taken into consideration as events which are abiogenic (Pacton et al. 2013). Nevertheless, research also suggested that microbes might play a prominent task in carbonate precipitation during the genesis of speleothem (Northup and Lavoie 2001; Baskar et al. 2005, 2006, 2007, 2009; Mulec et al. 2007; Jones 2010; Banerjee and Joshi 2013, 2014, 2016a, 2016b; Banerjee et al. 2019).

A range of geochemical events in caves including dissolution and precipitation processes result not only in the settlement of carbonates, iron, manganese oxides, silicates, nitrates, and sulfur compounds but also aid in the interruption of the host limestone rock (Pacton et al. 2013). Aptly, it can rightly be said that the cave microbes intervene in between a comprehensive array of constructive and destructive events that together can affect the crystal fabric development of speleothems (Jones 2010). Positive practices comprise binding and trapping by filamentous microorganisms, microbe calcification, and precipitation of minerals (Cañaveras et al. 2001), while the destructive practices include dissolution of mineral surfaces or microbially influenced corrosion that can ensue over discharge of mineral and organic acids (sulfuric acid) and exoenzymes, mechanical attack, and a host of additional tools. Reactions involving bacteria capable of oxidizing sulfur-, iron-, and manganese are of specific concern in cave dissolution techniques (Northup and Lavoie 2001). Investigations have also ascertained the bacterial relationship with iron deposition in caves. But substantiation for a dynamic microbial character in the creation of such cave settlements is scarce till date (Pacton et al. 2013). However, the existence of microbes does not instantly entail that they are involved with the development of minerals, since they might perhaps merely have been submerged during the precipitation of minerals; thereby an evaluation of the precise act that the microbes might have enacted in the mineral precipitation is more often than not quite an unattainable one (Jones 2010). Yet, by having a critical outlook on microscopical and geochemical data accrued so far, scientists intend to confine exactly the microbial role in stalactite formation (Pacton et al. 2013).

In view of the fact that ^{13}C and ^{18}O ratios are customarily used as paleoclimate proxies, investigators are constantly on the lookout for microbial activity, and they are quite hopeful that these miniature microbial factories might have a prominent control on the carbonate composition. This in turn complicates the understanding of isotope time sequences with respect to climatic changes (Pacton et al. 2013). Variations in long-term speleothem ^{13}C values can highlight a range of variations in the carbon content, for example, modifications in the pattern of surface vegetation blanketing the cave, alterations in degassing associated with low or high drip rates, or it can be differences in host rock dissolution (Johnson et al. 2006; Pacton et al. 2013). In a study on forestation by Pacton et al. (2013), the investigators have witnessed that lesser drip rates and long-lasting CO_2 degassing coupled with an alteration from taiga forest to open tundra during the utmost dry season resulted in enriched ^{13}C values. They are firm believers that further negative ^{13}C values are due to the outcome of microbially mediated reactions. They have provided robust evidence of microbial origin by experiments which proved solid carbon isotope diminution inside the crack and EPS-like morphological evidence. But no microbial effect can be featured in the case of ^{18}O which is largely managed by abiotic course of actions such as temperature. Speleothem calcite development through purely abiogenic progressions has previously been confronted, and it is quite likely that frequent microbial-calcite alliance is established in an extensive range of climatic settings in caves (Jones 2010, 2011).

4 Seeking Out Microbial Presence in the Cosmos

A significant aspect of astrobiology is to expand our concept of life on Earth, particularly restrictions to life on Earth and life in extreme settings (Kommedal 2017). The information gathered from surveying the extreme surroundings as possible microbial environments assists us in appreciating which set of environmental parameters is proficient in assisting microbial growth (Fig. 1). The elementary prerequisites for life are water as a solvent, nutrients, and energy extracted from a chemical or the sun. These three utmost necessities for life are the foundation for the quest for life in extraterrestrial space (Kommedal 2017).

4.1 Habitability

The goal of astrobiology is to outline the elementary prerequisites for life which can withstand metabolic movements and multiplication in terrestrial settings over ages (Cockell et al. 2016). Cockell et al. (2016) describe habitability as the capability of an atmosphere to upkeep the action of minimum one identified organism.

Fig. 1 Seeking out microbial life in extreme environments of Krem Phyllut cave in Meghalaya, India, and correlating the same with the Martian environment

4.2 *Biosignatures*

Biosignatures are defined as chemical (elemental, organic, and/or inorganic), morphological, and isotopic touches of life conserved in sediments, rocks, and minerals (Westall and Cavalazzi 2011). They signify the physical existence of the organisms apart from signs of their metabolic actions (Fig. 2). These are firmly of biotic origin (Marais et al. 2008). In the quest for life away from Earth, it is necessary to differentiate between abiotic chemistry and biosignatures to prevent false negatives and false positives. Abiotic signatures look like life parameters under these conditions (Horneck et al. 2016).

4.3 *Approaches to Search for Life in Extraterrestrial Region*

Our knowledge of the possibility of discovering life in the Solar System and habitability has been considerably amplified since the past decade. A shift is observed as we have moved from seeing Mars–Earth–Venus as inhabitable to seeing the entire solar system as possibly fit to live in (Kommedal 2017). The freezing biospheres in the external portions of the solar system were one time considered inhabitable. But at present, they are vital objects in the quest for life in extraterrestrial space (Marion et al. 2003). The crucial controlling standard in the quest for life further than Earth

Fig. 2 Biotic signatures depicting role of microorganisms in speleothem formations in hypogean environment

is to monitor the presence of water to estimate the prospective for inhabitability. Earlier astrobiology operations have mainly stressed on primary examination, but undertakings in the present era are more engrossed astrobiology undertakings armed with probing the presence of liquid water (McKay et al. 2014; Shapiro and Schulze-Makuch 2009).

5 Quest for Microbial Life in the Rock-Strewn Planets: Venus, Mars, and Mercury

The probable occurrence of water in the subsurfaces and surfaces of Venus, Mars, and Mercury marks them fascinating objects where biosignatures and/or microbial life can be established (Kommedal 2017). Though frost has been established nearby to the extremes on Mercury, the constraints are not so favorable for life that whatsoever argument regarding foundation cannot be assumed an accurate probability (Schulze-Makuch et al. 2005). The external parameters on Venus are enormously dehydrating. Due to this condition, the absence of water and high temperature are the regulating elements for life (Schulze-Makuch et al. 2005). Whereas life at the subsurface or surface is considered doubtful, life can be considered possible in the Venusian atmosphere (Schulze-Makuch et al. 2005). Prominent among these planets is Mars where the subsurface has been reflected as settings that have the potential to upkeep chemotrophic organisms (Kommedal 2017).

6 Astrobiological Importance of Lava Tube Caves

Much before the recognition that most of the microbial life on Earth exists in the dark hypogean zone, Gold (1992) put forward his views on the exhaustive existence of microorganisms in the subsurface of the Earth in total darkness, irrespective of the surface world driven by photosynthesis. Understanding that the subsurface of Earth harboring innumerable types of bioconstructions has astrobiological implications (Kommedal 2017), the subsurface is assumed to afford inhabitable settings on planetary objects where superficial parameters are unreceptive for the existence of life, for example, in the case of Mars, the Red Planet (Kommedal 2017).

Lava tube caves are known to exist where basaltic lava is prevalent on Earth. These form due to the vigorous flow of lava (Northup et al. 2011). These caves have an oligotrophic, extreme atmosphere as a result of no photosynthesis taking place deep inside (the gloomy region). Nevertheless, records have shown that the lava tube caves are an abode of widespread mineral formations of biogenic sources comprising various colorations and microbial mats (Northup et al. 2011). In these caves, skylights are prevalent where the roof has caved in and present a normal means to enter the subsurface (Lavoie et al. 2017). On the Martian surface, the unearthing of lava tube caves approves the theory of employing these caves as astrobiology prototypes (Kommedal 2017). The biogeochemistry of lava tube caves remains rather uninvestigated in spite of their occurrence (Northup et al. 2011). Investigations carried out by Lavoie et al. (2017) summed up the sole microbiological study within these caves which correlated to the exterior populations. Focusing on these types of caves in seeking out Martian life calls for an improved grasp of the geomicrobiology of these caves prevalent on Earth (Kommedal 2017). Orbital imagery and remote-sensing have pointed cave-like features and lava tube caves on the Red Planet (Leveille and Datta 2010). This recognition has steered to the remarks of these caves as objects to study in astrobiology. Steady physicochemical parameters in these caves render them an improved microbial growth and additional mineral precipitation (Leveille and Datta 2010).

7 Mars: An Astrobiological Target

Mars is utmost related to Earth in the midst of the entire range of planets in the Solar System (Kommedal 2017). Mars and Earth have similarities in their geological origin history. The occurrence of water close to the surface has played a pivotal part in determining the present-day planets. As Mars and Earth were formed from almost identical mass, i.e., carbonaceous chondrites, they are hypothesized to harbor similar elements (Kommedal 2017). These carbonaceous chondrites were generated during the primordial Solar System in an oxygen-affluent region. It consisted primarily of complex hydrocarbons and hydrated silicates (McSween and Richardson 1977). On accepting that extreme habitats on Earth might support microorganisms, Boston et al. (1992) proposed the chances of searching chemosynthetic environments in the subsurface of Mars.

7.1 Early Mars

Signals of rivers, high erosion rates, and presence of valley networks, among other features in environment since the initial Noachian have suggested a warm and wet early Martian climate (4.1–3.7 billion years ago). These environmental conditions endorsed huge waterbodies over time (Kommedal 2017). But the majority of this water disappeared 3.1–3.7 billion years earlier, in the Hesperian (Cockell 2014). Noachian documents the finest proof that earlier parameters were dissimilar from present circumstances. It is frequently proposed that previous ecological settings on the Red Planet were analogous to those on Earth (Haberle 1998). The geographical views connected to the Noachian advocate a wetter and warmer one.

7.2 The Likelihood of Life Being Discovered on Mars

Hecht et al. (2009) studied statistics involved in the Phoenix lander operation and gathered signs regarding the existence of perchlorates in soil. The discovery was further reinforced by information from Curiosity and Mars Reconnaissance Orbiter. The incidence of soil perchlorates has insinuations for life as a result of heating, where they transform into highly volatile compounds (Kerr 2013). Water has earlier been known as a necessity for sustaining life. The composition of water in the top-most surface of the Martian subsurface has been recorded by the Neutron Spectrometer (MONS) and Mars Odyssey Gamma-Ray Spectrometer (GRS), which demonstrates extensive dissemination (Kommedal 2017). The prevalence of water as snow and ice is stipulated to be present at a depth of 35 m (Christensen 2006).

The six main elements, CHNOPS, essential for life on Earth have been established on Mars (Wackett et al. 2004). Carbon dioxide comprises 95% of the atmospheric gases, and as a result of exchange at the atmospheric level, carbon is still considered to be present near the Martian subsurface (Cockell 2014). There can also be water molecules which can radiolytically split to generate hydrogen atoms (Lin et al. 2005). One more likely source of hydrogen on Mars can be the serpentinization reactions (Ehlmann et al. 2010). In spite of nitrogen being found on Mars, it is anonymous whether the nitrogen could be secured by life in the subterranean environment (Cockell 2014). Volcanic activities, impact happenings, and lightning are considered abiotic phenomena which might trap nitrogen. This can convert nitrogen to valuable states, but the depth at which this nitrogen is found is unclear (Cockell 2014). Consequently, nitrogen can be a regulating element in the Martian biosphere (Cockell 2014). Oxygen is fixed in place to most of the further elements, such as H_2O, CO_2, perchlorates, sulfates, etc. (Cockell 2014). Mars is chiefly comprised of basaltic rocks. Evidence advocates the presence of phosphate in the secondary mineral called apatite (Cockell 2014). Varied arrangements of sulfate and sulfuric compounds are extensive on Mars, which is exhibited by their extensive prevalence (Cockell 2014). Orbital observations and in situ measurements have proved that the geosystems on Mars are intensely affected by the sulfur cycle. These outcomes advocate that the key factor in facilitating aqueous procedures in Martian history has been sulfur (Kommedal 2017).

7.3 *Life in Deep Underground Caves on Mars*

The Mars Exploration Program of NASA studies Mars as a planetary system with the aim of gathering data and information on the evolution of Mars as a planet, the geological processes, and the possibility of Mars to have supported or to support life. The Mars exploration strategy has evolved from "Follow the Water" to "Explore habitability" to "Seeks Signs of Life." The Program has four goals, namely, Goal 1: Life, Goal 2: Climate, Goal 3: Geology, and Goal 4: Humans. What is important from the perspective of Life on Mars is Goal 1, which is based on the idea that the planets Mars and Earth may have had similar history during their early evolution, as life arose early on Earth. The key question asked in scientific explorations is whether life ever had arose in Mars. The key elements of this goal are centered on two strategic points. The first is determining if environments contain evidence of earlier life in the form of potential habitability and conserved biosignatures of life. The second key point of this goal is to determine if the environments have high potential for current habitability and if biosignatures have evidence of extant life.

The Martian underground life theory hypothesis and the Jet Propulsion Laboratory research have provided basic evidence such as the presence of water which led NASA to claim the presence of life in the subterranean caves on Mars. The space research scientist of NASA, Vlada Stamenković elaborated the Mars hypogean life theories at a planetary occasion in 2020 (https://www.republicworld.com/technology-news/science/nasa-says-there-could-be-life-in-deep-underground-caves-of-mars.html). According to Stamenković, the Martian surface was very cold and dry, apart from being oxidizing. He proposed sending a rover to Mars for the detection of underground water and other chemical substances which can indicate the signs of life. He, however, underlined the fact that the liquid available on the Martian surface might not be constant for a long period of time as the extended existence would have indicated the possibility of extant life forms on Mars. The US Geological Survey's (USGS) Astrogeology Science Center has been credited with discovering and mapping more than 1000 potential cave sites on Mars.

NASA space study program has sent a warm electronics box (WEB) rover in 2020 to study the underground cave surfaces of Mars. The aim is to collect Martian samples related to life events in the underground caves which could be alive or extinct and direct them to Earth. Such sample gathering exercise is planned with the use of the 2260-pound space probe. This development is the result of the discovery of oxygen adjacent to the constellation of Ursa Major. It is a cluster about 560 million light years away, named Markarian 231.

There have been constant questions raised by scientists over the years on whether life could hide inside Mars'caves. Mars rover has been capturing views of caves.

Photos and data drawn from the Orbiter probe of NASA mission have provided assumptions regarding what seem to be uncommon structures in cave and water like patterns of drainage captured in uninterrupted pictures of the identical zone over a decade. The drainage channels evident in the captured photos of the Mars surface had raised hopes of researchers finding proof of flow of liquid in the present time on the surface of the planet resembling run-off from mountain gullies. The gullies' patterns have been seen to evolve as observed in the photographs dating back to the mission of 1999. The Mars Reconnaissance Orbiter (MRO) provides the most recent evidence which indicates the flows were loose and dry material on the Mars surface. This observation does not totally crush the water theory of Mars as other gullies offer strong evidence within the last few million years of water flowing on Mars phasing to climate cycles repeats. The tilts and orbits of Mars cause different phases of repeating climate cycles like in the Earth. This causes some eras to become warmer than others during these climate cycles. The pictures from the MRO's high-resolution camera showcase flawless signs of liquid flows in the form of channels within the gullies. Photos and reports from Mars show circles on the mountains of Mars as analyzed by the Odyssey probe. The circles are concluded as cave skylights or deep holes by researchers. The evidence, photos, and reports gathered from Mars so far indicate that the caves on Mars could offer a safe place for current or past life.

Roboticists from NASA's Jet Propulsion Laboratory have been working on caves on earth with robots designed to climb the porous walls of caves on Mars. One such robot is named LEMUR, a climbing field test robot tried in the caves of Mexico, located at Grants. The place is surrounded by volcanic lava of ancient times spread throughout the site. There is a web of subterranean passages called lava tubes: winding, long caves of sufficient width. The tubes found there are like the ones on Mars, although they are bigger on Mars due to weaker gravity. These caves in Mexico are described as quite a handful of the finest regions in the solar system to explore for celestial life.

The designated robot, LEMUR, climbs the wall similar to cockroaches by gripping the surface with tiny microspines, hundreds in number, fitted in the feet, the design of which was inspired by inputs from entomologists. LEMUR setting out to climb the cave wall was a feat in planetary exploration, especially with reference to exploring life in underground caves on Mars. Planetary geologists have had close-ups of the Mars surfaces way back in the 1960s which showed the moons of Mars and Earth having lava tubes created by flowing lava during volcanism. The main entrance to one of these caves was established by the Mars Reconnaissance Orbiter of NASA in 2007, with a distinct "skylight," created by the partial collapsing of the roof of a tube, which looks like a giant pit from above. SELENE, the lunar orbiter of Japan, by 2009, had recorded skylights on the surface of the Moon along with other holes and pits of several geological origins. Ever since the discovery of these structures on the Moon, space speleology has emerged as a new field of space science.

The fundamental principle on which the search for biosignatures in Mars' caves is the fact that the planetary surfaces are exposed to exceeding limits of intense cosmic radiation, regular planetary events, and higher thermal extremes which render preservation of biosignatures a difficult task over a period of geologic time. The underground caves prevalent on Mars, however, offer a shield from cosmic rays and lesser-scale impact measures thus providing a comparatively more constant atmosphere for the preservation of biosignatures over a long period of time. Thus these caves become prospective targets for astrobiological studies and search for biosignatures beyond Earth. A rover or penetrator launched from orbit finds direct access to the cave with natural openings to deeper subsurface materials without the need for drilling. Artificial intelligence and machine learning algorithms applied to analyses of Mars orbital data are identifying numerous putative cave access points to the underground Martian environments. The advanced planetary cave identification and access provides a strong base for future exploratory astrobiology missions to Mars and subsequent search for biosignatures of life in the Martian caves.

Flowing lava with molten rock can chew the path into the ground creating caves that might extend to a few kilometers. Such caves on the Moon are of immense interest to planetary geologists, astrobiologists, and space explorers. These caves or tubes created by the lava are relatively more hospitable and stable than the ones found on the surface of the planet. Such comparatively hospitable conditions appeal for search of life forms from microbes to other colonizers (Perkins 2020). If Mars at all hosted life, then it must have progressed into such hospitable refugia with the surface becoming harsh to habitat over a period of geologic time.

Data collected and analyzed for decades by the researchers led to speculation that lava tubes on the moon might present a vital seat for putting up living space even for colonists (Coombs and Hawke 1992). Laura Kerber, a geologist from NASA's Jet Propulsion Laboratory, explains that the environment in such lava tubes is protected from cosmic ray radiations, ultraviolet light, and active elements from the sun. Further, the rock would provide shield and protection from micrometeorites. The underground environment on the Moon, as on Earth, would be buffered from extreme temperature variations. Between 2009 and 2013, the temperature swings on an extrapolated form. The data collected by the Lunar Reconnaissance Orbiter indicate the temperatures near the lunar equator reach 120 °C in the morning and dive to −130 °C throughout the two-week-extensive darkness (Kerber 2020). The bioindicators, in preserved or fossilized form in the Martian cave, can provide a deeper insight not only into the history and evolution of Mars and life forms but also can lead to connecting Earth with other habitable planets.

8 Sulfur Caves as Analogs of the Martian Surface

The subsurfaces of Earth till long ago were thought of as devoid of life (Fendrihan 2010). With the passage of time, microorganisms have been identified in hypogean habitats, such as caves and mines (Fig. 3). This overwhelming biodiversity involved a subtle interplay of the cycles of phosphorous, carbon, sulfur, nitrogen, manganese, and further elements (Fliermans and Balkwill 1989).

Fig. 3 Robust paleoclimatic propositions deep inside a cave environment

Cave ecosystems are generally regarded as dark, oligotrophic environments with steady low temperatures (Gounot 1999). Quite a number of caves can harbor organic material supplied by water channels or from bats (Fendrihan 2010). Scientific reports have confirmed that caves can be an abode to a multifaceted group of microorganisms; some are biofilm formers on cave walls, while others remain associated with stalactites, stalagmites, and numerous other karst bioconstructions (Mulec 2008). These microorganisms have the ability to form areas of unusual colorations and corrosion and even showed promising interactions with minerals (Fendrihan 2010).

With the emergence of the sulfuric-acid speleogenetic concept, microbiologists observed critically that typical living systems which thrived the sulfur caves formed as a result of sulfuric acid speleogenesis (Boston et al. 2006). The list of caves belonging to this category worldwide is limited but growing at a fast pace, providing researchers with prospective analogs for a better grasp of these sites from a much broader perspective. This chance interaction between water, sulfide, and life in the hypogean environments might also showcase an impending replica of extraterrestrial bionetworks for the Red Planet (Fig. 4). This can support us in the quest for astrobiological exploration for searching life (Boston et al. 1992, 2001).

Boston (2003) has mentioned the presence of volcanic lava-tube caves on the Red Planet, but the quest remains on the possibility of any other means of bioconstruction formations on the Red Planet. On this aspect, it is worth mentioning that the topography of Mars and Earth are poles apart; probably biospeleogenesis can be hypothesized based on our knowledge about the planet (Boston et al. 2006). This curiosity on Martian caves extends beyond the biogeochemistry of cave formation to the probability that the caves might serve as storehouses of paleoclimatic indicators, geochemical facts, and significantly, footprints of earlier period life (Boston et al. 1992, 2001; Boston 1999, 2000).

The Mars Pathfinder Mission in 1997 has thrown light on subsurface ice (Mitrofanov et al. 2002). The Mission experts have given a firm conclusion regarding the presence of water zone underneath the surface based on an estimate of the flow of geothermal heat on Mars (Boston et al. 1992). An exciting breakthrough was revealed by the landing of the Mars Exploration Rover (MER) Spirit in Gusev

Fig. 4 Black sooty depositions on the cave wall showcase a fine interplay of sulfuric acid speleogenesis on earth

Crater which gathered substantial proof that the depression might be a superficial pond or a miniature ocean at some point of stretch (Squyres et al. 2004). Baker et al. (1991) propagated this impression of aquatic presence during the 1990s. Subsequent MER operations during 2004–2005 have also proved the proliferant combinations of sulfur on Mars (Rieder et al. 1997; Gellert et al. 2004; Boston et al. 2006). The most up-to-date eye-opener that shed light on the presence of microbes was the noticeable discovery of chemically reduced gases, namely, methane and ammonia, in the Red Planet (Wong et al. 2004). It has been proposed that the presence of this reduced gaseous envelope could serve as a source of energy for the sustenance of subterranean microbial biota (Boston et al. 1992, 2006; Boston 2010).

9 Ice-Cold Moons: Enceladus, Europa, and Titan

Enceladus, Europa (moon of Jupiter), and Titan (moon of Saturn) are the most gifted icy world objects in the quest for extraterrestrial life (Russell et al. 2014). Enceladus, being one of the moons of Saturn, is of importance from an astrobiological point of view. The trails emanating from the southern pole provinces have known configuration. This highlights the region as inhabitable to life similar to Earth (Kommedal 2017). Moreover, the freezing exterior shields a universal subterranean sea that is most probably created by tidal forces resulting in geothermal heating (McKay et al. 2014). At the rocky core regions of Enceladus, the existence of hydrothermal vents might sustain microorganisms below its freezing exterior (Russell et al. 2014). Another crucial object in our pursuit for possibly habitable domains is the Jovian moon, Europa (Kommedal 2017). Marion et al. (2003) argued prospective Europan environments for the occurrence of life. It was resolved that Europa embraces ample nutrients and energy to upkeep a slight biomass. Based on their strategies, Shapiro and Schulze-Makuch (2009) proposed Saturn's biggest moon Titan to be the topmost significance in the pursuit of life owing to the probability of searching for life. It was found that Titan comprises organic chemicals which are confirmed by a cycle relying on the gas methane (Lunine and Lorenz 2009).

10 Conclusion

A voyage to Mars or another celestial body is a multifaceted one and an expensive affair, owing to which environmental facts from other planets are limited (Kommedal 2017). Mars remains to be the topmost important object. It is also the planet for which there is maximum data regarding present and past environmental settings (Cockell 2014). Earth is thought of as a harbor for specific habitats. As our understanding regarding environmental parameters and life on the subterranean surface progresses, the more we are able to better grasp our concept of these locations as impending environments away from Earth. The metabolism of microorganisms in hypogean environments is exceedingly different; quite a handful of them can be chemolithotrophs, carrying out Fe III, Mg reduction, sulfide reduction, or others, while the remaining ones can serve as heterotrophs (Fendrihan 2010). Innumerable metabolic pathways can be present in order to thrive in such inhospitable circumstances. The evidence that these subsurface microorganisms are viable for indefinite periods in rocks (for example amber, halite), permafrost, and ancient ice is actually a challenge for the scientific society (Fendrihan 2010). Future investigations like the mission EXOMARS might capture the lead from explorations in probing for microbial life in the extreme environments of the Red Planet (Fendrihan 2010).

Even though carbonate formation can be facilitated by microorganisms under diverse environmental circumstances, comprising dry and cold climates mainly related to permafrost incidence, the calcite growth commencement and cessation timings are not fixed with accuracy at this phase. Calcite ^{13}C has been understood to reflect the shift in vegetation patterns, even though up-to-date investigations indicate the influence of cave aeration (Tremaine et al. 2011; Pacton et al. 2013). Supplemented to that, it has been proved that degassing of CO_2 from the drip water had an effect on the pace of calcite precipitation (Kowalczk and Froelich 2010; Lambert and Aharon 2011). These analyses showed that an accurate and patient evaluation might showcase the usefulness of ^{13}C as a paleoclimatic pointer. Nonetheless, the impending influence of microbial action on the fractionation of carbon isotope is by and large neglected in paleoclimatic surveys (Pacton et al. 2013). As highlighted by Pacton et al. (2013), hiatuses generated as a result of permafrost buildup are associated with ^{13}C exhaustion and diligently connected to porous cracks. Hence, these biosignatures might be a marker of environmental adjustments in diverse climatic settings.

Competing Interests All the authors declare that they have no competing interests.

References

Azúa-Bustos A, González-Silva C, Mancilla RA, Salas L, Palma RE, Wynne JJ, McKay CP, Vicuña R (2009) Ancient photosynthetic eukaryote biofilms in an Atacama Desert coastal cave. Microb Ecol 58(3):485–496

Baker VR, Strom RG, Gulick VC, Kargel JS, Komatsu G, Kale VS (1991) Ancient oceans and ice sheets and the hydrological cycle on Mars. Nature 352:589–594

Banerjee S, Joshi SR (2013) Insights into cave architecture and the role of bacterial biofilm. P Natl A Sci India B 83(3):277–290

Banerjee S, Joshi SR (2014) Ultrastructural analysis of calcite crystal patterns formed by biofilm bacteria associated with cave speleothems. J Microsc Ultrastruct 2(4):217–223

Banerjee S, Joshi SR (2016a) Mineralogical footprints of bacterial biofilms associated with Labit cave, a part of the longest cave system in India. Geomicrobiol J 33(8):699–708

Banerjee S, Joshi SR (2016b) Culturable bacteria associated with the caves of Meghalaya in India contribute to speleogenesis. J Cave Karst Stud 78(3):144–157

Banerjee S, Jha DK, Joshi SR (2019) Cave microbiome for human welfare. In: Satyanarayana T, Das S, Johri B (eds) Microbial diversity in ecosystem sustainability and biotechnological applications. Springer, Singapore, pp 3–30

Baskar S, Baskar R, Mauclaire L, McKenzie JA (2005) Role of microbial community in stalactite formation, Sahastradhara caves, Dehradun, India. Curr Sci India 88:1305–1308

Baskar S, Baskar R, Mauclaire L, McKenzie JA (2006) Microbially induced calcite precipitation by culture experiments-possible origin for stalactites in Sahastradhara, Dehradun, India. Curr Sci India 90:58–64

Baskar S, Baskar R, Kaushik A (2007) Evidences for microbial involvement in the genesis of speleothem carbonates, Borra Caves, Visakhapatnam, India. Curr Sci India 92:350–355

Baskar S, Baskar R, Lee N, Theophilus PK (2009) Speleothems from Mawsmai and Krem Phyllut caves, Meghalaya, India: some evidences on biogenic activities. Environ Geol 57:1169–1186

Boston PJ (1999) The search for extremophiles on Earth and beyond: what is extreme here may be just business-as-usual elsewhere. Ad Astra 11(1)

Boston PJ (2000) Life below and life 'out there'. Geotimes 45(8):14–17

Boston PJ (2003) Extraterrestrial caves. In: Gunn J (ed) Encyclopedia of cave and karst science. Fitzroy-Dearborn Publishers Ltd., London, pp 355–358

Boston PJ (2010) Location, location, location! Lava caves on Mars for habitat, resources, and the search for life. J Cosmol 12:3957–3979

Boston PJ, Ivanov MV, McKay CP (1992) On the possibility of chemosynthetic ecosystems in subsurface habitats on Mars. ICARUS 95:300–308

Boston PJ, Spilde MN, Northup DE, Melim LA, Soroka DS, Kleina LG, Lavoie KH, Hose LD, Mallory LM, Dahm CN, Crossey LJ, Schelble RT (2001) Cave biosignature suites: microbes, minerals and Mars. Astrobiology 1(1):25–55

Boston PJ, Hose LD, Northup DE, Spilde MN (2006) The microbial communities of sulfur caves: a newly appreciated geologically driven system on Earth and potential model for Mars. Geol Soc Am Bull 404:331

Cañaveras JC, Sanchez-Moral S, Soler V, Saiz-Jimenez C (2001) Microorganisms and microbially induced fabrics in cave walls. Geomicrobiol J 18:223–240

Chapelle FH, O'Neill K, Bradley PM, Methe BA, Ciufo SA, Knobel LL, Lovley DR (2002) A hydrogen-based subsurface microbial community dominated by methanogens. Nature 415:312–315

Christensen PR (2006) Water at the poles and in permafrost regions of Mars. Elements 2:151–155

Cockell CS (2014) The subsurface habitability of terrestrial rocky planets: Mars. In: Kallmeyer J, Wagner D (eds) Microbial life in the deep biosphere. Walter de Gruyter GmbH, Berlin

Cockell CS, Bush T, Bryce C, Direito S, Fox-Powell M, Harrison JP, Lammer H, Landenmark H, Martin-Torres J, Nicholson N, Noack L, O'Malley-James J, Payler SJ, Rushby A, Samuels

T, Schwendner P, Wadsworth J, Zorzano MP (2016) Habitability: a review. Astrobiology 16:89–117
Coombs CR, Hawke BR (1992) A search for intact lava tubes on the moon: possible lunar base habitats. In: Proceedings of the second conference on lunar bases and space activities of the 21st century. NASA, Johnson Space Center, Houston, p 219
Corliss JB, Dymond J, Gordon LI, Edmond JM, Von Herzen RP, Ballard RD, Green K, Williams D, Bainbridge A, Crane K, Van Andel TH (1979) Submarine thermal sprirngs on the galapagos rift. Science 203:1073–1083
Edwards KJ, Becker K, Colwell F (2012) The deep, dark energy biosphere: Intraterrestrial life on earth. Annu Rev Earth Planet Sci 40(40):551–568
Ehlmann BL, Mustard JF, Murchie SL (2010) Geologic setting of serpentine deposits on Mars. Geophys Res Lett 37
Fendrihan S (2010) Microorganisms isolated from subsurface environments and their importance for astrobiology and theoretical biology. ELBA Bioflux 2(1):23–36
Fliermans CB, Balkwill DL (1989) Microbial life in deep terrestrial subsurfaces. Bioscience 39:370
Gellert R, Rieder R, Anderson RC, Bruckner J, Clark BC, Dreibus G, Economou T, Klingelhofer G, Lugmair GW, Ming DW, Squyres SW, d'Uston C, Wanke H, Yen A, Zipfel J (2004) Chemistry of rocks and soils in Gusev crater from the alpha particle X-ray spectrometer. Science 305:829–832
Gold T (1992) The deep, hot biosphere. Proc Natl Acad Sci USA 89:6045–6049
Gounot E (1999) Microbial life in permanent cold soils. In: Margesin R, Shinner F (eds) Cold adapted organisms. Springer, Berlin, pp 3–15
Haberle RM (1998) Early Mars climate models. J Geophys Res E Planet 103:28467–28479
Hartley A, Chong G, Houston J, Mather A (2005) 150 million years of climatic stability: evidence from the Atacama Desert, northern Chile. J Geol Soc 162:421–424
Hecht MH, Kounaves SP, Quinn RC, West SJ, Young SMM, Ming DW, Catling DC, Clark BC, Boynton WV, Hoffman J, Deflores LP, Gospodinova K, Kapit J, Smith PH (2009) Detection of perchlorate and the soluble chemistry of Martian soil at the Phoenix Lander Site. Science 325:64–67
Horneck G, Walter N, Westall F, Grenfell JL, Martin WF, Gomez F, Leuko S, Lee N, Onofri S, Tsiganis K, Saladino R, Pilat-Lohinger E, Palomba E, Harrison J, Rull F, Muller C, Strazzulla G, Brucato JR, Rettberg P, Capria MT (2016) AstRoMap European astrobiology roadmap. Astrobiology 16:201–243
Johnson KR, Hu CY, Belshaw NS, Henderson GM (2006) Seasonal trace-element and stable-isotope variations in a Chinese speleothem: the potential for high-resolution paleomonsoon reconstruction. Earth Planet Sci Lett 244:394–407
Jones B (2010) Microbes in caves: agents of calcite corrosion and precipitation. In: Pedley HM, Rogerson M (eds) Tufas and speleothems: unravelling the microbial and physical controls, vol 336. Geological Society, Special Publications, London, pp 7–30
Jones B (2011) Stalactite growth mediated by biofilms: example from Nani Cave, Cayman Brac, British West Indies. J Sediment Res 81:322–338
Kerber L (2020) The state of extraterrestrial cave science and exploration. Plenary lecture at the third international planetary caves conference. https://www.hou.usra.edu/meetings/3rdcaves2020/pdf/3rdcaves2020_program.htm#sess101. Accessed 17 October 2020
Kerr RA (2013) Pesky perchlorates all over Mars. Science 340:138–138
Kommedal EG (2017) Microbiology of lava tube skylights on earth: implications for astrobiology. Master's thesis, University of Stavanger, Norway
Kowalczk A, Froelich P (2010) Cave air ventilation and CO_2 outgassing by radon-222 modeling: how fast do caves breathe? Earth Planet Sci Lett 289:209–219
Lambert W, Aharon P (2011) Controls on dissolved inorganic carbon and ^{13}C in cave waters from DeSoto Caverns: implications for speleothem ^{13}C assessments. Geochim Cosmochim Acta 75:53–768

Lavoie KH, Winter AS, Read KJ, Hughes EM, Spilde MN, Northup DE (2017) Comparison of bacterial communities from lava cave microbial mats to overlying surface soils from Lava Beds National Monument, USA. PLoS One 12:e0169339
Leveille RJ, Datta S (2010) Lava tubes and basaltic caves as astrobiological targets on earth and Mars: a review. Planet Space Sci 58:592–598
Leverington D (2007) A history of solar system studies. AAS/Div Ext Solar Syst Abst 1:53–70
Lin LH, Hall J, Lippmann-Pipke J, Ward JA, Lollar BS, Deflaun M, Rothmel R, Moser D, Gihring TM, Mislowack B, Onstott TC (2005) Radiolytic H_2 in continental crust: nuclear power for deep subsurface microbial communities. Geochem Geophys Geosyst 6(7):Q07003
Lunine JI, Lorenz RD (2009) Rivers, lakes, dunes, and rain: crustal processes in Titan's methane cycle. Annu Rev Earth Planet Sci 37:299–320
Madigan MT, Clark DP, Stahl D, Martinko JM (2010) Brock biology of microorganisms, 13th edn. Benjamin Cummings
Marais DJD, Nuth JA, Allamandola LJ, Boss AP, Farmer JD, Hoehler TM, Jakosky BM, Meadows VS, Pohorille A, Runnegar B, Spormann AM (2008) The NASA astrobiology roadmap. Astrobiology 8:715–730
Marion GM, Fritsen CH, Eicken H, Payne MC (2003) The search for life on Europa: limiting environmental factors, potential habitats, and Earth analogues. Astrobiology 3:785–811
McKay CP (2004) What is life—and how do we search for it in other worlds? PLoS Biol 2:1260–1263
McKay CP, Friedmann EI, Gómez-Silva B, Cáceres-Villanueva L, Andersen DT, Landheim R (2003) Temperature and moisture conditions for life in the extreme arid region of the Atacama desert: four years of observations including the El Niño of 1997–1998. Astrobiology 3(2):393–406
McKay CP, Anbar AD, Porco C, Tsou P (2014) Follow the plume: the habitability of Enceladus. Astrobiology 14:352–355
McSween HY, Richardson SM (1977) Composition of carbonaceous chondrite matrix. Geochim Cosmochim Acta 41:1145–1161
Mitrofanov I, Anfimov D, Kozyrev A, Litvak M, Sanin A, Tret'yakov V, Krylov A, Shvetsov V, Boynton W, Shinohara C, Hamara D, Saunders RS (2002) Maps of subsurface hydrogen from the high energy neutron detector. Mars Odys Sci 297:78–81
Mulec J (2008) Microorganisms in hypogeon: examples from Slovenian karst caves. Acta Carsol 37(1):153–160
Mulec J, Kosi G, Vrhovsek D (2007) Algae promote growth of stalagmites and stalactites in karst caves (Skocjanske jame, Slovenia). Carbonates Evaporites 22:6–9
Navarro-González R, Rainey FA, Molina P, Bagaley DR, Hollen BJ, de la Rosa J, Small AM, Quinn RC, Grunthaner FJ, Cáceres L, Gomez-Silva B, McKay CP (2003) Mars-like soils in the Atacama Desert, Chile, and the dry limit of microbial life. Science 302:1018–1021
Northup DE, Lavoie K (2001) Geomicrobiology of caves: a review. Geomicrobiol J 18:199–222
Northup DE, Melim LA, Spilde MN, Hathaway JJM, Garcia MG, Moya M, Stone FD, Boston PJ, Dapkevicius MLNE, Riquelme C (2011) Lava cave microbial communities within mats and secondary mineral deposits: implications for life detection on other planets. Astrobiology 11:601–618
Pacton M, Breitenbach SF, Lechleitner FA, Vaks A, Rollion-Bard C, Gutareva OS, Osintcev AV, Vasconcelos C (2013) The role of microorganisms in the formation of a stalactite in Botovskaya cave, Siberia: paleoenvironmental implications. Biogeosciences 10(9):6115–6130
Parnell J, McMahon S (2016) Physical and chemical controls on habitats for life in the deep subsurface beneath continents and ice. Philos Trans R Soc London, Ser A 374(2059):20140293
Pedersen K (2000) Exploration of deep intraterrestrial microbial life: current perspectives. FEMS Microbiol Lett 185:9–16
Perkins S (2020) Lava tubes may be havens for ancient alien life and future human explorers. PNAS 117(30):17461–17464

Rieder R, Economou T, Wanke H, Turkevich A, Crisp J, Bruckner J, Dreibus G, McSween HY (1997) The chemical composition of Martian soil and rocks returned by the mobile alpha proton X-ray spectrometer: preliminary results from the X-ray mode. Science 278(5344):1771–1774

Russell MJ, Barge LM, Bhartia R, Bocanegra D, Bracher PJ, Branscomb E, Kidd R, McGlynn S, Meier DH, Nitschke W, Shibuya T, Vance S, White L, Kanik I (2014) The drive to life on wet and icy worlds. Astrobiology 14:308–343

Schulze-Makuch D, Dohm JM, Fairen AG, Baker VR, Fink W, Strom RG (2005) Venus, Mars, and the ices on Mercury and the Moon: Astrobiological implications and proposed mission designs. Astrobiology 5:778–795

Shapiro R, Schulze-Makuch D (2009) The search for alien life in our solar system: strategies and priorities. Astrobiology 9:335–343

Squyres SW, Arvidson RE, Bell JF, Bruckner J, Cabrol NA, Calvin W, Carr MH, Christensen PR, Clark BC, Crumpler L, Des Marais DJ, d'Uston C et al (2004) The Spirit Rover's Athena science investigation at Gusev Crater, Mars. Science 305:794–799

Stevens TO, McKinley JP (2000) Abiotic controls on H_2 production from basaltwater reactions and implications for aquifer biogeochemistry. Environ Sci Technol 34:826–831

Tremaine DM, Froelich PN, Wang Y (2011) Speleothem calcite farmed in situ: modern calibration of ^{18}O and ^{13}C paleoclimate proxies in a continuously-monitored natural cave system. Geochim Cosmochim Acta 75:4929–4950

Wackett LP, Dodge AG, Ellis LBM (2004) Microbial genomics and the periodic table. Appl Environ Microbiol 70:647–655

Westall F (2012) The early earth. In: Impey C, Lunine JI, Funes JG (eds) Frontiers of astrobiology. Cambridge University Press, Cambridge, UK, pp 89–114

Westall F, Cavalazzi B (2011) Biosignatures in rocks. In: Reitner J, Thiel V (eds) Encyclopedia of geobiology. Springer, Dordrecht, pp 189–201

Westall F, Foucher F, Bost N, Bertrand M, Loizeau D, Vago JL, Kminek G, Gaboyer F, Campbell KA, Breheret JG, Gautret P, Cockell CS (2015) Biosignatures on Mars: what, where, and how? Implications for the search for Martian life. Astrobiology 15:998–1029

Whitman WB, Coleman DC, Wiebe WJ (1998) Prokaryotes: the unseen majority. Proc Natl Acad Sci USA 95:6578–6583

Wilkins MJ, Daly RA, Mouser PJ, Trexler R, Sharma S, Cole DR, Wrighton KC, Biddle JF, Denis EH, Fredrickson JK, Kieft TL, Onstott TC, Peterson L, Pfiffner SM, Phelps TJ, Schrenk MO (2014) Trends and future challenges in sampling the deep terrestrial biosphere. Front Microbiol 5:481

Wong AS, Atreya SK, Formisano V, Encrenaz T, Ignatiev NI (2004) Atmospheric photochemistry above possible Martian hot spots. Adv Space Res 33(12):2236–2239

Ecological Restorations of Wetlands by Geomicrobes

Prasann Kumar and Komal Agrawal

Contents

Abstract Wetlands are ecosystems of incredible biodiversity renowned for purifying water and sequestering carbon. But humanity's more recent activities and a changing climate have endangered them. As a result, it is necessary to reform the traditional restoration methods, mainly based on physical and chemical treatment, to restore the wetland ecological system more comprehensively. Here, we describe the ability of geomicrobes, which are essential for many geological and chemical

P. Kumar (✉)
Department of Agronomy, School of Agriculture, Lovely Professional University, Phagwara, Punjab, India

K. Agrawal
Department of Microbiology, School of BioEngineering and Biosciences, Lovely Professional University, Phagwara, Punjab, India

S. R. Joshi et al. (eds.), *Mineral Transformation and Bioremediation by Geo-Microbes*, Interdisciplinary Biotechnological Advances,
https://doi.org/10.1007/978-981-96-3033-2_11

processes, to contribute to environmental recovery in a wetland system. Geomicrobes provide an environmentally friendly ecosystem service to restore wetlands by improving nutrient cycling, promoting plant growth, and stabilising sediments. It emphasises research on the roles of geomicrobes in wetlands, outlines their ecological functions, and proposes a strategy for incorporating microbial-based restoration technologies into wetland restoration works.

Keywords Wetlands · Microbes · Restoration · Remediation · Climate

Abbreviations

CH_4	Methane
CO_2	Carbon dioxide
EPS	Extracellular polymeric substances

1 Introduction

Wetlands are intricate and varied ecosystems distinguished by water, either in a permanent or seasonal state, which floods the land, establishing optimal circumstances for aquatic flora and a broad range of fauna. The distinctiveness of these ecosystems lies in their capacity to operate as both terrestrial and aquatic environments, rendering them crucial transitional zones between these two types of habitats. Wetlands' morphology is influenced by geographical location, hydrology, soil composition, and the specific vegetation they sustain (Kumar et al. 2023a, 2024j; Sharma et al. 2024). There are four main categories of wetlands: marshes, swamps, bogs, and fens, each characterised by unique habitats and ecological roles. Marshes are distinguished by the presence of herbaceous vegetation, including grasses, reeds, and sedges, and they frequently exhibit shallow, elevated bodies of water. Conversely, swamps provide habitat for arboreal vegetation, such as trees and shrubs, commonly located in floodplains or near rivers. The acidic wetlands known as bogs are characterised by the accumulation of peat, a form of organic material, and are commonly inhabited by mosses, notably *Sphagnum*. Fens, similar to bogs, are wetlands that produce peat. However, they obtain nutrients from groundwater and typically have lower acidity levels, allowing them to sustain a more varied plant community than bogs. Wetlands confer crucial ecological services necessary for the preservation of environmental health and the welfare of humans. Water purification is classified as one of the most essential services provided by wetlands. Wetland vegetation and microbiota effectively remove contaminants such as nitrates, phosphates, heavy metals, and sediments from surface water before entering larger water bodies such as rivers, lakes, or oceans. This method aids in preserving water quality and mitigates the potential for waterborne illnesses and detrimental algal blooms.

Furthermore, wetlands play a crucial role in managing floods. Wetlands function as inherent reservoirs, assimilating surplus precipitation and gradually releasing it, alleviating the consequences of flooding and diminishing the likelihood of harm to adjacent human communities and infrastructure (Devi et al. 2024e, f; Kumar and Singh 2024a, b, c, d). Moreover, wetlands function as crucial sites for carbon sequestration, so assuming great importance in the battle against climate change. Wetland soils, especially those found in peatlands, accumulate substantial carbon. The liberation of stored carbon into the atmosphere, such as carbon dioxide, during the drainage or degradation of wetlands, contributes to global warming. Additionally, wetlands serve as biodiversity hotspots, offering habitat for various species, such as birds, fish, amphibians, and plants. Numerous endangered species' survival depends on wetlands, so these ecosystems are crucial for preserving biodiversity. Notwithstanding their significance, wetlands face substantial jeopardy from human activities. Urbanisation has devastated wetlands as land is transformed into residential space, commercial establishments, and infrastructure. This phenomenon not only results in the depletion of biodiversity but also reduces the capacity of wetlands to carry out vital ecological functions such as water purification and erosion control. Fluvial agricultural runoff, containing fertilisers and pesticides, contaminates wetlands and disturbs the inherent equilibrium of nutrient cycling, resulting in eutrophication, a condition in which an excess of nutrients triggers algal blooms that exhaust oxygen and cause the death of aquatic organisms. Industrial pollution, defined as the release of heavy metals and hazardous chemicals, exacerbates the deterioration of wetland water quality, causing damage to both wildlife and individuals dependent on these ecosystems. Moreover, climate change presents a persistent danger to wetlands through escalating sea levels, modified precipitation patterns, and elevated temperatures, all of which can disturb the hydrology of these delicate ecosystems. Wetland ecological restoration refers to restoring wetlands that have been degraded or destroyed to a state that closely resembles their initial operational condition. The principles of environmental restoration underscore the need to remediate impaired ecosystems to reinstate their inherent processes and functions. This paradigm is founded on comprehending healthy ecosystems' structure, composition, and function and using that knowledge to restore deteriorated environments. Restoration efforts in wetlands seek to restore hydrology, soil structure, and plant communities to establish conditions that support the sustainability of wetland biodiversity and ecosystem services. Due to their substantial ecological, economic, and social advantages, the restoration of wetlands is of utmost importance. The capacity of degraded wetlands to carry out vital functions such as water filtration, flood control, and carbon sequestration is diminished (Devi and Kumar 2024a; Kumar et al. 2024f, e). The effects of this phenomenon extend beyond local ecosystems and have wider consequences for the quality of water, preservation of biodiversity, and regulation of climate. The restoration of wetlands can effectively address these concerns by reinstating the inherent mechanisms that sustain the provision of clean water, diminish the likelihood of flooding, and sequester carbon. Furthermore, restored wetlands can function as crucial habitats for wildlife, conserving species variety and bolstering ecological resilience in response to climate change. The objectives of wetland restoration are diverse, aiming to improve ecological services and restore biodiversity. An essential goal is to reinstate the hydrological functions of wetlands, so ensuring that water levels and flow patterns replicate natural conditions. This often entails the removal of obstacles to water flow, such as dikes or drainage systems, and reintroducing indigenous plant species that regulate water movement and enhance water quality. An additional crucial objective is reinstating the wetland's capacity to function as a carbon sink by restoring the requisite conditions for producing peat or other organic material accumulation. Furthermore, wetland restoration often addresses social and economic objectives, including enhancing water quality for human consumption, improving recreational prospects, and bolstering community resilience to natural calamities such as floods and droughts. Microorganisms, known as geomicrobes, are essential in geochemical processes in the periodic exchange of elements such as carbon, nitrogen, sulfur, and phosphorus (Devi and Kumar 2024b; Kumar et al. 2024a, d, g). These microorganisms play a crucial role in the operation of wetlands, facilitating essential biogeochemical processes that support the ecosystem. Understanding the function of geomicrobes is crucial in wetland restoration as these small organisms often play a dominant role in the ecological processes that contribute to the value of wetlands, such as nutrient cycling, organic matter breakdown, and pollutant breakdown. The microbial ecology in wetlands is characterised by tremendous diversity, as many species of bacteria, archaea, fungi, and protists inhabit different ecological niches. Through interactions with one another and their surroundings, these microorganisms establish intricate networks that control the accessibility of nutrients and the decomposition of organic matter. Wetland soils are generally characterised by anaerobic conditions, indicating low oxygen levels, which provide specific microbial species, including anaerobic bacteria and methanogens, with favourable environments. Under oxygen-deprived conditions, these microorganisms play a vital role in the wetland carbon cycle by facilitating processes such as methanogenesis and methane generation. Geomicrobes exert influences on biogeochemical cycles in wetlands through various mechanisms. Within the carbon cycle, they chemically decompose organic matter and emit either carbon dioxide or methane, contingent upon oxygen. Methanogenic bacteria are crucial in anaerobic wetlands as they transform organic carbon into methane, a highly potent greenhouse gas. Additional microorganisms, known as methanotrophs, can metabolise methane, thus mitigating the total greenhouse gas emissions originating from wetlands. Within the nitrogen cycle, nitrifying and denitrifying bacteria convert nitrogen compounds into gases such as nitrogen or nitrous oxide, which are then released into the atmosphere or rendered available to plants. Similarly, sulfur-reducing bacteria in wetlands contribute to the sulfur cycle by transforming sulfate into sulfide, which can be utilised by other microorganisms or integrated into plant constituents. Microbial activity significantly affects phosphorus cycling in wetlands by facilitating the solubilisation of phosphorus, enabling its availability for plant absorption and preventing its buildup in a state that may cause eutrophication (Choudhury et al. 2024; Kumar et al. 2024h, k; Kumar and Choudhury 2024a; Pandey et al. 2024).

2 Wetland Degradation and the Need for Restoration

2.1 *Causes of Wetland Degradation*

Wetlands are substantially endangered by various human activities, notably pollution from agriculture, industry, and urbanisation. Agricultural activities frequently include the application of fertilisers and pesticides, which have the potential to cascade into adjacent wetlands, transporting substantial quantities of nutrients, namely nitrogen and phosphorus. The excessive accumulation of nutrients results in eutrophication, which expedites the proliferation of algae and aquatic plants, disturbing the inherent equilibrium of wetland ecosystems. As algae blooms perish and undergo decomposition, they metabolise oxygen from the water, resulting in oxygen depletion or hypoxia. This condition adversely affects aquatic life, resulting in extreme mortality where only a few organisms can survive. Likewise, industrial pollution, including heavy metals, chemicals, and untreated waste, pollutes the water and soil of wetlands, causing harm to plants and animals and deteriorating the ecosystem's overall health. Urbanisation exacerbates these issues, as the physical development of buildings, roads, and infrastructure results in heightened stormwater runoff and transports contaminants such as oil, garbage, and sediments into wetlands. This introduces pollutants and modifies the inherent hydrological dynamics, exacerbating the wetland ecosystem strain. Habitat fragmentation poses a significant risk to wetlands, as the conversion of land for agricultural or urban use segregates wetland habitats into comparatively smaller areas (Devi et al. 2024a; Dey et al. 2024; Kumar et al. 2024i; Kumar and Choudhury 2024b; Kumar and Raut 2024; Sharma and Kumar 2024). This phenomenon hinders the ecological processes of species migration, diminishes genetic variability, and increases ecosystems' vulnerability to changes. Furthermore, it impairs wetlands' capacity to fulfil their role as uninterrupted ecological pathways, reducing their efficacy in sustaining biodiversity and delivering ecological services. The phenomenon of climate change presents a persistent and fundamental danger to wetlands. Elevated sea levels resulting from global warming threaten coastal wetlands, resulting in the submersion of these ecosystems and the erosion of land by the ocean. Variations in temperature and modified precipitation patterns also disturb the hydrological structure of wetlands, impacting water availability and the timing of seasonal water flows. Droughts or excessive flooding can significantly damage wetland ecosystems, impairing their capacity to sustain biodiversity, sequester carbon, and maintain water quality.

2.2 *Consequences of Wetland Degradation*

The deterioration of wetlands, these invaluable centres of biodiversity, has significant ramifications for our planet. Wetlands are habitats for many plant, animal, and microbial species, making them one of Earth's most biologically diverse ecosystems. When wetlands undergo drainage, pollution, or fragmentation, the habitats they provide are eradicated, substantially reducing biodiversity. Numerous avian, piscine, amphibian, and invertebrate species depend on wetlands for sustenance, reproduction, and thermal refuge. The depletion of these crucial habitats leads to decreased species numbers and even complete extinctions, especially for species that are specialised or exclusive to wetland ecology. This decline in biodiversity undermines the whole ecosystem, rendering it more susceptible to environmental fluctuations and less robust in the presence of pressures such as climate change. Furthermore, the degradation of wetlands causes species extinction and disturbs the inherent water and nutrient cycling process, which is a crucial ecological role of these ecosystems. Wetlands operate as essential intermediaries between land and water environments, controlling water movement and eliminating contaminants before entering rivers, lakes, and oceans. Modifying or destroying wetlands undermines their capacity to control water flow and ensure water purity. Consequently, this results in heightened runoff, erosion, and sedimentation in adjacent water bodies. Nutrient cycling, namely the nitrogen and phosphorus cycles, significantly relies on wetland ecosystems. Microbial populations in healthy wetlands are crucial in transforming nutrients into forms capable of being utilised by plants or safely eliminated from the ecosystem. The depletion of wetlands diminishes these inherent chemical reactions, resulting in imbalances of nutrients and exacerbating problems such as eutrophication in neighbouring aquatic systems (Devi et al. 2024d; Kumar et al. 2024c; Kumar 2024a; Saurabh et al. 2024). Wetland degradation has wide-ranging consequences that extend well beyond their boundaries, impacting the overall health of our planet. Yet another vital role of wetlands is the sequestration of carbon. Wetland soils, particularly those found in peatlands, have a substantial capacity to retain carbon, rendering them crucial in mitigating climate change. Nevertheless, when wetlands undergo drainage or degradation, the organic material lodged in the soils of the wetlands becomes exposed to oxygen, resulting in its breakdown and carbon dioxide (CO_2) emission into the atmosphere. This process not only reduces the ability of wetlands to function as carbon sinks but also adds to the overall greenhouse gas emissions on a global scale. Furthermore, the absence of oxygen in numerous wetlands also promotes the generation of methane (CH_4), a potent greenhouse gas. The regulation of methane emissions in intact wetlands is facilitated by microbial processes that consume methane. However, degraded wetlands tend to release elevated levels of both methane and CO_2. This shift intensifies global climate change by augmenting the total concentration of greenhouse gases in the atmosphere. Therefore, the depletion of wetlands weakens the functioning of local ecosystems and has extensive consequences on global climate regulation.

2.3 Need for Ecological Restoration

Restoring degraded wetlands is paramount to preserving and reinstating these distinct habitats' ecological services. With their profound biodiversity, wetlands are essential in managing water flow, mitigating floods, and purifying contaminants. The degradation of wetlands results in the loss or significant reduction of these functions, causing adverse effects on the surrounding ecosystems and human populations. Degraded wetlands, for instance, have reduced capacity to absorb surplus rainfall, heightening adjacent regions' vulnerability to flooding. Additionally, they acquire a diminished capacity to effectively remove contaminants from surface water, resulting in substandard water quality that can adversely affect downstream ecosystems and human water sources. The restoration of wetlands enables the recovery of the essential services they provide, enhancing both the ecological health and the welfare of the communities reliant on them (Devi et al. 2024b; Kumar 2024b, c; Kumar et al. 2024c). Moreover, restored wetlands contribute to mitigating climate change by augmenting carbon sequestration. This is because well-functioning wetlands can store substantial quantities of carbon in their soils, diminishing the carbon dioxide concentration in the atmosphere. A crucial determinant in operating both natural and restored wetlands is the microbial populations that reside within them. Microbial organisms, such as bacteria, archaea, fungi, and protists, are crucial in driving several biogeochemical processes in wetlands. They perform a pivotal function in nutrient cycling, which involves the conversion of carbon, nitrogen, phosphorus, and sulfur, essential for preserving ecological equilibrium. For instance, specific bacteria are accountable for the processes of nitrification and denitrification in the nitrogen cycle, which govern the availability of nitrogen in forms capable of being utilised by plants. Microbial activity in the carbon cycle involves the breakdown of organic matter and the subsequent generation of gases, including carbon dioxide and methane. Some microbes, known as methanotrophs, even consume methane, contributing to regulating greenhouse gas emissions. Microbiological communities play a crucial role in water purification in wetlands. The degradation of pollutants and organic matter converts harmful substances into less toxic or inert forms, enhancing water quality. The diversity and composition of these microbial communities are essential, as various species carry out specialised functions that enhance the ecosystem's overall fitness. Strategies aimed at increasing microbial diversity and function in restored wetlands can significantly improve the effectiveness of rehabilitation projects. Hence, it is crucial to integrate knowledge of microbial ecology into wetland restoration approaches since these microorganisms are fundamental to the restoration and enduring viability of wetland ecosystems (Jagadesh et al. 2024; Karam et al. 2023; Kumar 2023; Kumar and Choudhury 2024c; Kumar and Kumar 2023; Raj et al. 2024).

3 Geomicrobes and Their Role in Wetland Restoration

3.1 Microbial Communities in Wetlands

The geomicrobes, comprising bacteria, archaea, fungi, and protists, are essential to wetland ecosystems. They facilitate essential biogeochemical processes and exert an impact on the health of ecosystems. Every individual microbial species has a distinct function within the wetland environment, making valuable contributions to the processes of nutrient cycling, organic matter breakdown, and the general operation of the ecosystem. Within wetlands, bacteria represent the most prevalent and varied category of geomicrobes. They participate in many biochemical processes, encompassing carbon, nitrogen, sulfur, and phosphorus cycling. Within anaerobic wetland soils, bacteria engage in metabolic activities such as methanogenesis, the breakdown of organic matter to generate methane, and denitrification, the conversion of nitrates into nitrogen gas. These processes effectively decrease nitrogen availability for plant growth and contribute to the prevention of eutrophication. Archaea, a distinct assemblage of unicellular microorganisms, are frequently encountered in diverse ecological settings and assume a crucial function in methane generation in wetlands. In saturated, anaerobic environments, methanogenic archaea generate substantial amounts of methane by transforming simple compounds like carbon dioxide and hydrogen into energy. Fungi play a crucial role as decomposers in wetlands by breaking down complex organic substances such as lignin and cellulose that are resistant to decomposition by other organisms. The impact of fungal activity on plant litter and peat decomposition is substantial, affecting the carbon equilibrium in wetlands. Protists, such as amoebas, ciliates, and flagellates, play a dual role in nutrient cycling and as predators of bacteria and other tiny microorganisms (Devi et al. 2023a; 2023c; Dey et al. 2023b; Kumar and Choudhury 2024d; Kumari et al. 2023; Sharma and Kumar 2023; Srivastav et al. 2024). They regulate microbial populations and the overarching microbial food web in wetlands. In wetlands, the distribution and diversity of microbial communities are influenced by several environmental factors such as soil type, water chemistry, plant communities, and hydrology. Water saturation in wetland soils often leads to anaerobic conditions, which create low-oxygen environments conducive to the growth of anaerobic microorganisms such as methanogens and sugar-reducing bacteria. Conversely, aerobic organisms, such as nitrifying bacteria, are more prevalent in regions with elevated oxygen concentrations, such as the soil surface or the vicinity of plant roots. Microbiological diversity in wetlands is often characterised by a high degree of specialisation, where various microorganisms occupy specific ecological niches. The composition of microbial communities is also affected by the particular categories of wetlands, namely marshes, bogs, and fens. The swamps each contain distinct microbial communities that have evolved to suit the specific conditions of each ecosystem. For example, acidic bogs characterised by elevated organic matter levels often exhibit unique combinations of fungus and methanogens. Simultaneously, marshes with higher nutrient content may house various bacteria that participate in nutrient cycling. The rhizosphere, the compact soil area immediately flanking plant roots, is a critical zone for microbial activity in wetlands. The rhizosphere microbiome consists of bacteria, fungi, and other microorganisms that strongly interact with plant roots, establishing symbiotic or mutualistic associations. Within wetland environments, plant roots frequently emit organic substances, including sugars, amino acids, and organic acids, that function as a nutritional resource for microorganisms. By decomposing organic matter and transforming nutrients into readily absorbable forms, these microorganisms reciprocally improve the availability of nutrients for plants. For instance, mycorrhizal fungi establish symbiotic relationships with plant roots, facilitating the uptake of vital nutrients like phosphorus from the soil (Devi and Kumar 2023b; Kumar and Dey 2024; Kumar and Singh 2024e; Sharma et al. 2023a). Concurrently, nitrogen-fixing bacteria transform atmospheric nitrogen into forms available for plant utilisation. The interactions within the rhizosphere play a critical role in maintaining plant health and promoting growth in nutrient-deficient wetland soils, enabling wetland plants to flourish in otherwise tricky environments. The microbiological communities in the rhizosphere assist plants in adapting to environmental stressors, such as salinity, drought, or flooding, by improving the absorption of water and nutrients and providing protection against fungi. In wetlands, the ecological functioning of these ecosystems relies heavily on the diversity, distribution, and interactions of geomicrobes. These microbes play a crucial role in preserving the health and resilience of wetland environments (Table 1).

3.2 *Role of Geomicrobes in Biogeochemical Cycles*

Geomicroorganisms are essential in controlling wetland biogeochemical cycles, affecting nutrient availability, greenhouse gas emissions, and water quality. Within the carbon cycle, methanogenic archaea, which are geomicrobes, facilitate methanogenesis, a process in anaerobic environments where organic material is broken down to generate methane (CH_4). Methane is a highly significant greenhouse gas, but its effects are mitigated by methane-oxidising bacteria (methanotrophs), which metabolise methane before its release into the atmosphere. These microorganisms play a vital role in carbon cycling by deciding whether carbon is sequestered in the soil or emitted back into the atmosphere. In the nitrogen cycle, geomicrobes play a crucial role in facilitating vital processes such as nitrification, the conversion of ammonia into nitrates, and denitrification, the conversion of nitrates into nitrogen gas, reducing nitrogen in the ecosystem. Additionally, some bacteria engage in nitrogen fixation, transforming atmospheric nitrogen into a usable form for plants and restoring nitrogen in nutrient-deficient wetland soils (Devi et al. 2023c; Kumar et al. 2024b; Kumar and Choudhury 2024e; Sinam et al. 2023). These microbiological processes control the availability of nitrogen and prevent excessive nutrient saturation, which can result in eutrophication. In the sulfur cycle, sulfate-reducing bacteria catalyse the conversion of sulfate into hydrogen sulfide in the absence of

Table 1 Microbes and their restoration application

Microbe	Type	Primary role	Restoration application	Key process involved
Azotobacter	Nitrogen-fixing bacteria	Fixes atmospheric nitrogen into usable forms	Enhances nitrogen availability for plant growth in nutrient-poor wetlands	Nitrogen fixation
Rhizobium	Nitrogen-fixing bacteria	Symbiotic nitrogen fixation with legumes	Supports growth of leguminous plants in restored wetlands	Nitrogen fixation
Frankia	Nitrogen-fixing bacteria	Fixes nitrogen in symbiosis with non-legumes	Useful in non-legume-associated plants for restoration	Nitrogen fixation
Anabaena	Cyanobacteria	Fixes nitrogen in aquatic environments	Promotes growth in aquatic wetland environments	Nitrogen fixation
Clostridium	Anaerobic bacteria	Decomposes organic material under anoxic conditions	Facilitates organic matter breakdown in waterlogged soils	Anaerobic decomposition
Methanobacterium	Methanogenic archaea	Produces methane in anaerobic conditions	Controls methane emissions in restored wetlands	Methanogenesis
Methanosarcina	Methanogenic archaea	Converts acetate to methane	Reduces methane production through the management of methanogens	Methanogenesis
Methanotrophs	Methanotrophic bacteria	Oxidises methane into CO_2	Mitigates methane emissions from wetland soils	Methane oxidation
Thiobacillus	Sulfur-oxidising bacteria	Oxidises sulfide to sulfate	Detoxifies sulfide-rich environments in salt marshes	Sulfur oxidation
Beggiatoa	Sulfur-oxidising bacteria	Oxidises sulfur in anoxic conditions	Important for sulfur cycling in salt marshes	Sulfur oxidation
Desulfovibrio	Sulfate-reducing bacteria	Reduces sulfate to sulfide	Balances sulfate and reduces sulfur toxicity in marsh soils	Sulfate reduction

(continued)

Table 1 (continued)

Microbe	Type	Primary role	Restoration application	Key process involved
Desulfobacter	Sulfate-reducing bacteria	Reduces sulfate and participates in organic matter decomposition	Reduces sulfate and assists in organic matter breakdown in anaerobic zones	Sulfate reduction, organic decomposition
Pseudomonas	Denitrifying bacteria	Converts nitrate to nitrogen gas	Controls nitrate pollution and combat eutrophication	Denitrification
Paracoccus	Denitrifying bacteria	Converts nitrate to nitrogen gas	Assists in reducing nitrogen levels in nutrient-rich freshwater wetlands	Denitrification
Nitrosomonas	Nitrifying bacteria	Converts ammonia to nitrite	It helps control nitrogen levels by oxidising ammonia in wetland systems	Nitrification
Nitrobacter	Nitrifying bacteria	Converts nitrite to nitrate	Completes the nitrification process in aerobic zones of wetlands	Nitrification
Bacillus subtilis	Soil bacteria	Decomposes organic matter and produces growth-promoting hormones	Facilitates plant growth and organic matter breakdown in wetlands	Organic matter decomposition
Actinobacteria	Filamentous bacteria	Breaks down complex organic compounds	Enhances soil structure and nutrient cycling in peatland restoration	Organic matter decomposition
Cellulomonas	Cellulose-degrading bacteria	Decomposes cellulose from plant material	Necessary for organic matter decomposition in plant-dominated wetlands	Cellulose degradation
Phanerochaete chrysosporium	White-rot fungus	Breaks down lignin in woody material	Accelerates breakdown of lignin-rich material in wetland restoration projects	Lignin degradation

(continued)

Table 1 (continued)

Microbe	Type	Primary role	Restoration application	Key process involved
Trichoderma	Fungi	Suppresses soil pathogens and enhances plant growth	Used in biocontrol and promotes plant health in restored wetlands	Disease suppression, plant growth promotion
Mycorrhizal fungi	Symbiotic fungi	Enhances nutrient uptake in plants	Improves phosphorus and nutrient uptake for plants in wetland environments	Nutrient absorption
Aspergillus	Fungi	Decomposes organic material and helps recycle nutrients	Aids in nutrient cycling and organic matter breakdown	Organic matter decomposition
Penicillium	Fungi	It breaks down organic material and inhibits pathogens	Necessary for organic decomposition and pathogen control in wetland soils	Organic matter decomposition, disease suppression
Phosphobacteria	Phosphate-solubilising bacteria	Converts insoluble phosphate to soluble forms	Increases phosphorus availability in nutrient-limited wetlands	Phosphate solubilisation
Brevibacillus	Phosphate-solubilising bacteria	Solubilises inorganic phosphorus	Enhances phosphorus availability, aiding in plant nutrition in restored wetlands	Phosphate solubilisation
Azospirillum	Nitrogen-fixing bacteria	It fixes nitrogen and promotes plant root growth	Used in conjunction with plant species to enhance root growth and nutrient availability	Nitrogen fixation, plant growth promotion
Paenibacillus	Phosphate-solubilising bacteria	Produces enzymes that release phosphorus from organic matter	Supports plant growth by enhancing phosphorus cycling	Phosphate solubilisation

(continued)

Table 1 (continued)

Microbe	Type	Primary role	Restoration application	Key process involved
Sphingomonas	Aromatic compound-degrading bacteria	Degrades recalcitrant organic pollutants	Degrades persistent pollutants such as aromatic compounds in contaminated wetlands	Pollutant degradation
Dehalococcoides	Dehalogenation bacteria	Reduces halogenated compounds such as chlorinated solvents	Used in remediating polluted wetlands with toxic chemicals	Pollutant degradation
Geobacter	Iron-reducing bacteria	Reduces Fe(III) to Fe(II)	Facilitates bioremediation of metal-contaminated soils in wetlands	Iron reduction
Shewanella	Iron-reducing bacteria	Reduces metals, including uranium and chromium	Used in bioremediation of metal-contaminated wetlands	Metal reduction
Siderophore-producing bacteria	Iron-chelating bacteria	Solubilises iron and improves plant growth	Enhances iron availability in wetland restoration by chelating iron	Iron solubilisation
Lactobacillus	Lactic acid bacteria	Improves soil structure and suppresses pathogens	Used to promote healthy soil and suppress disease-causing organisms in wetland soils	Soil structure improvement, disease suppression
Arbuscular mycorrhizal fungi	Fungi	Symbiotic association with plant roots	Improves nutrient uptake, particularly phosphorus, and plant stress tolerance	Nutrient absorption, plant-microbe symbiosis
Vibrio	Sulfur-reducing bacteria	Reduces sulfur compounds	Important for sulfur cycling in anaerobic conditions of salt marshes	Sulfur reduction

(continued)

Table 1 (continued)

Microbe	Type	Primary role	Restoration application	Key process involved
Micrococcus	Hydrocarbon-degrading bacteria	Breaks down hydrocarbons and other pollutants	Biodegradation of oil spills and organic contaminants in wetlands	Pollutant degradation
Alcaligenes	Pollutant-degrading bacteria	Degrades a wide range of organic pollutants	Used in the cleanup of contaminated freshwater wetlands	Pollutant degradation
Pseudomonas fluorescens	Biocontrol agent	Inhibits pathogens and promotes plant health	Protects plants in restored wetlands from pathogenic microbes	Disease suppression, plant growth promotion
Bacillus thuringiensis	Biocontrol agent	Produces toxins that target insect pests	Biological control of insect pests in wetland restoration projects	Pest control
Acinetobacter	Hydrocarbon-degrading bacteria	Breaks down hydrocarbons in oil-contaminated wetlands	Important in the bioremediation of oil spills in coastal wetland restoration	Hydrocarbon degradation
Halobacterium	Halophilic archaea	Thrives in high-salinity environments	Important for microbial diversity and stability in salt marsh restoration	Halotolerance, microbial stability

oxygen while sulfur-oxidising bacteria reverse this process in the presence of oxygen. These microorganisms play a crucial role in the process of sulfur cycling, which impacts the absorption of nutrients by plants and the chemistry of soil. Microbiological communities in the phosphorus cycle increase phosphorus availability by solubilising phosphate, enabling plants to access it. Although wetland soils typically retain phosphorus in insoluble forms, specific bacteria and fungi can liberate this phosphorus by enzymatic activity, facilitating nutrient cycling. These microbial processes jointly impact the availability of nutrients and the quality of water in wetlands. Geomicrobes play a crucial role in maintaining the ecological equilibrium by controlling the cycling of vital elements. They prevent nutrient imbalances that may reduce water quality and promote the growth of plants and other essential organisms for wetland ecosystems.

3.3 Geomicrobes in Restoration Practices

Through their impact on soil stabilisation, organic matter decomposition, nutrient cycling, and pollutant degradation, geomicrobes play a crucial role in wetland restoration practices, contributing to the recovery and sustainability of degraded ecosystems. Soil stabilisation is a vital function of microbes in ecological restoration. Wetlands frequently experience erosion and sediment displacement, especially during restoration activities when the ecosystem is susceptible to damage. Microbial organisms, mainly bacteria and fungi, facilitate soil particle aggregation by forming biofilms and extracellular polymeric substances (EPS). These compounds create an adhesive matrix that retains soil, suppresses erosion, and enhances the structural stability of wetland soils. This microbiological activity improves the ability of restored wetlands to withstand environmental pressures such as flooding and intense rainfall. Biodegradation of organic matter is another crucial role geomicrobes perform in wetland restoration. Wetlands contain significant organic sediment, and cycling this material is essential for preserving the ecological equilibrium. Microbiological communities, comprising bacteria and fungi, decompose intricate organic compounds into more basic forms, such as deceased plant material and debris. This process not only facilitates the recycling of nutrients into the ecosystem but also inhibits the buildup of organic matter, thus averting the potential development of anaerobic conditions and methane generation. Promoting microbial decomposition is crucial in restoration projects to guarantee the efficient turnover of organic matter, support plant growth, and preserve soil health. Microorganisms are essential to soil fertility and cycling to restore nutrient-deficient or degraded soils. Wetlands frequently exhibit phytonutrient levels of vital elements such as nitrogen and phosphorus, which are indispensable for establishing and developing plants. Nitrogen-fixing bacteria transform atmospheric nitrogen into ammonia, facilitating its accessibility to plants, whereas phosphate-solubilising microorganisms liberate bound phosphorus from the soil, augmenting its availability (Dey et al. 2023c; Hidangmayum et al. 2023; Kumar et al. 2023c; Saini et al. 2023; Sharma et al. 2023). Microbial processes enhance soil fertility, facilitating indigenous plant proliferation and optimising restoration initiatives' overall efficacy. Through the facilitation of nutrient cycling, geomicrobes contribute to the restoration of the wetland ecosystem, assuring its long-term sustainability and resilience. A highly significant application of geomicrobes in restoration is their utilisation in the breakdown of pollutants and bioremediation. Pollutants such as heavy metals, pesticides, and industrial waste frequently contaminate wetlands, significantly damaging ecological health. Bioremediation is how bacteria, fungi, and archaea hydrolyse or neutralise these pollutants. Specifically, metal-reducing bacteria can transform toxic heavy metals into less detrimental forms, whereas hydrocarbon-degrading microorganisms can decompose oil and other organic contaminants. Introducing or promoting these microorganisms in wetland restoration can significantly enhance the ecosystem's capacity to recuperate from pollution, improving water quality and overall ecological health. By harnessing the many abilities of geomicrobes, restoration protocols can enhance their efficacy in stabilising soils, facilitating nutrient cycling, controlling organic matter, and breaking down pollutants. This, in turn, promotes deteriorated wetlands' recovery and long-term viability.

4 Microbial Mechanisms in Wetland Restoration

4.1 Phytoremediation and Microbial Synergies

The interplay between plants and microbes is crucial for restoring wetland ecosystems. Wetlands are distinctive ecological systems that offer essential functions such as water purification, carbon sequestration, and providing habitat for biodiversity. Nevertheless, the deterioration of wetlands caused by human activities has mandated the implementation of restoration initiatives. Within these ecological settings, plants and microbes collaborate in a mutually beneficial manner to establish ideal circumstances that facilitate ecosystem restoration. Microbial organisms, especially those in the rhizosphere of plants and endophytic microorganisms living in plant tissues, play a vital role in promoting plant growth and enhancing stress tolerance, which is essential for successful restoration. Rhizosphere refers to the limited soil area affected by root secretions and the corresponding microbial activity. This region is abundant with bacteria, fungi, and other microorganisms that facilitate the uptake of nutrients by plants, defend them against diseases, and improve their capacity to withstand environmental pressures such as salinity, waterlogging, and heavy metal pollution, which are prevalent in deteriorated wetlands. Microorganisms in the rhizosphere can fix atmospheric nitrogen, solubilise phosphates, and decompose organic matter, enhancing plant nutrient availability (Devi et al. 2023b; Dey et al. 2023a; Kumar et al. 2018a, d; 2023b; Kumar and Dwivedi 2018; Sharma et al. 2023b, c). Through enhancing soil health and its capacity to cycle nutrients, these microorganisms establish a favourable environment for the establishment and growth of plants, a critical aspect of wetland restoration endeavours. Endophytic microorganisms, which reside within the tissues of plants without inducing damage, also function crucially in plant development and ability to withstand stress. These microorganisms can synthesise hormones, such as auxins and cytokinins, stimulating plant growth. They also assist plants in adaptively regulating their physiological and biochemical processes in response to environmental stress. Furthermore, endophytes can confer resistance to pathogens by the synthesis of antimicrobial substances or by stimulating the plant's immune system. Regarding wetland restoration, endophytic microorganisms can improve the survival and productivity of wetland plants under challenging environments, so facilitating the development of robust plant communities that are crucial for the restoration process. Optimising the interactions between plants and microbes presents potential approaches to enhance the success of restoration in wetlands. Restoration practitioners can introduce advantageous rhizosphere and endophytic microorganisms to enhance wetland plants' establishment, growth, and survival rates. Furthermore, choosing plants that support strong microbial communities can improve the probability of successful restoration. Facilitating these interactions can develop more robust ecosystems that endure environmental pressures, guaranteeing long-term recovery and sustainability of wetland habitats.

4.2 Bioaugmentation and Biostimulation

Bioaugmentation systematically introduces particular advantageous microbial strains into an environment to improve its capacity to degrade or neutralise pollutants. These microbial strains are often selected based on their capacity to metabolise or detoxify specific contaminants, such as hydrocarbons, heavy metals, or organic pollutants. The fundamental concept underlying bioaugmentation is that introducing microorganisms with valuable traits into an environment can significantly enhance the overall rate of contaminant breakdown. Bioaugmentation improves this process since these microorganisms may not naturally occur in the polluted area or in inadequate amounts to remove the contaminants adequately. Bioaugmentation is especially beneficial in settings where the indigenous microbial populations are either too sluggish or inefficient in remediating pollutants. In oil spill remediation, introducing hydrocarbon-degrading bacteria can accelerate the biological decomposition of oil in the surrounding environment. This approach directly enhances the population of microorganisms explicitly specialised to break down contaminants, enabling more effective biodegradation and decreased pollutant levels within shorter periods (Dwivedi and Kumar 2011; Kumar et al. 2011, 2017, 2022a; Naik and Kumar 2020; Pankaj et al. 2012). In contrast, biostimulation refers to the process of enhancing the development and functionality of native microbial populations by the introduction of nutrient amendments, such as nitrogen, phosphorus, or carbon sources. In contrast to bioaugmentation, which involves the introduction of novel microbial species, biostimulation seeks to augment the native microbial population's ability to degrade pollutants. The underlying concept of biostimulation is that numerous microorganisms can break down pollutants. Still, their capacity may be constrained by insufficient nutrients or other crucial elements in their surroundings. By supplying the essential nutrients, oxygen, or electron acceptors, the native microorganisms can enhance their performance, resulting in higher rates of contaminant breakdown. Both bioaugmentation and biostimulation are imperative in environmental remediation, particularly in mitigating pollutants such as heavy metals and organic contaminants. For example, bioaugmentation can introduce metal-resistant microbial strains that transform toxic metal ions into less detrimental forms in regions of heavy metal contamination, such as restoring lead or cadmium. Biostimulation can augment microbial populations responsible for degrading organic contaminants, such as petroleum hydrocarbons or chlorinated solvents. Both approaches are commonly integrated to eliminate soil, water, and sediment pollutants. Bioaugmentation and biostimulation are becoming increasingly vital in bioremediation as environmentally friendly and economically efficient alternatives to conventional chemical treatments to restore contaminated ecosystems (Kotia et al. 2021; Kumar et al. 2019, 2022c; Kumar and Naik 2020; Kumar and Pathak 2019; Pathak et al. 2017).

4.3 Methanotrophs and Wetland Restoration

Methanotrophic bacteria are crucial in reducing methane emissions from wetlands, especially in rehabilitated ecosystems where methane production may rise due to anaerobic breakdown of organic material. These microorganisms, which employ methane as their primary energy source and carbon, function as a biological filter by metabolising methane before it can be released into the atmosphere. Methanotrophs predominantly reside in the aerobic regions of wetland habitats, where there is an abundance of oxygen to facilitate the process of methane oxidation. Metabolic methane oxidation significantly decreases total methane emissions, a crucial factor as methane is a potent greenhouse gas that substantially contributes to climate change. Within rehabilitated wetlands, where the processes of carbon cycling and microbial activities are actively reinstated, methanotrophic bacteria play a crucial role in maintaining a balance between methane production and oxidation. This thus helps to minimise the environmental consequences of these ecosystems. Wetlands are dynamic ecosystems characterised by both anaerobic (oxygen-deprived) and aerobic (oxygen-rich) zones, which interact to facilitate methane cycling (Das et al. 2022; Kumar et al. 2018b, c, 2022b; Kumar and Dwivedi 2020; Kumar and Mistri 2020; V. Kumar et al. 2021). Within the anaerobic zones, commonly located in deeper soils or submerged sediments, methane is generated by methanogenic archaea by organic matter decomposition. Next, this methane diffuses towards the oxygenated, aerobic regions near the water's surface. Within the aerobic zones, methanotrophic bacteria enzymatically convert methane into carbon dioxide and water by oxidising it chemically. These bacteria employ methane monooxygenase (MMO) enzymes to commence the methane oxidation process, necessitating oxygen as an electron acceptor. Nevertheless, recent studies have demonstrated that methane oxidation can occur in inert environments, enabled by methanotrophic archaea and sulfate-reducing bacteria groups. To further reduce methane emissions, these consortia employ alternative electron acceptors, such as sulfate or nitrate, to facilitate methane oxidation without oxygen. Through their operation in aerobic and anaerobic zones, methanotrophic processes significantly decrease wetland methane emissions (Table 2).

Table 2 Microbial mechanism involved and its restoration application

Microbial mechanism	Microbe involved	Primary function	Restoration application	Key process involved
Nitrogen fixation	*Azotobacter, rhizobium, Frankia*	Converts atmospheric nitrogen into bioavailable forms	Enhances nitrogen availability for plants in nutrient-poor wetlands	Nitrogen fixation
Symbiotic nitrogen fixation	*Rhizobium*	Nitrogen fixation with legume roots	Supports growth of leguminous plants in restored wetlands	Nitrogen fixation
Free-living nitrogen fixation	*Azotobacter*	Fixes nitrogen independently of plant roots	Increases nitrogen levels in soils without plant associations	Nitrogen fixation
Ammonia oxidation (nitrification)	*Nitrosomonas*	Converts ammonia to nitrite	Regulates nitrogen availability in wetlands	Nitrification
Nitrite oxidation (nitrification)	*Nitrobacter*	Converts nitrite to nitrate	Completes the nitrification process in aerobic wetland soils	Nitrification
Denitrification	*Pseudomonas, Paracoccus*	Converts nitrate to nitrogen gas	Reduces nitrate pollution and eutrophication	Denitrification
Methanogenesis	*Methanobacterium, Methanosarcina*	Produces methane in anaerobic environments	Methane control in wetland soils	Methanogenesis
Methane oxidation	*Methanotrophs*	Converts methane to CO_2	Reduces methane emissions from wetlands	Methane oxidation
Sulfur oxidation	*Thiobacillus, Beggiatoa*	Converts sulfide to sulfate	Reduces sulfur toxicity in salt marsh soils	Sulfur oxidation
Sulfate reduction	*Desulfovibrio, Desulfobacter*	Reduces sulfate to sulfide	Balances sulfur cycling in anaerobic wetland environments	Sulfate reduction

(continued)

Table 2 (continued)

Microbial mechanism	Microbe involved	Primary function	Restoration application	Key process involved
Phosphate solubilisation	*Brevibacillus, Paenibacillus*	Converts insoluble phosphorus into bioavailable forms	Increases phosphorus availability for plants	Phosphorus cycling
Lignin degradation	*Phanerochaete chrysosporium*	Breaks down lignin in woody materials	Enhances decomposition of plant material in wetland soils	Lignin degradation
Cellulose degradation	*Cellulomonas*	Decomposes cellulose from plant material	Facilitates breakdown of organic matter in wetland ecosystems	Cellulose degradation
Organic matter decomposition	*Clostridium, Bacillus subtilis*	Breaks down organic material in anaerobic environments	Supports nutrient cycling and soil health in waterlogged wetlands	Anaerobic decomposition
Chitin degradation	*Trichoderma*	Breaks down chitin and inhibits pathogens	Disease suppression and organic matter decomposition	Chitin degradation
Humus formation	*Actinobacteria*	Breaks down complex organic compounds	Improves soil structure and nutrient cycling in wetlands	Organic matter transformation
Polysaccharide degradation	*Pseudomonas fluorescens*	Breaks down polysaccharides from plant debris	Enhances carbon cycling and nutrient availability	Organic matter decomposition
Hydrocarbon degradation	*Sphingomonas, Acinetobacter*	Breaks down hydrocarbons and petroleum-based pollutants	Remediates oil spills and hydrocarbon-contaminated wetlands	Pollutant degradation
Halogenated compound degradation	*Dehalococcoides*	Reduces halogenated pollutants	Removes chlorinated solvents and similar contaminants	Pollutant degradation
Iron reduction	*Geobacter, Shewanella*	Reduces Fe(III) to Fe(II)	Assists in bioremediation of metal-contaminated wetlands	Metal reduction

(continued)

Table 2 (continued)

Microbial mechanism	Microbe involved	Primary function	Restoration application	Key process involved
Siderophore production	*Siderophore-producing bacteria*	Chelates iron to increase availability for plants	Improves iron uptake and plant health in nutrient-poor wetlands	Iron solubilisation
Pathogen suppression	*Lactobacillus, bacillus thuringiensis*	Inhibits pathogenic microbes in the soil	Promotes plant health by controlling soil pathogens	Disease suppression
Mycorrhizal symbiosis	*Arbuscular mycorrhizal fungi*	Enhances nutrient uptake (especially phosphorus)	Promotes plant growth and stress tolerance in wetland restoration	Symbiotic nutrient absorption
Anaerobic carbon decomposition	*Clostridium, Methanosarcina*	Breaks down organic carbon in anoxic environments	Reduces organic matter in waterlogged, low-oxygen wetlands	Anaerobic decomposition
Nitrate reduction	*Denitrifying bacteria*	Reduces nitrate to nitrogen gas	Helps combat nitrate accumulation and eutrophication	Nitrate reduction
Sulfur reduction	*Desulfovibrio*	Converts sulfate to sulfide	Plays a key role in sulfur cycling and balancing marsh ecosystems	Sulfur cycling
Phosphate uptake enhancement	*Mycorrhizal fungi*	Improves phosphorus uptake through symbiotic relationships	Increases phosphorus availability in nutrient-limited soils	Phosphorus cycling
Anaerobic respiration	*Shewanella*	Utilises metal ions and sulfur compounds as electron acceptors	Enhances respiration in anaerobic conditions in restored wetlands	Anaerobic respiration
Carbon dioxide fixation	*Cyanobacteria*	Converts CO_2 into organic matter	Enhances carbon sequestration in aquatic wetland ecosystems	Photosynthesis, carbon fixation

(continued)

Table 2 (continued)

Microbial mechanism	Microbe involved	Primary function	Restoration application	Key process involved
Heavy metal chelation	*Pseudomonas*	Chelates heavy metals to reduce toxicity	Important for remediating metal-contaminated wetlands	Metal chelation
Polycyclic aromatic hydrocarbon (PAH) degradation	*Pseudomonas, Sphingomonas*	Breaks down PAHs, which are persistent organic pollutants	Restores wetlands contaminated with organic pollutants	Pollutant degradation
Reduction of chromium and uranium	*Geobacter, Shewanella*	Reduces toxic heavy metals, such as uranium and chromium	Used for bioremediation of metal-contaminated wetlands	Metal reduction
Sulfide detoxification	*Thiobacillus*	Oxidises sulfide to less harmful sulfate	Mitigates sulfide toxicity in anaerobic wetland sediments	Sulfide oxidation
Halotolerance and osmoregulation	*Halobacterium*	Maintains cellular functions in high-salinity environments	Supports microbial diversity in saline wetland restoration	Osmoregulation
Biocontrol of pathogens	*Trichoderma, bacillus thuringiensis*	Inhibits soil-borne pathogens and insect pests	Biological control of pests and diseases in restored wetlands	Disease suppression, pest control
Phosphorus cycling	*Paenibacillus, Phosphobacteria*	Solubilises organic and inorganic phosphorus	Improves phosphorus availability in soils for plant growth	Phosphate solubilisation
Organic pollutant biodegradation	*Pseudomonas, Alcaligenes*	Breaks down organic pollutants and contaminants	Cleans up organic pollutants in wetland environments	Pollutant degradation
Hydrogen sulfide oxidation	*Beggiatoa*	Converts hydrogen sulfide into sulfate	Critical for sulfur cycling and detoxification in salt marshes	Sulfide oxidation

(continued)

Table 2 (continued)

Microbial mechanism	Microbe involved	Primary function	Restoration application	Key process involved
Metal oxidation and reduction	*Shewanella*, *Geobacter*	Facilitates the oxidation and reduction of metals	Used for bioremediation of metal-polluted wetlands	Metal cycling
Carbon sequestration	*Methanotrophic bacteria*, *cyanobacteria*	Fixes carbon through methane oxidation and photosynthesis	Enhances carbon sequestration in wetland soils and water bodies	Methane oxidation, photosynthesis
Methane suppression	*Methanotrophs*	Converts methane into less harmful CO_2	Controls methane emissions in wetlands, reducing greenhouse gases	Methane oxidation
Hydrogen oxidation	*Hydrogen-oxidising bacteria*	Uses hydrogen as an energy source under anoxic conditions	Assists in nutrient cycling and anaerobic respiration	Anaerobic respiration

4.4 *Redox Processes and Microbial Mediation*

Redox gradients play a crucial role in the biogeochemical processes that underlie the operation of wetland ecosystems. This gradient arises from the fluctuating oxygen availability in the soil and water columns of the wetland, resulting in well-defined zones that impact microbial activity, nutrient cycling, and pollutant breakdown. Aqueous wetlands typically have aerobic surface layers characterised by abundant oxygen, while deeper layers gradually become anaerobic as oxygen levels decrease. The redox gradient, or stratification, arises from the sluggish rate of oxygen diffusion into the saturated soils and the significant biological need for oxygen in these systems (Aley et al. 2022; Chakraborty et al. 2021; Goud et al. 2022; Kumar et al. 2021b; Upadhyay et al. 2023). These redox gradients enable a wide range of microbial biological processes that depend on various electron acceptors, including nitrate, sulfate, iron, and carbon dioxide, in the absence of oxygen. The presence of this variety is crucial for preserving the healthy state of the ecosystem since it enables several routes for the decomposition of organic matter, cycling of nutrients, and breakdown of pollutants. The spatial heterogeneity of redox conditions in wetlands gives rise to a diverse array of microenvironments, each characterised by unique biochemical processes that contribute to the overall functioning of the ecosystem. Microbes primarily mediate redox transformations in wetlands. In aerobic zones, microorganisms employ oxygen as an electron acceptor to decompose organic tissue, liberating carbon dioxide and facilitating nutrient cycling. Yet, in oxygen-deprived anaerobic environments, microorganisms transition to other electron acceptors like nitrate, sulfate, and iron to sustain their metabolic activities. For instance, denitrifying bacteria convert nitrate into nitrogen gas, sulfate-reducing bacteria utilise sulfate to generate hydrogen sulfide, and iron-reducing bacteria catalyse the conversion of ferric iron to ferrous iron. Microbiological processes play a vital role in maintaining the redox equilibrium in wetlands and guaranteeing the uninterrupted progression of biogeochemical cycles. The electron donors in these redox processes frequently originate from the organic material in wetlands, whereas the electron acceptors fluctuate depending on the redox potential of the surrounding environment. As environmental conditions decrease, microorganisms systematically access the most energetically advantageous electron acceptors, beginning with oxygen and progressing to nitrate manganese, iron, sulfate, and carbon dioxide. A systematic reduction is crucial for identifying the prevailing biogeochemical processes in particular wetland areas. Redox conditions significantly influence the availability of nutrients and the breakdown of pollutants in wetlands. Reducing nitrate to nitrogen gas eliminates surplus nitrogen from the system in anaerobic zones, thus mitigating eutrophication. Redox gradients also impact phosphorus cycling by allowing reduced iron to liberate bound phosphorus into the water column in the absence of oxygen, so enhancing its bioavailability. Moreover, redox gradients play a crucial role in the degradation of pollutants. Microbial degradation of complex organic pollutants, such as petroleum hydrocarbons and chlorinated solvents, can be achieved in anaerobic environments by utilising sulfate or nitrate as electron acceptors. The capacity to use alternative electron acceptors to break down pollutants renders wetlands very efficient natural bioremediation systems for polluted environments (Kumar et al. 2018, 2020, 2021a, Kumari et al. 2022; Reddy et al. 2022). Microbial populations in wetlands perform the role of mediating redox transformations, contributing to the regulation of nutrient availability, maintenance of ecosystem health, and improvement of the degradation of environmental pollutants.

5 Case Studies in Wetland Restoration Using Geomicrobes

Coastal wetlands, such as salt marshes and mangroves, are essential ecosystems that provide several ecological advantages, including functioning as barriers against storm surges, offering habitats for various species, and serving as critical carbon deposition sites. Nevertheless, these ecosystems face growing risks from human activities such as coastal development, pollution, and climate change. Restoration of coastal wetlands, namely salt marshes and mangroves, is now a top priority for environmental conservation initiatives. The use of microbial amendments has been shown to greatly improve restoration results. Microbial amendments in salt marsh restoration tend to prioritise the augmentation of indigenous microbial populations, which play a vital role in nutrient cycling, soil stabilisation, and plant growth

(Dwivedi and Kumar 2011; Kumar et al. 2018c, 2022c; Kumar and Dwivedi 2018; Kumar and Pathak 2019). Nitrogen-fixing bacteria enhance nitrogen availability in the nutrient-deficient soils commonly present in salt marshes, facilitating the growth and development of salt-tolerant plants such as Spartina. Furthermore, these plants play a crucial role in stabilising the marsh substrate and establishing favourable circumstances for restoring indigenous fauna. Furthermore, sulfur-oxidising bacteria have a crucial function in restoring salt marshes. These bacteria play a crucial role in sulfur cycling, especially in salt marshes, where sulfides can build up due to the anaerobic decomposition of organic material. Sulfur-oxidising bacteria, such as *Thiobacillus*, transform poisonous sulfides into sulfates, threatening plant and microbial life. Detoxification is crucial for rehabilitating salt marshes since it establishes a more favourable environment for plant roots and other organisms, resulting in a more robust and operational ecosystem. Furthermore, the process of mangrove restoration has also derived advantages from using microbial amendments. Microbial communities play a crucial role in supporting the growth and nutrient acquisition of mangroves, which flourish in saline and damp environments. Often, nitrogen-fixing bacteria, mycorrhizal fungus, and phosphate-solubilising bacteria are deliberately introduced or promoted in mangrove restoration interventions. These microorganisms augment the accessibility of vital nutrients such as nitrogen and phosphorus, often scarce in mangrove ecosystems. Moreover, applying these microbial amendments can enhance the ability of mangroves to tolerate stressors such as salinity, waterlogging, and heavy metal pollution, increasing their resilience to environmental fluctuations. Microbiological interventions are vital in expediting the recovery, improving plant growth, and fostering long-term ecological stability in salt marsh and mangrove restoration applications. Inland freshwater wetlands, encompassing bogs, marshes, and swamps, are ecologically heterogeneous systems that offer essential ecological functions such as water purification, flood management, and habitat for a rich array of species. Nevertheless, similar to coastal wetlands, they are susceptible to deterioration caused by agricultural runoff, changes in land use, and pollution. Microbiological interventions play a crucial role in restoring freshwater wetlands by restoring important ecological processes, namely the cycling of nutrients and the breakdown of contaminants. Microbiological amendments can be employed in bogs and marshes to enhance the breakdown of organic material, facilitate nutrient cycling, and stimulate soil formation. Microbial activity in these wetlands is particularly challenged by the saturated and anoxic conditions, which result in limited oxygen availability and dominance of anaerobic processes. The regulation of methane emissions, a powerful greenhouse gas generated in anaerobic wetland soils, heavily depends on methanotrophic bacteria. Facilitating the activity of these bacteria can effectively reduce the emission of methane, thus enhancing the climate-friendliness of restored wetlands. Moreover, introducing microorganisms that enhance organic matter decomposition, such as cellulolytic bacteria and fungi, can expedite the degradation of plant material, thus fostering soil advancement and the liberation of nutrients (Devi et al. 2023c; Devi and Kumar 2023a, b; Kumar et al. 2023c, 2024b; Kumar and Choudhury 2024d, e; Sharma et al. 2023a). Consequently, this facilitates the promotion of plant growth and improves the overall restoration of the wetland. Eutrophication is a significant problem in freshwater wetlands, especially in agricultural regions, caused by excessive nutrient concentrations, mainly nitrogen and phosphorus. Eutrophication results in detrimental algal blooms, hypoxia, and a decline in biodiversity. To address this issue, microbial interventions frequently prioritise the augmentation of denitrifying bacteria populations. These bacteria enzymatically transform nitrate, a prevalent agricultural contaminant, into nitrogen gas by denitrification; surplus nitrogen from the water is efficiently eliminated. This microbiological intervention is particularly beneficial in mitigating nutrient contamination in rehabilitated freshwater wetlands, aiding in the prevention of eutrophication and restoring ecological equilibrium. Constructed wetlands are artificial systems specifically created to replicate the inherent processes of wetlands for wastewater treatment. Their growing popularity in water treatment stems from their ability to offer an environmentally friendly and economically efficient alternative to traditional wastewater treatment techniques. The capacity of built wetlands to effectively filter contaminants, eliminate nutrients, and break down pollutants is mostly determined by the activity of microbial communities in the system. Microbial populations in artificially created wetlands have a pivotal function in improving the effectiveness of water treatment. Within these systems, microbial biofilms form on plant roots, soil particles, and other surfaces, enzymatically decomposing organic matter and contaminants by distinct metabolic pathways. Aerobic bacteria decompose organic compounds in the presence of oxygen, but anaerobic bacteria degrade more complex pollutants in deeper, oxygen-poor zones. To enhance microbial activity and maximise the removal of contaminants, constructed wetlands are often planned with alternating zones of aerobic and anaerobic conditions. Nitrogen removal in constructed wetlands is accomplished by the biochemical processes of nitrification (converting ammonia to nitrate by aerobic bacteria) and denitrification (reducing nitrate to nitrogen gas by anaerobic bacteria). Moreover, the microbial populations in modified wetlands can eliminate heavy metals, pathogens, and other pollutants from wastewater. Specialised bacteria, such as sulfate-reducing bacteria, can immobilise heavy metals by transforming them into less soluble forms, preventing their contamination of groundwater or surface water. In addition, wetland plants facilitate the growth of microbial communities by supplying oxygen to the rhizosphere (root zone) and providing surfaces for the attachment of biofilms. The symbiotic relationship between plants and microbes in built wetlands significantly improves the overall treatment efficacy, rendering these systems very successful in the remediation of diverse contaminants, ranging from nutrients to hazardous pollutants. Peatlands are highly distinctive wetland ecosystems that accumulate substantial carbon as partially broken-down organic material. These ecosystems serve as crucial carbon sinks, facilitating more carbon storage per unit area than other terrestrial ecosystems. Nevertheless, when peatlands are drained or degraded, they transform into substantial reservoirs of carbon dioxide (CO_2) and methane (CH_4), potent greenhouse gases. Thus, restoring peatlands is essential to addressing climate change by reinstating their role as carbon sinks and decreasing greenhouse gas emissions. Microbial communities are crucial regulators of carbon sequestration and greenhouse gas emissions in peatland restoration. The highly saturated and anaerobic environment of well-maintained peatlands retards the breakdown of organic material, so facilitating the gradual accumulation of carbon. Nevertheless, methanogenic archaea can generate methane without oxygen, raising concerns about methane emissions in revitalised peatlands. To achieve a balance between carbon sequestration and reducing methane emissions, restoration efforts often prioritise the restoration of the natural hydrology of peatlands, which subsequently affects microbial activity. Methanotrophic bacteria metabolise methane before its release into the atmosphere and play a vital role in controlling methane emissions in rehabilitated peatlands. These bacteria exhibit their highest activity level in the peat's oxygenated regions near plant roots, where they enzymatically convert methane into carbon dioxide. Restoration initiatives can mitigate methane emissions and preserve the peatland's carbon storage capacity by stimulating the development and activity of methanotrophs. Microbiological communities in restored peatlands are crucial in methane regulation and other vital processes, including nutrient cycling and soil formation (Devi et al. 2024a, c, d; Kumar et al. 2024c, i; Saurabh et al. 2024). Efficient management of the equilibrium between the microbial breakdown of organic material and carbon sequestration is crucial for the sustained effectiveness of peatland restoration endeavours. Managing water levels and promoting diverse microbial communities in restored peatlands can maintain their role as efficient carbon sinks, thus helping mitigate climate change's global impacts.

6 Challenges and Future Prospects

While microbial inoculants and bioaugmentation have gained traction in wetland restoration efforts, some significant challenges and limitations must be addressed. One of the primary limitations of current microbial inoculants is their inconsistent performance in the field. Inoculated microbes often struggle to establish themselves in natural ecosystems, where native microbial communities are already well-adapted to local environmental conditions. The introduced microorganisms may face competition from indigenous microbes, limiting their ability to perform the desired functions, such as nutrient cycling or contaminant degradation. Furthermore, environmental conditions such as pH, temperature, and nutrient availability can significantly influence the survival and activity of inoculated microbes, making it difficult to predict how they will perform across different restoration sites. Another major challenge is the potential for unintended consequences, such as introducing invasive microbial species. When microbial inoculants are introduced into a wetland, there is a risk that non-native microorganisms may outcompete native species, disrupting the ecological balance and potentially causing harm to the ecosystem. Invasive microbes can alter nutrient cycling, outcompete beneficial native organisms, and even contribute to the degradation of the very ecosystems they were intended to restore. The lack of long-term studies on the ecological impacts of microbial inoculants further complicates the understanding of potential risks, raising concerns about the sustainability and safety of these approaches. Scaling up microbial applications for large-scale restoration projects presents yet another set of challenges. While microbial inoculation may be feasible for small, controlled environments, applying these techniques across vast wetland landscapes is significantly more complex. Large-scale restoration projects require the development of cost-effective production and delivery systems for microbial inoculants, which can be challenging to achieve. Additionally, the logistics of applying microbial treatments uniformly across large areas are complicated by the inherent heterogeneity of natural ecosystems. Factors such as varying soil types, water levels, and plant communities can create uneven microbial growth and activity conditions, making it challenging to ensure consistent results. Furthermore, monitoring the success of microbial applications on a large scale requires advanced technologies and methodologies, which are not always readily available or affordable for many restoration initiatives. Despite these challenges, recent advances in microbial ecology and technology offer new tools and approaches for overcoming the limitations of microbial applications in wetland restoration. Emerging technologies such as metagenomics and metatranscriptomics are revolutionising how scientists monitor microbial activity in ecosystems. Metagenomics involves the analysis of genetic material recovered directly from environmental samples, allowing researchers to identify and quantify microbial communities without the need for traditional culturing methods. Metatranscriptomics goes a step further by analysing the gene expression profiles of microbial communities, providing insights into which genes are actively being expressed in response to environmental conditions. These technologies offer a more detailed understanding of microbial dynamics in restored wetlands, enabling scientists to monitor the success of microbial inoculants and assess their functional contributions to ecosystem restoration. In addition to advances in monitoring, microbial bioengineering and synthetic biology are opening up new possibilities for designing and developing microbial strains tailored explicitly for restoration purposes. Bioengineering techniques can enhance microorganisms' abilities to degrade pollutants, cycle nutrients, or tolerate extreme environmental conditions. For example, researchers are exploring ways to engineer microbes with enhanced nitrogen-fixing capabilities to support plant growth in nutrient-poor wetlands or with improved abilities to degrade complex organic pollutants. Synthetic biology approaches also allow for microbial consortia—combinations of different microorganisms that work together to perform multiple functions. These consortia can be designed to optimise nutrient cycling, pollutant degradation, and soil stabilisation, offering a multi-functional approach to ecosystem restoration. Restoration efforts can become more resilient and effective by engineering microbial consortia that mimic the complex interactions found in natural ecosystems. Microbial consortia are particularly promising for wetland restoration, as they can potentially address multiple ecological challenges simultaneously. For example, a consortium of nitrogen-fixing, sulfate-reducing, and methane-oxidising bacteria could enhance plant growth, reduce sulfate toxicity, and mitigate methane emissions in restored wetlands. The use of microbial consortia also aligns with the principles of ecological restoration, which emphasise the importance of restoring not just individual species but the complex interactions that sustain ecosystem function. As our understanding of microbial ecology grows, engineered microbial consortia in wetland restoration can potentially become a powerful tool for achieving long-term ecosystem recovery. The impact of global climate change on microbial-mediated restoration processes is a growing concern. Wetlands are particularly vulnerable to climate change due to their sensitivity to changes in temperature, precipitation patterns, and sea levels. These environmental changes can alter the hydrology, nutrient availability, and plant communities in wetlands, affecting microbial communities' composition and activity. For instance, rising temperatures can accelerate microbial decomposition rates, increasing greenhouse gas emissions from wetlands, mainly methane and carbon dioxide. Changes in water availability, such as prolonged droughts or flooding, can also disrupt the redox conditions in wetlands, affecting the balance between aerobic and anaerobic microbial processes. These shifts in microbial activity can compromise the ability of restored wetlands to sequester carbon and mitigate climate change. Adaptive strategies are needed to ensure the success of wetland restoration under changing environmental conditions. One approach is to focus on restoring wetlands with diverse microbial communities capable of adapting to various environmental conditions. By promoting microbial diversity, restoration projects can increase the resilience of wetlands to climate change, ensuring that critical functions such as nutrient cycling, pollutant degradation, and carbon sequestration are maintained even as environmental conditions shift. Additionally, restoration efforts may need to incorporate climate-resilient plant species that can support the activity of beneficial microbial communities under changing conditions. For example, salt-tolerant plant species could be introduced into coastal wetlands vulnerable to sea-level rise, helping maintain plant-microbe interactions essential for ecosystem stability. Climate change also presents opportunities for advancing our understanding of microbial ecology in wetland ecosystems. Scientists can study how microbial communities respond to stressors such as temperature fluctuations, water availability, and nutrient changes as environmental conditions change. These insights can inform the development of new microbial inoculants and restoration strategies better suited to the challenges of a changing climate. By integrating climate science with microbial ecology, restoration efforts can become more adaptive and forward-thinking, ensuring that wetlands continue to provide critical ecosystem services in the face of global environmental change. The role of policy in promoting microbial approaches to wetland restoration cannot be overstated. Effective policies are needed to support research, development, and implementation of microbial-based restoration techniques. Governments and regulatory agencies can play a crucial role by funding microbial research, establishing guidelines for the safe use of microbial inoculants, and encouraging the integration of microbial approaches into national and regional restoration initiatives. In many countries, wetland restoration is already a priority due to these ecosystems' ecological and economic benefits, such as flood protection, water filtration, and carbon sequestration. However, policy frameworks must evolve to recognise microbial interventions' potential to enhance these restoration efforts' success. Public awareness and community involvement are also crucial for the success of microbial-based wetland restoration projects. Many restoration initiatives rely on the participation of local communities, who can play a crucial role in monitoring wetland health, planting vegetation, and managing restoration sites. Educating the public about the importance of microbial communities in wetland ecosystems can foster more significant support for restoration efforts and encourage more sustainable land-use practices. Additionally, involving local communities in restoration projects can help ensure that restoration goals align with the needs and priorities of those who depend on these ecosystems for their livelihoods. Microbial-based restoration projects can become more sustainable and impactful by fostering a sense of ownership and stewardship. The economic and ecological benefits of microbial interventions in wetland restoration are significant. Wetlands provide valuable ecosystem services, such as water purification, flood mitigation, and carbon storage, with direct and indirect economic value. By enhancing the efficiency of restoration efforts, microbial inoculants and bioaugmentation techniques can help accelerate the recovery of these ecosystems, resulting in faster returns on investment for restoration projects. Moreover, microbial approaches can reduce the need for expensive chemical treatments or engineering solutions, making restoration more cost-effective. Ecologically, microbial interventions can improve the resilience of wetlands to environmental stressors, ensuring that these ecosystems continue to provide critical services in the face of climate change and other challenges. While there are significant challenges to applying microbial inoculants and bioaugmentation in wetland restoration, advances in microbial ecology and technology are opening up new possibilities for overcoming these limitations. Emerging technologies such as metagenomics, synthetic biology, and microbial consortia offer promising tools for enhancing the success of restoration projects. However, it is essential to consider the potential risks, such as invasive microbial species and the challenges of scaling up these techniques for large-scale restoration efforts. Climate change adds another layer of complexity, as shifting environmental conditions can alter microbial dynamics in wetlands. Adaptive strategies that promote microbial diversity and resilience are needed to ensure the long-term success of restoration efforts. Finally, policy support, public awareness, and community involvement are critical for advancing microbial approaches to wetland restoration, as these interventions offer significant economic and ecological benefits. As our understanding of microbial ecology grows, the potential for microbes to play a central role in ecosystem restoration becomes increasingly apparent.

7 Conclusion

In summary, geomicrobes are of utmost importance in restoring wetlands as they facilitate vital processes, including nutrient cycling, degradation of pollutants, and regulation of greenhouse gas emissions. Whether naturally occurring or introduced through bioaugmentation, these microorganisms significantly restore wetlands' recovery and long-term stability. Their participation in essential biogeochemical processes, such as nitrogen, sulfur, and carbon cycling, guarantees the preservation of ecological parameters that promote plant development, improve soil health, and control water quality. The capacity of geomicrobes to metabolise pollutants and correct nutrient imbalances makes them highly promising in tackling urgent environmental issues encountered by degraded wetlands, including eutrophication and pollution. The involvement of microorganisms such as methanotrophic bacteria in controlling methane emissions emphasises their crucial contribution to preserving the climate-regulating capabilities of wetlands. The ongoing progress in microbial ecology presents numerous excellent prospects for future research. A highly promising exploration avenue involves advancing microbial consortia, which integrate several species of microorganisms to generate versatile solutions for restoring ecosystems. By exploiting the synergistic interactions among different organisms, it is potentially feasible to tackle several ecological issues concurrently, ranging from nutrient insufficiencies to the degradation of pollutants and the sequestration of carbon. Furthermore, progress in metagenomics, metatranscriptomics, and microbial bioengineering provides promising prospects for a deeper understanding and control of microbial communities for improved restoration results. These technologies can offer more profound knowledge of the particular roles of microbial populations in rehabilitated wetlands, facilitating more focused and efficient interventions. Furthermore, synthetic biology can potentially engineer microorganisms adapted to the specific conditions of various wetland types, enhancing the adaptability and resilience of restoration efforts in response to environmental changes. To fully harness the advantages of geomicrobes in the restoration of wetlands, it is crucial to include microbial interventions in more comprehensive strategies for managing and conserving wetlands. Restoration of wetlands necessitates a comprehensive approach considering the interaction among microbial communities, plant species, hydrology, and external environmental criteria, as wetlands are dynamic ecosystems. Integrating microbial interventions into current wetland conservation endeavours can augment the efficacy of these programmes, thus fostering the development of more sustainable and resilient vegetation. Furthermore, incorporating microbial approaches into policy frameworks and management plans can guarantee that restoration endeavours are backed by scientific research and can be adjusted to forthcoming environmental circumstances, such as climate change. Utilising geomicrobial techniques improves ecological restoration and supports the overall objectives of wetland conservation, safeguarding these critical ecosystems for future generations.

The continuous study of microbial ecology will provide further tools and knowledge that can be used to restore wetlands, so ensuring the continued functionality, biodiversity, and resilience of these vital ecosystems in an ever more rapidly changing world.

References

Aley P, Singh J, Kumar P (2022) Chapter 23 – Adapting the changing environment: microbial way of life. In: Kumar A, Singh J, Ferreira LFRBT-MUCC (eds) Microbiome under changing climate. Woodhead Publishing, pp 507–525. https://doi.org/10.1016/B978-0-323-90571-8.00023-7

Chakraborty S, Kumar P, Sanyal R, Mane AB, Arvind Prasanth D, Patil M, Dey A (2021) Unravelling the regulatory role of miRNAs in secondary metabolite production in medicinal crops. Plant Gene 27:100303. https://doi.org/10.1016/j.plgene.2021.100303

Choudhury D, Kumar P, Zhimo VY, Sahoo J (2024) Crop rotation patterns and soil health management. In: Bioremediation of emerging contaminants from soils: soil health conservation for improved ecology and food security. Elsevier, Department of Plant Pathology, School of Agriculture, Lovely Professional University, Phagwara, India, pp 565–589. https://doi.org/10.1016/B978-0-443-13993-2.00025-6

Das T, Saha SC, Sunita K, Majumder M, Ghorai M, Mane AB, Prasanth DA, Kumar P, Pandey DK, Al-Tawaha AR, Batiha GE-S, Shekhawat MS, Ghosh A, Sharifi-Rad J, Dey A (2022) Promising botanical-derived monoamine oxidase (MAO) inhibitors: pharmacological aspects and structure-activity studies. S Afr J Bot 146:127–145. https://doi.org/10.1016/j.sajb.2021.09.019

Devi P, Kumar P (2023a) Role of oxidative enzymes in the alleviation of oxidative stress in plant system. In: Microbial oxidative enzymes: biotechnological applications. De Gruyter, Department of Agronomy, School of Agriculture, Lovely Professional University, Phagwara, India, pp 109–128. https://doi.org/10.1515/9783111062235-006

Devi P, Kumar P (2023b) Effect of microbial oxidative enzymes on soil health. In: Microbial oxidative enzymes: biotechnological applications. De Gruyter, Department of Agronomy, School of Agriculture, Lovely Professional University, Phagwara, India, pp 37–61. https://doi.org/10.1515/9783111062235-003

Devi P, Kumar P (2024a) Challenges in the storage of herbal medicine products and strategies for sustainable management. In: Reference series in phytochemistry. Springer Science and Business Media B.V., Department of Agronomy, School of Agriculture, Lovely Professional University, Phagwara, India, pp 1739–1767. https://doi.org/10.1007/978-3-031-43199-9_37

Devi P, Kumar P (2024b) Herbal medicine and pregnancy. In: Reference series in phytochemistry. Springer Science and Business Media B.V., Department of Agronomy, School of Agriculture, Lovely Professional University, Phagwara, India, pp 693–722. https://doi.org/10.1007/978-3-031-43199-9_25

Devi P, Kumar P, Sharma K, Singh J (2023a) Chapter 10 – Is there a relationship between myconanotechnology and sustainable development? In: Kuddus M, Ahmad IZ, Hussain CMBT-M, A. of N. in B (eds) Progress in biochemistry and biotechnology. Academic Press, pp 175–207. https://doi.org/10.1016/B978-0-443-15262-7.00009-7

Devi P, Dey SR, Kumar P (2023c) Integration of water resources management in rural areas. In: Water resources management for rural development: challenges and mitigation. Elsevier, Department of Agronomy, School of Agriculture, Lovely Professional University, Phagwara, India, pp 139–152. https://doi.org/10.1016/B978-0-443-18778-0.00013-1

Devi P, Dey SR, Saini L, Kumar P, Panigrahi S, Dwivedi P (2023b) Toward sustainable agriculture: strategies involving phytoprotectants against reactive oxygen species. In: Reactive oxygen species: prospects in plant metabolism. Springer Nature, Department of Agronomy,

School of Agriculture, Lovely Professional University, Phagwara, India, pp 229–247. https://doi.org/10.1007/978-981-19-9794-5_13

Devi P, Dey SR, Sharma K, Kumar P, Singh J (2024a) Waste management as an opportunity for the inclusion of vulnerable groups. In: Waste management in the circular economy. Springer International Publishing, Department of Agronomy, School of Agriculture, Lovely Professional University, Phagwara, India, pp 207–230. https://doi.org/10.1007/978-3-031-42426-7_11

Devi P, Dogra P, Kumar P, Singh J (2024b) Global status of biomass energy programmes-challenges and roadmap. In: Biomass energy for sustainable development. CRC Press, Lovely Professional University, Phagwara, India, pp 394–414. https://doi.org/10.1201/9781003406501-20

Devi P, Kapoor P, Kumar P (2024c) Biomarkers for toxicity detection. In: Biomarkers in environmental and human health biomonitoring: an integrated perspective. Elsevier, Department of Agronomy, School of Agriculture, Lovely Professional University, Phagwara, India, pp 313–329. https://doi.org/10.1016/B978-0-443-13860-7.00004-5

Devi P, Kaur A, Kumar P (2024d) Phytoremediation of metal contaminated soil using energy crops: soil health maintenance along with biofuel production. In: Bioremediation of emerging contaminants from soils: soil health conservation for improved ecology and food security. Elsevier, Department of Agronomy, School of Agriculture, Lovely Professional University, Phagwara, India, pp 307–333. https://doi.org/10.1016/B978-0-443-13993-2.00015-3

Devi P, Kumar P, Singh J (2024e) Biotechnological approaches as clean technology for the development of a sustainable environment. In: Development in waste water treatment research and processes: role of environmental microbiology in industrial wastewater research. Elsevier, Department of Agronomy, School of Agriculture, Lovely Professional University, Phagwara, India, pp 381–403. https://doi.org/10.1016/B978-0-443-13609-2.00016-1

Devi P, Kumar P, Singh J (2024f) Application of potential microbes in bioremediation. In: Development in waste water treatment research and processes: role of environmental microbiology in industrial wastewater research. Elsevier, Department of Agronomy, School of Agriculture, Lovely Professional University, Phagwara, India, pp 201–225. https://doi.org/10.1016/B978-0-443-13609-2.00021-5

Dey SR, Devi P, Kumar P (2023a) Organic contaminants in aquatic environments: sources and impact assessment. In: Metal organic frameworks for wastewater contaminant removal. Wiley, Department of Agronomy, School of Agriculture, Lovely Professional University, Phagwara, India, pp 299–317

Dey SR, Kumar P, Singh J (2023b) Growth of mustard under open field conditions with salicylic acid and antimicrobial thiourea. Biopestic Int 19:183–192. https://doi.org/10.59467/BI.2023.19.183

Dey SR, Sharma M, Kumar P (2023c) Effects and responses of chromium on plants. In: Environmental science and engineering. Springer Science and Business Media Deutschland, GmbH, Department of Agronomy, School of Agriculture, Lovely Professional University, Phagwara, India, pp 385–427. https://doi.org/10.1007/978-3-031-44029-8_14

Dey SR, Sharma M, Kumar P (2024) Phytoremediation of phenolic compounds from soil. In: Bioremediation of emerging contaminants from soils: soil health conservation for improved ecology and food security. Elsevier, Department of Agronomy, School of Agriculture, Lovely Professional University, Phagwara, India, pp 335–358. https://doi.org/10.1016/B978-0-443-13993-2.00016-5

Dwivedi P, Kumar P (2011) Anti-diabetic medicinal plants and their conservation: waging green war on diabetes. Med Plants 3:181–189. https://doi.org/10.5958/j.0975-4261.3.3.031

Goud EL, Singh J, Kumar P (2022) Climate change and their impact on global food production. In: Microbiome under changing climate. Elsevier, pp 415–436

Hidangmayum A, Dwivedi P, Kumar P, Upadhyay SK (2023) Seed Priming and foliar application of chitosan ameliorate drought stress responses in mungbean genotypes through modulation of morpho-physiological attributes and increased antioxidative defense mechanism. J Plant Growth Regul 42:6137–6154. https://doi.org/10.1007/s00344-022-10792-1

Jagadesh M, Dash M, Kumari A, Singh SK, Verma KK, Kumar P, Bhatt R, Sharma SK (2024) Revealing the hidden world of soil microbes: metagenomic insights into plant, bacteria, and fungi interactions for sustainable agriculture and ecosystem restoration. Microbiol Res 285:127764. https://doi.org/10.1016/j.micres.2024.127764

Karam N, Kumar P, Anand YR, Choudhury D (2023) Management of rice blast—a review of the strategies available. Biopestic Int 19:107–116. https://doi.org/10.59467/BI.2023.19.107

Kotia A, Rutu P, Singh V, Kumar A, Dhoke S, Kumar P, Singh DK (2021) Rheological analysis of rice husk-starch suspended in water for sustainable agriculture application. In: 2nd international conference on functional materials, manufacturing and performances, ICFMMP 2021. Elsevier Ltd, School of Mechanical Engineering, Lovely Professional Unievrsity, Phagwara, India, pp 1962–1966. https://doi.org/10.1016/j.matpr.2021.09.325

Kumar P (2023) Degradation of xenobiotics by bacteria and fungi: an overview. Biopestic Int 19:117–132. https://doi.org/10.59467/BI.2023.19.117

Kumar P (2024a) Biochar production from marine algae and its application in the treatment of industrial effluent. In: Marine biomass: biorefinery, bioproducts and environmental bioremediation. De Gruyter, Department of Agronomy, School of Agriculture, Lovely Professional University, Phagwara, India, pp 57–93. https://doi.org/10.1515/9783111353951-004

Kumar P (2024b) Application of marine biomass for carbon neutrality, sustainable environment, and economy. In: Marine biomass: biorefinery, bioproducts and environmental bioremediation. De Gruyter, Department of Agronomy, School of Agriculture, Lovely Professional University, Phagwara, India, pp 95–128. https://doi.org/10.1515/9783111353951-005

Kumar P (2024c) Soil conservation for global sustainability. In: World sustainability series. Springer Science and Business Media Deutschland. GmbH, Department of Agronomy, School of Agriculture, Lovely Professional University, Phagwara, India, pp 103–128. https://doi.org/10.1007/978-3-031-57456-6_6

Kumar P, Choudhury D (2024a) Optimized phytoremediation process for the sustainable management radionuclides. In: Bioremediation of emerging contaminants from soils: soil health conservation for improved ecology and food security. Elsevier, Department of Agronomy, School of Agriculture, Lovely Professional University, Phagwara, India, pp 443–464. https://doi.org/10.1016/B978-0-443-13993-2.00020-7

Kumar P, Choudhury D (2024b) Role of indigenous knowledge in agricultural soil reclamation without disturbing other ecosystems. In: Bioremediation of emerging contaminants from soils: soil health conservation for improved ecology and food security. Elsevier, Department of Agronomy, School of Agriculture, Lovely Professional University, Phagwara, India, pp 465–488. https://doi.org/10.1016/B978-0-443-13993-2.00021-9

Kumar P, Choudhury D (2024c) Microbial native soil bacteria against cadmium toxicity. In: Springer Water. Springer Nature, Department of Agronomy, School of Agriculture, Lovely Professional University, Phagwara, India, pp 193–226. https://doi.org/10.1007/978-3-031-54005-9_9

Kumar P, Choudhury D (2024d) Toxicity of Rhizospheric cadmium contaminated soil and its phytoremediation. In: Springer Water. Springer Nature, Department of Agronomy, School of Agriculture, Lovely Professional University, Phagwara, India, pp 227–264. https://doi.org/10.1007/978-3-031-54005-9_10

Kumar P, Choudhury D (2024e) Hydroponic removal of organic contaminants from water. In: Springer Water. Springer Nature, Department of Agronomy, School of Agriculture, Lovely Professional University, Phagwara, India, pp 143–164. https://doi.org/10.1007/978-3-031-53258-0_6

Kumar P, Dey SR (2024) Hydroponics phytoremediation: an overview. In: Springer Water. Springer Nature, Department of Agronomy, School of Agriculture, Lovely Professional University, Phagwara, India, pp 361–396. https://doi.org/10.1007/978-3-031-53258-0_13

Kumar P, Dwivedi P (2018) Putrescine and glomus mycorrhiza moderate cadmium actuated stress reactions in zea mays l. By means of extraordinary reference to sugar and protein. Vegetos 31:74–77. https://doi.org/10.5958/2229-4473.2018.00076.9

Kumar P, Dwivedi P (2020) Lignin estimation in sorghum leaves grown under hazardous waste site. Plant Arch 20:2558–2561

Kumar VA, Kumar P (2023) Revolutionizing agriculture with satellite technology for farmers: a review. Biopestic Int 19:97–106. https://doi.org/10.59467/BI.2023.19.97

Kumar P, Mistri TK (2020) Transcription factors in SOX family: potent regulators for cancer initiation and development in the human body. Semin Cancer Biol 67:105–113. https://doi.org/10.1016/j.semcancer.2019.06.016

Kumar P, Naik M (2020) Biotic symbiosis and plant growth regulators as a strategy against cadmium and lead stress in chickpea. Plant Arch. 20:2495–2500

Kumar P, Pathak S (2019) Responsiveness index of sorghum (Sorghum bicolor (L.) Moench) grown under cadmium contaminated soil treated with putrescine and mycorrhiza. Bangladesh J Bot 48:139–143

Kumar P, Raut AM (2024) Microbes-assisted bioaugmentation process in the reduction of emerging industrial pollutants from soil. In: Bioremediation of emerging contaminants from soils: soil health conservation for improved ecology and food security. Department of Agronomy, School of Agriculture, Lovely Professional UniversityElsevier, Phagwara, India, pp 519–540. https://doi.org/10.1016/B978-0-443-13993-2.00023-2

Kumar P, Singh J (2024a) Exploring enzyme biotechnology's role in green chemistry and advancements in pharmaceutical technologies. In: Enzyme biotechnology for environmental sustainability. Elsevier, Department of Agronomy, School of Agriculture, Lovely Professional University, Phagwara, India, pp 465–495. https://doi.org/10.1016/B978-0-443-22072-2.00015-2

Kumar P, Singh J (2024b) Environmental and health risk assessment of chemical pollutants in drinking water and wastewaters. In: Trends in biological processes in industrial wastewater treatment. Institute of Physics Publishing, Department of Agronomy, School of Agriculture, School of Bioengineering and Biosciences, Lovely Professional University, Phagwara, India, pp 5.1–5.30. https://doi.org/10.1088/978-0-7503-5678-7ch5

Kumar P, Singh J (2024c) Bioprocesses in industrial wastewater treatment: trends and prospects. In: Trends in biological processes in industrial wastewater treatment. Institute of Physics Publishing, Department of Agronomy, School of Agriculture, School of Bioengineering and Biosciences, Lovely Professional University, Phagwara, India, pp 15.1–15.50. https://doi.org/10.1088/978-0-7503-5678-7ch15

Kumar P, Singh J (2024d) Emerging contaminants and ways to reduce them. In: Trends in biological processes in industrial wastewater treatment. Institute of Physics Publishing, Department of Agronomy, School of Agriculture, School of Bioengineering and Biosciences, Lovely Professional University, Phagwara, India, pp 4.1–4.34. https://doi.org/10.1088/978-0-7503-5678-7ch4

Kumar P, Singh J (2024e) Hydroponic: an eco-friendly future. In: Springer Water. Springer Nature, Department of Agronomy, School of Agriculture, Lovely Professional University, Phagwara, India, pp 223–267. https://doi.org/10.1007/978-3-031-53258-0_9

Kumar P, Mandal B, Dwivedi P (2011) Heavy metal scavenging capacity of Mentha spicata and Allium cepa. Med Plants 3:315–318. https://doi.org/10.5958/j.0975-4261.3.4.053

Kumar P, Singh BN, Dwivedi P (2017) Plant growth regulators, plant adaptability and plant productivity: a review on abscisic acid (ABA) signaling in plants under emerging environmental stresses. In: Sustaining future food security in changing environments. Nova Science Publishers, Inc., Department of Plant Physiology, Institute of Agricultural Sciences, Banaras Hindu University, Varanasi, India, pp 81–97

Kumar P, Pathak S, Kumar M, Dwivedi P (2018) Role of secondary metabolites for the mitigation of cadmium toxicity in Sorghum Grown under mycorrhizal inoculated hazardous waste site. In: Kumar N (ed) Biotechnological approaches for medicinal and aromatic plants. Springer, Singapore, pp 199–212

Kumar P, Harshavardhan M, Kumar PS, Yumnam J, Jyoti N, Naik M, Misao L, Purnima K (2018a) Effect on chlorophyll a/b ratio in cadmium contaminated maize leaves treated with putrescine and mycorrhiza. Ann Biol 34:281–283

Kumar P, Kumar S, Harshavardhan M, Naik M, Yumnam J, Kumar PS, Jyoti N, Misao L, Purnima (2018b) Evaluation of plant height and leaf length of sorghum grown under different sources of nutrition. Ann Biol 34:284–286

Kumar P, Pathak S, Amarnath KS, Veerendra Brahma Teja P, Dileep B, Kumar K, Singh M, Siddique A (2018c) Effect of growth regulator on morpho-physiological attributes of chilli: a case study. Plant Arch. 18:1771–1776

Kumar P, Yumnam J, Kumar PS, Misao L, Jyoti N, Naik M, Purnima, Kumar S, Harshavardhan M (2018d) Cadmium induced changes in germination of maize seed treated with mycorrhiza. Ann Agri Bio Res 23:169–170

Kumar D, Rameshwar SD, Kumar P (2019) Effect of intergated application of inorganic and organic fertilizers on the roots of chickpea. Plant Arch. 19:857–860

Kumar P, Kumar T, Singh S, Tuteja N, Prasad R, Singh J (2020) Potassium: a key modulator for cell homeostasis. J Biotechnol 324:198–210

Kumar V, Dwivedi P, Kumar P, Singh BN, Pandey DK, Kumar V, Bose B (2021) Mitigation of heat stress responses in crops using nitrate primed seeds. South African J Bot 140:25–36. https://doi.org/10.1016/j.sajb.2021.03.024

Kumar P, Devi P, Dey SR (2021a) Fungal volatile compounds: a source of novel in plant protection agents. In: Volatiles and metabolites of microbes. Elsevier, pp 83–104

Kumar P, Sharma K, Saini L, Dey SR (2021b) Chapter 8 – Role and behavior of microbial volatile organic compounds in mitigating stress. In: Kumar A, Singh J, Samuel JBT-V, M. of M (eds) Volatiles and metabolites of microbes. Academic Press, pp 143–161. https://doi.org/10.1016/B978-0-12-824523-1.00010-9

Kumar P, Goud EL, Devi P, Dey SR, Dwivedi P (2022a) Heavy metals: transport in plants and their physiological and toxicological effects. In: Plant metal and metalloid transporters. Springer Nature, Department of Agronomy, School of Agriculture, Lovely Professional University, Phagwara, India, pp 23–54. https://doi.org/10.1007/978-981-19-6103-8_2

Kumar P, Goud EL, Devi P, Koul B (2022b) Metal pollutants in the environment. In: Environmental microbiology: emerging technologies. De Gruyter, Department of Agronomy, School of Agriculture, Lovely Professional University, Jalandhar, India, pp 291–323. https://doi.org/10.1515/9783110727227-012

Kumar P, Koul B, Sharma M (2022c) Phytoremediation of heavy metals. In: Heavy metals in plants: physiological to molecular approach. CRC Press, Department of Agronomy, School of Agriculture, Lovely Professional University, Jalandhar, India, pp 369–388. https://doi.org/10.1201/9781003110576-17

Kumar P, Choudhury D, Sahoo J, Devi MI (2023a) Exploring the potential for extended investigation: a preliminary examination of Liriomyza trifolii and its bibliometric foundation for analytical pursuits. J Food Chem Nanotechnol 9:S359–S366. https://doi.org/10.17756/jfcn.2023-s1-045

Kumar P, Latha DH, Singh J (2023b) Rhizoremediation of organic emerging soil contaminants: green technology. In: Clean technologies toward a sustainable future: physicochemical, biochemical and biotechnological approaches. IWA Publishing, Department of Agronomy, School of Agriculture, Lovely Professional University, Phagwara, India, pp 211–229. https://doi.org/10.2166/9781789063783_0211

Kumar P, Saini L, Sharma M (2023c) Assessment of the phytochemical constituents and metabolites in the medicinal plants and herbal medicine used in the treatment and management of respiratory diseases. In: Reference series in phytochemistry. Springer Science and Business Media B.V., Department of Agronomy, School of Agronomy, Lovely Professional University, Phagwara, India. https://doi.org/10.1007/978-3-031-21973-3_19-1

Kumar P, Choudhury D, Dey SR, Sharma M (2024a) Application of genetically modified crops against metallic contaminations. In: Bioremediation of emerging contaminants from soils: soil health conservation for improved ecology and food security. Elsevier, Department of Agronomy, School of Agriculture, Lovely Professional University, Phagwara, India, pp 541–564. https://doi.org/10.1016/B978-0-443-13993-2.00024-4

Kumar P, Dey SR, Choudhury D (2024b) Effectiveness of cadmium on biochemical shift of pea plant treated with mycorrhiza and putrescine. Nat Environ Pollut Technol 23:275–285. https://doi.org/10.46488/NEPT.2024.v23i01.023

Kumar P, Dwivedi P, Upadhyay SK (2024c) Optimization of polyamine and mycorrhiza in sorghum plant for removal of hazardous cadmium. Plant Physiol Biochem 214:108846. https://doi.org/10.1016/j.plaphy.2024.108846

Kumar P, Karam N, Choudhury D (2024d) Assessment of the phytochemical constituents and metabolites in the medicinal plants and herbal medicine used in the treatment and Management of Skin Diseases. In: Reference series in phytochemistry. Springer Science and Business Media B.V., Department of Agronomy, School of Agriculture, Lovely Professional University, Phagwara, India, pp 533–573. https://doi.org/10.1007/978-3-031-43199-9_21

Kumar P, Khokhar A, Dey SR (2024e) Evidence from the use of herbal medicines in the management and prevention of common eye diseases. In: Reference series in phytochemistry. Springer Science and Business Media B.V., Department of Agronomy, School of Agriculture, Lovely Professional University, Phagwara, India, pp 763–795. https://doi.org/10.1007/978-3-031-43199-9_27

Kumar P, Saini L, Sharma M (2024f) Assessment of the phytochemical constituents and metabolites in the medicinal plants and herbal medicine used in the treatment and Management of Respiratory Diseases. In: Reference series in Phytochemistry. Springer Science and Business Media B.V., Department of Agronomy, School of Agronomy, Lovely Professional University, Phagwara, India, pp 481–510. https://doi.org/10.1007/978-3-031-43199-9_19

Kumar P, Saurabh S, Sharma K (2024g) Assessment of the phytochemical constituents and metabolites in medicinal plants and herbal remedies used in the treatment and management of reproductive diseases: polycystic ovary syndrome. In: Reference series in phytochemistry. Springer Science and Business Media B.V., Department of Agronomy, School of Agriculture, Lovely Professional University, Phagwara, India, pp 511–532. https://doi.org/10.1007/978-3-031-43199-9_20

Kumar P, Sharma K, Singh J (2024h) Analysis of complex microbial communities in soil and wastewater treatment processes. In: Microbial metagenomics in effluent treatment plant. Elsevier, Department of Agronomy, School of Agriculture, Lovely Professional University, Phagwara, India, pp 25–41. https://doi.org/10.1016/B978-0-443-13531-6.00002-1

Kumar P, Sharma K, Singh J (2024i) Response of microbial community to environment changes. In: Microbial metagenomics in effluent treatment plant. Elsevier, Department of Agronomy, School of Agriculture, Lovely Professional University, Phagwara, India, pp 43–61. https://doi.org/10.1016/B978-0-443-13531-6.00006-9

Kumar P, Singh J, Singh G (2024j) Renewable hydrogen opportunities and challenges. In: Renewable hydrogen: opportunities and challenges in commercial success. Elsevier, Department of Agronomy, School of Agriculture, Lovely Professional University, Phagwara, India, pp 193–220. https://doi.org/10.1016/B978-0-323-95379-5.00010-9

Kumar P, Srivastav AL, Chaudhary V, van Hullebusch ED, Busquets R (2024k) Bioremediation of emerging contaminants from soils: soil health conservation for improved ecology and food security. Elsevier, Department of Agronomy, School of Agriculture, Lovely Professional University, Phagwara, India. https://doi.org/10.1016/C2022-0-02758-6

Kumari P, Singh J, Kumar P (2022) Impact of bioenergy for the diminution of an ascending global variability and change in the climate. In: Microbiome under changing climate. Elsevier, pp 469–487

Kumari P, Kumar P, Singh J (2023) Toxic contamination of aquatic habitat and its remediation. In: Spatial modeling of environmental pollution and ecological risk. Elsevier, Department of Agronomy, School of Agriculture, Lovely Professional University, Phagwara, India, pp 209–220. https://doi.org/10.1016/B978-0-323-95282-8.00021-3

Naik M, Kumar P (2020) Role of growth regulators and microbes for metal detoxification in plants and soil. Plant Arch 20:2820–2824

Pandey SK, Kumar P, Dey SR (2024) Xenobiotic stress management by microbial endophytes. In: Bioremediation of emerging contaminants from soils: soil health conservation for improved ecology and food security. Elsevier, College of Agriculture, Powarkheda Narmadapuram, Jawaharlal Nehru Agriculture University, Jabalpur, India, pp 241–259. https://doi.org/10.1016/B978-0-443-13993-2.00012-8

Pankaj PK, Kumar P, Nigam RC, Mishra PK (2012) Monitoring and surveillance of synthetic pyrethroids and organophosphate in different brands of soft drinks. J Chem Pharm Res 4:3939–3943

Pathak S, Kumar P, Mishra PK, Kumar M (2017) Mycorrhiza assisted approach for bioremediation with special reference to biosorption. Pollut Res 36:329–332

Raj A, Goutam E, Kumar P, Singh J (2024) Polyphenol oxidase: an enzyme of plants. In: Polyphenol oxidases: function, wastewater remediation, and biosensors. De Gruyter, Department of Agronomy, School of Agriculture, Lovely Professional University, Phagwara, India, pp 105–128. https://doi.org/10.1515/9783111033525-004

Reddy SS, Kumar P, Dwivedi P (2022) Heavy metal transporters, phytoremediation potential, and biofortification. In: Plant metal and metalloid transporters. Springer, pp 387–405

Saini L, Sharma K, Kumar P (2023) Water efficiency in agriculture as a vital approach toward water management in rural areas. In: Water resources management for rural development: challenges and mitigation. Elsevier, Department of Agronomy, School of Agriculture, Lovely Professional University, Phagwara, India, pp 127–138. https://doi.org/10.1016/B978-0-443-18778-0.00020-9

Saurabh K, Meena RK, Kumar P, Upadhyaya A, Das A (2024) Role of hyperaccumulators in the reduction of emerging industrial pollutants from soil. In: Bioremediation of emerging contaminants from soils: soil health conservation for improved ecology and food security. Elsevier, ICAR Research Complex for Eastern Region, Patna, India, pp 489–517. https://doi.org/10.1016/B978-0-443-13993-2.00022-0

Sharma K, Kumar P (2023) Microbe-mediated impact on cell membrane stability and leaf pigmentation in pearl millet under toxic mercury stress in soil. Biopestic Int 19:217–228. https://doi.org/10.59467/BI.2023.19.217

Sharma K, Kumar P (2024) Environmental threats posed by xenobiotics. In: Bioremediation of emerging contaminants from soils: soil health conservation for improved ecology and food security. Elsevier, Department of Agronomy, School of Agriculture, Lovely Professional University, Phagwara, India, pp 183–201. https://doi.org/10.1016/B978-0-443-13993-2.00009-8

Sharma M, Dey SR, Kumar P (2023) Antioxidant defence: a key mechanism of chromium tolerance. In: Environmental science and engineering. Springer Science and Business Media Deutschland GmbH, Department of Agronomy, School of Agriculture, Lovely Professional University, Phagwara, India, pp 91–116. https://doi.org/10.1007/978-3-031-44029-8_5

Sharma K, Devi P, Dey SR, Kumar P (2023a) Mercury phytovolatilization: an overview of the mechanism and mitigation. In: Role of green chemistry in ecosystem restoration to achieve environmental sustainability. Elsevier, Department of Agronomy, School of Agriculture, Lovely Professional University, Phagwara, India, pp 325–331. https://doi.org/10.1016/B978-0-443-15291-7.00031-6

Sharma K, Devi P, Kumar P (2023b) Metal-organic frameworks for wastewater contaminants removal. In: Metal organic frameworks for wastewater contaminant removal. Wiley, Department of Agronomy, School of Agriculture, Lovely Professional University, Phagwara, India, pp 95–117

Sharma K, Devi P, Kumar P, Dey A, Dwivedi P (2023c) Hazardous phytotoxic nature of reactive oxygen species in agriculture. In: Reactive oxygen species: prospects in plant metabolism. Springer Nature, Department of Agronomy, School of Agriculture, Lovely Professional University, Phagwara, India, pp 135–146. https://doi.org/10.1007/978-981-19-9794-5_8

Sharma K, Kumar P, Singh J (2024) Ecology and diversity of microbial communities involved in the removal of priority contaminants and micropollutants in waste. In: Development in waste water treatment research and processes: role of environmental microbiology in industrial wastewater research. Elsevier, Department of Agronomy, School of Agriculture, Lovely

Professional University, Phagwara, India, pp 181–199. https://doi.org/10.1016/B978-0-443-13609-2.00014-8

Sinam V, Kumar P, Singh J (2023) Bioremediation and ecorestoration strategies of aquatic environment. In: Spatial modeling of environmental pollution and ecological risk. Elsevier, Department of Agronomy, School of Agriculture, Lovely Professional University, Phagwara, India, pp 483–499. https://doi.org/10.1016/B978-0-323-95282-8.00025-0

Srivastav AL, Patel N, Rani L, Kumar P, Dutt I, Maddodi BS, Chaudhary VK (2024) Sustainable options for fertilizer management in agriculture to prevent water contamination: a review. Environ Dev Sustain 26:8303–8327. https://doi.org/10.1007/s10668-023-03117-z

Upadhyay SK, Devi P, Kumar V, Pathak HK, Kumar P, Rajput VD, Dwivedi P (2023) Efficient removal of total arsenic (As3+/5+) from contaminated water by novel strategies mediated iron and plant extract activated waste flowers of marigold. Chemosphere 313:137551. https://doi.org/10.1016/j.chemosphere.2022.137551

Geomicrobiological Strategies in Rehabilitation of Coal Mine Dump Ecosystems

Lily Shylla and Santa Ram Joshi

Contents

Abstract Mining, particularly coal and mineral exploration, serves not only as a cornerstone of global economic development but also brings significant environmental consequences, most notably the accumulation of mine wastes. Poor waste management practices lead to indiscriminate dumping of various mining residues, exacerbating environmental degradation. Key issues like soil contamination, water pollution, and air quality deterioration are common at mining sites worldwide, worsened by the conventional dumping of coal mining waste, including waste rock or overburden (OB) material, on valuable land. To address this urgent environmental challenge, the use of bioremediation and phytoremediation techniques offers promising solutions by leveraging the natural abilities of microorganisms, particularly geomicrobes. Geo microorganisms, which play a crucial role in Earth's biogeochemical cycles, are significantly affected by coal mining activities, leading to altered soil physicochemical properties and changes in microbial community struc-

L. Shylla (✉)
North East Centre for Technology Application and Reach (NECTAR), Shillong, Meghalaya, India

S. R. Joshi
Microbiology Laboratory, Department of Biotechnology and Bioinformatics, North-Eastern Hill University, Shillong, Meghalaya, India

S. R. Joshi et al. (eds.), *Mineral Transformation and Bioremediation by Geo-Microbes*, Interdisciplinary Biotechnological Advances,
https://doi.org/10.1007/978-981-96-3033-2_12

tures. However, despite the disruptions caused by mining, these microorganisms play an essential role in rehabilitating degraded land. They contribute to soil restoration by converting toxic contaminants into less harmful substances and supporting plant growth through metabolic regulation. This chapter explores the importance of studying geomicrobes and their pivotal role in restoring ecological balance in coal mine dumps, offering a pathway to more sustainable mining practices.

Keywords Coal mining · Overburden dumps · Geomicrobes · Rehabilitation · Humic substances · Bioremediation · Phytoremediation

Abbreviations

ACC	1-Aminocyclopropane-1-carboxylate
AMD	Acid mine drainage
AMF	Arbuscular mycorrhizal fungi
C/N Ratio	Carbon to nitrogen ratio
EPS	Extracellular polymeric substances
IAA	Indole-3-acetic acid
OB	Overburden
OBMs	Mine overburden materials
PGP	Plant-growth-promoting
SOC	Soil organic carbon
SOM	Soil organic matter
SRB	Sulfate-reducing bacteria

1 Introduction

Coal is an organic, combustible rock that is mostly made up of carbon with varying amounts of various other elements, primarily hydrogen, sulfur, oxygen, and nitrogen (Jovanovski et al. 2023), interlayered with other sedimentary rocks, formed from built-up plant matter that has been changed over millions of years by decay and extreme changes in temperature and pressure. The thickness of its beds can range from under a millimeter to several meters deep. Coal is prized for its high carbon content, which makes it an important resource for energy and heat production (Voncken 2020). It has been used for almost 3000 years and was known to the Chinese before the Christian era. The usage of coal was known to the Greeks in the fourth century A.D. In the ninth century, it was utilized in England as a residential fuel. Coal mining was greatly aided by the eighteenth-century industrial revolution that followed England's creation of the steam engine. Coke produced from bituminous coal started to replace charcoal in the iron ore smelting sectors, which further increased demand for coal (Brown and Dey 1975). The "Statistical Review of World

Energy 2020" estimates that 167.58 million tons of coal were produced worldwide in 2019. These days, the main uses of coal are in the production of electricity and, to a lesser degree, in heavy industries like the steel and iron sectors (Yılmaz and Uslu 2007). Pyrites, a type of ferrous sulfate, are found in considerable amounts in coal. When pyrite is exposed to atmospheric oxygen during mining operations, it undergoes an oxidation process that, in the presence of bacteria, turns it into sulfuric acid and ferrous sulfate. The resulting sulfuric acid decreases the pH of the water and soil in aquatic and terrestrial habitats, respectively, which has an impact on the activity and population of organisms that live there (Brown and Dey 1975).

Coal can be extracted by open cast and underground mining methods (Shylla et al. 2021). Open-cast mining is the most prevalent mining method in the globe, characterized by high labor efficiency, large mining capacity, quick construction, low production costs, and high resource recovery rates (Tian et al. 2014). In India, nearly 80% of coal production comes from open-cast mining compared to 20% from underground mining. There are numerous environmental issues associated with both mining methods (Gupta and Paul 2015). However, compared to underground mining, massive amounts of waste rock are produced by open-pit mining, and this excess rock is piled up in overburden dumps (Ghosh 2002). Along with other related mining operations, extraction of coal, particularly via opencast mining (also known as open-pit mining), creates unwanted tailing dams and dumps that disrupt communities of microbes and alter the soil's nutritional status in mined areas due to excessive stockpiling and leaching (Gupta and Paul 2015). One of the biggest streams of liquid and solid waste on Earth is generated by the mining and extraction of energy resources, industrial minerals, and metals. This includes rock, sediment, gangue, refuse, tailings, roasted ore, processing chemicals, and other non-economic by-products, all considered mine dump wastes (Hudson-Edwards 2016). Over the next century, it is anticipated that the volumes of these "mine wastes" will at least double as a result of rising energy and mineral needs as well as declining ore grades (Gupta and Paul 2015). Overburden (OB) is a common word used to describe this waste product produced during open-pit mining (Prashant et al. 2010).

According to data from the Indian Bureau of Mines, 7,55,440 hectares are under mining leases, while 37,772 hectares, or 5%, are actively mined (Juwarkar and Jambhulkar 2008). In India, 623 Mt. of coal was produced by removing 1870 Mt. of overburden during mining operations (IBM 2001). According to earlier research, the overburden generated by open-pit mining is around 4 m^3 for every ton of mined coal; therefore, the extraction of 250 million tons of coal results in 1000 m^3 of overburden. According to reports, mining activities damage 60 km^2 of land annually for this amount of mined coal, and around 75 km^2 is utilized for the disposal of overburden, external rock, and soil (Kundu and Ghose 1997; Juwarkar and Jambhulkar 2008).

The mining process entails clearing the site of overburden materials, such as vegetation and topsoil, and then disposing of them all at once in adjacent hollows or plains. As a result, it totally shifts the mountain from one location to another, resulting in massive, artificial mountains or unnatural wastelands. The natural succession process may require centuries to restore vegetation to the mining overburden materials (OBMs), which are not conducive to plant growth (Sheoran et al. 2010). The high sulfur content and potentially hazardous components in the overburden cause very acidic leachates to occur (Dutta et al. 2017). This dump causes soil erosion, degradation, and pollution of the environment by changing the natural topography of the soil, creating acid mine drainage, and impeding the natural succession of plant development. Mine spoils' chemical characteristics can hinder seed germination, and their incredibly low water-holding capacity can restrict the growth of seedlings (Sarkar et al. 2017). Many studies have indicated that overburden lacks essential minerals and nutrients, making it unsuitable for supporting plant growth (Dutta et al. 2017). The depth and area of coal mine overburden dumps typically affect their physical and chemical characteristics (Singh and Narzary 2021a, b). The minerals and heavy metal content, along with other chemical properties of the soil, are derived from the parent rock materials, such as metal-rich rocks or shales (He et al. 2005). The physicochemical properties of overburden vary by region due to differences in the geological composition of rocks (Juwarkar and Jambhulkar 2008). Additionally, these wastes are extremely vulnerable to erosion by water and wind (Straupnik 2022). Furthermore, the waste's fine-grained percentage may cause soil crusts and fractures to form, which would be detrimental to the soil structure (Hossner and Hons 1992).

Two primary environmental issues associated with coal mining are:

(i) **Acid Mine Drainage (AMD),** is a significant environmental concern characterized by elevated concentrations of toxic metalloids and metals such as arsenic, lead, sulfate and iron and have low pH. This phenomenon arises when sulfide minerals, particularly pyrite, present in mining waste come into contact with water and oxygen, leading to a series of oxidation reactions that generate sulfuric acid. The resulting acidic drainage is often laden with dissolved heavy metals, rendering it highly corrosive and toxic to both aquatic ecosystems and human populations (Johnson and Hallberg 2005). The environmental implications of AMD are profound and enduring. Once introduced into the water cycle, AMD can contaminate groundwater, streams, lakes, and estuaries for decades, if not centuries, after mining operations have ceased. The persistence of AMD is attributed to a self-sustaining oxidation process driven by the presence of sulfide minerals, water, and oxygen. This process not only generates sulfuric acid but also enhances the solubility of toxic metals, exacerbating the potential for environmental damage (Kuyucak 1999; Johnson 2003). AMD is visually distinctive, often manifesting as green or reddish water. The coloration depends on the metal composition of the drainage: green AMD typically indicates high ferrous iron content, while reddish water usually signals elevated levels of ferric iron. This visual aspect underscores the ecological impact of AMD, which can lead to severe disruption of aquatic habitats, posing risks to biodiversity and water quality (Méndez-García et al. 2014). Effective management and remediation of AMD are critical to mitigating its long-term impacts. Treatment methods can include neutralization of acidity using alkaline materials or the capture of dissolved metals through filtration and precipitation techniques. However, due to the complex nature of AMD and its persistent effects, proactive strategies are vital. These strategies involve minimizing the exposure of sulfide-rich waste to oxygen and water, thereby reducing future AMD generation and its environmental footprint (Johnson and Hallberg 2005).

(ii) **Erosion of mine tailings and coal mine dumps** presents significant environmental challenges, primarily due to the substantial amounts of fine particles that contain toxic metals and metalloids. These waste materials are particularly vulnerable to erosion driven by wind and water, leading to a variety of public health and ecological concerns (Bruneel et al. 2022). Dust generated from coal mine dumps can create hazardous conditions for nearby communities, especially on windy days, posing serious health risks associated with airborne contaminants such as lead, arsenic, and other heavy metals (Fahey and Newson 1997). Erosion in semi-arid and arid climates is exacerbated by episodic rainfall and intense wind, which can wash away soil and create gullies on the slopes of these dumps. This erosion threatens the stability of the dumps and results in significant maintenance challenges, often requiring costly repairs (Blight and Steffen 1979). The cumulative effects of erosion not only degrade soil quality but also lead to nutrient depletion, which undermines the potential for natural vegetative recovery and complicates future rehabilitation efforts (Bruneel et al. 2022). In tropical, temperate, and equatorial climates, the leaching of toxic metals from coal mine dumps caused acid mine drainage (AMD) formation, further complicating environmental remediation. This leaching occurs when precipitation mobilizes contaminants, resulting in significant degradation of water quality in nearby streams and groundwater systems (Mendez and Maier 2008; Nirola et al. 2016). To mitigate these pressing issues, establishing vegetative cover has emerged as a crucial strategy for controlling erosion in coal mine dumps, particularly in arid regions. Vegetation can significantly reduce wind and water erosion, stabilize slopes, and improve the aesthetic value of disturbed landscapes, which is vital for community acceptance of mining operations. However, coal mine dumps often present inhospitable conditions for plant growth due to factors such as high acidity, the presence of heavy metals, nutrient deficiencies, and extreme moisture variability (Gorber et al. 1978).

Mining wastes contain hazardous substances including lead, arsenic, mercury, and radioactive materials like uranium that can pose serious environmental risks if site management is not strictly enforced. These contaminants pose major risks to both human health and the environment since they are mutagenic, cytotoxic, and carcinogenic (Nordberg et al. 2014). According to Bruneel et al. (2022), residents are at risk by drinking tainted water, breathing in contaminant dust, or eating crops and animals raised on contaminated soil (Fig. 1).

The reclamation of mining sites, which involves the restoration of land resources, has become a global priority. The "Surface Mining Control and Reclamation Act," passed by the US Congress in August 1977, established uniform land reclamation practices in US mining regions. In 1950, Germany's "Prussian Mining Law" required the reclamation of mining landscapes. In 1989, China implemented the

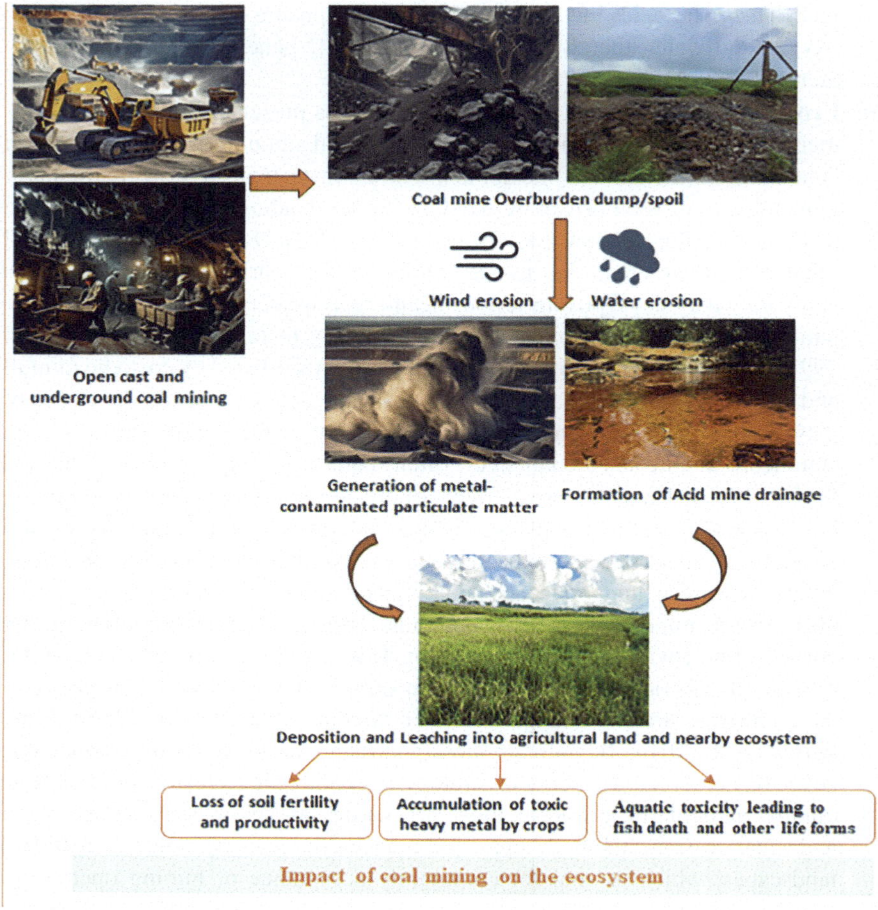

Fig. 1 Graphical representation of coal mining's impact on the ecosystem

"Land Reclamation Regulations," which resulted in notable advancements in ecological restoration and land reclamation (Li et al. 2022).

Reclamation expenses are high, and numerous research has been conducted globally (Aznar-Sánchez et al. 2018). Since metals and metalloids can't be degraded, they can only be converted into less toxic or less soluble forms or usage of precipitation processes for easier elimination (Hallberg 2010; Bolan et al. 2014). Chemical precipitation, soil replacement, solidification/stabilization, filtering for AMD, and soil cleaning for coal mine dumps and tailings are frequently prohibitively expensive or ineffectual, particularly at abandoned mining sites. Thus, establishing cost-effective and long-term remediation methods is crucial, especially in developing nations (Bruneel et al. 2022).

An often-overlooked aspect of traditional remediation approaches is the role of soil microorganisms and their effects on soil chemistry. While plant cover regeneration is typically emphasized, interactions between plants and soil and their effects on plant growth are important (Rivera et al. 2014). Microbes are central to soil formation and contribute to nutrient and organic matter cycles, along with metal and mineral cycling (da Silva et al. 2023). Microorganisms influence redox transformations of metals, mineral formation or dissolution, and the formation of organometallic complexes, affecting the toxicity and mobility of metallic elements (Gadd 2010). Prokaryotic and eukaryotic organisms have evolved mechanisms to resist and metabolize metals and metalloids, potentially playing a crucial role in site remediation. Methods such as microbial iron oxidation and sulfate reduction are used to precipitate metals from AMD solutions (Johnson and Hallberg 2005). Moreover, phyto-stabilization aims to stabilize tailings via vegetative cover that can prevent pollutant dispersion, though plant growth is a hindrance in harsh mine tailing and waste environments. Microorganisms can stimulate plant growth in coal mine dumps and polluted soils or immobilize pollutants (Rivera et al. 2014). Recent research highlights the potential of microbes in soil bioengineering (Graf et al. 2015; Vergani and Graf 2016). Bioengineering involves biologically driven influences that modify soil profile characteristics, including the role of microorganisms and major plant taxa in nutrient uptake and soil modification. Understanding soil chemistry and microbiological activity is crucial for mine spoil rehabilitation, and bioengineering approaches could enhance this process (Verboom and Pate 2006; Li et al. 2006). Monitoring microbial communities, particularly filamentous microbes, can assist in improving soil storage and management, thereby facilitating successful soil rehabilitation and ecosystem recovery (Wildman 2009) (Fig. 2).

The effects of these approaches on rehabilitation of the contaminated mine dump using geomicrobiological strategies can be monitored using the following methods:

2 Soil Aggregation by Geomicrobes

In mining operations, the elimination of overlying soil and the subsequent deposition of fragmented rock in the form of tailing or overburden dumps lead to significant environmental challenges (Rai et al. 2011). These dumps, which occupy large areas, result in the degradation of land and loss of soil quality. The loose nature of these dump materials makes them highly susceptible to wind erosion, which disperses fine particles over surrounding fertile land, negatively impacting plant growth and soil health (Barapanda et al. 2001).

Microbial communities play a crucial role in mitigating these issues through the enhancement of soil aggregation. Recent studies have identified two primary mechanisms by which microbes contribute to soil aggregation: (1) the formation of soil aggregates via fungal hyphae, and (2) the secretion/production of extracellular polymeric substances (EPS) (Blankinship et al. 2016). Plant roots, in association with fungal hyphae, are instrumental in developing and stabilizing soil aggregates by interlacing with soil particles, which aids in the compartmentalization of soil and the formation of microbial niches (Oades 1993; da Silva et al. 2023).

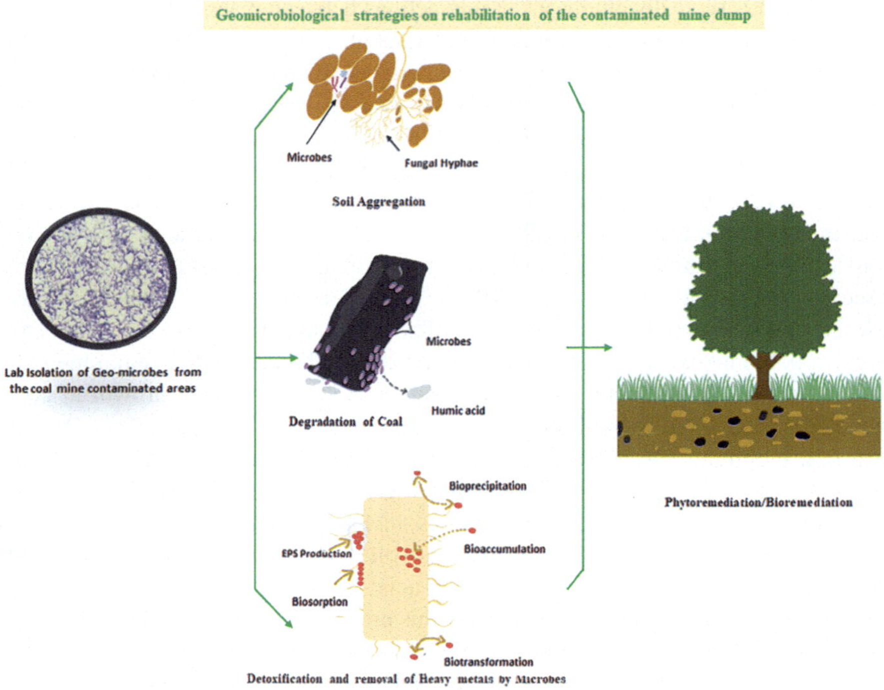

Fig. 2 Graphical representation of rehabilitation strategies for coal mine dumps using geomicrobes

The mycorrhizal fungi influence on soil structure has been well reviewed by Rillig and Mummey (2006), noting that arbuscular mycorrhizal fungi (AMF)—which form arbuscules by penetrating the cortical cells of plant roots—significantly enhance soil aggregation. The significant growth of AMF's extraradical hyphal is a key factor in the stabilization of soil aggregates (Wilson et al. 2009; Luna et al. 2016). Nevertheless, the degree to which AMF promotes soil aggregation varies with species, as different AMF families produce hyphae at different rates (Hart and Reader 2002). Further research is needed to elucidate how species diversity and environmental factors—such as nutrient levels, water content, and soil pH—affect hyphal growth and soil structure (da Silva et al. 2023).

In addition to fungal contributions, fungi and bacteria secrete EPS, as secondary metabolites function as binding agents for soil aggregates. These biopolymers play a critical role in creating a stable soil environment and accessing distant resources (Tisdall et al. 2012; Wang et al. 2014). EPS along with other biopolymers are particularly sensitive to changes in hydration, which influences its effectiveness in soil aggregation (Billings et al. 2015; Flemming et al. 2016). The EPS matrix is crucial both structurally and physiologically, as it protects microbial cells from desiccation, strengthens nutrient diffusion under dry conditions, and has a larger effect on soil water (Tecon and Or 2017).

A comprehensive understanding of the roles of fungal hyphae and bacterial EPS in soil aggregation is essential for the long-term preservation of post-mining sites (da Silva et al. 2023). Future research should focus on microbial interactions, soil ecology, and the specific functions of these microorganisms in various agricultural environments to optimize their function and usage in the field (Mitter et al. 2019).

3 Degradation of Coal by Geomicrobes

Coal, a complex and heterogeneous carbonaceous material interspersed with various mineral inclusions, provides a challenging substrate for microbial degradation. Geomicrobes, which include both bacteria and fungi, exhibit significant potential for breaking down coal through a range of mechanisms. The process of microbial coal degradation can broadly be classified into depolymerization, solubilization, and utilization (Akimbekov et al. 2022). Solubilization, a nonenzymatic process, occurs in alkaline environments where microorganisms produce alkaline metabolites, chelators, and surfactants that facilitate the dissolution of coal into a black liquid. Depolymerization involves enzymatic activity where lignin-degrading oxidoreductases (such as manganese peroxidase, lignin peroxidase, and laccase) and others cause hydrolases to cleave off the covalent bonds within the coal macromolecule, resulting in the release of lower molecular weight substances. This enzymatic breakdown typically occurs at pH levels below 6 (Weng et al. 2021). Several microbes such as *Trichoderma* sp., *Penicillium* sp., *Mycobacterium* sp., *Bacillus* sp., *Enterobacter* sp., *Acinetobacter* sp., and *Rhodococcus* sp. have been shown to release ligninolytic enzymes in a culture medium that contains coal. Ligninolytic microorganisms and saprotrophic fungi could serve as biocatalysts for coal transformation (Sekhohola et al. 2013). Utilization involves the biodegradation of these solubilized and depolymerized coal components by numerous bacteria and fungi, which use these components as carbon sources for growth (Akimbekov et al. 2022). Low-rank coals, such as lignite, are particularly susceptible to microbial attack than higher-rank coals due to their increased oxygen content. This increased oxygen content offers “biological doorways” for microbial interaction, resulting in greater water solubility and improved bioavailability of coal components. Additionally, the structural similarity of low-rank coals to lignin enhances their vulnerability to microbial degradation, as lignin-degrading microorganisms can target these similarities effectively (Akimbekov et al. 2022; Haider et al. 2018).

Among bacterial species, those isolated from coal environments often show superior coal-degrading capabilities compared to exogenous microorganisms due to their adaptation to the complex conditions of coal environments. These bacteria facilitate coal degradation through mechanisms such as biotransformation, biosolubilization, and bioconversion. For example, strains of *Cupriavidus*, *Pseudomonas*, and *Alcaligenes* have demonstrated effective depolymerization of aromatic compounds present in coal, such as phenol and toluene (David et al. 2017). The use of bacterial consortia, or “microbial cocktails,” has proven beneficial as these mixed communities can work synergistically to enhance the degradation process (Haider et al. 2018).

Fungal decomposition of coal differs notably from bacterial reactions. Fungi primarily release a range of organics including single-ring aromatics, polyaromatic hydrocarbons, and aromatic nitrogen compounds, with a smaller proportion of aliphatics (Haider et al. 2013). The fungal mechanisms of coal degradation involve both enzymatic and nonenzymatic agents. Enzymes such as peroxidases, hydrolases (including lignin peroxidase and manganese peroxidase), and phenol oxidases (like laccases) play crucial roles in breaking down coal's complex structure. Nonenzymatic agents produced by fungi, such as alkaline metabolites, surfactants, and chelators, also contribute to the modification of the coal matrix and the production of smaller, water-soluble molecules (Hofrichter and Fakoussa 2001; Akimbekov et al. 2022). Indigenous fungal species, well-adapted to the coal environment, have demonstrated effectiveness in coal modification. Examples include *Fusarium oxysporum*, *Paecilomyces farinosus*, and *Phanerochaete chrysosporium* (David et al. 2017).

The application of these microbial processes extends to bioremediation efforts aimed at restoring environments contaminated by coal mining. Mining activities lead to significant environmental challenges such as acid mine drainage, heavy metal contamination, and soil erosion (David et al. 2017). Traditional chemical approaches for site rehabilitation have shown limited effectiveness and cost-efficiency, leading to increased interest in bioremediation strategies. Microbial solubilization, oxidation, and liquefaction have been highlighted as effective techniques for rehabilitating coal-affected sites (Akimbekov et al. 2022). For instance, studies have demonstrated that autochthonous bacteria from coal mine spoil and fungal strains capable of colonizing and metabolizing discarded coal can significantly aid in environmental restoration. Moreover, microbial consortia that include various microorganisms working together have shown promise in enhancing coal degradation and improving rehabilitation outcomes (Hamidovic et al. 2016; David et al. 2017).

The degradation of coal by geomicrobes involves a complex interplay of microbial processes that transform coal into less harmful forms. Both bacteria and fungi contribute to this process through solubilization, depolymerization, and utilization, each utilizing different mechanisms to achieve coal modification. The application of these microbial processes in bioremediation offers a promising approach to mitigating the environmental impacts of coal mining and supporting the recovery of contaminated sites (Akimbekov et al. 2022).

4 Soil Geomicrobes and Soil Organic Matter

Soil organic matter (SOM) is a vital component of soil, impacting a variety of biological, chemical, and physical properties necessary for soil health and productivity. SOM is essential for the production and stability of soil aggregates, water retention, nutrient cycling, disease suppression, cation exchange capacity, and pH buffering (Celestina et al. 2019). These functions are critical from an agronomic context, considering they have the capacity to directly influence the productivity of crops

(Oelofse et al. 2015). However, coal mining activities can severely degrade soil quality, significantly altering the soil organic carbon (SOC) pool, which, in turn, affects the regional and global environment, crop productivity, and soil fertility (Xu et al. 2019). The removal of vegetation and topsoil during mining operations, along with the disruption of topography and hydrological regimes, contributes to a marked reduction in soil carbon stocks (Maharaj et al. 2007).

The effect of coal mining on soil extends to the microbial communities that are essential for the breakdown of organic matter and the cycling of nutrients (da Silva et al. 2023). The physicochemical changes induced by mining can lead to a significant reduction in microbial diversity and abundance, including key groups such as bacteria, fungi, and archaea (Ram and Masto 2010). Despite this disturbance, the remaining soil microbes are essential for the process of soil reconstruction and land restoration. Bacteria, for example, are instrumental in the decomposition of organic matter, while fungi are involved in breaking down complex macromolecules and forming symbiotic relationships with other organisms. Archaea are crucial in the nitrogen cycle, contributing to the overall stability of the ecosystem. The diverse microbial community in soil facilitates the cycling of SOM through various pathways, with fungi favoring the decomposition of SOM with high lignin and cellulose content via slower pathways, and bacteria dominating the decomposition of moist, nitrogen-rich tissues through faster pathways (Heuer et al. 1997; Ferris et al. 2004). The stability of soil carbon fractions produced by fungi is generally higher than those mediated by bacteria, making fungi a prominent group in studies of soil structure formation (Oades 1993).

Humic compounds, the most complicated and biologically active organic molecules in soil, serve an important role in strengthening microbial community composition, enhancing nutrient availability, and maintaining soil physicochemical properties (Akimbekov et al. 2022). These substances also enhance plant resilience against diseases and extreme environmental conditions. Interestingly, humic substances derived from low-rank coals share similar properties with those found in soil. Coal may possess humic acid from 25% to 85%, in contrast to 1–5% in most of the soils, providing a robust and novel source of humic acids for soil amendments (Dong et al. 2009). Coal-degrading microorganisms, particularly fungi and bacteria, have shown potential in converting coal to value-added products such as fulvic acids and humic acids (Akimbekov et al. 2022). Studies have shown that certain microbial consortia, including strains of *Serratia* and *Exiguobacterium*, can effectively degrade coal, leading to the production of humified organics. Furthermore, fungi such as *Mucor circinelloides* and *Aspergillus tubingensis* have been identified as efficient producers of humic acid, while others like *Penicillium daleae* and *Trichoderma koningiopsis* are noted for their production of fulvic acid (Nsa et al. 2022; Valero et al. 2016). These microbial processes not only aid in the biodegradation of coal but also contribute to the reclamation of soils affected by mining, enhancing their organic matter content and overall fertility (Akimbekov et al. 2022).

5 Geomicrobial Rehabilitation of Saline-Sodic Soil

Salinity is a widespread issue in coal mining areas worldwide, presenting significant challenges for soil rehabilitation. Coal spoils, the residual material left after coal extraction and coal mine dump where the spoils are dumped, often exhibit elevated salinity levels, particularly sodicity, which complicates the reclamation and restoration of these disturbed lands (Szczepanska and Twardowska 1999; Gozzard et al. 2009; Li et al. 2014). Sodicity in soils, characterized by an excess of sodium ions relative to other essential cations, can critically impair soil structural stability, resulting in various forms of degradation. Sodic soils are prone to surface crusting, reduced infiltration, low hydraulic conductivity, and the development of hard, dense subsoil layers, all of which impede root growth, seedling establishment, and the flow of air and water throughout the soil profile (Cubillos-Hinojosa et al. 2017). Additionally, the dispersion of clay particles upon wetting, driven by the high sodium ion content, exacerbates soil compaction and increases susceptibility to erosion and other forms of physical degradation (da Silva et al. 2023). The severe sodicity observed in coal spoils and dumps, such as those in Central Queensland, Australia, where electrical conductivity (EC) values can reach up to 25 mS/cm, underscores the extent of this problem and its implications for soil management and plant growth (Madsen and Mulligan 2006; Park et al. 2013).

Microbial communities, especially those adapted to extreme environments, offer a promising approach for mitigating the salinity issues associated with coal mine spoils. While microorganisms cannot directly alter the sodic or saline nature of soils, certain salt-tolerant microbes can thrive under these harsh conditions and contribute to improvements in soil structure and nutrient cycling, thus facilitating the establishment of salt-tolerant plant species (da Silva et al. 2023). For instance, extracellular polysaccharide (EPS)-producing bacteria have shown potential in alleviating salinity stress in plants. These bacteria can reduce sodium ion (Na+) influx into plant roots, mitigating the toxic effects of salinity (Ashraf et al. 2004). Additionally, EPS produced by bacteria can bind cations, including Na+, potentially lowering the sodium content accessible to uptake by plants and the reduction of salt stress (Geddie and Sutherland 1993). Bacteria such as *Aeromonas hydrophila/caviae* (strains MAS765), *Bacillus* sp. (strains MAS617, MAS620, MAS820), and *Bacillus insolitus* (strains MAS17) have shown positive results in alleviating salinity stress and survival rate in salt-sensitive plants (Ashraf et al. 2004).

Furthermore, certain microorganisms can convert or solubilize coal into humified organic matter, which can act as a soil amendment to enhance the physical and chemical properties of saline-sodic soils. Low-rank coals are considered effective amendments for enhancing soil quality and productivity due to their importance as a nutritive substrate for many soil microbes (Giannouli et al. 2009). Bacteria such as *Streptomyces* sp., *Pseudomonas* sp., *Bacillus* sp., *Rhodococcus* sp., and *Escherichia* sp., isolated from coal samples, have demonstrated the ability to solubilize coal and generate humic substances that enhance soil structure and nutrient availability (Sharma et al. 2015; Egamberdieva et al. 2019). Soil amendments containing 1%

coal and coal-solubilizing bacteria have been demonstrated to promote microbial activity, increase soil respiration, and enhance hydrolytic enzyme function, resulting in improved soil cation exchange capacity (Cubillos-Hinojosa et al. 2017).

Field studies have confirmed the effectiveness of using coal and coal-solubilizing bacteria for rehabilitating saline-sodic soils. The incorporation of these amendments has been associated with significant reductions in sodium adsorption ratio, electrical conductivity, and exchangeable sodium percentage, along with increases in soil respiration and microbiological activity (Cubillos-Hinojosa et al. 2017; Little et al. 2014). These changes contribute to the stabilization of soil aggregates and the reduction of erosion risks, which are crucial for the successful restoration of coal mining areas. However, despite these promising results, the inherent chemical variability and functional variety of coal as a modifier present hurdles (da Silva et al. 2023). Additional research is required to elucidate the chemical mechanisms involved in its use as a slow-release fertilizer. Additionally, potential soil contamination due to elevated levels of chlorides and inorganic constituents in certain types of coal must be carefully managed to avoid adverse environmental impacts (Akimbekov et al. 2022; Akimbekov et al. 2021; Cubillos-Hinojosa et al. 2017).

6 Bioremediation of Heavy Metals by Geomicrobes

Waste generated by mining, in the form of overburden or waste rock, is typically disposed of on mining sites, leading to significant environmental degradation (Barapanda et al. 2001). To mitigate these adverse effects, various techniques have been developed, including mine waste management, waste land management, and the reutilization of mine waste in civic domains. Restoration of abandoned mine sites is still a significant environmental issue (Dowarah et al. 2009). Soil restoration at mining sites tainted with coal or metal residues usually consisted of removing contaminated soil to secure landfills and then filling the higher strata. However, this strategy is expensive and necessitates considerable site rehabilitation activities. Alternatively, bioremediation strategies have been used to repair contaminated soils (Salt et al. 1995; Cunningham and Ow 1996).

Coal mining is a major global contributor to heavy metal toxicity, both geogenic and anthropogenic, primarily through acid mine drainage (AMD) (Akimbekov et al. 2022). Acid mine drainage, originating from abandoned coal mines, poses a persistent environmental threat that may stay detrimental for several centuries or even decades post-mine closure, continuously contributing to unfavorable environmental conditions (Johnson and Hallberg 2005). Studies on the assessment, management, and impact of AMD have garnered considerable global attention, with a focus on sustainable remediation strategies. Significant efforts have been focused on decreasing heavy metal concentrations and promoting microbial activity (Kothe and Büchel 2014). The use of indigenous microbes is a promising biotic technique for addressing the challenges posed by acid mine drainage (AMD) (Akimbekov et al. 2022). Bioremediation exploits the natural metabolic processes of microorganisms, including bacteria and fungi, to detoxify and eliminate harmful heavy metals from the contaminated ecosystem (Karnwal et al. 2024; Kulshreshtha et al. 2014). This approach involves two key processes: biotransformation and biodegradation, during which contaminants are converted into less harmful or nonhazardous substances. Bioremediation is defined as the enhancement of microbial activity to facilitate the rapid degradation of harmful organic pollutants to acceptable levels in groundwater, soil, materials, and sediments (Fulekar 2009; Singh et al. 2014). Recent developments in this field also focus on biological methods that can either immobilize or precipitation of inorganic pollutants such as heavy metals (Akimbekov et al. 2022). To stimulate microbial activity, various growth substances, nutrients, and terminal electron acceptors or donors may be added, resulting in enhanced degradation and biotransformation of organic pollutants. The energy and carbon necessary for these processes are derived from the breakdown of organic compounds by microbes (Fulekar 2009; Singh et al. 2014). Often, these microorganisms convert contaminants into methane or carbon dioxide, along with water and biomass. Biodegradation thus involves the breakdown of organic materials and the transformation of inorganic substances into environmentally safe forms (Kulshreshtha et al. 2014).

The role of microorganisms in remediation of heavy metal present in mine waste dumps has been extensively reported. Sulfate-reducing bacteria (SRB), often species such as *Desulfitobacterium*, *Desulfosporosinus*, and certain members of the *Actinobacteria* and *Firmicutes* phyla, have been extensively isolated from mining sites and are known for their ability to attenuate toxic metal/metalloid concentrations (Cohen 2006). The treatment of acid mine drainage (AMD) using sulfate-reducing bacteria (SRB) entails the conversion of sulfates into hydrogen sulfide, which subsequently reacts with metals, effectively eliminating them from the solution. Additionally, SRB help neutralize the acidity of AMD by producing alkalinity (Martins et al. 2010; Hard et al. 1997). A study by Luptakova and Kusnierova (2005) found that a mixed culture of *Desulfovibrio* and *Desulfotomaculum* was capable of effectively removing Cu^{2+} from model solutions in a batch reactor over a period of 5–6 days. Moreover, Dong et al. (2020) showed that when SRB were used alongside coal gangue, they achieved significant treatment efficiency in AMD remediation, particularly for the removal of chemical oxygen demand (COD), sulfate (SO_4^{2-}), ferrous iron (Fe^{2+}), and manganese (Mn^{2+}). Similarly, *Thiobacillus* sp., a sulfate-reducing bacterium, has been predominantly utilized for AMD remediation (Gupta and Paul 2015).

The capacity for bioremediation of heavy metal-tolerant bacteria native to rat-hole coal mine environments has been extensively documented. Shylla et al. (2021) reported that among 12 isolates, three (i.e., *Bacillus siamensis* KH-12A, *Serratia marcescens* KH-CC, and *Bacillus altitudinis* KH-16F) demonstrated a high maximum tolerable concentration (MTC) for Fe (500 ppm), Mn (830 ppm), and for Pb (1400 ppm), Among the isolates, *Serratia marcescens* exhibited the greatest remediation capabilities for Pb and Mn achieving removal rates of 83% and 72.5%, respectively. This bioremediation process not only facilitates the precipitation of toxic metals but also contributes to neutralizing the acidic conditions characteristic of AMD-impacted sites (Shylla et al. 2021).

Additionally, Zheng et al. (2019) identified 23 bacterial genera that were highly sensitive to Cd^{2+} and Hg^{2+} in agricultural soils impacted by coal mining, including *Acidobacteria*, *Actinobacteria*, *Chloroflexi*, *Candidate division WS3*, *Gemmatimonadetes*, *Thermotogae*, and *Proteobacteria*. They found that certain bacterial populations can thrive in the mineral-depleted and highly toxic environments of coal-overburdened strata. Singh et al. (2014) assessed the bacterial diversity in active opencast coal mining areas and discovered that isolates from Actinobacteria, Firmicutes, and Proteobacteria demonstrated significant tolerance (5–12 mM) to heavy metals such as Ni^{2+}, Cu^{2+}, Cr^{6+}, As^{3+}, and Cd^{2+}, suggesting their potential for bioremediation of contaminated sites. Furthermore, Gandhi et al. (2015) selected *Enterobacter* spp., *Klebsiella* spp., and *Acinetobacter gyllenbergii* due to their high resistance to heavy metals (Cd^{2+}, Pb^{2+}, Fe^{2+}, Mn^{2+}, and Cu^{2+}) and their biochemical characteristics, noting a correlation between elevated metal resistance and multi-antibiotic resistance.

Fungi also play an important role in heavy metal bioremediation. Species such as *Penicillium simplicissimum*, *Aspergillus niger*, and *Trichoderma* spp. have demonstrated the ability to precipitate metals as insoluble oxalates or adsorb them onto their cell walls, effectively removing metals from the geochemical cycle (Rao et al. 2005; Ren et al. 2009). Other fungal species like *Rhizopus* sp., *Mucor* sp., and *Aspergillus* spp. have been utilized for the remediation of AMD and metal-polluted soils (Gupta and Paul 2015). Specifically, thermophilic/thermotolerant micromycete cultures of *Aspergillus* spp. isolated from coal seam spoil have been shown to bind Cu^{2+} as low-solubility crystalline moolooite (Sokolyanskaya et al. 2020). These fungi are particularly effective in remediating metal-contaminated soils, including those associated with mining activities (Bruneel et al. 2022).

7 Geomicrobes and Their Role in Phytoremediation

The growing demand for metals and raw materials, combined with advancements in extraction methods, has highlighted the potential for reusing tailings when economically feasible (Dino et al. 2018). Traditional soil remediation techniques for metal and metalloid contamination include methods such as vitrification, soil replacement, soil washing, solidification/stabilization, electrokinetic remediation, and in situ flushing (Bolan et al. 2014; Peng et al. 2018). However, these approaches have significant drawbacks, including the use of toxic chemicals, alterations to soil physical, chemical, and biological properties, extended treatment times for large areas, and prohibitive costs. Additionally, mine tailings often contain high concentrations of metals and metalloids, making these methods generally unsuitable for their remediation (Bruneel et al. 2022).

In contrast, phytoremediation offers a sustainable, environmentally friendly, and cost-effective solution. This solar-driven technology not only rehabilitates landscapes but also enhances soil fertility, controls erosion, and is socially accepted due to its contribution to CO_2 fixation (Novo et al. 2018; Karaca et al. 2018). Numerous experiments have addressed phytoremediation of metal- and metalloid-contaminated soils with a specific focus on soils and tailings from mine (Ernst 2005; Mendez and Maier 2008; Karaca et al. 2018; Novo et al. 2018).

Research has increasingly explored the use of PGP microbes to enhance phytoremediation efficiency in heavily polluted and contaminated ecosystems such as mining areas. The association of plants with microbial communities can improve plant resistance to metals and metalloids and stimulate growth (Glick 2010; Rajkumar et al. 2012; Ahemad and Kibret 2014; Ma et al. 2016; Novo et al. 2018).

Plant-growth-promoting (PGP) bacteria are recognized for their broad range of beneficial activities in sustainable agronomy and soil restoration and remediation. They influence plant growth, stress resistance, hormone balance, immunity, and soil nutrient availability (Ciftci et al. 2021; Goswami et al. 2016). PGP bacteria enhance plant growth through multiple mechanisms, including the synthesis of phytohormones like indole-3-acetic acid (IAA), ethylene, cytokinins, abscisic acid, and gibberellins, all of which play key roles in physiological processes, like defense, germination, reproduction, and stress protection (Goswami et al. 2016). Microbes also enhance nutrient availability, such as iron, by producing siderophores that solubilize insoluble iron compounds (Ma et al. 2011; Glick 2012). Nitrogen-fixing bacteria which are symbiotic in nature, such as *Rhizobium*, *Bradyrhizobium*, and *Mesorhizobium*, promote legume growth by fixing atmospheric nitrogen (Wani et al. 2008; Ahemad and Kibret 2014). Microorganisms can also convert insoluble phosphates into accessible forms by mechanisms such as organic acid release, chelation, or phosphatase activity (Ma et al. 2011; Ullah et al. 2015). Some PGP bacteria defend plants from illnesses by producing antibiotics, synthesizing siderophores, or inducing systemic resistance (Goswami et al. 2016; Ma et al. 2016). Coal-solubilizing or degrading bacteria isolated from coal slurry in discard dumps often exhibit PGP characteristics (Akimbekov et al. 2022). Titilawo et al. (2020) investigated the genetic relatedness of coal-degrading rhizosphere bacteria (*Bacillus*, *Escherichia*, *Citrobacter*, *Serratia*, *Exiguobacterium*, and *Microbacterium*) to PGP bacteria sequences in the NCBI GenBank database, revealing PGP traits in 31 of 47 strains. Strains like *Pseudomonas* sp. NU36 and *Acinetobacter* sp. NU25 demonstrated significant plant growth promotion effects, such as IAA production, siderophore synthesis, and inorganic phosphate solubilization, alongside stress tolerance (Upadhyay et al. 2016). Barman et al. (2022) isolated and characterized ten bacterial isolates from coal dumping areas, among which *Bacillus* spp. exhibited tolerance to multiple heavy metals and demonstrated capabilities such as siderophore production, phosphate solubilization, IAA production, and ACC deaminase activity.

Revegetation with fast-growing leguminous species presents an alternative approach for rehabilitating degraded coal mining areas. Legumes enhance nutrient cycling, increase soil carbon content, and reduce erosion (Akimbekov et al. 2022). They form symbiotic relationships with rhizobia, which increase nitrogen deposition, lower the soil C/N ratio, and enrich soil organic matter, thus promoting nutrient cycling (Marra et al. 2012). Moura et al. (2016) found that native rhizobia from coal mining sites effectively nodulated *Mimosa* plants, improving shoot dry matter and enhancing growth in *Marica* species. Xia et al. (2018) isolated and identified 307 bacterial strains from switchgrass in coal fields, with 58% of these strains increasing lamina length compared to controls. *Delftia* spp. has been noted for its PGP characteristics, such as IAA and siderophore production, as well as growth promotion in *Brassica juncea* (Roy and Roy 2019). Actinomycetes like *Amycolatopsis* sp. and *Streptomyces* sp. isolated from birch rhizospheres in coal mine dumps showed siderophore and antibacterial compound production, alongside increased survival under heavy metal conditions (María et al. 2016).

Endophytic fungi also play a crucial role in plant health. In switchgrass studies, fungi isolated from reclaimed coal mining sites enhanced shoot height, fresh weight, and dry weight, demonstrating their functional benefits (Xia et al. 2018). Arbuscular mycorrhizal fungi (AMF), such as *Funneliformis mosseae*, significantly increased plant survival and growth in sea buckthorn (Zhang et al. 2020). The amalgamation of AMF and PGP bacteria into coal mine spoil banks has been shown to improve the growth of species like high-biomass hemp and reed canary grass while reducing the need for organic amendments (Gryndler et al. 2008). Despite technological breakthroughs and significant experience with PGP bacteria, building a commercially viable integrated bioprocess utilizing native coal PGP bacteria remains challenging. Recognizing native microbes with plant-growth-promoting properties as key "bioengineers" is essential for developing effective rhizosphere strategies in disturbed soils (Akimbekov et al. 2022).

8 Conclusion

Geomicrobes are essential for transforming nutrient-poor mine dumps, spoil, and tailing into productive soils. These microbes drive crucial soil processes such as carbon, nitrogen, and phosphorus cycling, while also enhancing soil stability and aggregation through their interactions in the rhizosphere. Hence integrating microbial strategies into rehabilitation of coal mining affected areas can significantly enhance soil quality and functionality. Geomicrobes facilitate soil aggregation and stability, mitigate coal contamination, and contribute to the management of heavy metal pollution. They also play a crucial part in enhancing soil organic matter and in addressing saline-sodic conditions as they have high salt-tolerant capacity and produce extracellular polymeric substances. Their ability to improve soil structure and support plant health is particularly valuable in addressing the challenges presented by saline, sodic, and dispersive mine spoils, dumps, and tailing which are prone to erosion. Furthermore, combining phytoremediation with plant-growth-promoting microorganisms offers a sustainable method for restoring metal-contaminated soils.

However, continued research and the development of advanced microbial inoculants are crucial for optimizing these strategies and ensuring effective and sustainable land restoration in areas impacted by mining.

Competing Interests The authors declare that they have no competing interests.

References

Ahemad M, Kibret M (2014) Mechanisms and applications of plant growth promoting rhizobacteria: current perspective. J King Saud Univ Sci 26(1):1–20

Akimbekov N, Digel I, Abdieva G, Ualieva P, Tastambek K (2021) Lignite biosolubilization and bioconversion by *Bacillus* sp. The collation of analytical data. Biofuels 12:247–258

Akimbekov NS, Digel I, Tastambek KT, Marat AK, Turaliyeva MA, Kaiyrmanova GK (2022) Biotechnology of microorganisms from coal environments: from environmental remediation to energy production. Biology 11:1306. https://doi.org/10.3390/biology11091306

Ashraf M, Hasnain S, Berge O, Mahmood T (2004) Inoculating wheat seedlings with exopolysaccharide-producing bacteria restricts sodium uptake and stimulates plant growth under salt stress. Biol Fertil Soils 40(3):157–162. https://doi.org/10.1007/s00374-004-0766-y

Aznar-Sánchez J, García-Gómez J, Velasco-Muñoz J, Carretero-Gómez A (2018) Mining waste and its sustainable management: advances in worldwide research. Fortschr Mineral 8(7):284. https://doi.org/10.3390/min8070284

Barapanda P, Singh SK, Pal BK (2001) Utilization of coal mining wastes: an overview. In: National seminar on environmental issues and waste management in mining and allied industries. Regional Engg. College, Rourkela, India, pp 177–182

Barman D, Dutta I, Jha DK (2022) Heavy metal resistant bacteria from coal dumping site with plant growth promoting potentials. Biologia 77:533–545

Billings SA, Tiemann LK, Ballantyne FIV, Lehmeier CA, Min K (2015) Investigating microbial transformations of soil organic matter: synthesizing knowledge from disparate fields to guide new experimentation. Soil 1(1):313–330. https://doi.org/10.5194/soil-1-313-2015

Blankinship JC, Fonte SJ, Six J, Schimel JP (2016) Plant versus microbial controls on soil aggregate stability in a seasonally dry ecosystem. Geoderma 272:39–50. https://doi.org/10.1016/j.geoderma.2016.03.008

Blight GE, Steffen KH (1979) Geotechnics of gold mining waste disposal. In: Current practice in mine waste disposal, Committee on Embankment Dams and Slope of the Geotechnical Engineering Division. ASCE, New York, pp 1–52

Bolan N, Kunhikrishnan A, Thangarajan R, Kumpiene J, Park J, Makino T, Kirkham MB, Scheckel K (2014) Remediation of heavy metal (loid) s contaminated soils–to mobilize or to immobilize. J Hazard Mater 266:141–166. https://doi.org/10.1016/j.jhazmat.2013.12.018

Brown CJ, Dey AK (1975) The mineral and nuclear fuels of the India subcontinent and Burma. Oxford University Press, Delhi, pp 1–170

Bruneel O, Mghazli N, Sbabou L, Héry M, Casiot C (2022) Role of microorganisms in rehabilitation of mining sites, focus on Sub Saharan African countries. J Geochem Explor 205:106327. https://doi.org/10.1016/j.gexplo.2019.06.009.hal-02615510

Celestina C, Hunt JR, Sale APWGE, Franks (2019) Attribution of crop yield responses to application of organic amendments: A critical review. Soil Tillage Res 186:135–145

Ciftci C, Tekdal D, Cetiner S (2021) 3 – The importance of plant growth–promoting rhizobacteria for plant productivity. In: Singh VP, Singh S, Tripathi DK, Prasad SM, Bhardwaj R, Chauhan DK (eds) Abiotic stress and legumes. Academic Press, Cambridge, MA, pp 69–80

Cohen RRH (2006) Use of microbes for cost reduction of metal removal from metals and mining industry waste streams. J Clean Prod 14:1146–1157

Cubillos-Hinojosa JG, Valero N, Peralta Castilla ADJ (2017) Effect of a low rank coal inoculated with coal solubilizing bacteria for the rehabilitation of a saline-sodic soil in field conditions. Revista Facultad Nacional de Agronomia Medellín 70:8271–8283

Cunningham SD, Ow DW (1996) Promises and prospects of phytoremediation. Plant Physiol 110:715–719

da Silva GOA, Southam AG, Gagen EJ (2023) Accelerating soil aggregate formation: a review on microbial processes as the critical step in a post-mining rehabilitation. Soil Res 61(3):209–223. https://doi.org/10.1071/SR22092

David Y, Baylon MG, Pamidimarri SDVN, Baritugo KA, Chae CG, Kim YJ, Kim TW, Kim MS, Na JG, Park SJ (2017) Screening of microorganisms able to degrade low-rank coal in aerobic conditions: potential coal biosolubilization mediators from coal to biochemicals. Biotechnol Bioprocess Eng 22:178–185

Dino GA, Mehta N, Rossetti P, Ajmone-Marsan F, De Luca DA (2018) Sustainable approach towards extractive waste management: two case studies from Italy. Res Policy 59:33–43

Dong L, Córdova-Kreylos AL, Yang J, Yuan H, Scow KM (2009) Humic acids buffer the effects of urea on soil ammonia oxidizers and potential nitrification. Soil Biol Biochem 41:1612–1621

Dong Y, Di J, Yang Z, Zhang Y, Wang X, Guo X, Li Z, Jiang G (2020) Study on the effectiveness of sulfate-reducing bacteria combined with coal gangue in repairing acid mine drainage containing Fe and Mn. Energies 13:995

Dowarah J, Deka Boruah HP, Gogoi J, Pathak N, Saikai N, Handique AK (2009) Eco-restoration of a high-sulphur coal mine overburden dumping site in Northeast India: a case study. J Earth Sys Sci 118:597–608

Dutta M, Saikia J, Taffarel SR, Waanders FB, de Medeiros D, Cutruneo CMNL, Silva LFO, Saikia BK, B.K. (2017) Environmental assessment and nano-mineralogical characterization of coal, overburden and sediment from Indian coal mining acid drainage. Geosci Front 8(6):1285–1297

Egamberdieva D, Wirth S, Bellingrath-Kimura SD, Mishra J, Arora NK (2019) Salt-tolerant plant growth promoting rhizobacteria for enhancing crop productivity of saline soils. Front Microbiol 10:2791

Ernst WHO (2005) Phytoextraction of mine wastes—options and impossibilities. Chemie der Erde Geochemistry 65:29–42

Fahey M, Newson TA (1997) Aspects of the geotechnics of mining wastes and tailings dams. In: Bouazza A, Kodikara J, Parker R (eds) Proceedings of the 1st Australia-New Zealand conference on environmental geotechnics—GeoEnvironment 97, Melbourne/Victoria, Australia, 26–28 November, pp 115–134

Ferris H, Venette RC, Scow KM (2004) Soil management to enhance bacterivore and fungivore nematode populations and their nitrogen mineralisation function. Appl Soil Ecol 25:19–35. https://doi.org/10.1016/j.apsoil.2003.07.001

Flemming HC, Wingender J, Szewzyk U, Steinberg P, Rice SA, Kjelleberg S (2016) Biofilms: an emergent form of bacterial life. Nat Rev Microbiol 14:563–575. https://doi.org/10.1038/nrmicro.2016.94

Fulekar MH (2009) Bioremediation of fenvalerate by *Pseudomonas aeruginosa* in a scale up bioreactor. Rom Biotechnol Lett 14(6):4900–4905

Gadd GM (2010) Metals, minerals and microbes: geomicrobiology and bioremediation. Microbiology 156:609–643. https://doi.org/10.1099/mic.0.037143-0

Gandhi V, Priya A, Priya S, Daiya V, Kesari J, Prakash K, Kumar Jha A, Kumar K, Kumar N (2015) Isolation and molecular characterization of bacteria to heavy metals isolated from soil samples in Bokaro Coal Mines, India. Pollution 1:287–295

Geddie JL, Sutherland IW (1993) Uptake of metals by bacterial polysaccharides. J Appl Bacteriol 74(4):467–472. https://doi.org/10.1111/j.1365-2672.1993.tb05155.x

Ghosh R (2002) Land use in mining areas of India. Envis Monograph No. 9, CME

Giannouli A, Kalaitzidis S, Siavalas G, Chatziapostolou A, Christanis K, Papazisimou S, Papanicolaou C, Foscolos A (2009) Evaluation of Greek low-rank coals as potential raw material for the production of soil amendments and organic fertilizers. Int J Coal Geol 77:383–393

Glick BR (2010) Using soil bacteria to facilitate phytoremediation. Biotechnol Adv 28(3):367–374

Glick BR (2012) Plant growth-promoting bacteria: mechanisms and applications. Scientifica 2012:963401. https://doi.org/10.6064/2012/963401

Gorber DM, Ibbotson BG, Knapp RA (1978) Trends in uranium mining waste management. In: Proceedings of the international symposium on waste treatment and utilisation, University of Waterloo, Waterloo, Ontario, Canada, July 5–7

Goswami D, Thakker JN, Dhandhukia PC (2016) Portraying mechanics of plant growth promoting rhizobacteria (PGPR): a review. Cogent Food Agric 2(1):1127500. https://doi.org/10.1080/23311932.2015.1127500

Gozzard E, Vink S, Nanjappa V, Moran CJ (2009) Salt dissolution dynamics on surface mine spoils. Min Technol 118:177–184. https://doi.org/10.1179/174328610X12682159815181

Graf F, te Kamp L, Auer M, Acharya MS, Wei W (2015) Soil aggregate stability in eco engineering: comparison of field and laboratory data with an outlook on a new modelling approach. In: Wu W (ed) Recent advances in modelling landslides and debris flows. Springer, pp 29–47

Gryndler M, Sudová R, Püschel D, Rydlová J, Janousková M, Vosátka M (2008) Cultivation of high-biomass crops on coal mine spoil banks: can microbial inoculation compensate for high doses of organic matter? Bioresour Technol 99:6391–6399

Gupta AK, Paul B (2015) Ecorestoration of coal mine overburden dump to prevent environmental degradation: a review. Res J Environ Sci 9(7):307–319

Haider R, Ghauri MA, SanFilipo JR, Jones EJ, Orem WH, Tatu CA, Akhtar K, Akhtar N (2013) Fungal degradation of coal as a pre-treatment for methane production. Fuel 104:717–725

Haider R, Ghauri MA, Rahim MU (2018) On comparison between fungal and bacterial pretreatment of coal for enhanced biogenic methane generation. J Geom 35:432–437

Hallberg KB (2010) New perspectives in acid mine drainage microbiology. Hydrometallurgy 104(3–4):448–453. https://doi.org/10.1016/j.hydromet.2009.12.013

Hamidovic S, Teodorovic S, Lalevic B, Jovicic Petrovic J, Jovic J, Kikovic D, Raicevic V (2016) Bioremediation potential assessment of plant growth-promoting autochthonous bacteria: a lignite mine case study. Pol J Environ Stud 25:113–119

Hard BC, Friedrich S, Babel W (1997) Bioremediation of acid mine water using facultatively methylotrophic metal-tolerant sulfate-reducing bacteria. Microbiol Res 152:65–73

Hart MM, Reader RJ (2002) Taxonomic basis for variation in the colonization strategy of arbuscular mycorrhizal fungi. New Phytol 153(2):335–344. https://doi.org/10.1046/j.0028-646X.2001.00312.x

He ZL, Yang XE, Stoffella PJ (2005) Trace elements in agroecosystems and impacts on the environment. J Trace Elem Med Biol 19:125–140. https://doi.org/10.1016/j.jtemb.2005.02.010

Heuer H, Krsek M, Baker P, Smalla K, Wellington EM (1997) Analysis of actinomycete communities by specificamplification of genes encoding 16S rRNA and gel-electrophoretic separation in denaturing gradients. Appl Environ Microbiol 63:3233–3241. https://doi.org/10.1128/aem.63.8.3233-3241.1997

Hofrichter M, Fakoussa RM (2001) Microbial degradation and modification of coal. Biopolymers 1:393–429

Hossner L, Hons F (1992) Reclamation of mine tailings. In: Lal R, Stewart BA (eds) Soil restoration. Springer, New York, pp 311–350. https://doi.org/10.1007/978-1-4612-2820-2_10

Hudson-Edwards K (2016) Tackling mine wastes. Science 352(6283):288–290. https://doi.org/10.1126/science.aaf3354

Indian Bureau of Mines (IBM) (2001) Annual report. Ministry of Steel and Mines, Government of India

Johnson DB (2003) Chemical and microbiological characteristics of mineral spoils and drainage waters at abandoned coal and metal mines. Water Air Soil Pollut 3:47–66

Johnson DB, Hallberg KB (2005) Acid mine drainage treatment options: a review. Sci Total Environ 338(1):3–14

Jovanovski G, Boev B, Makreski P (2023) Chemistry and geology of coal: nature, composition, coking, gasification, liquefaction, production of chemicals, formation, peatification, coalification, coal types, and ranks. ChemTexts 9(2)

Juwarkar AA, Jambhulkar HP (2008) Phytoremediation of coal mine spoil dump through integrated biotechnological approach. Bioresour Technol 99(11):4732–4741

Karaca O, Cameselle C, Reddy KR (2018) Mine tailing disposal sites: contamination problems, remedial options and phytocaps for sustainable remediation. Rev Environ Sci Biotechnol 17:205. https://doi.org/10.1007/s11157-017-9453-y

Karnwal A, Martolia S, Dohroo A, Al-Tawaha ARMS, Malik T (2024) Exploring bioremediation strategies for heavy metals and POPs pollution: the role of microbes, plants, and nanotechnology. Front Environ Sci 12(2024). https://doi.org/10.3389/fenvs.2024.1397850

Kothe E, Büchel G (2014) UMBRELLA: using MicroBes for the REgulation of heavy metaL mobiLity at ecosystem and landscape scAle. Environ Sci Pollut Res 21:6761–6764

Kulshreshtha A, Agrawal R, Barar M, Saxena S (2014) A review on bioremediation of heavy metals in contaminated water. IOSR J Environ Sci Toxicol Food Technol 8:44–50. https://doi.org/10.9790/2402-08714450

Kundu NK, Ghose MK (1997) Shelf life of stockpiled topsoil of an opencast coal mine. Environ Conserv 24(1):24–30

Kuyucak N (1999) Acid mine drainage prevention and control options. International Mine Water Association Proceedings

Li X, Zhang L, Zhang Z (2006) Soil bioengineering and the ecological restoration of riverbanks at the Airport Town, Shanghai, China. Ecol Eng 26:304–314. https://doi.org/10.1016/j.ecoleng.2005.10.011

Li X, Park JH, Edraki M, Baumgartl T (2014) Understanding the salinity issue of coal mine spoils in the context of salt cycle. Environ Geochem Health 36(3):453–465. https://doi.org/10.1007/s10653-013-9573-4

Li Y, Zhou W, Jing M, Wang S, Huang Y, Geng B, Cao Y (2022) Changes in reconstructed soil physicochemical properties in an opencast mine dump in the Loess Plateau Area of China. Int J Environ Res Public Health 19(2):706. https://doi.org/10.3390/ijerph19020706

Little KR, Rose MT, Jackson WR, Cavagnaro TR, Patti AF (2014) Do lignite-derived organic amendments improve early-stage pasture growth and key soil biological and physicochemical properties? Crop Pasture Sci 65:899–910

Luna L, Miralles I, Andrenelli MC, Gispert M, Pellegrini S, Vignozzi N, Solé-Beneta A (2016) Restoration techniques affect soil organic carbon, glomalin and aggregate stability in degraded soils of a semiarid Mediterranean region. Catena 143:256–264. https://doi.org/10.1016/j.catena.2016.04.013

Luptakova A, Kusnierova M (2005) Bioremediation of acid mine drainage contaminated by SRB. Hydrometallurgy 77:97–102

Ma Y, Prasad MNV, Rajkumar M, Freitas H (2011) Plant growth promoting rhizobacteria and endophytes accelerate phytoremediation of metalliferous soils. Biotechnol Adv 29(2):248–258

Ma Y, Rajkumar M, Zhang C, Freitas H (2016) Beneficial role of bacterial endophytes in heavy metal phytoremediation. J Environ Manag 174:14–25

Madsen PA, Mulligan DR (2006) Effect of NaCl on emergence and growth of a range of provenances of *Eucalyptus citriodora, Eucalyptus populnea, Eucalyptus camaldulensis* and *Acacia salicina*. For Ecol Manag 228(1):152–159. https://doi.org/10.1016/j.foreco.2006.02.044

Maharaj S, Barton CD, Karathanasis TA, Rowe HD, Rimmer SM (2007) Distinguishing "new" from "old" organic carbon on reclaimed coal mine sites using thermogravimetry: I. Method development. Soil Sci 172:292–301

María AM, Andrés I, Eugenio J, Héctor M, Susana CS (2016) Revealing the biotechnological potential of Delftia sp. JD2 by a genomic approach. AIMS Bioeng 3:156–175

Marra ML, Fonsêca Sousa Soares CR, de Oliveira SM, Avelar Ferreira PA, Lima Soares B, de Fráguas Carvalho R, de Lima JM, de Souza Moreira FM (2012) Biological nitrogen fixation and phosphate solubilization by bacteria isolated from tropical soils. Plant Soil 357:289–307

Martins M, Santos ES, Pires C, Barros RJ, Costa MC (2010) Production of irrigation water from bioremediation of acid mine drainage: comparing the performance of two representative systems. J Clean Prod 18:248–253

Mendez MO, Maier RM (2008) Phytoremediation of mine tailings in temperate and arid environments. Rev Environ Sci Biotechnol 7(1):47–59. https://doi.org/10.1007/s11157-007-9125-4

Méndez-García C, Mesa V, Sprenger RR, Richter M, María Diez MS, Solano J, Bargiela R et al (2014) Microbial stratification in low pH oxic and suboxic macroscopic growths along an acid mine drainage. ISME J 8(6):1259–1274

Mitter B, Brader G, Pfaffenbichler N, Sessitsch A (2019) Next generation microbiome applications for crop production—limitations and the need of knowledge-based solutions. Curr Opin Microbiol 49:59–65. https://doi.org/10.1016/j.mib.2019.10.006

Moura GGD, Armas RDD, Meyer E, Giachini AJ, Rossi MJ, Soares CRFS (2016) Rhizobia isolated from coal mining areas in the nodulation and growth of leguminous trees. Revista Brasileira de Ciencia do Solo 40:1–10

Nirola R, Megharaj M, Beecham S, Aryal R, Thavamani P, Vankateswarlu K, Saint C (2016) Remediation of metalliferous mines, revegetation challenges and emerging prospects in semi-arid and arid conditions. Environ Sci Pollut Res 23(20):20131–20150. https://doi.org/10.1007/s11356-016-7372-z

Nordberg GF, Fowler BA, Nordberg M (eds) (2014) Handbook on the toxicology of metals. Academic Press, p 1381

Novo LA, Castro PM, Alvarenga P, da Silva EF (2018) Plant growth–promoting rhizobacteria-assisted phytoremediation of mine soils. In: Bio-geotechnologies for mine site rehabilitation. Elsevier, pp 281–295. https://doi.org/10.1016/B978-0-12-812986-9.00016-6

Nsa IY, Akinyemi BT, Bello-Akinosho M, Ezechukwu SN, Bayode TB, Igbinigie EE, Adeleke RA (2022) Development of a saprophytic fungal inoculum for the biodegradation of sub-bituminous coal. SN Appl Sci 4:53

Oades JM (1993) The role of biology in the formation, stabilization and degradation of soil structure. Geoderma 56(1–4):377–400. https://doi.org/10.1016/0016-7061(93)90123-3

Oelofse M, Markussen B, Knudsen L, Schelde K, Olesen JE, Jensen LS, Sander Bruun S (2015) Do soil organic carbon levels affect potential yields and nitrogen use efficiency? An analysis of winter wheat and spring barley field trials. Eur J Agron 66:62–73

Park JH, Li X, Edraki M, Baumgartl T, Kirsch B (2013) Geochemical assessments and classification of coal mine spoils for better understanding of potential salinity issues at closure. Environ Sci: Processes Impacts 15:1235–1244. https://doi.org/10.1039/c3em30672k

Peng W, Li X, Xiao S, Fan W (2018) Review of remediation technologies for sediments contaminated by heavy metals. J Soils Sediments 18(4):1701–1719

Prashant MCN, Ghosh R, Mandal PK (2010) Use of crushed and washed overburden for stowing in underground mines: a case study. J Mines Met Fuels 58(1–2):7–12

Rai AK, Paul B, Singh G (2011) A study on physico chemical properties of overburden dump materials from selected coal mining areas of Jharia coalfields, Jharkhand, India. Int J Environ Sci 1(6):1350–1360

Rajkumar M, Sandhya S, Prasad MNV, Freitas H (2012) Perspectives of plant-associated microbes in heavy metal phytoremediation. Biotechnol Adv 30(6):1562–1574

Ram LC, Masto RE (2010) An appraisal of the potential use of fly ash for reclaiming coal mine spoil. J Environ Manag 91(3):603–617. https://doi.org/10.1016/j.jenvman.2009.10.004

Rao KR, Rashmi K, Latha JNL, Mohan PM (2005) Bioremediation of toxic metal ions using biomass of *Aspergillus fumigatus* from fermentative waste. Indian J Biotechnol 4:139–143

Ren WX, Li PJ, Geng Y, Li XL (2009) Biological leaching of heavy metals from a contaminated soil by *Aspergillus Niger*. J Hazard. Mat 167:164–169

Rillig MC, Mummey DL (2006) Mycorrhizas and soil structure. New Phytol 171(1):41–53. https://doi.org/10.1111/j.1469-8137.2006.01750.x

Rivera D, Mejías V, Jáuregui BM, Costa-Tenorio M, L'opez-Archilla AI, Peco B (2014) Spreading topsoil encourages ecological restoration on embankments soil fertility, microbial activity and vegetation cover. PLoS One 9(7):e101413. https://doi.org/10.1371/journal.pone.0101413

Roy S, Roy M (2019) Characterization of plant growth promoting feature of a neutromesophilic, facultatively chemolithoautotrophic, sulphur oxidizing bacterium Delftia sp. strain SR4 isolated from coal mine spoil. Int J Phytoremediation 21:531–540

Salt DE, Blaylock M, Kumar NPBA, Dushenkov V, Ensley BD, Chet I, Raskin I (1995) Phytoremediation: A novel strategy for the removal of toxic metals from the environment using plants. Biotechnology 13:468–474

Sarkar B, Wijesekara H, Mandal S, Singh M, Bolan NS (2017) Characterization and improvement in physical, chemical, and biological properties of mine waste. In: Bolan NS, Kirkham MB, Ok YS (eds) Spoil to soil mine site rehabilitation and revegetation. USA, CRC Press/Taylor and Francis, pp 3–16

Sekhohola LM, Igbinigie EE, Cowan AK (2013) Biological degradation and solubilisation of coal. Biodegradation 24:305–318

Sharma A, Singh P, Kumar S, Kashyap PL, Srivastava AK, Chakdar H, Singh RN, Kaushik R, Saxena AK, Sharma AK (2015) Deciphering diversity of salt-tolerant *Bacilli* from saline soils of eastern indo-gangetic plains of India. Geomicrobiol J 32:170–180

Sheoran V, Sheoran AS, Poonia P (2010) Soil reclamation of abandoned mine land by revegetation: a review. Int J Soil Sediment Water 3(2):13

Shylla L, Barik SK, Joshi SR (2021) Characterization and bioremediation potential of native heavy-metal tolerant bacteria isolated from rat-hole coal mine environment. Arch Microbiol 203:2379–2392

Singh KN, Narzary D (2021a) Geochemical characterization of mine overburden strata for strategic overburden-spoil management in an opencast coal mine. Environ Chall 3(2021):100060. https://doi.org/10.1016/j.envc.2021.100060

Singh KN, Narzary D (2021b) Heavy metal tolerance of bacterial isolates associated with overburden strata of an opencast coal mine of Assam (India). Environ Sci Pollut Res 28:63111–63126

Singh R, Singh P, Sharma R (2014) Microorganism as a tool of bioremediation technology for cleaning environment: a review. Int Acad Ecol Environ Sci 4(1):1–6

Sokolyanskaya LO, Ivanov MV, Ikkert OP, Kalinina AE, Evseev VA, Glukhova LB, Karnachuk OV (2020) Copper precipitation as insoluble oxalates by thermotolerant *Aspergillus* spp. from burning wastes of coal mining. Microbiology 89:498–501

Statistical Review of World Energy. https://www.bp.com/en/global/corporate/energy-economics/statistical-review-of-world-energy.html

Straupnik A (2022) An overview of the environmental impact of coal industry waste. IOP Conf Ser Earth Environ Sci 1070(012004):012004

Szczepanska J, Twardowska I (1999) Distribution and environmental impact of coal-mining wastes in Upper Silesia, Poland. Environ Geol 38:249–258. https://doi.org/10.1007/s002540050422

Tecon R, Or D (2017) Biophysical processes supporting the diversity of microbial life in soil. FEMS Microbiol Rev 41(5):599–623. https://doi.org/10.1093/femsre/fux039

Tian H, Cai Q, Zhen X (2014) Development prospects of surface coal mining industry in China. Coal Eng 46:11–14

Tisdall JM, Nelson SE, Wilkinson KG, Smith SE, McKenzie BM (2012) Stabilisation of soil against wind erosion by six saprotrophic fungi. Soil Biol Biochem 50:134–141. https://doi.org/10.1016/j.soilbio.2012.02.035

Titilawo Y, Masudi WL, Olawale JT, Sekhohola-Dlamini LM, Cowan AK (2020) Coal-degrading bacteria display characteristics typical of plant growth promoting rhizobacteria. Processes 8:1111

Ullah A, Heng S, Munis MFH, Fahad S, Yang X (2015) Phytoremediation of heavy metals assisted by plant growth promoting (PGP) bacteria: a review. Environ Exp Bot 117:28–40

Upadhyay N, Verma S, Pratap Singh A, Devi S, Vishwakarma K, Kumar N, Pandey A, Dubey K et al (2016) Soil ecophysiological and microbiological indices of soil health: a study of coal mining site in Sonbhadra, Uttar Pradesh. J Soil Sci Plant Nutr 16:778–800

Valero N, Melgarejo LM, Ramírez R (2016) Effect of low-rank coal inoculated with coal solubilizing bacteria on edaphic materials used in post-coal-mining land reclamation: A greenhouse trial. Chem Biol Technol Agric 3:20

Verboom WH, Pate JS (2006) Bioengineering of soil profiles in semiarid ecosystems: the 'phytotarium' concept: a review. Plant Soil 289:71–102. https://doi.org/10.1007/s11104-006-9073-8

Vergani C, Graf F (2016) Soil permeability, aggregate stability and root growth: a pot experiment from a soil bioengineering perspective. Ecohydrology 9(5):830–842. https://doi.org/10.1002/eco.1686

Voncken JHL (2020) The origin and classification of coal. In: Voncken JHL (ed) Geology of coal deposits of South Limburg, The Netherlands: including adjacent German and Belgian areas. Springer International Publishing, Cham, pp 25–40

Wang S, Srivastava AK, Wu QS, Fokom R (2014) The effect of mycorrhizal inoculation on the rhizosphere properties of trifoliate orange (*Poncirus trifoliata L. Raf.*). Sci Hortic 170:137–142. https://doi.org/10.1016/j.scienta.2014.03.003

Wani PA, Khan MS, Zaidi A (2008) Effect of metal-tolerant plant growth-promoting rhizobium on the performance of pea grown in metal-amended soil. Arch Environ Contam Toxicol 55(1):33–42

Weng C, Peng X, Han Y (2021) Depolymerization and conversion of lignin to value-added bioproducts by microbial and enzymatic catalysis. Biotechnol Biofuels 14:84

Wildman HG (2009) Managing microorganisms to improve mine site rehabilitation success. ACARP report C16034. Microbial Management Systems, Lapstone

Wilson GWT, Rice CW, Rillig MC, Springer A, Hartnett DC (2009) Soil aggregation and carbon sequestration are tightly correlated with the abundance of arbuscular mycorrhizal fungi: results from long-term field experiments. Ecol Lett 12:452–461. https://doi.org/10.1111/j.1461-0248.2009.01303.x

Xia Y, Amna A, Opiyo SO (2018) The culturable endophytic fungal communities of switchgrass grown on a coal-mining site and their effects on plant growth. PLoS One 13:e0198994

Xu Z, Zhang Y, Yang J, Liu F, Bi R, Zhu H, Lv C, Yu J (2019) Effect of underground coal mining on the regional soil organic carbon pool in farmland in a mining subsidence area. Sustain For 11(18):4961. https://doi.org/10.3390/su11184961

Yılmaz AO, Uslu T (2007) The role of coal in energy production—consumption and sustainable development of Turkey. Energy Policy 35(2):1117–1128

Zhang Y, Bi Y, Shen H, Zhang L (2020) Arbuscular Mycorrhizal Fungi Enhance Sea Buckthorn Growth in Coal Mining Subsidence Areas in Northwest China. J Microbiol Biotechnol 30:848–855

Zheng L, Li Y, Shang W, Dong X, Tang Q, Cheng H (2019) The inhibitory effect of cadmium and/or mercury on soil enzyme activity, basal respiration, and microbial community structure in coal mine–affected agricultural soil. Ann Microbiol 69:849–859

Geomicrobiology of Arsenic-Contaminated Groundwater

Prasann Kumar and Komal Agrawal

Contents

Abstract Arsenic-contaminated groundwater, a global health threat, poses significant health risks for millions worldwide. The complex biogeochemical processes that control arsenic mobility and toxicity often render traditional remediation methods inadequate. However, the field of geomicrobiology, a study of microbial–mineral interactions, offers a beacon of hope in the form of potential solutions for addressing arsenic contamination. This study delves into the processes that occur during the cycling of arsenic in groundwaters, investigating how microorganisms influence the mobilization and immobilization of arsenic and its speciation. By

P. Kumar (✉)
Department of Agronomy, School of Agriculture, Lovely Professional University, Phagwara, Punjab, India

K. Agrawal
Department of Microbiology, School of BioEngineering and Biosciences, Lovely Professional University, Phagwara, Punjab, India

S. R. Joshi et al. (eds.), *Mineral Transformation and Bioremediation by Geo-Microbes*, Interdisciplinary Biotechnological Advances,
https://doi.org/10.1007/978-981-96-3033-2_13

understanding these microbial agents, we can devise more effective methods of monitoring and remediating arsenic-contaminated groundwater. The chapter presents a comprehensive survey of current research, identifies critical microbes, and proposes innovative strategies for harnessing geomicrobes to eliminate arsenic from water systems. This extensive framework aims to enhance our understanding of geomicrobiological mechanisms and leverage them to develop more efficient and sustainable techniques for managing arsenic-contaminated water supplies. It unveils new paradigms and practical solutions to address one of the most pressing global environmental health issues related to mineral-microbial relationships.

Keywords Arsenic · Groundwater · Microbes · Immobilization · Remediation

Abbreviations

CCA	Chromated copper arsenate
COPD	Chronic obstructive pulmonary disease
GBM	Ganges-Brahmaputra-Meghna
IARC	International Agency for Research on Cancer
WHO	World Health Organization

1 Introduction

The contamination of groundwater with arsenic is a worldwide environmental and public health challenge that impacts millions of people in different areas. This naturally occurring metalloid in groundwater is attributed to geological processes, but human activities have intensified its occurrence in many regions (P. Kumar 2024c; P. Kumar, Devi et al. 2021a; V. Kumar et al. 2021b; M. Sharma et al. 2023d). The pervasive presence of arsenic in groundwater presents significant hazards to drinking water supplies, especially in areas where groundwater is a main source of drinkable water. The geographical dispersion of groundwater contaminated with arsenic is uneven, with some areas experiencing more severe contamination due to particular geological and climatic factors. The most impacted areas encompass substantial portions of South and Southeast Asia and certain regions in South and North America. Undoubtedly, the Ganges-Brahmaputra-Meghna (GBM) delta in Bangladesh and India is the most renowned epicentre of arsenic pollution. Within this area, a significant number of people depend on groundwater that frequently contains levels of arsenic that surpass the recommended threshold of 10 μg/L set by the World Health Organization (WHO). The problem in Bangladesh is of such magnitude that it has been characterized as the most substantial instance of widespread poisoning of a population in recorded history. Southeast Asian countries, including Vietnam and Cambodia, also encounter substantial issues concerning arsenic contamination in groundwater. There is extensive documentation of elevated arsenic levels in the Red River Delta and the Mekong Delta, which has significantly impacted millions of individuals. The contamination in these areas is mostly linked to the inherent sedimentary basins' geochemical conditions, where the reductive breakdown of iron oxyhydroxides results in the release of arsenic into the groundwater. In South America, the Pampa plains of Argentina and the Altiplano of Bolivia and Peru are characterized by extensive arsenic contamination. Here, volcanic activity and geothermal processes are responsible for the elevated arsenic concentrations in the groundwater (Devi, Kumar, et al. 2024a; P. Kumar 2024a, b; P. Kumar et al. 2024a, b, c, d, e, f, g, h, i, j, k; P. Kumar and Mistri 2020; P. Kumar and Singh 2024a; K. Sharma et al. 2024). The issue is especially severe in rural regions with limited alternative water sources. Environmental pollution caused by arsenic is a significant issue in the western states of the United States, namely in certain regions of Nevada, California, and Arizona. The contamination in these regions is frequently associated with mining operations and the inherent process of arsenic leaching from sedimentary rocks and soils. Arid conditions worsen the issue by causing reduced rates of groundwater recharge, which in turn concentrates arsenic levels in the aquifers. Additional regions, including some of China, Mexico, and several African countries, also encounter arsenic pollution, although to different degrees. In China, the Xinjiang and Inner Mongolia regions are renowned for their elevated concentrations of arsenic in groundwater, mostly attributed to inherent geological factors. The Lagunera region in Mexico experiences notable arsenic problems, principally caused by excessive groundwater extraction that disrupts the geochemical balance and releases arsenic. Another factor complicating the worldwide occurrence of arsenic in groundwater is the inadequate infrastructure and resources in many affected areas to monitor and control contamination adequately. Due to insufficient water quality testing and treatment facilities in developing countries, millions unintentionally ingest water contaminated with arsenic daily. The health consequences of arsenic exposure are extensive, rendering it one of the foremost public health hazards linked to groundwater pollution. Arsenic is a highly toxic and carcinogenic substance. Prolonged exposure to water contaminated with arsenic can result in a medical condition called arsenicosis or chronic arsenic poisoning. Manifestations of arsenicosis encompass cutaneous lesions, excessive pigmentation, and keratosis, affecting the palms and soles. As prolonged exposure persists, the likelihood of experiencing more serious health consequences escalates. A paramount health consequence of prolonged exposure to arsenic is the heightened susceptibility to several types of cancer. Arsenic is categorized as a Group 1 carcinogen by the International Agency for Research on Cancer (IARC), indicating that it is a recognized carcinogen specific to humans. The predominant malignancies linked to arsenic exposure include cutaneous cancer, nephron tumour, and pulmonary carcinoma. Moreover, there is compelling evidence establishing a correlation between arsenic exposure and the development of liver, kidney, and prostate cancers. The latency period for arsenic-related malignancies can be of considerable duration, frequently extending over several decades. Consequently, persons who are exposed to arsenic in the present may not manifest cancer until several years later. In addition to cancer, exposure to arsenic has been associated with various other health-related issues. The prevalence of cardiovascular diseases, such as hypertension and ischemic heart disease, is higher in populations that are exposed to elevated levels of arsenic. Arsenic also presents hazards to the respiratory system, as evidenced by research indicating a higher prevalence of chronic obstructive pulmonary disease (COPD) and bronchitis throughout populations exposed to it. Moreover, arsenic has the potential to compromise the immune system, thus increasing the vulnerability of persons to infections and other illnesses. Particularly problematic are the effects of arsenic on the health of mothers and children. Pregnant women who are exposed to water contaminated with arsenic face an increased likelihood of experiencing problematic pregnancy outcomes, such as miscarriages, stillbirths, and low birth weight. Furthermore, empirical data indicates that exposure to arsenic during pregnancy can result in developmental complications in offspring, such as compromised cognitive abilities and heightened vulnerability to a range of health issues in the future (Devi et al. 2024b; P. Kumar and Singh 2024d, b; Kumari et al. 2022). The ramifications of arsenic contamination on society and the economy are equally substantial. In numerous afflicted areas, the illness linked to arsenic exposure diminishes the population's productivity, resulting in economic losses at both the individual and communal levels. Patients afflicted with persistent health disorders resulting from exposure to arsenic frequently encounter exorbitant medical expenses, which can be financially debilitating, particularly for families with limited financial resources. In rural regions, where healthcare availability is restricted, the economic consequences are even more substantial, as those affected may be unable to work or provide for their families. Furthermore, the existence of arsenic in groundwater has wider economic consequences. In several impacted areas, agriculture serves as the main means of sustenance. Nevertheless, the contamination of groundwater used for irrigation with arsenic can lead to the accumulation of arsenic in crops, so diminishing agricultural output and presenting supplementary health hazards to consumers. Furthermore, this pollution can harm agricultural products' marketability, resulting in financial losses for farmers and compromising food security. In regions afflicted by extensive arsenic contamination, the absence of reliable potable water sources can result in escalated water provision and treatment expenses. Implementing costly water treatment technologies or alternative water supply systems may compel governments and communities to allocate resources away from other crucial domains of development. Management of arsenic contamination can impose a significant economic burden on already constrained public health and infrastructure budgets, especially in developing nations. Moreover, it is crucial not to disregard the social consequences of arsenic pollution. The social isolation and discrimination seen in many affected communities can be attributed to the stigma surrounding arsenicosis and other health problems linked to arsenic. Individuals displaying apparent symptoms of arsenicosis may experience social exclusion or marginalization, which worsens the social and economic difficulties they encounter. Globally, the widespread occurrence and dispersion of groundwater contaminated with arsenic pose a major obstacle to public health and socio-economic progress. Adverse health effects of arsenic exposure are profound, with enduring repercussions that can impact whole generations (Devi et al. 2024a, b, c, d, e, f; Goud et al. 2022; Kumar et al. 2024a, b, c, d, e, f, g, h, i, j, k; Reddy et al. 2022). The socio-economic ramifications are equally significant, as arsenic compromises economic output, escalates health-care expenditures, and intensifies poverty and inequality. Effective resolution of this problem necessitates a holistic strategy encompassing the surveillance and control of groundwater quality, creating cost-effective and efficient water treatment methods, and promoting knowledge regarding the hazards associated with arsenic exposure. Successful mitigation of the consequences of this global crisis and safeguarding the health and well-being of affected populations can only be achieved by synchronized efforts at the local, national, and international levels. The interdisciplinary discipline of geomicrobiology investigates the intricate interactions between microorganisms and geological processes. Situated at the convergence of microbiology, geology, chemistry, and environmental science, this field of study specifically examines the reciprocal relationship between microorganisms and the geochemical and mineralogical processes of the Earth. Geomicrobiology investigates the functions of microorganisms in biogeochemical cycles, the process of mineral formation and transformation, and the consequences of microbial activity on the environment, particularly the subsurface ecological system. In addition to soils and sediments, geomicrobiology encompasses many environments, including deep-sea hydrothermal vents and subsurface aquifers. Within such habitats, microorganisms can greatly modify the chemical composition of their surroundings, frequently enabling or influencing crucial geological processes. These processes include recycling crucial elements such as carbon, nitrogen, sulfur, and metals, indispensable for preserving the Earth's ecosystems. Furthermore, geomicrobiology studies biomineral formation, in which microorganisms play a role in the precipitation of minerals by their metabolic interactions. These biominerals can influence the physical and chemical characteristics of sediments and rocks, affecting the porosity and permeability of geological units. This discipline also encompasses the investigation of extremophiles, microorganisms that flourish in extremely harsh environments, such as elevated temperatures, pressures, salinities, or acidities. These organisms offer valuable insights into the boundaries of life on Earth and maybe on other planets, providing indications about the beginnings of life and the potential existence of life beyond our present planet. Another crucial field in geomicrobiology is the investigation of microbial metabolisms that can impact the movement and concentration of metals and metalloids, such as arsenic. Comprehending these microbial processes is crucial for tackling environmental issues, such as pollution and contamination, in which microbes can either worsen or alleviate the destructive impacts of harmful substances. Geomicrobiology is critically important in bioremediation, using microbial processes to purify polluted environments. By comprehending the inherent mechanisms by which microorganisms effectively eliminate or immobilize pollutants, scientists can devise methods to augment these processes and implement them in contaminated areas. Geomicrobiology is the study of the interaction between microbial communities and geological materials in groundwater systems and how this interaction affects the mineral composition of the groundwater. Biological transformations investigate the microbial processes in reducing or oxidizing metals, metalloids, and organic compounds. Transformations of this nature can have a substantial effect on the quality of groundwater, making geomicrobiology an essential discipline for the management of water resources and the safeguarding of public health. The problem of arsenic contamination in groundwater is multifaceted, encompassing geological, hydrological, and biological processes. The field of geomicrobiology plays a crucial role in comprehending the biogeochemical mechanisms that control the movement, distribution, and impact of arsenic in the subsurface environment (Aley et al. 2022; Das et al. 2022; Devi et al. 2024a, b, c, d, e, f; Devi and Kumar 2024a; Kumar, Khokhar et al. 2024e; Kumar et al. 2018a, b, c, d, e). The impact of arsenic contamination can be either mitigated or exacerbated by microorganisms, depending on the particular environmental conditions and microbial communities. Redox processes are a primary mechanism by which microorganisms influence arsenic contamination. The predominant oxidation states of arsenic in groundwater are arsenate (As(V)) and arsenite (As(III)). The mobility and toxicity of arsenic are strongly influenced by its oxidation state, whereby arsenite exhibits greater mobility and potential toxicity compared to arsenate. Microorganisms can actively facilitate the conversion of arsenate to arsenite and the oxidation of arsenite to arsenate, so exerting a direct influence on the spatial distribution and characteristics of arsenic in groundwater. In anaerobic settings, such as deep aquifers or sediments, specific bacteria can utilize arsenate as a terminal electron acceptor during their metabolic activities, chemically reducing it to arsenite. Microbiological reduction of arsenate can result in elevated levels of arsenite in groundwater, exacerbating the contamination levels. By contrast, in aerobic or microaerophilic conditions, other microorganisms can convert arsenite into arsenate, which may result in the immobilization of arsenic as it adheres to mineral surfaces or forms solid phases. Microbiological processes play a crucial role in determining the destiny of arsenic in groundwater systems and, as a result, the hazards linked to arsenic pollution. Iron and sulfur cycling are intricately connected to the movement of arsenic in groundwater, and the study of geomicrobiology offers a valuable understanding of the interactions between these cycles and arsenic. In numerous aquifers, arsenic is linked to iron oxides or sulfides, and these minerals' microbial reduction or oxidation can either release or immobilize arsenic. By reducing ferric iron (Fe(III)) to ferrous iron (Fe(II)), iron-reducing bacteria can cause the breakdown of iron oxides and the subsequent liberation of arsenic that was previously attached to or integrated into these minerals. Sulfate-reducing bacteria can convert sulfate into sulfide, forming arsenic sulfides and limiting its movement in groundwater. Nevertheless, these processes are not simple, as the formation of secondary minerals can influence microbial activity outputs, competing ions, and particular environmental conditions. Furthermore, the function of microbial communities in natural attenuation and bioremediation is a crucial component of geomicrobiology in arsenic contamination. By facilitating the precipitation of arsenic-containing minerals or converting arsenic into less mobile or less toxic forms, natural microbial processes can sometimes aid in mitigating arsenic contamination. Understanding these processes enables the development of bioremediation techniques that can augment the inherent ability of microbial communities to purify polluted groundwater. For instance, introducing particular electron donors or acceptors can prompt the proliferation of microorganisms that facilitate the formation of arsenic precipitates or their conversion into less detrimental forms. Investigating microbial interactions with arsenic also encompasses examining the genetic and enzymatic processes that facilitate the metabolism of arsenic by microorganisms at the molecular level. This encompasses identifying genes implicated in the resistance to and metabolism of arsenic, such as those that encode enzymes for arsenate production or arsenite oxidation. Enhanced comprehension of these molecular processes enables scientists to more accurately forecast the reactions of microbial communities to alterations in environmental conditions and the subsequent effects on the movement and toxicity of arsenic. Furthermore, geomicrobiology plays a role in developing biosensors and bioindicators to monitor arsenic contamination. Developing techniques for detecting and quantifying arsenic in groundwater, utilizing microorganisms that exhibit specific responses to arsenic, offers significant insights for water resource management and public health protection. A comprehensive understanding of the intricate mechanisms governing the behaviour of arsenic in groundwater systems is essential in the field of geomicrobiology. Microbiologists can elucidate the processes that control arsenic's movement, speciation, and toxicity by studying the interactions between microorganisms and their geological environment. Acquiring this information is crucial for formulating efficient approaches to control arsenic pollution, whether by natural reduction, bioremediation, or systematic monitoring. Given that groundwater is an essential resource for millions of people globally, the knowledge acquired from geomicrobiology will be crucial in guaranteeing this resource's safety and long-term viability in the presence of environmental issues (Kumar et al. 2018a, b, c, d, e, 2024a, b, c, d, e, f, g, h, i, j, k; Kumar and Choudhury 2024c; Pandey et al. 2024).

2 Sources and Geochemistry of Arsenic in Groundwater

2.1 Natural Sources of Arsenic

As a naturally occurring element, arsenic is present in the Earth's crust at different concentration levels. The occurrence of this substance in groundwater is intricately linked to the geological and mineralogical properties of the zones it impacts. A comprehensive knowledge of arsenic's geological origins and mineral sources is essential to determine the elements contributing to its movement and discharge into groundwater, presenting substantial environmental and public health concerns. Arsenic has complex geological origins linked to a broad spectrum of rock types and mineral deposits. The element arsenic is commonly present in minute quantities in sulfide minerals, such arsenopyrite (FeAsS), realgar (As4S4), and orpiment (As2S3), as well as in oxide minerals such as iron oxides, including goethite and hematite. Minerals of this nature are often linked to hydrothermal systems, in which fluids at high temperatures come into contact with the adjacent rock, forming metal-rich mineral deposits, including arsenic. Hydrothermal veins, particularly present in volcanic and geothermal areas, are significant reservoirs of minerals containing arsenic. Under these conditions, arsenic can be deposited with other metals, such as gold, silver, copper, and lead, becoming a prevalent secondary product of mining operations. Furthermore, sedimentary rocks substantially impact the geological dispersion of arsenic. The incorporation of arsenic into sedimentary formations occurs by weathering and erosion of rocks that contain arsenic, followed by the deposition of these sediments in different geological environments. Specifically, arsenic is present in clays, shales, and sandstones, especially in regions with a past of volcanic eruptions or where ancient hydrothermal systems were active. In such instances, arsenic is frequently bonded to iron oxides, clays, or organic substances present in the sediments, where it maintains a considerable level of stability until there are changes in environmental conditions (Devi et al. 2023a, b, c; Devi and Kumar 2023b; Dey et al. 2023a, b, c, 2024; Kumar et al. 2024a; Kumar and Choudhury 2024e; Saurabh et al. 2024). Volcanic ash and tuff are significant geological sources of arsenic. Volcanic activity triggers the emission of arsenic into the atmosphere, where it is subsequently integrated into ash particles that ultimately accumulate on the Earth's surface. In due course, these ash layers can integrate into the sedimentary record, augmenting the arsenic concentration in the surrounding geology. In areas characterized by a past of volcanic eruptions, such as the Andes in South America or the Pacific Ring of Fire, the predominant sources of arsenic are volcanic rocks and ash deposits. Furthermore, arsenic can be present in secondary minerals that arise from the weathering and modification of primary arsenic-containing minerals and the aforementioned primary sources. For example, the process of oxidative weathering of sulfide minerals can stimulate the mobilization of arsenic, resulting in the creation of secondary arsenate minerals like scorodite ($FeAsO_4{}^32H_2O$) and pharmacosiderite ($KFe_4(AsO_4)3(OH)4^36\text{-}7H_2O$). Although these secondary minerals can act as transient reservoirs for arsenic, they can dissolve and subsequently release arsenic into the environment, especially in groundwater systems, under specific circumstances. The introduction of arsenic into groundwater is an intricate phenomenon controlled by many geological, geochemical, and biological parameters. These processes may be classified into two main categories: natural mechanisms, which involve the movement of arsenic from its geological origins, and anthropogenic mechanisms, in which human activities such as mining, agriculture, and industrial processes have a substantial impact. The reductive dissolution of iron oxides is an extensively researched natural process by which arsenic is released into groundwater. Arsenic is adsorbed onto the surfaces of iron oxides, such as goethite and hematite, which are plentiful in sediments and soils in numerous aquifers. Under oxidation conditions, these iron oxides exhibit stability while the arsenic remains immobilized. Once the redox conditions of the aquifer change to reducing conditions, typically caused by the presence of organic matter that supports microbial respiration, iron-reducing bacteria can utilize ferric iron (Fe(III)) as an electron acceptor to convert it into ferrous iron (Fe(II)). The reduction process induces the disintegration of iron oxides and the resulting liberation of arsenic into the surrounding groundwater. In regions such as the Ganges-Brahmaputra-Meghna delta in South Asia, where high levels of organic matter and reducing conditions contribute to widespread arsenic contamination, this mechanism is especially pervasive in sedimentary aquifers. An additional vital natural process for the release of arsenic is the oxidative weathering of sulfide minerals that contain arsenic. The process involves the interaction between oxygenated water, typically derived from rainfall or surface water infiltration, and sulfide minerals such as arsenopyrite, which contain arsenic. Electron oxidation of these minerals results in the liberation of arsenic and other elements, including sulfur and iron, into the adjacent environment. One example of a by-product of the oxidation of arsenopyrite is the formation of arsenate (As(V)), which can subsequently be transported in the groundwater under specific circumstances. The mechanism described is especially pertinent in regions with a mining background, where exposed sulfide minerals undergo weathering processes that expedite the release of arsenic. An additional important source of arsenic in volcanic and geothermal areas is the interaction between groundwater and volcanic rocks and hydrothermal fluids rich in arsenic. In these settings, the high temperatures and acidic conditions resulting from geothermal activity can extract arsenic from volcanic glass and other arsenic minerals. The transportation of arsenic into groundwater systems by these fluids can result in its establishment as a persistent contaminant. The Andes region, renowned for its geothermal activity and volcanic past, exemplifies these processes that contribute to increased concentrations of arsenic in groundwater. pH fluctuations can also induce the breakdown of minerals containing arsenic. The acidity or alkalinity of the adjacent water affects the solubility of minerals containing arsenic in many aquifers. In alkaline environments, the adsorption of arsenic onto mineral surfaces can undergo desorption, resulting in enhanced mobility and a subsequent rise in its concentration within groundwater. Similarly, under acidic conditions, the corrosion of minerals containing arsenic can be intensified, resulting in the liberation of arsenic. Fluctuations in pH can occur due to natural phenomena, such as the infiltration of acidic precipitation, and human activities, such as acid mine drainage or fertilizer use (Devi and Kumar 2024b; Karam et al. 2023; Kumar 2023; Kumar and Singh 2024c, e; Kumar and Kumar 2023). Industrial activities can greatly intensify the discharge of arsenic into underground water sources. A widely recognized example is the influence of mining, in which the extraction and treatment of ores containing arsenic subject these materials to weathering mechanisms. In the presence of oxygen and water, tailings and waste rock generated by mining activities have the potential to leach arsenic into adjacent water sources. Acid mine drainage, caused by the oxidation of sulfide minerals in mine waste, can expedite the discharge of arsenic and other hazardous species into the groundwater. Furthermore, agricultural practices can exacerbate the problem of arsenic pollution in groundwater. Arsenic-containing pesticides, herbicides, and fertilizers have traditionally been a major contributor to the presence of arsenic in agricultural soils. Nevertheless, these chemicals have been mostly eliminated in numerous countries, and their impact continues, as arsenic remains in the soil and may enter the groundwater through leaching. Moreover, excessive extraction of groundwater for irrigation can contribute to the decline of the water table and establish circumstances that facilitate the liberation of arsenic from sediments, especially in areas where groundwater serves as the main water supply for agriculture. Excessive groundwater extraction can result in subsidence and aquifer compaction, distorting the geochemical environment and facilitating the release of arsenic. Under certain circumstances, the extraction of substantial amounts of groundwater can attract water from nearby regions, particularly areas with elevated arsenic concentrations, resulting in a general rise in arsenic levels in the extracted water. The contamination of wells with arsenic in areas where groundwater is used for drinking water presents substantial health hazards. Excavating wells into aquifers abundant in arsenic can unintentionally access water contaminated with arsenic, exposing significant populations to this hazardous element. The extensive arsenic exposure in regions such as Bangladesh and West Bengal can be attributed to the heavy dependence on shallow tube wells for drinking water since these wells frequently come into contact with arsenic sediments. The processes by which arsenic is released into groundwater are intricate and affected by natural and human-induced elements. A comprehensive grasp of these processes is crucial for formulating effective approaches to reduce arsenic pollution and safeguard groundwater resources. To effectively manage arsenic in groundwater, it is necessary to adopt a comprehensive strategy that considers the geological, geochemical, and biological mechanisms that control the movement of arsenic and the influence of human activities on these mechanisms. The mitigation of both natural and anthropogenic causes that contribute to the release of arsenic can effectively decrease the likelihood of arsenic contamination and ensure the protection of the health of affected communities.

2.2 Anthropogenic Sources of Arsenic

Industrial activities have long been recognized as significant contributors to environmental arsenic contamination, particularly in water systems. The use, production, and disposal of arsenic and arsenic-containing compounds in various industries have led to the widespread distribution of this toxic element, often with devastating environmental and public health consequences. Mining and mineral processing are primary industrial sources of arsenic contamination. Arsenic is commonly associated with metal ores, particularly those containing copper, gold, lead, and zinc. During the extraction and processing of these ores, arsenic is often released as a by-product. The smelting of these ores, which involves heating the materials to extract the desired metals, can release significant amounts of arsenic into the atmosphere through gaseous emissions or particulate matter. These emissions can settle onto the land or water bodies, leading to localized contamination. For instance, in areas with a history of extensive mining and smelting activities, such as the Copperbelt region in Zambia or the historic mining districts of the Western United States, arsenic contamination of soil, water, and air is a persistent issue. The disposal of mining waste, including tailings and slag, is another primary source of arsenic contamination. These waste materials often contain high concentrations of arsenic, which can leach into groundwater and surface water, particularly in the presence of oxygen and water. Acid mine drainage, a common problem in mining regions, exacerbates the release of arsenic from mine waste. The acidic conditions generated by the oxidation of sulfide minerals can increase the solubility of arsenic, leading to higher concentrations in nearby water sources. This issue is particularly acute in areas where abandoned or poorly managed mines are left exposed to the elements, with no remediation efforts to contain the spread of contaminants. The metallurgical industry processes and refines metals and contributes to arsenic contamination. During the smelting and refining of metals, arsenic can be released into the environment through air emissions and wastewater discharges (Devi et al. 2024a, b, c, d, e, f; Jagadesh et al. 2024; Kumar et al. 2023a, b, c, 2024a, b, c, d, e, f, g, h, i, j, k; Kumar and Dey 2024; Kumar and Pathak 2019; Pathak et al. 2017; Raj et al. 2024). In producing non-ferrous metals, such as copper and lead, arsenic is often volatilized and emitted as a gas during smelting. These emissions can travel long distances and settle far from the source, leading to widespread contamination of land and water bodies. In addition, the treatment and disposal of metallurgical waste, including slag and sludge, can lead to arsenic leaching into groundwater and surface water. The electronics industry is another significant source of arsenic contamination. Arsenic is used to produce semiconductors, which are employed to manufacture integrated circuits and other electronic components. The processing of these materials can generate arsenic-containing waste, which, if not properly managed, can lead to environmental contamination. Wastewater from semiconductor manufacturing facilities often contains arsenic, which can be released into water bodies if not adequately treated. Furthermore, the disposal of electronic waste (e-waste), such as discarded computers and mobile phones, can contribute to arsenic contamination, particularly in regions where e-waste recycling is poorly regulated or conducted under unsafe conditions. Historically, the wood preservation industry has been a significant source of arsenic contamination, mainly using chromated copper arsenate (CCA) as a wood preservative. CCA-treated wood was widely used for constructing outdoor structures such as decks, playgrounds, and utility poles. The arsenic in CCA-treated wood can leach into the soil and groundwater over time, leading to localized contamination. Although the use of CCA has been restricted or banned in many countries, legacy contamination from previously treated wood remains a concern. Sites where CCA-treated wood was produced, used, or disposed of are often contaminated with arsenic, posing risks to human health and the environment. The production and use of glass and ceramics also contribute to arsenic contamination. Arsenic is a fining agent in the glass industry to remove bubbles and impurities during manufacturing. In ceramics, arsenic compounds are sometimes used in glazes and pigments. Disposing of waste materials from these industries, including spent glass and ceramics, can lead to the release of arsenic into the environment. Wastewater from glass and ceramics manufacturing facilities can also contain arsenic, contributing to water pollution if not properly managed. Other industrial activities contributing to arsenic contamination include the production of pesticides, herbicides, and fungicides, which historically contain arsenic as an active ingredient. Although arsenic-based pesticides have been largely phased out in many parts of the world, their legacy persists in soils and groundwater, particularly in agricultural regions where these chemicals were heavily applied. Agricultural practices have significantly contributed to arsenic contamination, mainly through the historical use of arsenic-based pesticides, herbicides, and fertilizers. Agriculture widely uses these chemicals to control pests, diseases, and weeds. Still, their long-term environmental impact has proven to be severe, with arsenic residues remaining in soils and water bodies for decades after their application. One of the most notorious arsenic-based pesticides was lead arsenate, extensively used in orchards, particularly in the United States, to control insect pests such as the codling moth. Lead arsenate was favoured for its effectiveness, but it also led to the accumulation of arsenic in soils where it was applied. Over time, the repeated application of lead arsenate resulted in the buildup of arsenic to levels that posed significant risks to the environment and human health. The arsenic in contaminated soils can leach into groundwater, particularly in regions with high rainfall or irrigation, contaminating drinking water sources. Although lead arsenate was banned in the 1970s, the legacy of its use remains a persistent problem in many former orchard areas, where arsenic levels in soil and groundwater continue to pose risks. Another significant agricultural source of arsenic is the use of arsenic-containing herbicides. One of the most common arsenic-based herbicides was monosodium methanearsonate (MSMA), used to control grasses and weeds in crops such as cotton, sugarcane, and turfgrass. Like lead arsenate, MSMA contributed to the accumulation of arsenic in soils, particularly in regions where it was applied repeatedly over multiple growing seasons. The arsenic from MSMA can be mobilized under certain soil conditions, particularly in high water content, leading to its leaching into groundwater. The use of MSMA has been restricted in many countries due to its environmental impact, but its legacy remains, particularly in agricultural regions where it was heavily applied. Arsenic contamination from agriculture is not limited to pesticides and herbicides; it also extends to fertilizers. In some regions, phosphate fertilizers commonly used to promote plant growth can contain trace amounts of arsenic. The application of these fertilizers can contribute to the gradual accumulation of arsenic in soils, particularly in areas where fertilizers are applied in large quantities. While the arsenic content in modern fertilizers is generally low, the cumulative effect of repeated applications over time can still result in elevated levels of arsenic in agricultural soils (Devi and Kumar 2023a; Dey et al. 2023a, b, c; Kumar et al. 2019; 2024a, b, c, d, e, f, g, h, i, j, k; Kumari et al. 2023; Sharma and Kumar 2023; Srivastav et al. 2024). This is particularly concerning in regions where groundwater is the primary source of irrigation water, as the mobilization of arsenic from soils into groundwater can lead to contamination of water supplies. In addition to chemical inputs, agricultural practices such as irrigation can also influence the mobility and distribution of arsenic in the environment. The over-extraction of groundwater for irrigation can lower the water table and create conditions that promote the release of arsenic from sediments into the groundwater. This is particularly problematic in regions where groundwater is already prone to arsenic contamination, such as the Ganges-Brahmaputra-Meghna delta in South Asia. In these areas, the widespread use of groundwater for irrigation has exacerbated the problem of arsenic contamination, as the drawdown of the water table can lead to the mobilization of arsenic from deeper sediments into the groundwater. Using groundwater for rice cultivation is another agricultural activity that can contribute to arsenic contamination. Rice is often grown in flooded fields, where the anaerobic conditions created by the standing water can promote the reduction of arsenic-bearing minerals, releasing arsenic into the soil and water. The uptake of arsenic by rice plants can result in the accumulation of arsenic in the edible parts of the plant, posing risks to human health. This is a significant concern in regions where rice is a staple food and where groundwater is contaminated with arsenic, as the consumption of arsenic-contaminated rice can lead to chronic exposure. In some regions, animal feed additives containing arsenic have also contributed to arsenic contamination. Arsenic-based compounds, such as roxarsone, were historically used as feed additives in poultry and swine production to promote growth and improve feed efficiency. The use of these additives resulted in the excretion of arsenic in animal waste, which, when applied to fields as fertilizer, contributed to the accumulation of arsenic in soils. The runoff from these fields can carry arsenic into nearby water bodies, contaminating surface water and groundwater. Although the use of arsenic-based feed additives has mainly been discontinued, the legacy of their use persists, particularly in regions with intensive livestock production. The impact of agricultural practices on arsenic contamination is a complex issue that requires careful management to mitigate the risks. Reducing the use of arsenic-containing chemicals, improving the management of irrigation practices, and monitoring the levels of arsenic in soils and water are essential steps in addressing the problem. In regions where arsenic contamination is a significant concern, implementing sustainable agricultural practices and using alternative water sources for irrigation can help reduce the risk of further contamination. Additionally, public awareness and education are crucial in promoting the adoption of safer agricultural practices and reducing the exposure of communities to arsenic-contaminated food and water. The anthropogenic sources of arsenic contamination are diverse and pervasive, encompassing various industrial and agricultural activities. The legacy of arsenic use in these sectors continues to pose significant environmental and public health challenges, particularly in regions with widespread contamination. Addressing these challenges requires a multifaceted approach that includes regulating and monitoring industrial and agricultural practices, remediating contaminated sites, and promoting sustainable alternatives. By taking these steps, we can mitigate the impact of arsenic contamination and protect the health of the environment and the communities that depend on it.

2.3 Arsenic Speciation and Mobility

The primary oxidation states of arsenic in groundwater are arsenite (As(III)) and arsenate (As(V)). Considering the substantial differences in chemical behaviour, toxicity, and mobility between these two species, it is essential to comprehend arsenic speciation to evaluate the hazards of arsenic pollution and devise efficient remediation methods. The toxicity and mobility of arsenite, also known as As(III), in groundwater are generally higher than arsenate. Typically, it occurs under reducing conditions, when the lack of oxygen or low oxygen levels enable the +3 oxidation state to remain stable. In its trivalent state, arsenite is primarily found in groundwater as H_3AsO_3 (arsenic acid). The solubility of this neutral variant of arsenic in water is substantial, hence posing challenges in its removal using traditional water treatment methods. Moreover, arsenite has a lower propensity to adhere to mineral surfaces than arsenate, enhancing its relative mobility within aquifers. Conversely, arsenate, or As(V), dominates in oxidizing environments. Typically, arsenate is present in groundwater as negatively charged oxyanions, namely $H_2AsO_4^+$ and $HAsO_4^{2+}$. These species exhibit lower water solubility compared to arsenite and have a greater tendency to adsorb onto iron oxides, clay minerals, and other particles present in the aquifer matrix. Consequently, arsenate is comparatively less mobile than arsenite. Nevertheless, the simplicity with which arsenate can be converted into arsenite under specific circumstances complicates the prediction of arsenic movement in groundwater systems. Several parameters, such as pH, redox potential, the presence of other chemical species, and the mineral composition of the aquifer, alter the speciation and mobility of arsenic in groundwater. The pH of groundwater substantially impacts the process of arsenic speciation. Under acidic conditions, both arsenite and arsenate can exhibit enhanced mobility as a result of an increase in solubility. However, arsenate forms less soluble complexes under neutral to alkaline conditions, decreasing mobility. The level of electron availability in the environment, known as the redox potential, is crucial in determining the presence of arsenic as either As(III) or As(V). Arsenite exhibits greater stability under reducing conditions commonly encountered in deeper or organic-rich groundwater, resulting in increased mobility. By contrast, oxidizing conditions promote the occurrence of arsenate, which can be immobilized by adsorbing onto minerals (Choudhury et al. 2024; Kumar and Choudhury 2024d; Kumar and Raut 2024; Sharma et al. 2023a, b, c, d; Sharma and Kumar 2024; Upadhyay et al. 2023). The mobility of arsenic can also be influenced by other chemical species present in the groundwater. For instance, dissolved iron can create iron oxyhydroxides, which have a high affinity for adsorbing arsenate, hence delimiting their ability to move. The displacement of competing anions, such as phosphate and bicarbonate, can enhance the mobility of arsenate on mineral surfaces. Through its interaction with the arsenic species and the redox conditions, organic matter can either increase or decrease the mobility of arsenic. The behaviour of arsenic in groundwater is fundamentally influenced by the geochemical parameters of an aquifer, such as pH, redox potential, and the concentrations of other chemical species. The speciation of arsenic and its adsorption onto minerals are directly influenced by pH. The likelihood of arsenic remaining in solution is higher in acidic environments, enhancing mobility. By contrast, arsenate can create stable complexes with minerals such as iron oxides at elevated pH levels, decreasing mobility. The redox potential, denoted as Eh, determines the oxidation state of arsenic and, by extension, its environmental behaviour. The predominant form of arsenic in shallow, oxygen-rich groundwater is arsenate under oxidizing conditions. By contrast, arsenite is more durable in reducing conditions, such as those in deeper, oxygen-deprived groundwater or sediments abundant in organic matter. Reducing iron and manganese oxides in these conditions can release previously absorbed arsenic into the groundwater, enhancing mobility. The presence of other geochemical factors, including sulfides, carbonates, and organic contaminants, influences arsenic behaviour. Sulfides can promote the creation of insoluble minerals composed of arsenic sulfide, immobilizing arsenic. Nevertheless, alterations in redox conditions can potentially remobilize these arsenic minerals. By buffering the pH of the groundwater, carbonates can indirectly influence the speciation and mobility of arsenic. Conversely, organic matter can bind arsenic, decreasing its mobility or facilitating reducing conditions, causing the formation of arsenate into the more easily moveable arsenite form. An aquifer's sediment and mineral composition plays a crucial role in regulating the movement and accessibility of arsenic in groundwater. Iron and manganese oxides are crucial minerals used for the adsorption and immobilization of arsenic, particularly in its arsenate state. These oxides function as inherent filters, capturing arsenic on their surfaces and impeding its translocation through the aquifer. Nevertheless, when exposed to reducing conditions, these oxides can dissolve, liberating adsorbed arsenic and returning it to the groundwater. The substantial surface area and cation exchange capacity of clay minerals contribute to their important involvement in the adsorption of arsenic. Nevertheless, the presence of other competing ions can influence the affinity of arsenic for these minerals, diminishing arsenic adsorption efficiency. By stabilizing arsenite or forming arsenic-sulfide complexes, sediments abundant in organic matter or sulfide minerals can generate reducing environments that enhance the mobility of arsenic. Furthermore, the pH and buffering capacity of the groundwater can be influenced by sediments consisting of carbonate minerals, indirectly impacting the speciation and mobility of arsenic. The interaction between geochemical conditions and sediment composition ultimately governs how arsenic is mobilized or immobilized in groundwater systems. Gaining insight into these interactions is essential for forecasting the destiny of arsenic in polluted aquifers and developing efficient strategies for remediation (Table 1).

3 Microbial Transformations of Arsenic

3.1 Arsenic-Transforming Microorganisms

Microbial species that respire and oxidize arsenic are essential in the biogeochemical cycling of arsenic in natural settings. In groundwater systems, these microorganisms play a crucial role in facilitating the conversion of arsenic between its different oxidation states, greatly influencing its mobility, toxicity, and bioavailability. Microbial species that respire arsenic, such as *Geobacter* and *Shewanella*, can utilize arsenate (As(V)) as a terminal electron acceptor in anaerobic respiration,

Table 1 Different sources of arsenic and its association with the environment

Source	Geochemical behaviour of arsenic	Associated environment	Primary arsenic species	Key mechanisms of release	Microbial involvement	Human influence
1. Natural geogenic (bedrock)	Arsenic released through weathering of arsenic-bearing minerals	Granitic and volcanic rocks	Arsenate (As(V)), Arsenite (As(III))	Weathering, dissolution	Limited microbial impact	Minimal human impact
2. Mining activities	Acid mine drainage increases arsenic mobility	Mining areas (especially gold and copper)	Arsenate (As(V)), Arsenite (As(III))	Oxidation of arsenopyrite	Microbial sulfide oxidation	Mining waste and tailings
3. Agricultural runoff	Pesticides containing arsenic leach into groundwater	Agricultural regions	Methylated arsenic species	Leaching from arsenic-based pesticides	Methylating bacteria	Excessive fertilizer use
4. Industrial effluents	Chemical waste increases arsenic levels	Urban/industrial zones	Inorganic arsenic species	Discharge of arsenic-containing chemicals	Possible industrial biofilters	Industrial wastewater
5. Coal combustion	Fly ash from coal burning releases arsenic	Power plants, coal-rich regions	Arsenite (As(III))	Deposition of fly ash in water sources	Limited microbial role	Industrial emissions
6. Volcanic activity	Arsenic released during volcanic eruptions	Volcanic regions	Arsenate (As(V)), Arsenite (As(III))	Direct release from volcanic gases	Sulfur-oxidizing bacteria	None
7. Weathering of sulfide minerals	Oxidation of arsenopyrite, galena, and others releases arsenic	Sulfide mineral-rich regions	Arsenate (As(V))	Sulfide oxidation	Sulfate-reducing bacteria	Minimal human influence
8. Thermal springs	Hot springs dissolve arsenic from geothermal rocks	Geothermal regions	Arsenite (As(III))	Dissolution in hot, acidic conditions	Thermophilic bacteria	None
9. Uranium mining	Associated with arsenic release in tailings and waste	Uranium-rich regions	Arsenate (As(V)), Arsenite (As(III))	Oxidation of tailings	Uranium-reducing microbes	Industrial mining

10. Groundwater irrigation	Over-extraction and drawdown of groundwater can mobilize arsenic	Agricultural regions	Arsenite (As(III)), Arsenate (As(V))	Redox changes from irrigation practices	Redox-cycling bacteria	Irrigation practices
11. Fossil fuel extraction	Arsenic released from oil and gas extraction	Oil and gas fields	Arsenate (As(V))	Extraction-related geochemical shifts	Sulfate-reducing bacteria	Fossil fuel extraction
12. Landfills and waste disposal	Leachate from waste increases arsenic in nearby groundwater	Landfills, waste dumps	Inorganic arsenic	Leaching from arsenic-containing waste	Nitrifying bacteria	Improper waste disposal
13. Water table fluctuations	Seasonal changes in water levels mobilize arsenic	Shallow aquifers, wetlands	Arsenite (As(III)), Arsenate (As(V))	Redox-driven release from sediments	Redox cycling bacteria	Groundwater extraction
14. Sedimentary aquifers	Arsenic released from pyrite-bearing sediments	Sedimentary basins	Arsenate (As(V)), Arsenite (As(III))	Dissolution and redox changes	Anaerobic bacteria	Minimal human impact
15. Desiccation of wetlands	Oxidation of arsenic in soils as wetlands dry up	Coastal and inland wetlands	Arsenate (As(V)), Methylated species	Oxidation of organic matter	Sulfate-reducing bacteria	Agricultural runoff
16. Irrigation canals	Mobilization of arsenic through artificial recharge	Agricultural regions	Arsenite (As(III)), Arsenate (As(V))	Dissolution of sediments due to water flow	Possible bacteria-mediated processes	Agricultural practices
17. Desalination brine	Concentrated arsenic in brine from desalination processes	Desalination plants	Arsenate (As(V))	Concentration through brine disposal	Limited microbial influence	Desalination discharge
18. Fertilizer use	Phosphate fertilizers release arsenic into groundwater	Agricultural areas	Arsenite (As(III)), Methylated arsenic	Leaching and chemical displacement	Arsenic methylating bacteria	Overuse of fertilizers

(continued)

Table 1 (continued)

Source	Geochemical behaviour of arsenic	Associated environment	Primary arsenic species	Key mechanisms of release	Microbial involvement	Human influence
19. Chemical manufacturing	Industrial chemicals with arsenic enter water bodies	Chemical plants	Inorganic arsenic	Discharge of waste	Industrial microbial treatment	Chemical waste disposal
20. Poultry waste	Arsenic-based feed additives in waste leach into groundwater	Poultry farms	Methylated arsenic species	Leaching from poultry waste	Microbial methylation of arsenic	Use of arsenic-based additives
21. Geothermal power plants	Release of arsenic-rich steam into water sources	Geothermal energy sites	Arsenite (As(III)), Arsenate (As(V))	Release from geothermal fluids	Thermophilic microorganisms	Energy production waste
22. Saltwater intrusion	Coastal aquifers have arsenic mobilized through saltwater mixing	Coastal regions	Arsenite (As(III)), Arsenate (As(V))	Ion exchange and displacement	Halophilic bacteria	Urbanization and over-extraction
23. Hard rock mining	Sulfide ores release arsenic during mining operations	Mineral-rich areas	Arsenate (As(V)), Arsenite (As(III))	Oxidation of arsenopyrite	Sulfate-reducing bacteria	Mining practices
24. Irrigation water recharge	Recharge systems can remobilize arsenic from sediments	Agricultural areas	Arsenite (As(III))	Changes in groundwater flow and redox	Redox-sensitive bacteria	Irrigation and water management
25. Rice paddy cultivation	Anaerobic conditions in paddy soils increase arsenic solubility	Southeast Asia	Arsenite (As(III)), Arsenate (As(V))	Anaerobic reduction and microbial activity	Methanogenic bacteria	Agricultural intensification
26. Oil refineries	Industrial discharge containing arsenic into nearby water bodies	Oil refining zones	Inorganic arsenic	Discharge of wastewater	Limited microbial role	Industrial activities
27. Aquaculture	Arsenic from feed and medications can enter water bodies	Coastal aquaculture farms	Methylated arsenic species	Leaching of feed additives	Arsenic-methylating bacteria	Excessive feed use

28. Industrial sediments	Sediments near industrial sites accumulate arsenic	Industrial zones	Arsenate (As(V))	Adsorption/desorption in sediments	Microbial respiration	Industrial waste management
29. Phosphate rock weathering	Natural phosphate rocks release arsenic into groundwater	Phosphate-rich regions	Arsenate (As(V))	Weathering and dissolution	Phosphate-solubilizing bacteria	Minimal human influence
30. Metal smelting	Arsenic released from smelting processes	Smelting areas	Arsenite (As(III))	Emissions and deposition from smelting	Limited microbial role	Industrial emissions
31. Urban runoff	Arsenic from industrial and household waste enters water systems	Urban areas	Inorganic arsenic	Runoff from roads, buildings, and waste	Limited microbial role	Urban pollution
32. Hydraulic fracturing	Arsenic mobilized from shale formations during fracking	Shale gas regions	Arsenate (As(V)), Arsenite (As(III))	Fracturing-induced release	Anaerobic bacteria	Fracking waste management
33. Seawater desalination	Reject brine concentrates arsenic in surrounding waters	Coastal regions	Arsenate (As(V))	Evaporation and concentration	Limited microbial involvement	Desalination processes
34. Evaporation basins	Evaporative concentration of arsenic in surface water	Arid and semi-arid regions	Arsenite (As(III))	Concentration through evaporation	Halophilic bacteria	Water management practices
35. Contaminated soils	Arsenic in contaminated soils leaches into groundwater	Industrial/agricultural areas	Arsenite (As(III))	Leaching from contaminated sites	Microbial arsenic methylation	Soil remediation challenges
36. Land reclamation	Disturbance of sediments during land reclamation mobilizes arsenic	Reclaimed coastal land	Arsenite (As(III))	Sediment oxidation and disturbance	Sulfur-oxidizing bacteria	Land development

(continued)

Table 1 (continued)

Source	Geochemical behaviour of arsenic	Associated environment	Primary arsenic species	Key mechanisms of release	Microbial involvement	Human influence
37. Ore processing	Tailings from ore processing leach arsenic into nearby water sources	Mining and processing areas	Arsenite (As(III))	Leaching from tailings	Sulfate-reducing bacteria	Industrial practices
38. Flooding	Inundation of arsenic-rich soils releases arsenic	Floodplain regions	Arsenate (As(V)), Arsenite (As(III))	Anaerobic reduction during flooding	Methanogens and anaerobes	Climate change-induced flooding
39. Fossil fuel combustion	Atmospheric deposition of arsenic from coal burning into groundwater	Regions near coal plants	Arsenate (As(V))	Atmospheric fallout and leaching	Limited microbial role	Industrial pollution
40. Lead smelting	Arsenic co-occurs with lead in smelting emissions	Lead mining areas	Arsenite (As(III))	Release during smelting	Sulfate-reducing bacteria	Industrial pollution control
41. Natural river erosion	Arsenic from eroded riverbanks enters water systems	Riverine environments	Arsenite (As(III))	Erosion and sediment transport	Limited microbial role	River management practices

converting it into arsenite (As(III)). This process frequently occurs in oxygen-deprived environments, where these microorganisms play a role in the movement of arsenic, so enhancing its susceptibility to contamination. Conversely, arsenic-oxidizing microorganisms, such as *Pseudomonas* species, convert arsenite into the less poisonous and less easily transported arsenate in the presence of oxygen. Microbiological activity is vital in immobilizing arsenic and can be utilized in bioremediation approaches to decrease arsenic pollution in groundwater. Specific microbial taxa and others are fundamental in the biotransformation of arsenic in the environment, underscoring the significance of microbial processes in managing arsenic pollution (Kumar et al. 2024a, b, c, d, e, f, g, h, i, j, k; Kumar et al. 2024b; Kumar and Choudhury 2024a; Sharma et al. 2023a, b, c, d; Sinam et al. 2023).

3.2 Mechanisms of Arsenic Reduction

An essential step in the biogeochemical cycling of arsenic, especially in anoxic environments like sediments and groundwater, is the microbial reduction of arsenate (As(V)) to arsenite (As(III)). A heterogeneous assemblage of anaerobic bacteria predominantly facilitates this reduction by employing arsenate as a terminal electron acceptor in their respiratory metabolism. Microbiological growth in oxygen-depleted environments is not only sustained by this process but also plays a crucial role in the mobility and toxicity of arsenic in natural systems. Arsenate reductase (Arr) is the specific enzyme that catalyses the conversion of As(V) to As(III). The cytoplasmic enzyme known as arsenate reductase is encoded by the arr genes, commonly present in the genomes of bacteria that respire arsenic, including species of *Geobacter*, *Shewanella*, and *Desulfotomaculum*. To produce energy, these bacteria utilize arsenate as an electron acceptor and combine its reduction with the oxidation of organic or inorganic substrates. The enzyme arsenate reductase catalyses redox reactions to convert arsenate into arsenite. The enzyme's active site is equipped with a molybdenum cofactor that efficiently transfers electrons from a donor molecule to arsenate, transforming it into arsenite. The reduced form, known as arsenite, is subsequently discharged into the adjacent environment, where it exhibits greater mobility and toxicity than arsenate, exacerbating the level of arsenic pollution in groundwater—environmental parameters impact arsenate reductase activity, including pH, redox potential, and electron donor availability. Comprehending the processes of As(V) reduction and the function of arsenate reductase in this transformation is crucial for developing bioremediation approaches that target regulating arsenic movement and reducing its effects on human health and the environment (Chakraborty et al. 2021; Hidangmayum et al. 2023; Kumar et al. 2018a, b, c, d, e; Kumar and Choudhury 2024b; Kumar and Naik 2020).

3.3 Mechanisms of Arsenic Oxidation

The biochemical process of microbial oxidation of arsenite [As(III)] to arsenate [As(V)] is critically important in the detoxification and biogeochemical cycling of arsenic in the environment. As a toxic metalloid, arsenic exists in several oxidation states, with arsenite [As(III)] being more mobile and hazardous than arsenate [As(V)]. Microbial species that can oxidize arsenite help decrease its toxicity and bioavailability, alleviating arsenic pollution in natural settings, especially in water and soil systems. The oxidation process is predominantly facilitated by a pivotal enzyme known as arsenite oxidase, which reduces the transformation of As(III) to As(V). This enzymatic reaction facilitates the survival of microorganisms in environments abundant in arsenic. It connects the process to their energy metabolism, particularly for microbes that use arsenic as an energy source during metabolism.

The function of arsenite oxidase in microbial metabolism is essential for the survival of specific bacteria and archaea in environments contaminated with arsenic. A molybdenum-containing enzyme called arsenite oxidase catalyses the oxidation of As(III) by transferring electrons from As(III) to an appropriate electron acceptor, usually oxygen or nitrate, depending on the metabolic pathway used by the microorganism. Microbes utilize the energy produced by this electron transfer chain to facilitate their growth and survival. In aerobic environments, oxygen acts as the terminal electron acceptor, directly oxidizing As(III) to As(V). Nevertheless, certain microorganisms utilize nitrate or other substitute electron acceptors without oxygen, so arsenite oxidation is a fundamental component of anaerobic respiration. Chemolithoautotrophic microorganisms often possess the capacity to oxidize As(III), enabling them to obtain energy from the oxidation of inorganic matter such as arsenite. These microorganisms employ arsenite oxidation to produce a proton motive force across their cellular membranes, facilitating the production of adenosine triphosphate (ATP), the cell's energy currency. This energy acquisition mechanism enables these microorganisms to flourish in habitats characterized by limited availability of organic carbon sources, such as deep subsurface waters, hot springs, or soils abundant in arsenic (Dwivedi and Kumar 2011; Kumar et al. 2017, 2020; 2022a, b, c; Kumar and Dwivedi 2020; Naik and Kumar 2020). Arsenite oxidation is advantageous for microbial energy metabolism. It has environmental consequences, as it facilitates the immobilization of arsenic by transforming it into the less poisonous and more readily precipitable arsenate moiety. Arsenate adsorption onto mineral surfaces, mainly iron oxides, decreases its mobility in groundwater and mitigates its capacity to pollute drinking water sources. Alongside its function in energy metabolism, arsenite oxidation is a component of a more comprehensive detoxification mechanism used by microorganisms residing in environments contaminated with arsenic. Certain bacteria have developed the capacity to oxidize As(III) not to acquire energy but to decrease the amount of the poisonous metalloid retained within their cells. The microorganisms frequently harbour genes that confer resistance to arsenic, such as those encoding arsenite oxidase. These genes allow them to survive and multiply in unfavourable conditions characterized by elevated levels of arsenic. Oxidation of arsenite to arsenate mitigates the harmful impact of arsenic on biological functions, enabling the survival of these microorganisms in habitats where other species may be hindered or eradicated by arsenic poisoning. Generally, the function of microbial oxidation of As(III) to As(V) and the enzyme arsenite oxidase is essential for microbial metabolism and the detoxification of the environment. This biogeochemical stabilization of arsenic in natural systems is crucial for the survival of specialized microorganisms that use arsenic oxidation as a metabolic adaptation. Deciphering these microbial processes significantly impacts environmental bioremediation endeavours to reduce arsenic pollution in water and soils. Analysis of biotransformation mechanisms involving methylated arsenic species is crucial for comprehending the movement and effects of arsenic on natural environments, especially groundwater systems. Inorganic arsenic (As) undergoes microbial methylation, converting arsenite [As(III)] and arsenate [As(V)] into methylated arsenic compounds, including monomethyl arsenic acid (MMA) and dimethylarsinic acid (DMA). Microbiological enzymes such as arsenite methyltransferase catalyse this conversion, which modifies the chemical characteristics of arsenic, so affecting its mobility, bioavailability, and toxicity. The process of methylation of arsenic was traditionally believed to promote its detoxification, as organic arsenic compounds were generally considered less poisonous than their inorganic equivalents. Nevertheless, scientific studies have demonstrated that certain types of methylated arsenic, namely trivalent methylated compounds, can possess significant toxicity and pose even greater risks than inorganic arsenic. The compounds can disrupt cellular processes such as DNA repair and enzyme activity, worsening the health hazards associated with arsenic in groundwater or other environments (Kotia et al. 2021; Kumar et al. 2011; 2022a, b, c; Kumar and Dwivedi 2018). The influence of methylation on the mobility and toxicity of arsenic is substantial since methylated arsenic species always exhibit higher solubility and mobility compared to inorganic arsenic forms. The increased mobility of arsenic amplifies the likelihood of its contamination spreading through groundwater systems, potentially impacting more densely populated regions. Methylation can decrease arsenic adsorption onto mineral surfaces, such as iron oxides, which usually serve to immobilize inorganic arsenic. Therefore, methylation processes can either reduce or worsen the severity of arsenic pollution, depending on the particular environmental circumstances and the types of arsenic present. The alteration of arsenic speciation in groundwater by microbial processes is intricately connected to other biogeochemical cycles, namely those comprising sulfur and iron. Precisely, bacteria that reduce sulfate can convert arsenate into arsenite, whereas bacteria that reduce iron can break down iron oxides, so liberating arsenic that was previously attached to mineral surfaces. The intricate interplay between arsenic and the sulfur and iron cycles can profoundly impact the destiny and conduct of arsenic in groundwater, influencing its toxicity and the possibility of human exposure. Understanding these microbial and geochemical interactions is essential for formulating efficient interventions to control arsenic pollution in impacted areas (Table 2).

Table 2 Different types of microbes and their role

Microbe	Type	Feature	Arsenic transformation	Key mechanisms	Optimal conditions	Notable example	Environmental role
1. *Geobacter sulfurreducens*	Bacterium	Geobacteraceae family, dissimilatory metal-reducing	Arsenate (As(V)) reduction	Reduction to arsenite (As(III))	Anaerobic, neutral to slightly acidic pH	Soil and sediment bioremediation	Key in reducing arsenate in contaminated environments
2. *Shewanella oneidensis*	Bacterium	Gamma-proteobacteria, versatile electron acceptor	Arsenate (As(V)) reduction	Reduction to arsenite (As(III))	Anaerobic, diverse electron donors	Bioremediation research	Involved in metal and radionuclide reduction
3. *Thermomonospora curvata*	Actinobacteria	Soil bacterium, thermophilic	Arsenate (As(V)) reduction	Reduction to arsenite (As(III))	Thermophilic conditions	Hot springs environments	Contributes to arsenic cycling in geothermal areas
4. *Desulfotomaculum ruminis*	Bacterium	Firmicutes, sulfate-reducing	Arsenate (As(V)) reduction	Reduction to arsenite (As(III))	Anaerobic, high-sulfate environments	Rumen of cattle	Plays a role in sulfur and arsenic cycling
5. *Pseudomonas putida*	Bacterium	Gamma-proteobacteria, versatile metabolism	Arsenite (As(III)) oxidation	Oxidation to arsenate (As(V))	Aerobic, diverse environments	Soil and industrial sites	Bioremediation and detoxification of arsenic
6. *Bacillus cereus*	Bacterium	Firmicutes, soil bacterium	Arsenite (As(III)) oxidation	Oxidation to arsenate (As(V))	Aerobic, neutral pH	Soil and plant rhizosphere	Detoxification and transformation in soil
7. *Ralstonia metallidurans*	Bacterium	Beta-proteobacteria, metal-resistant	Arsenite (As(III)) oxidation	Oxidation to arsenate (As(V))	Aerobic, metal-rich environments	Contaminated soil	Metal resistance and detoxification
8. *Sulfolobus solfataricus*	Archaea	Sulfolobaceae family, extremophile	Arsenite (As(III)) oxidation	Oxidation to arsenate (As(V))	Acidic, high-temperature environments	Geothermal habitats	Role in extreme environments and arsenic cycling

9. *Clostridium* sp.	Bacterium	Firmicutes, anaerobic	Arsenate (As(V)) reduction	Reduction to arsenite (As(III))	Anaerobic, diverse environments	Anaerobic digesters	Biotransformation in anaerobic conditions
10. *Methylobacterium extorquens*	Bacterium	Alpha-proteobacteria, methanol-utilizing	Arsenic methylation	Methylation to various organic arsenic species	Aerobic, methanol presence	Soil and water environments	Organic arsenic compound production
11. *Arthrobacter* spp.	Bacterium	Actinobacteria, versatile metabolism	Arsenate (As(V)) reduction	Reduction to arsenite (As(III))	Aerobic and anaerobic environments	Soil	Important for bioremediation
12. *Corynebacterium glutamicum*	Bacterium	Actinobacteria, industrial strain	Arsenate (As(V)) reduction	Reduction to arsenite (As(III))	Aerobic and fermentative conditions	Industrial bioreactors	Biotechnological applications in arsenic detoxification
13. *Pseudomonas fluorescens*	Bacterium	Gamma-proteobacteria, plant-associated	Arsenite (As(III)) oxidation	Oxidation to arsenate (As(V))	Aerobic, soil	Plant rhizosphere	Enhances plant health through arsenic detoxification
14. *Burkholderia cepacia*	Bacterium	Beta-proteobacteria, soil bacterium	Arsenite (As(III)) oxidation	Oxidation to arsenate (As(V))	Aerobic, diverse environments	Contaminated soil	Plays a role in bioremediation
15. *Sulfurospirillum barnesii*	Bacterium	Delta-proteobacteria, sulfate-reducing	Arsenate (As(V)) reduction	Reduction to arsenite (As(III))	Anaerobic, sulfate-rich environments	Sediments	Sulfate reduction and arsenic transformation
16. *Desulfovibrio vulgaris*	Bacterium	Delta-proteobacteria, sulfate-reducing	Arsenate (As(V)) reduction	Reduction to arsenite (As(III))	Anaerobic, high-sulfate environments	Soil, sediments	Important in arsenic and sulfur cycles

(continued)

Table 2 (continued)

Microbe	Type	Feature	Arsenic transformation	Key mechanisms	Optimal conditions	Notable example	Environmental role
17. *Anaeromyxobacter dehalogenans*	Bacterium	Delta-proteobacteria, anaerobic	Arsenate (As(V)) reduction	Reduction to arsenite (As(III))	Anaerobic, diverse organic substrates	Contaminated environments	Arsenic reduction in contaminated sites
18. *Clostridium acetobutylicum*	Bacterium	Firmicutes, fermentative	Arsenate (As(V)) reduction	Reduction to arsenite (As(III))	Anaerobic, fermentative conditions	Fermentation processes	Arsenic reduction in industrial fermentations
19. *Geothrix fermentans*	Bacterium	Delta-proteobacteria, iron-reducing	Arsenate (As(V)) reduction	Reduction to arsenite (As(III))	Anaerobic, iron-rich environments	Soil and sediments	Role in iron and arsenic cycling
20. *Methanosarcina acetivorans*	Archaea	Methanosarcinaceae, methane-producing	Arsenic methylation	Methylation to various organic arsenic species	Anaerobic, methane-rich environments	Anaerobic digesters	Organic arsenic compound production
21. *Sulfurimonas denitrificans*	Bacterium	Epsilon-proteobacteria, sulfur-oxidizing	Arsenite (As(III)) oxidation	Oxidation to arsenate (As(V))	Microaerophilic, sulfur-rich environments	Marine sediments	Sulfur and arsenic oxidation
22. *Bacillus thuringiensis*	Bacterium	Firmicutes, insect pathogen	Arsenate (As(V)) reduction	Reduction to arsenite (As(III))	Aerobic soil environments	Soil	Potential bioremediation applications
23. *Azoarcus* sp.	Bacterium	Beta-proteobacteria, denitrifying	Arsenate (As(V)) reduction	Reduction to arsenite (As(III))	Anaerobic, nitrogen-rich environments	Soil and sediments	Role in nitrogen and arsenic cycles
24. *Eubacterium oxidoreducens*	Bacterium	Firmicutes, iron-reducing	Arsenate (As(V)) reduction	Reduction to arsenite (As(III))	Anaerobic, iron-rich conditions	Soil and sediments	Important for arsenic and iron reduction

25. *Lactobacillus plantarum*	Bacterium	Firmicutes, lactic acid bacterium	Arsenate (As(V)) reduction	Reduction to arsenite (As(III))	Anaerobic, fermentative conditions	Fermentation processes	Arsenic reduction in food processing
26. *Thermodesulfovibrio Yellowstone*	Bacterium	Delta-proteobacteria, thermophilic	Arsenate (As(V)) reduction	Reduction to arsenite (As(III))	Thermophilic, sulfate-rich environments	Hot springs	Arsenic reduction in geothermal environments
27. *Xanthobacter autotrophicus*	Bacterium	Alpha-proteobacteria, autotrophic	Arsenite (As(III)) oxidation	Oxidation to arsenate (As(V))	Aerobic, carbon dioxide fixation	Soil	Bioremediation and environmental cleanup
28. *Nocardioides* sp.	Bacterium	Actinobacteria, soil bacterium	Arsenate (As(V)) reduction	Reduction to arsenite (As(III))	Aerobic and anaerobic	Soil	Arsenic reduction and bioremediation
29. *Streptomyces coelicolor*	Bacterium	Actinobacteria, antibiotic producer	Arsenite (As(III)) oxidation	Oxidation to arsenate (As(V))	Aerobic soil environments	Soil	Role in soil arsenic cycling

4 Microbial Interactions with Geological and Geochemical Processes

Iron and sulfur cycling is of essential importance in the movement and dispersion of arsenic in the environment, especially in groundwater systems. Iron-reducing and sulfur-reducing bacteria play a crucial role in these processes by actively affecting the conversion of minerals that either adsorb or release arsenic. Iron-reducing bacteria, such as species of *Geobacter*, catalyse the reduction of ferric iron [Fe(III)] in iron oxides to ferrous iron [Fe(II)], so inducing the disintegration of iron minerals. This process can liberate previously absorbed arsenic onto iron oxide surfaces, enhancing its mobility and capacity to pollute potable water. Furthermore, sulfur-reducing bacteria, such as *Desulfovibrio* species, catalyse sulfate reduction to sulfide, facilitating the formation of sulfide minerals such as pyrite (FeS_2). The mobility of arsenic is influenced by the adsorption or release of these sulfide minerals, which is contingent upon environmental conditions. The interplay among iron, sulfur, and arsenic cycling is intricate, and the existence of these microorganisms can either augment or impede the movement of arsenic in the subsurface. The impact of iron and sulfur minerals on the movement of arsenic is substantial due to their frequent dual role as both sinks and sources of arsenic. Under oxygen-rich environments, iron oxides such as goethite and hematite bind to arsenic, immobilizing it and inhibiting its mobility (Devi et al. 2023a, b, c; Dey et al. 2023a, b, c; Kumar et al. 2018a, b, c, d, e; 2023a, b, c; Pankaj et al. 2012; Saini et al. 2023). However, when iron-reducing bacteria are active under reducing conditions, these iron oxides dissolve, liberating arsenic into the adjacent water. Similarly, sulfur minerals can capture arsenic by creating arsenic sulfides or facilitate its release when sulfide levels drop or sulfur-reducing bacteria break down sulfide minerals. The continuous and dynamic chemical changes in these minerals, caused by microbial activity, directly affect the behaviour of arsenic in groundwater, influencing its distribution and toxicity. The microbial impact on the sorption and desorption of arsenic is intricately linked to how micromolecules modify mineral surfaces. Specific microorganisms can generate organic acids or other metabolic by-products that alter mineral surfaces' electrical charge and geometry, influencing their capacity to adsorb arsenic. For example, bacterial-generated biofilms can envelop mineral surfaces, obstructing sites for arsenic adsorption or establishing novel microenvironments that promote desorption rates. Microbial metabolism can alter the oxidation state of arsenic, converting arsenate [As(V)] into arsenite [As(III)], which is less strongly bound and hence more mobile, resulting in the reduced adsorption of arsenate onto mineral surfaces. Furthermore, microorganisms can detoxify iron and sulfur minerals, as previously mentioned, facilitating the breakdown of these minerals containing arsenic and releasing arsenic into the underground water. The alterations in mineral surfaces caused by microorganisms and the process of arsenic speciation establish a dynamic system in which arsenic is consistently adsorbed and desorbed depending on microbial activity and environmental conditions. This emphasizes the ongoing impact of these processes on the movement of arsenic. Microbes influence groundwater flow and arsenic distribution beyond chemical processes through biofilm development and changes in hydrological patterns. Biofilms, intricate assemblages of cells encased in a matrix of extracellular polymeric substances (EPS), substantially impact changing hydrological parameters. They can block pores in soil and sediment, decreasing permeability and modifying groundwater flow patterns. Biofilms can also generate confined anaerobic environments, which promote the activity of bacteria that reduce iron and sulfur, exerting additional influence on the speciation and movement of arsenic. The physical impediment resulting from biofilms can cause the diversion of groundwater flow, concentrating arsenic in specific regions and generating concentrated foci of pollution. Furthermore, microbial communities actively modify hydrological patterns by causing alterations in soil structure by synthesizing extracellular polymeric substances (EPS), which bind soil particles together. This phenomenon can significantly decrease the velocity of water flow and generate stagnant areas conducive to the accumulation of arsenic. Overall, the microbial activity-driven cycling of iron and sulfur is crucial in determining the mobility and distribution of arsenic in groundwater systems. Bacteriological reduction and dissolution of iron and sulfur minerals impact the release of arsenic, while alterations in mineral surfaces caused by microbes affect the processes of arsenic sorption and desorption. Thus, the development of biofilms and the dynamics of microbial communities significantly influence groundwater movement, determining the geographical spread of arsenic pollution. A comprehensive understanding of these microbial processes is crucial for effectively controlling arsenic contamination and safeguarding water resources in impacted areas.

5 Geomicrobiological Arsenic Contamination

The contamination of groundwater with arsenic is a pressing worldwide health concern, with the movement and dispersion of arsenic in various environmental conditions heavily influenced by geomicrobiological processes. Two regions that have been thoroughly investigated for the presence of arsenic pollution are the Ganges-Brahmaputra Delta in South Asia and the dry regions of the Western United States. The geomicrobiological dynamics in both areas manifest distinct characteristics that impact the concentrations of arsenic in groundwater, contributing to extensive contamination. The Ganges-Brahmaputra Delta, which traverses portions of India and Bangladesh, is among the most severely affected areas by arsenic contamination globally. Many individuals in these regions depend on groundwater as their primary source of drinking water, and the concentration of arsenic in the groundwater frequently surpasses the established threshold of 10 μg/L set by the World Health Organization (WHO). The contamination is mainly caused by the natural migration of arsenic from sedimentary deposits rather than pollution from industrial activities. Sediments deposited by rivers over thousands of years contain arsenic, which, under specific circumstances, can be mobilized into groundwater, resulting in extensive arsenic poisoning among local communities. Geomicrobiological investigations conducted in the Ganges-Brahmaputra Delta have uncovered the crucial significance of microbial processes in the movement of arsenic (Kumar et al. 2011, 2017,

2023a, b, c, 2024a, b, c, d, e, f, g, h, i, j, k; Kumar, Sharma et al. 2021c; Kumar and Dwivedi 2018; Naik and Kumar 2020). Iron oxide-rich sediments in the delta effectively condense arsenic, rendering it immobile. Iron-reducing bacteria, including *Geobacter* and *Shewanella* species, catalyse the reduction of ferric iron [Fe(III)] to ferrous iron [Fe(II)], resulting in the fragmentation of iron oxides and the liberation of arsenic into groundwater. Furthermore, the sediment contains sulfate-reducing bacteria, which, by their metabolic processes, can either capture arsenic by producing sulfide minerals or release it, depending on the specific geochemical conditions in the area. Various factors, including organic matter, influence these microbial processes to support microbial metabolism and the subsurface environment's redox state. Identifying these microbial processes has redirected the attention of remediation endeavours towards regulating microbial activity and the redox conditions in groundwater to alleviate arsenic pollution. Arsenic pollution in groundwater is a significant environmental issue in the dry areas of the Western United States, especially in states such as Nevada, Arizona, and California. In contrast to the Ganges-Brahmaputra Delta, where arsenic is released to alleviate conditions, arsenic pollution in the Western United States is typically linked to oxidizing conditions in volcanic and geothermal regions and sedimentary basins. The groundwater in these areas predominantly contains elevated levels of dissolved arsenic due to the erosion of rocks abundant in arsenic, volcanic deposits, and geothermal processes. These sources discharge arsenic into the water in the presence of natural, oxidizing conditions. Water scarcity worsens the issue in arid areas since the high evaporation rates effectively accumulate arsenic in the groundwater. Geomicrobiological processes at the local level in the Western United States also impact the movement of arsenic. Microorganisms in certain regions exert an influence on the speciation of arsenic by converting arsenate [As(V)] into arsenite [As(III)] by biochemical respiration. Of its higher toxicity and mobility, arsenite presents a more significant threat of contamination. Geothermal springs and volcanic communities frequently support thermophilic and chemolithotrophic microorganisms that flourish in high-temperature conditions, impacting the process of arsenic cycling via redox changes. An example is when sulfur-oxidizing bacteria generate acidic conditions that facilitate the breakdown of arsenic minerals, liberating arsenic into the groundwater. Conversely, iron-reducing bacteria can potentially cause the precipitation of arsenic by facilitating the creation of iron-arsenic complexes under certain circumstances. A thorough knowledge of geomicrobiology is necessary to effectively control arsenic pollution in arid environments, as these microbial processes play a significant role in determining the distribution of arsenic. Geomorphology, hydrology, and microbiology significantly influence the magnitude and intensity of arsenic pollution in groundwater in both the Ganges-Brahmaputra Delta and the Western United States. Microbial organisms, especially those engaged in iron and sulfur cycling, play a crucial role in mobilizing or sequestering arsenic, depending on the area's redox conditions and mineral composition. Despite variations in environmental conditions, understanding geomicrobiology is essential for devising efficient remediation approaches in both areas. An investigation into the microbial mechanisms that impact the movement of arsenic presents promising strategies for reducing pollution and safeguarding public health in impacted regions (Choudhury et al. 2024;

Devi and Kumar 2023a; Jagadesh et al. 2024; Kumar et al. 2019; Kumar and Raut 2024; Sharma and Kumar 2023, 2024; Srivastav et al. 2024).

6 Bioremediation of Arsenic-Contaminated Groundwater

Bioremediation efficiently reduces arsenic contamination in groundwater and soil by using naturally occurring or engineered microorganisms to break down, detoxify, or immobilize contaminants. The core concept of bioremediation is to use microbial metabolic processes to convert arsenic from toxic and mobile forms, such as arsenite [As(III)], into less pathogenic and less mobile forms, such as arsenate [As(V)], or to facilitate the sequestration of arsenic in stable mineral phases. Biologic remediation techniques are categorized into in situ and ex situ approaches, each possessing distinct benefits and constraints. Microbiological methods offer distinct benefits because of their environmental sustainability, cost-efficiency, and capacity to selectively target certain pollutants without significant chemical processes. Nevertheless, constraints arise from the reliance on environmental factors such as pH, temperature, and redox potential, which can impede microbial metabolic activity or the effectiveness of bioremediation endeavours. Moreover, the sluggish rate of microbial dynamics and possible difficulties in managing or expanding microbial populations pose significant obstacles. In situ, bioremediation mechanisms are implemented directly at the contamination site without excavation or removal of contaminated materials, minimizing disruption and increasing cost-effectiveness for large-scale contamination (Dwivedi and Kumar 2011; P. Kumar et al. 2011; P. Kumar, Choudhury et al. 2023a; P. Kumar, Goud, Devi, Dey et al. 2022a; P. Kumar, Yumnam, Kumar, Misao et al. 2018e; P. Kumar and Dwivedi 2018; Naik and Kumar 2020; Saini et al. 2023). The enhancement of groundwater microbial arsenic reduction or oxidation is crucially dependent on biostimulation and bioaugmentation. Biostimulation is the introduction of nutrients or electron acceptors to augment the development and functionality of indigenous microorganisms capable of either oxidizing arsenite to arsenate, which can be more easily absorbed onto mineral surfaces, or reducing arsenate to arsenite for storage in arsenic-sulfide minerals. Bioaugmentation is the implementation of targeted bacterial strains, such as iron-reducing or sulfur-reducing species, which exhibit high efficiency in the conversion of arsenic within specific environmental settings. Efficacious in situ bioremediation initiatives have been implemented in regions such as Bangladesh, where the injection of organic substrates like molasses into the subsurface has effectively triggered microbial activity to decrease the concentration of arsenic in groundwater. Furthermore, acetate biostimulation has been used in the Western United States to augment iron-reducing bacteria and immobilize arsenic by adsorbing it onto iron oxides. Ex situ bioremediation methods entail extracting polluted water or soil from its initial site for targeted treatment under controlled circumstances. This approach enables more precise manipulation of environmental parameters that impact microbial activity, possibly accelerating remediation rates. Microbial systems, such as constructed wetlands or bioreactors, are commonly employed to treat water contaminated with arsenic. These systems generate ideal conditions that enable microorganisms to eliminate arsenic using bioaccumulation, precipitation, or oxidation-reduction processes. One possible approach is to develop microbial bioreactors that facilitate arsenite oxidation to arsenate and then employ filtration to eliminate the arsenate waste. Nevertheless, ex-situ bioremediation encounters obstacles such as exorbitant operational expenses, logistical complexities associated with the transportation of polluted substances, and the requirement for ongoing surveillance and upkeep of microbial metabolism in artificial systems. The potential of emerging bioremediation technologies, such as genetic engineering and synthetic biology, to improve the effectiveness and selectivity of microbial systems for arsenic remediation is considerable. Genomic engineering methods can be employed to alter microorganisms to increase the expression of genes associated with arsenic oxidation, reduction, or sequestration. This, in turn, enhances their ability to detoxify arsenic in different environmental circumstances. Arsenite oxidase genes can be introduced into bacteria to enhance their capacity to facilitate arsenite oxidation to arsenate in oxidative conditions. Likewise, it is possible to modify sulfate-reducing bacteria to increase their ability to produce stable arsenic-sulfide minerals under favourable reducing conditions. Synthetic biology can create novel metabolic pathways that integrate arsenic detoxification with other biogeochemical processes, enhancing the resilience and adaptability of microbial systems to dynamic environmental contexts. Although currently in the experimental phase, these technologies are promising as a future bioremediation strategy. They can potentially address the constraints of conventional microbial methods and provide more accurate regulation of arsenic biogeochemical cycles. Bioremediation generally provides a potent array of methods for dealing with arsenic pollution, in which microbial processes exert a significant and noteworthy influence. In situ techniques provide a cost-efficient and minimally intrusive approach, whereas ex situ methods enhance control and efficiency in particular situations. Technological progress in genetic engineering and synthetic biology is expected to improve the ability of microorganisms to break down arsenic, making bioremediation a more feasible approach for controlling arsenic pollution in both natural and artificial settings (Devi et al. 2024a, b, c, d, e, f; Dey, Devi, et al. 2023a; P. Kumar et al. 2024a, b, c, d, e, f, g, h, i, j, k; P. Kumar, Latha et al. 2023b; P. Kumar and Dey 2024; P. Kumar and Singh 2024e; Pathak et al. 2017; K. Sharma and Kumar 2023).

7 Challenges and Future Directions

Microbial arsenic transformations influence the mobility and toxicity of arsenic in natural environments. Notwithstanding substantial progress, there are still considerable deficiencies in our comprehension of these changes. A primary constraint is the inadequate identification and characterization of microbial species capable of transforming arsenic. Although arsenate-reducing and arsenite-oxidizing bacteria are well-documented, the range of microbial communities capable of these chemical reactions has not been thoroughly investigated. Numerous investigations depend on laboratory-based models, which frequently fall short of capturing the intricacy of natural ecosystems where several microbial interactions can impact the metabolism of arsenic. Furthermore, the scientific understanding of the involvement of anaerobic microorganisms in arsenic conversion, especially in underground settings, is limited because of challenges in replicating these conditions through experimental means. One notable area of insufficient understanding is microorganisms' biochemical mechanisms to convert arsenic. For instance, although the processes of detoxifying arsenic are extensively researched, there is limited understanding of how microorganisms carry out arsenic metabolism for energy generation in the presence of environmental stressors like fluctuating pH or salinity. The arsenic cycling will likely involve numerous unidentified microbial pathways and biochemical processes. The present study centres on two primary processes: the reduction of arsenate and the oxidation of arsenite. Unexplored pathways, such as arsenic methylation and demethylation, may be crucial in arsenic's movement, toxicity, and sequestration. Furthermore, the interactions between microorganisms that process arsenic and other microbial communities in soil or sediment ecosystems have not been thoroughly investigated. Such interactions can potentially impact the rates and routes of arsenic transformation. Gaining insight into the interspecies communication processes and the function of microbial consortia in arsenic metabolism is crucial for a comprehensive understanding of the cycling of arsenic in different ecosystems. These gaps impede our capacity to forecast the movement of arsenic in various environmental conditions, presenting difficulties in developing appropriate remediation measures. An inherent technological obstacle in geomicrobiology is the complexity of investigating subsurface living populations. Microbial communities play a crucial role in the cycling of arsenic in subsurface aquifers, where most arsenic contamination occurs. The complexity and limited accessibility of these environments pose challenges in sampling and analysing microbial populations and activities. Conventional microbiological methods, such as culturing, are often insufficient for investigating subsurface communities because of the challenge of reproducing real-life conditions such as low oxygen, pressure, and temperature gradients. Furthermore, subsurface microorganisms' sluggish growth rates and metabolic activities challenge their containment within laboratory environments. The emergence of sophisticated molecular methodologies, such as metagenomics and single-cell genomics, has facilitated examining microbial variety in numerous ecological settings. However, these approaches still encounter constraints in connecting microbial activity to particular environmental conditions. An important obstacle in geomicrobiology is the direct correlation between microbial activity and geochemical measures. Although metagenomic and proteomic techniques can detect microbial species and identify their metabolic capacity, linking this information to real-time biogeochemical processes is still challenging. Microbiological processes often operate at temporal and spatial scales significantly smaller than those usually encompassed by geochemical measurements. Moreover, the intricate interplay among microbial communities, minerals, and organic matter in subsurface environments introduces additional complexity. Identifying microbial genes and enzymes associated with arsenic transformations does not provide a direct connection to the wider environmental factors that regulate the movement of arsenic. The influence of geochemical gradients, particularly those related to redox conditions, on microbial arsenic transformations is substantial. However, accurately quantifying these gradients at the precise level of microbial activity is often challenging. Focused research efforts will be necessary to address the knowledge gaps in microbial arsenic transformations and overcome technological challenges. Key future objectives include thoroughly characterizing the microbial communities engaged in arsenic cycling, particularly in subsurface settings. Enhancing methodologies for real-time quantification of microbial activity, such as microfluidics or bioelectrochemical sensors, may yield more precise evaluations of microbial involvement in arsenic transformations. Furthermore, it is crucial to prioritize integrating molecular, geochemical, and ecological data to develop more comprehensive models of arsenic cycling. The models must consider the multi-scale interactions between microorganisms and environmental variables. Advancements in the field of geomicrobiology, especially in the understanding and mitigation of arsenic contamination, will heavily rely on interdisciplinary research. Collaborations among microbiologists, geochemists, hydrologists, and environmental engineers can potentially enhance the efficacy of arsenic bioremediation methods. One potential application of combining microbial genomics with geochemical modelling is the ability to forecast the effects of alterations in land use, climate, or water management practices on the movement of arsenic in groundwater systems. Advancements in environmental sensing and data analytics can enable real-time monitoring of microbial and geochemical processes, enhancing the adaptive management of polluted environments. Through the integration of microbial ecology and geochemistry, interdisciplinary research has the potential to revolutionize the management of sites contaminated with arsenic.

8 Conclusion

The present chapter has examined the pivotal significance of geomicrobiology in comprehending the process of arsenic cycling and its consequences for environmental systems, namely groundwater management. A crucial discovery is that microbial processes substantially impact arsenic's movement, distribution, and harmful effects in natural settings. The transformation of arsenic by microorganisms, namely bacteria, through reduction, oxidation, methylation, and demethylation significantly influences the movement of arsenic in soils, sediments, and deep-water sources. Arsenic-reducing microorganisms, for instance, promote the transformation of less toxic arsenate (As(V)) into the more mobile and toxic arsenite (As(III)), influencing groundwater contamination dynamics. In addition, arsenic-oxidizing microorganisms can reverse this process by immobilizing arsenic under specific circumstances. However, their functioning in subsurface environments is not well understood. Identifying several knowledge gaps has revealed a restricted comprehension of the variety and ecological functions of microorganisms engaged in arsenic conversion, particularly in anaerobic and subsurface settings. The intricate nature of microbial communities and their interplay with other environmental parameters, such as redox potential, pH, and organic matter, poses difficulties in forecasting the behaviour of arsenic in natural systems. Moreover, although considerable progress has been achieved by employing molecular analysis to detect microbial genes and enzymes associated with arsenic conversion, difficulties remain in connecting these discoveries to actual biogeochemical processes. Significant technological obstacles have been emphasized in the investigation of subsurface microbial ecology. Accessing and reproducing in-situ conditions in laboratory experiments is challenging, and the sluggish growth rates of many subsurface microorganisms further complicate research endeavours. In conclusion, the chapter underscores the need to promote interdisciplinary research to enhance our understanding of arsenic cycling and devise efficient remediation methods. Analysis of geomicrobiological factors has great potential to enhance groundwater management strategies, especially in areas affected by arsenic pollution. A comprehensive understanding of the impact of microbial communities on arsenic mobility and speciation can enable environmental managers to devise more efficient approaches for the prevention and mitigation of groundwater contamination. An immediate consequence is the potential for bioremediation approaches that exploit microbial processes to immobilize or detoxify arsenic. As an illustration, activating arsenic-oxidizing bacteria in polluted aquifers might facilitate the transformation of arsenite into arsenate. This state is more prone to precipitation and immobilization under specific environmental circumstances. Similarly, controlling the redox conditions in polluted areas to suppress arsenic-reducing microorganisms could restrict the production of hazardous, mobile arsenite. Furthermore, incorporating geomicrobiological knowledge into predictive models of arsenic behaviour can augment the precision of risk evaluations and enhance decision-making in managing groundwater resources. This could be especially advantageous in developing early warning systems for arsenic contamination in regions where microbial activity is vital in disseminating arsenic. Consequently, interventions could be more finely directed, reducing superfluous remediation efforts and concentrating on microbial processes that impact arsenic cycling the most. Geomicrobiological analyses can provide valuable information for implementing sustainable water management strategies in areas where human activities, such as agriculture or mining, worsen groundwater pollution. By comprehending the microbial mechanisms that control the release of arsenic from sediments, managers can adopt land-use strategies that minimize disturbances to these processes, thus decreasing the likelihood of contamination. Ensuring the incorporation of geomicrobiology into environmental management and research is essential for effectively tackling the worldwide issues posed by arsenic pollution. Addressing arsenic pollution becomes considerably more pressing, given the world's increasing dependence on groundwater for drinking water, agriculture, and industry. Conventional methods for managing groundwater, particularly those that emphasize physical and chemical processes, are inadequate for dealing with the intricate nature of arsenic cycling. Given the crucial role that microbiological processes play in determining the fate of arsenic in the environment, it is necessary to include geomicrobiological insights into scientific research and practical management strategies. Specifically, the need for interdisciplinary partnerships cannot be overstated in enhancing our knowledge of arsenic pollution and formulating effective remedies. Integrating microbiologists, geochemists, hydrologists, and engineers will facilitate the development of novel methodologies that connect microbial activity with comprehensive environmental processes. The collaboration has the potential to generate innovative bioremediation techniques, enhance predictive models, and promote more sustainable groundwater management practices. The overwhelming magnitude of arsenic pollution worldwide, especially in Southeast Asia, highlights the necessity for comprehensive strategies that include geomicrobiology. To enhance the protection of crucial water resources and ensure public health, it is necessary to pursue further research in microbial arsenic transformations and implement these findings to address practical environmental issues. The incorporation of geomicrobiology into groundwater management has the potential to fundamentally revolutionize our approach to one of the most urgent global environmental pollution challenges.

References

Aley P, Singh J, Kumar P (2022) Adapting the changing environment: microbial way of life. In: Kumar A, Singh J, Ferreira LFRBT-MUCC (eds) Microbiome under changing climate. Woodhead Publishing, pp 507–525. https://doi.org/10.1016/B978-0-323-90571-8.00023-7

Chakraborty S, Kumar P, Sanyal R, Mane AB, Arvind Prasanth D, Patil M, Dey A (2021) Unravelling the regulatory role of miRNAs in secondary metabolite production in medicinal crops. Plant Gene 27:100303. https://doi.org/10.1016/j.plgene.2021.100303

Choudhury D, Kumar P, Zhimo VY, Sahoo J (2024) Crop rotation patterns and soil health management. In: Bioremediation of emerging contaminants from soils: soil health conservation for improved ecology and food security. Elsevier, pp 565–589. https://doi.org/10.1016/B978-0-443-13993-2.00025-6

Das T, Saha SC, Sunita K, Majumder M, Ghorai M, Mane AB, Prasanth DA, Kumar P, Pandey DK, Al-Tawaha AR, Batiha GE-S, Shekhawat MS, Ghosh A, Sharifi-Rad J, Dey A (2022) Promising botanical-derived monoamine oxidase (MAO) inhibitors: pharmacological aspects and structure-activity studies. S Afr J Bot 146:127–145. https://doi.org/10.1016/j.sajb.2021.09.019

Devi P, Kumar P (2023a) Effect of microbial oxidative enzymes on soil health. In: Microbial oxidative enzymes: biotechnological applications. De Gruyter, pp 37–61. https://doi.org/10.1515/9783111062235-003

Devi P, Kumar P (2023b) Role of oxidative enzymes in the alleviation of oxidative stress in plant system. In: Microbial oxidative enzymes: biotechnological applications. De Gruyter, pp 109–128. https://doi.org/10.1515/9783111062235-006

Devi P, Kumar P (2024a) Challenges in the storage of herbal medicine products and strategies for sustainable management. In: Reference series in Phytochemistry: Vol. Part F3164. Springer Science and Business Media B.V, pp 1739–1767. https://doi.org/10.1007/978-3-031-43199-9_37

Devi P, Kumar P (2024b) Herbal medicine and pregnancy. In: Reference series in phytochemistry: Vol. Part F3164. Springer Science and Business Media B.V, pp 693–722. https://doi.org/10.1007/978-3-031-43199-9_25

Devi P, Dey SR, Kumar P (2023a) Integration of water resources management in rural areas. In: Water resources management for rural development: challenges and mitigation. Elsevier, pp 139–152. https://doi.org/10.1016/B978-0-443-18778-0.00013-1

Devi P, Dey SR, Saini L, Kumar P, Panigrahi S, Dwivedi P (2023b) Toward sustainable agriculture: strategies involving phytoprotectants against reactive oxygen species. In: Reactive oxygen species: prospects in plant metabolism. Springer Nature, pp 229–247. https://doi.org/10.1007/978-981-19-9794-5_13

Devi P, Kumar P, Sharma K, Singh J (2023c) Is there a relationship between myconanotechnology and sustainable development? In: Kuddus M, Ahmad IZ, C. M. B. T.-M. and A. of N. in

B. Hussain (eds) Progress in biochemistry and biotechnology. Academic Press, pp 175–207. https://doi.org/10.1016/B978-0-443-15262-7.00009-7
Devi P, Dey SR, Sharma K, Kumar P, Singh J (2024a) Waste management as an opportunity for the inclusion of vulnerable groups. In: Waste management in the circular economy. Springer International Publishing, pp 207–230. https://doi.org/10.1007/978-3-031-42426-7_11
Devi P, Dogra P, Kumar P, Singh J (2024b) Global status of biomass energy programmes-challenges and roadmap. In: Biomass energy for sustainable development. CRC Press, pp 394–414. https://doi.org/10.1201/9781003406501-20
Devi P, Kapoor P, Kumar P (2024c) Biomarkers for toxicity detection. In: Biomarkers in environmental and human health biomonitoring: an integrated perspective. Elsevier, pp 313–329. https://doi.org/10.1016/B978-0-443-13860-7.00004-5
Devi P, Kaur A, Kumar P (2024d) Phytoremediation of metal contaminated soil using energy crops: soil health maintenance along with biofuel production. In: Bioremediation of emerging contaminants from soils: soil health conservation for improved ecology and food security. Elsevier, pp 307–333. https://doi.org/10.1016/B978-0-443-13993-2.00015-3
Devi P, Kumar P, Singh J (2024e) Application of potential microbes in bioremediation. In: Development in waste water treatment research and processes: role of environmental microbiology in industrial wastewater research. Elsevier, pp 201–225. https://doi.org/10.1016/B978-0-443-13609-2.00021-5
Devi P, Kumar P, Singh J (2024f) Biotechnological approaches as clean technology for the development of a sustainable environment. In: Development in waste water treatment research and processes: role of environmental microbiology in industrial wastewater research. Elsevier, pp 381–403. https://doi.org/10.1016/B978-0-443-13609-2.00016-1
Dey SR, Devi P, Kumar P (2023a) Organic contaminants in aquatic environments: sources and impact assessment. In: Metal organic frameworks for wastewater contaminant removal. Wiley, pp 299–317. https://www.scopus.com/inward/record.uri?eid=2-s2.0-85174103623&partnerID=40&md5=dbfb41dc5474b67aaaca5dec64c73718
Dey SR, Kumar P, Singh J (2023b) Growth of mustard under open field conditions with salicylic acid and antimicrobial thiourea. Biopesticides International 19(2):183–192. https://doi.org/10.59467/BI.2023.19.183
Dey SR, Sharma M, Kumar P (2023c) Effects and responses of chromium on plants. In: Environmental science and engineering: Vol. Part F1975. Springer Science and Business Media Deutschland GmbH, pp 385–427. https://doi.org/10.1007/978-3-031-44029-8_14
Dey SR, Sharma M, Kumar P (2024) Phytoremediation of phenolic compounds from soil. In: Bioremediation of emerging contaminants from soils: soil health conservation for improved ecology and food security. Elsevier, pp 335–358. https://doi.org/10.1016/B978-0-443-13993-2.00016-5
Dwivedi P, Kumar P (2011) Anti-diabetic medicinal plants and their conservation: waging green war on diabetes. Medicinal Plants 3(3):181–189. https://doi.org/10.5958/j.0975-4261.3.3.031
Goud EL, Singh J, Kumar P (2022) Climate change and their impact on global food production. In: Microbiome under changing climate. Elsevier, pp 415–436
Hidangmayum A, Dwivedi P, Kumar P, Upadhyay SK (2023) Seed priming and foliar application of chitosan ameliorate drought stress responses in mungbean genotypes through modulation of morpho-physiological attributes and increased antioxidative defense mechanism. J Plant Growth Regul 42(10):6137–6154. https://doi.org/10.1007/s00344-022-10792-1
Jagadesh M, Dash M, Kumari A, Singh SK, Verma KK, Kumar P, Bhatt R, Sharma SK (2024) Revealing the hidden world of soil microbes: metagenomic insights into plant, bacteria, and fungi interactions for sustainable agriculture and ecosystem restoration. Microbiol Res 285:127764. https://doi.org/10.1016/j.micres.2024.127764
Karam N, Kumar P, Anand YR, Choudhury D (2023) Management of rice blast—a review of the strategies available. Biopesticides International 19(2):107–116. https://doi.org/10.59467/BI.2023.19.107
Kotia A, Rutu P, Singh V, Kumar A, Dhoke S, Kumar P, Singh DK (2021) Rheological analysis of rice husk-starch suspended in water for sustainable agriculture application. In: 2nd interna-

tional conference on functional materials, manufacturing and performances, ICFMMP 2021, vol 50, pp 1962–1966. https://doi.org/10.1016/j.matpr.2021.09.325
Kumar P (2023) Degradation of xenobiotics by bacteria and fungi: an overview. Biopesticides International 19(2):117–132. https://doi.org/10.59467/BI.2023.19.117
Kumar P (2024a) Application of marine biomass for carbon neutrality, sustainable environment, and economy. In: Marine biomass: biorefinery, bioproducts and environmental bioremediation. De Gruyter, pp 95–128. https://doi.org/10.1515/9783111353951-005
Kumar P (2024b) Biochar production from marine algae and its application in the treatment of industrial effluent. In: Marine biomass: biorefinery, bioproducts and environmental bioremediation. De Gruyter, pp 57–93. https://doi.org/10.1515/9783111353951-004
Kumar P (2024c) Soil conservation for global sustainability. In: World sustainability series: Vol. Part F2887. Springer Science and Business Media Deutschland GmbH, pp 103–128. https://doi.org/10.1007/978-3-031-57456-6_6
Kumar P, Choudhury D (2024a) Hydroponic removal of organic contaminants from water. In: Springer water: Vol. Part F2309. Springer Nature, pp 143–164. https://doi.org/10.1007/978-3-031-53258-0_6
Kumar P, Choudhury D (2024b) Microbial native soil bacteria against cadmium toxicity. In: Springer water: Vol. Part F2532. Springer Nature, pp 193–226. https://doi.org/10.1007/978-3-031-54005-9_9
Kumar P, Choudhury D (2024c) Optimized phytoremediation process for the sustainable management radionuclides. In: Bioremediation of emerging contaminants from soils: soil health conservation for improved ecology and food security. Elsevier, pp 443–464. https://doi.org/10.1016/B978-0-443-13993-2.00020-7
Kumar P, Choudhury D (2024d) Role of indigenous knowledge in agricultural soil reclamation without disturbing other ecosystems. In: Bioremediation of emerging contaminants from soils: soil health conservation for improved ecology and food security. Elsevier, pp 465–488. https://doi.org/10.1016/B978-0-443-13993-2.00021-9
Kumar P, Choudhury D (2024e) Toxicity of rhizospheric cadmium contaminated soil and its phytoremediation. In: Springer water: Vol. Part F2532. Springer Nature, pp 227–264. https://doi.org/10.1007/978-3-031-54005-9_10
Kumar P, Dey SR (2024) Hydroponics phytoremediation: an overview. In: Springer water: Vol. Part F2309. Springer Nature, pp 361–396. https://doi.org/10.1007/978-3-031-53258-0_13
Kumar P, Dwivedi P (2018) Putrescine and glomus mycorrhiza moderate cadmium actuated stress reactions in zea mays l. By means of extraordinary reference to sugar and protein. Vegetos 31(3):74–77. https://doi.org/10.5958/2229-4473.2018.00076.9
Kumar P, Dwivedi P (2020) Lignin estimation in sorghum leaves grown under hazardous waste site. Plant Arch 20:2558–2561
Kumar VA, Kumar P (2023) Revolutionizing agriculture with satellite technology for farmers: a review. Biopestic Int 19(2):97–106. https://doi.org/10.59467/BI.2023.19.97
Kumar P, Mistri TK (2020) Transcription factors in SOX family: potent regulators for cancer initiation and development in the human body. Semin Cancer Biol 67:105–113. https://doi.org/10.1016/j.semcancer.2019.06.016
Kumar P, Naik M (2020) Biotic symbiosis and plant growth regulators as a strategy against cadmium and lead stress in chickpea. Plant Arch 20:2495–2500
Kumar P, Pathak S (2019) Responsiveness index of sorghum (Sorghum bicolor (L.) Moench) grown under cadmium contaminated soil treated with putrescine and mycorrhiza. Bangladesh J Bot 48(1):139–143
Kumar P, Raut AM (2024) Microbes-assisted bioaugmentation process in the reduction of emerging industrial pollutants from soil. In: Bioremediation of emerging contaminants from soils: soil health conservation for improved ecology and food security. Elsevier, pp 519–540. https://doi.org/10.1016/B978-0-443-13993-2.00023-2
Kumar P, Singh J (2024a) Bioprocesses in industrial wastewater treatment: trends and prospects. In: Trends in biological processes in industrial wastewater treatment. Institute of Physics Publishing, pp 15.1–15.50. https://doi.org/10.1088/978-0-7503-5678-7ch15

Kumar P, Singh J (2024b) Emerging contaminants and ways to reduce them. In: Trends in biological processes in industrial wastewater treatment. Institute of Physics Publishing, pp 4.1–4.34. https://doi.org/10.1088/978-0-7503-5678-7ch4

Kumar P, Singh J (2024c) Environmental and health risk assessment of chemical pollutants in drinking water and wastewaters. In: Trends in biological processes in industrial wastewater treatment. Institute of Physics Publishing, pp 5.1–5.30. https://doi.org/10.1088/978-0-7503-5678-7ch5

Kumar P, Singh J (2024d) Exploring enzyme biotechnology's role in green chemistry and advancements in pharmaceutical technologies. In: Enzyme biotechnology for environmental sustainability. Elsevier, pp 465–495. https://doi.org/10.1016/B978-0-443-22072-2.00015-2

Kumar P, Singh J (2024e) Hydroponic: an eco-friendly future. In: Springer Water: Vol. Part F2309. Springer Nature, pp 223–267. https://doi.org/10.1007/978-3-031-53258-0_9

Kumar P, Mandal B, Dwivedi P (2011) Heavy metal scavenging capacity of Mentha spicata and Allium cepa. Medicinal Plants 3(4):315–318. https://doi.org/10.5958/j.0975-4261.3.4.053

Kumar P, Singh BN, Dwivedi P (2017) Plant growth regulators, plant adaptability and plant productivity: A review on abscisic acid (ABA) signaling in plants under emerging environmental stresses. In: Sustaining future food security in changing environments. Nova Science Publishers, Inc, pp 81–97

Kumar P, Harshavardhan M, Kumar PS, Yumnam J, Jyoti N, Naik M, Misao L, Purnima, Kumar S (2018a) Effect on chlorophyll a/b ratio in cadmium contaminated maize leaves treated with putrescine and mycorrhiza. Ann Biol 34(3):281–283

Kumar P, Kumar S, Harshavardhan M, Naik M, Yumnam J, Kumar PS, Jyoti N, Misao L, Purnima. (2018b) Evaluation of plant height and leaf length of sorghum grown under different sources of nutrition. Ann Biol 34(3):284–286

Kumar P, Pathak S, Amarnath KS, Veerendra Brahma Teja P, Dileep B, Kumar K, Singh M, Siddique A (2018c) Effect of growth regulator on morpho-physiological attributes of chilli: a case study. Plant Arch 18(2):1771–1776

Kumar P, Pathak S, Kumar M, Dwivedi P (2018d) Role of secondary metabolites for the mitigation of cadmium toxicity in sorghum grown under mycorrhizal inoculated hazardous waste site. In: Biotechnological approaches for medicinal and aromatic plants: conservation, genetic improvement and utilization. Springer, Singapore, pp 199–212

Kumar P, Yumnam J, Kumar PS, Misao L, Jyoti N, Naik M, Purnima, Kumar S, Harshavardhan M (2018e) Cadmium induced changes in germination of maize seed treated with mycorrhiza. Ann Agri Bio Res 23(2):169–170

Kumar D, Rameshwar SD, Kumar P (2019) Effect of intergated application of inorganic and organic fertilizers on the roots of chickpea. Plant Archives 19(1):857–860

Kumar P, Kumar T, Singh S, Tuteja N, Prasad R, Singh J (2020) Potassium: A key modulator for cell homeostasis. J Biotechnol 324:198–210

Kumar P, Devi P, Dey SR (2021a) Fungal volatile compounds: a source of novel in plant protection agents. In: Volatiles and metabolites of microbes. Elsevier, pp 83–104

Kumar V, Dwivedi P, Kumar P, Singh BN, Pandey DK, Kumar V, Bose B (2021b) Mitigation of heat stress responses in crops using nitrate primed seeds. S Afr J Bot 140:25–36. https://doi.org/10.1016/j.sajb.2021.03.024

Kumar P, Sharma K, Saini L, Dey SR (2021c) Chapter 8 – Role and behavior of microbial volatile organic compounds in mitigating stress. In: Kumar A, Singh J, J. B. T.-V. and M. of M. Samuel (eds) Volatiles and metabolites of microbes. Academic Press, pp 143–161. https://doi.org/10.1016/B978-0-12-824523-1.00010-9

Kumar P, Goud EL, Devi P, Dey SR, Dwivedi P (2022a) Heavy metals: transport in plants and their physiological and toxicological effects. In: Plant metal and metalloid transporters. Springer Nature, pp 23–54. https://doi.org/10.1007/978-981-19-6103-8_2

Kumar P, Goud EL, Devi P, Koul B (2022b) Metal pollutants in the environment. In: Environmental microbiology: emerging technologies. De Gruyter, pp 291–323. https://doi.org/10.1515/9783110727227-012

Kumar P, Koul B, Sharma M (2022c) Phytoremediation of heavy metals. In: Heavy metals in plants: physiological to molecular approach. CRC Press, pp 369–388. https://doi.org/10.1201/9781003110576-17

Kumar P, Choudhury D, Sahoo J, Devi MI (2023a) Exploring the potential for extended investigation: a preliminary examination of Liriomyza trifolii and its bibliometric foundation for analytical pursuits. J Food Chem Nanotechnol 9(Special Issue 1):S359–S366. https://doi.org/10.17756/jfcn.2023-s1-045

Kumar P, Latha DH, Singh J (2023b) Rhizoremediation of organic emerging soil contaminants: green technology. In: Clean technologies toward a sustainable future: physicochemical, biochemical and biotechnological approaches. IWA Publishing, pp 211–229. https://doi.org/10.2166/9781789063783_0211

Kumar P, Saini L, Sharma M (2023c) Assessment of the phytochemical constituents and metabolites in the medicinal plants and herbal medicine used in the treatment and management of respiratory diseases. In: Reference series in Phytochemistry, vol 2023. Springer Science and Business Media B.V. https://doi.org/10.1007/978-3-031-21973-3_19-1

Kumar P, Choudhury D, Dey SR, Sharma M (2024a) Application of genetically modified crops against metallic contaminations. In: Bioremediation of emerging contaminants from soils: soil health conservation for improved ecology and food security. Elsevier, pp 541–564. https://doi.org/10.1016/B978-0-443-13993-2.00024-4

Kumar P, Dey SR, Choudhury D (2024b) Effectiveness of cadmium on biochemical shift of pea plant treated with mycorrhiza and putrescine. Nat Environ Pollut Technol 23(1):275–285. https://doi.org/10.46488/NEPT.2024.v23i01.023

Kumar P, Dwivedi P, Upadhyay SK (2024c) Optimization of polyamine and mycorrhiza in sorghum plant for removal of hazardous cadmium. Plant Physiol Biochem 214:108846. https://doi.org/10.1016/j.plaphy.2024.108846

Kumar P, Karam N, Choudhury D (2024d) Assessment of the phytochemical constituents and metabolites in the medicinal plants and herbal medicine used in the treatment and management of skin diseases. In: Reference series in phytochemistry: Vol. Part F3164. Springer Science and Business Media B.V, pp 533–573. https://doi.org/10.1007/978-3-031-43199-9_21

Kumar P, Khokhar A, Dey SR (2024e) Evidence from the use of herbal medicines in the management and prevention of common eye diseases. In: Reference series in phytochemistry: Vol. Part F3164. Springer Science and Business Media B.V, pp 763–795. https://doi.org/10.1007/978-3-031-43199-9_27

Kumar P, Saini L, Sharma M (2024f) Assessment of the phytochemical constituents and metabolites in the medicinal plants and herbal medicine used in the treatment and management of respiratory diseases. In: Reference series in phytochemistry: Vol. Part F3164. Springer Science and Business Media B.V, pp 481–510. https://doi.org/10.1007/978-3-031-43199-9_19

Kumar P, Saurabh S, Sharma K (2024g) Assessment of the phytochemical constituents and metabolites in medicinal plants and herbal remedies used in the treatment and management of reproductive diseases: polycystic ovary syndrome. In: Reference series in Phytochemistry: Vol. Part F3164. Springer Science and Business Media B.V, pp 511–532. https://doi.org/10.1007/978-3-031-43199-9_20

Kumar P, Singh J, Singh G (2024h) Renewable hydrogen opportunities and challenges. In: Renewable hydrogen: opportunities and challenges in commercial success. Elsevier, pp 193–220. https://doi.org/10.1016/B978-0-323-95379-5.00010-9

Kumar P, Srivastav AL, Chaudhary V, van Hullebusch ED, Busquets R (2024i) Bioremediation of emerging contaminants from soils: soil health conservation for improved ecology and food security. In: Bioremediation of emerging contaminants from soils: soil health conservation for improved ecology and food security. Elsevier. https://doi.org/10.1016/C2022-0-02758-6

Kumar P, Sharma K, Singh J (2024j) Analysis of complex microbial communities in soil and wastewater treatment processes. In: Microbial metagenomics in effluent treatment plant. Elsevier, pp 25–41. https://doi.org/10.1016/B978-0-443-13531-6.00002-1

Kumar P, Sharma K, Singh J (2024k) Response of microbial community to environment changes. In: Microbial metagenomics in effluent treatment plant. Elsevier, pp 43–61. https://doi.org/10.1016/B978-0-443-13531-6.00006-9

Kumari P, Singh J, Kumar P (2022) Impact of bioenergy for the diminution of an ascending global variability and change in the climate. In: Microbiome under changing climate. Elsevier, pp 469–487

Kumari P, Kumar P, Singh J (2023) Toxic contamination of aquatic habitat and its remediation. In: Spatial modeling of environmental pollution and ecological risk. Elsevier, pp 209–220. https://doi.org/10.1016/B978-0-323-95282-8.00021-3

Naik M, Kumar P (2020) Role of growth regulators and microbes for metal detoxification in plants and soil. Plant Arch 20:2820–2824

Pandey SK, Kumar P, Dey SR (2024) Xenobiotic stress management by microbial endophytes. In: Bioremediation of emerging contaminants from soils: soil health conservation for improved ecology and food security. Elsevier, pp 241–259. https://doi.org/10.1016/B978-0-443-13993-2.00012-8

Pankaj PK, Kumar P, Nigam RC, Mishra PK (2012) Monitoring and Surveillance of synthetic pyrethroids and organophosphate in different brands of soft drinks. J Chem Pharm Res 4(8):3939–3943

Pathak S, Kumar P, Mishra PK, Kumar M (2017) Mycorrhiza assisted approach for bioremediation with special reference to biosorption. Pollut Res 36(2):329–332. https://www.scopus.com/inward/record.uri?eid=2-s2.0-85028617468&partnerID=40&md5=da296311c0f3d29ea94d21631fe01d69

Raj A, Goutam E, Kumar P, Singh J (2024) Polyphenol oxidase: an enzyme of plants. In: Polyphenol oxidases: function, wastewater remediation, and biosensors. De Gruyter, pp 105–128. https://doi.org/10.1515/9783111033525-004

Reddy SS, Kumar P, Dwivedi P (2022) Heavy metal transporters, phytoremediation potential, and biofortification. In: Plant metal and metalloid transporters. Springer, pp 387–405

Saini L, Sharma K, Kumar P (2023) Water efficiency in agriculture as a vital approach toward water management in rural areas. In: Water resources management for rural development: challenges and mitigation. Elsevier, pp 127–138. https://doi.org/10.1016/B978-0-443-18778-0.00020-9

Saurabh K, Meena RK, Kumar P, Upadhyaya A, Das A (2024) Role of hyperaccumulators in the reduction of emerging industrial pollutants from soil. In: Bioremediation of emerging contaminants from soils: soil health conservation for improved ecology and food security. Elsevier, pp 489–517. https://doi.org/10.1016/B978-0-443-13993-2.00022-0

Sharma K, Kumar P (2023) Microbe-mediated impact on cell membrane stability and leaf pigmentation in pearl millet under toxic mercury stress in soil. Biopesticides International 19(2):217–228. https://doi.org/10.59467/BI.2023.19.217

Sharma K, Kumar P (2024) Environmental threats posed by xenobiotics. In: Bioremediation of emerging contaminants from soils: soil health conservation for improved ecology and food security. Elsevier, pp 183–201. https://doi.org/10.1016/B978-0-443-13993-2.00009-8

Sharma K, Devi P, Dey SR, Kumar P (2023a) Mercury phytovolatilization: an overview of the mechanism and mitigation. In: Role of green chemistry in ecosystem restoration to achieve environmental sustainability. Elsevier, pp 325–331. https://doi.org/10.1016/B978-0-443-15291-7.00031-6

Sharma K, Devi P, Kumar P (2023b) Metal-organic frameworks for wastewater contaminants removal. In: Metal organic frameworks for wastewater contaminant removal. wiley, pp 95–117

Sharma K, Devi P, Kumar P, Dey A, Dwivedi P (2023c) Hazardous phytotoxic nature of reactive oxygen species in agriculture. In: Reactive oxygen species: prospects in plant metabolism. Springer Nature, pp 135–146. https://doi.org/10.1007/978-981-19-9794-5_8

Sharma M, Dey SR, Kumar P (2023d) Antioxidant defence: A key mechanism of chromium tolerance. In: Environmental science and engineering: Vol. Part F1975. Springer Science and Business Media Deutschland GmbH, pp 91–116. https://doi.org/10.1007/978-3-031-44029-8_5

Sharma K, Kumar P, Singh J (2024) Ecology and diversity of microbial communities involved in the removal of priority contaminants and micropollutants in waste. In: Development in waste water treatment research and processes: role of environmental microbiology in industrial wastewater research. Elsevier, pp 181–199. https://doi.org/10.1016/B978-0-443-13609-2.00014-8

Sinam V, Kumar P, Singh J (2023) Bioremediation and ecorestoration strategies of aquatic environment. In: Spatial modeling of environmental pollution and ecological risk. Elsevier, pp 483–499. https://doi.org/10.1016/B978-0-323-95282-8.00025-0

Srivastav AL, Patel N, Rani L, Kumar P, Dutt I, Maddodi BS, Chaudhary VK (2024) Sustainable options for fertilizer management in agriculture to prevent water contamination: a review. Environ Dev Sustain 26(4):8303–8327. https://doi.org/10.1007/s10668-023-03117-z

Upadhyay SK, Devi P, Kumar V, Pathak HK, Kumar P, Rajput VD, Dwivedi P (2023) Efficient removal of total arsenic (As3+/5+) from contaminated water by novel strategies mediated iron and plant extract activated waste flowers of marigold. Chemosphere 313:137551. https://doi.org/10.1016/j.chemosphere.2022.137551

Geobiomineralization at the Interface of Plant–Microbe Interaction and the Role of Mycorrhizal Fungi

Vishal Kumar Mohan and Santa Ram Joshi

Contents

Abstract Biomineralization is a process in which living organisms catalyze the formation of minerals, this has drawn interest for its possible uses in environment remediation. In order to promote biomineralization, microbes use a variety of strategies, such as sorption and sequestration, mineral precursor synthesis, metabolism and transformation, and cooperative relationships among microbial communities. Mycorrhizal fungi enhance nutrient uptake, alter the soil's biogeochemical composition, and establish symbiotic relationships with plants. These fungi increase plant roots' surface area for nutrient uptake through their vast hyphal networks, speeding up the minerals' weathering and dissolution. Mycorrhizal fungi produce organic acids and enzymes that aid in the transformation of minerals liberating vital minerals substrates and affecting soil fertility. Mycorrhizal fungi also help in the immobilization and detoxification of contaminated soil hence it becomes crucial to

V. K. Mohan · S. R. Joshi (✉)
Microbiology Laboratory, Department of Biotechnology and Bioinformatics, North-Eastern Hill University, Shillong, Meghalaya, India

S. R. Joshi et al. (eds.), *Mineral Transformation and Bioremediation by Geo-Microbes*, Interdisciplinary Biotechnological Advances,
https://doi.org/10.1007/978-981-96-3033-2_14

understand the process underlying plants, microbes, and minerals to fully investigate the potential of biomineralization as a sustainable strategy for environmental management; more investigation into this area is necessary.

Keywords Mycorrhiza · Plant–microbe · Rhizobacteria · Symbiosis · Nitrogen · Phosphorus

Abbreviations

AMF	Arbuscular mycorrhiza Fungi
AOB	Ammonia oxidation bacteria
Cd	Cadmium
EcM	Ectomycorrhiza
ErM	Ericoid mycorrhiza
K	Potassium
N	Nitrogen
OM	Orchid Mycorrhiza
P	Phosphorus
PGPR	Plant growth promoting rhizobacteria
QQ	Quorum quenching
Qs	Quorum sensing
RPEs	Rhizosphere priming effects
SLB	Southern leaf blight
SOM	Soil organic matter
Zn	Zinc

1 Introduction

The process in which mineral deposits occur inside or outside the variety of an organism's cells is known as biomineralization. These minerals consist of magnitites and other deposits that contain iron, gold deposits, calcium carbonate, and calcium phosphate in nonvertebrate shells (Boskey 2003). The compound microscope was invented in the late seventeenth century, leading to biomineralization's foundation. Van Leeuwenhoek used his cutting-edge microscope to study a variety of objects, including mineralized tissues. Van Leeuwenhoek (1693) identified the osteons found in many bones by magnifying 400 times, and Havers (1691) also identified the lamella; each lamella being only three microns thick (Weiner 2007). According to phylogenetic data, biominerals first appeared in the fossil record long after the various phyla had split apart. Since none of the biomineralizing animals in different phyla shared a common ancestor that was also biomineralizing, the species had to evolve ways to make carbonate biominerals individually (Gilbert et al. 2022).

Metalloid and metal pollutants can build up throughout the food chain and cause multiple health issues. They are of the utmost importance to the environment. Significantly abundant mineral phases can release or sorb metal pollutants based on the chemical conditions of the surrounding environment. They have the ability to catalyze redox reactions that alter the speciation of organic pollutants and as a result their mobility (Benzerara et al. 2010).

Our terrestrial ecosystem is fundamentally based on interaction between plants and microbes. Plant–microbe interaction can take many forms, including mutualism, competition parasitism, and commensalism. Commensalism and mutualism are the most prevalent types of interaction, wherein one or both species gain from the partnership, respectively (Wu et al. 2009). The range of microorganisms that live alongside plants, such as bacteria, protists, archaea, and fungi is referred to as the "plant microbiota." The term "plant Microbiota" describes all of the plant microbiota's individuals, genomes, habitats, and environmental conditions (Pantigoso et al. 2022). Understanding the microbial population and the root architecture is crucial to understanding how these interactome associate, given the microbial speciation, microbial diversification, structural complexity, and interactions surrounding the root system. High levels of interactions between microbes and plants have led to their classification as metaorganisms or holobionts (Nadarajah and Abdul Rahman 2021). Mineral deposition occurs when metabolites secreted by microbes interact with ions or other substances in the surrounding environment (Frankel and Bazylinski 2003). Bacteria break down organic matter below the surface, while fungi degrade material above it. As a result, the abundance of bacteria, fungi, and actinomycetes indicates soil biological activity (Bai et al. 2012). Microorganisms are vital to the soil phosphorus cycle and regulate the uptake of phosphorus by plants (Richardson and Simpson 2011). Inorganic nitrogen present in the soil is mostly immobilized by microorganisms present in the soil. The primary mechanisms that make up the nitrogen cycle in soil include denitrification, ammonium reduction, nitrification, and ammonification (Xiong et al. 2021). In the rhizosphere, soil microbes are essential to the cycling and transformation of phosphorus. They mineralize organic P using extracellular enzymes (Wang et al. 2023).

2 Plant–Microbe Interaction

Numerous above- and belowground microorganisms coexist in dense populations within plants for mutualistic purposes. The microorganisms that infect plants can be divided into four groups: phyllospheric microbes grow on surfaces of leaves, endophytic microbes live inside tissues, epiphytes live on the surface, and rhizospheric microbes live in the soil near the roots (Fig. 1). The rhizosphere is thought to be the most dynamic among them, having a major impact status on plants (Vishwakarma et al. 2020). Even though a wide variety of microbes frequently surround plants, most microorganisms are unable to colonize plants, and this is mostly explained by the fact that plants have developed active defense layers over time that shield them

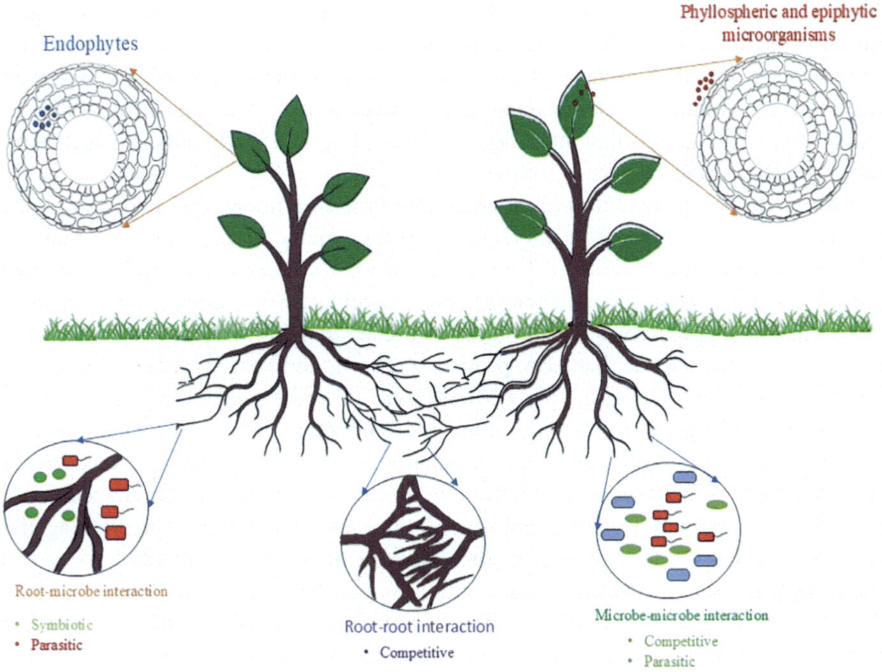

Fig. 1 Aboveground and belowground plant–microbe interaction

from the majority of microorganisms (Aung et al. 2018). Endophytes are mostly non-pathogenic fungi or bacteria that invade the interior spaces of plant tissues, such as the seeds, flowers, leaves, roots, or stems at some stage in their life cycles. Endophytes can be localized in plant tissues in an almost dormant stage, or they can be systemic affecting several different plant tissues in terms of their distribution and metabolic activity (Bacon and White 2016). Numerous studies have shown that endophytes and epiphytes don't hurt plants, rather they result in the synthesis of some essential substances such as hormones. Auxin, which is produced by certain microorganisms, is important for the development and growth of plants. Microbes and their host plants can interact in ways that are advantageous or detrimental to the microbe or the plant, and depending on the type of association, these types of association are categorized as parasitism, synergism, competition, neutralism, amensalism, or mutualism. It is also possible for some plant-associated microbes to merely develop saprophytically from the nutrients the host plant releases (Kumar et al. 2017). External environment conditions significantly impact the interactions between microbes and plants. The well-known "disease triangle" theory explains how a plant disease develops: the pathogen must be genetically virulent, the plant must be genetically susceptible, and the environment must support both the pathogen's virulence and the plant's susceptibility (Cheng et al. 2019).

3 Phyllosphere Microorganism

The phyllosphere, a microenvironment, stretches from the leaf surface outward to the outer edge of the boundary layer enclosing the leaf and inward into the leaf tissues. However, the phyllosphere is divided into two separate but connected niches called "phylloplane" and "phyllotelma," which comprise the leaf surface landscape and waterscape (Thapa and Prasanna 2018). The aerial or aboveground parts of plants, such as flowers, fruits, leaves, and stems, are collectively referred to as the phyllosphere, with leaves being the most prevalent component. The plant's well-organized, multilayered leaves are shielded from various biotic and abiotic stresses by a thin, waxy cuticle layer that encircles the upper and lower epidermis. This creates an unfavorable environment for microorganisms (Bashir et al. 2022).

Common inhabitants of the phyllosphere include the taxonomic groups of algae, fungi, bacteria, and yeast that can exist as endophytes or epiphytes, and free-living microorganisms. Members of the nematodes, protozoa, and archaea represent the less common species. It has been observed that organisms and yeasts that sporulate frequently and quickly colonize the surroundings more actively. Filamentous fungi, which are mostly spores, are usually ephemeral residents of leaf surfaces, along with many plant pathogenic microbes and oomycetes. They frequently make up a sizable amount of the phyllosphere, which hurts the vigor and health of plants (Thapa and Prasanna 2018). The foliar existence and numerous amounts of these microorganisms, whether parasitic, commensal, or mutualistic, are the consequence of multiple concurrent processes, such as microbial movement to and from the phyllosphere, as well as microbial multiplication, dispersal, and death within it (Leveau 2019).

Understanding the phyllosphere microbiota can help one understand the mechanism that controls interactions between microorganisms, plants, and the atmosphere in both natural and artificial settings. The phyllosphere is an extreme and unstable habitat for epiphytic microorganisms, which reside on the surface of plant tissues. It has oligotrophy-related traits, such as limiting carbon and nitrogen, and a multitude of highly fluctuating physiochemical constraints (Bringel and Couée 2015). Phyllosphere, endophytic, and epiphytic fungi inhabit different microenvironments. Plants have more control over fungal colonization within their own cells than they do over exterior surfaces because endophytic fungi interact with the plant's internal surroundings and absorb nourishment from the host cells, while epiphytic fungi interact with the outer world and depend on nutrients that accumulate on leaves from the environment or released from leaves (Yao et al. 2019). Epiphytes encounter numerous external environmental stressors, including temperature, humidity, and sun radiation. Endophytes, on the other hand, live in a more protected environment but still face threats from plant defense mechanisms. The various stressors that endophytes and epiphytes encounter may have a significant impact on the makeup of their communities. The few studies that have compared the epiphyte and endophytic fungal communities on the leaves of woody plants like *Camellia japonica* and *Coffea arabica* have shown that they car distinct communities (Gomes et al. 2018).

These investigations have concentrated on the bacteria connected to a particular host species phyllosphere. The environment that the epiphyte community lives in is nutrient-poor and unstable, with constant fluctuations in radiation, humidity, and temperature. Contrarily, the endophytic community lives in a more stable environment than epiphytes do with the host plant's defense mechanism likely to be their biggest obstacle (Mina et al. 2020). A few decades of studying rhizobacteria and microbial endophytes have given us a partial understanding of the vital function of the plant microbiome in maintaining the development and existence of plants, hence it is the evidence that epiphytic and endophytic microorganisms are essential elements of the plant surrounding, despite their incompleteness. Three main functional groups of endophytic microbes have been identified: (1) microorganisms that lessen biotic stress in the host plant; (2) microorganisms that shield the host plant from biological stressors; and (3) microorganisms that boost the host's levels of iron, nitrogen, vitamins, or phosphorus to support nutrition. This functional characteristic of harmless endophytic and epiphytic plants enables the design and creation of microbiomes for crop plants to enhance cultivation using fewer agrochemical inputs and at less expense. We must first comprehend the functioning of microbiomes to synthesize them (Bacon and White 2016).

4 Root–Root Interaction

Underlying a multitude of interesting interactions, between plants, chemical signaling plays a vital role as an intermediary of interactions among plants, and they can recognize and react to chemical signals that are released by their neighbors, both intraspecific and interspecific, both above and below ground, airborne chemicals mediate aboveground signaling interaction, whereas belowground plant signals are driven by root exudates (Wang et al. 2020).

Plants that coexist depend on a limited set of basic resources, including light, water, and nutrients. Since scarce resources are typically found at their best during particular times and at particular places, interactions among plants therefore carry a high risk of competition and may cause "competitive exclusion under limiting similarity" in a community of plants, meaning that species are more likely to become extinct format he community as a result of competition with other species that share similar traits, according to ecological theory. A great diversity of plant species is thought to be able to coexist in a particular habitat because of the wide range of features that different plant species have evolved that enable species to occupy distinct niches over time and space known as niche complementary theory (Faget et al. 2013).

Thousands of distinct compounds, broadly classified as soluble sugar, amino acids, fatty acids, proteins, organic acids, and more, are found in root exudates and serve a variety of purposes in the rhizosphere. Given that several root exudates have been linked to nutrient availability, they may act as a mediator in resource competition. For instance, the exudation of protons, carboxylates, and acid phosphatases raises the concentration of ions, such as P and Zn, that can be absorbed by plants. Root exudates have a function beyond just making nutrients more accessible to plants. In the rhizosphere, exudates also have a direct impact on how root–root interaction turns out. Alloelophathy in which plants release phytotoxins that hinder the establishment, growth, or survival of their neighbors is possibly the most well-known of these, and this reduces competition and increases the availability of resources (Mommer et al. 2016a, b).

5 Microbe–Microbe Interaction

Interactions between microbes associated with roots are comparatively more intricate than those between microorganisms in bilk soil. This is likely because microbial cells are more numerous and diverse in the rhizosphere with the majority of beneficial interaction, suggesting that mutualistic relationships are more likely to occur in the rhizosphere. Additionally, through signals secreted by the roots, which are typically beneficial to the plants themselves, plant roots mediate inverse interactions among harmful and helpful microbes. According to research, soil-borne pathogens attack crops after roots have recruited and promoted antagonistic microbes. This implies that plants might defend themselves against infections by using their roots (Zhang et al. 2017).

There are four ways that microbes can communicate with each other: (1) between pathogen and endophyte; (2) between pathogen microbes; (3) change in lifestyle in the environment; and (4) succession by microbes. The microbial community in the soil and on plant surfaces can be impacted by pathogens. It was found that the local microbiota in maize with Southern leaf blight (SLB) infection had decreased in richness (Nadarajah and Abdul Rahman 2021).

When bacteria are sufficiently densely populated, they can carry out a variety of ecologically significant activities that are not seen on less concentrated cells, such as the synthesis of extracellular enzymes, exopolysaccharides, biofilm formulation, and the expression of virulent traits, and bacterial quorum sensing (QS) is a type of population behavior that depends on cell density. It involves the regulation of gene expression through the production, detection, and response to particular small molecules (Deangelis et al. 2008). Many bacteria use quorum sensing to obtain the greatest competitive advantages; consequently, quorum quenching (QQ) is used by a variety of organisms to offset the advantages of QS. The QQ mechanism is extensively present in both prokaryotes and eukaryotes, and it is crucial for interactions between microbes and hosts (Chong et al. 2012).

6 Microbial Interaction at the Rhizosphere

For millions of years, prokaryotes and eukaryotes have coexisted on Earth. Including the endogenous bacterial flora, humans are thought to contain 10^{14} bacterial cells and 10^{13} human cells. Because of this long-standing relationship, prokaryotes and eukaryotes have developed both advantageous and disadvantageous relationships (Badri et al. 2009). The soil around plant roots, or rhizosphere, provides a unique environment for microbial colonization. Lorenz Hiltner was the one who originally coined the term "rhizosphere." It is a complex matrix that is home to a wide variety of microbiological species (Haldar and Sengupta 2015). Higher bacterial redundancy and activity but lower diversification are associated with the rhizosphere compared to non-rhizosphere soil. Numerous abiotic and biotic factors, such as the physiochemical characteristics of the soil, signals and chemicals produced by the plant, and interacting microorganisms, influence the structure of the rhizosphere microbiome. In the rhizosphere zone, plant roots release diverse compounds that play crucial roles as chemoattractants and repellents (Jain et al. 2020). Key ecosystem processes like productivity, biogeochemical cycling, and stress tolerance are influenced by below-ground interaction between soil microbes, plant roots, and the abiotic environment (Mommer et al. 2016a, b). The weathering of rocks and minerals produces soil, which varies in composition depending on the parent material's origin, climate, and vegetation. Most soil carbon is derived, directly or indirectly, from plants. Although weathering can occur due to physical and chemical factors, most weathering processes involve plants, mainly their roots, or microbial activities that rely on root-derived carbon (Lambers et al. 2009). The primary source of organic carbon in soil is plant roots. Even though the soil around a plant's root may only make up 1–2% of the entire soil volume, this zone is a hub for microbial carbon transformation. It can contribute 30–40% of the total soil organic carbon input; compared to the surrounding bulk soil, microbial densities and activities are often up to ten times higher in the rhizosphere (Pett-Ridge et al. 2021). Because biofertilizers contain specialized microbes that can fix, solubilize, decompose, or mobilize nutrition sources, applying them through seed or soil improves plant uptake of nutrients. Research on biofertilizers began in 1896 when Nobbe and Hiltner (1896) developed "Nitragin," the first commercially grown and patented Rhizobium culture (Mohanram and Kumar 2019).

Based on our current understanding, plant–microbe interaction has been divided into three categories: (i) negative interaction, i.e., pathogenic, (ii) positive interaction, where either one or both partners benefit from a close relationship (symbiosis) from a loose relationship, either both partner benefit or one partner benefits without the other's harm (associative), and (iii) neutral interaction, where neither party is harmed and neither party receive a direct benefit from the interaction (Singh et al. 2004). Most rhizobacteria are commensals that form mutual relationships with plants using root exudates as nourishment. Certain bacteria known as plant growth-promoting rhizobacteria (PGPR) can favorably influence plant growth and fitness when connected to roots (Pii et al. 2015). Plant health is also impacted by interactions among microbes, plants, and soil regarding iron dynamics. This is demonstrated by certain soils naturally suppressing a significant class of soil-borne diseases, including plant wilting brought on by the pathogenic *Fusarium oxyporum* (Lemanceausoils et al. 2009). Both in vitro and in vivo arbuscular mycorrhizal fungi and other fungi, along with bacteria associated with the roots, are essential to the plant-root system's operation (Atkinson and Watson 2000). Microbial interactions in the rhizosphere can be controlled as low-input biotechnology to support environmentally friendly and sustainable agrotechnological practices, as shown by strategic and applied research (Barea et al. 2013).

Rhizosphere microbes, such as phosphate-solubilizing bacteria, increase nutrient availability of soil for root uptake and enhance plant resistance to abiotic and biotic stresses by producing phytohormones. Plant roots secrete exudates from photosynthesis, a significant carbon source for microorganism growth, as shown in Fig. 2 (Pathan et al. 2020). The main elements influencing the attraction of microbial proliferation in the rhizosphere are exudates from plant roots. Smaller molecules weight substances such as amino acids, mucilage, and other cell constituents make up these exudates. Terpenes, flavonoids, glucosinolates, and other secondary metabolites are among the many substances that plants release (Velmourougane et al. 2017). Plant roots and PGPR establish a mutually beneficial relationship through chemical communication. The intricate relationship between plant roots and PGPR in the root surrounding preconditions plants for a more robust defense reaction in response to external stimuli. The trigger events happen more intensely in response to secondary stimuli. The metabolomes alter due to these natural, primed, and triggered physiological states. Targeted and untargeted metabolomic activities were

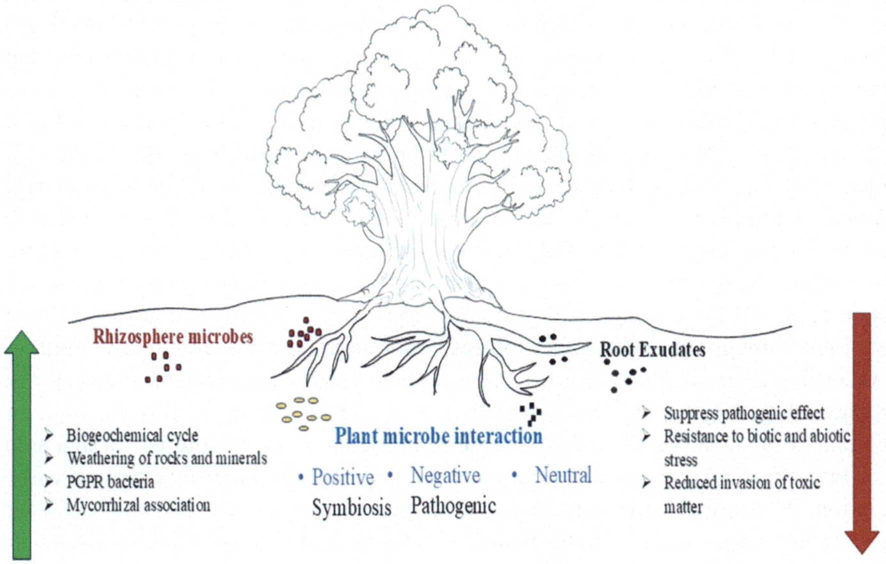

Fig. 2 Plant and microbe association under biotic and abiotic influences

studied (Rehman 2020). One of plant roots' most amazing metabolic characterizations is their capacity to secrete various compounds into the rhizosphere. Root exudates carry about 5–21% of all photosynthesis retained carbon into rhizosphere. The type of plant, the age of the individual plants, and environmental biotic and abiotic variables all affect the amount and quality of root exudates (Haichar et al. 2014). The main focus of plant–plant interaction research in the past few decades has been chemically mediated plant signaling and response mechanisms. Chemicals released by conspecific and interspecific neighbors may be utilized by coexisting plants. Plants, in particular, can recognize chemicals as signals and react by causing an allelopathic interface, which controls interactions between or within species (Wang et al. 2020). A step in the rhizodeposition process is root exudation. Plant species, plant age, and external variables like biotic and abiotic stressors all affect the amount and quality of root exudates. Young seedlings usually exude between 30 and 40% of their carbon as root exudates. This indicates that root exudates represent a significant carbon cost to the plant (Badri and Vivanco 2009).

7 Mycorrhizal Fungi: Diversity and Function

Mycorrhizal fungi are the most prevalent and functionally significant soil microorganisms in various natural and semi-natural ecosystems. Arbuscular mycorrhizal fungi and almost all plants in temperate grasslands form mutualistic relationships (Johnson et al. 2005). Of all the interactions that plants participate in, the arbuscular mycorrhizal (AM) symbiosis is the most widespread due to its prevalence from both geographical (found on every continent and in most biomes) and evolutionary (the capacity to participate in the symbiosis is lost, not gained) viewpoint (Powell and Rillig 2018). *Glomarales* fungi and most vascular plants have a relationship that dates back to 600 Ma, and it was likely a significant factor in the establishment of plants on land (Munkvold et al. 2004). According to molecular evidence and fossil records, AM symbiosis is an ancient evolutionary process that started with the first terrestrial plants. Though molecular research has indicated a significantly more significant number of species, the species diversity of AMF is low, with only about 240 recognized species. Low diversity in AMF has been proposed as a defining feature of mutualistic symbiosis due to low selection pressure for fungal partner speciation (Lee et al. 2013). Although the mycelial morphologies of AMF are closely related to their functioning, different species' morphologies are frequently similar. According to fossil evidence, some have probably not changed much since early in their evolutionary history (Hart and Klironomos 2003). The mycorrhizal mycelium enhances the host plant's ability to acquire nutrients by increasing its surface area for nutrient uptake, particularly ectomycorrhizal and arbuscular-mycorrhizal association. Additionally, the hyphae can enter tiny microsites that the much coarser plant roots cannot reach. The formation of nutrient-depleted soil volumes surrounding roots is caused by the active uptake of poorly mobile nutrients, such as

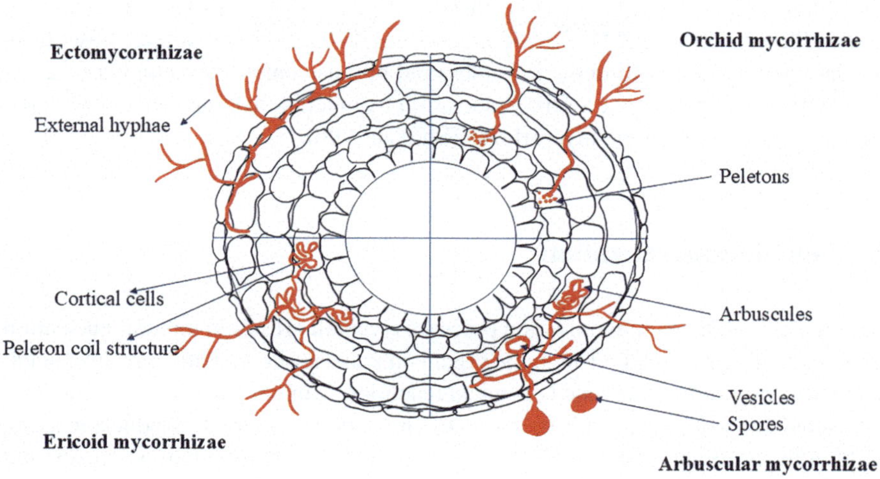

Fig. 3 Types of mycorrhizae

phosphorus. Mycorrhizal hyphae can bridge this gap by supplying nutrients from farther away soil (Finlay 2004).

Mycorrhizas do not always have a mutualistic relationship because the benefits to fungi in myco-heterotrophic plant association are implausible. Mutualistic relationships encompass a broad spectrum of symbiosis and nonsymbiosis, as well as direct and indirect relationships, many of which work through mechanisms other than the transfer of nutrients (Brundrett 2004). Four main types of mycorrhizae are identified based on anatomy and partner identity as shown in Fig. 3: ericoid mycorrhiza (ErM), arbuscular mycorrhiza (AM), orchid mycorrhiza (OM), and ectomycorrhiza (EcM) (Mohan and Joshi 2024). Even though the majority of vascular plant species only form one type of mycorrhiza symbiosis, occasionally, EcM and ErM fungi will also co-colonize AM roots (Soudzilovskaia et al. 2020).

Arbuscular mycorrhiza develops their fungal hyphae intracellularly within the plant cortex cells, while ectomycorrhizas develop their fungal hyphae intercellularly around the cortex cells. Arbuscular mycorrhiza-producing fungi are classified as belonging to the division Glomeromycota, whereas fungi that form ectomycorrhizas can be classified as either Ascomycota or Basidiomycota (Bejerano et al. 2014). Orchid mycorrhiza with an estimated 25,000 species in the Orchidaceae family is the most prominent plant family in the world. Fungal partners provide chemicals to orchids that are not just carbon. Mycorrhiza G. repens was known to have acquired 100 times more P than nonmycorrhizal control (Dearnaley 2007). Ectomycorrhiza EcM is the fungal mycorrhiza association that occurs primarily in woody trees and is caused by fungi that infect about two vascular plant roots. A hyphal mantle is formed by ectomycorrhiza around the tips of the root, and a Hartig net of hyphae can pass through the cortical root cells. Through the vigorous hyphae, the fungi take photosynthates from their host plant, giving it extra nutrients and water (Shrestha et al. 2021). Ericoid mycorrhiza, because it only affects hosts from a select few families within the Eri-cales, and because only a tiny number of ascomycetous fungi participate in the association as mycobionts, has been considered the most specific of all the mycorrhizas (Straker 1996).

8 Geobiomineralization

Biologically induced mineralization processes typically result in mineral nucleation and extracellular growth through the organism's metabolic activity and subsequent metabolic by-product-mediated chemical reactions. Mineral particles are frequently deposited due to an organism's secretion of one or more metabolic products reacting with ions or other compounds in the surrounding environment (Frankel and Bazylinski 2003). Since soil microorganisms are primary players in the transformation and circulation of soil organic matter, they are crucial to the nitrogen cycle of the ecosystem. At the same time, bacteria are crucial to the breakdown of organic matter underground, and fungi contribute more to the degradation of organic matter on the surface. Consequently, it is generally accepted that the abundance of bacteria, fungi, and actinomycetes is a crucial indicator of the biological activity levels in soil. Numerous microbes' species are breaking down the protein and its derivatives (Bai et al. 2012).

Biomineralization is the process of mineral crystals deposited in the living organism's matrix. During development, this process forms organic-based skeletal structures, like bone, which are dynamic and complex and have structural and metabolic functions (Hu et al. 2011). Plants use calcium oxalate for a variety of purposes. Calcium oxalate crystals assist plants in distributing light evenly to the chloroplasts in the palisade cells. These crystals function as agents that retard combustion and make trees less appetizing to termites in woody tissues. They offer defense against grazing animals when found in leaves. Furthermore, the accumulation of metal oxalate minerals is a significant carbon source for diazotrophic bacteria in the tree and the surrounding litter (Braissant et al. 2004). When compared to other heavy metals, cadmium (Cd) ranks as one of the most common and hazardous environmental pollutants. A significant amount of hazardous Cd is released into the environment through human activities and natural processes. Adsorption, electrodeposits, biological techniques, membrane separation, and chemical precipitation have been proposed to remediate Cd-polluted water and soils. Microorganism-mediated remediation and bioremediation are two biological treatments that have drawn much interest in the suggested techniques (Yin et al. 2021).

9 Mineralization of Phosphorus

Remobilizing soil-existing phosphorus (P) stocks is crucial to mitigating the depletion of rock phosphate resources suitable for fertilizer production. Topsoil (0–15 cm) usually has total P contents between 50 and 500 mg kg^{-1}. However, phosphate ions' high sorption capacity makes only a tiny portion of the P bioavailable. A significant amount of P in temperate zine soils is found in organic forms and must be mineralized for the plant to use (Spohn and Kuzyakov 2013). Since microbes are essential for the phosphorus (P) cycle in soil, they are crucial in regulating the amount of P plants can access. Thus, for many years, there has been much interest in comprehending microbes' role in P nutrition for plants and the possibility of altering specific microbes to boost soil P availability. Because P deficiency is prevalent in tropical and weathered soils across the globe, the demand has increased. Despite the fact that many soils include a relatively large amount of P that is only accessible to plants, P fertilizer costs are increasing and plant absorption efficacy from soil and fertilizer sources is frequently low (Richardson and Simpson 2011). Various microbial organisms, such as algae, fungi, bacteria, and actinomycetes, can solubilize and mineralize P. There have been reports of *Pseudomonas* spp., *Bacillus circulans*, and *Agrobacterium* spp. They are mobilizing poorly available phosphorus through solubilization and mineralization. Various strains of *Azotobacter*, *Burkholderia*, *Enterobacter*, *Erwinia*, *Bacillus*, *Kushner*, *Paenibacillus*, etc., are among the other bacteria that solubilized and mineralized phosphorus (Fig. 4) (Alori et al. 2017). The ability of microorganisms to liberate metabolites that chelate the cation attached to the phosphate, like organic acids, and convert it into soluble forms is what determines that phosphate solubilizing occurs via a variety of microbial mechanisms and processes, such as proton extrusion and organic acid production. Numerous microbial P solubilizing mechanisms can be found in nature, and fungi and bacteria are mainly responsible for the global cycling of inorganic soil phosphates (Khan et al. 2009). PSMs can dissolve soil P into orthophosphate forms that are soluble and available to plants, primarily PO_{43}-HPO_{42}-, and H_2PO_4. Thus, there has been much interest in identifying the contribution of PSMs to plant nutrition acquisition

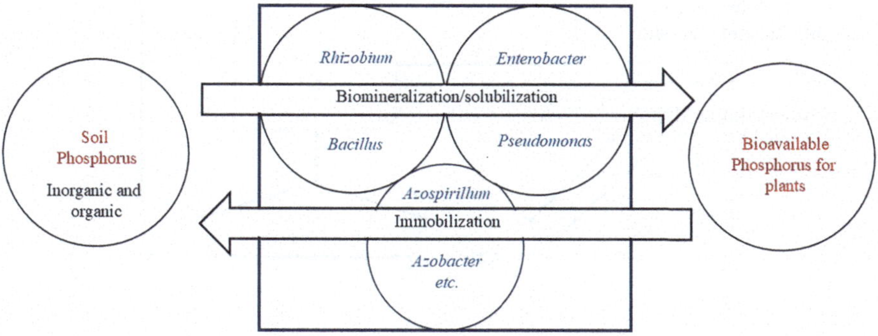

Fig. 4 Diagram illustrating some processes in the phosphorus cycle in soil

comprehending the potential for modifying PSMs to increase soil P availability to restore other soil health, even though PSM inoculation is a generally acknowledged environmentally friendly method of raising soil soluble P concentration and agriculture productivity, there has not been enough focus on understanding the role that PSM play in P geochemical processes, such as sorption-desorption, weathering, and dissolution-precipitation (Tian et al. 2021).

10 Mineralization of Nitrogen

All organisms require nitrogen (N) as an essential nutrient. There is fierce competition between microorganisms and plants in terrestrial ecosystems because N is typically scarce for plants. Microorganisms can use various organic and mineral compounds because they have evolved distinct mechanisms for taking up and assimilating both minerals and organic forms of nitrogen as shown in Fig. 5 (Geisseler et al. 2010). Most processes generating and immobilizing inorganic N depend on the microbial biomass. The production of the enzyme necessary for ammonification and N assimilation for microbial growth is catalyzed by microbial biomass. Even though the microbial biomass aline only makes up 2–5% of the organic N in the earth's humus pool, which is maintained by components of the cell walls and gives the soil organic N through turnover and death (Myrold and Bottomley 2008). While organic N sources show an even inorganic emit pattern due to N mineralization, ammonium fertilizers offer a substantial but short-lived flush of inorganic N when applied. Soil microbes may be affected differently by minerals and organic N fertilizers (Ouyang and Norton 2020). Figure 5 shows that the interaction

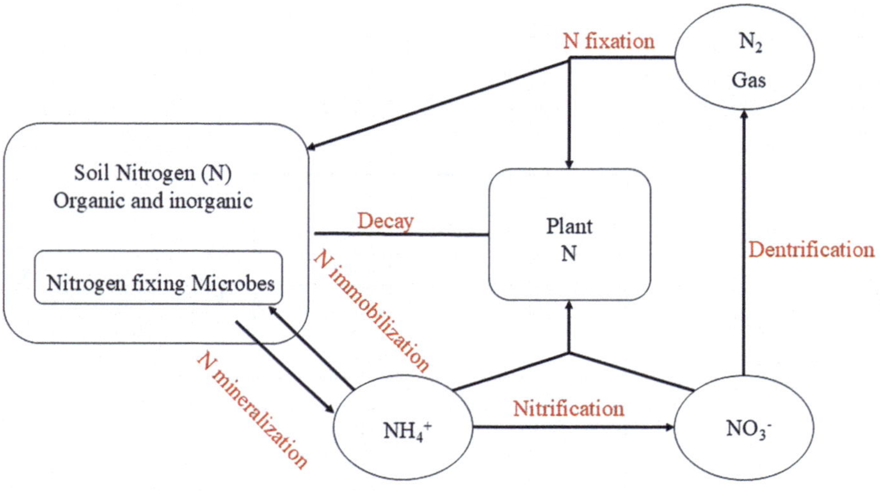

Fig. 5 Diagram illustration of nitrogen cycle

between roots, microorganisms, and soil organic matter (SOM) boosts activity of microbes and biomass and produces rhizosphere priming effects (RPEs). Compared to bare soil, the SOM decomposition can increase to 380%, decrease by 50%, or remain constant when growing plants. Beneficial RPEs may boost organic nitrogen (N) mineralization by up to 437% because microorganisms produce more extracellular enzymes to mine nitrogen (N) from SOM, which is then accessible for uptake by plants and microbes (Jiang et al. 2021). The soil nitrogen cycle mainly consists of nitrification, denitrification, ammonification, and ammonium reduction. These processes are primarily fueled by nitrogen-fixing microorganisms, dentrifying and nitrifying useful bacteria, and nitrate dissimilatory reduction to ammonium process functional microorganisms (Xiong et al. 2021). Microbial functional genes control the processes involved in the cycling of N in soil in terrestrial ecosystems. The nirK and nirS genes contribute to NO_3 consumption and N2O production, the nosZ gene mediates the conversion of N_2O to N_2 during the denitrification process, and the amoA gene in ammonia oxidation bacteria (AOB) and archaea (AOA) controls inorganic N availability and N_2O production via nitrification (Dai et al. 2020). Fungi can depolymerize litter and break down organic matter in the soil; certain fungi can even help plants directly absorb organic nitrogen from the soil. Furthermore, by stimulating N mineralization through the microbial loop, bacteria can increase plant productivity in addition to mineralization of N from soil organic matter to support their growth (Xiao et al. 2022).

11 Mineralization of Potassium

In the rhizosphere, soil microbes play a vital role in the cycling and transformation of P. Microorganisms produce and utilize extracellular enzymes to mineralize organic P because they can use phosphorylated compounds as carbon sources. Acid and alkaline phosphatase activities are used as stand-ins for gaining the mineralization of organic P to accessible inorganic P because monoester P is believed to be the primary organic P form (Wang et al. 2023). One essential and necessary nutrient for plant growth is potassium (K). As most agronomic crops require massive amounts of absorption from the root zone for growth and production, it is categorized as a macronutrient. This enzyme is essential for photosynthesis, enzyme activation, and protein synthesis. The soil's potassium levels have decreased due to crop uptake, leaching, water runoff, and soil erosion, which has resulted in an ever-increasing interest in intensive and rigorous agriculture (Teotia et al. 2016). The K is released by silicate bacteria from the feldspar and illite minerals (Fig. 6). K and other silicate minerals elements could be released with the help of silicate bacteria. The K release was caused by bacterial strains in an acidic pH and oxygen-rich environment (Masood and Bano 2016). P, including Pi and Po, is readily bound to Ca, Al, Mg, Mn, and Fe to form insoluble salts. As a result, most P in soils is present in inaccessible forms, such as organic soluble/insoluble forms. Utilizing phosphorus (P),

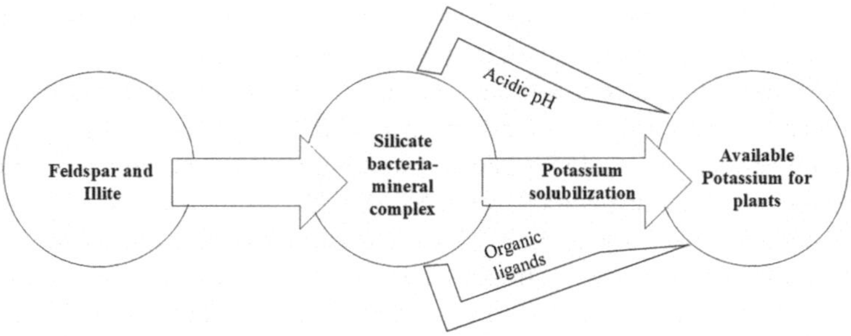

Fig. 6 Schematic depiction of potassium solubilization

which includes Po mineralization and Pi solubilization, is an essential process in soils that gives plants and microbes access to available P (Wan et al. 2020).

12 Mycorrhizal Fungi in Mineral Weathering

The enormous quantity of nutrients trapped in soil minerals almost a century ago raised the question of whether plants could actively access this possible source of nutrients. Studies on the function of microorganisms, such as mycorrhizal fungi, in mineral weathering fungi have emerged five decades later. A provocative article titled "Rock Eating Fungi" was published in the journal *Nature*. Evidence of fungal hyphae, which are likely mycorrhizal, physically or chemically penetrating feldspar grains, was reported in this publication (Smits and Wallander 2017). Mycorrhizal hyphae reportedly use strongly complexing organic acid exudates to penetrate aluminum silicate minerals. The mycorrhiza can absorb minerals released by weathering in these pores and transfer them to the host plant's roots, avoiding competition from other organisms for uptake (Van Breemen et al. 2000).

According to several studies, ectomycorrhizal fungi can only dissolve minerals in order to obtain nutrients, such as phosphorus (P), which is an essential nutrient found in the mineral apatite; there have been reports of "tunnels" in mineral grains, which are likely the result of ectomycorrhizal fungus excreting organic acids that dissolve mineral structure and absorb nutrients. Ectomycorrhiza roots have been shown to absorb nutrients directly from the minerals that form rocks. This suggests that the global success of the ectomycorrhizal tree may be attributed to a direct result of ectomycorrhizal mineral dissolution (Koele et al. 2014). In order to support a vast hyphae network with a high absorption area for nutrient absorption, mycorrhizal types use their host photosynthetic organisms. The nutrient-foraging abilities of the fungus enhance root functioning in trees so that AM and EM fungi fully encase tree root tips to absorb the soil-root interface. By excluding low molecular weight organic compounds, EM fungi can improve mineral weathering and regulate the movement of elements from soil to trees (Quirk et al. 2012).

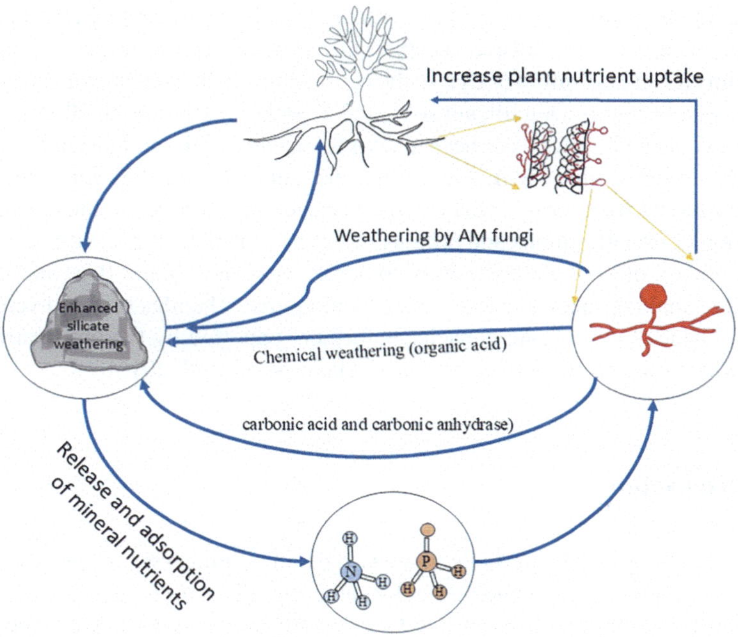

Fig. 7 AM fungi and weathering phenomena in soil

Arbuscular mycorrhiza attach themselves to roots and spread out into the soil, where they feed on the carbon and energy of their host plant and scavenge for nutrients like PO_{42}-, NH_4^+, and K and Zn as shown in Fig. 7. The root-depletion zone of a plant is greatly expanded when its carbon is invested in fungal biomass rather than roots because fungi have branched hyphae (1–10 μm) that significantly improve the bioactive surface area compared to roots. These characteristics of AM fungi probably affect how rocks weather (Verbruggen et al. 2020).

Recent reviews of the mechanisms of silicate weathering by EM and AM fungi have examined how minerals weather. There are three main ways that mycorrhizal fungi weather rocks, with a particular focus on their interaction with carbonate rocks: (1) chemical weathering caused by complicated metal cations, binding cations, and released organic acids that acidify the rhizosphere, (2) physical weathering brought on by minerals being penetrated or trenched by mycorrhiza hyphae, and (3) action of carbonic acid produced from respiratory CO_2 and the enzyme carbonic anhydrase, which catalyzes this acid's synthesis (Thorley et al. 2015).

It is thought that mycorrhizal fungi have considerably higher rates of mineral dissolution in the current (underground) rock weathering timescales. It is well known that the introduction of early land plants greatly accelerated terrestrial weathering. It is spread more widely because of their partnership with mycorrhizal fungi (Verbruggen et al. 2020).

Feldspar grains from podzol E horizons were found to have Tunnel-like features. These tunnels were thought to have formed due to organic anions released at the tips of fungal hyphae, most likely ectomycorrhizal, dissolving minerals. Fungal feldspar tunneling would aid in feldspar weathering and result in the influx of dissolved potassium (K) and cadmium (Cd) into the ecosystem. In the majority of ecosystems, feldspar weathering is a significant source of K and Ca (Smits et al. 2005).

Ectomycorrhizal fungi have been shown to actively seek out nutrients and preferentially colonize areas with higher nutrient content, and they are believed to enhance mineral weathering. This is large because they produce organic acids. The production of organic anions and protons by ectomycorrhizal hyphae tips can facilitate the release of vital elements from minerals, such as calcium (Ca) and magnesium (Mg), making them accessible for fungal uptake. The idea of ectomycorrhizal fungi's enhanced weathering capacity is further supported by the observation that they produce mineral tunneling features (Remiszewski et al. 2016).

13 Discussion

Mycorrhizal fungi and bacteria facilitate two essential processes in Earth's ecosystem, i.e., biomineralization and mineral weathering. Plant roots' exudates send signals that draw microorganisms and start various advantageous processes. The ability of bacteria to enhance plant growth (rhizobacteria) and their part in the mineralization of essential nutrients, especially phosphorus, nitrogen, and potassium, are well known. Mycorrhiza fungi, which coexist symbiotically with plant roots, release organic acids and enzymes that dissolve minerals, enabling plants to absorb vital nutrients like calcium and phosphorus. They play a significant part in the weathering of minerals. By precipitating minerals, these fungi also aid in the biomineralization process, which helps to form and stabilize soil. Similarly, bacteria impact these processes through their metabolic activities; they produce compounds such as organic acids that improve the dissolution of minerals and aid in creating biogenic minerals. These microorganisms work together to maintain the productivity and health of terrestrial ecosystems by accelerating the cycle of elements.

References

Alori ET, Glick BR, Babalola OO (2017) Microbial phosphorus solubilization and its potential for use in sustainable agriculture. Front Microbiol 8:971. https://doi.org/10.3389/fmicb.2017.00971

Atkinson D, Watson CA (2000) The beneficial rhizosphere: a dynamic entity. Appl Soil Ecol 15(2):99–104. https://doi.org/10.1016/S0929-1393(00)00084-6

Aung K, Jiang Y, He SY (2018) The role of water in plant–microbe interactions. Plant J 93(4):771–780. https://doi.org/10.1111/tpj.13795

Bacon CW, White JF (2016) Functions, mechanisms and regulation of endophytic and epiphytic microbial communities of plants. Symbiosis 68(1–3):87–98. https://doi.org/10.1007/s13199-015-0350-2

Badri DV, Vivanco JM (2009) Regulation and function of root exudates. Plant Cell Environ 32(6):666–681. https://doi.org/10.1111/j.1365-3040.2009.01926.x

Badri DV, Weir TL, van der Lelie D, Vivanco JM (2009) Rhizosphere chemical dialogues: plant-microbe interactions. Curr Opin Biotechnol 20(6):642–650. https://doi.org/10.1016/j.copbio.2009.09.014

Bai J, Gao H, Xiao R, Wang J, Huang C (2012) A review of soil nitrogen mineralization as affected by water and salt in coastal wetlands: issues and methods. Clean (Weinh) 40(10):1099–1105. https://doi.org/10.1002/clen.201200055

Barea J-M, Pozo M-J, Azcón R, Azcón-Aguilar C (2013) Microbial interactions in the rhizosphere. Mol Microbial Ecol Rhizosphere. https://doi.org/10.1002/9781118297674ch4

Bashir I, War AF, Rafiq I, Reshi ZA, Rashid I, Shouche YS (2022) Phyllosphere microbiome: diversity and functions. Microbiol Res 254:126888. https://doi.org/10.1016/j.micres.2021.126888

Bejerano RN, Navarro-Ródenas A, Gutiérrez A (2014) Types of Mycorrhizal Association. Soil Biology 69–80. https://doi.org/10.1007/978-3-642-40096-4_5

Benzerara K, Miot J, Morin G, Ona-Nguema G, Skouri-Panet F, Férard C (2010) Importance, mécanismes et implications environnementales de la biominéralisation par les microorganismes. Comptes Rendus – Geoscience 343(2):160–167. https://doi.org/10.1016/j.crte.2010.09.002

Boskey AL (2003) Biomineralization: an overview. Connect Tissue Res 44(1):5–9. https://doi.org/10.1080/03008200390152007

Braissant O, Cailleau G, Aragno M, Verrecchia EP (2004) Biologically induced mineralization in the tree Milicia excelsa (Moraceae): its causes and consequences to the environment. Geobiology 2(1):59–66. https://doi.org/10.1111/j.1472-4677.2004.00019.x

Bringel F, Couée I (2015) Pivotal roles of phyllosphere microorganisms at the interface between plant functioning and atmospheric trace gas dynamics. Front Microbiol 6. https://doi.org/10.3389/fmicb.2015.00486

Brundrett M (2004) Diversity and classification of mycorrhizal associations. Biol Rev Camb Philos Soc 79(3):473–495. https://doi.org/10.1017/S1464793103006316

Cheng YT, Zhang L, He SY (2019) Plant-microbe interactions facing environmental challenge. Cell Host Microbe 26(2):183–192. https://doi.org/10.1016/j.chom.2019.07.009

Chong TM, Koh CL, Sam CK, Choo YM, Yin WF, Chan KG (2012) Characterization of quorum sensing and quorum quenching soil bacteria isolated from Malaysian tropical montane forest. Sensors 12(4):4846–4859. https://doi.org/10.3390/s120404846

Dai Z, Yu M, Chen H, Zhao H, Huang Y, Su W, Xia F, Chang SX, Brookes PC, Dahlgren RA, Xu J (2020) Elevated temperature shifts soil N cycling from microbial immobilization to enhanced mineralization, nitrification and denitrification across global terrestrial ecosystems. Glob Chang Biol 26(9):5267–5276. https://doi.org/10.1111/gcb.15211

Deangelis KM, Lindow SE, Firestone MK (2008) Bacterial quorum sensing and nitrogen cycling in rhizosphere soil. FEMS Microbiol Ecol 66(2):197–207. https://doi.org/10.1111/j.1574-6941.2008.00550.x

Dearnaley JDW (2007) Further advances in orchid mycorrhizal research. Mycorrhiza 17(6):475–486. https://doi.org/10.1007/s00572-007-0138-1

Faget M, Nagel KA, Walter A, Herrera JM, Jahnke S, Schurr U, Temperton VM (2013) Root-root interactions: extending our perspective to be more inclusive of the range of theories in ecology and agriculture using in-vivo analyses. Ann Bot 112(2):253–266. https://doi.org/10.1093/aob/mcs296

Finlay, R.D., 2004. Mycorrhizal fungi and their multifunctional roles, in: Mycologist. Cambridge University Press 18(2), 91–96. doi:https://doi.org/10.1017/S0269915X04002058

Frankel RB, Bazylinski DA (2003) Biologically induced mineralization by bacteria. Rev Mineral Geochem 54:95–114

Geisseler D, Horwath WR, Joergensen RG, Ludwig B (2010) Pathways of nitrogen utilization by soil microorganisms – a review. Soil Biol Biochem 42(12):2058–2067. https://doi.org/10.1016/j.soilbio.2010.08.021

Gilbert PUPA, Bergmann KD, Boekelheide N, Tambutté S, Mass T, Marin F, Adkins JF, Erez J, Gilbert B, Knutson V, Cantine M, Ortega Hernández J, Knoll AH (2022) Biomineralization. Integrating mechanism and evolutionary history. Sci Adv 8:eabl9653. https://www.science.org
Gomes T, Pereira JA, Benhadi J, Lino-Neto T, Baptista P (2018) Endophytic and epiphytic phyllosphere fungal communities are shaped by different environmental factors in a Mediterranean ecosystem. Microb Ecol 76(3):668–679. https://doi.org/10.1007/s00248-018-1161-9
Haichar F. el Z, Santaella C, Heulin T, Achouak W (2014) Root exudates mediated interactions belowground. Soil Biology and Biochemistry 77:69–80. https://doi.org/10.1016/j.soilbio.2014.06.017
Haldar S, Sengupta S (2015) Plant-microbe cross-talk in the rhizosphere: insight and biotechnological potential. Open Microbiol J 9:1–7
Hart MM, Klironomos JN (2003) Diversity of arbuscular mycorrhizal fungi and ecosystem functioning. In: Mycorrhizal ecology, vol 157. Springer
Havers C (1691) Osteologia nova. Samuel Smith, London
Hu J, Liu X, Ma PX (2011) Biomineralization and bone regeneration: principles of regenerative medicine, 2nd edn. Elsevier, pp 733–745. https://doi.org/10.1016/B978-0-12-381422-7.10040-9
Jain A, Chakraborty J, Das S (2020) Underlying mechanism of plant–microbe crosstalk in shaping microbial ecology of the rhizosphere. Acta Physiol Plant 42(1). https://doi.org/10.1007/s11738-019-3000-0
Jiang Z, Liu Y, Yang J, Brookes PC, Gunina A (2021) Rhizosphere priming regulates soil organic carbon and nitrogen mineralization: the significance of abiotic mechanisms. Geoderma 385:114877. https://doi.org/10.1016/j.geoderma.2020.114877
Johnson D, IJdo M, Genney DR, Anderson IC, Alexander IJ (2005) How do plants regulate the function, community structure, and diversity of mycorrhizal fungi. J Exp Bot 56(417):1751–1760. https://doi.org/10.1093/jxb/eri192
Khan AL, Jilani G, Akhtar MS, Naqvi SM, Rasheed M, Ali Khan A, Jilani G, Saleem Akhtar M, Muhammad Saqlan Naqvi S, Rasheed M (2009) A phosphorus solubilizing bacteria: occurrence, mechanisms and their role in crop production. J Agric Biol Sci 1(1):48–58
Koele N, Dickie IA, Blum JD, Gleason JD, de Graaf L (2014) Ecological significance of mineral weathering in ectomycorrhizal and arbuscular mycorrhizal ecosystems from a field-based comparison. Soil Biol Biochem 69:63–70. https://doi.org/10.1016/j.soilbio.2013.10.041
Kumar J, Singh D, Ghosh P, Kumar A (2017) Endophytic and epiphytic modes of microbial interactions and benefits. In: Plant-microbe interactions in agro-ecological perspectives, vol 1. Springer Singapore, pp 255–771. https://doi.org/10.1007/978-981-10-5813-4_12
Lambers H, Mougel C, Jaillard B, Hinsinger P (2009) Plant-microbe-soil interactions in the rhizosphere: an evolutionary perspective. Plant Soil 321(1–2):83–115. https://doi.org/10.1007/s11104-009-0042-x
Lee EH, Eo JK, Ka KH, Eom AH (2013) Diversity of arbuscular mycorrhizal fungi and their roles in ecosystems. Mycobiology 41(3):121–125. https://doi.org/10.5941/MYCO.2013.41.3.121
Lemanceausoils P, Bauer P, Kraemer S, Briat JF (2009) Iron dynamics in the rhizosphere as a case study for analyzing interactions between soils, plants and microbes. Plant Soil 321(1–2):513–535. https://doi.org/10.1007/s11104-009-0039-5
Leveau JH (2019) A brief from the leaf: latest research to inform our understanding of the phyllosphere microbiome. Curr Opin Microbiol 49:41–49. https://doi.org/10.1016/j.mib.2019.10.002
Masood S, Bano A (2016) Mechanism of potassium solubilization in the agricultural soils by the help of soil microorganisms: Potassium Solubilizing Microorganisms for Sustainable Agriculture. Springer India, pp 137–147. https://doi.org/10.1007/978-81-322-2776-2_10
Mina D, Pereira JA, Lino-Neto T, Baptista P (2020) Epiphytic and endophytic bacteria on olive tree phyllosphere: exploring tissue and cultivar effect. Microb Ecol 80(1):145–157. https://doi.org/10.1007/s00248-020-01488-8

Mohan VK, Joshi SR (2024) Mycorrhizal fungi as ecofriendly interventions for crop productivity: industrial microbiology and biotechnology. Springer Nature Singapore, Singapore, pp 663–677. https://doi.org/10.1007/978-981-97-1912-9_26
Mohanram S, Kumar P (2019) Rhizosphere microbiome: revisiting the synergy of plant-microbe interactions. Ann Microbiol 69(4):307–320. https://doi.org/10.1007/s13213-019-01448-9
Mommer L, Hinsinger P, Prigent-Combaret C, Visser EJW (2016a) Advances in the rhizosphere: stretching the interface of life. Plant Soil 407(1–2):1. https://doi.org/10.1007/s11104-016-3040-9
Mommer L, Kirkegaard J, van Ruijven J (2016b) Root-root interactions: towards a rhizosphere framework. Trends Plant Sci 21(3):209–217. https://doi.org/10.1016/j.tplants.2016.01.009
Munkvold L, Kjøller R, Vestberg M, Rosendahl S, Jakobsen I (2004) High functional diversity within species of arbuscular mycorrhizal fungi. New Phytol 164(2):357–364. https://doi.org/10.1111/j.1469-8137.2004.01169.x
Myrold DD, Bottomley PJ (2008) Nitrogen mineralization and immobilization. American Society of Agronomy
Nadarajah K, Abdul Rahman NSN (2021) Plant-microbe interaction: aboveground to belowground, from the good to the bad. Int J Mol Sci 22(19). https://doi.org/10.3390/ijms221910388
Nobbe F, Hiltner L (1896) Inoculation of the soil for cultivating leguminous plants. U.S. patent 570 813
Ouyang Y, Norton JM (2020) Short-term nitrogen fertilization affects microbial community composition and nitrogen mineralization functions in an agricultural soil. Appl Environ Microbiol 86(5). https://doi.org/10.1128/AEM.02278-19
Pantigoso HA, Newberger D, Vivanco JM (2022) The rhizosphere microbiome: plant–microbial interactions for resource acquisition. J Appl Microbiol 133(5):2864–2876. https://doi.org/10.1111/jam.15686
Pathan SI, Ceccherini MT, Sunseri F, Lupini A (2020) Rhizosphere as hostspot for plant-soil microbe interaction: Carbon and Nitrogen Cycling in Soil. Springer Singapore, pp 17–44. https://doi.org/10.1007/978-981-13-7264-3
Pett-Ridge J, Shi S, Estera-Molina K, Nuccio E, Yuan M, Rijkers R, Swenson T, Zhalnina K, Northen T, Zhou J, Firestone MK (2021) Rhizosphere biology: interactions between microbes and plants. In: Rhizosphere carbon turnover from cradle to grave: the role of microbe–plant interactions. Springer, Singapore, pp 51–73. https://doi.org/10.1007/978-981-15-6125-2_2
Pii Y, Mimmo T, Tomasi N, Terzano R, Cesco S, Crecchio C (2015) Microbial interactions in the rhizosphere: beneficial influences of plant growth-promoting rhizobacteria on nutrient acquisition process. A review. Biol Fertil Soils 51(4):403–415. https://doi.org/10.1007/s00374-015-0996-1
Powell JR, Rillig MC (2018) Biodiversity of arbuscular mycorrhizal fungi and ecosystem function. New Phytol 220(4):1059–1075. https://doi.org/10.1111/nph.15119
Quirk J, Beerling DJ, Banwart SA, Kakonyi G, Romero-Gonzalez ME, Leake JR (2012) Evolution of trees and mycorrhizal fungi intensifies silicate mineral weathering. Biol Lett 8(6):1006–1011. https://doi.org/10.1098/rsbl.2012.0503
Rehman F ur 2020. Chemistry of plant–microbe interactions in rhizosphere and rhizoplane. Indian J Pure Appl Biosci 8(5), 11–19. doi:https://doi.org/10.18782/2582-2845.8350
Remiszewski KA, Bryce JG, Fahnestock MF, Pettitt EA, Blichert-Toft J, Vadeboncoeur MA, Bailey SW, Lyon CB (2016) Elemental and isotopic perspectives on the impact of arbuscular mycorrhizal and ectomycorrhizal fungi on mineral weathering across imposed geologic gradients. Chem Geol 445(16):164–171. https://doi.org/10.1016/j.chemgeo.2016.05.005
Richardson AE, Simpson RJ (2011) Soil microorganisms mediating phosphorus availability. Plant Physiol 156(3):989–996. https://doi.org/10.1104/pp.111.175448
Shrestha S, Gautam TP, Mandal TN, Aryal HP (2021) Ecology and diversity of ectomycorrhiza in moist Tropical Forest of Sunsari District, Eastern Nepal. J Inst Sci Technol 26(1):35–42. https://doi.org/10.3126/jist.v26i1.37815

Singh BK, Millard P, Whiteley AS, Murrell JC (2004) Unravelling rhizosphere-microbial interactions: opportunities and limitations. Trends Microbiol 12(8):386–393. https://doi.org/10.1016/j.tim.2004.06.008
Smits MM, Wallander H (2017) Role of Mycorrhizal Symbiosis in Mineral Weathering and Nutrient Mining from Soil Parent Material: Mycorrhizal Mediation of Soil: Fertility, Structure, and Carbon Storage. Elsevier Inc., pp 35–46. https://doi.org/10.1016/B978-0-12-804312-7.00003-6
Smits MM, Hoffland E, Jongmans AG, Van Breemen N (2005) Contribution of mineral tunneling to total feldspar weathering. Geoderma 125(1–2):59–69. https://doi.org/10.1016/j.geoderma.2004.06.005
Soudzilovskaia NA, Vaessen S, Barcelo M, He J, Rahimlou S, Abarenkov K, Brundrett MC, Gomes SIF, Merckx V, Tedersoo L (2020) FungalRoot: global online database of plant mycorrhizal associations. New Phytol 227(3):955–966. https://doi.org/10.1111/nph.16569
Spohn M, Kuzyakov Y (2013) Phosphorus mineralization can be driven by microbial need for carbon. Soil Biol Biochem 61:69–75. https://doi.org/10.1016/j.soilbio.2013.02.013
Straker CJ (1996) Ericoid mycorrhiza: ecological and host specificity. Mycorrhiza 6(4):215–225. https://doi.org/10.1007/s005720050129
Teotia P, Kumar V, Kumar M, Shrivastava N, Varma A (2016) Rhizosphere microbes: potassium solubilization and crop productivity—present and future aspects: potassium solubilizing microorganisms for sustainable agriculture. Springer India, pp 315–325. https://doi.org/10.1007/978-81-322-2776-2_22
Thapa S, Prasanna R (2018) Prospecting the characteristics and significance of the phyllosphere microbiome. Ann Microbiol 68(5):229–245. https://doi.org/10.1007/s13213-018-1331-5
Thorley RMS, Taylor LL, Banwart SA, Leake JR, Beerling DJ (2015) The role of forest trees and their mycorrhizal fungi in carbonate rock weathering and its significance for global carbon cycling. Plant Cell Environ 38(9):1947–1961. https://doi.org/10.1111/pce.12444
Tian J, Ge F, Zhang D, Deng S, Liu X (2021) Roles of phosphate solubilizing microorganisms from managing soil phosphorus deficiency to mediating biogeochemical p cycle. Biology 10(2):1–19. https://doi.org/10.3390/biology10020158
Van Breemen N, Finlay R, Lundström U, Jongmans AG, Giesler R, Olsson M (2000) Mycorrhizal weathering: a true case of mineral plant nutrition. Biogeochemistry 49:53
Van Leeuwenhoek A (1693) An extract of a letter from Mr. Anth. Van. Leeuwenhoek, cointaining several observations on the texture of the bones of animals compared with that of wood: on the bark of tree: on the little scales found on the cuticula, etc. J R Soc 17:838–843
Velmourougane K, Saxena G, Prasanna R (2017) Plant-microbe interactions in the rhizosphere: mechanisms and their ecological benefits: plant-microbe interactions in agro-ecological perspectives. Springer Singapore, pp 193–219. https://doi.org/10.1007/978-981-10-6593-4_7
Verbruggen E, Struyf E, Vicca S (2020) Can arbuscular mycorrhizal fungi speed up carbon sequestration by enhanced weathering? Plants People Planet 3(5):445–453. https://doi.org/10.1002/ppp3.10179
Vishwakarma K, Kumar N, Shandilya C, Mohapatra S, Bhayana S, Varma A (2020) Revisiting plant–microbe interactions and microbial consortia application for enhancing sustainable agriculture: a review. Front Microbiol 11. https://doi.org/10.3389/fmicb.2020.560406
Wan W, Hao X, Xing Y, Liu S, Zhang X, Li X, Chen W, Huang Q (2020) Spatial differences in soil microbial diversity caused by pH-driven organic phosphorus mineralization. Land Degrad Dev 32(2):766–776. https://doi.org/10.1002/ldr.3734
Wang NQ, Kong CH, Wang P, Meiners SJ (2020) Root exudate signals in plant–plant interactions. Plant Cell Environ 44(4):1044–1058. https://doi.org/10.1111/pce.13892
Wang Y, Luo D, Xiong Z, Wang Z, Gao M (2023) Changes in rhizosphere phosphorus fractions and phosphate-mineralizing microbial populations in acid soil as influenced by organic acid exudation. Soil Tillage Res 225:105543. https://doi.org/10.1016/j.still.2022.105543
Weiner S (2007) Biomineralization: A structural perspective. J Struct Biol 163(3):229–234. https://doi.org/10.1016/j.jsb.2008.02.001

Wu CH, Bernard SM, Andersen GL, Chen W (2009) Developing microbe-plant interactions for applications in plant-growth promotion and disease control, production of useful compounds, remediation and carbon sequestration. Microb Biotechnol 2(4):428–440. https://doi.org/10.1111/j.1751-7915.2009.00109.x

Xiao R, Man X, Duan B, Cai T, Ge Z, Li X, Vesala T (2022) Changes in soil bacterial communities and nitrogen mineralization with understory vegetation in boreal larch forests. Soil Biol Biochem 166:108572. https://doi.org/10.1016/j.soilbio.2022.108572

Xiong Q, Hu J, Wei H, Zhang H, Zhu J (2021) Relationship between plant roots, rhizosphere microorganisms, and nitrogen and its special focus on rice. Agriculture (Switzerland) 11(3). https://doi.org/10.3390/agriculture11030234

Yao H, Sun X, He C, Maitra P, Li XC, Guo LD (2019) Phyllosphere epiphytic and endophytic fungal community and network structures differ in a tropical mangrove ecosystem. Microbiome 7(1):57. https://doi.org/10.1186/s40168-019-0671-0

Yin T, Lin H, Dong Y, Li B, He Y, Liu C, Chen X (2021) A novel constructed carbonate-mineralized functional bacterial consortium for high-efficiency cadmium biomineralization. J Hazard Mater 401:123269. https://doi.org/10.1016/j.jhazmat.2020.123269

Zhang R, Vivanco JM, Shen Q (2017) The unseen rhizosphere root–soil–microbe interactions for crop production. Curr Opin Microbiol 37:8–14. https://doi.org/10.1016/j.mib.2017.03.008

Bioremediation of Hydrocarbon-Contaminated Sites: A Geomicrobiological Approach

Deepshikha, Pradeep Verma, Prasann Kumar, and Komal Agrawal

Contents

Abstract The natural ecology and human health have been adversely affected by environmental pollution resulting from organic pollutants, heavy metals, and potentially hazardous waste products. While soil naturally contains heavy metals, rapid industrialization and human activities like uncontrolled use of agrochemicals have significantly increased their concentrations globally. Bioremediation provides an environmentally acceptable and effective approach to recovering areas damaged with toxic compounds by utilizing the intrinsic biological techniques of microbes

Deepshikha · K. Agrawal (✉)
Department of Microbiology, School of Bio Engineering and Biosciences, Lovely Professional University, Phagwara, Punjab, India

P. Verma
Bioprocess and Bioenergy Laboratory (BPBEL), Department of Microbiology, Central University of Rajasthan, Bandarsindri, Rajasthan, India
e-mail: pradeepverma@curaj.ac.in

P. Kumar
Department of Agronomy, School of Agriculture, Lovely Professional University, Phagwara, Punjab, India

S. R. Joshi et al. (eds.), *Mineral Transformation and Bioremediation by Geo-Microbes*, Interdisciplinary Biotechnological Advances,
https://doi.org/10.1007/978-981-96-3033-2_15

and plants to eliminate unstable chemicals. Recent research is enhancing the ability of naturally occurring microbes to cleave detrimental compounds that are released in the ecosystem. Among distinct bioremediation methods, microbe-based technique stands out for their effectiveness, lack of toxic end-products, and the natural presence of microbes that can biodegrade nutrients from surface contaminants. These microbes facilitate both in situ and ex situ bioremediation processes effectively. The advancement of bioremediation technology depends on comprehending the complexities of biodegradation mechanisms. The toxic petroleum hydrocarbons and their associated by-products are effectively removed economically and effectively by using an integrated approach that combines biostimulation and bioaugmentation. Thus, the current chapter will elaborate on the microbe-based remediation of the hydrocarbons along with its limitations and prospects.

Keywords Hydrocarbon · Remediation · Pollutants · Biodegradation · Bioaugmentation

Abbreviations

PAH	Polycyclic aromatic hydrocarbons
PCB	Polychlorinated biphenyls
PH	Petroleum hydrocarbon
PMUA	Poly (ethylene) glycol-modified urethane acrylate
POP	Persistent organic pollutants
TPH	Total petroleum hydrocarbons
T-RFLP	Terminal restriction fragment length polymorphism

1 Introduction

The rising community and rapid economic expansion in developing nations significantly contribute to the expanding necessity for energy. Since 1990, the world's energy consumption has risen by >50%, and between 2015 and 2040, it is predicted to rise by ~28%. Fossil fuels continue to be the predominant energy source among all major sectors, with industries and transportation together accounting for ~75% of the cumulative energy consumption by 2040 (Imam et al. 2019). The use of petroleum hydrocarbons (PHs) as energy sources presents numerous environmental threats. PHs are pollutants generated from extensive production, transportation, shipping activities, off-shore oil production, coastal oil refining, and accidental spills (Gao et al. 2022). Human operations, including industrial liquid discharges and municipal run-offs, contribute to PH pollution as well and it adversely affects the environment and poses health risks to various life forms. This categorization of PHs as toxic compounds places them among priority pollutants, which also include a significant number of organic substances and by-products (Ambaye et al. 2022).

Among these, persistent organic pollutants (POPs) and polycyclic aromatic hydrocarbons (PAHs) are priority pollutants because of their long-lasting presence in the environment and their resistance to natural degradation processes (Gennadiev et al. 2015). Typically, the average quantities of PH include approximately 57%, 29%, and 14% of aliphatic hydrocarbons, aromatic hydrocarbons, and asphaltenes respectively, containing various polar compounds (Auti et al. 2019). Aliphatic hydrocarbons are either unsaturated or saturated and can occur in branched as well as in linear chain configurations. Polycyclic structures and stable monocyclic rings, like benzene, are characteristics of aromatic hydrocarbons (Alotaibi et al. 2021). The extensive use of PH is a significant contributor to soil pollution, a global environmental concern. PAHs are particularly notable for their high toxicity, functioning as mutagens and carcinogens, which ultimately present significant health threats to humans (Brzeszcz and Kaszycki 2018). Thus, the elimination of PH from the environment becomes imperative for long-term sustainability. To address PH pollutants in soils a variety of conventional approaches are used, involving physical, chemical, and biological procedures (Quiza et al. 2015). However, bioremediation, which uses microorganisms to treat contaminated soil, emerges as an intriguing approach and can be conducted under both anaerobic and aerobic conditions. These methods are effective in degrading heavy metals and organic pollutants, potentially by changing microbial metabolism through processes such as transformation/sequestration ratios, biotransformation, bioaccumulation, and bioleaching exercises (Aydin et al. 2017). Thus, the current chapter emphasizes the influence of hydrocarbon on the environment, and how it can be bio-remediated along with the factors that affect the biodegradation efficiency. In addition, the chapter will also provide an insight into the mechanism and the roles of microbes in the biodegradation process.

2 Environmental Impact of Hydrocarbon Contaminants

The physical characteristics of soil are affected by its pH, which can change its surface, compaction, rigidity, resistance to penetration, and saturated hydraulic conductivity. Additionally, these hydrocarbons can impact the soil's chemical attributes, such as the concentration and content of minerals and heavy metals (Zhang et al. 2024). Furthermore, petroleum-derived substances lead to water (H_2O) and oxygen (O_2) deficiency, as well as reduced availability of nitrogen (N_2) and phosphorus (P_4) (Mohammadi et al. 2020). This contamination results in organic pollution of groundwater, affecting its usability and causing economic loss, decreased agricultural productivity, and environmental challenges. Additionally, the presence of PH poses risks to microorganisms, plants, animals, and humans (Ahmed and Fakhruddin 2018).

PH contaminations can originate from multiple sources, with crude oil being a notable example. Crude oil is predominantly composed of short-chain hydrocarbons, including a complex array of paraffinic, alicyclic, and aromatic hydrocarbons, along with small quantities of non-hydrocarbon substances such as phenol (Abbasian et al. 2015). The carbon (C) content generally falls between 83% and 87%, and the hydrogen content ranges from 10% to 14%. Additionally, small amounts of N_2, sulfur (S), O_2, and metals like nickel (Ni) may also be present (Sjöblom et al. 2003). Furthermore, hydrocarbon pollutants can cause both immediate and long-term effects, including genetic mutations, teratogenicity, immunotoxicity, carcinogenesis, and high bioaccumulation potential, resulting in degradation of animals, plants, and overall ecosystem (Varjani and Upasani 2019). These compounds can accumulate in various organisms, such as fish, phytoplankton, and other aquatic species, and subsequently enter the food chain, causing significant disturbances to the exposed organisms (Sakshi et al. 2019). The biomagnification and bioaccumulation of hydrocarbons can lead to reproductive, neurological, immune system, and overall development problems, including teratogenic effects, and are also linked to cancer cells in lungs, skin, liver, bladder, and gastrointestinal system (Pattabhiramaiah et al. 2018). Consequently, exposure to hydrocarbon contaminants poses numerous health risks. Furthermore, hydrocarbon pollutants and its associated toxins reduce species evenness, richness, and phylogenetic diversity, thereby altering community structure in contaminated soil environments (Xu et al. 2018).

3 Biodegradation and Bioremediation of Hydrocarbon Contaminates

Bioremediation encompasses various processes that employ biological systems to rehabilitate or cleanse polluted environments (Saadoun and Al-Ghzawi 2005). The role of microbial communities is frequently highlighted in bioremediation efforts. Many native microorganisms possess the capacity to facilitate environmental recovery through the oxidation, immobilization, or transformation of contaminants (Vishwakarma et al. 2020). They use the resultant metabolites to grow by a variety of methods, including biosorption, biostimulation, bio-attenuation, and bioaugmentation (Coelho et al. 2015) some of which are discussed in Table 1. A figurative representation of various bioremediation techniques is demonstrated in Fig. 1.

Biosorption represents a practical application within sustainable development and biotechnology. It relies on the interactions between biomass and pollutants, often involving multiple simultaneous interactions. Factors affecting these interactions consequently impact the process's efficiency (Fomina and Gadd 2014). The interactions that occur in a biosorption procedure include complexation/coordination, ion exchange, microprecipitation, chemisorption, physisorption, electrostatic interactions, and reduction (de Freitas et al. 2019). A study comparing the biodegradation and biosorption capabilities of various percentages of culture supernatants from *Aspergillus parasiticus* NRRL:3386, as well as varied quantities of living (fungal pellets) and dead biomasses (Filter-dried fungal pellets, autoclaved), was carried out. According to the study, after 4 days of incubation, 1 g of living biomass reduced the amount of petroleum by almost 80%. On the other hand, living biomass

Table 1 Overview of strategies applied for bioremediation

Strategy	Pollutants	Microorganism	Testing method	Removal efficiency (%)	Duration	Reference
BA	Crude oil	*Pseudomonas aeruginosa* PP3 and *Pseudomonas aeruginosa* PP4	SM	52–67	–	Muthukumar et al. (2023)
BA+BS	Petroleum refinery sludge	*Dietzia lutea* (IRB191), *Dietzia lutea* (IRB192), *Stenotrophomonas pavanii* (IRB19) and *Staphylococcus warneri* (BSM19)	–	90.3 ± 3.7	21 days	Behera et al. (2022)
BA	Oily waste	*P. aeruginosa* NCIM 5514	SM	76.14 ± 0.85	56 days	Varjani and Upasani (2021)
BA	Crude oil	FC1 (*P. chrysogenum and C. brachyspora*), FC2 (*S. brevicaulis and S. botryosum*), and FC3 (*S. brevicaulis, S. botryosum,* and *C. sphaerospermum*)	SM	77.62–97	–	Hamad et al. (2021)
BA	Diesel oil	*Sphingobacteriia Actinobacteria*	SM	88	64.5 weeks	Woźniak-Karczewska et al. (2019)
BA	Crude oil	*Rhodococcus erythropolis*	SM	30–38	182 days	Pacwa-Płociniczak et al. (2019)
BA+BS	Crude oil	*Eichhornia crassipes and genus Burkholderia*	SM	36–51	30 days	Tao et al. (2019)
Surfactants-assisted biodegradation	Phenanthrene	*Massilia*, *Bacillus*, *Lysobacter*, *Archrobacter*, and *Phenylobacterium* dominant genera for degradation	SM	72	Upto 35 days	Lu et al. (2019)
	Fluoranthene			64		
	Pyrene			58		
NA + BA+BS	Engine oil	*P. putida* TPHK-1 and *P. aeruginosa* TPHK-4	SM	31–75	210 days	Ramadass et al. (2018)
NA + BA+BS	Petroleum refinery waste	Methanogenic microbes (*Bacillus*, *Coprothermobacter*, *Rhodobacter*, *Pseudomonas*, etc.)	VM	57–75	120 days	Roy et al. (2018a, b)

BA Bioaugmentation; *BS* Biostimulation; *NS* Natural attenuation; *FC* Fungal consortia; *SM* Soil microcosm; *VM* Vial microcosm

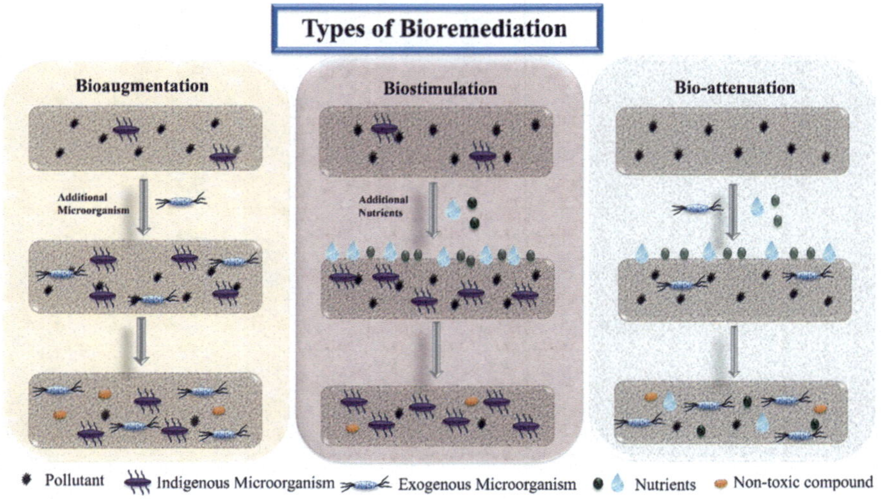

Fig. 1 Illustration of different types of bioremediation initiatives for the biodegradation of petroleum hydrocarbons

outperformed cell-free culture supernatant in terms of petroleum degradation. It's biosorption was obtained at over 50% with 1 g and 2.5 g of dead biomass, over 70% with 5 g and 7.5 g, and over 80% with 10 g of dead biomass. Additionally, the petroleum elimination efficiencies of 10 g of dead biomass and 2 g of live biomass both exceeded 80% (Ozyurek and Avcioglu 2020).

Biostimulation, another method for biodegradation of PHs, typically involves injecting specific nutrients into contaminated sites. Fertilizers, growth supplements, and trace minerals are added to the environment to enhance the activity and metabolic rate of indigenous microorganisms. Additionally, certain environmental factors such as pH, temperature, and O_2 levels are optimized (Kumar et al. 2011). One of the primary benefits of implementing indigenous bacteria in the biostimulation method is their natural compatibility and vast distribution in the subsurface, enabling them to avoid potential issues linked to environmental incompatibility (Di Gregorio et al. 2013). A study investigated biostimulation by microbial response to nitrate or compost during hydrocarbon biodegradation in petroleum-contaminated soil, as well as the isotopic impacts of δ13C and δ15N. The addition of compost and KNO_3 facilitated the elimination of total petroleum hydrocarbons (TPH), which was associated with a substantial increase in Actinobacteria and Firmicutes phyla. After 90 days of biostimulation, soil alpha diversity decreased. The potassium nitrate (KNO_3) supplementation resulted in a strong significant N isotope effect ($\varepsilon N = -24.20 \pm 9.54‰$) and a significant inverse C isotope effect ($\varepsilon c = 16.6 \pm 0.8‰$) (Liu et al. 2022).

Another method of enhancing the degradation of pollutants involves increasing the concentration of microbes with degradative abilities, these can be natural (Safdari et al. 2018), exotic (Qiao et al. 2014), or engineered microorganisms (Liu et al. 2021), to the contaminated site (Dehnavi and Ebrahimipour 2022). This strategy is based on the assumption that local microbial populations lack the ability to digest complex mixtures such as petroleum or are stressed as a result of recent spill exposure. Therefore, to increase the degradation rates exogenous microorganisms are introduced in the environment such a bioremediation procedure is called bioaugmentation (Adams et al. 2015). A study investigated the impact of bioaugmentation by isolating bacteria from oily waste sludge and assessing their oil degradation and biomass production capabilities in a liquid mineral medium. The bacterial strain was introduced twice into composting experiments with varying levels of final compost: once during the initial 56 days as the primary compost and again during the last 56 days as secondary composting stage. TPHs removal achieved ranged from 67.64% to 89.56% over the two-stage process (Poorsoleiman et al. 2020).

3.1 Factors Influencing Bioremediation Efficacy

Microbial remediation is affected by various elements, involving characteristics of microorganisms, physicochemical properties of contaminants, and environmental factors (pH, temperature, and nutrient availability) (Clarkson and Abubakar 2015). Bacteria that degrade hydrocarbons are prevalent and widespread at contaminated sites because they can rapidly adapt to polluted environments and use toxins as sources of C and energy for their metabolism and development (Chaudhary and Kim 2019). The amount of hydrocarbon biodegradation is in proportion to the presence and accessibility of naturally occurring hydrocarbon-degrading microbes in the polluted environment (Das and Chandran 2011). Bacterial communities that degrade hydrocarbons interact in hydrocarbon-contaminated environments via both synergistic and antagonistic processes. These interactions, whether between different bacterial species or between bacteria and fungi, influence the efficiency of biodegradation (Ying and Wei 2019). Research indicates that the introduction of exogenous bacterial inoculum often fails to enhance biodegradation due to its inability to outcompete native bacteria and cope with predation and environmental factors (Zawierucha et al. 2014). *Cycloclasticus* and *Alcanivorax borkumensis*, in particular, have a synergistic connection that improves polyaromatic hydrocarbon decomposition. *A. borkumensis* uses multiple hydrocarbon pollutants as metabolic sources, which avoid direct competition and produce biosurfactants that aid in PAH emulsification making it accessible to *Cycloclasticus*. Conversely, *Thalassolituus* produces metabolites that inhibit the growth of other oleophilic bacteria, reducing biodegradation rates (Macaulay 2015). Additionally, the interaction between indigenous and exogenous microbes often results in antagonistic effects that hinder their metabolic activities, further limiting the overall effectiveness of hydrocarbon biodegradation (Liu et al. 2020a, b). The favorable application of bioremediation capitalizes significantly on the interdependence of microbial populations, characterized by synergistic interactions. Research has demonstrated that certain microbial consortia involved in hydrocarbon degradation establish cooperative relationships that facilitate the complete breakdown of contaminants (Omrani et al. 2018).

4 Mechanism of the Degradation of Petroleum Hydrocarbons

The biodegradation of PHs is a multifaceted process influenced by presence and concentration of hydrocarbons. There are distinct types of hydrocarbons: saturated, aromatic, fatty acids, phenols, ketones, and esters, as well as resins (including quinolines, pyridines, sulfoxides, amides, and carbazoles) (Atlas and Bragg 2009). In general, hydrocarbons are susceptible to microbial degradation in the order such as linear alkanes > branched alkanes > tiny aromatic hydrocarbons > cyclic alkanes. Biodegradation of hydrocarbons in ecosystems is primarily carried out by bacteria, yeast, and fungi (Mangla et al. 2021). In natural habitats, these microorganisms break down hydrocarbons to different degrees through both aerobic and anaerobic mechanisms. However, metabolic limitations vary depending on the absence or presence of O_2 in mixed microbial populations (consortia), which are essential for degrading complex petroleum hydrocarbon mixtures in environments (Guarino et al. 2017). Generally, the presence of O_2 facilitates rapid and complete biodegradation of most organic pollutants. Through central metabolic pathways like the tricarboxylic acid cycle, these environmental biodegradation processes ultimately break down organic pollutants into smaller fragments. Key metabolic intermediates such as pyruvate, succinate, and acetyl-CoA primarily contribute to the biosynthesis of cellular mass. Additionally, the gluconeogenesis process synthesizes sugars necessary for various biological syntheses (Javaid et al. 2016).

4.1 Aerobic Degradation

Under aerobic conditions, most organic contaminants are rapidly and completely degraded. The initial intracellular attack on these pollutants involves oxidation and activation, with the incorporation of O_2 being facilitated by key enzymatic catalysts such as peroxidases and oxygenases. Specific enzyme systems enable the degradation of petroleum hydrocarbons. Furthermore, other mechanisms involved include microbial cell adhesion to substrates and the development of biosurfactants (Rahman et al. 2003). The microbial degradation of chlorinated oils, PHs, and other substances is facilitated by cytochrome P450 hydroxylases. These enzymes have been isolated from *Candida* species such as *C. apicola*, *C. tropicalis*, and *C. maltosa* (Van Beilen and Funhoff 2007). Alkane oxygenases, which are found in both prokaryotes and eukaryotes, are essential for alkane degradation in aerobic environments. They include integrated membrane di-iron alkane hydroxylases, copper-containing methane monooxygenases, and readily soluble di-iron methane monooxygenases. The efficiency and extent of pollutant biodegradation by fungal enzymes depend on variables such as nutrient availability, O_2, and enzymatic conditions, such as chemical structure, temperature, pH, and cellular transport properties (Zhang et al. 2016). Fungi have evolved mechanisms to mineralize and degrade various organic contaminates, largely due to complex structure of lignin. The extracellular peroxidases generated by these fungi are entrusted for the preliminary oxidation of PHs. Manganese peroxidases passively co-oxidize many PHs by enzyme-mediated lignin peroxidation, while fungal lignin peroxidases actively oxidize a number of PHs. Other enzymes such as lignin peroxidase, cytochrome P450 monooxygenase, laccase, manganese peroxidase, lipases, and proteases, have been extensively studied for their capability to degrade PAHs (Balaji et al. 2014).

4.2 Anaerobic Degradation

In the absence of oxygen, aromatic chemicals like benzoate, halo-benzoate, and chlorophenol are broken down by microbes. In such anaerobic conditions, molecular O_2 is substituted with alternative electron acceptors like manganese, nitrate, sulfate, and iron to degrade hydrocarbons present in groundwater oil plumes. Anaerobic bacteria are separated into two categories: facultative and obligate anaerobes. Obligate anaerobes include most sulfate-reducing bacteria, which strictly require an anaerobic environment. In contrast, facultative anaerobes can utilize either O_2 or hydrocarbons for degradation (Mangla et al. 2021). The two major metabolic processes engaged in the anaerobic breakdown of n-alkanes are distinct from those in aerobic hydrocarbon metabolism. Fumarate is added in the first method, activating the alkane at the subterminal C. This process is analogous to production of benzylsuccinate through the anaerobic breakdown of toluene. However, subsequent reactions diverge, comprising processes such as dehydrogenation and hydration (Widdel and Grundmann 2010). The second mechanism involves carboxylation and is primarily observed in the growth patterns of the sulfate-reducing strain. An alkane-degrading sulfate reducing strain Hxd3 was identified as *Desulfococcus oleovorans*. The microbe is unique among alkane degraders because it converts even-numbered alkanes into odd-numbered cellular fatty acids, while growth of odd-numbered alkanes results in even-numbered cellular fatty acids (So et al. 2003). The most well-understood and widespread anaerobic mechanism is radical-catalyzed incorporation of fumarate to hydrocarbon, which produces substituted succinate derivatives. Numerous alkyl-substituted benzenes and n-alkanes are known to be activated by this reaction (Kube et al. 2004). An investigation identified and employed a quinone-respiring *Bacillus* sp., which uses oil-contaminated soil to degrade crude oil. This strain achieved a maximum degradation of 76.7% of TPH with the assistance of a biosurfactant and the electron mediator anthraquinone-2,6-disulfonate. qPCR analysis revealed that anthraquinone-2,6-disulfonate enhanced electron transfer, reducing nitrite accumulation and improving TPH degradation. This research introduced a bacterial strain effective in degrading TPH from soil and proposed an enhanced denitrification mechanism for petroleum hydrocarbon degradation facilitated by the electron mediator (Liu et al. 2020a, b).

5 Geomicrobiology and Its Scope

The multidisciplinary field of geomicrobiology investigates the role of microbes in the earth's formation and geological processes, as well as their perspectives on the prospects for humanity (Muksy et al. 2023). Geobiology places a lot of emphasis on understanding how biological activities affect an element's cycle. This comprises the disintegration and breakdown of geological materials, the absorption of vital macro- and micronutrients required for life, and their long-term incorporation into the sedimentary record (Wani and Mir 2020).

6 Microbial Communities in Hydrocarbon-Contaminated Sites

Gaining an insight into the metabolic activities and composition of microbes involved in hydrocarbon degradation requires an understanding of their structures, which entails both in situ and ex situ processes. Research employing omics approaches, including metabarcoding, metagenomics, and meta-transcriptomics, across diverse locations has revealed distinct patterns of hydrocarbon contamination (Roy et al. 2018a, b). The most common bacterial genera identified in soil samples collected from contaminated sites were *Anaeromyxobacter*, *Bdellovibrio*, *Megamonas*, *Bacillus*, *Aquicella*, *Pseudomonas*, *Chromobacterium*, *Xanthomonas*, and *Acinetobacter*, analyzed by Illumina MiSeq sequencing platform (Dasgupta et al. 2018). In agricultural soils contaminated with PAHs, the most abundant fungal phyla were Ascomycota and Euryarchaeota (Zhou et al. 2017). Furthermore, analysis of sludge specimens from three different petroleum refineries identified hydrocarbon-degrading genera including *Mycobacterium*, *Geobacter*, and *Pseudomonas*, as well as methanogenic genera such as *Methanosaeta* and *Methanobacterium*, and members of the order Thermoplasmatales (Roy et al. 2018a, b). Studies on bacterial population in polluted sediments from various global coastline revealed a high relative abundance of bacterial families (Geobacteraceae and Desulfobacteracea) and genera (*Thalassospira*, *Alteromonas*, and *Marinobacter*) (Neethu et al. 2019). In sediment samples heavily contaminated with PAHs from three different mangrove regions along the Brazilian coast, members of the family Desulfobacteraceae, including *Desulfatibacillum* and *Desulfococcus*, were identified (Cabral et al. 2018). Additionally, genera associated with detoxification processes were present in the contaminated mangrove samples. These genera included *Desulfitobacterium*, *Vibrio*, *Rhodopseudomonas*, *Burkholderia*, *Pseudomonas*, and *Roseobacter* (de Sousa et al. 2020). Furthermore, such microbial interaction has resulted in the progress of bioremediation procedures. In the context of biodegradation and bioremediation, these microbial associations are known as microbial consortia.

6.1 Microbial Consortia in Bioremediation

A number of factors determine whether a microbial consortium is considered natural, artificial, or synthetic, including the methods used to develop them, the degree of human influence on their composition (involving genetic modifications), and the presence of microbial interactions (Massot et al. 2022). Natural consortia are microbial groups that arise without explicit human intervention. They exist naturally or may appear in the environment, originate from a single source (such as a specific soil), and can be obtained relatively easily through enrichment methods. These methods enhance the growth of certain members within a natural microbial community and are widely used to obtain natural degrading consortia (Ma et al. 2021). These uses contaminate as the singular C and energy source, selective pressure is applied to the microbial population, resulting in an enrichment culture composed of cells capable of thriving under the desired conditions. Conditions of enrichment culture have the ability to foster new interactions and abilities that are not present in the initial population (Table 2). Soil harbors inactive microbes that do not proliferate due to C-limitation (Joergensen and Wichern 2018). These microorganisms possess metabolic characteristics that are either inactive or operating at a minimal level. The enrichment culture's modifications in physical or chemical settings might awaken such latent microorganisms, resulting in the development of novel relationships and functionalities. M10, a naturally occurring PAH-degrading consortia from Antarctic soils, was discovered. This consortium was formed by four sequential enrichment cultures employing anthracene, benzo[a]pyrene, fluorene, and phenanthrene as C and energy sources. A detailed investigation of its composition, using a polyphasic technique that incorporated combined culture-dependent and culture-independent methodologies, discovered that the consortium was mostly composed of 17 separate bacterial populations. These populations belonged to genera *Sphingomonas*, *Pseudomonas*, *Stenotrophomonas*, and *Pedobacter/Brevibacterium* (Vázquez et al. 2013).

Artificial consortia refer to a broad category of microbial associations made up of selected strains in specific proportions. In a study, microbes were combined to form a consortium according to their ability to degrade, these strains were isolated from distinct plastic scrape disposition locations. This resulted in a consortium of four bacterial strains: *Acanthopleuribacter pedis*, *Bacillus aerius*, *Pseudomonas otitidis*, and *Bacillus cereus* (Anwar et al. 2013). A further strategy to create a strong bacterial consortium for the breakdown of crude oil included establishing 14 consortia capable of breaking down hydrocarbons. Five bacterial isolates from enrichment cultures obtained from three distinct contaminated soil locations were combined to generate these, which were selected from a collection of 23 microbes based on their capacity to degrade crude oil. After 5 weeks of incubating the conjunction of biosurfactant-producing and non-producing bacteria achieved 84% TPH degradation. The consortium comprised two biosurfactant-producing *Bacillus* spp., *B. cereus* and *B. pumilus*, isolated from soils of distinct locations. The intentional addition of microorganisms that produce surfactants raises the design's complexity

Table 2 Microorganisms for bioremediation of petroleum hydrocarbons

Microorganism		Source	Substrate	Incubation	Degradation (%)	Reference
Bacteria	*Achromobacter* sp. HZ01	PC seawater	Anthracene	28 °C for 30 days	29.8	Deng et al. (2014)
			Phenanthrene		50.8	
			Pyrene		34.4	
	Bacillus subtilis BMT4i (MTCC 9447)	AC soil	Benzo[a]Pyrene	37 °C for 28 days	84.66	Lily et al. (2009)
	Enterobacter sp. (MM087)	EOC soil	Phenanthrene	37 ± 0.5 °C for 24 h	80.2	Darma et al. (2016)
			Pyrene		59.7	
	Rhodococcus sp. P14	PC sediments	Phenanthrene	30 °C for 30 days	34	Song et al. (2011)
			Pyrene		43	
			Benzo[a]pyrene		30	
	Pseudoxanthomonas sp. DMVP2	PC sediment	Phenanthrene	37 °C for 72 h	86	Patel et al. (2012)
	Streptomyces sp. (ERI-CPDA-1)	PC soil	Naphthalene (0.1%)	30 °C for 7 days	99.14	Balachandran et al. (2012)
			Phenanthrene (0.1%)		17.5	

Fungus	*Aspergillus* sp. RFC-1	Rumaila oil field (SPS)	Naphthalene	30 °C for 7 days	84.6	Al-Hawash et al. (2019)
			Phenanthrene		50.3	
			Pyrene		55.1	
	Nocardia sp. H17–1	PC soil	Aliphatic	30 °C for 6 days	99.0 ± 0.1	Baek et al. (2006)
			Aromatic		23.8 ± 0.8	
	Penicillium sp. CHY-2	Soil	Butylbenzene	20 °C for 28 days	42	Govarthanan et al. (2017)
			Naphthalene		15	
			Acenaphthene		10	
			Ethylbenzene		4	
			Benzo[a]pyrene		2	
	Trichoderma sp.	–	Pyrene	240 h	63	Wang et al. (2008)
			Benzo(a)pyrene		37	
	Penicillium sp. CHY-2	Soil	Decane	20 °C for 28 days	49	Govarthanan et al. (2017)
			Dodecane		33	
			Octane		8	
	Fusarium sp.	–	Pyrene	240 h	69	Wang et al. (2008)
			Benzo(a)pyrene		37	

AC Automobile contaminated; *PC* Petroleum contaminated; *EOC* Engine oil contaminated; *SPS* Surface polluted sludge and makes consortium development more methodical (Patowary et al. 2016). Among the various combinations that result in artificial microbial consortia, a particular subgroup known as synthetic microbial consortia emerges. These are advanced variants of artificial consortia, necessitating presence of variants for specific, established activities or responsibilities, typically complimentary to other microbes (Kim et al. 2011). This classification of consortia encompasses organisms, which have been genetically modified to perform degradation or cooperate in procedures more efficiently than their wild-type equivalents. However, this genetic modification is not a mandatory criterion for inclusion in the classification (Mao et al. 2017).

6.2 Recent Advancements in Bioremediation

To advance the development of more effective environmental cleanup technologies, numerous bioinformatics tools are employed to analyze various data sources in the study of bioremediation. However, there is a significant shortage of information concerning the parameters that govern the development and metabolism of bacteria with the potential for bioremediation, resulting in a limited spectrum of bioremediation techniques. Bioinformatics has been employed to analyze microorganisms with bioremediation capabilities and to map their mineralization pathways and mechanisms (Kumari and Kumar 2021). Proteomic techniques, including 2-D polyacrylamide gel electrophoresis, mass spectrometry, and microarrays, are essential in investigating bioremediation methods, substantially improving the structure of microbial proteins with contaminant-degrading abilities. The integration of genomics, transcriptomics, metabolomics, and proteomics provides valuable insights into bioremediation studies (Sar and Islam 2012).

Organic pollutants, such as PH and PAHs, are absorbed in soil, reducing their solubility and mobility and thereby minimizing their environmental effect. The introduction of polymeric nano-network particles improves both the solubility and release of phenanthrene from polluted aquifer materials (Tripathi et al. 2022). Polymeric nanoparticles are made from source chain of poly(ethylene) glycol-modified urethane acrylate (PMUA). PMUA nanoparticles were developed to preserve their characteristics in the presence of varied bacterial populations (Nwuche et al. 2021).

Bacteria that exhibit tolerance to pollutants are ideal candidates for genetic modification and biochemical pathway engineering, as they are adapted to endure and accumulate various toxic, refractory, and nondegradable compounds under extreme conditions. Moreover, understanding biochemical functions is essential for evaluating microbial bioremediation processes (Gupta and Singh 2017). For instance, haloalkane dehalogenases have a pivotal role in the degradation of harmful pollutants, while fenpropathrin anaerobic decomposition pathway in *Bacillus* sp. DG-02 is critical for the elimination of pyrethrins from soil (Chen and Zhan 2019). Genetically engineered microbes have shown success in bioremediating hexachlorocyclohexane and methyl parathion. For instance, genetically manipulated *Pseudomonas putida* KT2440 has been studied, focusing on the bioremediation of organophosphates and pyrethroids (Gong et al. 2018). The advent of metabolic engineering has led to extensive documentation of the catabolism and degradation of various persistent compounds. Notably, *Pseudomonas* sp. WBC-3 and *Sphingobium japonicum* have exhibited effective pathways for the bioremediation of methyl parathion and the degradation of hexachloro-cyclohexane (Siddavattam et al. 2019).

7 Case Studies

An investigation was conducted to evaluate the soil organic matter impact on the biodegradation of TPH at various levels. Besides assessing the efficiency of TPH (fuel oil) degradation, the study also employed molecular methods, including terminal restriction fragment length polymorphism analysis (T-RFLP) and oligonucleotide microarray, to analyze TPH-degrading microbial population comprehensively. Results showed that TPH levels decreased from 10,000 mg/kg to a range of 1849–4352 mg/kg in dry-weight soil. Higher soil organic contents were associated with increased biodegradation effectiveness and kinetic rate constants. The resulting growth patterns demonstrated that different types of contaminated oil altered microbial diversification over time. Both T-RFLP and microarray profiles identified *Gordonia desulfuricans*, *G. alkanivorans*, and *Rhodococcus erythoropolis* as the predominant bacteria, while *Aspergillus versicolor* and *Fusarium oxysporum* were the prevailing fungi (Chen et al. 2020).

Research conducted a comprehensive analysis of soil samples collected from a contaminated industrial area. The study identified a predominance of microorganisms from the phylum Proteobacteria in these samples. The soil samples with the highest creosote concentrations exhibited limited microbial biodiversity. Prolonged exposure to hydrophobic hydrocarbons led to bacterial adaptation, enabling them to utilize these compounds as sources of C and energy, particularly in the genera *Pseudomonas*, *Sphingomonas*, and *Serratia*. After 28 days, the indigenous microbial consortia achieved approximately 30% biodegradation of creosote across all samples. Among the isolated strains, *Brevundimonas olei* and *Pseudomonas mendocina* demonstrated the foremost biodegradation capabilities, degrading over 60% of the total PAHs. The presence of the AT7 dispersant notably enhanced the total PAH biodegradation by *Serratia marcescens* and *Pseudomonas moseii* (Smułek et al. 2020).

Another study investigated the aerobic degeneration of polychlorinated biphenyls (PCBs) using bacterial strains, *Rhodococcus* sp. IN306, *Rhodococcus erythropolis* IN129, *Mycolicibacterium frederiksbergense* IN53, and a mixed culture M1 (composed of strains in equal ratios). Respirometric tests and chromatographic studies were used to evaluate the efficiency of both the individual strains and the combined culture. *Rhodococcus* sp. IN306 was the most effective, achieving a 54.4% degradation rate of 18 PCB congeners. The rate of biodegradation reduced as the amount of chlorine atoms increased. *M. frederiksbergense* IN53 degraded TPH with a 37.2% degradation rate. In a soil sample polluted with TPH and PCBs, the mixed culture M1 achieved the highest biodegradation efficiency, reducing PCB content by 51.8% and TPH content by 34.6%. The mixed culture M1's biodegradation ability was validated ex situ using the prism technique in nonsterile soil samples contaminated with wasted transformer oil and old petroleum hydrocarbons. After 6 months of inoculation with M1, TPH levels were reduced by 70.8% and PCB levels by 84.5% (Steliga et al. 2020).

8 Conclusion

Bioremediation is a valuable approach to waste management, particularly effective in removing waste from contaminated sites by employing organisms to neutralize or consume pollutants. The primary objective of bioremediation is to use data from various biological databases—including those detailing chemical composition and structure, organic compounds RNA/protein expression, pathway of microbial degradation, catalytic enzymes, and comparative genomics for interpretation of the specific reduction mechanisms utilized by specific organisms that targeted pollutants. The fulfillment of synthetic microbial consortia heavily relies on advancements in systematic and synthetic biology, and modeling and analytical techniques. Biotechnology innovations enable the engineering of microbial consortia with specific bioremediation functions. Top-down engineering involves altering environmental variables to prompt a natural microbial community to perform desired functions, while bottom-up engineering concentrates on redesigning metabolic pathways and reprogramming microbial interactions. To acquire precise oversight of microbial consortia and guarantee the achievement of specified functions and consistent outcomes, future research should prioritize researching metabolic interactions between species.

Conflict of Interest All the authors declare that they have no competing interests.

References

Abbasian F, Lockington R, Mallavarapu M, Naidu R (2015) A comprehensive review of aliphatic hydrocarbon biodegradation by bacteria. Appl Biochem Biotechnol 176:670–699. https://doi.org/10.1007/s12010-015-1603-5

Adams GO, Fufeyin PT, Okoro SE, Ehinomen I (2015) Bioremediation, biostimulation and bioaugmention: a review. Int J Environ Bioremediat Biodegrad 3(1):28–39. https://doi.org/10.12691/ijebb-3-1-5

Ahmed F, Fakhruddin ANM (2018) A review on environmental contamination of PHand its biodegradation. Int J Environ Sci Nat Res 11(3):1–7. https://doi.org/10.19080/IJESNR.2018.11.555811

Al-Hawash AB, Zhang X, Ma F (2019) Removal and biodegradation of different petroleum hydrocarbons using the filamentous fungus *Aspergillus* sp. RFC-1. Microbiologyopen 8(1):e00619. https://doi.org/10.1002/mbo3.619

Alotaibi F, Hijri M, St-Arnaud M (2021) Overview of approaches to improve rhizoremediation of petroleum hydrocarbon-contaminated soils. Appl Microbiol 1(2):329–351. https://doi.org/10.3390/applmicrobiol1020023

Ambaye TG, Chebbi A, Formicola F, Prasad S, Gomez FH, Franzetti A, Vaccari M (2022) Remediation of soil polluted with petroleum hydrocarbons and its reuse for agriculture: recent progress, challenges, and perspectives. Chemosphere 293:133572. https://doi.org/10.1016/j.chemosphere.2022.133572

Anwar M, Negi H, Zaidi M, Haider G, Gupta S, Goel R (2013) Biodeterioration studies of thermoplastics in nature using indigenous bacterial consortium. Braz Arch Biol Technol 56:475–484. https://doi.org/10.1590/S1516-89132013000300016

Atlas R, Bragg J (2009) Bioremediation of marine oil spills: when and when not–the Exxon Valdez experience. Microb Biotechnol 2(2):213–221. https://doi.org/10.1111/j.1751-7915.2008.00079.x

Auti AM, Narwade NP, Deshpande NM, Dhotre DP (2019) Microbiome and imputed metagenome study of crude and refined petroleum-oil-contaminated soils: potential for hydrocarbon degradation and plant-growth promotion. J Biosci 44:1–16. https://doi.org/10.1007/s12038-019-9936-9

Aydin S, Karaçay HA, Shahi A, Gökçe S, Ince B, Ince O (2017) Aerobic and anaerobic fungal metabolism and omics insights for increasing polycyclic aromatic hydrocarbons biodegradation. Fungal Biol Rev 31(2):61–72. https://doi.org/10.1016/j.fbr.2016.12.001

Baek KH, Yoon BD, Oh HM, Kim HS, Lee IS (2006) Biodegradation of aliphatic and aromatic hydrocarbons by Nocardia sp. H17-1. Geomicrobiol J 23(5):253–259. https://doi.org/10.1080/01490450600760633

Balachandran C, Duraipandiyan V, Balakrishna K, Ignacimuthu S (2012) Petroleum and polycyclic aromatic hydrocarbons (PAHs) degradation and naphthalene metabolism in Streptomyces sp.(ERI-CPDA-1) isolated from oil contaminated soil. Bioresour Technol 112:83–90. https://doi.org/10.1016/j.biortech.2012.02.059

Balaji V, Arulazhagan P, Ebenezer P (2014) Enzymatic bioremediation of polyaromatic hydrocarbons by fungal consortia enriched from petroleum contaminated soil and oil seeds. J Environ Biol 35(3):521–529

Behera ID, Nayak M, Mishra A, Meikap BC, Sen R (2022) Strategic implementation of integrated bioaugmentation and biostimulation for efficient mitigation of petroleum hydrocarbon pollutants from terrestrial and aquatic environment. Mar Pollut Bull 177:113492. https://doi.org/10.1016/j.marpolbul.2022.113492

Brzeszcz J, Kaszycki P (2018) Aerobic bacteria degrading both n-alkanes and aromatic hydrocarbons: an undervalued strategy for metabolic diversity and flexibility. Biodegradation 29:359–407. https://doi.org/10.1007/s10532-018-9837-x

Cabral L, de Sousa STP, Júnior GVL, Hawley E, Andreote FD, Hess M, de Oliveira VM (2018) Microbial functional responses to long-term anthropogenic impact in mangrove soils. Ecotoxicol Environ Saf 160:231–239. https://doi.org/10.1016/j.ecoenv.2018.04.050

Chaudhary DK, Kim J (2019) New insights into bioremediation strategies for oil-contaminated soil in cold environments. Int Biodeterior Biodegradation 142:58–72. https://doi.org/10.1016/j.ibiod.2019.05.001

Chen S, Zhan H (2019) Biodegradation of synthetic pyrethroid insecticides. In: Microbial metabolism of xenobiotic compounds. Springer Singapore, pp 229–244

Chen YA, Liu PWG, Whang LM, Wu YJ, Cheng SS (2020) Effect of soil organic matter on petroleum hydrocarbon degradation in diesel/fuel oil-contaminated soil. J Biosci Bioeng 129(5):603–612. https://doi.org/10.1016/j.jbiosc.2019.12.001

Clarkson MA, Abubakar SI (2015) Bioremediation and biodegradation of hydrocarbon contaminated soils: a review. IOSR J Environ Sci Toxicol Food Technol 9(11):38–45. https://doi.org/10.9790/2402-091113845

Coelho LM, Rezende HC, Coelho LM, De Sousa PA, Melo DF, Coelho NM (2015) Bioremediation of polluted waters using microorganisms. Adv Bioremediat Wastewater Pollut Soil 10(10.5772):60770

Darma UZ, Aziz NAA, Zulkefli SZ, Mustafa M (2016) Identification of phenanthrene and pyrene degrading bacteria from used engine oil contaminated soil. Int J Sci Eng Res 7:680–686

Das N, Chandran P (2011) Microbial degradation of PHcontaminants: an overview. Biotechnol Res Int 2011(1):941810. https://doi.org/10.4061/2011/941810

Dasgupta A, Saikia R, Handique PJ (2018) Mapping the bacterial community in Digboi Oil Refinery, India by high-throughput sequencing approach. Curr Microbiol 75:1441–1446. https://doi.org/10.1007/s00284-018-1541-x

de Freitas GR, da Silva MGC, Vieira MGA (2019) Biosorption technology for removal of toxic metals: a review of commercial biosorbents and patents. Environ Sci Pollut Res 26:19097–19118. https://doi.org/10.1007/s11356-019-05330-8

de Sousa STP, Cabral L, Lacerda-Júnior GV, Noronha MF, Ottoni JR, Sartoratto A, de Oliveira VM (2020) Exploring the genetic potential of a fosmid metagenomic library from an oil-impacted mangrove sediment for metabolism of aromatic compounds. Ecotoxicol Environ Saf 189:109974. https://doi.org/10.1016/j.ecoenv.2019.109974

Dehnavi SM, Ebrahimipour G (2022) Comparative remediation rate of biostimulation, bioaugmentation, and phytoremediation in hydrocarbon contaminants. Int J Environ Sci Technol 19(11):11561–11586. https://doi.org/10.1007/s13762-022-04343-0

Deng MC, Li J, Liang FR, Yi M, Xu XM, Yuan JP, Peng J, Wu CF, Wang JH (2014) Isolation and characterization of a novel hydrocarbon-degrading bacterium Achromobacter sp. HZ01 from the crude oil-contaminated seawater at the Daya Bay, southern China. Mar Pollut Bull 83(1):79–86. https://doi.org/10.1016/j.marpolbul.2014.04.018

Di Gregorio S, Azaizeh H, Lorenzi R (2013) Biostimulation of the autochthonous microbial community for the depletion of polychlorinated biphenyls (PCBs) in contaminated sediments. Environ Sci Pollut Res 20:3989–3999. https://doi.org/10.1007/s11356-012-1350-x

Fomina M, Gadd GM (2014) Biosorption: current perspectives on concept, definition and application. Bioresour Technol 160:3–14. https://doi.org/10.1016/j.biortech.2013.12.102

Gao H, Wu M, Liu H, Xu Y, Liu Z (2022) Effect of petroleum hydrocarbon pollution levels on the soil microecosystem and ecological function. Environ Pollut 293:118511. https://doi.org/10.1016/j.envpol.2021.118511

Gennadiev AN, Pikovskii YI, Tsibart AS, Smirnova MA (2015) Hydrocarbons in soils: origin, composition, and behavior. Eurasian Soil Sci 48:1076–1089. https://doi.org/10.1134/S1064229315100026

Gong T, Xu X, Dang Y, Kong A, Wu Y, Liang P, Wang S, Yu H, Xu P, Yang C (2018) An engineered Pseudomonas putida can simultaneously degrade organophosphates, pyrethroids and carbamates. Sci Total Environ 628:1258–1265. https://doi.org/10.1016/j.scitotenv.2018.02.143

Govarthanan M, Fuzisawa S, Hosogai T, Chang YC (2017) Biodegradation of aliphatic and aromatic hydrocarbons using the filamentous fungus Penicillium sp. CHY-2 and characterization of its manganese peroxidase activity. RSC Adv 7(34):20716–20723. https://doi.org/10.1039/C6RA28687A

Guarino C, Spada V, Sciarrillo R (2017) Assessment of three approaches of bioremediation (Natural attenuation, landfarming and bioagumentation–assistited landfarming) for a petroleum hydrocarbons contaminated soil. Chemosphere 170:10–16. https://doi.org/10.1016/j.chemosphere.2016.11.165

Gupta S, Singh D (2017) Role of genetically modified microorganisms in heavy metal bioremediation. In: Advances in environmental biotechnology. Springer Singapore, pp 197–214. https://doi.org/10.1007/978-981-10-4041-2_12

Hamad AA, Moubasher HA, Moustafa YM, Mohamed NH, Abd-El Rhim EH (2021) Petroleum hydrocarbon bioremediation using native fungal isolates and consortia. Sci World J 2021(1):6641533. https://doi.org/10.1155/2021/6641533

Imam A, Suman SK, Ghosh D, Kanaujia PK (2019) Analytical approaches used in monitoring the bioremediation of hydrocarbons in petroleum-contaminated soil and sludge. TrAC Trends Anal Chem 118:50–64. https://doi.org/10.1016/j.trac.2019.05.023

Javaid MK, Ashiq M, Tahir M (2016) Potential of biological agents in decontamination of agricultural soil. Scientifica 2016(1):1598325. https://doi.org/10.1155/2016/1598325

Joergensen RG, Wichern F (2018) Alive and kicking: why dormant soil microorganisms matter. Soil Biol Biochem 116:419–430. https://doi.org/10.1016/j.soilbio.2017.10.022

Kim HJ, Du W, Ismagilov RF (2011) Complex function by design using spatially pre-structured synthetic microbial communities: degradation of pentachlorophenol in the presence of Hg (II). Integr Biol 3(2):126–133. https://doi.org/10.1039/c0ib00019a

Kube M, Heider J, Amann J, Hufnagel P, Kühner S, Beck A, Reinhardt R, Rabus R (2004) Genes involved in the anaerobic degradation of toluene in a denitrifying bacterium, strain EbN1. Arch Microbiol 181:182–194. https://doi.org/10.1007/s00203-003-0627-3

Kumar A, Bisht BS, Joshi VD, Dhewa T (2011) Review on bioremediation of polluted environment: a management tool. Int J Environ Sci 1(6):1079–1093. https://doi.org/10.1007/s00203-003-0627-3

Kumari P, Kumar Y (2021) Bioinformatics and computational tools in bioremediation and biodegradation of environmental pollutants. In: Bioremediation for environmental sustainability. Elsevier, pp 421–444. https://doi.org/10.1016/B978-0-12-820318-7.00019-8

Lily MK, Bahuguna A, Dangwal K, Garg V (2009) Degradation of benzo [a] pyrene by a novel strain Bacillus subtilis BMT4i (MTCC 9447). Braz J Microbiol 40:884–892. https://doi.org/10.1590/S1517-83822009000400020

Liu X, Li Z, Zhang C, Tan X, Yang X, Wan C, Lee DJ (2020a) Enhancement of anaerobic degradation of petroleum hydrocarbons by electron intermediate: performance and mechanism. Bioresour Technol 295:122305. https://doi.org/10.1016/j.biortech.2019.122305

Liu H, Tan X, Guo J, Liang X, Xie Q, Chen S (2020b) Bioremediation of oil-contaminated soil by combination of soil conditioner and microorganism. J Soils Sediments 20:2121–2129. https://doi.org/10.1007/s11368-020-02591-6

Liu Y, Zhang H, He X, Liu J (2021) Genetically engineered methanotroph as a platform for bioaugmentation of chemical pesticide contaminated soil. ACS Synth Biol 10(3):487–494. https://doi.org/10.1021/acssynbio.0c00532

Liu H, Wu M, Guo X, Gao H, Xu Y (2022) Isotope fractionation (δ 13 C, δ 15 N) and microbial community response in degradation of petroleum hydrocarbons by biostimulation in contaminated soil. In: Environmental science and pollution research. Springer Nature, pp 1–10. https://doi.org/10.1007/s11356-021-16055-y

Lu L, Zhang J, Peng C (2019) Shift of Soil PAHs (PAHs) dissipation pattern and microbial community composition due to rhamnolipid supplementation. Water Air Soil Pollut 230:1–14. https://doi.org/10.1007/s11270-019-4118-9

Ma M, Gao W, Li Q, Han B, Zhu A, Yang H, Zheng L (2021) Biodiversity and oil degradation capacity of oil-degrading bacteria isolated from deep-sea hydrothermal sediments of the South Mid-Atlantic Ridge. Mar Pollut Bull 171:112770. https://doi.org/10.1016/j.marpolbul.2021.112770

Macaulay BM (2015) Understanding the behaviour of oil-degrading micro-organisms to enhance the microbial remediation of spilled petroleum. Appl Ecol Environ Res 13(1):247–262. https://doi.org/10.15666/aeer/1301_247262

Mangla H, Sharma H, Dave S, Sudan J, Pathak H (2021) Microbial mechanism of petroleum hydrocarbons degradation: an environmental perspective. Appl Environ Biotechnol 6(2):32–42. https://doi.org/10.26789/AEB.2021.02.005

Mao X, Oremland RS, Liu T, Gushgari S, Landers AA, Baesman SM, Alvarez-Cohen L (2017) Acetylene fuels TCE reductive dechlorination by defined Dehalococcoides/Pelobacter consortia. Environ Sci Technol 51(4):2366–2372. https://doi.org/10.1021/acs.est.6b05770

Massot F, Bernard N, Alvarez LMM, Martorell MM, Mac Cormack WP, Ruberto LA (2022) Microbial associations for bioremediation. What does "microbial consortia" mean? Appl Microbiol Biotechnol 106(7):2283–2297. https://doi.org/10.1007/s00253-022-11864-8

Mohammadi L, Rahdar A, Bazrafshan E, Hamid Dahmardeh M, Susan ABH, Kyzas GZ (2020) PHRemoval from wastewaters: a review. PRO 8(4):447. https://doi.org/10.3390/pr8040447

Muksy R, Kolo K, Abdullah SM (2023) Bacterial and fungal-mineral interactions and their application in bioremediation–a review. J Ecol Eng 24(5):1

Muthukumar B, Surya S, Sivakumar K, AlSalhi MS, Rao TN, Devanesan S, Arunkumar P, Rajasekar A (2023) Influence of bioaugmentation in crude oil contaminated soil by Pseudomonas species on the removal of total petroleum hydrocarbon. Chemosphere 310:136826. https://doi.org/10.1016/j.chemosphere.2022.136826

Neethu CS, Saravanakumar C, Purvaja R, Robin RS, Ramesh R (2019) Oil-spill triggered shift in indigenous microbial structure and functional dynamics in different marine environmental matrices. Sci Rep 9(1):1354. https://doi.org/10.1038/s41598-018-37903-x

Nwuche CO, Igbokwe VC, Ajagbe DD, Onwosi CO (2021) Nanoparticles, biosurfactants and microbes in bioremediation. In: Rhizomicrobiome dynamics in bioremediation. CRC Press, pp 162–179

Omrani R, Spini G, Puglisi E, Saidane D (2018) Modulation of microbial consortia enriched from different polluted environments during petroleum biodegradation. Biodegradation 29:187–209. https://doi.org/10.1007/s10532-018-9823-3

Ozyurek SB, Avcioglu NH (2020) The role of Aspergillus parasiticus NRRL: 3386 strain on petroleum biodegradation and biosorption processes. Eur J Biol 79(2):144–150

Pacwa-Płociniczak M, Czapla J, Płociniczak T, Piotrowska-Seget Z (2019) The effect of bioaugmentation of petroleum-contaminated soil with Rhodococcus erythropolis strains on removal of petroleum from soil. Ecotoxicol Environ Saf 169:615–622. https://doi.org/10.1016/j.ecoenv.2018.11.081

Patel V, Cheturvedula S, Madamwar D (2012) Phenanthrene degradation by Pseudoxanthomonas sp. DMVP2 isolated from hydrocarbon contaminated sediment of Amlakhadi canal, Gujarat, India. J Hazard Mater 201:43–51. https://doi.org/10.1016/j.jhazmat.2011.11.002

Patowary K, Patowary R, Kalita MC, Deka S (2016) Development of an efficient bacterial consortium for the potential remediation of hydrocarbons from contaminated sites. Front Microbiol 7:1092. https://doi.org/10.3389/fmicb.2016.01092

Pattabhiramaiah M, Shanthala M, Rajashekara S, Sheikh F, Naik S (2018) Biodegradation of PAHs (PAHs) by microbes isolated from the marine sponge biemna fortis (topsent 1897). In: Microbial action on hydrocarbons. Springer Singapore, pp 199–222. https://doi.org/10.1007/978-981-13-1840-5_9

Poorsoleiman MS, Hosseini SA, Etminan A, Abtahi H, Koolivand A (2020) Effect of two-step bioaugmentation of an indigenous bacterial strain isolated from oily waste sludge on petroleum hydrocarbons biodegradation: scaling-up from a liquid mineral medium to a two-stage composting process. Environ Technol Innov 17:100558. https://doi.org/10.1016/j.eti.2019.100558

Qiao J, Zhang C, Luo S, Chen W (2014) Bioremediation of highly contaminated oilfield soil: bioaugmentation for enhancing aromatic compounds removal. Front Environ Sci Eng 8:293–304. https://doi.org/10.1007/s11783-013-0561-9

Quiza L, St-Arnaud M, Yergeau E (2015) Harnessing phytomicrobiome signaling for rhizosphere microbiome engineering. Front Plant Sci 6:507. https://doi.org/10.3389/fpls.2015.00507

Rahman KS, Rahman TJ, Kourkoutas Y, Petsas I, Marchant R, Banat IM (2003) Enhanced bioremediation of n-alkane in petroleum sludge using bacterial consortium amended with rhamnolipid and micronutrients. Bioresour Technol 90(2):159–168. https://doi.org/10.1016/S0960-8524(03)00114-7)

Ramadass K, Megharaj M, Venkateswarlu K, Naidu R (2018) Bioavailability of weathered hydrocarbons in engine oil-contaminated soil: impact of bioaugmentation mediated by Pseudomonas spp. on bioremediation. Sci Total Environ 636:968–974. https://doi.org/10.1016/j.scitotenv.2018.04.379

Roy A, Dutta A, Pal S, Gupta A, Sarkar J, Chatterjee A, Saha A, Sarkar P, Sar P, Kazy SK (2018a) Biostimulation and bioaugmentation of native microbial community accelerated bioremediation of oil refinery sludge. Bioresour Technol 253:22–32. https://doi.org/10.1016/j.biortech.2018.01.004

Roy A, Sar P, Sarkar J, Dutta A, Sarkar P, Gupta A, Mohapatra B, Pal S, Kazy SK (2018b) Petroleum hydrocarbon rich oil refinery sludge of North-East India harbours anaerobic, fermentative, sulfate-reducing, syntrophic and methanogenic microbial populations. BMC Microbiol 18:1–22. https://doi.org/10.1186/s12866-018-1275-8

Saadoun IM, Al-Ghzawi ZD (2005) Bioremediation of petroleum contamination. Biorem Aquat Terr Ecosyst 132:173–212

Safdari MS, Kariminia HR, Rahmati M, Fazlollahi F, Polasko A, Mahendra S, Wilding WV, Fletcher TH (2018) Development of bioreactors for comparative study of natural attenuation, biostimulation, and bioaugmentation of petroleum-hydrocarbon contaminated soil. J Hazard Mater 342:270–278. https://doi.org/10.1016/j.jhazmat.2017.08.044

Sakshi, Singh SK, Haritash AK (2019) PAHs: soil pollution and remediation. Int J Environ Sci Technol 16:6489–6512. https://doi.org/10.1007/s13762-019-02414-3

Sar P, Islam E (2012) Metagenomic approaches in microbial bioremediation of metals and radionuclides. In: Microorganisms in environmental management: microbes and environment. Springer, Dordrecht, pp 525–546. https://doi.org/10.1007/978-94-007-2229-3_23

Siddavattam D, Yakkala H, Samantarrai D (2019) Lateral transfer of organophosphate degradation (*opd*) genes among soil bacteria: mode of transfer and contributions to organismal fitness. J Genet 98:1–10. https://doi.org/10.1007/s12041-019-1068-3

Sjöblom J, Aske N, Auflem IH, Brandal Ø, Havre TE, Sæther Ø, Westvik A, Johnsen EE, Kallevik H (2003) Our current understanding of water-in-crude oil emulsions: recent characterization techniques and high-pressure performance. Adv Colloid Interf Sci 100:399–473. https://doi.org/10.1016/S0001-8686(02)00066-0

Smułek W, Sydow M, Zabielska-Matejuk J, Kaczorek E (2020) Bacteria involved in biodegradation of creosote PAH–A case study of long-term contaminated industrial area. Ecotoxicol Environ Saf 187:109843. https://doi.org/10.1016/j.ecoenv.2019.109843

So CM, Phelps CD, Young LY (2003) Anaerobic transformation of alkanes to fatty acids by a sulfate-reducing bacterium, strain Hxd3. Appl Environ Microbiol 69(7):3892–3900. https://doi.org/10.1128/AEM.69.7.3892-3900.2003

Song X, Xu Y, Li G, Zhang Y, Huang T, Hu Z (2011) Isolation, characterization of Rhodococcus sp. P14 capable of degrading high-molecular-weight polycyclic aromatic hydrocarbons and aliphatic hydrocarbons. Mar Pollut Bull 62(10):2122–2128. https://doi.org/10.1016/j.marpolbul.2011.07.013

Steliga T, Wojtowicz K, Kapusta P, Brzeszcz J (2020) Assessment of biodegradation efficiency of polychlorinated biphenyls (PCBs) and petroleum hydrocarbons (TPH) in soil using three individual bacterial strains and their mixed culture. Molecules 25(3):709. https://doi.org/10.3390/molecules25030709

Tao K, Zhang X, Chen X, Liu X, Hu X, Yuan X (2019) Response of soil bacterial community to bioaugmentation with a plant residue-immobilized bacterial consortium for crude oil removal. Chemosphere 222:831–838. https://doi.org/10.1016/j.chemosphere.2019.01.133

Tripathi S, Sanjeevi R, Anuradha J, Chauhan DS, Rathoure AK (2022) Nano-bioremediation: nanotechnology and bioremediation. In: Research anthology on emerging techniques in environmental remediation. IGI Global, pp 135–149

Van Beilen JB, Funhoff EG (2007) Alkane hydroxylases involved in microbial alkane degradation. Appl Microbiol Biotechnol 74:13–21. https://doi.org/10.1007/s00253-006-0748-0

Varjani S, Upasani VN (2019) Influence of abiotic factors, natural attenuation, bioaugmentation and nutrient supplementation on bioremediation of petroleum crude contaminated agricultural soil. J Environ Manag 245:358–366. https://doi.org/10.1016/j.jenvman.2019.05.070

Varjani S, Upasani VN (2021) Bioaugmentation of Pseudomonas aeruginosa NCIM 5514–a novel oily waste degrader for treatment of petroleum hydrocarbons. Bioresour Technol 319:124240. https://doi.org/10.1016/j.biortech.2020.124240
Vázquez S, Nogales B, Ruberto L, Mestre C, Christie-Oleza J, Ferrero M, Bosch R, Mac Cormack WP (2013) Characterization of bacterial consortia from diesel-contaminated Antarctic soils: towards the design of tailored formulas for bioaugmentation. Int Biodeterior Biodegradation 77:22–30. https://doi.org/10.1016/j.ibiod.2012.11.002
Vishwakarma GS, Bhattacharjee G, Gohil N, Singh V (2020) Current status, challenges and future of bioremediation. In: Bioremediation of pollutants. Elsevier, pp 403–415. https://doi.org/10.1016/B978-0-12-819025-8.00020-X
Wang X, Gong Z, Li P, Zhang L, Hu X (2008) Degradation of pyrene and benzo (a) pyrene in contaminated soil by immobilized fungi. Environ Eng Sci 25(5):677–684. https://doi.org/10.1089/ees.2007.0075
Wani KS, Mir BA (2020) Microbial geo-technology in ground improvement techniques: a comprehensive review. Innov Infrastruct Solut 5(3):82. https://doi.org/10.1007/s41062-020-00335-6
Widdel F, Grundmann O (2010) Biochemistry of the anaerobic degradation of non-methane alkanes. In: Handbook of hydrocarbon and lipid microbiology. Springer, Berlin, Heidelberg. https://doi.org/10.1007/978-3-540-77587-4_64
Woźniak-Karczewska M, Lisiecki P, Białas W, Owsianiak M, Piotrowska-Cyplik A, Wolko Ł, Ławniczak Ł, Heipieper HJ, Gutierrez T, Chrzanowski Ł (2019) Effect of bioaugmentation on long-term biodegradation of diesel/biodiesel blends in soil microcosms. Sci Total Environ 671:948–958. https://doi.org/10.1016/j.scitotenv.2019.03.431
Xu X, Liu W, Tian S, Wang W, Qi Q, Jiang P, Gao X, Li F, Li H, Yu H (2018) Petroleum hydrocarbon-degrading bacteria for the remediation of oil pollution under aerobic conditions: a perspective analysis. Front Microbiol 9:2885. https://doi.org/10.3389/fmicb.2018.02885
Ying TENG, Wei CHEN (2019) Soil microbiomes—a promising strategy for contaminated soil remediation: a review. Pedosphere 29(3):283–297. https://doi.org/10.1016/S1002-0160(18)60061-X
Zawierucha I, Malina G, Ciesielski W, Rychter P (2014) Effectiveness of intrinsic biodegradation enhancement in oil hydrocarbons contaminated soil. Arch Environ Protect 40(1):101–113. https://doi.org/10.2478/aep-2014-0010
Zhang JH, Xue QH, Gao H, Ma X, Wang P (2016) Degradation of crude oil by fungal enzyme preparations from Aspergillus spp. for potential use in enhanced oil recovery. J Chem Technol Biotechnol 91(4):865–875. https://doi.org/10.1002/jctb.4650
Zhang G, Liu T, Li H, Wang Z, Huang X, Yi X, Yan D (2024) Experimental Study on the Effects of Heavy Metal Pollution on Soil Physical Properties and Microstructure Evolution. Appl Sci, 14(5):2022. https://doi.org/10.3390/app14052022
Zhou ZF, Wang MX, Zuo XH, Yao YH (2017) Comparative investigation of bacterial, fungal, and archaeal community structures in soils in a typical oilfield in Jianghan, China. Arch Environ Contam Toxicol 72:65–77. https://doi.org/10.1007/s00244-016-0333-1

Geomicrobes in Environmental Remediation of Polyaromatic Hydrocarbons

Stevenson Thabah, Aawaj Kuloong Rai, and Santa Ram Joshi

Contents

Abstract Hydrocarbon contamination poses a serious environmental threat, and bioremediation of hydrocarbon-contaminated sites using the geomicrobiology approach involves utilizing microorganisms to degrade, transform, or stabilize organic and inorganic contaminants, particularly hydrocarbons present in soil, groundwater, or other geological materials. This approach utilizes the understand-

S. Thabah · A. K. Rai · S. R. Joshi (✉)
Microbiology Laboratory, Department of Biotechnology & Bioinformatics, North-Eastern Hill University, Shillong, Meghalaya, India
e-mail: srjoshi2006@nehu.ac.in

S. R. Joshi et al. (eds.), *Mineral Transformation and Bioremediation by Geo-Microbes*, Interdisciplinary Biotechnological Advances,
https://doi.org/10.1007/978-981-96-3033-2_16

ing of microbial communities and their interaction with geological conditions to optimize bioremediation strategies. Geomicrobiology helps identify native hydrocarbon-degrading microbes, optimize environmental conditions for their activity, and understand degradation pathways. Based on this knowledge, bioremediation strategies like biostimulation, bioaugmentation, and bioventing can be tailored. Bioremediation offers advantages like cost-effectiveness, sustainability, and in situ treatment. Challenges include the complexities of the subsurface environment and recalcitrant hydrocarbons. Future directions involve optimizing bioremediation for diverse geological settings and engineering microbes to tackle persistent contaminants.

Keywords Bioremediation · Geomicrobiology · Hydrocarbons biostimulation · Bioaugmentation

Abbreviations

HMW PAHs	High-molecular-weight PAHs
LMW PAHs	Low-molecular-weight PAHs
PAHs	Polyaromatic hydrocarbons

1 Introduction

Polyaromatic hydrocarbons (PAHs) belong to a class of organic aromatic pollutants having fused benzene rings. PAHs are discharged by natural processes, such as volcanic activity and fires in the forest in the environment. However, major sources of environmental contamination arise through anthropogenic activities, such as partial combustion of organic matter for instance coal, crude oil, wood, and gasoline. Oil spills, industrial discharges, or accidental leakage from storage tanks also significantly contribute to the environmental prevalence of PAHs (Patel et al. 2020). Once released into the environment, they cause significant damage to the environment including human health, since they remain hydrophobic which makes them resistant to degradation. In case of aquatic environments, PAHs from oil spills can greatly affect marine life, reduce oxygen levels, and disrupt food chains. In soil, PAHs can greatly reduce soil fertility, reduce microbial diversity, and contaminate groundwater. They pose serious risks to human health such as respiratory problems, skin irritation, and in some cases, even cancer (Truskewycz et al. 2019). Given the widespread occurrence and persistence of PAH contamination, there is a critical need for effective remediation strategies to control and reduce hydrocarbon pollution to protect both the environment and the general public's health. Traditional methods of remediation of PAH-contaminated sites involve both physical and chemical processes, such as excavation, incineration, and chemical oxidation. These methods, however, tend to be expensive, energy-intensive, and environmentally harmful. Bioremediation can be seen as an alternative promising approach to degrade these organic pollutants thereby restoring the affected area with minimal ecological disturbances (Yuniati 2018). Because it does not involve the use of harsh chemical treatments and can effectively harness the biodegradative capabilities of natural microorganisms, bioremediation has gained a lot more traction as an alternative to cleaning up petroleum-contaminated sites in recent years (Truskewycz et al. 2019). Bioremediation is achieved using either in situ methods or ex situ methods, depending on the site conditions, and has the potential to achieve complete remediation of the PAH-contaminated sites. One approach to bioremediation is using geomicrobiology, which is an interdisciplinary field that explores the interactions between microorganisms and geological processes. It combines microbiology, geology, chemistry, and environmental science to understand how microorganisms influence the abiotic environment and how these environmental conditions, in turn, affect microbial activity. Geomicrobiology can help us understand the role of geomicrobes in degrading PAHs and how the environment affects these microbial activities. Understanding these interactions will not only enable the identification of native PAH-degrading microbes but will also help optimize bioremediation strategies for PAH-contaminated sites.

2 PAHs Contamination in the Environment and Their Biodegradation

PAHs contamination in the environment primarily results from petroleum oil spills, industrial discharges, vehicle emissions, incomplete combustion of fossil fuels, and use of pesticides in agriculture (Patel et al. 2020). These hydrocarbons are highly toxic and pose long-term environmental hazards. In soils, PAHs often bind tightly to particles, making them difficult to remove and resulting in toxicity to plants, which diminishes agricultural productivity and harms soil-dwelling organisms, disrupting the ecological balance. In water bodies, PAHs can create surface slicks that block sunlight and reduce oxygen levels, adversely affecting aquatic life. Groundwater contamination is particularly alarming as it can compromise drinking water sources and pose significant risks to human health (Abha and Swaranjit 2012). PAHs can have harmful effects on human health through inhalation of contaminated air, dermal absorption, and ingestion of contaminated food or water. Severe exposure can cause problems such as irritation in the eyes and skin, while certain chronic cases that involve prolonged exposure to PAHs have been linked to various types of cancer including skin, lung, skin, and gastrointestinal cancers (Kim et al. 2013). Given the severe human health implications of PAH contamination, it is crucial to develop effective remediation strategies to mitigate environmental PAH pollution. Based on the number of aromatic rings, PAHs are categorized into two types, low-molecular-weight (LMW) and high-molecular-weight (HMW) PAHs (Patel et al. 2020). LMW PAHs have 2–3 fused aromatic rings (e.g., anthracene, naphthalene,

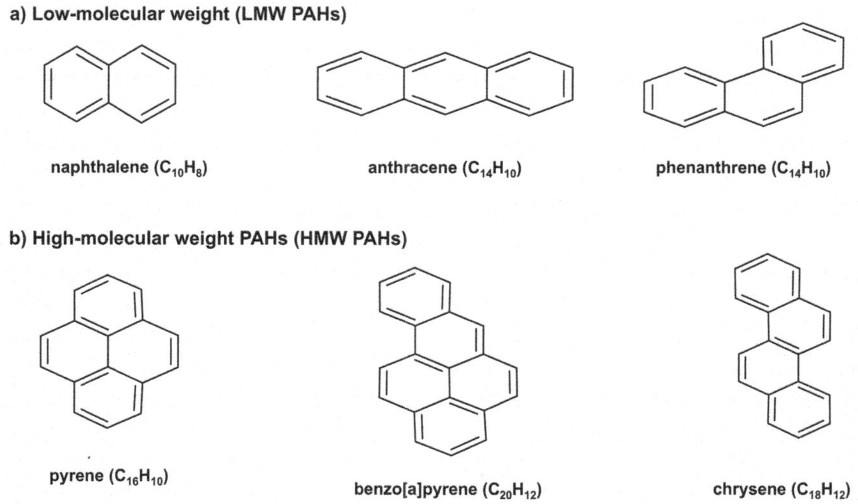

Fig. 1 Structural representation of LMW and HMW PAHs

and phenanthrene) and tend to be more volatile, soluble in water, and more susceptible to microbial biodegradation. In contrast, HMW PAHs have four or more fused aromatic rings (e.g., pyrene, benzo[a]pyrene, and chrysene) are more hydrophobic, less volatile, and more resistant to microbial degradation (Fig. 1). The increased stability and lower bioavailability of HMW PAHs make them more likely to persist in the environment and hence tend to accumulate in the long term (Sui et al. 2021). It is essential to understand the involvement of specific types of hydrocarbons for predicting the fate and transport in soil and water as well as their biodegradation potential. The hydrocarbons can be alkanes that are simple to aromatic compounds, which are more complex, and these significantly influence their physical and chemical properties such as solubility, volatility, and reactivity, all of which play a significant role in their potential for biodegradation in the environment (Hidalgo et al. 2020).

Other factors that can influence the biodegradation of PAHs include several factors, which include temperature, moisture, oxygen, nutrient availability, concentration of the hydrocarbon, and the composition of the microorganisms and their community (Sui et al. 2021). These factors are crucial in determining the efficiency and effectiveness of biodegradation processes.

3 Geomicrobes in Environmental Remediation of PAHs

3.1 Role of Microorganisms in PAH Biodegradation

Geomicrobes are microbes that inhabit the Earth's subsurface and interact with geological processes, influencing the movement, bioavailability, and transformation of organic and inorganic components. Understanding the interactions between geomicrobes and PAHs can help us identify native PAH-degrading microbes and optimize bioremediation strategies for contaminated sites. Different microorganisms like bacteria, archaea, fungi, and algae contribute to bioremediation of PAHs. They have evolved various metabolic pathways and mechanisms to utilize these hydrocarbons as carbon and energy sources and break down the complex hydrocarbons into simpler, less toxic compounds (Ławniczak et al. 2020). Among the bacterial genera, *Pseudomonas*, *Acinetobacter*, *Rhodococcus*, *Alcanivorax*, *Bacillus*, and *Mycobacterium* are widely capable of degrading PAHs and have been elaborately profiled for their potential for applications in bioremediation due to their efficiency and adaptability to different geological conditions (Mahjoubi et al. 2018; Soare et al. 2019; Cai et al. 2021). Members of the *Alcanivorax* genus are considered "hydrocarbonoclastic" bacteria that are highly specialized in utilizing alkanes as their only source of carbon and energy and are known to proliferate in response to oil spills in marine environments. Their abundance increases significantly following an oil spill event, indicating their important role in bioremediation of oil contamination (Wang and Shao 2013). Among fungal species, *Aspergillus*, *Penicillium*, and *Trichoderma* have also demonstrated their potential to degrade an array of PAHs, and through their interactions with bacteria, they expand the range of PAHs that they can target. However, they can collaboratively maximize the degradation potential of the microbial consortia of bacteria and fungi (Matúš et al. 2023). Archaea bacteria such as *Methanobacterium*, *Methanosaeta*, and *Methanosarcina* also contribute to the eco-restoration and remediation of PAHs contaminated sites, especially under extremely harsh conditions of high salt concentration, temperature, or pH (Jiménez et al. 2016). Their adaptability and metabolic capabilities make them thrive and effectively degrade PAHs, especially in anaerobic conditions through methanogenesis. Algae such as cyanobacteria (*Agmenellum quadruplicatum*) and microalgae (*Oscillatoria* sp.) have shown promising potential for PAH-contaminated sites bioremediation through their ability to degrade various petroleum hydrocarbons (Ghosal et al. 2016; Hamouda et al. 2023) but also enhance the surrounding ecosystem by providing oxygen and serving food for other organisms benefiting the microbial community as a whole. Fungi and algae are capable of breaking down PAHs in aerobic conditions, while bacteria and archaea can degrade these hydrocarbons in both presence and absence of oxygen (Truskewycz et al. 2019). So, depending upon the availability of oxygen PAHs degradation can either be aerobic or anaerobic.

3.2 Aerobic Biodegradation of PAHs

Degradation of hydrocarbons including PAHs under aerobic conditions is the fastest and most effective way for bioremediation of hydrocarbons contaminating the environment. In aerobic conditions, bacteria such as *Pseudomonas*, *Bacillus*, *Sphingomonas*, *Rhodococcus*, and *Mycobacterium* that possess monooxygenase or dioxygenase enzymes initiate the process by adding oxygen atoms to the aromatic rings of PAH molecules. The aromatic rings of the PAH are first hydroxylated by dioxygenase enzymes producing cis-dihydrodiols. These cis-dihydrodiols are subsequently rearomatized into diol intermediate through the action of dehydrogenase enzymes. The diol intermediates are then cleaved by either intradiol or extradiol dioxygenases following either an ortho- or meta-ring cleavage pathway. This cleavage results in simpler compounds such as catechols, which are further broken down into intermediates of the TCA cycle, leading to the final production of CO_2 and H_2O (Ghosal et al. 2016; Patel et al. 2020). The initial oxidation of the aromatic ring is the rate-limiting step in PAHs biodegradation. Given the important role of oxygenases and dioxygenases in aerobic biodegradation of PAHs, research studies have concentrated on isolation, characterization, and engineering of these enzymes to enhance their activity and substrate range. Recent studies have employed metagenomic and directed evolution approaches to discover and engineer novel aromatic ring-cleaving dioxygenase enzymes with improved properties for potential bioremediation applications (Wang et al. 2023).

Ligninolytic fungi, especially *Phanerochaete chrysosporium*, *Pleurotus ostreatus*, and *Bjerkandera adusta*, the group of white rot fungi, are reported to secrete various extracellular non-specific oxidative enzymes like lignin peroxidase, laccase, and manganese peroxidase that can oxidize and degrade a wide range of PAHs especially HMW PAHs (Ghosal et al. 2016). The peroxidase enzymes act on the aromatic rings of PAHs, initiating its oxidative cleavage to produce PAH quinones which further metabolize the ring structure through fission to produce less complex and more bioavailable intermediates that are further degraded by the fungi (Shahsavari et al. 2019). Non-ligninolytic fungi such as *Aspergillus* and *Penicillium* rely on intracellular enzymes like cytochrome P450 monooxygenases for PAH degradation. These enzymes oxidize PAHs by incorporating oxygen atoms into the aromatic ring, converting PAHs to arene oxides. Arene oxides are highly reactive intermediates that can undergo nonenzymatic rearrangement converting into phenols. Arene oxides can also be enzymatically converted to trans-dihydrodiols via epoxide hydrolase enzymes. Phenols and trans-dihydrodiols can be conjugated into more soluble forms such as O-glucosides or other intermediates, making them easier to further metabolize (Aydin et al. 2017). Both ligninolytic and non-ligninolytic fungi are capable of degrading PAHs, with ligninolytic fungi relying on extracellular enzymes to break down the PAHs into smaller intermediates and non-ligninolytic fungi using intracellular mechanisms to oxidize PAHs into more manageable forms. These processes lead to the breakdown of the complex, recalcitrant structures of PAHs, facilitating further degradation.

Table 1 Comparison of ligninolytic and non-ligninolytic fungal PAH degradation

Aspect	Ligninolytic fungi	Non-ligninolytic fungi
Main enzymes	Laccases, lignin peroxidase, Mn peroxidase	Cytochrome P450 monooxygenase, epoxide hydrolase
Mode of action	Extracellular oxidative enzymes degrade PAHs	Intracellular oxidation via cytochrome P450
Degradation pathway	Oxidative ring cleavage of PAHs	Oxidation to arene oxides, rearrangement to phenols or dihydrodiols
End products	Soluble ring-opened intermediates	Phenols, O-glucosides, trans-dihydrodiols
Fungal sp.	*Phanerochaete chrysosporium, Bjerkandera adusta, Pleurotus ostreatus*	*Aspergillus, Penicillium*

In marine environments, bacteria dominate the PAH degradation process, while fungi are more prominent in freshwater and terrestrial environments. Bacteria are particularly efficient at degrading LWM PAHs, whereas fungi are adept at breaking down HWM PAHs. Therefore, the combination of bacterial and fungal biodegradation can provide a synergistic approach to bioremediation of a mixture of PAHs (Table 1).

3.3 Anaerobic Biodegradation of PAHs

While aerobic biodegradation of PAHs is the most effective pathway for degradation of PAHs, anaerobic biodegradation can play an important role in oxygen-limited environments such as deep subsurface, aquatic sediments, and oil reservoirs (Boopathy 2017). Many diverse anaerobic bacteria such as *Desulfobacterium*, *Geobacter*, *Dechloromonas*, *Rhodopseudomonas*, and methanogenic archaea have been identified to degrade PAHs under anaerobic conditions but their mechanism has not been fully understood (Nzila 2018; Wartell et al. 2021). In anaerobic conditions, these microbes involve sulfate, nitrate, or iron instead of oxygen as alternative terminal electron acceptors and hence play a key role in geochemical cycling.

For instance, microbes such as sulfate-reducing bacteria (SRB) have the ability to degrade PAHs under anaerobic conditions using sulfate as an electron acceptor (Rabus et al. 2016). Nitrate-reducing bacteria are known to degrade both LWM and HWM PAHs under anaerobic conditions. Nitrate-reducing bacteria were found to degrade naphthalene and phenanthrene in the presence of nitrate (Dhar et al. 2019). These nitrate-reducing bacteria act as denitrifiers, where nitrate is reduced to nitrogen gas and PAH molecules are oxidized enabling their breakdown. Methanogenic archaea are another group of anaerobes that can utilize PAHs and convert them to methane as the final end-product, which has been observed in oil reservoirs and contaminated groundwater (Gieg et al. 2008). Though degradation of PAHs in the absence of oxygen is slow compared to its degradation in aerobic conditions, they can contribute significantly to the overall removal of hydrocarbons, especially through syntrophic interactions between microbial groups. In petroleum-contaminated environments, the anaerobic degradation of hydrocarbons is commonly associated with methanogenesis. Fermentation bacteria break down hydrocarbons into alcohols, hydrogen, and volatile fatty acids. These intermediates are then utilized by methanogens such as *Methanocoocus* to produce methane and CO_2 (Jiménez et al. 2016). These intermediates can also be used by other anaerobic bacteria like sulfate-reducers, denitrifiers, and iron-reducers for their growth (Wartell et al. 2021).

Though the mechanisms of anaerobic PAH degradation are not yet fully understood, they generally involve the initial activation of the PAH molecules. This activation can be achieved through methylation, addition of fumarate, carboxylation, and hydroxylation using alternative oxygen sources (Joye et al. 2016).

One of the primary mechanisms for activation of PAHs such as naphthalene is methylation followed by the addition of fumarate. Naphthalene is methylated to 2-methylnaphthalene, which gets converted to naphthyl-2-methylsuccinate by naphthyl-2-methyl succinate synthase enzyme. Once naphthyl-2-methylsuccinate is formed, it is activated by coenzyme-A to form naphthyl-2-methyl-succinyl-CoA, which further undergoes a series of reactions, breaking down into smaller intermediates that can enter the β-oxidation pathway, a key central metabolic pathway resulting in the production of acetyl-CoA. Acetyl-CoA is a critical metabolite that, in aerobic conditions, typically feeds into the TCA (tricarboxylic acid) cycle. However, in anaerobic environments where oxygen is unavailable, alternative electron acceptors like sulfate (SO_4^{2-}), nitrate (NO_3^-), or ferric iron (Fe^{3+}) take the role of the terminal electron acceptor. These anaerobic processes enable energy generation and further degradation of the hydrocarbons, with final products such as sulfides, nitrogen gas, or methane being produced, depending on the microbial pathways involved (Meckenstock et al. 2016; Rabus et al. 2016).

Carboxylation of aromatic hydrocarbons like naphthalene leads to addition of a carboxyl group (-COOH) to the aromatic compound resulting in the production of 2-naphthoic acid. This compound facilitates the first step in breaking down the complex aromatic hydrocarbons. 2-naphthoic acid through a series of reactions gets converted to 2-naphthoyl-CoA which undergoes further ring cleavage, carbon rearrangement, and finally β-oxidation pathway (Kümmel et al. 2015).

Hydroxylation using alternative electron acceptors such as sulfate (SO_4^{2-}), nitrate (NO_3^-), or iron (Fe^{3+}) instead of oxygen to add hydroxyl group (-OH) to the hydrocarbon, making it more reactive and allow further degradation. For example, in nitrate-reducing conditions, nitrate (NO_3^-) can be reduced to nitrite (NO_2^-) or other reduced forms, and the oxygen atoms from nitrate participate in hydroxylation reactions (Wartell et al. 2021). After hydroxylation, the hydroxylated intermediates like alcohols, aldehydes, or acids can be broken down into molecules like acetyl-CoA, which then enter the central metabolic pathways such as for methanogenesis or sulfate reduction depending on the microbial community and environment.

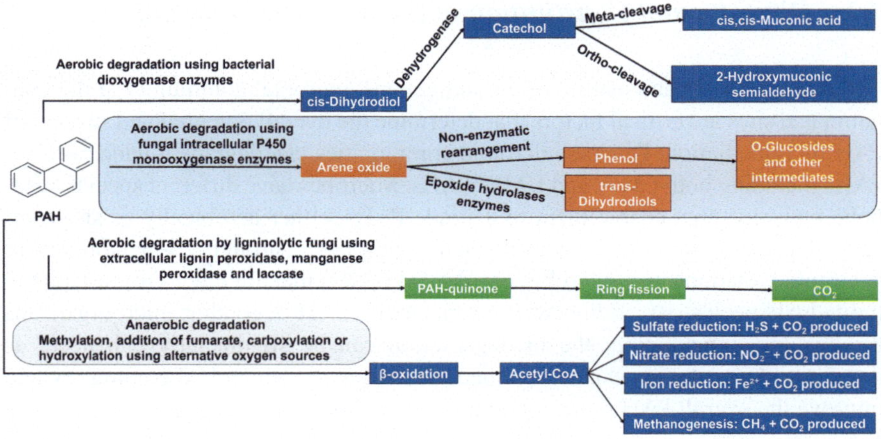

Fig. 2 Pathways for aerobic and anaerobic biodegradation of PAHs

Through both aerobic and anaerobic degradation of PAHs, geomicrobes not only clean up the environment but also help to restore the biogeochemical cycles such as the carbon and nitrogen cycles in disturbed environments. By producing methane or carbon dioxide during their metabolic processes, these microbes contribute to carbon recycling. By reducing sulfate (SO_4^{2-}) and ferric (Fe^{3+}) during PAH degradation, geomicrobes contribute to the mobilization of these elements, making them more bioavailable to other organisms in the ecosystem. Figure 2 represents the summary of biodegradation of PAHs.

4 Factors Influencing Bioremediation of PAHs

Geomicrobial biodegradation of PAHs is influenced by a variety of environmental and biological factors. One of the critical factors for bioremediation of PAHs is the nature and bioavailability of PAHs to the microbial community. Some of the PAHs are easily degraded but some PAHs are highly resistant to biodegradation. PAHs tend to be adsorbed tightly to soil particles or dissolved in an aqueous solution, limiting their accessibility to microbes capable of degrading them. The potential of the microbial community for degrading PAHs depends upon the diverseness and composition of the microbial community present in the contaminated environment. Environmental factors like temperature, oxygen availability, pH, and nutrient availability (e.g., nitrogen, phosphorus) play a vital role in the activity of microbes and the rate at which biodegradation takes place in the environment (Ghosal et al. 2016).

4.1 *Diversity and Constitution of the Microbial Community*

The diversity and composition of the indigenous microbial community at the contaminated sites are critical factors that determine the overall potential and success of PAH bioremediation. Diverse microbial communities can degrade a wide range of PAHs including both LMW and HMW PAHs. Microbes have different special metabolic pathways that enable them to degrade PAHs, either aerobically or anaerobically depending upon the environmental conditions. Through co-metabolism or syntrophic relationships, microbes can degrade PAHs that are otherwise resistant to individual species. By adding well-characterized PAH-degrading microorganisms through bioaugmentation, the biodegradation can be significantly enhanced, as these introduced strains can complement the indigenous microbial community and improve the overall PAH remediation efficiency (Shahsavari et al. 2019).

4.2 *Temperature*

Temperature is a key environmental factor that can significantly affect the microbial activity and biodegradation potential of geomicrobes involved in the remediation of PAHs. Most PAH-degrading microbes are mesophilic and work best between 20 and 40 °C. Higher temperatures not only increase the solubility and bioavailability of PAHs but also increase the microbial and enzymatic activities of PAH-degrading microbes, leading to faster biodegradation rates. But under extreme conditions, it can inhibit the growth and activities of these microbes, lowering the bioremediation efficiency. Under such high temperatures as those seen in oil reservoirs, composting piles, and geothermal sites, thermophilic bacteria, namely, *Bacillus* and *Thermus* spp., play an important role in the biodegradation of PAHs because they can survive and are better adapted to work under these conditions (Ghosal et al. 2016). In cold environments (below 15 °C) seen in polar regions, the rate of PAH degradation is slower due to both reduced microbial activity and the lower solubility of PAHs in aquatic environments. However, over a long period, biodegradation can still occur through the action of low-temperature tolerant and low-temperature loving microorganisms such as *Pseudomonas* and *Rhodococcus* (Margesin and Schinner 2001; Ghosal et al. 2016).

4.3 *pH*

In addition to temperature, pH is another critical environmental factor that can significantly influence the biodegradation of PAHs by microbes. Most PAH-degrading microbes prefer a neutral pH range between 6.5 and 7.5 and any deviations from the optimal range, either toward acidic or alkaline conditions, can disrupt the biodegradation rates. In acidic or highly alkaline conditions, the degradation of PAHs can be slower due to reduced microbial activities, lower enzyme efficiency, and decreased PAH bioavailability. However, specialized acidophilic and alkaliphilic microbes may still contribute to biodegradation in extreme pH environments, albeit at slower rates (Kebede et al. 2021).

4.4 *Oxygen*

As discussed earlier the rate and efficiency of PAH biodegradation are dependent on the availability of oxygen. Aerobic degradation is the most efficient method of degrading PAHs, with oxygen serving as the acceptor of electrons in the involved process. In oxygen-limited environments, anaerobic degradation of PAHs takes place, which is much slower, and instead of oxygen, alternative electron acceptors like sulfate, nitrate, and iron take part.

4.5 *Nutrient Availability*

Essential nutrients like phosphorus (P), nitrogen (N), and trace elements like potassium (K), magnesium (Mg), iron (Fe), and sulfur (S) and their availability are important for microbial growth and the metabolic activities of PAH-degrading microbes. An optimum ratio of carbon/nitrogen/phosphorus (C/N/P) for hydrocarbon degradation ranges from 100:10:1 to 100:20:1 (Kebede et al. 2021). Low nitrogen levels, especially in petroleum spills and industrial waste areas, lead to slower microbial growth and decreased enzyme activity. High nitrogen levels can favor the growth of non-degrading microbes over PAH-degrading ones, leading to competition for available nutrients. Trace elements such as Mg, K, S, and Fe are crucial since they can act as cofactors/coenzymes for PAH-degrading enzymes. In nutrient-poor environments, addition of NPK fertilizers or biostimulation-adding nutrients can stimulate microbial activity to enhance PAHs degradation. Optimizing nutrient levels is a key strategy in successful bioremediation of PAH-contaminated sites.

4.6 *Soil Type*

In terrestrial PAH-contaminated sites, factors such as soil texture, organic matter content, porosity, water retention, and nutrient availability greatly influence the biodegradation process. Sandy and loamy soils, with good aeration and water balance, generally support faster PAHs degradation, while clay-rich and organic soils with their strong sorption of PAHs and reduced oxygen flow support slower PAH degradation (Kebede et al. 2021). Soils with higher redox potential (well-oxygenated soils) promote aerobic biodegradation, whereas low redox soils such as waterlogged soils exhibit anaerobic degradation. Availability of $SO4^{2-}$, NO^{3-}, or Fe^{3+} as alternative electron acceptors in oxygen-limited soils can enable anaerobic degradation pathways for certain recalcitrant PAHs that may not be easily degraded by aerobic degradation, thereby expanding the range of PAHs that can be effectively mineralized by the indigenous microbial community.

5 Bioremediation Strategies

Three kinds of microbial bioremediation techniques are used for mitigation of PAHs contaminated soils: biostimulation, bioaugmentation, and natural attenuation (Fig. 3). These techniques are cost-effective and eco-friendly and hence are most preferred for bioremediation of pollutants such as PAHs (Bento et al. 2005).

5.1 Biostimulation

It involves enhancing the growth of hydrocarbon-degrading indigenous bacteria by stimulating their growth factors such as electron acceptors and nutrients (Tyagi et al. 2011). For effective remediation of hydrocarbons, optimization of surrounding environments of contaminated spots is necessary as the hydrocarbon-degrading microbes may not be efficient in normal conditions. Addition of stimulants to the subsurface is done through injection wells (Banerjee et al. 2016). Microbial

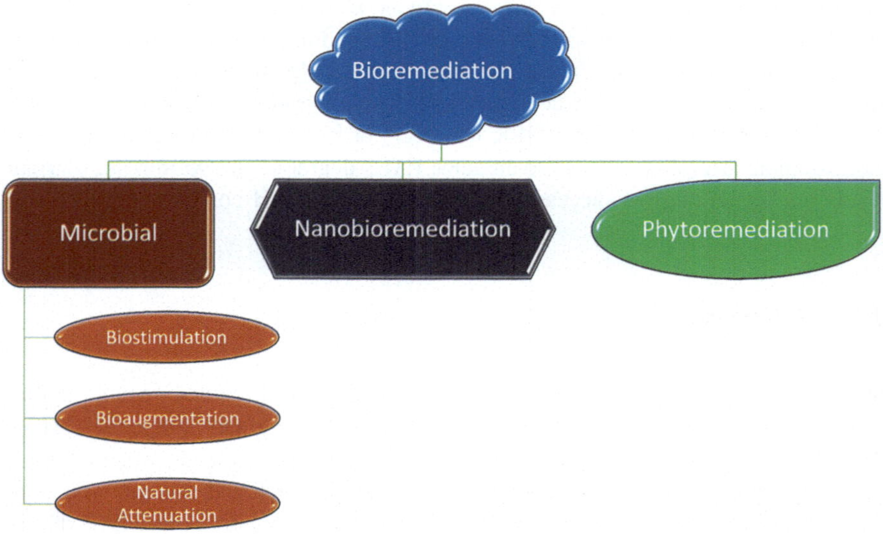

Fig. 3 Bioremediation strategies

degradation can be increased by adding rate-limiting nutrients in the environment. For obtaining optimum microbial degradation, environmental conditions like temperature, moisture content and pH can be optimized.

5.2 *Bioaugmentation*

Bioaugmentation is a process of introducing hydrocarbon-degrading bacteria to supplement or augment the preexisting microbial population for degradation of hydrocarbons (Tyagi et al. 2011). This method involves adding of bioinoculant that is beneficial to microbe which can be either indigenous or genetically modified microorganisms at the contaminated sites for bioremediation. The method is used mostly in oil-contaminated sites where the native soil microorganism is inadequate in degrading the pollutant hydrocarbons (Muter 2023).

5.3 *Natural Attenuation*

Bio-attenuation or intrinsic bioremediation known as natural attenuation is a process of biodegradation that relies on the degradation capability of naturally inhabiting native microorganisms without the interference of humans (Bisht et al. 2015). It is the simplest form of bioremediation where the process of degradation is natural and followed in environments where conditions for activity of native microorganisms are favorable. Human health and environment are at low risk in this process (Yu et al. 2005). Remediation by Enhanced Natural Attenuation (RENA) is an enhanced way of natural attenuation through which bioremediation is achieved more efficiently. Details such as microbial proliferation, aeration content of the soil, nutrient, and moisture content are some of the factors that are taken care of during RENA of pollutants (Ayilara and Babalola 2023).

5.4 *Nanobioremediation*

Nanobioremediation refers to the use of nanoparticles for remediation of the contaminated soil and water (El-Ramady et al. 2020). Titanium, gold zinc, aluminum, iron, silver, and carbon nanoparticles are mostly used in bioremediation (Alazaiza et al. 2021). Copper pollutants from acid mine drainage have been removed by using diatomaceous nanoparticles in conjunction with polyethyleneimine nanoparticles (Ji et al. 2023). Iron nanoparticles have shown an effective role in soil bioremediation through process of catalysis, adsorption, photodegradation, and ultimately filtration. Nanoremediation is one of the most effective remediation processes in the current state of the art due to the high reactivity of the nanomaterials (NMs) (Mukhopadhyay et al. 2022).

6 Challenges and Future Directions

Due to high specificity, RENA technology is incapable of degrading all the wastes as the pollution sites contain different types of wastes and pollutants. In addition, RENA-degraded pollutants are sometimes converted to other harmful substances and the process as a whole is time-consuming. The external nutrients that are supplied to the soil for enhancement of microbial activity may lead to alteration of natural eco-balance and the soil texture can also be destroyed. In cases of severe pollution and when rapid remediation is needed, RENA being a time-consuming process is less preferred. Co-inoculation is more efficient in microbial degradation of pollutants as suggested by Rahman et al. (2002). The co-inoculated bacteria will have a catabolic pathway. One such bacteria to be included in co-culture is *Mycobacterium* sp., which showed tremendous ability to degrade low as well as high molecular weight PAH, which was done by hydroxylation and oxidative mechanisms. Strain selection, contaminant types, introduction and inoculation of cultures, ecology of microbes, and environmental constraints are some of the factors that should be considered while designing strategies for bioremediation (Tyagi et al. 2011). Challenges of bioaugmentation include the viability of introduced microorganisms for bioremediation over time. Due to various biotic and abiotic stresses, such as fluctuations in pH, temperature, nutrient availability, and concentration of toxic pollutants, the introduced microorganism may not survive in the contaminated soil. In addition, non-indigenous microorganisms, antimicrobial compounds secreted by other microorganisms, predator protozoa, competition for limiting food and nutrition, bacteriophages, and fungi metabolites are always a threat to the added microorganism (Bisht et al. 2015). Laboratory and on-field test scenarios are different due to the change in the factors and conditions that are optimum for the normal growth and bioremediation by pollutant-degrading microbes. Thus, standardization is to be done for more accurate results.

7 Conclusion

Geomicrobes have a vital part in the environmental remediation of PAHs, providing a natural and sustainable solution to address these persistent compounds. By utilizing the metabolic capabilities of indigenous microbial communities, geomicrobes can be used to target PAHs in the environment under both aerobic and anaerobic conditions. These microbes break down the complex PAH compounds into simpler ones, adapting to various environmental conditions, which enhances their potential for bioremediation. By leveraging the natural metabolic capabilities of microorganisms and understanding their interactions with the environment, it is possible to develop effective strategies for the degradation of even the most recalcitrant hydrocarbons. However, challenges such as contaminant complexity, bioavailability, and the need for precise monitoring remain. Advances in genomic technologies, synthetic biology, and nanotechnology, along with a deeper integration of geomicrobiology into bioremediation practices, hold the potential to overcome these challenges and make pathways for more efficient and effective remediation efforts in the coming days.

Competing Interests All the authors declare that they have no competing interests.

References

Abha S, Swaranjit C (2012) Hydrocarbon pollution: effects on living organisms, remediation of contaminated environments, and effects of heavy metals co-contamination on bioremediation. In: Romero-Zeron L (ed) Introduction to Enhanced Oil Recovery (EOR) processes and bioremediation of oil-contaminated sites. IntechOpen, Rijeka, pp 186–206. https://doi.org/10.5772/48014

Alazaiza MYD, Albahnasawi A, Ali GAM, Bashir MJK, Copty NK, Amr SSA, Abushammala MFM, Al Maskari T (2021) Recent advances of nanoremediation technologies for soil and groundwater remediation: a review. Water (Switzerland) 13:2186. https://doi.org/10.3390/w13162186

Aydin S, Karaçay HA, Shahi A, Gökçe S, Ince B, Ince O (2017) Aerobic and anaerobic fungal metabolism and Omics insights for increasing polycyclic aromatic hydrocarbons biodegradation. Fungal Biol Rev 31:61–72. https://doi.org/10.1016/j.fbr.2016.12.001

Ayilara MS, Babalola OO (2023) Bioremediation of environmental wastes: the role of microorganisms. Front Agron 5:1183691. https://doi.org/10.3389/fagro.2023.1183691

Banerjee A, Dutta S, Mondal S, Roy A (2016) Bioremediation of hydrocarbon—a review. Int J Adv Res 4:1303–1313. https://doi.org/10.21474/IJAR01/734

Bento FM, Camargo FAO, Okeke BC, Frankenberger WT (2005) Comparative bioremediation of soils contaminated with diesel oil by natural attenuation, biostimulation and bioaugmentation. Bioresour Technol 96:1049–1055. https://doi.org/10.1016/j.biortech.2004.09.008

Bisht S, Pandey P, Bhargava B, Sharma S, Kumar V, Sharma KD (2015) Bioremediation of polyaromatic hydrocarbons (PAHs) using rhizosphere technology. Braz J Microbiol 46:7–21. https://doi.org/10.1590/S1517-838246120131354

Boopathy R (2017) Anaerobic degradation of petroleum hydrocarbons in sediments. In: Jonathan W, Wong C, Rajeshwar D, Tyagi D, Pandey A (eds) Current developments in biotechnology and bioengineering. Elsevier, Amsterdam, pp 475–490. https://doi.org/10.1016/B978-0-444-63664-5.00020-4

Cai Y, Wang R, Rao P, Wu B, Yan L, Hu L, Park S, Ryu M, Zhou X (2021) Bioremediation of petroleum hydrocarbons using *Acinetobacter sp.* SCYY-5 isolated from contaminated oil sludge: strategy and effectiveness study. Int J Environ Res Public Health 18:819. https://doi.org/10.3390/ijerph18020819

Dhar K, Subashchandrabose SR, Venkateswarlu K, Krishnan K, Megharaj M (2019) Anaerobic microbial degradation of polycyclic aromatic hydrocarbons: a comprehensive review. In: de Voogt P (ed) Reviews of environmental contamination and toxicology, vol 251. Springer, Cham, pp 25–108. https://doi.org/10.1007/398_2019_29

El-Ramady H, El-Henawy A, Amer M, Omara AE-D, Elsakhawy T, Salama A-M, Ezzat A, Elsherif A, Elmahrouk M, Shalaby T (2020) Agro-pollutants and their nano-remediation from soil and water: a mini-review. Environ Biodivers Soil Secur 4:361–375. https://doi.org/10.21608/jenvbs.2020.47751.1111

Ghosal D, Ghosh S, Dutta TK, Ahn Y (2016) Current state of knowledge in microbial degradation of Polycyclic Aromatic Hydrocarbons (PAHs): a review. Front Microbiol 7. https://doi.org/10.3389/fmicb.2016.01369

Gieg LM, Duncan KE, Suflita JM (2008) Bioenergy production via microbial conversion of residual oil to natural gas. Appl Environ Microbiol 74:3022–3029. https://doi.org/10.1128/AEM.00119-08

Hamouda RA, Alhumairi AM, Saddiq AA (2023) Simultaneous bioremediation of petroleum hydrocarbons and production of biofuels by the micro-green alga, cyanobacteria, and its consortium. Heliyon 9:e16656. https://doi.org/10.1016/j.heliyon.2023.e16656

Hidalgo KJ, Sierra-Garcia IN, Dellagnezze BM, de Oliveira VM (2020) Metagenomic insights into the mechanisms for biodegradation of polycyclic aromatic hydrocarbons in the oil supply chain. Front Microbiol 11. https://doi.org/10.3389/fmicb.2020.561506

Ji M, Li B, Majdi A, Alkhalifah T, Alturise F, Ali HE (2023) Application of nano remediation of mine polluted in acid mine drainage water using machine learning model. Chemosphere 311:136926. https://doi.org/10.1016/j.chemosphere.2022.136926

Jiménez N, Richnow HH, Vogt C, Treude T, Krüger M (2016) Methanogenic hydrocarbon degradation: evidence from field and laboratory studies. Microb Physiol 26:227–242. https://doi.org/10.1159/000441679

Joye S, Kleindienst S, Gilbert J, Handley K, Weisenhorn P, Overholt W, Kostka J (2016) Responses of microbial communities to hydrocarbon exposures. Oceanography 29:136–149. https://doi.org/10.5670/oceanog.2016.78

Kebede G, Tafese T, Abda EM, Kamaraj M, Assefa F (2021) Factors influencing the bacterial bioremediation of hydrocarbon contaminants in the soil: mechanisms and impacts. J Chem 2021:1–17. https://doi.org/10.1155/2021/9823362

Kim K-H, Jahan SA, Kabir E, Brown RJC (2013) A review of airborne polycyclic aromatic hydrocarbons (PAHs) and their human health effects. Environ Int 60:71–80. https://doi.org/10.1016/j.envint.2013.07.019

Kümmel S, Herbst F-A, Bahr A, Duarte M, Pieper DH, Jehmlich N, Seifert J, von Bergen M, Bombach P, Richnow HH, Vogt C (2015) Anaerobic naphthalene degradation by sulfate-reducing Desulfobacteraceae from various anoxic aquifers. FEMS Microbiol Ecol 91. https://doi.org/10.1093/femsec/fiv006

Ławniczak Ł, Woźniak-Karczewska M, Loibner AP, Heipieper HJ, Chrzanowski Ł (2020) Microbial degradation of hydrocarbons—basic principles for bioremediation: a review. Molecules 25:856. https://doi.org/10.3390/molecules25040856

Mahjoubi M, Cappello S, Souissi Y, Jaouani A, Cherif A (2018) Microbial bioremediation of petroleum hydrocarbon – contaminated marine environments. In: Zoveidavianpoor M (ed) Recent insights in petroleum science and engineering. InTechOpen, Rijeka. https://doi.org/10.5772/intechopen.72207

Margesin R, Schinner F (2001) Biodegradation and bioremediation of hydrocarbons in extreme environments. Appl Microbiol Biotechnol 56:650–663. https://doi.org/10.1007/s002530100701

Matúš P, Littera P, Farkas B, Urík M (2023) Review on performance of Aspergillus and Penicillium species in biodegradation of organochlorine and organophosphorus pesticides. Microorganisms 11:1485. https://doi.org/10.3390/microorganisms11061485

Meckenstock RU, Boll M, Mouttaki H, Koelschbach JS, Cunha Tarouco P, Weyrauch P, Dong X, Himmelberg AM (2016) Anaerobic degradation of benzene and polycyclic aromatic hydrocarbons. Microb Physiol 26:92–118. https://doi.org/10.1159/000441358

Mukhopadhyay R, Sarkar B, Khan E, Alessi DS, Biswas JK, Manjaiah KM, Eguchi M, Wu KCW, Yamauchi Y, Ok YS (2022) Nanomaterials for sustainable remediation of chemical contaminants in water and soil. Crit Rev Environ Sci Technol 52:2611. https://doi.org/10.1080/10643389.2021.1886891

Muter O (2023) Current trends in bioaugmentation tools for bioremediation: a critical review of advances and knowledge gaps. Microorganisms 11:710. https://doi.org/10.3390/microorganisms11030710

Nzila A (2018) Biodegradation of high-molecular-weight polycyclic aromatic hydrocarbons under anaerobic conditions: overview of studies, proposed pathways and future perspectives. Environ Pollut 239:788–802. https://doi.org/10.1016/j.envpol.2018.04.074

Patel AB, Shaikh S, Jain KR, Desai C, Madamwar D (2020) Polycyclic aromatic hydrocarbons: sources, toxicity, and remediation approaches. Front Microbiol 11. https://doi.org/10.3389/fmicb.2020.562813

Rabus R, Boll M, Heider J, Meckenstock RU, Buckel W, Einsle O, Ermler U, Golding BT, Gunsalus RP, Kroneck PMH, Krüger M, Lueders T, Martins BM, Musat F, Richnow HH, Schink B, Seifert J, Szaleniec M, Treude T, Ullmann GM, Vogt C, von Bergen M, Wilkes H (2016) Anaerobic microbial degradation of hydrocarbons: from enzymatic reactions to the environment. Microb Physiol 26:5–28. https://doi.org/10.1159/000443997

Rahman KSM, Thahira-Rahman J, Lakshmanaperumalsamy P, Banat IM (2002) Towards efficient crude oil degradation by a mixed bacterial consortium. Bioresour Technol 85:257–261. https://doi.org/10.1016/S0960-8524(02)00119-0

Shahsavari E, Schwarz A, Aburto-Medina A, Ball AS (2019) Biological degradation of Polycyclic Aromatic Compounds (PAHs) in soil: a current perspective. Curr Pollut Rep 5:84–92. https://doi.org/10.1007/s40726-019-00113-8

Soare M-G, Lakatos ES, Ene N, Malo (Dalanaj) N, Popa O, Babeanu N (2019) The potential applications of Bacillus sp. and Pseudomonas sp. strains with antimicrobial activity against phytopathogens, in waste oils and the bioremediation of hydrocarbons. Catalysts 9:959. https://doi.org/10.3390/catal9110959

Sui X, Wang X, Li Y, Ji H (2021) Remediation of petroleum-contaminated soils with microbial and microbial combined methods: advances, mechanisms, and challenges. Sustain For 13:9267. https://doi.org/10.3390/su13169267

Truskewycz A, Gundry TD, Khudur LS, Kolobaric A, Taha M, Aburto-Medina A, Ball AS, Shahsavari E (2019) Petroleum hydrocarbon contamination in terrestrial ecosystems—fate and microbial responses. Molecules 24:3400. https://doi.org/10.3390/molecules24183400

Tyagi M, da Fonseca MMR, de Carvalho CCCR (2011) Bioaugmentation and biostimulation strategies to improve the effectiveness of bioremediation processes. Biodegradation 22:231–241. https://doi.org/10.1007/s10532-010-9394-4

Wang W, Shao Z (2013) Enzymes and genes involved in aerobic alkane degradation. Front Microbiol 4. https://doi.org/10.3389/fmicb.2013.00116

Wang M, Ding M, Yuan Y (2023) Bioengineering for the microbial degradation of petroleum hydrocarbon contaminants. Bioengineering 10:347. https://doi.org/10.3390/bioengineering10030347

Wartell B, Boufadel M, Rodriguez-Freire L (2021) An effort to understand and improve the anaerobic biodegradation of petroleum hydrocarbons: a literature review. Int Biodeterior Biodegradation 157:105156. https://doi.org/10.1016/j.ibiod.2020.105156

Yu KSH, Wong AHY, Yau KWY, Wong YS, Tam NFY (2005) Natural attenuation, biostimulation and bioaugmentation on biodegradation of polycyclic aromatic hydrocarbons (PAHs) in mangrove sediments. Mar Pollut Bull 51:1071–1077. https://doi.org/10.1016/j.marpolbul.2005.06.006

Yuniati MD (2018) Bioremediation of petroleum-contaminated soil: a review. IOP Conf Ser: Earth Environ Sci 118:012063. https://doi.org/10.1088/1755-1315/118/1/012063

Microbial Life Within the Deep, Hot Subsurface Biosphere and Its Implications in Astrobiology

Sourav Mukhopadhyay, Swatilekha Sarkar, and Pinaki Sar

Contents

Abstract The deep, hot subsurface biosphere represents one of the largest biomes of our planet. Recent scientific explorations have confirmed that in spite of a multitude of physical and chemical extremities, the deep biosphere could harbor up to 19% of Earth's total biomass. Evolution, adaptation, and metabolic abilities of these extremophilic microorganisms intrigued us to answer the most fundamental ques-

S. Mukhopadhyay
Department of Bioscience and Biotechnology, Indian Institute of Technology Kharagpur, Kharagpur, West Bengal, India

Department of Biological Sciences, National University of Singapore, Singapore, Singapore
e-mail: sourav_mukhopadhyay@u.nus.edu

S. Sarkar · P. Sar (✉)
Department of Bioscience and Biotechnology, Indian Institute of Technology Kharagpur, Kharagpur, West Bengal, India

S. R. Joshi et al. (eds.), *Mineral Transformation and Bioremediation by Geo-Microbes*, Interdisciplinary Biotechnological Advances,
https://doi.org/10.1007/978-981-96-3033-2_17

tions about origin, evolution, limits, and mechanisms of adaptation of life on this planet and their exploration in other planetary bodies. This chapter presents a comprehensive review of: (a) general attributes of extremophiles; (b) deep subsurface as the hotspot for extremophiles and astrobiological research; and (c) astrobiological targets with subsurface analogues on Earth and role of bioenergetics model for predicting habitability. Overall, this review seeks to synthesize the existing state of knowledge on the topic and provides a direction for future research and applications in astrobiology.

Keywords Deep biosphere · Extremophilic microbes · Stress adaptation · Habitability · Astrobiology

Abbreviations

°C	Degree celsius
Atm	Atmospheric
MPa	Mega pascal
NaCl	Sodium chloride
O_2	Oxygen
CO_2	Carbon dioxide

1 Introduction

The term "extremophile" was coined over 30 years ago to describe microorganisms capable of thriving under extreme environmental conditions, such as high or low temperature, pH, salinity, and pressure—conditions that are hostile to humans and most known organisms (Arias et al. 2023). The "extreme" in "extremophile" is inherently anthropocentric, as it is based on environmental conditions that are unsuitable for human life and most microorganisms (Rothschild and Mancinelli 2001). In contrast, most life forms on Earth grow and reproduce under what are considered "physiological" conditions—such as a temperature of 37 °C, 1 atm pressure, 0.9% to 3% salinity, and a neutral pH of 7.4—ideal for human life and the growth of organisms like *Escherichia coli*. However, over the past century, research has revealed that numerous organisms not only survive but also require extreme conditions for their existence (Caro-Astorga et al. 2024). As a result, the boundaries of what constitutes "extreme" environments have been broadened, with studies expanding the range of temperature, pH, pressure, radiation, salinity, and nutrient availability that life can endure (Borsodi 2024).

While many organisms thrive under "physiological" conditions, a large portion of Earth's crust is subject to extreme environments, such as temperatures ranging from −89 °C in the Arctic to 400 °C on the deep seafloor. In response to these harsh conditions, microbes residing in these environments have developed unique physiological strategies to survive extreme variations in pH, temperature, pressure, and nutrient availability. Though extremophiles encompass a range of life forms, including protozoa, algae, and fungi, the majority are prokaryotic (Borsodi 2024). To better understand their stress tolerance mechanisms, distribution, and diversity, extremophilic prokaryotes are classified according to their specific adaptations to these extreme environmental parameters.

Studies show that extremophiles have evolved specialized adaptations to withstand single or multiple extreme conditions. Of particular interest is the deep subsurface biosphere, where high temperatures, nutrient limitations, and extreme pressures coexist, creating a unique environment for microbial life. This habitat is increasingly viewed as a valuable astrobiological analogue, offering insights into the potential for life beyond Earth. Astrobiological analogues can be classified based on the properties they replicate, such as compositional, electrochemical, or physical characteristics, and can also be categorized by the specific planetary conditions they simulate. Recent advancements in bioenergetics have led to the development of models that quantify key parameters like nutrient and energy turnover, along with growth rates, thereby providing a more detailed framework for understanding possible life processes in exoplanetary environments.

As extremophiles continue to intrigue researchers due to their unique adaptations to extreme environments, this study focuses on providing a deeper understanding of these organisms, specifically their stress tolerance mechanisms and the wide range of extreme habitats they occupy on Earth. A particular emphasis is placed on the deep subsurface biosphere, which supports a variety of extremophiles under high temperature, pressure, and nutrient limitations. This environment is considered a key astrobiological analog, offering insights into potential extraterrestrial life. Additionally, the study will explore other astrobiological analogues and examine recent advances in bioenergetic models, which help quantify microbial processes, to assess their relevance for life on exoplanets. This knowledge is essential for evaluating the potential for microbial life in exoplanetary environments and informing the design of targeted astrobiological missions and advancing the search for life on planets and moons with analogous extreme conditions.

2 Environmental Extremities and Adaptation of Extremophiles to Stress Conditions

2.1 Classification of the Extremophiles

Based on the propensity to survive in different, distinct extreme environmental conditions, the prokaryotic life forms can be categorized into a few classes:

1. Thermophiles: Thermophiles are defined as microorganisms that live at 41–122 °C. Their optimum growth temperature lies between 60 and

108 °C. Thermophiles are generally categorized into three categories: (a). moderate thermophiles ($T_{optimum}$: 50–60 °C), (b). extreme thermophiles ($T_{optimum}$: 60–80 °C), and (c). hyperthermophiles ($T_{optimum}$: 80–110 °C) (Kumar et al. 2019). The most thermophilic bacteria reported till date is *Geogemma barossii* which has the ability to grow at 129 °C (Kashefi and Lovley 2003).

2. Psychrophiles: Psychrophiles are referred to as microbes that grow fastest at a temperature of 15 °C or lower. However, they do not grow above 20 °C. Another category of cold-adapted microorganisms exhibit growth at temperatures close to the freezing point of water with a growth optima above 20 °C. They are known as psychrotrophic bacteria. The known limit of low temperature growth is −12 °C, and the microorganism that exhibits growth at such a low temperature is *Psychromonas ingrahamii* (Coker 2019).
3. Acidophiles: Acidophiles are microbes that tend to grow well in acidic conditions. Acidophilic microorganisms grow at pH1–5, and their growth optima is below pH3. The most acidophilic microorganism reported till date is *Picrophilus torridus,* which has the ability to grow at pH around 0 (Coker 2019).
4. Alkaliphiles: Alkaliphiles represent a category of extremophiles that are able to grow in alkaline environments. Alkaliphilic microbes show prominent growth at pH8.5 to 11, and their growth optima is generally around pH9. *Plectonema nostocorum* and *Hydrogenophaga* sp., both tend to grow at the known highest limit of pH (i.e., pH13) (Coker 2019).
5. Halophiles: The microorganisms that survive extreme hypersaline conditions and require at least 1.16% (w/v) salinity are termed halophiles. However, microorganisms that grow well even in the absence of salt but tolerate varying salt concentrations are defined as halotolerants. Halophiles are categorized into three groups: (a) slight halophiles (marine bacteria): halophiles which grow well at 1.16% to 3% (w/v) NaCl; (b) moderate halophiles: halophiles which grow well at 3% to 14.6% (w/v) NaCl; and (c) extreme halophiles: halophiles which exhibit optimal growth at 14.6% to 30.37% (saturated) (w/v) NaCl. *Haloferax volcanii* has been reported to grow at saturated NaCl concentrations (Hartman et al. 2010).
6. Barophiles: Barophiles and piezophiles are referred to as a class of microorganisms that have shorter generation times at higher pressure than 1 atm pressure (Bartlett 2002). *Thermococcus piezophilus* has been observed to survive at elevated pressure, as high as 125 MPa, and is the known most barophilic microorganism. Despite the small number of culturable strict piezophiles, environmental studies reveal that there is possibility of life at extreme high pressures. A report on piezotolerant strains has deciphered that life can tolerate transient exposures till 2000 MPa.

   Also, (7) radiation-resistant microbes, (8) xerophiles that prefer dry environments, and (9) desiccation-resistant microorganisms represent different classes of extremophiles.

It has been possible to delineate prokaryotic diversity in a framework of extreme physicochemical conditions, which is routinely represented by thermophilic,

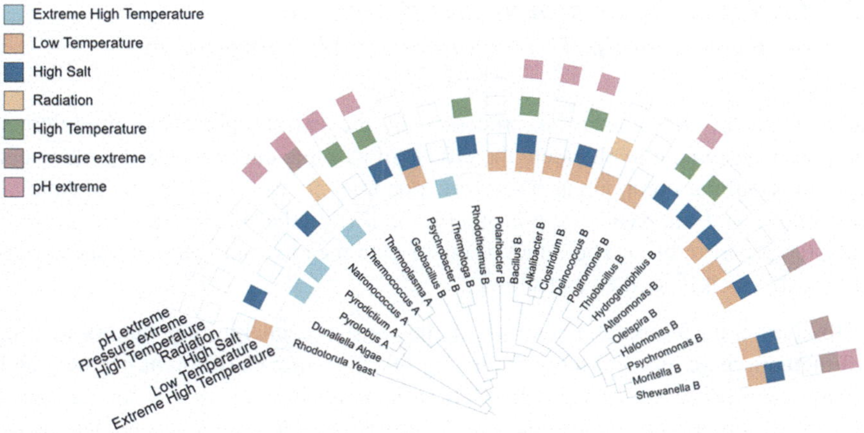

Fig. 1 Phylogenetic distribution of different extremophilic microbes (both prokaryotic and eukaryotic) reported from various extreme environments

psychrophilic, acidophilic, alkaliphilic, halophilic, barophilic, and radioresistant species. The phylogenetic distribution of different extremophiles is presented in Fig. 1. This systematics appears to be paradoxical, as one organism might unite specialized extremotolerant properties of several extreme groups (Shen et al. 2024). In the backdrop of extremophilic microbes, it is important to note that they represent a class of highly adapted microbes for the conditions considered and are able to metabolically and biochemically operate under a specific extreme environment. Consequently, while defining the limits of life at extremes, it is of little interest whether the organism is extremophilic or extremotolerant in nature as long as it manifests active growth under the specific extreme environmental precondition. Hence another modern school of systematics for extreme-adapted microorganisms is proposed (Rothschild and Mancinelli 2001). Subsequently, depending on the competence of adaptation to single or multiple environmental stress conditions, extremophiles are differentiated into two classes:

1. Specialist: Organisms that are adapted to single extremity and survive therein.
2. Generalist: Organisms that are adapted to multiple environmental stresses, including stressors that are non-indigenous to their habitats (Rothschild and Mancinelli 2001).

While notable effort has gone into studies of specialists, such as hyperthermophiles and radioresistant species, generalists remain of significant interest in relation to a broader array of their applicability in production of renewable energy, second-generation (2G) biofuel, in strategizing resilient storage and disposal of nuclear wastes, etc. (Sun et al. 2024). To utilize the potential of extremophilic/extremotolerant microbes for the enhancement of quality of human life, it is crucial to have clearer conceptions about how they are impacted by extreme environmental parameters and subsequently adapt to extreme environmental conditions.

2.2 Impact of the Environmental Extremities on Extremophilic/Extremotolerant Microorganisms

For analysis of the adaptation processes of the extremophiles/extremotolerant microbes, it is crucial to elucidate the effects of extreme environmental parameters on various cellular compounds. Therefore, one must focus primarily on the cellular morphology and the basic components of a cell (i.e., the membranes, the nucleic acids, the proteins, etc.). The effect of major extreme environmental parameters on microbial cells is described as follows:

1. Temperature: Temperature is one of the most crucial limiting factors to life. Fluctuations in ambient temperature significantly impact different physical and biological processes of the cell. With an increase in temperature, the enhancement of membrane fluidity and the denaturation of proteins and nucleic acids take place. The solubility of gases, important for vital biochemical functions (O_2 and CO_2) of the microbes, varies with an increase in temperature. Under low temperatures, essentially the enzyme reaction and solute uptake rates are decreased, membrane fluidity is reduced, and inhibitory nucleic acid structures and intracellular crystalline ice are formed that interfere with the viability of the cells. The most significant biological process sensitive to temperature is translation, where the ribosome and associated proteins aid in temperature sensing (Rothschild and Mancinelli 2001).
2. pH: It is well accepted that a wide array of biological processes relies on ambient pH values. Under acidic conditions, high differences in cellular concentration of protons with the external environment induce inflow of protons. The incoming protons, thereby, dramatically change the internal cellular pH, which leads to cell death. Under high pH, alkaliphiles do overcome an imbalance of H^+ ions too. Else, the cellular H^+ ions move out of the cells that significantly impact the integrity of the cellular membrane. Also, external environment-induced significant fluctuations of cellular pH greatly affect the protein folding and enzyme catalysis inside the cells (Mougi 2024).
3. Salinity: High salt concentrations result in water losses in the course of osmotic processes. High ambient/external osmotic pressure affects the permeability of the cell membrane and interferes with the integrity of the cells. High concentrations of salt also impact the proper folding of the protein molecules and might destabilize the nucleic acids (Lonescu et al. 2024).
4. Pressure: When subjected to elevated pressure conditions, the progress of chemical reactions at per with the increase in volume becomes impossible. Also, the spatial organization of lipids are subjected to densification under high ambient pressure. This phenomenon causes damage to the cell wall and cell membrane. High pressure might result in a decrease of cellular growth rate and production of biomass. This is due to the suppression of interplay of proteins with different substrates and subsequent inactivation of cellular enzymes (Bartlett 2002).

Apart from the discussed environmental parameters, (5) radiation and (6) desiccation are known to impact the cellular DNA. Structural damage and nicks in the DNA are the major effects of these two extremes that interfere with the viability of the cells (Coker 2019). All the discussed environmental parameters impact the microbes, firstly at the level of cellular morphology and consequently at the level of the membrane and basic macromolecules (Merino et al. 2019). Therefore, to further elucidate their extremophilic properties, it is important to understand how they circumvent the changes in the cellular components so as to ensure viability under the effect of the extreme environmental conditions.

2.3 *Stress Adaptations of the Extremophilic Bacteria*

For sustenance, in the presence of multiple stressors, the inhabitants of such extreme environments are presumed to have developed unique adaptations, especially at the level of their membranes and macromolecules, which might be refracted through their different stress responses.

1. Adaptations to extreme temperatures: All classes of thermophiles and psychrophiles, are exposed to external temperature constantly. Consequently, they have undergone adaptations at the level of macromolecules (i.e., DNA, proteins, and lipids) and cellular complexes (cell surface, RNA polymerase, ribosomes, etc.) so that these cellular organizations stay functional under a specific temperature range. Another important adaptation is at the level of lipid composition of cell membranes. Majorly, increasing the saturated fatty acids, short-chain and branched-chain fatty acids, helps counterbalance the high temperature-induced increased fluidity of the cell membrane. To combat the lethal effects of low temperatures, microbes have accumulated different physiological adaptations, such as regulation of the membrane fluidity, synthesis of chaperone molecules, and production of organic solute/cryoprotectant molecules. Compatible solutes are used as antifreeze for reduction in the cytoplasmic freezing point and inhibition of aggregation/denaturation of proteins, stabilization of membranes, and further scavenging of free radicles at low temperature conditions (Kashefi and Lovley 2003).
2. Adaptation to extreme pH: To circumvent the negative effects of high environmental pH conditions, alkaliphilic microbes have developed a cell wall with a significant negative charge on it that aids in reducing the pH of the external environment, which is just outside the cell. They might also develop a secondary cell wall, which is essentially acidic in nature and is constituted of teichuronopeptide, teichuronic acid, or polyglutamic acid. These acids have an affinity for H^+ and repulsion for OH^-, possibly aiding in the generation of the proton motive force needed to drive ATP synthesis (Amangeldina et al. 2024).

   To cope with the low pH conditions, acidophilic microbes have developed certain physiological mechanisms. One of those mechanisms is developing a proton-impermeable cell membrane. Another important physiological adaptation is reduction in the pore size of membrane channels. Acidophilic microbes possess a net positive potential charge inside the cell that counterbalances the high concentration of H^+ ions in the external environment (Coker 2019).

3. Adaptation to extreme salinity: For survival in the high-salinity environment, halophilic microbes have developed physiological mechanisms to strike the osmotic balance between the intracellular and the extracellular environment to minimize water loss. An important physiological mechanism is the intracellular synthesis of organic solute molecules (also known as compatible solutes/osmolytes), such as ectoine, hydroxyectoine, glycine-betaine, dimethylsulfoniopropionate, K-glutamate, trehalose, and proline (Zhu et al. 2010). For example, ectoine-mediated osomoadaptation provides resistance to survive in the hypersaline environment (Zhu et al. 2010). Halophilic microbes might attain the osmotic equilibrium by another physiological strategy via accumulation of K^+ in the cytoplasm as a "salt-in-cytoplasm" strategy (Wu et al. 2024). Increased intracellular KCl level might contribute to osmotic homeostasis in an external hypertonic environment (Zhu et al. 2010).
4. Adaptation to extreme pressure: To cope up with the high-pressure environment, bacterial piezophiles synthesize polyunsaturated, monounsaturated fatty acids, phosphatidylglycerol, and phosphatidylcholine in place of phosphatidylethanolamine. Other mechanisms of piezophilic adaptation are not yet known. However, a few investigations suggest that impact on the viability is not due to a specific group of genes/enzymes but rather is an outcome of a global change in metabolism (Bartlett et al. 2007).

   Adaptations to (5) extreme radiation and (6) desiccation require a very robust DNA repair orchestra. Having multiple copies of the genome has also been proposed to be an evolution to circumvent radiation–/desiccation-induced DNA damage (Coker 2019; Merino et al. 2019).

The specialists need to have a specific stress adaptation repertoire to counteract the specific level of environmental extremity. However, for the polyextremotolerant/polyextremophilic/generalist mode of sustenance, a very adaptive membrane framework and a robust metabolic architecture are required. Also, organic solute molecules might be a good strategy to rely on as they give protection from multiple stress conditions like temperature extremity, pressure extremity, and osmotic stress. Stress-induced metabolites (e.g., compatible solutes, exopolysaccharides, etc.) and enzymes are of prime industrial interest. For utilization of the enormous bioresource potentials, extremophilic and extremotolerant microbes have been isolated from different extreme habitats.

2.4 Distribution of Different Extreme Habitats on Earth

There are myriads of environments on Earth that involve different environmental conditions (Fig. 2). Based on the specific environmental extremity involved and the potential to harbor specialist microbes, different habitats on Earth are categorized into a few classes.

1. High-temperature habitats: Natural high-temperature habitats where thermophiles and hyperthermophiles are present include geothermal waters (Fig. 2), hot springs, mud pots, continental subsurface, volcanoes, fumaroles, geysers, deep-sea hydrothermal vents, and also anthropogenic environments, such as compost facilities and anaerobic reactors.
2. Low-temperature habitats: Natural low-temperature ecosystems that harbor psychrophiles/psychrotolerant microbes are the Southern Ice Field, Southern extreme zones, such as the Patagonia, the high mountains of Los Andes, vast ice regions, such as the Northern and the Antarctic continent, etc.
3. High pH/alkaline habitats: Natural alkaline ecosystems include highly alkaline hyper-saline lakes (e.g., Lake Natron, Tanzania) and alkaline soda lakes (e.g., Lake Mono, CA, United States). Engineered alkaline habitats that harbor alkaliphiles include highly alkaline enrichments generated by industrial activities (e.g., indigo dye plants) or soils with high alkalinity (e.g., estuaries with long

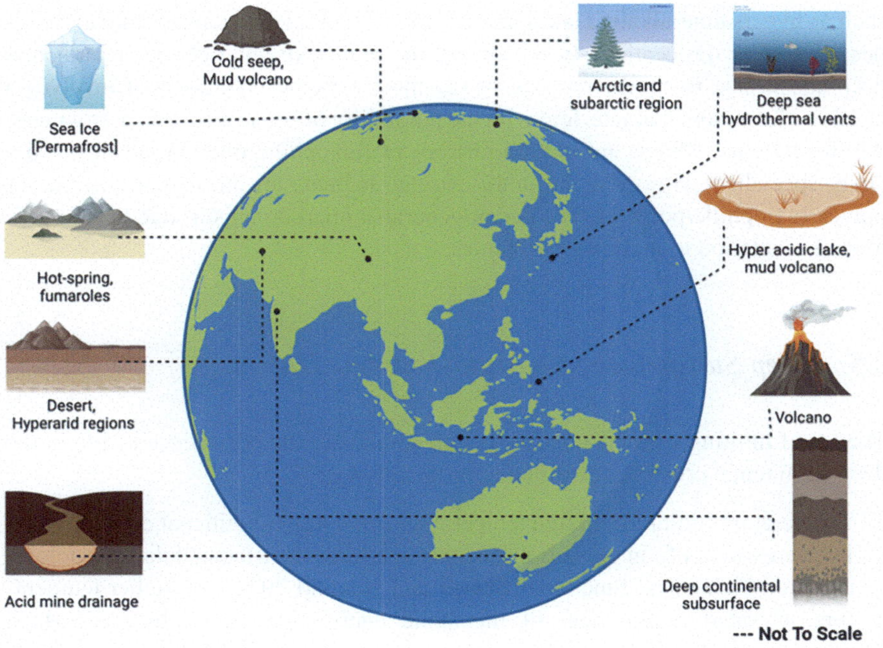

Fig. 2 Distribution of different extreme environments on Earth so far studied for the inhabiting life

periods of evaporation, clay particles with highly abundant alkaline crevices) (Sorokin et al. 2014).

4. Low pH/acidic habitats: Low pH natural sites where acidophilic microorganisms are present include acidic rivers (e.g., Rio Tinto), solfataric fields, and geothermal sulfur-rich sites. Acid mine drainage is an anthropogenic low-pH habitat.
5. High salinity environments: High saline environments include salars, saline lakes, oceans, polar ice, and coastal areas. Some noteworthy saline lakes are located in the Central region (Cáhuil Lagoon), the Atacama Salar, and the Southern Patagonia.
6. High pressure environments: Natural high-pressure habitats include deep lakes such as Baikal (1.6 km), Tanganyika (1.5 km), the ocean, and continental subsurface (e.g., Deccan subsurface). It is crucial to mention that hydrostatic pressure keeps increasing at a rate of 10.5 kPa per meter of depth, and for lithostatic (overburden) pressure, the rate of increase is 22.6 kPa per meter. This implies that microbes that grow at the bottom of the Mariana Trench (10.9 km below the ocean surface) realize 114.4 kPa of pressure, but microbes in the South African gold mines (3.5 km below Earth's surface) encounter 79.1 kPa (Bartlett et al. 2007; Coker 2019).

   Also, environments of exoplanets like the Martian surface and the surface of Titan involve exposure to (7) high radiation. Continental subsurface and desert environments like Kalahari and Atacama represent naturally occurring (8) desiccated environments.

However, there are habitats that involve more than one environmental extremity, such as hypersaline-alkaline lakes and desert soil that involve temperature extremities as well as desiccation stress. Among the multi-extreme habitats, continental deep subsurface is considered one of the most extreme habitats as it involves a number of stressors that interfere with the possibility of life in such an environment. All the extreme habitats are likely to harbor extremophilic microorganisms, especially specialists. However, the multi-extreme habitats are most likely to harbor polyextremophilic/polyextremotolerant/generalist microorganisms due to their prolonged adaptation to multiple extremities.

2.5 Deep Subsurface as a Multi-Extreme Habitat

The set of limiting extreme conditions that administer the occurrence of life in the deep continental crust are identified to be as follows:

1. Temperature limitations: High temperature is the most significant extremity that interferes with life in the subsurface through impacting the cellular physiological and biochemical functions (Beaver and Neufeld 2024). At higher temperature, elevated requirement of metabolic energy has been observed which compensates for degraded biomolecules. Therefore, high temperature stress, compounded with oligotrophy in the subsurface, makes it uninhabitable (Dutta

et al. 2018). In subsurface oceanic crust and continental crust, the temperature increases following a gradient of 15 and 25 °C/km subsequently with depth. Recent investigations on the deep continental subsurface of the Deccan traps indicate a temperature gradient of 15 °C/km in the granitic crust and 25 °C/km in the basaltic crust (Dutta et al. 2018). The putative temperature limit to life is 113 °C.

2. Energy limitations: Markedly low energy flux (i.e., around 103-fold lower than energy flux in surface environment) is another extremity that interferes with life within the deep, aphotic, and oligotrophic biosphere. Photosynthesis is absent from aphotic deep biosphere. Subsurface microbes completely depend on the buried organic matter and reduced complexes (Mn (II), Fe (II), ammonia, sulfur/sulfide, etc.), which, upon oxidation, release energy. It leads to a very slow catabolism rate among the subsurface organisms (thousand to million-fold lower than that of a model microbe that grows on a nutrient-rich medium). Sometimes the biomass turn-over in sub-seafloor deep biosphere is a few hundred to a few thousand years, 100-1000 times slower than in surface sediment (Hoehler and Jørgensen 2013). On the contrary, that of the model microorganisms remains in the order of hours or days.
3. Void Space: Void space/porosity is a significant physico-chemical parameter that delimits life in the subsurface biosphere. While considering a specific rock/sediment type that has an identical geologic history, void space decreases with depth exponentially (Parnell and McMahon 2016). In a deeper horizon, small particle size and high pressure delimit void space and fluid movement. This phenomenon exerts control over subsurface cell density.
4. Oxygen limitation: The subsurface environment is described as strictly anaerobic, eliminating the possibilities of radiolysis or surface recharge of water. Even following radiolysis or surface recharge, the deeper flow becomes anoxic within a short retention distance due to the presence of reduced organic matter or aerobic metabolism (Govil et al. 2018).
5. Pressure limitations: Hydrostatic and lithostatic pressure are another set of physico-chemical parameters that delimit life and shape the subsurface microbial community composition (Lauro and Bartlett 2008; Dutta et al. 2019). Biological processes, especially the formation of macromolecules, crucial to viability, are restricted in high pressure. In the deep continental crust, the pressure increases at a gradient of 15–28 MPa/km. A recent geomicrobiological investigation has shown a lithostatic pressure gradient of 26.7 MPa/km in Deccan traps (Dutta et al. 2018).
6. Ionic limitations: One of the understudied but crucial environmental parameters is the effect of dissolved ions. Fluctuations in dissolved ion concentration might interfere with microbial replication, which results in restriction of microbial growth. Dissolved ions might delimit life in the continental subsurface through impacting water activity, ionic strength, and chaotropicity (Payler et al. 2019).

The deep biosphere is further classified into several provinces (i.e., hydrothermal vents, continental subsurface, deep mines, etc.) that involve multiple environmental limitations to sustenance of life. All the deep biosphere extreme provinces involve multiple environmental limitations, including temperate, pH, pressure, oligotrophy, and so on. The organisms inhabiting the diverse subterrestrial deep biosphere provinces belong to Proteobacteria, Firmicutes, Actinobacteria, Archaea, etc. Isolation of extremophilic bacteria has been done from different deep biosphere habitats (Govil et al. 2018).

2.6 Deep Subsurface Habitats as the Hotspot for Extremophilic Bacteria

Major extremophilic bacteria that have been isolated from several deep subsurface environments are summarized in Table 1. The quest of exploring the extremophiles of deep biosphere started when Garland described the limit of prokaryotic life as up to 4 km beneath Earth's surface in 1971. Followed by various investigations to isolate and identify thermophilic bacteria from deep subsurface. Works on deep subsurface piezophilic communities have been carried out by Fang et al. (2017) and Dutta et al. (2019). Halophilic microbes have also been isolated by Finster et al. (2009), Gales et al. (2011), and Dong et al. (2014). For subsurface saline environments, halophiles have been isolated. However, Chang et al. (2007) and Zhang et al. (2014) observed a degree of halotolerance in subsurface isolates that do not belong to a saline subsurface site. Cason et al. (2012) have assessed radionuclide tolerance abilities of a subsurface isolate. All of these extremotolerant properties of the subsurface isolates allude to a generalist/polyextremotolerant mode of sustenance in the subsurface microbes. However, all the past works on extreme-adapted microorganisms of the deep subsurface have primarily concentrated on the isolation of thermophilic bacteria, piezophilic bacteria, and also the characterization of piezophilic community dynamics in deep subsurface. There has been a persistent knowledge gap regarding the nature and diverse environmental limits of extremophiles of deep subsurface due to technological hindrances and lack of access to subsurface. Also, there is a dearth of investigations on the characterization of the stress responses of the subsurface extremophilic bacteria under different extreme conditions. The study of deep subsurface extremophiles provides critical insights for astrobiology.

These microorganisms, which survive in environments characterized by extreme temperature, pressure, and energy limitations, offer valuable models for understanding the potential for life in extraterrestrial environments. Research into the metabolic strategies and survival mechanisms of these organisms under extreme conditions can inform predictions about the types of microbial life that might exist on other planets. In this regard, deep biosphere research plays a key role in advancing our understanding of microbial life processes that could potentially occur in similar extreme environments beyond Earth.

Table 1 Major extremophilic microbes, isolated from different deep subsurface habitats, and their various stress tolerant properties

Sl. No.	Name of the isolate	Subsurface site of isolation	Depth (m)	Brief description	References
1	*Paenibacillus* sp. and *Brevibacillus* sp.	Deep subsurface, Homestake gold mine	1500	Moderately thermophilic bacteria with ability to degrade cellulose and derivatives	Rastogi et al. (2009)
2	*Thermoanaerobacter ethanolicus*	Deep subsurface, Piceance Basin	Multiple depths	Thermophilic anaerobic bacterium with abilities to reduce various metals such as Fe (III), cobalt (III), chromium (VI), manganese (IV), and uranium (VI)	Roh et al. (2002)
3	*Desulfuromonas soudanensis* WTL	Soudan Underground Iron Mine	713	Electrochemically enriched novel halophilic iron-reducing bacterium	Badalamenti et al. (2016)
4	*Desulfotomaculum thermosubterraneum*	Geothermally active underground mine	250	Gram-positive, sulfate-reducing bacterium, thermophilic in nature	Kaksonen et al. (2006)
5	*Bacillus subterraneus*	Australian thermal deep aquifer	–	Facultative anaerobic bacterium with salinity tolerance 0–9%	Kanso et al. (2002)
6	*Geosporobacter subterraneus*	Subsurface deep aquifer, Paris basin, France	800 m	Strictly anaerobic, gram-positive bacterium, moderately thermophilic and slightly halotolerant	Klouche et al. (2007)
7	*Desulfovibrio bastinii* sp. nov. and *Desulfovibrio gracilis* sp. nov.	Emeraude oilfield, deep subsurface	–	Moderately halophilic, mesophilic, moderately acidophilic	Magot et al. (2004)
8	*Halobacteroides acetoethylicus* sp. nov.	Gas-bearing sandstone and hypersaline deep subsurface	–	Gram-negative, anaerobic rod, with mesophilic, halophilic nature (5–22% NaCl concentration)	Rengpipat et al. (1988)

(continued)

Table 1 (continued)

Sl. No.	Name of the isolate	Subsurface site of isolation	Depth (m)	Brief description	References
9	*Desulfovibrio putealis* sp. nov.	Paris basin, France	430	Sulfate-reducing bacterium with utilization ability of lactate, pyruvate, malate, fumarate, ethanol, butanol, acetate/H_2	Basso et al. (2005)
10	*Desulfocurvus vexinensis* gen. nov.	Paris Basin, France	830	Anaerobic, chemo-organotrophic bacterium with prominent halotolerant nature (till 20% NaCl concentration)	Klouche et al. (2007)
11	*Hydrogenobacter subterraneus*	Deep subsurface geothermal water	1500 m	A novel thermophilic (60–85 °C) isolate with strictly aerobic nature and heterotrophic life style	Takai et al. (2001a)
12	*Arthrobacter subterraneus* sp. nov.	Deep subsurface water, Korea	–	Mesophilic, moderately alkaliphilic. Halotolerant bacterium (till 13% NaCl concentration)	Chang et al. (2007)
13	*Microvirga subterranea* gen. nov., sp. nov.	Deep subsurface thermal aquifer	–	Aerobic, non-sporulating, gram-negative, rod-shaped bacterium; moderate thermophilic	Kanso and Patel (2003)
14	*Petrothermobacter organivorans* gen. nov., sp. nov.	Yabasi oil field, subsurface, Japan	–	A novel thermophilic, anaerobic, chemoheterotrophic, acetate-oxidizing and Fe (III)-, Mn (IV)-, nitrate-, and sulfate-reducing bacterium	Tamazawa et al. (2017)
15	*Thermoanaerobacter brockii*	Deep subsurface French oil well	2100	A strictly anaerobic, thermophilic, gram-positive, spore-forming eubacterium	Cayol et al. (1995)
16	*Bacillus flexus*	Deep subsurface oil reservoir	–	Mesophilic, multidrug-resistant halotolerant bacterium (up to 10% NaCl concentration)	Zhang et al. (2014)

(continued)

Table 1 (continued)

Sl. No.	Name of the isolate	Subsurface site of isolation	Depth (m)	Brief description	References
17	*Koleobacter methoxysyntrophicus gen.* nov., sp. nov.	Deep subsurface oil field formation water	1100	An anaerobic thermophilic, moderately alkaliphilic, rod-shaped bacterium	Sakamoto et al. (2021)
18	*Geobacillus thermodenitrificans* NG80-2	Deep subsurface oil reservoir, Northern China	–	Gram-positive, denitrifying, long-chain alkane-degrading facultative soil bacteria	Feng et al. (2007)
19	*Tessaracoccus profundi* sp. nov.	Deep subsurface cheapsake drill core	940	Aerobic, rod-shaped halophilic bacterium (20% NaCl concentration)	Finster et al. (2009)
20	*Halanaerocella petrolearia* gen. nov., sp. nov.	Deep subsurface hypersaline oil reservoir	1153	Anaerobic, halophilic (6–26% NaCl concentration) fermentative bacterium	Gales et al. (2011)
21	*Orenia metallireducens* sp. nov.	Deep subsurface saline environment, Illinois	2020	A novel moderately thermophilic, halophilic (up to 21% NaCl concentration) and metal-reducing bacterium	Dong et al. (2014)
22	*Thermus scotoductus* SA-01	Subsurface groundwater, Witeatersrand Basin	3200	Uranium-reducing thermophilic subsurface bacterium	Cason et al. (2012)
23	*Bacillus cereus* strain Q1	Daqing oil field, China	–	Biosurfactant-producing bacterium, thermophilic	Xiong et al. (2009)
24	*Alkaliphilus transvaalensis* gen. nov., sp. nov.	South African Gold mine	3200	An extremely alkaliphilic bacterium (pH10), with temperature range 20–50 °C	Takai et al. (2001b)
25	*Thermincola ferriacetica* sp. nov.	Subsurface hydrothermal spring	–	A new anaerobic, thermophilic, facultatively chemolithoautotrophic bacterium capable of dissimilatory Fe(III) reduction	Zavarzina et al. (2007)
26	*Tepidibacillus fermentans* gen. nov., sp. nov.	Subsurface gas storage	–	A moderately thermophilic anaerobic and microaerophilic bacterium	Slobodkina et al. (2013)

(continued)

Table 1 (continued)

Sl. No.	Name of the isolate	Subsurface site of isolation	Depth (m)	Brief description	References
27	*Thermodesulfovibrio* sp.	Subsurface aquifer, Western Siberia	2000	Facultatively alkaliphilic (pH8.5), moderately thermophilic (65 °C) bacterium	Frank et al. (2016)
28	*Anaerotalea alkaliphila* gen. nov., sp. nov.	Subsurface aquifer, Western Siberia	2287	An alkaliphilic, anaerobic, moderately thermophilic, fermentative bacterium	Frolova et al. (2021)
29	*Thermanaeromonas toyohensis* gen. nov., sp. nov.,	Subsurface geothermal aquifer, Japan	550	A novel thermophilic, strictly anaerobic, thiosulfate-reducing bacterium	Mori et al. (2002)
30	*Tepidibacillus infernus* sp. nov.,	Subsurface, TauTona Gold Mine	–	A moderately thermophilic, selenate- and arsenate-respiring hydrolytic bacterium	Podoskorskaya et al. (2016)
31	*Metallumcola ferriviriculae* strain MK1	Soudan Underground Mine State Park	713	Cytochrome-rich, mesophilic, Gram-positive, spore-forming Fe (III)-reducing bacterium	Hsu et al. (2024)
32	*Atrimonas thermophila* gen. nov., sp. nov.	Nakajo gas field in Niigata Prefecture	>1000	A novel anaerobic thermophilic bacterium of the phylum Atribacterota isolated	Kawamoto et al. (2024)
33	*Cytobacillus pseudoceanisediminis* sp. nov.	Deep subsurface saline spring at the Baksan Neutrino Observatory INR RAS (Kabardino-Balkaria, Russia)	–	A novel facultative, methylotrophic bacterium with high heavy metal resistance	Tarasov et al. (2023)
34	*Methylomonas aurea* sp. nov. and *Methylomonas subterranea* sp. nov.	Sanford Underground Research Facility (SURF)	–	Type 1 methane-oxidizing bacteria	Abin et al. (2024)

(continued)

Table 1 (continued)

Sl. No.	Name of the isolate	Subsurface site of isolation	Depth (m)	Brief description	References
35	*Candidatus Desulforudis audaxviator* dominates	Ground-water bearing fracture, the Fennoscandian Shield, Sweden	975	Primary producers in the oligotrophic continental subsurface	Westmeijer et al. (2024)
36	*Geochorda subterranea* gen. nov. sp. nov. and *Carboxydochorda subterranea* gen. nov. sp. nov.	Subsurface borehole 5P, Western Siberia	–	Polymer-degrading, facultative bacteria also capable of fermentative growth	Karnachuk et al. (2024)
37	*Anaerosoma tenue* gen. nov., sp. nov., and *Parvivirga hydrogeniphila* gen. nov., sp. nov.	Subsurface mud volcano Karabetova Gora	–	First cultured representative of uncultured clade OPB41	Khomyakova et al. (2022)
38	*Rhodothermus bifroesti* sp. nov.	Subsurface of the volcanic island Surtsey off the south coast of Iceland	70	Thermophilic bacterium	Bergsten et al. (2022)

3 Astrobiological Implications of the Extremophiles in General and Deep Life

3.1 Deep Biosphere Research and Astrobiology

The deep biosphere is an ecosystem of organisms and their living space that extends from a few meters to several kilometers deep beneath Earth's surface (Purkamo 2023). To better understand life under extreme energy-limiting conditions, harsh radiation, temperature and pressure fluctuation, deep sea, and subsurface would be the primary targets, as these low-energy settings offer conditions closely resembling those found on extraterrestrial planets, making them prime analogues for studying potential life beyond Earth (Takai 2011; Jones et al. 2018). Chemolithotrophic bacteria are primary colonizers of these habitats that are too hostile for other microorganisms, living at an interface where active biology must wrestle with the thermodynamic constraints on cellular biochemistry and structures. Extremophilic chemolithotrophic bacteria are crucial for astrobiological exploration within our solar systems. Microbial activities are a major player for stabilization (both thermal and chemical) of these extreme environments, making the conditions favorable for habitability. Sampling of the cryptic subsurface microbial communities from the basement of the Río Tinto headwaters indicates that the acidic environment results from the microbial oxidation of pyritic ores, a process that is consistently exothermic under both aerobic and anaerobic conditions, thereby releasing heat into the environment. Subsurface temperatures can be controlled by microbes through heat storage until they surpass the maximum growth temperature (Fernández-Remolar et al. 2008).

3.2 Extraterrestrial Astrobiological Targets with Subsurface Analogues on Earth

Understanding of life is derived from the observable and quantifiable occurrences that take place on Earth, and as a result, it is confined to this particular manifestation of life. Nevertheless, the laws of chemistry and physics encompass fundamental principles that enable us to extrapolate the conditions under which life could survive elsewhere. These principles indicate that life requires a liquid solvent, energy source, and building blocks to exist (Schwieterman et al. 2018). The key factor controlling the dimension of biospace for life on Earth is liquid water due to its cosmic abundance and physicochemical properties (Merino et al. 2019). A significant study on astrobiology focuses on analyzing the environment on Earth analogues to terrestrial planets (Martian & Venus environment), Jovian moons (Europa), and Saturn moons (Titan and Enceladus) (Preston and Dartnell 2014). These "analogue sites" inform researchers not only about the limits of life (habitability) but also long-term preservation of signs of past life and selection of most potential biosignatures to target with appropriate analytical methods. Field-testing or in situ validation of instruments responsible for biosignature detection, including maneuverability and traction of rover designs, is another function offered by terrestrial analogues. Selected terrestrial analogues for study should mimic the environment of those solar system locations, like Martian or Venus environment. A single analogue site on Earth cannot imitate all characteristics of the extraterrestrial environment; some sites are compositional (mineralogy, water, and organic contents) or electrochemical (redox potential, pH, water activity) analogues, while others are environmental (temperature, wind, and radiation) or physical analogues (thermo-physical, porosity) to the targeted environment (Marlow et al. 2008, 2011; Martins et al. 2017). In the context of planetary analogue studies, the environment on Earth is classified into different tiers based on the degree to which they replicate the conditions found on other planets, either through direct or indirect evidence. In case of Mars, robust scientific data such as images, spectral analysis, and other observational methods are there to certify the presence of volcanoes, ice caps, and impact craters. Therefore, basaltic and ice-covered environments on Earth are first-order Martian analogues. On the other hand, analogues of second order are based on indirect or speculative evidence that requires further validation. Lake Vostok serves as a second-order analogue of the brine ocean of Europa as direct evidence of Europa's Ocean remains elusive, abundant indirect indicators, such as the detection of surface features consistent with ice-tectonic activity and variations in its magnetic field, suggesting the presence of a conductive liquid beneath its icy shell, strongly support its existence (Preston and Dartnell 2014).

3.3 Terrestrial Planets

3.3.1 Venus

The surface of Venus experiences extreme temperature, averaging around 460 °C, which precludes the existence of liquid water and the stability of organic molecules essential for life as we understand it (Schulze-Makuch and Irwin 2002). However, in the upper atmosphere of Venus, a potential habitable zone has been identified within the lower and middle cloud layers where temperatures range from freezing to about 120 °C, and conditions may allow for the presence of liquid water droplets within the clouds, although droplets are highly acidic with dissolved H_2SO_4 (Cockell 1999). Analogue sites on Earth completely replicating conditions of Venus' atmosphere are limited, as presence of highly acidic conditions with long-lived aerosols are absent on earth. However, the presence of life in Earth's high-altitude atmosphere is well documented, offering valuable insights into potential Venusian atmospheric habitability. Unlike surface-based analogues, potential Venusian atmospheric analogues are transient and primarily associated with high-altitude atmospheric layers. For instance, Sattler et al. (2001) demonstrated bacterial growth in supercooled cloud droplets sampled from a meteorological station on an Alpine Mountain. Viable bacteria have also been detected at much higher altitudes, beyond the tropopause in the stratosphere, through balloon flights and high-altitude research aircraft (Smith et al. 2010; Wainwright 2015). It is crucial to underscore that the sulfuric acid clouds of Venus are far more acidic than even the most extreme acidic environments on Earth, such as the Dallol Geothermal Area in the Danakil Depression, Northern Afar (Ethiopia) (Cavalazzi et al. 2019). The Dallol pools are hypersaline, high-temperature (108 °C), anoxic hydrothermal systems with NaCl supersaturation and iron concentrations reaching 150 g/L, characterized by the lowest recorded environmental pH, ranging between −1.7 and 0 (Kotopoulou et al. 2019). Preliminary investigations indicate that life may not only survive but potentially thrive in these polyextreme hydrothermal conditions. There is evidence suggesting the presence of microorganisms, such as the diminutive (*0.2 µm) *Nanohaloarchaea* and other archaea phyla, within the Dallol acidic pools (Gómez et al. 2019). Another recently launched mission by NASA is AVENGERS (THE ANALOGS FOR VENUS' GEOLOGICALLY RECENT SURFACES (AVENGERS) INITIATIVE, which will monitor the recent volcanic activity of Venus through SAR image analysis, age estimation of volcanic activity, geological mapping, radiometric and interferometry analysis of active volcanoes, and thickness comparison of volcanic structures of Venus and terrestrial analogues (D'Incecco et al. 2024). Currently selected terrestrial analogues based on volcanic and tectonic activity after Mount Etna through AVENGERS initiative are as follows:

1. Kīlauea, an active shield volcano in Hawaii, can used as a terrestrial model hotspot mimicking the volcanic activities on Venus.
2. The East African Rift System (EARS) (10°N, 40°E), an active rift zone in East Africa comprising of many volcanoes, can be served as another terrestrial analogues showing similarity with formation's most active volcanic rift system of Venus, such as the Olapa Chasma-Idunn Mons system (45°S, 215°E).
3. Mount Merapi (7°S; 110°E) in Indonesia represents another terrestrial analogue site that requires investigation. Laboratory analysis of its volcanic products can give valuable insights for comparison of future spectral data obtained from Venus volcanism.
4. Bolivian volcanoes are selected as terrestrial analogues of Venus for their explosive characteristics that resemble some volcanic events that occurred at Venus.

As research advances in understanding the limits of life and the conditions that may exist in Venus' cloud layers, it becomes increasingly clear that more focused terrestrial analogue studies are necessary. These should target environments that simulate the key characteristics of the Venusian atmosphere, such as acidic, nutrient-poor conditions and the potential for life to persist within aerosolized droplets. Such studies will be critical in informing our understanding of the habitability potential in Venus' atmosphere, where extremophilic life forms could survive.

3.3.2 Mars

Mars is the most Earth-like planet in the solar system, believed to have offered a considerably warmer and wetter environment in the past (Preston and Dartnell 2014). Mars, however, witnessed substantial global environmental transformations in its history. Consequently, terrestrial analogue sites are employed not just to replicate circumstances characteristic of particular Martian locations but also to model various phases of its evolutionary history, specifically similar places emulating the past, present, and future environments of Mars. Terrestrial analogues of Mars can be categorized based on their resemblance to early, middle, and present Martian evolutionary conditions, providing insights into ancient aqueous environments, transitional phases of volcanic and climatic activity, and the current cold, arid, and oxidizing surface (Colehin and Delgado-Baquerizo. 2022). The recent exploration of Mars-analogous environments within the Indian subcontinent has yielded significant advancements, securing India as a key player in the global astrobiology landscape. Indian astrobiological sites hold significant importance in the field of planetary science and astrobiology, despite the presence of numerous well-known sites globally. The unique geological, environmental, and microbial diversity of Indian sites, such as the Lonar Crater, volcanic field of Dhinodhar hill, Kachchh, valleys developed on the basaltic rocks of Deccan trap in the upper reaches of the Tapi river (Alaldari drainage), the high-altitude Himalayan regions (Nubra-Shyok Valley, Satluj river basin), and the alkaline Sambhar Lake offer distinct analogues for extra-terrestrial environments (Pandey et al. 2020; Chavan et al. 2022). These sites provide valuable opportunities to study extremophiles and geochemical processes under conditions such as high radiation, salinity, and extreme temperatures—factors that closely mirror those on planets like Mars. Moreover, the distinct evolutionary history of these sites adds a different dimension to our understanding of biosignatures, planetary habitability, and the resilience of life in extreme environments. Indian astrobiological sites thus complement existing global sites by offering unique ecosystems that enrich the comparative study of potential extraterrestrial life. A key factor in evaluating the planet's habitability is the presence of evidence for water, temperature, pressure, and radiation. The current Martian surface is highly inhospitable for Earth-like life. In the absence of a protective magnetosphere, Mars is subjected to continuous exposure to DNA-damaging radiation, along with a lack of liquid water, extremely low average temperatures, low atmospheric pressure, and the presence of surface perchlorates—which pose further challenges (Fornaro et al. 2018; Coleine and Delgado-Baquerizo 2022). Perchlorates, in particular, can induce oxidative stress in potential microorganisms. However, certain regions on Earth are inhabited by polyextremophiles (e.g., xerophiles, psychrophiles, halophiles, and radioresistant organisms), which are remarkably adapted to conditions resembling those of modern Mars (Schultz et al. 2023). Fungi, yeasts, and many bacterial species belonging to *Debaryomyces hansenii, Vibrio, Bacillus, Salinovibrio, Staphylococcus,* and *Nesiotobacter* isolated from extreme environment of the earth showed remarkable tolerance up to 2.4 M perchlorate concentration and reduction ability (Acevedo-Barrios et al. 2019; Heinz et al. 2019). Polyextremophilic bacteria, yeasts, and fungi can be used as model organisms to offer valuable insights into the potential survival strategies of life in extraterrestrial environments, particularly on planets like Mars and Icy moons such as Europa. Perchlorate-tolerant bacteria isolated from Mars-analogue Qaidam Basin soils, which are exposed for 4 h in a near-space environment, not only show tolerance to Mars-like harsh environmental conditions but also have a high tolerance to perchlorate solutions. These polyextremophilic bacterial strains are promising candidates for future astrobiological research, such as exposure experiments on space stations or in laboratory-based planetary environment simulation chambers.

3.3.3 Icy Moons

Some of the most significant astrobiological targets within the solar system are located on planetary moons. Among these, the icy satellites of Jupiter and Saturn—Europa, Enceladus, and Titan—have emerged as highly promising candidates for hosting potentially habitable environments, as evidenced by recent space missions. These moons have been revealed to be geologically active, with subsurface oceans and dynamic processes that could support life. However, the challenge for astrobiological validation lies in the extreme nature of these environments, which far surpasses those found on Earth, making terrestrial analogues scarce. Moreover, our understanding of the actual conditions on these Icy moons are largely inferential, as it is primarily derived from indirect observations rather than direct empirical data.

The environment of Europa is constantly bombarded with ionizing radiation and temperature ranges from −187 to −147 °C (Dougherty et al. 2019), surpassing the lowest limit of temperature for microbial growth. However, the thick icy layer could provide protection to life from harsh radiation of the environment. Indirect evidence indicated the presence of a brine ocean beneath the ice layer of Europa (Chivers et al. 2023). Though the hydrating mineral composition is unknown till date, this potential habitable zone is gaining much attention for its astrobiological implications.

Titan possesses a nitrogen-rich atmosphere with a fluvial system, as well as lakes and seas comprised of a blend of heavier hydrocarbons, methane and ethane, dissolved nitriles, and/or atmospheric gases (Mitri et al. 2007). Methane-ethane mixtures are appropriate as a biosolvent; for this reason, the surface environment of Titan may be deemed habitable, albeit for unconventional life forms with biochemistry distinct from our own. Enceladus, one of Saturn's most intriguing moons, presents an environment that appears remarkably suitable for life. The discovery of plumes of water vapor and organic compounds erupting from fissures in its south polar region, known as the "tiger stripes," further suggests a dynamic and active environment conducive to life (Berne et al. 2024). The presence of these organic materials, along with evidence of high salinity and the potential for a stable energy source from hydrothermal vents, indicates that Enceladus possesses many of the fundamental conditions necessary for sustaining microbial ecosystems (Neveu et al. 2020). Moreover, the moon's relatively young surface suggests ongoing geological processes that may continually refresh its habitability. Together, these factors position Enceladus as a compelling candidate in the search for extraterrestrial life within our solar system. A map of representative locations studied as terrestrial analogues of Venus, Mars, and Icy moons is presented in Fig. 3, modified after Preston and Dartnell (2014), Coleine and Delgado-Baquerizo (2022).

3.4 Role of Bioenergetics Model for Predicting Habitability in Astrobiology and Extreme Habitats

For progression and maintenance of life, supply of energy and nutrients is required. Major challenges of astrobiology are predicting habitability of the extraterrestrial environment by studying poorly characterized extreme environments on Earth. Bioenergetic models play an important role in understanding the energetics and nutrient availability of these poorly characterized extreme environments, thereby predicting potential biological productivity. Major inputs required for the predictability of how much biomass an environment could host are the following: NutMEG (nutrition, maintenance, energy, and growth) and some knowledge on the local environment. Higgins and Cockell (2020) developed a bioenergetic model for dynamic monitoring of the organism and the environment it interacts with to predict growth rate, biomass production, biosignature, and other parameters. Finally, they have investigated, by matching model prediction along with laboratory observation, how

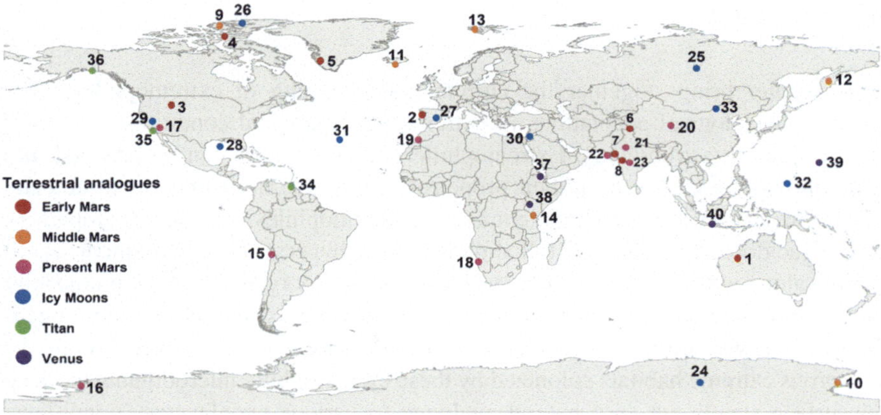

Fig. 3 A map of representative locations studied as terrestrial analogues of Venus, Mars, and icy moons. (Map created using the Free and Open Source QGIS 2023) [Early Mars: (1) The Pilbara region, West Australia, (2) Rio Tinto, SW Spain, (3) Yellowstone National Park, USA, (4) Haughton Impact Structure, Devon Island, (5) Dongwanzi Ophiolite Complex, China, (6) Nubra-Shyok Valley, India, (7) Dinodhar hills, India, (8) Alaldari drainage system, India. Middle Mars: (9) Axel Heiberg Island, Canadian High Arctic, (10) Beacon Valley, Antarctica; (11) Sub-glacial Volcanism, Iceland, (12) Kamchatka, Russian Federation, (13) Bockfjord Volcanic Complex, Svalbard, (14) Kilimanjaro, Tanzania, Africa. Present Mars: (15) The Atacama Desert, South Africa, (16) The Antarctic Dry Valleys, Antarctica, (17) The Mojave Desert, USA, (18) The Namib desert, Namibia, Africa, (19) Ibn Battuta Centre Sites, Morocco, (20) Qaidam Basin, Tibetan Plateau, (21) Sambhar Lake, India, (22) Kachch, India, (23) Lonar Crater, India. Icy Moons: (24) Lake Vostok, Antarctica, (25) Permafrost, Multiple sites, (26) The Borup Fiord Pass, Ellesmere Island, (27) Lake Tirez, Spain (28) Orca Basin, Gulf of Mexico, (29) Mono Lake, California, USA, (30) The Dead Sea, Israel, (31) The Lost City, Mid-Atlantic Ridge, (32) The Mariana Trench, USA, (33) Lidy Hot Springs, USA. Titan: (34) Pitch Lake, Trinidad and Tobago, (35) Rancho La Brea Tar Pits, California, USA, (36) Alaskan Oil Field, USA. Venus Environment: (37) Dallol Geothermal area, Ethiopia, 38. East African Rift, Africa (39) Kilauea Mountain, Hawaii (40) Mount Merapi, Indonesia]

placing a theoretical methanogen under energy or nutrient limitation can affect the growth rate and biosignature production (gas), total biomass production in a closed-well mixed gas system. Existing studies on modeling of extraterrestrial environments infer possible energetic habitability from free energy alone, while the NutMEG model offers the addition of dynamic calculation of kinetics for different metabolisms, maintenance costs, and variations in free energy yield based on local composition change with the course of time. Model predictability can be improved by comparing its first-ordered reactions with empirical evidence obtained from the real extreme environment of terrestrial analogues on Earth. In summary, bioenergetic models are foundational to astrobiology because they provide a theoretical and practical approach to understanding life's potential in alien environments. By assessing how energy sources and metabolic pathways may operate in diverse planetary conditions, bioenergetic models inform the search for extraterrestrial life and help shape exploration strategies that maximize the chances of detecting signs of life.

4 Summary

This chapter brings together the current understanding of extremophiles, their unique physiological adaptations to extreme environmental conditions, and their implications in astrobiology. Extremophilic microorganisms can be grouped into different categories based on their propensity to survive in distinct environmental conditions like tolerance to high temperature (thermophiles), pressure (barophiles), acidic (acidophiles), alkaline (alkaliphiles), and hypersaline environment (halophiles) along with desiccation and radioresistant microorganisms. Macromolecular and cellular-level modification strategies employed by extremophilic microorganisms to cope with different stressors have been discussed in this chapter. Among the numerous extreme habitats colonized by these extremophilic microorganisms, deep subsurface stands out as a natural analogue to various exoplanetary conditions, making it an ideal model for astrobiological research. This chapter presents the knowledge on the extremophilic life processes and contributes toward betterment of our understanding of potential life on other planets. Advancing theoretical frameworks and bioenergetic models have equipped us to better predict the extent of life processes considering factors such as free energy, nutrient availability, and maintenance costs occurring in extraterrestrial environments. Integrating our knowledge of extremophilic survival with these predictive models provides valuable insights, identifying habitable exoplanetary environments. Furthermore, this enhances our understanding of the limits of life, informs the search for biosignatures, and helps shape future missions aimed at detecting life beyond Earth.

Acknowledgments The authors gratefully acknowledge the support provided by the Ministry of Earth Sciences, Govt. of India (MoES/P.O.(Seismo)/1(383)/2020) and Science and Engineering Research Board (CRG/2021/007102), Core Research Grant scheme, Govt. of India for allowing the exploration of the extreme world of deep terrestrial biosphere and functionalities of various microorganisms living there. SM and SS gratefully acknowledge the financial assistance provided by the Ministry of Earth Sciences, Govt. of India, and University Grants Commission, Govt. of India, respectively.

Competing Interest The authors declare no competing interest.

References

Abin CA, Garner CT, Sankaranarayanan K, Sindelar RA, Kotary KF, Garner RM et al (2024) Methylomonas rivi sp. nov., Methylomonas rosea sp. nov., Methylomonas aurea sp. nov., and Methylomonas subterranea sp. nov., type I methane-oxidizing bacteria isolated from a freshwater creek and the deep terrestrial subsurface. Int J Syst Evol Microbiol 74:006506

Acevedo-Barrios R, Bertel-Sevilla A, Alonso-Molina J, Olivero-Verbel J (2019) Perchlorate-reducing bacteria from hypersaline soils of the Colombian Caribbean. Int J Microbiol 2019:6981865. https://doi.org/10.1155/2019/6981865

Amangeldina A, Tan ZW, Berezovsky IN (2024) Living in trinity of extremes: genomic and proteomic signatures of halophilic, thermophilic, and pH adaptation. Curr Res Struct Biol 7:100129

Arias PM, Butler J, Randhawa GS, Soltysiak MP, Hill KA, Kari L (2023) Environment and taxonomy shape the genomic signature of prokaryotic extremophiles. Sci Rep 13(1):16105

Badalamenti JP, Summers ZM, Chan CH, Gralnick JA, Bond DR (2016) Isolation and genomic characterization of *Desulfuromonas soudanensis* T, a metal- and electroderespiring bacterium from anoxic deep subsurface brine. Front Microbiol 7:913

Bartlett DH (2002) Pressure effects on in vivo microbial processes. Biochim Biophys Acta 1595:367–381

Bartlett, D. H., Lauro, F. M., & Eloe, E. A. (2007). Microbial adaptation to high pressure. Physiology and biochemistry of extremophiles, 331–348

Basso O, Caumette P, Magot M (2005) *Desulfovibrio putealis* sp. nov., a novel sulfate-reducing bacterium isolated from a deep subsurface aquifer. Int J Syst Evol Microbiol 55(1):101–104

Beaver RC, Neufeld JD (2024) Microbial ecology of the deep terrestrial subsurface. ISME J 18(1):wrae091

Bergsten P, Vannier P, Mougeolle A, Rigaud L, Marteinsson VT (2022) Rhodothermus bifroesti sp. nov., a thermophilic bacterium isolated from the basaltic subsurface of the volcanic Island Surtsey. Int J Syst Evol Microbiol 72:005214

Berne A, Simons M, Keane JT, Leonard EJ, Park RS (2024) Jet activity on Enceladus linked to tidally driven strike-slip motion along tiger stripes. Nat Geosci 17(5):385–391. https://doi.org/10.1038/s41561-024-01418-0

Borsodi AK (2024) Taxonomic diversity of extremophilic prokaryotes adapted to special environmental parameters in Hungary: a review. Biol Futur 75:183–192

Caro-Astorga J, Meyerowitz JT, Stork DA, Nattermann U, Piszkiewicz S, Vimercati L, DeBenedictis E (2024) Polyextremophile engineering: a review of organisms that push the limits of life. Front Microbiol 15:1341701

Cason ED, Piater LA, van Heerden E (2012) Reduction of U(VI) by the deep subsurface bacterium, *Thermus scotoductus* SA-01, and the involvement of the ABC transporter protein. Chemosphere 86(6):572–577

Cavalazzi B, Barbieri R, Gómez F, Capaccioni B, Olsson-Francis K, Pondrelli M, Rossi AP, Hickman-Lewis K, Agangi A, Gasparotto G, Glamoclija M, Ori GG, Rodriguez N, Hagos M (2019) The Dallol geothermal area, Northern Afar (Ethiopia)—an exceptional planetary field analog on earth. Astrobiology 19(4):553–578. https://doi.org/10.1089/ast.2018.1926

Cayol JL, Ollivier B, Patel BKC, Ravot G, Magot M, Ageron E et al (1995) Description of *Thermoanaerobacter brockii* subsp. *lactiethylicus* subsp. nov., isolated from a deep subsurface French oil well, a proposal to reclassify *Thermoanaerobacter finnii* as *Thermoanaerobacter brockii* subsp. *finnii* comb. nov., and an emended description of *Thermoanaerobacter brockii*. Int J Syst Bacteriol 45(4):783–789

Chang HW, Bae JW, Nam YD, Kwon HY, Park JR, Shin KS et al (2007) *Arthrobacter subterraneus* sp. nov., isolated from deep subsurface water of the South Coast of Korea. J Microbiol Biotechnol 17(11):1875–1879

Chavan A, Bhore V, Bhandari S (2022) Terrestrial Martian analogues from the Indian subcontinent: implications for hydrological activity on Mars. Icarus 385:115118. https://doi.org/10.1016/j.icarus.2022.115118

Chivers CJ, Buffo JJ, Schmidt BE (2023) Stable brine layers beneath Europa's chaos. Planet Sci J 4(9):159. https://doi.org/10.3847/PSJ/acea75

Cockell CS (1999) Life on Venus. Planet Space Sci 47(12):1487–1501. https://doi.org/10.1016/S0032-0633(99)00036-7

Coker JA (2019) Recent advancement in understanding extremophiles. F1000 Faculty Rev 8:F1000

Coleine C, Delgado-Baquerizo M (2022) Unearthing terrestrial extreme microbiomes for searching terrestrial-like life in the solar system. Trends Microbiol 30(11):1101–1115. https://doi.org/10.1016/j.tim.2022.04.002

D'Incecco, P and Filiberto, J and Garvin, JB and Arney, GN and Getty, SA and Zelenyi, LM and Zasova, LV and Korablev, O and Ivanov, MA and Head, JW and others (2024). The Analogs for VENus' GEologically Recent Surfaces (AVENGERS) Initiative: Updates and New Target Areas. 55th Lunar and Planetary Science Conference (LPSC)

Dong Y, Kumar CG, Chia N, Kim PJ, Miller PA, Price ND et al (2014) *Halomonas sulfidaeris*-dominated microbial community inhabits a 1.8 km-deep subsurface Cambrian sandstone reservoir. Environ Microbiol 16(6):1695–1708

Dougherty MK, Esposito LE, Krimigis MS (2019) Staurn from Cassini-Huygens. Springer

Dutta A, Dutta Gupta S, Gupta A, Sarkar J, Roy S, Mukherjee A, Sar P (2018) Exploration of deep terrestrial subsurface microbiome in Late Cretaceous Deccan traps and underlying Archean basement, India. Sci Rep 8:17459

Dutta A, Peoples LM, Gupta A, Bartlett DH, Sar P (2019) Exploring the piezotolerant/piezophilic microbial community and genomic basis of piezotolerance within the deep subsurface Deccan traps. Extremophiles 23(3):421–433

Fang J, Kato C, Runko GM, Nogi Y, Hori T, Li J et al (2017) Predominance of viable spore-forming piezophilic bacteria in high-pressure enrichment cultures from ~1.5 to 2.4 km deep coal-bearing sediments below the ocean floor. Front Microbiol 8:137

Feng L, Wang W, Cheng J, Ren Y, Zhao G, Gao C et al (2007) Genome and proteome of long-chain alkane degrading *Geobacillus thermodenitrificans* NG80-2 isolated from a deep subsurface oil reservoir. Proc Natl Acad Sci U S A 104(14):5602–5607

Fernández-Remolar DC, Gómez F, Prieto-Ballesteros O, Schelble RT, Rodríguez N, Amils R (2008) Some ecological mechanisms to generate habitability in planetary subsurface areas by chemolithotrophic communities: the Ro Tinto subsurface ecosystem as a model system. Astrobiology 8(1):157–173. https://doi.org/10.1089/ast.2006.0022

Finster KW, Cockell CS, Voytek MA, Gronstal AL, Kjeldsen KU (2009) Description of *Tessaracoccus profundi* sp. nov., a deep-subsurface actinobacterium isolated from a Chesapeake impact crater drill core (940 m depth). Antonie Van Leeuwenhoek 96(4):515–526

Fornaro T, Steele A, Brucato JR (2018) Catalytic/protective properties of martian minerals and implications for possible origin of life on Mars. Life 8(4):1–41. https://doi.org/10.3390/life8040056

Frank YA, Kadnikov VV, Lukina AP, Banks D, Beletsky AV, Mardanov AV et al (2016) Characterization and genome analysis of the first facultatively alkaliphilic Thermodesulfovibrio isolated from the deep terrestrial subsurface. Front Microbiol 7:2146

Frolova A, Merkel AY, Novikov AA, Bonch-Osmolovskaya EA, Slobodkin AI (2021) Anaerotalea alkaliphila gen. nov., sp. nov., an alkaliphilic, anaerobic, fermentative bacterium isolated from a terrestrial mud volcano. Extremophiles 25:301–309

Gales G, Chehider N, Joulian C, Battaglia-Brunet F, Cayol JL, Postec A et al (2011) Characterization of *Halanaerocella petrolearia* gen. nov., sp. nov., a new anaerobic moderately halophilic fermentative bacterium isolated from a deep subsurface hypersaline oil reservoir. Extremophiles 15(5):565–571

Gómez F, Cavalazzi B, Rodríguez N, Amils R, Ori GG, Olsson-Francis K, Escudero C, Martínez JM, Miruts H (2019) Ultra-small microorganisms in the polyextreme conditions of the Dallol volcano, Northern Afar, Ethiopia. Sci Rep 9(1):1–9. https://doi.org/10.1038/s41598-019-44440-8

Govil T, Rathinam NK, Salem DR, Sani RK (2018) Taxonomical diversity of extremophiles in the deep biosphere. Elsevier

Hartman AL, Norais C, Badger JH, Delmas S, Haldenby S, Madupu R et al (2010) The complete genome sequence of *Haloferax volcanii* DS2, a model archaeon. PLoS One 5:e9605

Heinz J, Waajen AC, Airo A, Alibrandi A, Schirmack J, Schulze-Makuch D (2019) Bacterial growth in chloride and perchlorate brines: halotolerances and salt stress responses of planococcus halocryophilus. Astrobiology 19(11):1377–1387. https://doi.org/10.1089/ast.2019.2069

Higgins PM, Cockell CS (2020) A bioenergetic model to predict habitability, biomass and biosignatures in astrobiology and extreme conditions: a bioenergetic model to predict habitability, biomass and biosignatures in astrobiology and extreme conditions. J R Soc Interface 17(171):20200588. https://doi.org/10.1098/rsif.2020.0588

Hoehler, T., Jørgensen, B. Microbial life under extreme energy limitation. Nat Rev Microbiol 11, 83–94 (2013)

Hsu D et al (2024) Isolation and genomic analysis of "Metallumcola ferriviriculae" MK1, a Gram-positive, Fe (III)-reducing bacterium from the Soudan Underground Mine, an iron-rich Martian analog site. Appl Environ Microbiol 90:e0004424

Jones RM, Goordial JM, Orcutt BN (2018) Low energy subsurface environments as extraterrestrial analogs. Front Microbiol 9:1–18. https://doi.org/10.3389/fmicb.2018.01605

Kaksonen AH, Spring S, Schumann P, Kroppenstedt RM, Puhakka JA (2006) *Desulfotomaculum thermosubterraneum* sp. nov., a thermophilic sulfate-reducer isolated from an underground mine located in a geothermally active area. Int J Syst Evol Microbiol 56(11):2603–2608

Kanso S, Patel BKC (2003) *Microvirga subterranea* gen. nov., sp. nov., a moderate thermophile from a deep subsurface Australian thermal aquifer. Int J Syst Evol Microbiol 53(2):401–406

Kanso S, Greene AC, Patel BKC (2002) *Bacillus subterraneus* sp. nov., an iron- and manganese-reducing bacterium from a deep subsurface Australian thermal aquifer. Int J Syst Evol Microbiol 52(3):869–874

Karnachuk OV et al (2024) Novel thermophilic genera Geochorda gen. nov. and Carboxydochorda gen. nov. from the deep terrestrial subsurface reveal the ecophysiological diversity in the class Limnochordia. Front Microbiol 15:1441865

Kashefi K, Lovley DR (2003) Extending the upper temperature limit for life. Science 301:934

Kawamoto H, Watanabe M, Mochimaru H, Nakahara N, Meng XY, Sakamoto S et al (2024) Atrimonas thermophila gen. nov., sp. nov., a novel anaerobic thermophilic bacterium of the phylum Atribacterota isolated from deep subsurface gas field and proposal of Atrimonadaceae fam. nov. within the class Atribacteria in the phylum Atribacterota. Syst Appl Microbiol 47:126515

Khomyakova MA, Zavarzina DG, Merkel AY, Klyukina AA, Pikhtereva VA, Gavrilov SN, Slobodkin AI (2022) The first cultivated representatives of the actinobacterial lineage OPB41 isolated from subsurface environments constitute a novel order Anaerosomatales. Front Microbiol 13:1047580

Klouche N, Fardeau ML, Lascourrèges JF, Cayol JL, Hacene H, Thomas P, Magot M (2007) *Geosporobacter subterraneus* gen. nov., sp. nov., a spore-forming bacterium isolated from a deep subsurface aquifer. Int J Syst Evol Microbiol 57(8):1757–1761

Kotopoulou E, Delgado Huertas A, Garcia-Ruiz JM, Dominguez-Vera JM, Lopez-Garcia JM, Guerra-Tschuschke I, Rull F (2019) A polyextreme hydrothermal system controlled by iron: the case of Dallol at the Afar Triangle. ACS Earth Space Chem 3(1):90–99. https://doi.org/10.1021/acsearthspacechem.8b00141

Kumar S, Dangi K. Arun, Shukla Pratyoosh, Baishya Debabrat, Khare K. Sunil et al (2019) Thermozymes:Adaptive strategies and tools for their biotechnological applications. Bioresour. Technol 278: 372–382

Lauro FM, Bartlett DH (2008) Prokaryotic lifestyles in deep sea habitats. Extremophiles 12(1):15–25

Lonescu, D., Zoccarato, L., Cabello-Yeves, P. J., & Tikochinski, Y. (2024). Extreme fluctuations in ambient salinity select for bacteria with a hybrid "salt-in"/" salt-out" osmoregulation strategy. Frontiers in Microbiomes, 2, 1329925

Magot M, Basso O, Tardy-Jacquenod C, Caumette P (2004) *Desulfovibrio bastinii* sp. nov. and *Desulfovibrio gracilis* sp. nov., moderately halophilic, sulfate-reducing bacteria isolated from deep subsurface oilfield water. Int J Syst Evol Microbiol 54(5):1693–1697

Marlow JJ, Martins Z, Sephton MA (2008) Mars on Earth: soil analogues for future Mars missions. Astron Geophys 49(2):2.20–2.23. https://doi.org/10.1111/j.1468-4004.2008.49220.x

Marlow JJ, Martins Z, Sephton MA (2011) Organic host analogues and the search for life on Mars. Int J Astrobiol 10(1):31–44. https://doi.org/10.1017/S1473550410000303

Martins Z, Cottin H, Kotler JM, Carrasco N, Cockell CS, dela Torre Noetzel R, Demets R, de Vera JP, d'Hendecourt L, Ehrenfreund P, Elsaesser A, Foing B, Onofri S, Quinn R, Rabbow E, Rettberg P, Ricco AJ, Slenzka K, Stalport F et al (2017) Earth as a tool for astrobiology—a European perspective. Space Sci Rev 209:43–81. https://doi.org/10.1007/s11214-017-0369-1

Merino N, Aronson HS, Bojanova DP, Feyhl-Buska J, Wong ML, Zhang S, Giovannelli D (2019) Living at the extremes: extremophiles and the limits of life in a planetary context. Front Microbiol 10:780

Mitri G, Showman AP, Lunine JI, Lorenz RD (2007) Hydrocarbon lakes on Titan. Icarus 186(2):385–394. https://doi.org/10.1016/j.icarus.2006.09.004

Mori K, Hanada S, Maruyama A, Marumo K (2002) Thermanaeromonas toyohensis gen. nov., sp. nov., a novel thermophilic anaerobe isolated from subterranean vein in the Toyoha mines. Int J Syst Evol Microbiol 52:1675–1680

Mougi A (2024) pH adaptation stabilizes bacterial communities. NPJ Biodivers 3(1):32

Neveu M, Anbar AD, Davila AF, Glavin DP, Mackenzie SM, Phillips-Lander CM, Sherwood B, Takano Y, Williams P, Yano H (2020) Returning samples from enceladus for life detection. Front Astron Space Sci 7:26. https://doi.org/10.3389/fspas.2020.00026

Pandey S, Clarke J, Nema P, Bonaccorsi R, Som S, Sharma M, Phartiyal B, Rajamani S, Mogul R, Martin-Torres J, Vaishampayan P, Blank J, Steller L, Srivastava A, Singh R, McGuirk S, Zorzano MP, Güttler JM, Mendaza T et al (2020) Ladakh: diverse, high-altitude extreme environments for off-earth analogue and astrobiology research. Int J Astrobiol 19(1):78–98. https://doi.org/10.1017/S1473550419000119

Parnell J, McMahon S (2016) Physical and chemical controls on habitats for life in the deep subsurface beneath continents and ice. Philos Trans R Soc A Math Phys Eng Sci 374:20140293

Payler SJ, Biddle JF, Lollar BS, Fox-Powell MG, Edwards T, Ngwenya BT et al (2019) An ionic limit to life in the deep subsurface. Front Microbiol 10:426

Preston LJ, Dartnell LR (2014) Planetary habitability: lessons learned from terrestrial analogues. Int J Astrobiol 13(1):81–98. https://doi.org/10.1017/S1473550413000396

Purkamo L (2023) Life in the deep subsurface and its implications for astrobiology. In: Oxford research encyclopedia of planetary science. Oxford University Press, pp 1–18. https://doi.org/10.1093/acrefore/9780190647926.013.274

QGIS.org, 2023. QGIS Geographic Information System. QGIS Association. http://www.qgis.org

Rastogi G, Muppidi GL, Gurram RN, Adhikari A, Bischoff KM, Hughes SR et al (2009) Isolation and characterization of cellulose-degrading bacteria from the deep subsurface of the Homestake gold mine, Lead, South Dakota, USA. J Ind Microbiol Biotechnol 36(4):585–593

Rengpipat S, Langworthy TA, Zeikus JG (1988) *Halobacteroides acetoethylicus* sp. nov., a new obligately anaerobic halophile isolated from deep subsurface hypersaline environments. Syst Appl Microbiol 11(1):28–35

Roh Y, Liu SV, Li G, Huang H, Phelps TJ, Zhou J (2002) Isolation and characterization of metal-reducing *Thermoanaerobacter* strains from deep subsurface environments of the Piceance Basin, Colorado. Appl Environ Microbiol 68(12):6013–6020

Rothschild LJ, Mancinelli RL (2001) Life in extreme environments. Nature 409:1092–1101

Sakamoto S, Nobu MK, Mayumi D, Tamazawa S, Kusada H, Yonebayashi H et al (2021) *Koleobacter methoxysyntrophicus* gen. nov., sp. nov., a novel anaerobic bacterium isolated from deep subsurface oil field and proposal of *Koleobacteraceae* fam. nov. and *Koleobacterales* ord. nov. within the class Clostridia of the phylum Firmicutes. Syst Appl Microbiol 44(1):126154

Sattler B, Puxbaum H, Psenner R (2001) Bacterial growth in supercooled cloud droplets. Geophys Res Lett 28(2):239–242. https://doi.org/10.1029/2000GL011684

Schulze-Makuch D, Irwin LN. Reassessing the possibility of life on venus: proposal for an astrobiology mission (2002). Astrobiology. 2(2): 197–202

Schultz J, dos Santos A, Patel N, Rosado AS (2023) Life on the edge: bioprospecting extremophiles for astrobiology. J Indian Inst Sci 103(3):721–737. https://doi.org/10.1007/s41745-023-00382-9

Schwieterman EW, Kiang NY, Parenteau MN, Harman CE, Dassarma S, Fisher TM, Arney GN, Hartnett HE, Reinhard CT, Olson SL, Meadows VS, Cockell CS, Walker SI, Grenfell JL,

Hegde S, Rugheimer S, Hu R, Lyons TW (2018) Exoplanet biosignatures: a review of remotely detectable signs of life. Astrobiology 18(6):663–708. https://doi.org/10.1089/ast.2017.1729

Shen L, Liu Y, Chen L, Lei T, Ren P, Ji M, Pucci F (2024) Genomic basis of environmental adaptation in the widespread poly-extremophilic *Exiguobacterium* group. ISME J 18(1):wrad020

Slobodkina GB, Panteleeva AN, Kostrikina NA, Kopitsyn DS, Bonch-Osmolovskaya EA, Slobodkin AI (2013) Tepidibacillus fermentans gen. nov., sp. nov.: a moderately thermophilic anaerobic and microaerophilic bacterium from an underground gas storage. Extremophiles 17:833–839

Smith DJ, Griffin DW, Schuerger AC (2010) Stratospheric microbiology at 20 km over the Pacific Ocean. Aerobiologia 26(1):35–46. https://doi.org/10.1007/s10453-009-9141-7

Sorokin DY, Berben T, Melton ED, Overmars L, Vavourakis CD, Muyzer G (2014) Microbial diversity and biogeochemical cycling in soda lakes. Extremophiles 18:791–809

Sun J, He X, Yilin LE, Al-Tohamy R, Ali SS (2024) Potential applications of extremophilic bacteria in the bioremediation of extreme environments contaminated with heavy metals. J Environ Manag 352:120081

Takai K (2011) Limits of life and the biosphere: lessons from the detection of microorganisms in the deep sea and deep subsurface of the Earth. In: Origins and evolution of life. Cambridge University Press, pp 469–486. https://doi.org/10.1017/cbo9780511933875.030

Takai K, Komatsu T, Horikoshi K (2001a) *Hydrogenobacter subterraneus* sp. nov., an extremely thermophilic, heterotrophic bacterium unable to grow on hydrogen gas, from deep subsurface geothermal water. Int J Syst Evol Microbiol 51(4):1425–1435

Takai K, Moser DP, Onstott TC, Spoelstra N, Pfiffner SM, Dohnalkova A, Fredrickson JK (2001b) Alkaliphilus transvaalensis gen. nov., sp. nov., an extremely alkaliphilic bacterium isolated from a deep South African gold mine. Int J Syst Evol Microbiol 51:1245–1256

Tamazawa S, Mayumi D, Mochimaru H, Sakata S, Maeda H, Wakayama T et al (2017) *Petrothermobacter organivorans* gen. nov., sp. nov., a thermophilic, strictly anaerobic bacterium of the phylum Deferribacteres isolated from a deep subsurface oil reservoir. Int J Syst Evol Microbiol 67(10):3982–3986

Tarasov K, Yakhnenko A, Zarubin M, Gangapshev A, Potekhina NV, Avtukh AN, Kravchenko E (2023) Cytobacillus pseudoceanisediminis sp. nov., a novel facultative methylotrophic bacterium with high heavy metal resistance isolated from the deep underground saline spring. Curr Microbiol 80:31

Wainwright M (2015) Biological entities and DNA-containing masses isolated from the stratosphere-evidence for a non-terrestrial origin. Astron Review 11(2):25–40. https://doi.org/10.1080/21672857.2015.1087751

Westmeijer G et al (2024) Candidatus Desulforudis audaxviator dominates a 975 m deep groundwater community in central Sweden. Commun Biol 7:1332

Wu Z, Li M, Qu L, Zhang C, Xie W (2024) Metagenomic insights into microbial adaptation to the salinity gradient of a typical short residence-time estuary. Microbiome 12(1):115

Xiong Z, Jiang Y, Qi D, Lu H, Yang F, Yang J et al (2009) Complete genome sequence of the extremophilic Bacillus cereus strain Q1 with industrial applications. J Bacteriol 191:1120–1121

Zavarzina DG, Sokolova TG, Tourova TP, Chernyh NA, Kostrikina NA, Bonch-Osmolovskaya EA (2007) Thermincola ferriacetica sp. nov., a new anaerobic, thermophilic, facultatively chemolithoautotrophic bacterium capable of dissimilatory Fe(III) reduction. Extremophiles 11:1–7

Zhang F, Jiang X, Chai L, She Y, Yu G, Shu F et al (2014) Permanent draft genome sequence of *Bacillus flexus* strain T6186-2, a multidrug-resistant bacterium isolated from a deep-subsurface oil reservoir. Mar Genomics 18:135–137

Zhu D, Zhang W, Zhang Q, Nagata S (2010) Accumulation and role of compatible solutes in fast-growing *Salinivibrio costicola* subsp. *yaniae*. Can J Microbiol 56:1020–1027

Geomicrobiological Factors That Control Bioremediation of Uranium-Debased Sites

Macmillan Nongkhlaw and Santa Ram Joshi

Contents

Abstract The growing demand for alternate source of energy has made uranium the most sought mineral in the world. Mining for this naturally occurring energy-rich element has increased in the last 60 years and thus has made uranium pollution an emerging environmental hazard. Previously, physical remediation and chemical remediation have been used in cleaning up of uranium-contaminated sites, but emerging technologies such as bioremediation have been utilized for the same. The use of microorganism for remediation of uranium-contaminated sites has received widespread attention for its cost-effectiveness and environmental-friendly approaches. Bioremediation of uranium-contaminated sites can be carried out by either employing the indigenous microbes from the contaminated sites or adding novel microbes that possess intrinsic bioremediation properties. Bioremediation of U-contaminated site depends on the speciation of uranium, nature of contaminated site, and their associated physicochemical properties of those sites. Furthermore, understanding the microbial community structure and their function is a crucial factor for successful bioremediation of uranium-contaminated sites.

M. Nongkhlaw · S. R. Joshi (✉)
Department of Biotechnology and Bioinformatics, North Eastern Hill University, Shillong, India

S. R. Joshi et al. (eds.), *Mineral Transformation and Bioremediation by Geo-Microbes*, Interdisciplinary Biotechnological Advances,
https://doi.org/10.1007/978-981-96-3033-2_18

Keywords Bioremediation · Uranium reduction · Uranium speciation · Bioprecipitation · Biosorption

Abbreviations

DO	Dissolved oxygen
EDTA	Ethylenediaminetetraacetate
NTA	Nitrilotriacetate
ORFRC	Oak Ridge Field Research Center

1 Introduction

The demand for energy is an ever-increasing dilemma that is directly influenced by the increase in population, economics growth, and technological advancement. This increase in demand for energy source is exacerbated by the need for clean energy sources. Nuclear energy which mainly involves controlled fission reaction has emerged as one of the alternate sources of energy generation because of its low input of raw material, reliability, and cleaner operation as compared to fossil fuels. Uranium is the energy source of nuclear energy, and mining for this naturally occurring radioactive element has been going on for the last 60 years. Adverse environmental impact of uranium mining and its effect on various life forms have been well documented (Gao et al. 2019). The process of bioremediation involves using microorganisms or their biomass to remove or transform (into less toxic) pollutants from contaminated environments (Saha et al. 2024; Shahina and Rafiq 2024). The interaction of microbes with geochemical processes is a relationship that dates back to the beginning of life on this planet (Dong et al. 2022). The scale of this interaction is both temporal and spatial, and significantly defines the habitability of the planet. The change in geological landscape can greatly influence the dynamics of a microbial community structure. Conversely, microbial metabolic activities affect solubility, transformation, and formation of minerals. Furthermore, the ability of microbes to alter the physicochemical properties of minerals has been exploited to clean up the environment from toxic metals that result from geological mishaps or anthropogenic causes.

Bioremediation is a process that involves the use of microorganisms, plants, and their biomass in the process of removing toxic chemicals including metals from a habitat or environment or the conversion of these chemicals into less toxic or non-toxic form. Bioremediation of toxic metals including uranium has been utilized as a standalone method or in conjunction with physical and chemical methods for cleaning up the environment.

2 Overview of Bioremediation of Uranium

Bioremediation of uranium-contaminated sites can be achieved by restricting the mobility of uranium either by bioprecipitation or bioaccumulation. Immobilization of uranium involves biologically reducing uranium [U(VI)] which is soluble to a less soluble uranium [U(IV)] that eventually precipitates out into the environment and effectively decreases its bioavailability (You et al. 2021). Generally, microbial reduction by members of *Shewanella* sp. and *Geobacter* sp. has been reported but other microbes have shown similar kind of activity (Chandler et al. 2010; Shelobolina et al. 2008; Yun-Juan et al. 2005). Uranium reduction can be carried in the outer membrane, cytoplasm, periplasm, and extracellularly through the enzymatic activity of various enzymes. Predominantly, bioremediation of uranium in an anaerobic environment is carried out by *Geobacter* species. At the molecular level, cytochrome proteins from *Geobacter* species are involved in uranium reduction as evident by their increased expression and prevalence in subsurface contaminated with uranium (Holmes et al. 2009). Similarly, *Shewanella oneidensis* lacking in cytochrome-c was unable to reduce uranium (Marshall et al. 2006). Usually, reduction of uranium in sulfur-reducing bacteria like *Desulfovibrio alaskensis* G20 can occur in presence of electron donor like hydrogen but can be bypassed by other donors like lactate and pyruvate. Extracellular reduction of uranium via extracellular pili has been reported in *Geobacter* species. These pili reduce uranium with the help of carbon ligands outside of the cell, and this mechanism acts as a protective barrier against the cytotoxic effect of uranium (Cologgi et al. 2011). Other methods of uranium reduction include the use of microbial metabolites capable of transferring electron to U(VI) without direct involvement of the cellular components. This mechanism was reported in *Shewanella* species that release flavin mononucleotide that act as an electron donor to mediate reduction of U(VI) (Yamasaki et al. 2017; Suzuki et al. 2010a, b). Generally, reduction of uranium occurs extracellularly or in the periplasm but cytoplasmic reduction has also been reported. In *D. aleskensis*, reduction of U(VI) in cytoplasm takes place with thioredoxins used as electron donor. The importance of thioredoxin (MreD) and thioredoxin reductase (MreE) encoded by *mre* operon with an additional oxidoreductase (MreG) in U(VI) reduction was elucidated by mutagenesis studies (Li et al. 2014; Li and Krumholz 2009).

Another mechanism for uranium precipitation involves the interaction of uranium with phosphates released by microorganism (Gudavalli et al. 2018; Zhang et al. 2021a, b). Phosphorous is a macronutrient that is required by microorganisms for their various metabolic processes (Smil 2000). Soluble phosphate, a form which is essential for microbes, is very scarce in the environment. Hence, microorganisms use phosphatase to release phosphate ions from minerals and organic sources. Uranyl ion interacts with phosphate ions leading to complexation and precipitation of the former. Irrespective of their localization, phosphatases are categorized as acid or alkaline depending upon their optimum pH for their activity. Phosphatase-mediated uranium precipitation utilizes phosphate that is released from various sources like glycerol-2-phosphate, polyphosphate, phytate, and glycerol-3-phosphate (Appukuttan et al. 2006; Powers et al. 2002; Li et al. 2019, 2017; Renninger et al. 2004). The role of acid phosphatase in uranium biomineralization was initially reported in *Serratia* sp. N14 (previously known as *Citrobacter* N14) (Macaskie et al. 1992). Following this, different *Serratia* spp. have been found to mediate uranium-phosphate precipitation in anaerobic condition and also in presence of gamma radiation (Newsome et al. 2015; Chandwadkar et al. 2018). Involvement of PhoN, a periplasmic acid phosphatase from *Serratia* sp. in uranium bioprecipitation, has been reported, but additional strain-specific properties like uranium nucleation sites were also necessary for uranium mineralization (Macaskie et al. 1994; Jeong et al. 1997). Recombinant bacterial strains like DH5alpha overexpressing PhoN from *S. typhi* showed increased uranium removal capacity as compared to *Serratia* sp. N14 (Basnakova et al. 1998). Similar observation was also seen with PhoN from *S. typhi* and its heterologous expression in *Deinococcus radiodurans* R1 (Appukuttan et al. 2006). Microorganisms isolated from uranium-contaminated site with harbored acid phosphatase have been studied for their uranium bioprecipitation property (Choudhary et al. 2012; Sanchez-Castro et al. 2017; Beazley et al. 2007; Martinez et al. 2007; Morrison et al. 2021). Phosphatase-positive bacteria grown in medium with organic phosphate and under dissimilar pH conditions showed uranium removal properties with organic phosphate supplementation. Uranium precipitation in alkaline environment like industrial effluent that is rich in carbonates is very rare because uranium-carbonate complex is stable and highly soluble at higher pH (Beazley et al. 2007; Martinez et al. 2007; Morrison et al. 2021). *Sphingomonas* sp. BSAR with alkaline phosphatase PhoK showed the ability to form uranium phosphate precipitate from the uranium-carbonate containing medium at pH 9 through the cleavage of glycerol-2-phosphate (Nilgiriwala et al. 2008). Higher precipitation of uranium was also observed when PhoK from *Sphingomonas* sp. BSAR was overexpressed in *Escherichia coli* and *Deinococcus radiodurans* R1 (Kulkarni et al. 2013). Besides PhoK, heterologous expression of periplasmic alkaline phosphatase PhoA from *E. coli* in *Pseudomonas* sp. was also found to effectively release Pi in sterilized soil slurries which in turn precipitate uranium from cell-free supernatant (Powers et al. 2002). Extracellular uranium-phosphate precipitation along with the upregulation of four alkaline phosphatases was also seen in *Stenotrophomonas bentonitica*. These findings showed that bacterial alkaline and acid phosphatases are potential candidates for use in immobilization of uranium in the environment (Pinel-Cabello et al. 2021). Furthermore, as observed in these studies, phosphate ion donors like glycerol-3-phosphate and glycerol-2-phosphate utilized for uranium precipitation are supplemented in the medium. Another mechanism for bioprecipitation of uranium through the formation of uranium phosphate is the use of naturally occurring organic phosphate like phytate and polyphosphates (Turner 2006). Phytate (inositol hexaphosphate) is stored in plants as phosphorous storage, and phytase enzymes are responsible for hydrolyzing phytate into lesser phosphorylated compound such as myo-inositol derivative and phosphate especially the inorganic category (Wyss et al. 1999). Phytase activity in acidic conditions was observed in microbial populations of uranium rich and contaminated site (Salome et al. 2017). Upregulation of a putative phytase in presence of uranium was also reported in *Caulobacter crescentus* indicating that this enzyme may play a role in detoxification of uranium which in turn can be exploited for bioremediation purposes (Hu et al. 2005). Another mechanism for phosphatase mediated uranium biomineralization is the interaction of naturally produced phosphate released by microorganism. Cellular polyphosphate from *Pseudomonas aeruginosa* was found to mediate the accumulation of uranium precipitate on the cell surface (Renninger et al. 2004). Accumulation of uranium phosphate in microbial cell surface has been observed in uraniferous soils indicating naturally occurring biomineralization. The use of biomineralization as a clean-up process for uranium does have its limitation that could result due to rapid precipitation that can hamper microbial metabolism and economic viability due to the use of external phosphate donors. However, other studies have shown that encrustation does not impact viability, and the use of phosphate donor could be reduced by finding other naturally occurring phosphate donor like polyphosphate or phytates. Uranium can also attach to microbial surfaces through a passive mechanism called biosorption. Bacterial surfaces are adorned with negatively charged ligands like hydroxyl, amine, carboxyl, phosphate, and sulfhydryl groups with the capacity to bind metals using the chemical sorption approach (Hufton et al. 2021). Dead biomass of bacterial cells has been found to possess better adsorption capacity because it is not affected by metal toxicity. Studies showed that microbial biosorption capacity for uranium in bacteria varied from concentrations of 45 mg g^{-1} to 615 mg g^{-1} dry weight of cells for uranium. Biosorption has been found to be more effective in treatment of effluents with low to medium concentrations as it leads to faster binding and removal of bound metals easily from a cell surface to regenerate the biosorbent for effective removal of the metal (Schiewer and Volesky 2000). Even though the bacteria showed promising potential to biosorb uranium, they do have their drawbacks which include competition for binding with other cations and rapid desorption from the cell surface. Biosorption does have its challenges but it can be employed as a "pump and treat" method for the treatment of effluents.

Bioremediation of uranium is an emerging field that is slowly gaining traction in research and funding because nuclear waste keeps on accumulating in the environment. Success in bioremediation of uranium-contaminated sites requires a deeper understanding of the various geomicrobiological factors that impact the previously described bioremediation strategies for uranium bioremediation (Fig. 1).

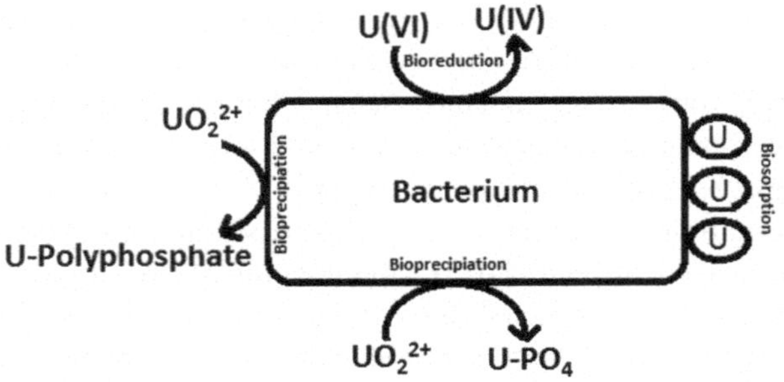

Fig. 1 Overview of uranium bioremediation

3 Geomicrobiological Factors That Impact Uranium Bioremediation

Microbial bioremediation is influenced by both biotic and abiotic factors. Microbial community structure and the diverse metabolic functions associated with their habitat play a very important role in the bioremediation process. Similarly, physicochemical properties of the targeted contaminant in the environment that it exists are a crucial factor that determines the path and outcome of any bioremediation strategies. Hence, bioremediation of uranium is also affected by these factors which in turn determine the practicability of the remediation processes.

3.1 Initial U(VI) Concentration

Like any other enzymatic-driven reaction, the rate of biological reduction of U(VI) will be faster if the initial concentration of U (VI) is higher. U(VI) on the other hand is toxic to the cells at higher concentration and that ultimately inhibits and retards the growth of the involved microorganism which in turn will terminate the reduction activity (Sani et al. 2006). Reactive oxygen species (ROS) formation when live cells are exposed to U(VI) is well documented (Gao et al. 2019). ROS are known to disrupt cellular membrane integrity and hinder the terminal oxidation chain in respiration by reducing ATP and NADH production. Genes involved in apoptotic cascades such as Bid, Bcl-2, Bax, and caspases, which lead to degradation of DNA degradation and cell death, are overexpressed in presence of U(VI) (Liu et al. 2015). The toxicity of uranium can be reduced by the presence of phosphate, carbonate, dissolved organic carbon (DOC), and sulfate that form complexes with UO_2 OH^+ and UO^{2+} or other cations such as Mg^{2+}, Ca^{2+}, and H^+ that compete for the biological ligands binding sites (Trenfield et al. 2011; Belli et al. 2015).

3.2 Uranium Speciation

Uranium speciation plays a crucial role in bioremediation strategies aimed at mitigating uranium contamination in soils and sediments. Understanding the chemical forms of uranium is essential for effective remediation, as different species exhibit varying mobility and bioavailability. Uranium exists in various oxidation state such as +III, +IV, +V, and +VI, but the two forms of uranium, namely, U(IV) and U(VI), are most predominant. However, U(IV) and U(VI) are stable and U(VI) is known to be more toxic than the other valencies in aqueous environment (Gao et al. 2019). The most easily reduced form of uranium is the bioavailable U(VI) which is represented by unstable free uranyl hydroxides, hydrated uranyl ions, and some uranyl complexes (Ulrich et al. 2011; Bone et al. 2017). Uranium speciation is also affected by the presence of carbonate, ethylenediaminetetraacetate (EDTA) nitrilotriacetate (NTA), citrate, and humic acid. The rate of reduction of U(VI) by *Shewanella oneidensis* MR-1 enhances with increasing concentration of EDTA. This could be the result of the formation of EDTA-uranium complexes that prevents the precipitation of the latter, thereby resulting in higher reduction of U(VI) (Sheng et al. 2011). Similarly, the presence of EDTA and NTA enhances U(VI) reduction rate by *Shewanella putrefaciens* (Suzuki et al. (2010b). However, the presence of citrate significantly decreases the reduction rate because citrate forms a multinuclear complex with U(VI), unlike NTA and EDTA that form mononuclear complexes with U(VI). This shows that biological availability of U(VI) is dependent on the structure it forms with organic complexes rather than the stability constant. Uranium reduction is also affected by the presence of humic acid as humic acid-uranyl complex structure interferes with the electron transfer process. Biological U(VI) reduction with humic acid mostly occurs at basic pH or at increased content of carbonates (Wang et al. 2019). Other studies also showed that the enzymatic U(VI) reduction gets better in the availability of humic acid and the end product is found to accumulate as a sheet of intracellular granules with nanoparticle crystals (Li et al. 2017).

Uranium is sensitive to the change in the redox environment, and it is influenced by the change in electrical conductivity and pH of the environment. Uranyl (UO_2^{2+}) forms highly soluble compounds under oxidative conditions, with the oxidation of U(IV) to stable uranyl oxide (UO_2) in reduced environment (Vodyanitskii 2011). Carbonate/phosphate concentration also plays an important role on U(VI) reduction to U(IV). The formation of U(VI)-carbonate/phosphate complexes is highly favored when there is an increase in carbonate/phosphate concentration in solution which leads to stabilization of the oxidation state of U(VI) (Papanicolaou et al. 2010; Liao et al. 2020). Furthermore, organic compounds can also act as the electron donors in the reduction of U(VI) to U(IV) (Gavrilescu et al. 2009; Vodyanitskii 2011). The speciation of uranium (U) is also significantly influenced by pH, affecting its solubility and mobility in various environments. Understanding these dynamics is crucial for environmental remediation and nuclear waste management. At low pH (around 2), uranium predominantly exists as soluble uranyl ions (UO_2^{2+}), which can lead to increased mobility in contaminated soils (Perdrial et al. 2022). As pH rises, uranium solubility decreases which enhance its retention in soils (Mühr-Ebert et al. 2019, Pan et al. 2022). U can precipitate as uranyl hydroxides or for instance, uranyl oxyhydroxides like metaschoepite form at higher pH levels, indicating a shift towards less soluble species (Pan et al. 2022). In contrast, while higher pH generally promotes precipitation and immobilization of uranium, the presence of complexing agents like phosphates can alter these dynamics, potentially enhancing solubility under certain conditions (Perdrial et al. 2022). This highlights the complexity of uranium behavior in environmental systems. As mentioned earlier, carbonates and bicarbonates affect uranium speciation. Studies showed that there is a negative correlation between carbonate or bicarbonate concentration with uranium bioreduction notably at pH around neutral and alkaline zone because of uranyl carbonate being strongly stable which in turn makes it less soluble in such conditions (Sheng and Fein 2014; Ulrich et al. 2011). Non-carbonate uranyl complex can be readily reduced as compared to carbonate uranyl complex (Belli et al. 2015). Due to thermodynamic constraints to obtain electrons in neutral to alkaline solutions, carbonate uranyl complexes have low redox potential in such environment (Xie et al. 2018). However, contradictory reports in some field tests concluded that bicarbonate-spiked reaction did not hinder the enzymatic U(VI) reduction, instead it was found to be helpful to the bioremediation of uranium in water bodies (Long et al. 2015). At the optimum rate, reduction of U(VI) was approximately 3.8 times higher when acetate and bicarbonate were added as compared to a test well where only acetate was injected. Furthermore, the concentration of U(VI) remained stable for an extensive time period without showing any rebound. In a separate field study, it was indicated that high U(VI) reduction rate is possible in groundwater where calcium-uranyl-carbonate complexes concentration is higher (>90%) (Williams et al. 2011). Hyun et al. (2009) also proposed that carbonate salts, especially the bicarbonates, can accelerate the dissolution of U(VI) from sediments which in turn promotes the growth of uranium-respiring microbes that help in increasing the reduction of U(VI) (Chandler Darrell et al. 2013). Hence, these conflicting observations in laboratory studies and field tests may lie in different environmental conditions. Therefore, prior baseline information on the source of uranium, its reactivity, and mobility is necessary before implementing bioremediation process.

Bioremediation of uranium (U) is significantly influenced by pH levels, which affect both the solubility of uranium and the metabolic activity of microorganisms involved. The redox potential and biosorption of uranium are affected by change in pH and that can influence the reduction rate. The U(VI) complex when combined with carbonate ions shows exceptionally low standard redox potential (e.g., $UO_2(CO_3)_3^{4-}/UO_2$), Eh° = −0.4221 V). The thermodynamic effect that drives electron transfer is hampered due to the relatively low redox potential which in turn impedes the reduction process. Acidic conditions can hinder microbial activity and U(VI) reduction, necessitating pH adjustments for effective bioremediation (Finneran et al. 2002). In low-pH environments, stimulating microbial U(VI) reduction for uranium bioremediation is challenging due to high nitrate levels and insufficient reduction rates, limiting effective immobilization of uranium (Shelobolina et al. 2007). The biosorption of uranium is highly pH-dependent, with increased uptake at higher pH levels (above 3.0) due to enhanced binding capacity of biomass (Yang and Volesky 1999). Similarly, biosorption of uranium by *Penicillium piscarium* and *Acidithiobacillus ferrooxidans* was effective at pH 3.5, removing 93.2–97.5% of uranium (Romero-Gonzalez et al. 2016; Coelho et al. 2020). Furthermore, phosphatase-mediated bioprecipitation of uranium is also affected by changes in pH. Uranium-tolerant phosphate solubilizing bacteria can potentially enhance bioremediation processes by altering pH levels conducive to uranium solubility and bioavailability.

As with all living cells, temperature and pH can have a direct impact on the growth and metabolic activity of microorganisms with U(VI) reduction potential. Microbes involved in U(VI) reduction displayed a bell-shaped curve in a reduction rate versus temperature studies which indicated a enzymatic type of a reaction under specific temperature ranges when the microbes are optimally active (Rao and Tian 2008). A study on mixed culture of *Clostridium*-like bacteria and *Desulfovibrio* by Boonchayaanant et al. (2009) found that an increase in temperature from 20 °C to 30 °C led to threefold increase of pseudo-second-order rate constant. Extracellular reduction of U(VI) in Geobacteraceae is promoted by pili expression and is generally dependent on temperature affecting the growth. For example, in *G. metallireducens*, pili assembly was promoted at 30 °C during the reduction of uranium whereas the assembly of pili in *G. sulfurreducens* was known to occur normally at 25 °C (Mahadevan and Zhao 2017).

The straight-through impact of pH on growth and metabolism of U(VI)-reducing microorganism is analogous to that of temperature. However, due to diverse tolerance of microorganism to acidic and alkaline conditions, it is difficult to stipulate the optimum and extreme conditions. The appropriate pH reported for the growth of *Desulfosporosinus* sp. strain GBSRB4.2, a sulfate-reducing bacteria, and certain *Clostridium* sp. ranged from 4.0 to 4.5 (Senko et al. 2009; Gilson et al. 2015; Gao and Francis 2008). A mixed culture of *Clostridium pasteurianum*, *C. sphenoides*, and *C. acetobutylicum*, and sulfate-reducing bacteria was found to reduce U(VI) optimally at proximal neutral pH (Yi et al. 2007; Gao and Francis 2008). Significant reduction of U(VI) at pH 10–10.5 by native microbial community in the sediments rich in lime has also been reported (Williamson et al. 2014).

pH significantly influences uranium bioremediation, as microbial processes can alter pH, affecting uranium solubility and mobility, with optimal conditions promoting the precipitation and biosorption of insoluble U(IV).

3.3 Competitive Electron Acceptors

Dissolved oxygen (DO) is a crucial player that affects the biological U(VI) reduction because of its strong redox potential property and its ability to re-oxidize the reduced uranium. DO can also disrupt the community structure of microorganism including the U(VI)-reducing bacteria. There is a delay in uranium reduction if DO is present even at small concentrations (Campbell et al. (2011). The population of anaerobic microorganisms involved in reduction of U(VI) would be reduced in low DO conditions while permitting facultative and aerobic microorganisms that lack U(VI) reduction potentials to populate the setup (Hu et al. 2016). Other favorable electron acceptors like Fe(III), nitrate, and sulfate are generally found to co-exist in a uranium-rich condition and are also beneficial for anaerobic microorganisms as electron acceptors. The strong redox potential of nitrate as compared to U(VI) makes the former more energy-efficient and the presence of nitrate makes it unfavorable for uranium reduction. A number of studies have shown that nitrate when completely utilized makes it possible to reduce U(VI) reduction, or even in a condition when nitrate concentration is very low (Finneran et al. 2002; Elias et al. 2003; Istok et al. 2004). Furthermore, the inhibition of metabolic activity and growth of SRB is observed in high concentration of nitrate, which ultimately affects the efficiency of U(VI) reduction (Hu et al. 2016). *Clostridium* bacteria with uranium reduction potential are not sensitive to the build-up of concentration of nitrate (Gao and Francis 2008). Fe(III) and sulfate presence do not affect the reduction of U(VI) significantly owing to similarity in redox potentials and due to the fact that most of sulfate- and Fe(III)-reducing bacteria are also efficient in reduction of U(VI). Furthermore, Fe(III) bioavailability in the environment is curbed by its insolubility, which diminishes its competitive ability for electrons (Komlos et al. 2008). According to a study, the amount of hydrated iron oxide is inversely related to the rate and extent of U(VI) reduction, but crystalline goethite (α-FeOOH) and hematite (Fe_2O_3) have no effect on these parameters (Wielinga et al. 2000). Other studies even suggested that the presence of Fe(III) is important to the sustainability of *Geobacter* activity and enhancement of biological reduction (Anderson et al. 2003; Moon et al. 2007; Zhuang et al. 2012). The complexed forms of Fe(III)/Fe(II) and U(VI)/U(IV) would likewise function as ESs in the availability of organic ligands (EDTA or citrate) to increase the rate of bioreduction of both U(VI) and Fe(III), according to a recent study (Zhang et al. 2021a, b).

3.4 Coexisting Metal Ions

Uranium ore leaching has Cu and Zn as typical coexisting heavy metals that are liberated in uranium-contaminated groundwater. Reduction of U(VI) is relatively not affected by lower concentrations of Zn ($\leq$20 mg/L) and Cu ($\leq$5 mg/L) (Yi et al. 2007); however, the harmful consequences of heavy metals on microbes can completely inhibit the reduction process with the concentrations of metals reaching around 25 mg/L and 15 mg/L respectively. It has also been discovered that alkaline earth elements, such as magnesium and calcium, affect the rate of reduction of U(VI). Because hematite and goethite can adsorb Ca^{2+} and lower its solubility, the negative impacts can be somewhat mitigated, but the rate of reduction of U(VI) falls as the concentration of Ca rises (Brooks et al. 2003; Stewart et al. 2007). Presence of Mg is unfavorable to reduction of U(VI) and similar effect is shown in presence of Ca. When Ca^{2+} (Mg^{2+}) combines with carbonate and uranyl ions to form a ternary complex, the result is a stearic obstruction or poor molecular orbital overlap between the U(VI) center and U(VI) terminal reductase. This reduces the U(VI) bioavailability and deters the transfer of electrons to U(VI) (Belli et al. 2015).

3.5 Role of Microbial Metabolism in Bioremediation of Uranium

To survive in an environment, microbes require the source of energy and carbon for the metabolism of cellular material and inorganic elements such as phosphorus, nitrogen, potassium, sulfur, magnesium, and calcium. The most common sources of carbon naturally available to microorganisms are organic carbon and carbon dioxide which usually are found as dissolved species in groundwater or total organic content in soil. In uranium bioremediation, different organic compounds like acetate, glucose, or ethanol are usually added to the subsurface environment to stimulate the growth microbial population native to that environment. The energy requirement for bacteria is achieved either from sunlight or by oxidation reaction of chemical nature. The oxidation of reduced inorganic or organic compounds— which can include ammonia, sulfide, and nitrite—is the typical metabolic process that microorganisms use in underground environments when light is scarce. The process of microbial oxidation involves the production of electrons that get released in a controlled and systematic way on to a terminal electron acceptor. In principle, the choice of the terminal electron acceptor will decide how much energy is generated from that exchange and that depends on the reduction potential of the oxidation/reduction couple. Generally, microbes will use a higher redox couple with higher reduction potential because more energy will be conserved as compared to a redox couple with lower reduction potential. During microbial uranium reduction, soluble U(VI) is used as a terminal electron acceptor for energy generation. This scenario will take place if other favorable redox couples (mostly nitrate, Fe(III) and sulfates) with better energy conservation are absent or diminished from the ecosystem. Previously, U(VI) reduction was thought to be primarily an abiological process; however, the discovery of different Fe(III)-reducing microorganisms that have the capacity to directly reduce U(VI) to U(IV) made a significant departure from that traditional view. These findings indicate the direct role that microbes can play in the reduction of uranium, rather than just facilitating the abiological reduction process through metabolic byproducts or surfaces. These findings suggest the presence of a novel metabolic pathway that enhances our understanding of microbial energy sources. The ability of reduction of U(VI) to U(IV) insoluble form by these microorganisms is vital for the process of immobilization of uranium in aquatic environment and sediments. U(VI) immobilization could reduce the spread of uranium contamination in environments such as aquifers and sediments, thereby contributing to environmental remediation efforts.

Microorganism involved in uranium reduction also requires electrons to carry out their metabolic processes. According to numerous reports, the most often utilized chemical in laboratory and field studies is acetate, which is followed by ethanol and lactate (Newsome et al. 2014). Furthermore, bacterial U(VI)-reducing activity can be stimulated by glycerol, methanol, pyruvate, glucose, formate, benzoate, butyrate, butanol, hydrogen, and electrodes (Gregory and Lovley 2005). According to a microcosm experiment conducted at Oak Ridge Base, ethanol reduces sediments at a faster pace than acetate (Luo et al. 2007). Other pilot studies carried out with the groundwater and sediments environment of the Oak Ridge showed varying degrees of reduction rate with different electron donors used; when methanol was used, the reduction process was delayed but was much greater than those with glucose or ethanol (Madden et al. 2009). Zhang et al. (2010) demonstrated that the impact of conversion of ethanol on U(VI) is significantly greater than that of oleate because of the production of more hydrogen intermediates by ethanol, suggesting that the rate at which organic substrate is consumed and intermediate hydrogen is produced can influence the rate at which U(VI) is reduced. Although there was a significant retardation in uranium removal, complex organic combinations like hydrogen-release compound (HRC) or vegetable oils have been demonstrated to effectively boost U(VI) biodegradation rates in addition to the usual simple organic compounds previously described (Barlett et al. 2012). More research is currently needed on the lowering of U(VI) by organic substrates, as this could be a major topic of study in the future.

Currently, differing observations are seen from various studies on different locations, and some findings are even contradictory. However, the choice of participating components must be in line with the biomass compositions and ambient circumstances in order to optimize not only the efficiency but also the rate of reduction of U(VI). In any case, U(VI) reduction may be improved by more complicated organic electron donors which act as sustained-release electron donors and simpler alcohols (methanol and ethanol). These experiments demonstrated a clear shortcoming in bioremediation as compared to chemical reduction. The overall indication is that a lengthy induction period for microbe formation is typically necessary before significant U(VI) reduction takes place in any environment.

4 Summary

The use of nuclear fuel as a source of energy is an unavoidable but rather an emergent requirement to support the burgeoning demand for an ever-increasing population on this planet. Concurrently this has made the release of spent nuclear fuel and its safe disposal a major headache for the world. This chapter highlights the various bioremediation techniques that can be used or have been tested in diverse geological settings and the geomicrobiological factors that can decide the usefulness of each method to remediate the environment. Furthermore, this chapter gave a brief outlook for the researcher on the various conditions that can play a major role in the feasibility and practicality of uranium bioremediation (Table 1).

Table 1 Selected microorganisms involved in bioremediation of uranium (Rogiers et al. 2022)

Bacteria	References
Uranium reduction	
Geobacter sp. M18	Yun-Juan et al. (2005)
Geobacter uraniireducens strain RF4	Holmes et al. (2009)
Geobacter sulfurreducens	Shelobolina et al. (2007), Orellana et al. (2013), Lloyd et al. (2003), Cologgi et al. (2011)
Shwanella oneidensis MR-1	Marshall et al. (2006), Bencheikh-Latmani et al. (2005)
Desulfovibrio alaskensis G20	Payne et al. (2002, 2004), Li and Krumholz (2009), Li et al. (2014)
Thermoterrabacterium ferrireducens	Khijniak et al. (2005)
Desulfovibrio desulfuricans	Lovley and Phillips (1992)
Anaeromyxobacter sp.	Sanford et al. (2007)
Clostridium sp.	Francis et al. (1994)
Bioprecipitation	
Serratia sp. N14	Macaskie et al. (1994), Jeong and Macaskie (1995), Jeong et al. (1997)
Escherichia coli DH5 α expressing PhoN from *Salmonella typhi*	Basnakova et al. (1998)
Escherichia coli DH5 α expressing PhoC from *Morganella morganii*	Basnakova et al. (1998)
Escherichia coli DH5 α expressing PhoN from a *Salmonella typhi* isolate	Appukuttan et al. (2006)
Deinococcus radiodurans R1 expressing PhoN from a *Salmonella typhi* isolate	Appukuttan et al. (2006)
Microbacterium oleivorans A9	Gallois et al. (2018)
Caulobacter OR37	Morrison et al. (2021)
Arthrobacter sp. X34	Beazley et al. (2007), Martinez et al. (2007)
Bacillus sp. Y9-2	Beazley et al. (2007), Martinez et al. (2007)
Rahnella sp. Y9602	Beazley et al. (2007), Martinez et al. (2007)
Serratia sp. strain OT II7	Chandwadkar et al. (2018)
Chryseobacterium sp. strain PMSZPI	Khare et al. (2020)
Sphingomonas sp. *BSAR-1*	Nilgiriwala et al. (2008)
Escherichia coli BL21 expressing PhoK from *Sphingomonas* sp. *BSAR-1*	Nilgiriwala et al. (2008)
Escherichia coli DH5 α expressing PhoN from *Salmonella typhi*	Kulkarni et al. (2016)
Deinococcus radiodurans R1expressing PhoK from *Sphingomonas* sp. BSAR-1	Kulkarni et al. (2016)

(continued)

Table 1 (continued)

Bacteria	References
Deinococcus radiodurans R1 expressing PhoK from *Sphingomonas* sp. BSAR-1	Kulkarni et al. (2016)
Caulobacter. crescentus NA1000	Yung and Jiao (2014)
Pseudomonas rhodesiae R1.2 + Pseudomonas veronii V1.2 expressing PhoA from *Escherichia coli*	Powers et al. (2002)
Stenotrophomonas bentonitica BII-R7	Pinel-Cabello et al. (2021)
Microbial communities from ORFRC	Salome et al. (2017)
Caulobacter crescentus CB15N	Hu et al. (2005), Yung and Jiao (2014)
Pseudomonas aeruginosa	Renninger et al. (2004)
Deinococcus radiodurans	Suzuki and Banfield (2004)
Allocasuarina torulosa	Acharya et al. (2017)
Paenibacillus sp. JG-TB8	Reitz et al. (2014)
Cuprividus metallidurans NA4	Rogiers et al. (2021)
Biosorption	
Bacillus sp. MRS-1	Sher and Li (2023)
Stenotrophomonas sp. *CICC 23833*	Hu et al. (2023)
Acidithiobacillus ferrooxidans	Kasra-Kermanshahi et al. (2019)
Bacillus cereus 12-2	Zhang et al. (2019)

References

Acharya C, Chandwadkar P, Nayak C (2017) Unusual versatility of the filamentous, diazotrophic cyanobacterium *Anabaena torulosa* revealed for its survival during prolonged uranium exposure. Appl Environ Microbiol 83:1–17. https://doi.org/10.1128/AEM.03356-16

Anderson RT, Vrionis HA, Ortiz-Bernad I, Resch CT, Long PE, Dayvault R, Karp K, Marutzky S, Metzler DR, Peacock A, White DC, Lowe M, Lovley DR (2003) Stimulating the in situ activity of geobacter species to remove uranium from the groundwater of a uranium-contaminated aquifer. Appl Environ Microbiol 69:5884–5891

Appukuttan D, Rao AS, Apte SK (2006) Engineering of *Deinococcus radiodurans* R1 for bioprecipitation of uranium from dilute nuclear waste. Appl Environ Microbiol 72:7873–7878. https://doi.org/10.1128/AEM.01362-06

Barlett M, Moon HS, Peacock AA, Hedrick DB, Williams KH, Long PE, Lovley D, Jaffe PR (2012) Uranium reduction and microbial community development in response to stimulation with different electron donors. Biodegradation 23:535–546

Basnakova G, Stephens ER, Thaller MC, Rossolini GM, Macaskie LE (1998) The use of *Escherichia coli* bearing a phoN gene for the removal of uranium and nickel from aqueous flows. Appl Microbiol Biotechnol 50:266–272. https://doi.org/10.1007/s002530051288

Beazley MJ, Martinez RJ, Sobecky PA, Webb SM, Taillefert M (2007) Uranium biomineralization as a result of bacterial phosphatase activity: insights from bacterial isolates from a contaminated subsurface. Environ Sci Technol 41:5701–5707. https://doi.org/10.1021/es070567g

Belli KM, DiChristina TJ, Van Cappellen P, Taillefert M (2015) Effects of aqueous uranyl speciation on the kinetics of microbial uranium reduction. Geochim Cosmochim Acta 157:109–124

Bencheikh-Latmani R, Williams SM, Haucke L, Criddle CS, Wu LY, Zhou JZ et al (2005) Global transcriptional pro?ling of shewanella oneidensis MR-1 during Cr(VI) and U(VI) reduction. Appl Environ Microbiol 71:7453–7460

Bone SE, Dynes JJ, Cliff J, Bargar JR (2017) Uranium(IV) adsorption by natural organic matter in anoxic sediments. Proc Natl Acad Sci USA 114:711–716

Boonchayaanant B, Nayak D, Du X, Criddle CS (2009) Uranium reduction and resistance to reoxidation under iron-reducing and sulfate-reducing conditions. Water Res 43:4652–4664

Brooks SC, Fredrickson JK, Carroll SL, Kennedy DW, Zachara JM, Plymale AE, Kelly SD, Kemner KM, Fendorf S (2003) Inhihition of bacterial U(VI) reduction by calcium. Environ Sci Technol 37:1850–1858

Campbell KM, Veeramani H, Ulrich K-U, Blue LY, Giammar DE, Bernier-Latmani R, Stubbs JE, Suvorova E, Yabusaki S, Lezama-Pacheco JS, Mehta A, Long PE, Bargar JR (2011) Oxidative dissolution of biogenic uraninite in groundwater at Old Rifle, CO. Environ Sci Technol 45:8748–8754

Chandler Darrell P, Knickerbocker C, Bryant L, Golova J, Wiles C, Williams Kenneth H, Peacock Aaron D, Long Philip E (2013) Profiling in situ microbial community structure with an amplification microarray. Appl Environ Microbiol 79:799–807

Chandler DP, Kukhtin A, Mokhiber R, Knickerbocker C, Ogles D, Rudy G et al (2010) Monitoring microbial community structure and dynamics during in situ U(VI) bioremediation with a field-portable microarray analysis system. Environ Sci Technol 44:5516–5522. https://doi.org/10.1021/es1006498

Chandwadkar P, Misra HS, Acharya C (2018) Uranium biomineralization induced by a metal tolerant Serratia strain under acid, alkaline and irradiated conditions. Metallomics 10:1078–1088. https://doi.org/10.1039/c8mt00061a

Choudhary S, Islam E, Kazy SK, Sar P (2012) Uranium and other heavy metal resistance and accumulation in bacteria isolated from uranium mine wastes. J Environ Sci Health Part a-Toxic/Hazardous Substances Environ Eng47:622–637. https://doi.org/10.1080/10934529.2012.650584

Coelho E, Reis TA, Cotrim M, Rizzutto M, Corrêa B (2020) Bioremediation of water contaminated with uranium using Penicillium piscarium. Biotechnol Prog 36(5):e30322. https://doi.org/10.1002/BTPR.3032

Cologgi DL, Lampa-Pastirk S, Speers AM, Kelly SD, Reguera G (2011) Extracellular reduction of uranium via Geobacter conductive pili as a protective cellular mechanism. Proc Natl Acad Sci USA 108:15248–15252. https://doi.org/10.1073/pnas.1108616108

Dong H, Huang L, Linduo Zhao Q, Zeng XL, Sheng Y, Shi L, Geng W, Jiang H, Li F, Zhang L, Guo D, Li G, Hou W, Chen H (2022) A critical review of mineral–microbe interaction and co-evolution: mechanisms and applications. Natl Sci Rev 9(10):128. https://doi.org/10.1093/nsr/nwac128

Elias DA, Krumholz LR, Wong D, Long PE, Suflita JM (2003) Characterization of microbial activities and U reduction in a shallow aquifer contaminated by uranium mill tailings. Microb Ecol 46:83–91

Finneran KT, Anderson RT, Nevin KP, Lovley DR (2002) Potential for bioremediation of uranium-contaminated aquifers with microbial U(VI) reduction. Soil Sediment Contam Int J 11:339–357

Francis AJ, Dodge CJ, Lu F, Halada GP, Clayton CR (1994) XPS and XANES studies of uranium reduction by Clostridium sp. Environ Sci Technol 28:636–639

Gallois N, Alpha-bazin B, Ortet P, Barakat M, Piette L, Long J et al (2018) Proteogenomic insights into uranium tolerance of a Chernobyl's Microbacterium bacterial isolate. J Proteome 177:148–157. https://doi.org/10.1016/j.jprot.2017.11.021

Gao WM, Francis AJ (2008) Reduction of uranium(VI) to uranium(IV) by clostridia. Appl Environ Microbiol 74:4580–4584

Gao N, Huang Z, Liu H, Hou J, Liu X (2019) Advances on the toxicity of uranium to different organisms. Chemosphere 237:124548

Gavrilescu M, Pavel LV, Cretescu I (2009) Characterization and remediation of soils contaminated with uranium. J Hazard Mater 163:475–510

Gilson ER, Huang S, Jaffe PR (2015) Biological reduction of uranium coupled with oxidation of ammonium by acidimicrobiaceae bacterium A6 under iron reducing conditions. Biodegradation 26:475–482

Gregory KB, Lovley DR (2005) Remediation and recovery of uranium from contaminated subsurface environments with electrodes. Environ Sci Technol 39:8943–8947

Gudavalli R, Katsenovich Y, Wellman D (2018) Quantification of kinetic rate law parameters for the dissolution of natural autunite in the presence of aqueous bicarbonate ions at high concentrations. J Environ Radioact 19:1–9. https://doi.org/10.1016/j.jenvrad.2018.04.007

Holmes DE, O'Neil RA, Chavan MA, N'Guessan LA, Vrionis HA, Perpetua LA et al (2009) Transcriptome of *Geobacter uraniireducens* growing in uranium- contaminated subsurface sediments. ISME J 3:216–230. https://doi.org/10.1038/ismej.2008.89

Hu P, Brodie EL, Suzuki Y, McAdams HH, Andersen GL (2005) Whole-genome transcriptional analysis of heavy metal stresses in *Caulobacter crescentus*. J Bacteriol 187:8437–8449. https://doi.org/10.1128/JB.187.24.8437

Hu N, Ding D-X, Li S-M, Tan X, Li G-Y, Wang Y-D, Xu F (2016) Bioreduction of U(VI) and stability of immobilized uranium under suboxic conditions. J Environ Radioact 154:60–67

Hu Z, Zhou Z et al (2023) Synergy of surface adsorption and intracellular accumulation for removal of uranium with Stenotrophomonas sp: performance and mechanisms. Environ Res 220:115093. https://doi.org/10.1016/j.envres.2022.115093

Hufton J, Harding J, Thomas S, Romero-Gonzalez ME (2021) The importance of the bacterial cell wall in uranium(VI) biosorption. Phys Chem Chem Phys 23(2):1566–1576. https://doi.org/10.1039/D0CP04067C

Hyun SP, Fox PM, Davis JA, Campbell KM, Hayes KF, Long PE (2009) Surface complexation modeling of U(VI) adsorption by aquifer sediments from a former mill tailings site at Ri?e, Colorado. Environ Sci Technol 43:9368–9373

Istok JD, Senko JM, Krumholz LR, Watson D, Bogle MA, Peacock A, Chang YJ, White DC (2004) In situ bioreduction of technetium and uranium in a nitrate-contaminated aquifer. Environ Sci Technol 38:468–475

Jeong BC, Macaskie LE (1995). PhoN-type acid phosphatases of a heavy metal-accumulating Citrobacter sp.: resistance to heavy metals and affinity towards phosphomonoester substrates. FEMS Microbiol Lett 130:211–214. https://doi.org/10.1111/j.1574-6968.1995.tb07722.x

Jeong BC, Hawes C, Bonthrone KM, Macaskie LE (1997) Localization of enzymically enhanced heavy metal accumulation by Citrobacter sp. and metal accumulation in vitro by liposomes containing entrapped enzyme. Microbiology 143:2497–2507. https://doi.org/10.1099/00221287-143-7-2497

Kasra-Kermanshahi R, Bahrami-Bavani M, Tajer-Mohammad-Ghazvini P (2019) Microbial clean-up of uranium in the presence of molybdenum using pretreated Acidithiobacillus ferrooxidans. J Radioanal Nucl Chem 322(2):1139–1149. https://doi.org/10.1007/s10967-019-06819-9

Khare D, Kumar R, Acharya C (2020) Genomic and functional insights into the adaptation and survival of *Chryseobacterium* sp. strain PMSZPI in uranium enriched environment. Ecotoxicol Environ Saf 191:110217. https://doi.org/10.1016/j.ecoenv.2020.110217

Khijniak TV, Slobodkin AI, Coker V, Renshaw JC, Livens FR, Bonch-Osmolovskaya, EA et al (2005) Reduction of uranium(VI) phosphate during growth of the thermophilic bacterium Thermoterrabacterium ferrireducens. Appl Environ Microbiol 71:6423–6426

Komlos J, Moon HS, Jaffe PR (2008) Effect of sulfate on the simultaneous bioreduction of iron and uranium. J Environ Qual 37:2058–2062

Kulkarni S, Ballal A, Apte SK (2013) Bioprecipitation of uranium from alkaline waste solutions using recombinant Deinococcus radiodurans. J Hazardous Mater 262:853–861. https://doi.org/10.1016/j.jhazmat.2013.09.057

Li R, Ibeanusi V, Hoyle-Gardner J, Crandall C, Jagoe C, Seaman J et al (2019) Bacterial-facilitated uranium transport in the presence of phytate at Savannah River Site. Chemosphere 223:351–357. https://doi.org/10.1016/j.chemosphere.2019.02.064

Li X, Krumholz LR (2009) Thioredoxin is involved in U(VI) and Cr(VI) reduction in Desulfovibrio desulfuricans G20. J. Bacteriol. 191:4924–4933. https://doi.org/10.1128/JB.00197-09

Li X, Zhang H, Ma Y, Liu P, and Krumholz LR (2014). Genes required for alleviation of uranium toxicity in sulfate reducing bacterium Desulfovibio alaskensis G20. Ecotoxicology 23:726–733. https://doi.org/10.1007/s10646-014-1201-2

Liu F, Du KJ, Fang Z, You Y, Wen GB, Lin YW (2015) Chemical and biological insights into uranium-induced apoptosis of rat hepatic cell line. Radiat Environ Biophys 54:207–216

Li XL, Ding CC, Liao JL, Du L, Sun Q, Yang JJ, Yang YY, Zhang D, Tang J, Liu N (2017) Microbial reduction of uranium (VI) by bacillus sp dwc-2: a macroscopic and spectroscopic study. J Environ Sci 53:9–15

Liao R, Shi Z, Chen Y, Wang X (2020) Redox potential and uranium sorption onto sediments: kinetic and thermodynamic characteristics. Chem Ecol 36:475–485

Lloyd JR, Leang C, Myerson ALH, Coppi MV, Cuifo S, Methe B et al (2003) Biochemical and genetic characterization of PpcA, a periplasmic c-type cytochrome in geobacter sulfurreducens. Biochem. J. 369:153–161

Long PE, Williams KH, Davis JA, Fox PM, Wilkins MJ, Yabusaki SB, Fang YL, Waichler SR, Berman ESF, Gupta M, Chandler DP, Murray C, Peacock AD, Giloteaux L, Handley KM, Lovley DR, Banfield JF (2015) Bicarbonate impact on U(VI) bioreduction in a shallow alluvial aquifer. Geochim Cosmochim Acta 150:106–124

Lovley DR, Phillips EJP (1992) Bioremediation of uranium contamination with enzymatic uranium reduction. Environ Sci Technol 26(11):2228–2234

Luo WS, Wu WM, Yan TF, Criddle CS, Jardine PM, Zhou JZ, Gu BH (2007) Influence of bicarbonate, sulfate, and electron donors on biological reduction of uranium and microbial community composition. Appl Microbiol Biotechnol 77:713–721

Macaskie LE, Empson RM, Cheetham AK, Grey CP, Skarnulis AJ (1992) Uranium bioaccumulation by a Citrobacter sp. as a result of enzymically mediated growth of polycrystalline HUO_2PO_4. Science 257:782–785. https://doi.org/10.1126/science.1496397

Macaskie LE, Bonthrone KM, Rouch DA (1994) Phosphatase-mediated heavy metal accumulation by a Citrobacter sp. and related enterobacteria. FEMS Microbiol Lett 121:141–146. https://doi.org/10.1111/j.1574-6968.1994.tb07090.x

Madden AS, Palumbo AV, Ravel B, Vishnivetskaya TA, Phelps TJ, Schadt CW, Brandt CC (2009) Donor-dependent extent of uranium reduction for bioremediation of contaminated sediment microcosms. J Environ Qual 38:53–60

Mahadevan GD, Zhao F (2017) A concise review on microbial remediation cells (MRCs) in soil and groundwater radionuclides remediation. J Radioanal Nucl Chem 314:1477–1485

Marshall MJ, Beliaev AS, Dohnalkova AC, Kennedy DW, Shi L, Wang Z et al (2006) c-Type cytochrome-dependent formation of U(IV) nanoparticles by *Shewanella oneidensis*. PLoS Biol 4:e268. https://doi.org/10.1371/journal.pbio.0040268

Martinez RJ, Beazley MJ, Taillefert M, Arakaki AK, Skolnick J, Sobecky PA (2007) Aerobic uranium (VI) bioprecipitation by metal-resistant bacteria isolated from radionuclide- and metal-contaminated subsurface soils. Environ Microbiol 9:3122–3133. https://doi.org/10.1111/j.1462-2920.2007.01422.x

Moon HS, Komlos J, Jaffe PR (2007) Uranium reoxidation in previously bioreduced sediment by dissolved oxygen and nitrate. Environ Sci Technol 41:4587–4592

Morrison KD, Zavarin M, Kersting AB, Begg JD, Mason HE, Balboni E et al (2021) Influence of Uranium concentration and pH on U-phosphate biomineralization by caulobacter OR37. Environ Sci Technol 55:1626–1636. https://doi.org/10.1021/acs.est.0c05437

Mühr-Ebert EL, Wagner F, Walther C (2019) Speciation of uranium: compilation of a thermodynamic database and its experimental evaluation using different analytical techniques. Appl. Geochem., 100, pp. 213–222. https://doi.org/10.1016/j.apgeochem.2018.10.006

Newsome L, Morris K, Lloyd JR (2014) The biogeochemistry and bioremediation of uranium and other priority radionuclides. Chem Geol 363:164–184. https://doi.org/10.1016/j.chemgeo.2013.10.034

Newsome L, Morris K, Lloyd JR (2015) Uranium biominerals precipitated by an environmental isolate of serratia under anaerobic conditions. PLoS One 10:e0132392. https://doi.org/10.1371/journal.pone.0132392

Nilgiriwala KS, Alahari A, Rao AS, Apte SK (2008) Cloning and overexpression of alkaline phosphatase PhoK from Sphingomonas sp. Strain BSAR-1 for bioprecipitation of uranium from alkaline solutions. Appl Environ Microbiol 74:5516–5523. https://doi.org/10.1128/AEM.00107-08

Orellana R, Leavitt JJ, Comolli LR, Csencsits R, Janot N, Flanagan KA et al (2013) U(VI) reduction by diverse outer surface c-type cytochromes of Geobacter sulfurreducens. Appl Environ Microbiol 79:6369–6374

Pan D, Zhao X, Wang P, Li P, Li Y, Wu W, Wang Z, Fan Q (2022) Insights into sorption speciation of uranium on phlogopite: Evidence from TRLFS and DFT calculation. J Hazard Mater Apr 5;427:128164. https://doi.org/10.1016/j.jhazmat.2021.128164. Epub 2021 Dec 28. PMID: 34991008

Papanicolaou F, Antoniou S, Pashalidis I (2010) Redox chemistry of sulphate and uranium in a phosphogypsum tailings dump. J Environ Radioact 101:601–605

Payne RB, Gentry DM, Rapp-Giles BJ, Casalot L, Wall JD (2002) Uranium reduction by Desulfovibrio desulfuricans strain G20 and a cytochrome c3 mutant. Appl Environ Microbiol 68:3129–3132. https://doi.org/10.1128/AEM.68.6.3129-3132.2002

Payne RB, Casalot L, Rivere T, Terry JH, Larsen L, Giles BJ et al (2004) Interaction between uranium and the cytochrome c3 of Desulfovibrio desulfuricans strain G20. Arch Microbiol 181:398–406

Perdrial N, Vázquez-Ortega A, Reinoso-Maset E, O'Day PA, Chorover J (2022) Effects of flow on uranium speciation in soils impacted by acidic waste fluids. J Environ Radioact 251:106955. https://doi.org/10.1016/j.jenvrad.2022.106955

Pinel-Cabello M, Jroundi F, López-Fernández M, Geffers R, Jarek M, Jauregui R et al (2021) Multisystem combined uranium resistance mechanisms and bioremediation potential of *Stenotrophomonas bentonitica* BII-R7: transcriptomics and microscopic study. J Hazard Mater 403:123858. https://doi.org/10.1016/j.jhazmat.2020.123858

Powers LG, Mills HJ, Palumbo VA, Zhang C, Delaney K, Sobecky PA (2002) Introduction of a plasmid-encoded phoA gene for constitutive overproduction of alkaline phosphatase in three subsurface Pseudomonas isolates. FEMS Microbiol Ecol 41:115–123. https://doi.org/10.1016/S0168-6496(02)00263-5

Rao L, Tian G (2008) The effect of temperature on the speciation of U(VI) in sulfate solutions. In: Merkel BJ, Hasche-Berger A (eds) Uranium, mining and hydrogeology. Springer, Berlin, pp 763–770

Reitz T, Rossberg A, Barkleit A, Selenska-Pobell S, Merroun ML (2014) Decrease of U(VI) immobilization capability of the facultative anaerobic strain *Paenibacillus* sp. JG-TB8 under anoxic conditions due to strongly reduced phosphatase activity. PLoS One 9:e102447. https://doi.org/10.1371/journal.pone.0102447

Renninger N, Knopp R, Nitsche H, Clark DS, Keasling JD (2004) Uranyl precipitation by *Pseudomonas aeruginosa* via controlled polyphosphate metabolism. Appl Environ Microbiol 70:7404–7412. https://doi.org/10.1128/AEM.70.12.7404-7412.2004

Rogiers T, Merroun ML, Williamson A, Leys N, Houdt RV, Boon N et al (2021) *Cupriavidus metallidurans* NA4 actively forms polyhydroxybutyrate-associated uranium-phosphate precipitates. J Hazard Mater 421:126737. https://doi.org/10.1016/j.jhazmat.2021.126737

Rogiers T, Van Houdt R, Williamson A, Leys N, Boon N, Mijnendonckx K (2022) Molecular Mechanisms Underlying Bacterial Uranium Resistance. Front Microbiol. 2022 Mar 10;13:822197. https://doi.org/10.3389/fmicb.2022.822197. PMID: 35359714; PMCID: PMC8963506

Romero-Gonzalez ME, Nwaobi, BC, Hufton J, Gilmour DJ (2016) Ex-situ Bioremediation of U(VI) from Contaminated Mine Water Using Acidithiobacillus ferrooxidans Strains. Front environ sci 4. https://doi.org/10.3389/FENVS.2016.00039

Saha M, Paromita Chakraborty S, Roy. (2024) Bioremediation: a remedy for environmental pollution. IIP, Chikkamagaluru, pp 276–305. https://doi.org/10.58532/v3bdbt7p2ch3

Salome KR, Beazley MJ, Webb SM, Sobecky PA, Taillefert M (2017) Biomineralization of U(VI) phosphate promoted by microbially-mediated phytate hydrolysis in contaminated soils. Geochim Cosmochim Acta 197:27–42. https://doi.org/10.1016/j.gca.2016.10.008

Sanchez-Castro I, Amador-Garcia A, Moreno-Romero C, Lopez-Fernandez M, Phrommavanh V, Nos J et al (2017) Screening of bacterial strains isolated from uranium mill tailings porewaters for bioremediation purposes. J Environ Radioact 166(Pt 1):130–141. https://doi.org/10.1016/j.jenvrad.2016.03.016

Sanford RA, Wu Q, Sung Y, Thomas SH, Amos BK, Prince EK et al (2007) Hexavalent uranium supports growth of Anaeromyxobacter dehalogenans and Geobacter spp. with lower than predicted biomass yields. Environ Microbiol 9:2885–2893

Sani RK, Peyton BM, Dohnalkova A (2006) Toxic effects of uranium on *Desulfovibrio desulfuricans* G20. Environ Toxicol Chem 25:1231–1238

Schiewer S, Volesky B (2000) Biosorption processes for heavy metal removal. In: D.R. Lovley (Editor), Environmental Microbe-Metal Interactions. ASM Press, Washington DC, pp. 329–362

Senko JM, Zhang GX, McDonough JT, Bruns MA, Burgos WD (2009) Metal reduction at low pH by a desulfosporosinus species: implications for the biological treatment of acidic mine drainage. Geomicrobiol J 26:71–82

Shahina SK, Rafiq S (2024) Bioremedial approach to the mitigation of environmental pollution. Bentham Science, Sharjah, pp 183–206. https://doi.org/10.2174/9789815196115124010011

Shelobolina ES, Coppi VM, Korenevsky AA, Didonato LN, Sullivan SA, Konishi H et al (2007) Importance of c-Type cytochromes for U(VI) reduction by *Geobacter sulfurreducens*. BMC Microbiol 7:16. https://doi.org/10.1186/1471-2180-7-16

Shelobolina ES, Vrionis HA, Findlay RH, Lovley DR (2008) *Geobacter uraniireducens* sp. nov., isolated from subsurface sediment undergoing uranium bioremediation. Int J Syst Evol Microbiol 58:1075–1078. https://doi.org/10.1099/ijs.0.65377-0

Sheng L, Szymanowski J, Fein JB (2011) The effects of uranium speciation on the rate of U(VI) reduction by shewanella oneidensis MR-1. Geochim Cosmochim Acta 75:3558–3567

Sheng L, Fein JB (2014) Uranium reduction by Shewanella oneidensis MR-1 as a function of NaHCO3 concentration: surface complexation control of reduction kinetics. Environ Sci Technol Apr 1;48(7):3768–75. https://doi.org/10.1021/es5003692. Epub 2014 Mar 12. PMID: 24576101

Sher S, Li R (2023) Bacillus sp. strain MRS-1: A potential candidate for uranyl biosorption from uranyl polluted sites. Saudi J Biol Sci 30. https://doi.org/10.1016/j.sjbs.2023.103873

Smil V (2000) Phosphorus in the environment: natural flows and human interferences. Annu Rev Energy Environ 25:53–88. https://doi.org/10.1146/annurev.energy.25.1.53

Stewart BD, Neiss J, Fendorf S (2007) Quantifying constraints imposed by calcium and iron on bacterial reduction of uranium(VI). J Environ Qual 36:363–372

Suzuki Y, Banfield JF (2004) Resistance to, and accumulation of, uranium by bacteria from a uranium-contaminated site. Geomicrobiol J 21:113–121. https://doi.org/10.1080/01490450490266361

Suzuki Y, Kitatsuji Y, Ohnuki T, Tsujimura S (2010a) Flavin mononucleotide mediated electron pathway for microbial U(VI) reduction. Phys Chem Chem Phys 12:10081–10087. https://doi.org/10.1039/c0cp00339e

Suzuki Y, Tanaka K, Kozai N, Ohnuki T (2010b) Effects of citrate, NTA, and EDTA on the reduction of U(VI) by shewanella putrefaciens. Geomicrobiol J 27:245–250

Trenfield MA, Ng JC, Noller BN, Markich SJ, van Dam RA (2011) Dissolved organic carbon reduces uranium bioavailability and toxicity. 2. Uranium VI speciation and toxicity to three tropical freshwater organisms. Environ Sci Technol 45:3082–3089

Turner BL (2006) Inositol phosphates: linking agriculture and the environment. In: Turner BL, Richardson AE, Mullaney EJ (eds) Inositol phosphates in soil: amounts, forms and significance of the phosphorylated inositol stereoisomers. CABI, Wallingford, pp 186–206. https://doi.org/10.1079/9781845931520.0186

Ulrich K-U, Veeramani H, Bernier-Latmani R, Giammar DE (2011) Speciation-dependent kinetics of Uranium(VI) bioreduction. Geomicrobiol J 28:396–409

Vodyanitskii YN (2011) Chemical aspects of uranium behavior in soils: a review. Eurasian Soil Sci 44:862–873

Wang Q, Zhu C, Huang X, Yang G (2019) Abiotic reduction of uranium(VI) with humic acid at mineral surfaces: competing mechanisms, ligand and substituent effects, and electronic structure and vibrational properties. Environ Pollut 254:113110

Wielinga B, Bostick B, Hansel CM, Rosenzweig RF, Fendorf S (2000) Inhibition of bacterially promoted uranium reduction: ferric (hydr)oxides as competitive electron acceptors. Environ Sci Technol 34:2190–2195

Williams KH, Long PE, Davis JA, Wilkins MJ, N'Guessan AL, Steefel CI, Yang L, Newcomer D, Spane FA, Kerkhof LJ, McGuinness L, Dayvault R, Lovley DR (2011) Acetate availability and its influence on sustainable bioremediation ofuranium-contaminated groundwater. Geomicrobiol J 28:519–539

Williamson AJ, Morris K, Law GTW, Rizoulis A, Charnock JM, Lloyd JR (2014) Microbial reduction of U(VI) under alkaline conditions: implications for radioactive waste geodisposal. Environ Sci Technol 48:13549–13556

Wyss M, Brugger R, Kronenberger A, Re R, Fimbel R, Oesterhelt G et al (1999) Biochemical characterization of fungal phytases (myo-inositol hexakisphosphate phosphohydrolases): catalytic properties. Appl Environ Microbiol 65:367–373. https://doi.org/10.1128/AEM.65.2.367-373

Xie JC, Wang JL, Lin JF, Zhou XH (2018) The dynamic role of pH in microbial reduction of uranium(VI) in the presence of bicarbonate. Environ Pollut 242:659–666

Yamasaki S, Tanaka K, Kozai N, Ohnuki T (2017) Effect of flavin compounds on uranium(VI) reduction-kinetic study using electrochemical methods with UV-vis spectroscopy. Appl Geochem 78:279–286. https://doi.org/10.1016/j.apgeochem.2017.01.014

Yang JB, Volesky B (1999) Modeling uranium-proton ion exchange in biosorption. Environ Sci Technol 33:4079–4085

Yi ZJ, Tan KX, Tan AL, Yu ZX, Wang SQ (2007) Influence of environmental factors on reductive bioprecipitation of uranium by sulfate reducing bacteria. Int Biodeterior Biodegrad 60:258–266

You W, Peng W, Tian Z, Zheng M (2021) Uranium bioremediation with U(VI)-reducing bacteria. Sci Total Environ 798:149107. https://doi.org/10.1016/j.scitotenv.2021.149107

Yung MC, Jiao Y (2014) Biomineralization of uranium by PhoY phosphatase activity aids cell survival in *Caulobacter crescentus*. Appl Environ Microbiol 80:4795–4804. https://doi.org/10.1128/AEM.01050-14

Yun-Juan C, Long PE, Geyer R, Peacock AD, Resch CT, Sublette K et al (2005) Microbial incorporation of 13C-labeled acetate at the field scale: detection of microbes responsible for reduction of U(VI). Environ Sci Technol 39:9039–9048. https://doi.org/10.1021/es051218u

Zhang F, Wu WM, Parker JC, Mehlhorn T, Kelly SD, Kemner KM et al (2010). Kinetic analysis and modeling of oleate and ethanol stimulated uranium (VI) bio-reduction in contaminated sediments under sulfate reduction conditions. J Hazard Mater 183:482–489

Zhang J, Chen H, Song J, Deng H, Chen Z, Lin Z (2019) Immobilization of Uranium at Nanoscale by Bacillus cereus 12-2 at Different U(VI) Concentration. J Nanosci Nanotechnol. Nov 1;19(11):7131–7138. https://doi.org/10.1166/jnn.2019.16661. PMID: 31039867

Zhang L, Chen Y, Xia Q, Kemner KM, Shen Y, O'Loughlin EJ, Pan Z, Wang Q, Huang Y, Dong H, Boyanov MI (2021a) Combined effects of Fe(III)-bearing clay minerals and organic ligands on U(VI) bioreduction and U(IV) speciation. Environ Sci Technol 55:5929–5938

Zhang K, Zhang D, Wu X, Xue Y (2021b) Continuous and efficient immobilization of heavy metals by phosphate-mineralized bacterial consortium. J Hazard Mater 416:125800. https://doi.org/10.1016/j.jhazmat.2021.125800

Zhuang K, Ma E, Lovley DR, Mahadevan R (2012) The design of long-term effective uranium bioremediation strategy using a community metabolic model. Biotechnol Bioeng 109:2475–2483

Unlocking Geomicrobiological Perspectives in Bioremediation of Hydrocarbons

Dina Barman and Kaushik Bhattacharjee

Contents

D. Barman
Microbial Ecology Laboratory, Department of Botany, Gauhati University, Guwahati, India

K. Bhattacharjee (✉)
CSIR-Central Drug Research Institute, Lucknow, Uttar Pradesh, India
e-mail: kaushik.b@cdri.res.in

S. R. Joshi et al. (eds.), *Mineral Transformation and Bioremediation by Geo-Microbes*, Interdisciplinary Biotechnological Advances,
https://doi.org/10.1007/978-981-96-3033-2_19

Abstract The increasing industries coupled with sustained population growth and the high demand for hydrocarbons have driven unparalleled economic progress and advancement. Nevertheless, these hydrocarbons have undeniably caused significant environmental challenges in recent decades. It pollutes soil, agricultural lands, and water bodies, which ultimately pose a serious threat to aquatic and marine ecosystems and human health. To mitigate these hazards, it is imperative to address these pollutants before their effects exacerbate. Bioremediation emerges as a sustainable, cost-effective, and efficient solution, employing bacteria, fungi, and algae to degrade complex organic pollutants into simpler compounds. Geomicrobiological approaches offer promising solutions for bioremediation, leveraging the metabolic capabilities of microorganisms to degrade hydrocarbons. This chapter explores the geomicrobiological aspects of bioremediation, elucidating the intricate interactions between microorganisms, hydrocarbons, and the environment. By highlighting key factors such as microbial diversity, metabolic pathways, and environmental conditions, it provides valuable insights for managing hydrocarbon contamination in an eco-friendly manner. This knowledge fosters the development of effective and sustainable bioremediation strategies, grounded in a comprehensive understanding of geomicrobiological perspectives.

Keywords Bioremediation · Hydrocarbon · Geomicrobiology · Strategy

Abbreviations

PAHs	Polycyclic aromatic hydrocarbons
PCBs	Polychlorobiphenyls
PCDDs	Polychlorinated dibenzo-p-dioxins
PCDFs	Polychlorinated dibenzofurans
TPHs	Total petroleum hydrocarbons

1 Introduction

Rapid industrialization and growing dependence on petroleum-based products have triggered substantial environmental issues, with hydrocarbon pollution being a primary concern. Total petroleum hydrocarbons (TPHs), which refer to chemical compounds derived from crude oil, are used to manufacture a variety of goods across industries. By 2030, the global consumption of TPHs is projected to reach 106.6 million barrels, up from 85 million barrels in 2016. Petroleum components are often introduced into the environment through spills, leaks, exploration activities, and improper waste disposal (Zoghi and Mafigholami 2023). Each year, natural crude oil seepage is estimated at approximately 600,000 metric tons, subject to an error margin of 200,000 metric tons. Whether through accidental or human-induced release, hydrocarbons contribute significantly to water and soil contamination (Das and Chandran 2011). Over time, these pollutants surpass natural levels, spreading through ecosystems and resulting in serious environmental, ecological, and social consequences globally. This underscores the urgent need for effective remediation measures (Zoghi and Mafigholami 2023).

Traditional soil remediation methods, which primarily rely on mechanical and chemical processes, often result in incomplete contaminant breakdown. Additionally, these methods are unable to degrade long- chain-containing hydrocarbons (Ravi et al. 2022). In contrast, bioremediation has gained recognition as an effective and environmentally friendly approach for hydrocarbon degradation, utilizing the metabolic capabilities of microorganisms (Ravi et al. 2022). Various microorganisms, including bacteria, fungi, and yeast, can break down hydrocarbons into simpler molecules, though the process is complex, involving multiple steps and nonlinear pathways (Radhakrishnan et al. 2023; Mohammed et al. 2023). This cost-efficient method can be widely applied to contaminated areas, leading to the complete mineralization of hydrocarbons into carbon dioxide and water (Mohammed et al. 2023).

Geomicrobiology, the intersection of microbiology and geology, offers a profound understanding of microbial processes in geological environments and their influence on biogeochemical cycles. It is within this interdisciplinary field that new insights into microbial communities and their roles in hydrocarbon degradation are being unlocked, paving the way for more efficient and sustainable bioremediation approaches. By integrating insights from geomicrobiology, this chapter contributes to innovative and environmentally sustainable bioremediation strategies. These strategies are not only effective in mitigating hydrocarbon pollution but also in preserving ecosystem health, making them vital tools in the management of contaminated sites. This comprehensive understanding of geomicrobiological perspectives lays the groundwork for future advancements in bioremediation technologies, fostering more efficient and eco-friendly solutions to address the global challenge of hydrocarbon contamination.

2 Understanding Hydrocarbon

Hydrocarbons consist entirely of hydrogen and carbon atoms in organic compound form. They are the primary constituents of fuels like crude oil, natural gas, coal, and petroleum products. Hydrocarbons can be classified into several types, including aliphatic and aromatic hydrocarbons, polycyclic aromatic hydrocarbons (PAHs), polychlorinated dibenzo-p-dioxins (PCDDs), and polychlorobiphenyls (PCBs) (Tursi et al. 2018).

The PAHs, consisting of two or more linked aromatic rings are one of the major environmental pollutants. The pollution caused by PAHs is broadly classified into two main categories, i.e., anthropogenic and natural emissions (Patel et al. 2020). Natural sources, including volcanic explosions and wildfires, are considered insignificant or minor contributors. In contrast, anthropogenic emissions are the primary drivers of pollution by PAH and are further divided into four types, i.e., emission from industries through the production of various industrial items like coal-tar, aluminum, iron, steel, rubber, power, combustion of waste, gas-fueled engines of industrial machinery, etc.; domestic release through burning of garbage and wood, cooking, smoking, etc.; vehicular emissions; and agricultural activities via uncontrolled biomass combustion, burning of agricultural left-over, use of pesticides (Patel et al. 2020). The PAHs are also categorized into petrogenic, pyrogenic, and biogenic based on their origin. The petrogenic ones are formed from fuel sources, pyrogenic ones are produced via incomplete combustion, and biogenic compounds are synthesized through the metabolism of material by plants, phytoplankton, and different microorganisms (Patel et al. 2020).

PCBs are another organic pollutant that represents a considerable threat to ecosystem and human health. The PCBs are aromatic hydrocarbons derived from biphenyl by substituting 1–10 hydrogen atoms with chlorine atoms. Due to their chemical properties, they are widely used in dielectric fluids (e.g., transformers, capacitors), as pesticide additives, insulators, etc. (Montano et al. 2022). PCBs are generally exposed to environment via thermal processes, primarily ignition of waste, smelting of steel, combustion of coal and wood of industrial and/or domestic origin, and inadequate dumping of PCB waste (Montano et al. 2022). They are highly carcinogenic, toxic, and prone to sluggish decomposition; hence they are categorized into the most hazardous pollutants (Meagher 2000).

Polychlorinated dibenzofurans (PCDFs), dioxin-like polychlorinated biphenyls (dl-PCBs), and polychlorinated dibenzo-p-dioxins (PCDDs) are the three main classes of persistent organic pollutants (POPs) owing to their carcinogenic, immunotoxic, and neurotoxic properties as well as their capacity of bioaccumulation and persistence for longer period in environment (Schuster et al. 2015). Dioxins are highly lipophilic and tend to accumulate in aquatic food webs, especially within epibenthic food chains. Due to their bioaccumulation and biomagnification properties, dioxins are detected as contaminants not only near their source but also in distant locations (Taylor et al. 2024). They are generally exposed by accidental releases as by-products from various sources, including chemical production, incineration, and other combustion processes (Schuster et al. 2015).

3 Hydrocarbons and Their Environmental Impact

As a finite and critical resource, hydrocarbons are vital in driving global economic growth, social advancement, and human well-being. With the unprecedented rise in global population and rapid industrialization, the demand for hydrocarbons has also surged dramatically (Daâssi and Almaghribi 2022). However, the excessive consumption of hydrocarbons significantly contributes to the contamination of water and soil ecosystems, with harmful effects on human health. Moreover, there is a persistent risk of environmental pollution from leaks during the storage, transportation, refining, processing, and disposal of oil. Due to their stability and long-lasting presence in the environment, hydrocarbons are recognized as major environmental pollutants. Their nature, chemical composition, and ability to be transported through the environment renders them very toxic to living organisms. The hydrocarbon, mainly petroleum products (PH) are primarily introduced into ecosystems through oil spills, industrial discharges, and improper waste disposal. It is estimated that approximately 8.8 million metric tons of oil are released into aquatic environments globally each year, with around 90% of this pollution directly attributable to human activities, including oil spills on land and in marine environments (Ambaye et al. 2022). Similarly, leaking of natural gas (NG) results in environmental risks of varying degrees of severity. In some of the rare cases, these leaks result in the accretion of NG and can create an explosion hazard, endangering both property and the lives of humans. Such incidents often draw national attention due to their sudden occurrence and potentially devastating impact on communities and infrastructure (Aradhi et al. 2023).

3.1 Soil and Agricultural Land Pollution

The hydrocarbons affect the soil's physio-chemical and enzymatic properties, including soil pH, bulk density, moisture content, electrical conductivity, soil total organic carbon, total nitrogen, and availability of phosphorus, potassium, and enzyme activities like phosphatase, urease, and dehydrogenase, and augment the heavy metal concentration (Ambaye et al. 2022; Barman et al. 2022). The persistence and toxicity of hydrocarbons also degrade the biological properties of soil and reduce microbial diversity and microbial biomass carbon (Daâssi and Almaghribi 2022). This in turn degrades fertility and health of soil and affects the overall crop productivity. Higher concentrations of PH are also toxic to crops and vegetation, impairing soil aeration and water infiltration, which hinders plant growth and development. Petroleum pollution also causes oxidative stress, leaf deformities, and cell death in plants. The pollution spreads through the food chain, ultimately posing serious risks to human health (Ambaye et al. 2022) (Fig. 1).

3.2 Water Bodies and Aquatic Ecosystems

The introduction of hydrocarbons into water bodies has even more severe consequences. Despite their low solubility in water, these compounds exhibit high toxicity, mutagenicity, and carcinogenicity by entering the food chain. It directly impacts the oxygenation process, destroys aquatic habitats, endangers marine species, and poses serious risks to human health (Fig. 1). The water bodies' contamination due to improper disposal of oil and oil spillage is relatively common practice. Due to their low solubility in water, they persist as a separate phase, leading to their

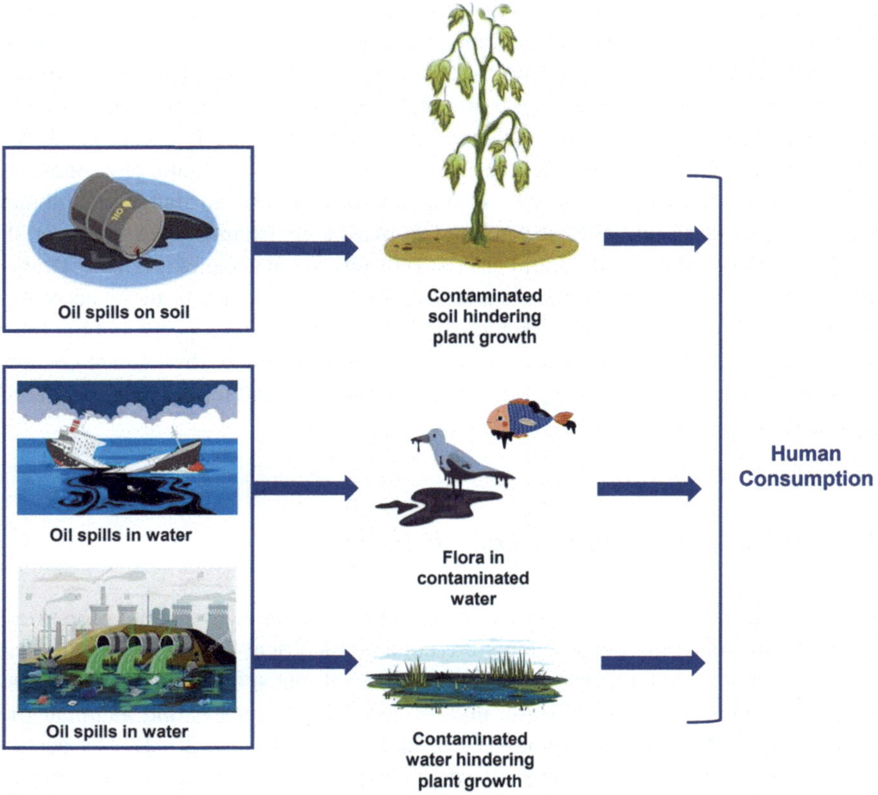

Fig. 1 Environmental impact of petroleum hydrocarbons on soil, water, and human health

presence as free-floating, dissolved in soil particles, emulsified, and adsorbed to soil particles, thus creating secondary contamination sources (Tursi et al. 2018).

In recent decades, countless oil spill incidents have taken place across the globe, releasing vast quantities of crude oil. One of the most notorious was the Deepwater Horizon spill that unleashed around 4.9 million barrels of oil into the Gulf of Mexico in 2010. This spill spread across a large area, severely impacting the ecosystem of the Gulf. After analysis, it was observed that those crude oils contained PAHs, which are toxic and cause various harmful effects such as immunotoxic, cardiotoxic, and embryonic defects to different benthic organisms, fish, and marine vertebrates dwelling in the ecosystem (Honda and Suzuki 2020). The primary concern regarding PAHs is their carcinogenic potential. Due to their hydrophobic nature, PAHs are readily absorbed into cells, leading to the activation of cytochrome P450 (CYP) enzyme expression. These enzymes metabolize PAHs into various byproducts, some of which attach to DNA potentially causing carcinogenic or mutagenic effects (Honda and Suzuki 2020). Another incident is related to Niger Delta, Nigeria. The region is rich in natural resources, especially gas and oil, which contribute over 90% of export earnings of Nigeria and around 80% of revenue generated. Since oil was first discovered there in 1956, the region has driven Nigeria's rise as Africa's largest oil producer and 12th globally. However, oil production has brought significant problems, related to pollution from frequent oil spills, industrial waste, and gas flaring. These issues have compounded over decades, severely affecting both the ecosystem and local communities. One of the most notable spills occurred on February 23, 2019, covering approximately 10.5 km^2 (Ewim et al. 2023).

3.3 Human Health Concerns

The hydrocarbons especially PAHs are recognized as mutagenic, carcinogenic, and teratogenic agents. They enter the human body through dermal contact, inhalation, and ingestion; however, their effects are based on their concentration, toxicity, duration, and route (Mallah et al. 2022). Short-term exposure to PAHs leads to acute symptoms, while prolonged exposure results in chronic health disorders. Inhalation of PAHs, especially in occupational settings, triggers respiratory issues, while certain PAHs, including anthracene and naphthalene, potentially cause allergic reactions. Exposure to high concentrations of PAHs leads to chronic health issues, like, respiratory problems (asthma, bronchitis), metabolic problems (diabetes, obesity), and mental issues (depression, anxiety). The exposure of PAH also has been recognized as contributing to the weakening of immune function, potentially increasing individuals' vulnerability to infections and other health-related issues (Venkatraman et al. 2024). The PAHs are generally lipophilic leading to their accumulation in adipose-rich organs, which increases the risk of tumors in areas like the pancreas, lungs, skin, bladder, and others. PAH exposure is linked to cardiovascular diseases (CVD) such as atherosclerosis and myocardial infarction (Mallah et al. 2022).

4 Geomicrobiology and Hydrocarbon

Geomicrobiology is an interdisciplinary field that bridges microbiology and geology to investigate the influence of microorganisms on geological and geochemical processes. A key area of its application lies in the exploration and biodegradation of hydrocarbons. Studying the interactions between microbes and hydrocarbons offers valuable insights with far-reaching implications for energy production and environmental remediation efforts. The field of geomicrobiology plays a pivotal role in advancing our understanding of Earth's subsurface ecosystems, including the biogeochemical cycling of elements, mineral formation and degradation, and microbial transformations of both organic and inorganic substrates. Among these processes, microbial transformations are particularly significant, as they not only influence geological landscapes but also provide practical solutions for mitigating environmental pollution (Gadd 2010).

4.1 Role of Geomicrobiology in Hydrocarbon Exploration

Geomicrobiology plays a pivotal role in hydrocarbon exploration by harnessing specific microorganisms that thrive in hydrocarbon-enriched environments. Through microbial prospecting for oil and gas (MPOG), geomicrobiologists identify abnormal concentrations of hydrocarbon-oxidizing bacteria, providing indirect yet reliable signals of the presence of subsurface oil and gas deposits. This method complements a combination of geochemical, geological, and geophysical techniques to assess the hydrocarbon potential of an area, thus improving success rates in the exploration of petroleum. By leveraging the natural interaction between microbes and hydrocarbons, this technique aids in pinpointing potential oil and gas reserves. Typically, gaseous hydrocarbons migrate from subsurface petroleum deposits. Hydrocarbon-oxidizing bacteria utilize these gases as their primary carbon and energy sources, predominantly colonizing surface soils above hydrocarbon-bearing formations. The isolation and quantification of these bacteria serve as indirect indicators in oil prospecting (Rasheed et al. 2012, 2015) (Fig. 2).

4.2 Role of Geomicrobiology in the Recovery of Hydrocarbon

Geomicrobiology plays a decisive role in the recovery of hydrocarbons, particularly in microbial-enhanced oil recovery (MEOR). The method is based on the natural metabolic processes of microorganisms to improve the extraction of oil from reservoirs. In this process, microorganisms particularly living bacteria along with indispensable nutrients are introduced into the injection well. Under suitable

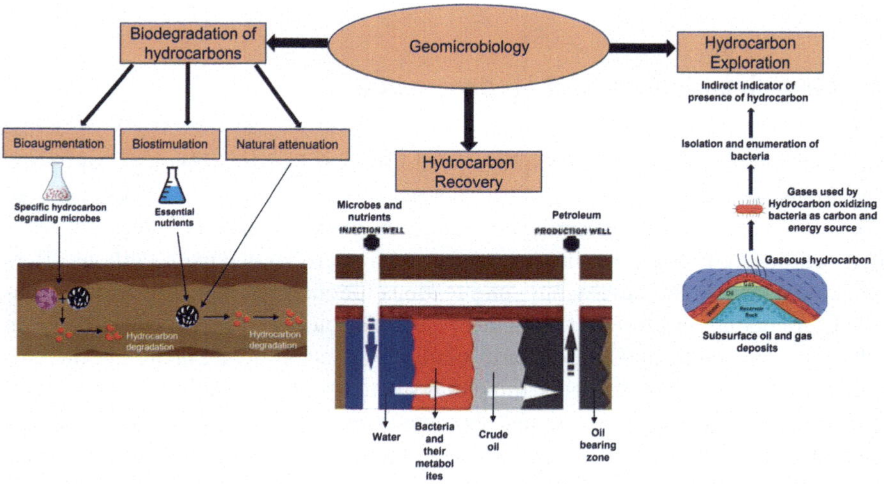

Fig. 2 The role of geomicrobiology in hydrocarbon exploration, recovery, and degradation

environmental conditions within the reservoir, the introduced microorganisms proliferate rapidly, and their metabolic by-products, like biogas (methane, hydrogen, carbon dioxide, nitrogen), bioacids (butyric acid, propionic acid), biosurfactants (rhamnolipid, trehalose lipids, lichenysin, viscosin), solvents (acetone, ethanol, butanol, propane-2-diol), and biopolymer, facilitate the mobilization of the residual oil (Fig. 2). However, the growth and impact of the microbes in MEOR are influenced by several factors, including type of microbial strain used, nutrient availability to the microbial strain, as well as various reservoir characteristics such as pH, temperature, permeability, pressure, salinity, porosity, etc. Through the MEOR process, nearly 50% of the residual oil can potentially be recovered (Yernazarova et al. 2024; Nikolova and Gutierrez 2020).

Among the different microbes used in MEOR, bacteria like *Clostridium* and *Enterobacter* and archaea *Methanobacterium* produce gases such as CH_4, H_2, H_2S, and N_2 as metabolic by-products. The generation of these gases can increase reservoir pressure, thus pushing oil toward production wells (Van Hamme et al. 2003). This process, called "gas-lift" in MEOR, enhances oil recovery by increasing the mobility of oil trapped in rock pores. In addition to gas production, some microbes like *Clostridium*, *Enterobacter*, *Bacillus*, *Zymomonas*, and *Klebsiella* produce organic acids including propionic acids, acetic acid, and solvents like 1-butanol, acetone, ethanol, and propan-2-diol as metabolic by-products. The organic acid can dissolve carbonate minerals in reservoir rocks and eventually increase the permeability and widen the pores. The solvents produced by microbes also dissolute rocks to increase the permeability and reduce the viscosity thus enhancing oil flow (Yernazarova et al. 2024). Some of the strains of *Bacillus*, *Pseudoxanthomonas taiwanensis*, *Pseudomonas*, *Brevibacillus*, and *Streptomyces are* involved in MEOR by producing biosurfactants that are biodegradable, nontoxic, and environmentally friendly. The released biosurfactants reduce the interfacial tension between oil and water. This reduction in tension allows for more efficient displacement of oil from the reservoir, thus enhancing recovery of oil (Purwasena et al. 2024; Nikolova and Gutierrez 2020). Additionally, biopolymers (pullulan, xanthan gum, curdlan, levan, and dextran scleroglucan) produced by *Bacillus* sp., *Pseudomonas* sp., *Rhizobium viscosum*, *Xanthomonas* sp., *Acinetobacter venetianus*, and *Sphingomonas* sp. contribute significantly to enhancing oil recovery. These biopolymers accomplish this by promoting selective plugging, enhancing the viscidity of the removing fluid, and lowering the mobility ratio of water-to-oil, thus improving the effectiveness of oil extraction (Yernazarova et al. 2024).

4.3 Role of Geomicrobiology in Hydrocarbon Biodegradation and Bioremediation

Geomicrobiology plays an important role in the biodegradation and bioremediation of hydrocarbon-contaminated environments, harnessing the metabolic capabilities of microorganisms to break down harmful hydrocarbon compounds. They do it by biostimulation, bioaugmentation, and natural attenuation. In natural attenuation, naturally occurring microbial communities can break down hydrocarbon pollutants without human intervention. This is a cost-effective, eco-friendly method for dealing with contamination. Biostimulation involves enhancing the activity of microbes through the addition of nutrients (nitrogen and phosphorus), oxygen, or electron acceptors to speed up the degradation of hydrocarbons. This method stimulates indigenous microbial populations to break down contaminants more effectively. In some cases where native microbes are insufficient for effective remediation, bioaugmentation introduces specialized, hydrocarbon-degrading microbes into contaminated sites to enhance degradation (Fig. 2).

5 Degradation of Hydrocarbon: Geomicrobiological Perspective

The levels of hydrocarbon continue to rise due to urbanization and industrialization, and addressing this escalating issue becomes increasingly critical. Microorganisms have long been recognized for their remarkable ability to transform both organic and inorganic substrates. They have been integral to Earth's geological and geochemical processes for billions of years. One of the most critical aspects of geomicrobiology in hydrocarbon systems is the degradation of hydrocarbons. This is particularly relevant in bioremediation, where microorganisms are employed to clean up environments contaminated by hydrocarbon pollutants such as petroleum.

Bioremediation is a reliable, promising, cost-effective, and environmentally sustainable method to remediate organic and inorganic contaminants (Gadd 2010). In the context of hydrocarbon degradation, geomicrobiology is concerned with how microorganisms transform these organic pollutants into less harmful substances through biochemical pathways. This transformation is key to bioremediation efforts, which rely on natural processes to cleanse environments contaminated with hydrocarbons. Effective bioremediation hinges on the optimization of environmental conditions to promote microbial activity and degradation rates. Two major strategies employed in bioremediation are biostimulation and bioaugmentation.

5.1 Biostimulation

Biostimulation is a bioremediation strategy that enhances microbial activity by adding essential nutrients to contaminated environments. This process encourages the growth of indigenous microorganisms, enabling them to effectively degrade hydrocarbons by addressing the depletion of essential nutrients like nitrogen and phosphorus, which often restrict microbial activity in contaminated environments. By introducing growth-limiting nutrients and adjusting the environmental factors (temperature, pH, moisture, redox potential), the microbial degradation process can be stimulated. Oxygen, which is crucial in oxygen-limited environments, can be also supplied via methods like bioventing or biosparging. These processes facilitate the metabolism of hydrocarbons, thus speeding up bioremediation (Kebede et al. 2021) (Fig. 2, Table 1).

5.2 *Bioaugmentation*

When the native microbial community is incapable of degrading certain hydrocarbons, bioaugmentation is employed. This approach introduces specific hydrocarbon-degrading microbial strains into the contaminated environment to enhance bioremediation. For successful biodegradation, the introduced microbes need to be capable of metabolizing a broad range of contaminants, resisting mutations, tolerating diverse biotic and abiotic conditions, and competing effectively with indigenous microorganisms and predators. Bioaugmentation is particularly effective in environments with increased levels of recalcitrant hydrocarbons, like PAHs or chlorinated compounds. Various bioaugmentation techniques, including phytoaugmentation, rhizosphere bioaugmentation, and gene bioaugmentation, are commonly applied to clean up contaminated sites (Kebede et al. 2021). To date, researchers have utilized various microbial strains isolated from polluted environments such as seawater and soil, including *Pseudomonas*, *Bacillus*, *Corynebacterium*, *Rhodococcus*, *Acinetobacter*, *Gordonia*, *Halomonas*, *Alcanivorax*, *Microbacterium*, *Marinobacter*, *Cobetia*, *Cycloclasticus*, *Citrobacter*, *Enterobacter*, *Ralstonia*, *Lactobacillus*, *Acidocella*, *Chlorella*, and their consortia, to degrade petroleum hydrocarbons (Sayed et al. 2021) (Table 1).

5.3 *Combined Effect of Biostimulation and Bioaugmentation*

Some of the researchers concluded that for the degradation of petroleum hydrocarbon, biostimulation with the addition of phosphorus and nitrogen is more efficient than bioaugmentation (Wu et al. 2019). However, combining bioaugmentation with biostimulation can significantly enhance the microbial activity. This involves introducing aeration for aerobic microorganisms, supplementing nutrients when they are insufficient, and adding surfactants to tackle hydrophobic pollutants. Such strategies are often employed in the remediation of hydrocarbon-contaminated soils, particularly in temperate climates (Al-Kaabi et al. 2022). Chaudhary et al. (2021) conducted a study isolating diesel-degrading bacteria from diesel-contaminated soil samples to remediate the pollution. The study also concluded that biostimulation alone is insufficient for effective remediation in soils lacking indigenous oil-degrading microbes. This highlights the importance of a comprehensive strategy incorporating both consortium bioaugmentation and biostimulation for optimal soil restoration (Table 1).

Table 1 Role of microbes in the degradation of hydrocarbons using biostimulation and bioaugmentation

Pollutant	Applied treatments	Stimulator	Microorganisms	Degraded (%)	Time	References
Weathered polluted soils	BS-BA	Carbon/nitrogen/phosphorus	*Pseudomonas aeruginosa* D7S1		16 weeks	Al-Kaabi et al. (2022)
Petroleum hydrocarbon	BS	NH_4NO_3 and KH_2PO_4	–	28.3%	12 weeks	Wu et al. (2019)
Petroleum hydrocarbon	BA	–	Petroleum-degrading microflora	13.9%	12 weeks	Wu et al. (2019)
Petroleum refinery sludge	BS	$NaNO_3$ and KH_2PO_4	–	>80%	90 days	Sarkar et al. (2016)
Petroleum oil	BA		*Pseudomonas*, *Rhodococcus*, and *Acinetobacter*	66%	15 days	Perdigão et al. (2020)

100 g/kg of crude oil	BS-BA	8 g/g of nitrogen-phosphorus ratio	600 cell/g bacteria isolated from crude oil-polluted soil	93.75%	60 days	Popoola et al. (2022)
Petroleum hydrocarbons	BS-BA	Lipophilic fertilizers (uric acid and lecithin) and biosurfactants (rhamnolipids)	Hydrocarbon-degrading consortia		90 days	Nikolopoulou et al. (2013)
PAH	BA		*Flavobacterium johnsoniae* and *Shewanella baltica*	93.32%		El-Sheshtawy et al. (2022)
PAH	BA		*Bacillus sorensis* isolated from oily polluted soil	32.4%		Oualha et al. (2019)
Diesel	BA		*Bacillus sorensis* isolated from oily polluted soil	39.2%		Oualha et al. (2019)
Crude oil	BA		*Bacillus subtilis* isolated from petroleum contaminated sediments and industrial wastewater	>94.0%		Oyetibo et al. (2017)
Phenanthrene	BA		*Bacillus subtilis* isolated from petroleum-contaminated sediments and industrial wastewater	>85.0%		Oyetibo et al. (2017)
Diesel oil	BS	2% (m/m) diesel oil, urea, K_2HPO_4		41.9%	90 days	Giovanella et al. (2021)
Total petroleum hydrocarbons	BA		Addition of native dominant degrading bacteria, i.e., *Bacillus albus*, *Bacillus havnesii*, *Pseudomonas balearica*	99.87% in surface soil, 96.04% in middle soil, and 71.63% in deep soils	300 days	Nong et al. (2023)
Total petroleum hydrocarbons	BS	NH_4NO_3 and KH_2PO_4		99.75% in surface soil, 99.73% in middle soil, and 99.84% in deep soils		Nong et al. (2023)
Petroleum hydrocarbon	BS	Kenaf-core and rhamnolipid		82.57%	90 days	Omenna et al. (2024)

BS biostimulation, *BA* bioaugmentation

6 Microbes in Remediation of Hydrocarbon-Contaminated Sites

Till now, researchers have isolated different microorganisms from various niches and explored their capacity in singly or in consortia for the bioremediation of hydrocarbon-contaminated sites.

It has been observed that numerous bacteria from various genera, including *Bacillus*, *Pseudomonas*, *Klebsiella*, *Vibrio*, *Corynebacterium*, *Nocardia*, *Brevibacterium*, *Acinetobacter*, *Flavobacterium*, and *Shewanella*, play crucial roles in the degradation of PHCs (Mekonnen et al. 2024). The symbiosis of *Bacillus cereus* D12, *Bacillus soreness* D11, and *Pseudomonas stutzeri* D13 was used to remove petroleum hydrocarbons present in the weathered soil (Alsayegh et al. 2021). Ravi et al. (2022) successfully isolated *Bacillus licheniformis* ARMP2 and *Pseudomonas aeruginosa* ARMP8 from petroleum hydrocarbon-contaminated soil, and both strains demonstrated strong potential for degrading petroleum compounds. In a similar vein, *Shewanella baltica* and *Flavobacterium johnsoniae* isolated from oil-contaminated soil at the Suez Oil Processing Company in Egypt demonstrated the ability to biodegrade a range of polycyclic aromatic hydrocarbons (El-Sheshtawy et al. 2022).

Similarly, different fungi like *Aspergillus*, *Penicillium*, *Trichoderma*, *Paecilomyces*, *Fusarium*, *Polyporus*, *Phomopsis*, *Ganoderma*, *Agaricus*, and *Pleurotus* were potent in the degradation of PHCs. Dhailappan et al. (2022) used *Pleurotus ostreatus* to degrade PHC-contaminated soil; Novianty et al. (2021) used consortia of *Aspergillus* sp. to degrade n-alkane contaminated site (Mekonnen et al. 2024).

It was also observed that different extreme ecological environments like saline regions are susceptible to pollution because of their proximity to oil industries. These harsh conditions necessitate the use of extremophiles like Archaea for effective bioremediation, as conventional microorganisms may not thrive in such environments. Researchers have used different archaea like *Natrialba*, *Haloferax*, *Halobacterium*, and *Haloarchaea* in the degradation of PAH (Patel et al. 2020).

Algae also play a vital role in the bioremediation of PAHs in aquatic environments, as they are the primary producers in the estuarine and coastal ecosystems. Microalgae including *Agardhiella subulate*, *Raphidocelis subcapitata*, *Scenedesmus acutus*, *Ankistrodesmus braunii*, *Selenastrum capricornutum*, *Scenedesmus quadricauda*, *Chlorella vulgaris*, *Prototheca zopfii*, *Anabaena fertilissima*, *Anabaena variabilis*, *Chlamydomonas*, *Synechocystis*, *Agmenellum quadruplicatum*, and *Phormidium corium* were reported to degrade PAHs (Satpati et al. 2023).

Bioremediation utilizing microbial consortium strategies, such as co-cultures of bacteria-fungi, and the synergistic ability of bacteria-algae and fungi-algae have been recognized to be among the utmost bioremediation methods in both laboratory and large-scale applications. Sharma et al. (2016) utilized a bacterial-fungal co-culture approach using *Streptomyces rochei* PAH-13, *Serratia marcescens* L-11, and *Phanerochaete chrysosporium* VV-18 to degrade PAHs. Similarly,

Subashchandrabose et al. (2019) applied a bacterial-algal synergistic approach employing *Chlorella* sp. MM3 and *Rhodococcus wratislaviensis* 9 for the bioremediation of PAH-contaminated soil.

Microbial enzyme-based bioremediation is another effective method utilizing enzymes extracted from microbes for the removal of PAHs. Microbial enzymes like lignin peroxidase, oxygenase, dehydrogenase, manganese peroxidase, laccases, and phenoloxidases play a vital role in the oxidation of PAH. These enzymes exhibit high efficiency and selectivity because of their rapid rate of reaction and their capability to operate across a wide spectrum of pH and temperature levels. However, the high cost of enzyme extraction, purification, and production remains a limitation (Patel et al. 2020).

7 Metabolic Pathways in Hydrocarbon Degradation in Microbial Systems

Hydrocarbon degradation is a critical process for reducing environmental pollution, especially in areas contaminated by petroleum products. Microorganisms like bacteria, fungi, and archaea are essential in decomposing hydrocarbons via their metabolic pathways. These metabolic processes are complex and involve a variety of enzymes that transform hydrocarbons into simpler compounds, eventually resulting in the formation of harmless substances like water and carbon dioxide.

7.1 Aerobic Pathways

Microorganisms employ aerobic metabolic pathways to break down hydrocarbons in the presence of oxygen. The key enzymes driving this process are oxygenase, specifically monooxygenases and dioxygenases. The initial step in aerobic metabolism involves the incorporation of an oxygen atom as a hydroxyl group, catalyzed by oxygenases. Monooxygenases facilitate the cleavage of the oxygen-oxygen bond in O_2, incorporating one oxygen atom into the hydrocarbon structure while reducing the other to H_2O. Conversely, dioxygenases insert both oxygen atoms into aromatic hydrocarbons, playing an essential role in activating aromatic molecules and opening aromatic rings (Pandolfo et al. 2023).

During the degradation of aliphatic hydrocarbons, microbes catalyze the inception of oxygen atoms into either the terminal or subterminal methyl group of hydrocarbon molecules through enzymatic oxidation. Oxidation in the subterminal methyl group leads to the formation of secondary alcohol, which is ultimately oxidized to ketone, and then to ester. Upon hydrolysis of the ester yielding an alcohol, which is further transformed into an aldehyde and eventually a fatty acid. When the terminal methyl group is oxidized, it produces an alcohol that undergoes a series of oxidations to convert it into an aldehyde and then a carboxyl group. In both pathways, the resulting fatty acids are metabolized via β-oxidation to generate acetyl-CoA, which enters the Krebs cycle for energy production (Pandolfo et al. 2023).

7.2 Anaerobic Pathways

In oxygen-limited environments, microorganisms employ anaerobic metabolic pathways for hydrocarbon degradation. These pathways do not rely on oxygen but instead utilize alternative electron acceptors (manganese, nitrate, sulfate, carbon dioxide). Microbes including denitrifying bacteria (*Pseudomonas*, *Thauera*, *Azoarcus*), sulfate-reducing bacteria (*Desulfovibrio*, *Desulfobacter*), iron-reducing bacteria (*Geobacter*, *Shewanella*), and methanogenic archaea (*Methanosarcina*, *Methanobacterium*) are mainly used in the pathway to degrade hydrocarbons. Anaerobic degradation is slower compared to aerobic processes but is vital in environments such as deep soils, sediments, or subsurface waters. During anaerobic degradation, substrates are carboxylated, followed by the breaking of the ring (Ambaye et al. 2022). Anaerobic hydrocarbon degradation involves distinct biochemical mechanisms that differ from those in aerobic pathways. The most common mechanisms include the following.

7.2.1 Anaerobic Hydroxylation

Anaerobic hydroxylation of hydrocarbons refers to the biochemical process in which hydrocarbons are hydroxylated in the absence of oxygen. This process is typically carried out by specialized anaerobic microorganisms, such as sulfate-reducing bacteria or nitrate-reducing bacteria. Key enzymes involved in anaerobic hydroxylation include oxygen-independent hydroxylases or reductases, which can activate the hydrocarbon molecule by transferring electrons, leading to the formation of a reactive intermediate that is hydroxylated. This step is often followed by further metabolic transformations that allow the microorganisms to utilize hydrocarbons as energy or carbon sources.

Anaerobic hydroxylation was initially discovered during the degradation of propylbenzene and ethylbenzene. This process is carried out by denitrifying bacteria and is facilitated by the key enzyme ethylbenzene dehydrogenase (EBDH). Furthermore, the sulfate-reducing bacterium *Desulfococcus oleovorans* participates in an alternative pathway for anaerobic alkane degradation, which is catalyzed by other enzymes, including EBDH. This pathway circumvents the usual activation of alkanes through the addition of fumarate, instead of hydroxylating alkanes at the subterminal methylene group, resulting in the formation of iso-alcohols. Anaerobic hydroxylation reaction also occurs in the preliminary steps of degradation of phenolic compounds like *p*-Cresol, and *p*-ethylphenol and terpenoid compounds like limonene (Boll et al. 2020).

7.2.2 Fumarate Addition

One of the most studied mechanisms in anaerobic hydrocarbon degradation is the addition of fumarate to hydrocarbons, particularly alkylated aromatic hydrocarbons like cresols, ethylbenzene, toluene, cyclohexane, and xylenes. It involved the activation of hydrocarbons by fumarate addition to the methyl group of hydrocarbons by the fumarate-adding enzymes (FAE) that create an intermediate and can be further utilized in β-oxidation (Boll et al. 2020).

7.2.3 Carboxylation

In certain pathways, hydrocarbons are carboxylated (i.e., a carboxyl group is added) before being further metabolized. This mechanism has been detected during the anaerobic degradation of aromatic hydrocarbons, such as benzene and naphthalene. One of the most common carboxylation reactions involves the direct incorporation of carbon dioxide (CO_2) into hydrocarbons. The hydrocarbon, typically in the form of an organometallic intermediate, reacts with CO_2 to yield a carboxylated product. For example, the incorporation of CO_2 into naphthalene produces 2-naphthoate, followed by 2-naphthoyl-CoA and finally tetrahydro-2-naphthoyl-CoA. Similarly, benzene gets converted into benzoyl-CoA by direct incorporation of CO_2 (Boll et al. 2020).

7.2.4 Hydrocarbon Activation via Reverse Methanogenesis

Certain methanogenic archaea utilize a reverse methanogenesis pathway, where hydrocarbons like methane are initially activated by enzymes that reverse the conventional methane production process, resulting in their breakdown. Hallam and colleagues discovered that the genomes of these methanotrophic archaea encode methyl-CoM reductase (Mcr), an enzyme that usually reduces the methyl-CoM to generate methane and heterodisulfide containing coenzyme B and coenzyme M. Under suitable circumstances, this reaction was reversed by Mcr, resulting in the production of methyl-CoM. Moreover, an alternative mechanism for butane activation by Mcr proteins in thermophilic archaea has been proposed, suggesting potential pathways for the activation of non-methane alkanes. Wang and colleagues reexamined 64 metagenomes and identified putative multicarbon oxidizing mcr genes across various archaeal phyla. Their analysis also indicated the horizontal transfer of mcr genes into Asgard archaea. Additionally, they suggested that Helarchaeota may oxidize short-chain alkanes via alkyl-CoM intermediates (Laczi et al. 2020).

8 Factors Influencing Geomicrobial Remediation

Geomicrobial remediation of hydrocarbon-contaminated sites includes utilizing microorganisms to eliminate or degrade hydrocarbons in the environment. Several factors influence the effectiveness of this bioremediation process.

8.1 *Type of Hydrocarbon*

The type of hydrocarbon present in a contaminated site serves an essential role in determining the efficiency and approach of bioremediation. Hydrocarbons vary in their chemical structure, which directly affects their biodegradability, toxicity, and interaction with microbial communities. These variations can significantly impact the success of bioremediation strategies. The hydrocarbon in higher concentration becomes more toxic to microorganisms, inhibiting their growth and thus reducing the efficiency of bioremediation (Mekonnen et al. 2024). Moreover, the chemical structure of the hydrocarbon influences its biodegradability. The linear hydrocarbons are more easily metabolized by microbes in comparison to cyclic hydrocarbons, polycyclic aromatic hydrocarbons (PAHs), and asphaltenes. Additionally, the biodegradability of hydrocarbons is inherently influenced by their composition, with simpler compounds being more readily utilized by hydrocarbon-degrading microorganisms. Crude oil and kerosene contain a mixture of different hydrocarbon types, including alkanes, aromatics, and complex molecules such as asphaltenes. The biodegradability of these mixtures varies widely, with lighter fractions being more easily degraded, while heavier, more complex compounds may persist in the environment. The asphaltenes, for instance, are highly resistant to microbial degradation and may require long-term or specialized treatment strategies (Mekonnen et al. 2024).

8.2 *Microbial Community*

Indigenous hydrocarbon-degrading bacteria found in hydrocarbon-contaminated sites are often better adapted to local environmental conditions, as they utilize hydrocarbons as the primary source of energy and carbon for their growth and development. These bacteria are highly efficient in biodegradation, having evolved to metabolize the specific hydrocarbons present in their environment. Equipped with enzymes like oxygenases and dehydrogenases, they break down complex hydrocarbons into simpler, less harmful compounds. Their presence accelerates the degradation process, significantly reducing the environmental impact of contamination. Hence, in areas where these bacteria are abundant, bioremediation becomes more efficient and rapid (Kebede et al. 2021). Hydrocarbon-degrading bacteria often engage in complex interactions, involving both competition and cooperation, which can significantly influence the efficiency of bioremediation. Competition occurs in two forms: intraspecific and interspecific. In intraspecific competition, bacteria produce metabolites that inhibit the growth of other bacterial species. For example, *Thalassolituus*, an oil-degrading bacterium, antagonizes the growth of other oleophilic bacteria by releasing inhibitory metabolites. In interspecific competition, bacteria and fungi compete for essential nutrients like nitrogen, phosphorus, and trace minerals, which are often scarce in contaminated soils or waters. Strong competitors may monopolize these nutrients, potentially slowing or inhibiting the activity of other key degraders.

On the other hand, cooperation, or a synergistic interaction, occurs when one species partially degrades a compound, and another species further breaks down the intermediate products. This cooperative metabolism broadens the range of hydrocarbons that can be degraded at contaminated sites. A well-documented example is the synergy between *Cycloclasticus* and *Alcanivorax borkumensis* in the degradation of polyaromatic hydrocarbons (Kebede et al. 2021). Such cooperation accelerates the degradation process, as bacteria work together to metabolize complex compounds more efficiently. Conversely, intense competition may hinder degradation if crucial bacterial species are outcompeted or inhibited.

8.3 Environmental Conditions

Various environmental factors like salinity, temperature, oxygen levels, pH, and nutrient availability significantly influence the bioremediation process. Among these factors, temperature is crucial in influencing the efficiency of microbial hydrocarbon degradation.

Microbes are capable of degrading hydrocarbons across a broad temperature range, with higher temperatures generally boosting microbial activity. This increase in temperature enhances enzymatic reactions and improves hydrocarbon bioavailability, leading to faster degradation rates. In contrast, lower temperatures reduce microbial growth and enzyme activity, slowing down the degradation process (Daâssi and Almaghribi 2022). Most of the hydrocarbon-degrading microorganisms perform best within particular temperature ranges, i.e., 15–20 °C in marine environments, 20–30 °C in freshwater habitats, and 30–40 °C in soil. To overcome the limitations of lower temperatures, recent research has explored the use of thermophilic microorganisms, which can function effectively at higher temperatures. These thermophilic microbes have shown remarkable adaptability, enabling them to regulate their metabolic mechanisms and thrive under harsh environmental conditions, offering a promising solution for improving bioremediation in extreme temperature settings (Mekonnen et al. 2024).

Soil and water pH have a significant impact on the activity of hydrocarbon-degrading microorganisms. Most microbes involved in bioremediation thrive in a neutral to mildly alkaline pH range, while fungi are more capable of interacting with hydrocarbons in slightly acidic conditions. Extreme pH levels can inhibit microbial growth, disrupt enzyme function, and reduce the efficiency of hydrocarbon degradation. Therefore, maintaining an optimal pH within this range is critical for effective bioremediation. In cases where soils are overly acidic or alkaline, pH adjustments can be made through amendments to create more favorable conditions for microbial activity (Mekonnen et al. 2024).

Salinity is a critical factor influencing the bioremediation of hydrocarbon-contaminated sites. High salinity levels can inhibit the growth of many hydrocarbon-degrading microorganisms due to osmotic stress (Mekonnen et al. 2024). However, certain halophilic (salt-tolerant) bacteria, such as *Marinobacter*, *Alcanivorax*, and *Halomonas*, have adapted to these conditions and are capable of degrading hydrocarbons in saline environments. These halophilic microbes are essential for bioremediation in saltwater or brackish environments, where standard hydrocarbon degraders would struggle to survive (Martins and Peixoto 2012). Conversely, in freshwater or low-salinity soils, non-halophilic hydrocarbon degraders tend to perform more effectively.

Hydrocarbon-degrading microorganisms need vital nutrients like carbon, nitrogen, and phosphorus to support their growth and metabolic functions. In many contaminated environments, hydrocarbons provide a rich carbon source, but nutrients like nitrogen and phosphorus are often limiting. This nutrient limitation can significantly slow down the biodegradation process. Maintaining the right C:N:P ratio is essential for supporting the growth of microbes. Research has demonstrated that a C:N:P ratio of 100:10:1 optimizes microbial degradation in soils contaminated with PAHs (Daâssi and Almaghribi 2022). However, Premnath et al. (2021) reported that *Klebsiella pneumoniae* degrade hydrocarbons more efficiently at 10:2 of C:N ratio. An overabundance of nutrients can also adversely affect biodegradation, as excessively high concentrations may inhibit the process and reduce its overall efficiency (Daâssi and Almaghribi 2022).

Oxygen is another crucial factor for the aerobic biodegradation of hydrocarbons, as most hydrocarbon-degrading bacteria depend on oxygen to metabolize contaminants through oxidative pathways (Mekonnen et al. 2024). Aerobic conditions enable faster and more efficient degradation processes. In polycyclic aromatic hydrocarbon contaminated sites, the aerobic microorganisms use molecular oxygen to initiate hydrocarbon oxidation by secreting oxygenase enzymes, i.e., monooxygenases or dioxygenases, which break down complex benzene rings into smaller, more manageable hydrocarbon structures. Through the activity of these degrading enzymes, various reactions such as oxidation and hydrogenation take place, ultimately resulting in the remediation of contaminated areas (Daâssi and Almaghribi 2022). Nonetheless, in oxygen-limited environments, adding oxygen can significantly increase the rate of oxygenase-catalyzed microbial degradation of PHCs (Mekonnen et al. 2024).

9 Conclusion and Future Prospects

Geomicrobiology provides a crucial framework for understanding and harnessing the metabolic capabilities of microorganisms in the bioremediation of hydrocarbons. By unlocking the potential of microbial communities and optimizing environmental conditions, bioremediation offers a powerful tool for mitigating the impact of hydrocarbon pollution on ecosystems. Continued research in this field promises to refine these techniques and expand their applications in addressing one of the most pressing environmental challenges of our time.

The future of bioremediation lies in the continued exploration of microbial diversity and the development of more targeted, efficient methods for enhancing hydrocarbon degradation. Advances in genomic technologies such as metagenomics and transcriptomics offer unprecedented insight into the composition and metabolic potential of microbial communities in contaminated environments. One exciting avenue for future research is the engineering of microbial strains with enhanced degradation capabilities through synthetic biology. These engineered microbes could be tailored to degrade specific hydrocarbon compounds, making bioremediation efforts more efficient and environmentally sustainable.

Competing Interests All the authors declare that they have no competing interests.

References

Al-Kaabi N, Disi ZA, Al-Ghouti MA, Solling TI, Zouari N (2022) Interaction between indigenous hydrocarbon-degrading bacteria in reconstituted mixtures for remediation of weathered oil in soil. Biotechnol Rep 36:e00767. https://doi.org/10.1016/j.btre.2022.e00767

Alsayegh SY, Al-Ghouti MA, Zouari N (2021) Study of bacterial interactions in reconstituted hydrocarbon-degrading bacterial consortia from a local collection, for the bioremediation of weathered oily-soils. Biotechnol Rep 29:e00598. https://doi.org/10.1016/j.btre.2021.e00598

Ambaye TG, Chebbi A, Formicola F, Prasad S, Gomez FH, Franzetti A, Vaccari M (2022) Remediation of soil polluted with petroleum hydrocarbons and its reuse for agriculture: recent progress, challenges, and perspectives. Chemosphere 293:133572. https://doi.org/10.1016/j.chemosphere.2022.133572

Aradhi KK, Dasari BM, Banothu D, Manavalan S (2023) Spatial distribution, sources and health risk assessment of heavy metals in topsoil around oil and natural gas drilling sites, Andhra Pradesh, India. Sci Rep 13:10614. https://doi.org/10.1038/s41598-023-36580-9

Barman D, Dutta I, Jha DK (2022) Heavy metal resistant bacteria from coal dumping site with plant growth promoting potentials. Biologia 77:533–545. https://doi.org/10.1007/s11756-021-00963-y

Boll M, Estelmann S, Heider J (2020) Anaerobic degradation of hydrocarbons: mechanisms of hydrocarbon activation in the absence of oxygen. In: Boll M (ed) Anaerobic utilization of hydrocarbons, oils, and lipids: handbook of hydrocarbon and lipid microbiology. Springer, Cham. https://doi.org/10.1007/978-3-319-50391-2_2

Chaudhary DK, Bajagain R, Woo S, Jaisoo J (2021) Effect of consortium bioaugmentation and biostimulation on remediation efficiency and bacterial diversity of diesel-contaminated aged soil. World J Microbiol Biotechnol 37:1–12. https://doi.org/10.1007/s11274-021-02999-3

Daâssi D, Almaghribi FQ (2022) Petroleum-contaminated soil: environmental occurrence and remediation strategies. 3 Biotech 12(6):139. https://doi.org/10.1007/s13205-022-03198-z

Das N, Chandran P (2011) Microbial degradation of petroleum hydrocarbon contaminants: an overview. Biotechnol Res Int 2011:941810. https://doi.org/10.4061/2011/941810

Dhailappan A, Arumugam A, Rajendran G, Ravichandran M (2022) A dualistic approach to investigate the remedial potential and edible property of Pleurotus ostreatus on hydrocarbon-contaminated soil. J Appl Biol Biotechnol 10:70–78. https://doi.org/10.7324/jabb.2022.100310

El-Sheshtawy HS, Aman D, Nassar HN (2022) A novel bioremediation technique for petroleum hydrocarbons by bacterial consortium immobilized on goethite-chitosan nanocomposite. Soil Sediment Contam 31:176–199. https://doi.org/10.1080/15320383.2021.1916737

Ewim DRE, Orikpete OF, Scott TO, Onyebuchi CN, Onukogu AO, Uzougbo CG, Onunka C (2023) Survey of wastewater issues due to oil spills and pollution in the Niger Delta area of Nigeria: a secondary data analysis. Bull Natl Res Cent 47:116. https://doi.org/10.1186/s42269-023-01090-1

Gadd GM (2010) Metals, minerals and microbes: geomicrobiology and bioremediation. Microbiology 156:609–643. https://doi.org/10.1099/mic.0.037143-0

Giovanella P, de Azevedo Duarte L, Kita DM, de Oliveira VM, Sette LD (2021) Effect of biostimulation and bioaugmentation on hydrocarbon degradation and detoxification of diesel-contaminated soil: a microcosm study. J Microbiol 59:634–643. https://doi.org/10.1007/s12275-021-0395-2

Honda M, Suzuki N (2020) Toxicities of polycyclic aromatic hydrocarbons for aquatic animals. Int J Environ Res Public Health 17(4):1363. https://doi.org/10.3390/ijerph17041363

Kebede G, Tafese T, Abda EM, Kamaraj M, Assefa F (2021) Factors influencing the bacterial bioremediation of hydrocarbon contaminants in the soil: mechanisms and impacts. J Chem 2021:1–17. https://doi.org/10.1155/2021/9823362

Laczi K, Erdeiné Kis Á, Szilágyi Á, Bounedjoum N, Bodor A, Vincze GE, Kovács T, Rákhely G, Perei K (2020) New frontiers of anaerobic hydrocarbon biodegradation in the multi-omics era. Front Microbiol 11:590049. https://doi.org/10.3389/fmicb.2020.590049

Mallah MA, Changxing L, Mallah MA, Noreen S, Liu Y, Saeed M, Xi H, Ahmed B, Feng F, Mirjat AA et al (2022) Polycyclic aromatic hydrocarbon and its effects on human health: an overeview. Chemosphere 296:133948. https://doi.org/10.1016/j.chemosphere.2022.133948

Martins LF, Peixoto RS (2012) Biodegradation of petroleum hydrocarbons in hypersaline environments. Braz J Microbiol 43(3):865–872. https://doi.org/10.1590/S1517-83822012000300003

Meagher RB (2000) Phytoremediation of toxic elemental and organic pollutants. Curr Opin Plant Biol 3:153–162. https://doi.org/10.1016/s1369-5266(99)00054-0

Mekonnen BA, Aragaw TA, Genet MB (2024) Bioremediation of petroleum hydrocarbon contaminated soil: a review on principles, degradation mechanisms, and advancements. Front Environ Sci 12:1354422. https://doi.org/10.3389/fenvs.2024.1354422

Mohammed SA, Omar Zrary TJ, Hasan AH (2023) Degradation of crude oil and the pure hydrocarbon fractions by indigenous soil microorganisms. Biologia 78:3637–3651. https://doi.org/10.1007/s11756-023-01513-4

Montano L, Pironti C, Pinto G, Ricciardi M, Buono A, Brogna C, Venier M, Piscopo M, Amoresano A, Motta O (2022) Polychlorinated Biphenyls (PCBs) in the environment: occupational and exposure events, effects on human health and fertility. Toxics 10:365. https://doi.org/10.3390/toxics10070365

Nikolopoulou M, Pasadakis N, Kalogerakis N (2013) Evaluation of autochthonous bioaugmentation and biostimulation during microcosm-simulated oil spills. Mar Pollut Bull 72(1):165–173. https://doi.org/10.1016/j.marpolbul.2013.04.007

Nikolova C, Gutierrez T (2020) Use of microorganisms in the recovery of oil from recalcitrant oil reservoirs: current state of knowledge, technological advances and future perspectives. Front Microbiol 10:2996. https://doi.org/10.3389/fmicb.2019.02996

Nong J, Peng P, Pan J, Shen T, Xie QL (2023) Effect of bioaugmentation and biostimulation on hydrocarbon degradation and bacterial community composition in different petroleum-contaminated soil layers. Water Air Soil Pollut 234:189. https://doi.org/10.1007/s11270-023-06161-7

Novianty R, Saryono A, Pratiwi NW, Hidayah A, Juliantari E (2021) The diversity of fungi consortium isolated from polluted soil for degrading petroleum hydrocarbon. Biodiversitas 22:5077–5084. https://doi.org/10.13057/BIODIV/D221145

Omenna EC, Omage K, Ezaka E, Azeke MA (2024) Bio-augmentation and bio-stimulation with kenaf core enhanced bacterial enzyme activities during bio-degradation of petroleum hydrocarbon in polluted soil. Sci Rep 14(1):8. https://doi.org/10.1038/s41598-023-50882-y

Oualha M, Al-Kaabi N, Al-Ghouti M, Zouari N (2019) Identification and overcome of limitations of weathered oil hydrocarbons bioremediation by an adapted *Bacillus sorensis* strain. J Environ Manag 250:109455. https://doi.org/10.1016/j.jenvman.2019.109455

Oyetibo GO, Chien MF, Ikeda-Ohtsubo W, Suzuki H, Obayori OS, Adebusoye SA, Ilori MO, Amund OO, Endo G (2017) Biodegradation of crude oil and phenanthrene by heavy metal resistant *Bacillus subtilis* isolated from a multi-polluted industrial wastewater creek. Int Biodeterior Biodegrad 120:143–151. https://doi.org/10.1016/j.ibiod.2017.02.021

Pandolfo E, Barra Caracciolo A, Rolando L (2023) Recent advances in bacterial degradation of hydrocarbons. Water 15:375. https://doi.org/10.3390/w15020375

Patel AB, Shaikh S, Jain KR, Desai C, Madamwar D (2020) Polycyclic aromatic hydrocarbons: sources, toxicity, and remediation approaches. Front Microbiol 11:562813. https://doi.org/10.3389/fmicb.2020.562813

Perdigão R, Almeida CMR, Santos F, Carvalho MF, Mucha AP (2020) Optimization of an autochthonous bacterial consortium obtained from beach sediments for bioremediation of petroleum hydrocarbons. Water 13:66. https://doi.org/10.3390/w13010066

Popoola LT, Yusuff AS, Adeyi AA, Omotara OO (2022) Bioaugmentation and biostimulation of crude oil contaminated soil: process parameters influence. South Afr J Chem Eng 39:12–18. https://doi.org/10.1016/j.sajce.2021.10.003

Premnath N, Mohanrasu K, Rao RGR, Dinesh GH, Prakash GS, Pugazhendhi A, Jeyakanthan J, Govarthanan M, Kumar P, Arun A (2021) Effect of C/N substrates for enhanced extracellular polymeric substances (EPS) production and poly cyclic aromatic hydrocarbons (PAHs) degradation. Environ Pollut 275:116035. https://doi.org/10.1016/j.envpol.2020.116035

Purwasena IA, Amaniyah M, Astuti DI, Firmansyah Y, Sugai Y (2024) Production, characterization, and application of *Pseudoxanthomonas taiwanensis* biosurfactant: a green chemical for microbial enhanced oil recovery (MEOR). Sci Rep 14:10270. https://doi.org/10.1038/s41598-024-61096-1

Radhakrishnan A, Balaganesh P, Vasudevan M, Natarajan N, Chauhan A, Arora J, Ranjan A, Rajput VD, Sushkova S, Minkina T et al (2023) Bioremediation of hydrocarbon pollutants: recent promising sustainable approaches, scope, and challenges. Sustainability 15:5847. https://doi.org/10.3390/su15075847

Rasheed MA, Kalpana MS, Prasanna MV, Lakshmi M, Madhavi T, Tiwari DM, Patil DJ, Dayal AM, Raju SV (2012) Geo-microbial and light gaseous hydrocarbon anomalies in the near surface soils of Deccan Syneclise Basin, India: implications to hydrocarbon resource potential. J Pet Sci Eng 84-85:33–41. https://doi.org/10.1016/j.petrol.2012.01.010

Rasheed MA, Hasan SZ, Rao PLS, Boruah A, Sudarshan V, Kumar B, Harinarayana T (2015) Application of geo-microbial prospecting method for finding oil and gas reservoirs. Front Earth Sci 9:40–50. https://doi.org/10.1007/s11707-014-0448-5

Ravi A, Ravuri M, Krishnan R, Narenkumar J, Anu K, Alsalhi MS, Devanesan S, Kamala-Kannan S, Rajasekar A (2022) Characterization of petroleum degrading bacteria and its optimization conditions on effective utilization of petroleum hydrocarbons. Microbiol Res 265:127184. https://doi.org/10.1016/j.micres.2022.127184

Sarkar J, Kazy SK, Gupta A, Dutta A, Mohapatra B, Roy A, Bera P, Mitra A, Sar P (2016) Biostimulation of indigenous microbial community for bioremediation of petroleum refinery sludge. Front Microbiol 7:1407. https://doi.org/10.3389/fmicb.2016.01407

Satpati GG, Gupta S, Biswas RK, Choudhury AK, Kim JW, Davoodbasha M (2023) Microalgae mediated bioremediation of polycyclic aromatic hydrocarbons: strategies, advancement and regulations. Chemosphere 344:140337. https://doi.org/10.1016/j.chemosphere.2023.140337

Sayed K, Baloo L, Sharma NK (2021) Bioremediation of total petroleum hydrocarbons (tph) by bioaugmentation and biostimulation in water with floating oil spill containment booms as bioreactor basin. Int J Environ Res Public Health 18:2226. https://doi.org/10.3390/ijerph18052226

Schuster JK, Harner T, Fillmann G, Ahrens L, Altamirano JC, Aristizábal B, Bastos W, Castillo LE, Cortés J, Fentanes O et al (2015) Assessing polychlorinated dibenzo-p-dioxins and polychlorinated dibenzofurans in air across Latin American countries using polyurethane foam disk passive air samplers. Environ Sci Technol 49(6):3680–3686. https://doi.org/10.1021/es506071n

Sharma A, Singh SB, Sharma R, Chaudhary P, Pandey AK, Ansari R, Vasudevan V, Arora A, Singh S, Saha S, Nain L (2016) Enhanced biodegradation of PAHs by microbial consortium with different amendment and their fate in in-situ condition. J Environ Manag 181:728–736. https://doi.org/10.1016/j.jenvman.2016.08.024

Subashchandrabose SR, Venkateswarlu K, Venkidusamy K, Palanisami T, Naidu R, Megharaj M (2019) Bioremediation of soil long-term contaminated with PAHs by algal- bacterial synergy of *Chlorella* sp. MM3 and *Rhodococcus wratislaviensis* strain 9 in slurry phase. Sci Total Environ 659:724–731. https://doi.org/10.1016/j.scitotenv.2018.12.453

Taylor MD, Langdon KA, Smith JA, Stevenson G, Edge K (2024) Polychlorinated dibenzodioxins/furans and dioxin-like polychlorinated biphenyls in fish and crustaceans of a recreationally fished estuary, following targeted remediation. Sci Total Environ 921:171089. https://doi.org/10.1016/j.scitotenv.2024.171089

Tursi A, Beneduci A, Chidichimo F, De Vietro N, Chidichimo G (2018) Remediation of hydrocarbons polluted water by hydrophobic functionalized cellulose. Chemosphere 201:530–539. https://doi.org/10.1016/j.chemosphere.2018.03.044

Van Hamme JD, Singh A, Ward OP (2003) Recent advances in petroleum microbiology. Microbiol Mol Biol Rev 67:503–549. https://doi.org/10.1128/mmbr.67.4.503-549.2003

Venkatraman G, Giribabu N, Mohan PS, Muttiah B, Govindarajan VK, Alagiri M, Abdul Rahman PS, Karsani SA (2024) Environmental impact and human health effects of polycyclic aromatic hydrocarbons and remedial strategies: a detailed review. Chemosphere 351:141227. https://doi.org/10.1016/j.chemosphere.2024.141227

Wu M, Wu J, Zhang X, Ye X (2019) Effect of bioaugmentation and biostimulation on hydrocarbon degradation and microbial community composition in petroleum-contaminated loessal soil. Chemosphere 237:124456. https://doi.org/10.1016/j.chemosphere.2019.124456

Yernazarova A, Shaimerdenova U, Akimbekov N, Kaiyrmanova G, Shaken M, Izmailova A (2024) Exploring the use of microbial enhanced oil recovery in Kazakhstan: a review. Front Microbiol 15:1394838. https://doi.org/10.3389/fmicb.2024.1394838

Zoghi P, Mafigholami R (2023) Optimisation of soil washing method for removal of petroleum hydrocarbons from contaminated soil around oil storage tanks using response surface methodology. Sci Rep 13:15457. https://doi.org/10.1038/s41598-023-42777-9

Native Geomicrobes and Their Role in Ore Deposits

Debaraty Chakravarty and Santa Ram Joshi

Contents

Abstract Microbes dwell and thrive under varied conditions. Certain microbes are found in ore deposits and remain exclusive to these mineral sites. In today's industrial era, effects of leaching of metals like arsenic from or nearby ore deposits can lead to excessive soil toxicity while polluting the nearby water bodies. Depletion of minerals from the ore deposits can lead to a habitat change of the residual microbiota, and it is yet to be confirmed if the new habitat thus formed is hostile to the original dwellers while supporting the growth of new microbes. Exploration of these native microbes can offer insights regarding their survival strategies in ore deposits and their possible role in both ore formation and leaching minerals from polluted sites. This chapter explores such geo microbes, the techniques employed, and also their relevant use in commerce and industry.

Keywords Microbes · Bioremediation · Bioprospecting · Mining · Leaching

D. Chakravarty · S. R. Joshi (✉)
Microbiology Laboratory, Department of Biotechnology and Bioinformatics, North-Eastern Hill University, Shillong, Meghalaya, India

S. R. Joshi et al. (eds.), *Mineral Transformation and Bioremediation by Geo-Microbes*, Interdisciplinary Biotechnological Advances,
https://doi.org/10.1007/978-981-96-3033-2_20

Abbreviations

Cd	Cadmium
Cu	Copper
Hg	Mercury
Mo	Molybdenum
Pb	Lead
U	Uranium
Zn	Zinc

1 Introduction

Microbes have a role in elementary processes like biogeochemical cycling, alterations in metals and minerals, breakdown of organic matter, bioweathering processes, and the creation of soil and sediment (Gadd 2010). These processes are integral components of natural biogeochemical cycles for metals and several other associated elements. These elements such as sulfur, phosphorous, metalloids, actinides, and metal radionuclides may be present in organic matter, soil, rocks, and minerals. Metal and mineral transformation thus influence the niche/habitat of the organisms. Role of microbes in ore formation has been recognized and broadly documented (Macqueen 1985; Southam and Saunders 2005; Hofmann 2011). Mining of ores leads to destruction of the natural flora, fauna, and microbial communities residing in the area (Newsome and Falagán 2021). Post-mining operations make the area vacant and desolate, while simultaneously posing a likely risk of becoming a garbage dump (Gross 2018). Constituents of mines along with garbage can leak into nearby water bodies affecting the new ecosystem. Likewise, contaminated running water may seep through different outlets and over a period of time erodes landscape of large areas and deposits the garbage material into subsequent downhill regions (Kumar and Kumar 2018). Metal and mineral transformation can thus have both beneficial and detrimental effects on the environment.

2 Mineral Ores

Valuable elements can be extracted from natural deposits in nature. These aggregations are called mineral ores (Heinrich and Candela 2014). Common examples of ores in deposits include iron ore, copper ore, aluminum ore, and rare earth element ores.

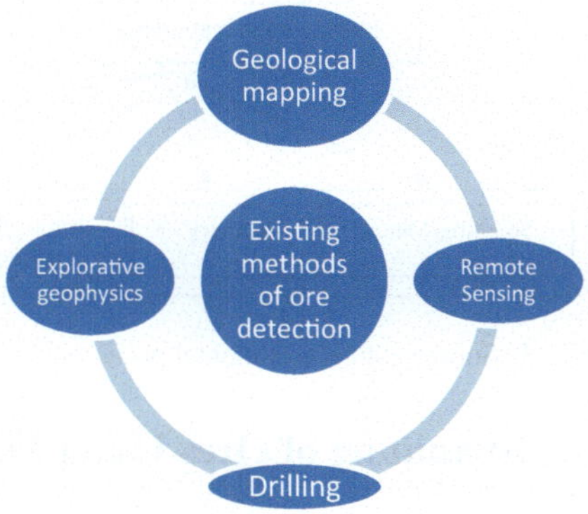

Fig. 1 Schematic representation of few techniques employed in ore refining

3 Methods in Ore Refining

Different methods that can be exploited in investigating the microbial ecology of bacteria native to ore deposits include surface sampling, geophysical, geochemical, and geostatistical techniques (Fig. 1).

4 Bioremediation

Biological systems are crucial in the remediation, immobilization, and detoxification of organic and inorganic pollutants (Sharma 2020). Minerals and elements deposited by microbes exhibit properties which can be of industrial importance (Gadd 2010). Likewise, such microbe-mediated alterations may lead to spoilage of natural properties of materials, and also to biocorrosion of metals and alloys, metal contamination, etc. Overall, such occurrences may have severe social, environmental, and economic consequences. Microbes utilize minerals by colonizing mineral surfaces, breaking the minerals to extract the energy from chemical component, and also forming their own minerals such as chalk and silica shells (Gross 2018). Microbes can serve as bioindicators for resources in unexplored regions. When water percolates through hillside dumps deposited by miners, brine present from the area contains dissolved metals. Dump bioleaching has mostly been used for extracting copper from ores. Microbes dwelling on mineral ores are stimulated by irrigation and conditioning. On completion of the washing stages, the metal is collected as effluent from the dump. The most commonly employed microbe is *Acidithiobacillus ferrooxidans* which oxidizes ferrous and sulfur minerals. It is resistant to arsenic and uranium. Other bacteria involved in bioleaching are also from *Acidithiobacillus*, *Sulfobacillus*, or *Leptospirillum* genus while archaea belong to the genus *Sulfolobus*, *Metallosphaera*, and *Acidianus* (Gross 2018) (Fig. 2).

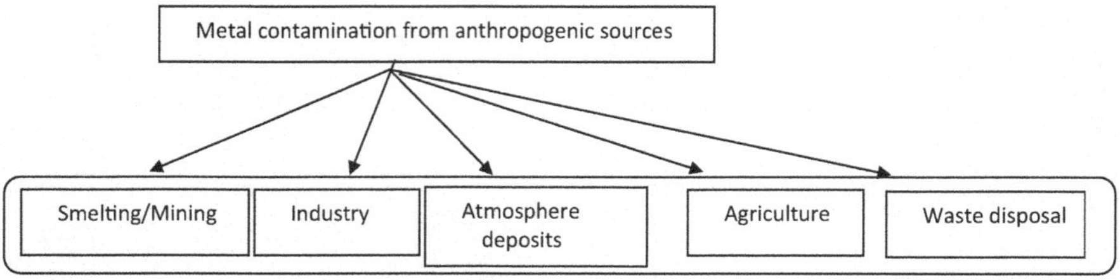

Fig. 2 Sources of environmental pollution

5 Biomining of Ores Using Microbes

Formation of carbonates and oxides may stabilize the ores formed by microbial biosorption and adhesion (Robles-Fernández et al. 2022). It is understood that microbes exhibit such variation in their properties and subsequent function at specific stages of microbial growth. Biofilm formation, biomineralization, quorum sensing, and fossilization are other geochemical processes involved in cycling, transformation, and distribution of metals in the environment (Gross 2018). Certain geomicrobes have a role in adhesion and absorption to hydrocarbons, thus being immaculate in bioremediation. These hydrocarbon-oxidizing microbes can serve as an indicator species and provide an insight into locations which have rich oil reserves (Rasheed et al. 2011). Recent times which are focusing on renewable and clean energy, discovery, and study of such indigenous microbiota may help in understanding the limitations faced, thereby offering opportunities to improve the existing mineral mining and extraction techniques (Atlas 1981; Klusman 1993). Such research offers scope to facilitate minimum waste generation on mining or reutilization of the generated waste.

Other less common methods of refining heavy metals involve mineral precipitation by ureolytic bacteria (Jalilvand et al. 2020). Urease-producing bacteria from contaminated soils were isolated and identified for assessing the potential of selected isolates for biomineralization of heavy metals. Subsequently, *Sporosarcina pasteurii* was used to precipitate Zn, Pb, and Cd from solutions of different metal ion concentration and the precipitates thus generated were compared. Certain bacteria were more resilient in elevated levels of metal concentrations and thereby more efficient in remediation of polluted sites (Jalilvand et al. 2020).

Studies have also revealed that sulfate-reducing bacteria can produce bicarbonate which thereby reduces acidity present in mine effluents, by utilizing the existing hydrogen ions (Römer and Schwartz 1965; Thompson and Ferris 1990; Douglas and Beveridge 1998). Thus, such process is useful in treating surface water and contaminated groundwater which are commonly seen having acidic conditions. Through geochemical analyses, it was understood that uranium ores are present in schist rocks that are metamorphosed with chlorite and biotite (Islam et al. 2014). It was also found that microbial communities dwelling in these subsurface uranium deposits consisted mainly of *Proteobacteria*, *Acidobacteria*, *Bacteroidetes*,

Firmicutes, etc. Other ore deposits which contain metalloids also pose a greater risk for soil contamination, and it becomes imminent to contain their spread by understanding the sequestration strategies employed by microbial community (Sun et al. 2017).

Microorganisms can extract minerals such as gold, copper, silver, cobalt, uranium, zinc, and nickel. The primary microorganisms used for mining include *Thiobacillus thioxidans, T. ferroxidans, T. thermophilica, Pseudomonas fluorescens, P. putida, Achromobacter, Bacillus licheniformis, and Sulfobus acidocaldarius* (Sajjad et al. 2019; Margaryan 2021). Extraction of uranium is mostly done by in situ bioleaching techniques by oxidation of uranium in presence of H_2SO_4-Fe_3^+ solution. It is an indirect mechanism where microbial action is on the oxidized product (Sajjad et al. 2019). However, heap bioleaching can lead to high uranium recovery. Other metals which can be recovered by bioleaching are nickel, gold, silver, cobalt, molybdenum, and antimony.

As an alternative to the mining techniques employed conventionally which are hazardous to the environment (Gavrilescu 2022), microbial mining involves two main methods. The first is the direct method where microbes degrade mineral ore via enzymes which separate ore and minerals and the other method is when microbes produce substances which interact and process the ore thereby consisting ultimately of a mineral mixture (Sarkodie et al. 2022). This biomining is preceded by biooxidation, which involves solubilization of metals such as gold via chemical and biological leaching processes. Lithotrophs and organotrophs facilitate these stages of metal recovery and concentration from these ores. Chemolithoautotrophs, on solubilization, decrease the pH of surrounding environment, which further intensifies metal solubilization by microbial leaching (Chen et al. 2022).

Some heterotrophs, however, improve biomining ability of autotrophs by enacting on metal-soluble metal sulfides, and others target non-sulfide rare earth metals. *Bacillus licheniformis*, *B. polymyxa*, and *Pseudomonas putida* are the most commonly used heterotrophic bacteria involved in metal solubilization (Sana et al. 2021). Among fungi, *Aspergillus* and *Penicillium* are the most predominant organisms (Bosecker 1997).

Studies have shown that *Ralstonia metallidurans* is present in biofilms on gold (Reith et al. 2006). This microbe may be exploited by industries for processing of gold containing ores by exploiting the biochemical and physiological pathways that enable the bacteria for gold extraction from ores.

6 Ore Deposition Effects on Microbiota

While deposition of heavy metals may affect vegetation growth, alkalinity, and salinity in an area (Bradshaw et al. 1965; Xie et al. 2016), they exert noticeable effect on microbiota as well (Giller et al. 1998). Diversity, richness, and bioactivity of microbes decrease with a surge in heavy metal deposition in an area (DeLeo and Ehrlich 1994). However, the diversity of metal resistant microbial population gradually increases over time (Xie et al. 2014). While microbes can accumulate metals in their cells, they also can release organic acids into the soil ensuring greater solubility of the heavy metals (Sánchez-Castro et al. 2023).

Monitoring of the degraded areas by assessing the impact of physical and chemical attributes of the ore and the bioavailabilty of metals in an area can give an insight on the soil microbial biomass of the region (Lado et al. 2008). By evaluating the enzyme activity, density of microbial groups, soil density, porosity, and texture, the effectiveness of the rehabilitation process can be analyzed (Hinojosa et al. 2008). More residual microbiota have a greater capacity for bioremediation potential.

7 Native Microbiota in Ores

Bacteria adsorb and accumulate metals like mercury, cadmium, lead, etc. *Bacillus circulans* and *Rhodospirillum* sp. can accumulate Cd, Pb, and Hg. Fungi such as *Rhizopus arrhizus*, *Pencillium lapidorum*, *P. chrysogenum*, *P. spimuiosum*, *Aspergillus oryzae*, *Mucor haemalis*, *Agaricus bisporus*, common yeast, and *Sporobolomyces salmonicolour* readily exhibit biosorption properties. Algal species include *Chlorella vulgaris* and *Chlorella regularis* and can adsorb metals like Pb, Hg, Cu, Mo, and U. Other bioaccumulators are marine organisms which include *Luminaria*, *Ulva*, *Codium*, and green algae *Hydrodictyon reticulatum*.

8 Mineral Transformations by Microbes in Ore Deposits (Fig. 3)

Organisms such as nitrate reducers and urea-degrading bacteria release bicarbonate and can create an alkaline environment and aid in mineral deposition. Essential elements required for growth of microbes are extracted from environment. Organism like *Betaproteobacteria* produce metabolites to enhance the dissolution of available minerals (Lepleux et al. 2012; Dong et al. 2022). Methanotrophs like methane

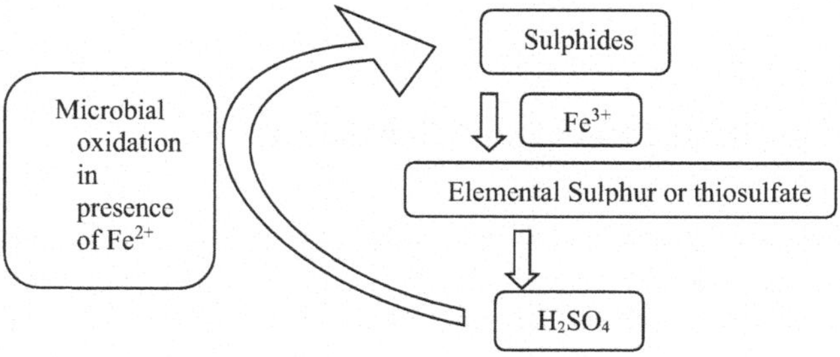

Fig. 3 Mode of action of *Acidithiobacillus ferroxidans*

oxidizing microbes regulate production of methanobactin based on availability of copper (Kraemer et al. 2015).

Bacteria, including dead bacterial strains, are the most efficient biosorbents due to abundance of active chemisorption sites and high surface to volume ratio. Dead strain of *Bacillus sphaericus* showed more biosorption of chromium compared to live strain (Velásquez and Dussan 2009; Pande et al. 2022). *Aspergillus niger* and *Monodictys pelagic* were found to be involved in bioaccumulation of lead and chromium (Sher and Rehman 2019). Bioleaching involves solubilization followed by purification such as ion exchange, selective precipitation, adsorption, and membrane separation (Rohwerder et al. 2003).

A crucial step in bioremediation involves biotransformation where a chemical compound is altered structurally so that the toxicity, in the relatively more polar form, is reduced by metal –microbe interaction (Pande et al. 2022). Examples include *Micrococcus* sp. and *Acinetobacter* sp. involved in oxidation of arsenic in its less toxic form (Nagvenkar and Ramaiah 2010; Thatoi et al. 2014). Such microbial transformation is developed through isomerization, oxidation, methylation, and attachment to new functional groups, forming new carbon bonds. These also aid in adaptation of microbes to their environment, as aerobic and anaerobic microbes thriving in different temperature and different optimal pH. Joint methods of phytoremediation and bioaugmentation effectively may lead to the cleaning up process known as rhizoremediation (Pande et al. 2022). Plant growth-promoting bacteria and those residing in the rhizosphere are an efficient and ecofriendly method for elimination of heavy metals from plants (Ali et al. 2013; Pande et al. 2022; Sati et al. 2023).

Microbes are mixed with nanomaterials to form a nanocomposite, e.g., immobilization of *Halomonas* sp., for the removal of lead and cadmium (Cao et al. 2020). Advanced molecular techniques and omics based techniques can be used for better ways to deal with metal pollution (Bosecker 1997; Pande et al. 2022; Sayqal and Ahmed 2021).

Production of siderophores by microbes can lead to formation of stable complexes (Neubauer et al. 2000; Rajkumar et al. 2010). Nanosorbents enhance chemical activity and adsorption capacity, amplify surface binding energy, and minimize internal diffusion resistance.

Community structure of microbes and biofilm formation depends on minerals and mineral composition on the surface (Zhou et al. 2021). The interior bulk of the metallic site may vary from the site to which the microbes attach themselves as the surface is exposed to weathering and episodes of weathering and leaching (Boeva et al. 2021).

Acid mine drainage is generated by metal sulfide oxidation (Brimhall et al. 2005). Wastes from acid mine drainage include sulfide minerals such as pyrite, marcasite, galena, chalcopyrite, and arsenopyrite (Kang et al. 2001). In sulfidic mines, fermentation and respiration are ways enabling carbon cycling. Microbes such as *Veillonella*, *Staphylococus*, and *Clostridium* spp. generate byproducts via fermentation. These organic substances can thereby be used as substrates by variety of sulfate-reducing microbes and methanogens. Acid mine waters can be a reservoir of iron-oxidizing bacteria such as *Ferrovum* sp. and *Gallionella* sp. (Hedrich et al. 2011; Tischler et al. 2013; Ortiz-Castillo et al. 2021). In presence of iron-oxidizing bacterium, schwertmannite formation may take place via three methods: Primary nucleation stage, formation of minerals ferrihydrite and schwertmannite, and their transformation into goethite (in presence of organic carbon) and lepidocrocite. Jarosite and alunite environments harbor microbial environments which play a key role in the acidic nature of systems and subsequent biochemical changes of microbes generate conditions that favor alunite formation (Potter-McIntyre and McCollom 2018). These acidic environments facilitate colonization within porous layers of iron heap precipitate. The reduction of iron sulfate minerals like jarosite helps in reduction of toxic heavy metals under natural conditions (Wray 2011; Nazari et al. 2014). Metal-reducing organisms also reduce Cr(VI) to Cr(III) as a detoxification mechanism (Zhou et al. 2012; Xu et al. 2013).

Carbonate materials are used by microbes to interfere neutralization processes as they attach to high energy sites and prevent dissolution. In biogeochemical weathering, microbial transformation of silica minerals such as vermiculite aids in meeting their nutritional requirements, as it is accompanied by the release of K, Mg, and Al (Ivanova and Chizhikova 2014).

9 Prospects and Benefits of Using Microbes in Ore Explorations

Biocontrol measures involve many approaches: bactericides which inhibit microbially mediated constituent release from ores, encapsulating minerals to prevent oxidation, enhancing biofilm growth of microbes by supplementing inorganic nutrients, and introducing organic amendments (Ortiz-Castillo et al. 2021). For bioremediation, organism like *Shewanella* sp. which utilize nitrate and take part in iron cycling can be used to address nitrate pollution (Liu et al. 2019; Lu et al. 2021). Additionally, engineered microbes in natural environments such as *Geobacter sulfurreducens* are electroactive and can potentially generate current in biochemical systems (Logan et al. 2019).

Microbes while aiding acid mine drainage and metal leaching also lead in metal immobilization such as conversion of sulfate to sulfide (Ayangbenro et al. 2018). Microbial consortia are involved in various enzymatic and nonenzymatic processes which influence and are also affected by the biogeochemical processes such as formation, degradation, alteration of secondary minerals, and mobilization of metals (Lybrand et al. 2019). Thus, they can be used to understand weathering levels in an ore site (Courchesne et al. 2021). Presence of *Acidithiobacillus ferrooxidans* and *Leptospirillum* spp. indicates an acidic environment while archaea such as *Ferroplasma acidiphilum* are seen in oxidized zones (Huang et al. 2011). Augmentation of iron- and sulfur-oxidizing microbes indicates acid drainage generation and loss of vegetation. Likewise, increase in neutrophilic heterotrophic microbes indicates successful plant establishment (Schippers et al. 2000; Londry and Sherriff 2005). This information is also helpful in developing effective strategies to manage microbially mediated contaminant release at the source.

Bio-desulfurization involves removal of pyrite, which contributes the maximum in acid mine drainage from coal by microbes, thus contributing to an environment-friendly and economical approach to handle mine waste (Joulian et al. 2020). Ore deposits have a concentrated amount of chemical elements that are valuable economically. However, as they exist in minimal amount, they do not add up sufficiently for extraction purposes (Levett et al. 2021). Bioleaching makes recovery easier from low-grade ores and can concentrate metals from wastes making the process environment-friendly and economical. Biosorption, as the name suggests, involves microbial cell surface adsorption of metals from mine wastes or dilute mixtures (Sánchez-Castro et al. 2023). This too involves removal of toxic metals from industrial effluents and recovery of valuable and toxic metals. Biosorption processes employ a variety of bacteria, algae, yeasts, and molds for effluent treatment and metal recovery (Levett et al. 2021). Microbial membranes being negatively charged enable metallic ions to adhere on the microbial surfaces.

10 Mineral Deposition in Bacterial Cell Envelope (Fig. 4)

Sulfur deposition originally believed to have occurred in the cytoplasm was eventually found to be occurring in the periplasm. Common examples include *Beggiatoa* and *Thiothrix* sp. and *Thiovulum* strains (Ehrlich 1999). When selenite and tellurite are enzymatically reduced, it leads to formation of the elemental Se and Te (Gerrard et al. 1974; Ehrlich 1999). Se deposition was observed in cell envelope of *E. coli* and Te in *Bacillus subtilis* and *Proteus vulgaris* (van Iterson and Leene 1964). The cyanobacterium *Synechococcus* is epicellularly involved in CaCo3 shedding

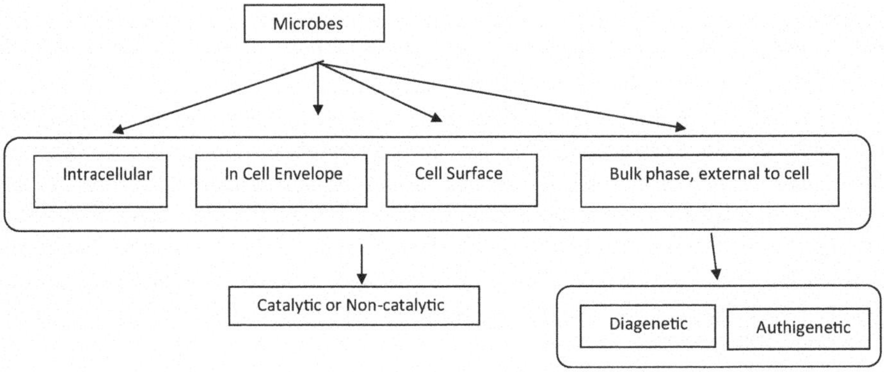

Fig. 4 Mineral deposition mediated by microbes in various locations of the cell

(Thompson 1998). This CaCo3 thus formed is utilized in forming bioherms (Thompson and Ferris 1990).Calcareous algae involved in precipitation include *Chara*, *Halimeda*, and *Emiliania huxleyi* (De Vrind-Jong and De Vrind 1997). In coccolithophores, such as *Pleurochrysis carterae*, it is the cytoplasm which is involved in the formation of calciferous scales (Van der Wal et al. 1983).

11 Microbial Mineral Nucleation

Diatoms have frustules where silica is deposited (de Vrind-Jong and de Vrind 1997). In mycelium of fungi, *Rhizopus arrhizus*, precipitation as uranyl hydroxide takes place (Tsezos and Volesky 1982). *Citrobacter* sp. can have the ability to adsorb uranyl ions and Cd^{2+}, Cu^{2+}, Mn^{2+}, Pb^{2+}, La^{3+}, Am^{3+}, Th^{4+}, Pu^{4+} ions followed by subsequent deposition as phosphate salts (Macaskie and Dean 1984; Macaskie et al. 1992). This bacteria produces an excess of acid phosphatase, which thereby on cleaving glycerol 2-phosphate deposits phosphate salts (Montgomery et al. 1995).

Oxidation of Fe^{2+} to Fe^{3+} by *Gallionella ferruginea* (Lütters and Hanert 1989; Hallbeck and Pedersen 1991) results in deposition of amorphous FeOH in the bacterial stalk (Ghiorse 1984), whereas *Leptothrix discophora* deposits in its sheath (De Vrind-Jong et al. 1990; Corstjens et al. 1992). The bacteria where ferric iron gets deposited in exopolysaccharide layer include *Siderocapsa*, *Naumanniella*, *Ochrobium*, *Siderococcus*, *Pedomicrobium*, and *Seliberia* (Ghiorse and Ehrlich 1992). Manganese oxide deposition can also be observed in budding bacteria such as *Hyphomicrobium* and *Pedomicrobium* (Aristovskaya 1963; Tyler and Marshall 1967). *Geobacter metallireducens*, present in river sediments, can produce magnetite in bulk phase on reduction of amorphous ferric oxide in presence of acetate (Lovley et al. 1987). *Thiobacillus thioparus* can oxidize sulfur under restricted supply of oxygen (Sokolova et al. 1968; van den Ende and van Gemerden 1993). *Chlorobium* produces sulfur which remains adhered to the cell surface (van Gemerden 1986). *Pseudomonas fluorescens* and *Bacillus* sp. can aerobically reduce chromium in bulk phase while aerobic process is followed by *Enterobacter cloacae* for chromium reduction (Bopp and Ehrlich 1988; DeLeo and Ehrlich 1994; Ohtake et al. 1990; Wang and Xiao 1995; Wang and Shen 1995).

For formation of uraninite ore, *Desulfovibrio desulfuricans*, *Geobacter metallireducens*, and *Shewanella putrefaciens* can enzymatically reduce metal oxides to metal ores (Lovley et al. 1991; Gorby and Lovley 1992; Lovley and Phillips 1992a, 1992b). Ore formation and mineralization may be biologically induced (Lowenstam and Weiner 1989) or biologically controlled (Mann 1995). The former happens without the aid of any cell structures or appendages whereas if mineral formation occurs inside vesicles, or structures it is known as biologically controlled mineral formation (Mann 1995).

12 Microbiota in Extremes: Volcanoes and Polar Regions

Edaphic factors and vegetative growth of the region also influence microbial growth and network formation (Chen et al. 2020). Volcanic gases have a huge amount of sulfur present as sulfur dioxide or hydrogen sulfide and water vapor (Garcia-Lopez et al. 2022). Thus, Arctic glaciers are a habitat of *Thiomonas thermosulfata*, *Thermodesulfovibrio thiophilus*, and *Desulfovibrio*. Disintegration and eroding of volcanic rocks and ion exchange between the pyroclasts and the glacier runoff are also aided by microbes. The basalt-laden rocks serve as an energy reservoir for chemolithotrophic bacteria as they contain a sizeable proportion of reduced iron and manganese. Antarctic glaciers also harbor iron-oxidizing bacteria such as *Ferrimicrobium acidiphilum*, simultaneously with iron-reducing bacteria such as *Rhodoferax ferrireducens* and *Deferribacter autotrophicus*. In glaciers, the commonly found genera include *Hymenobacter*, *Actinoallomurus*, *Glaciimonas*, *Frankia*, *Pontibacter*, *Symploca*, and *Cyanobacterium*. In the Arctic region, those identified also include *Deinococcus*, *Polaromonas*, *Metylobacterium*, *Dinitrobacterium*, and *Nitriliruptor*. Those exclusive to Antarctic glaciers are *Salinibacterium*, *Flavobacterium*, and *Singulisphaera*. Submarine volcanic environments are said to be a rich source of many iron oxide deposits detected on the sea floor (Jiang and Wang 2005). Certain microbes present in volcanic areas such as those belonging to *Conexibacter*, Agricales, and Gaiellales possess an efficient adaptability to environmental fluctuations and reabsorption of minerals like sodium and calcium (Wang et al. 2024). Bacteria experience greater effect than fungal communities due to changes caused by volcanic disturbances. A case study in South China revealed organic shale rock loaded with minerals and organic matter was the result of volcanic eruptions. It was suggested that mild volcanic activity can be a source of nutrients like Fe, Si, Mn, N, P, and organic matter. An ecosystem exhibiting volcanic activity can have potent effects on growth of microbiota in the area (Wang et al. 2024). In an excess of ash leachate from volcanic eruption, it could be seen that the increased nutrient content present in the ash enhanced microbial growth, e.g., increased synthesis of chlorophyll-a by *Anabaena pseudoichthyoides*. This study also showed the four stages of growth including acclimatization, flourishing, growth decline, and reaching an equilibrium threshold. The treatment group which received volcanic ash and granite had a faster growth rate than the control. Bacteria possess diversity in molecular and metabolic aspects such that they modify the chemistry of their environment in different ways that benefits them.

In an environment comprising of heavy metals, microbes may develop resistance to heavy metals using mechanisms such as biosorption, bioaccumulation, and transfer (Dexter-Dyer et al. 1984). Bacterial plasmids can confer heavy metal resistance to the bacteria as it may transmit genes for arsenic, bismuth, cobalt, lead, nickel, silver, and mercury. Thus, a survival is ensured for such microbes in these heavy metal-contaminated environments. Likewise, microalgae can also uptake metals like zinc and mercury, and is influenced positively in presence of light. Affinity of metal accumulation may vary in microalgae such as Hg > Cd > Pb > Zn as observed in *Chlamydomonas.* Certain anaerobic bacteria such as *Serratia marcescens* and *Agrobacterium tumefaciens* highlight the possibility of microbes solubilizing gold. Bacteria like *Bacillus megaterium, B. mesentericus*, *Pseudomonas liquefaciens*, and *Bacterium nitrificans* can dissolve gold by producing amino acids. Their mutant strains produce amino acids in excess leading to greater effectiveness. *Thiobacillus ferroxidans* may also be involved in gold dissolution. A study by Korobushkina et al. (1976) showed that gold is soluble in certain amino acids as gold and nitrogen bonds can be formed in alkaline media. Those amino acids with strong tendency for dissolving gold include histidine, asparagine, serine, aspartic acid, alanine, and glycine.

A study by Reimer (1975) explains that sedimentary Archean gold deposits, as seen in Witwatersrand, are a result of moderate natural processes such as bacteria metabolizing gold under either aerobic and anaerobic conditions or environmental factors. The biological activity on gold can induce reactions even on an inert metal like gold. These may involve processes such as adsorption, dissolution of gold from minerals, organic matter degradation, concentration in oxidizing and reducing environments, and redepositing near fossil remains. A possibility remains that large humic complexes could have accumulated colloidal gold particles from weathering of conglomerate rocks (Bertrand 1972; Nagy 1981).

Iron and manganese metabolism and precipitation can take place in a solitary microbial mat although the metals may be present in microbial mats in oxidizing and reducing layers, in an alternate manner (Krumbein 1979) as revealed from Sabkha-type sediments of Precambrian rocks, known commonly as stromatolites. These laminated rocks thus have been known to be at a marginal border between marine environment and weathering. The former environment provides appropriate environment for pH and nutrients while the latter supplies the dissolved metals. Additionally, lagoon environments too can serve as a reservoir for material from both terrestrial and marine origin. Fungal stromatolites have been discovered additionally (Kretzschmar 1982), which accumulate large amounts of iron and manganese, similar to pre-Cambrian banded iron information.

Horikoshi et al. (1981), observed that accumulation tendency of uranium decreased in the order actinomycete, bacteria, yeasts, fungi. Strong chelators such as EDTA could facilitate release of metals and ligand adsorption and has a major role in the uptake process. Dead cyanobacteria and algae are also involved in heavy metal adsorption (Ferguson and Bubela 1974). Metal ion accumulation may occur while being in contact with previously formed ion rather than biosorbent (Beveridge 1978).

References

Ali H, Khan E, Sajad MA (2013) Phytoremediation of heavy metals—concepts and applications. Chemosphere 91(7):869–881

Aristovskaya TV (1963) The natural forms of existence of soil bacteria

Atlas RM (1981) Microbial degradation of petroleum hydrocarbons: an environmental perspective. Microbiol Rev 45(1):180–209

Ayangbenro AS, Olanrewaju OS, Babalola OO (2018) Sulfate-reducing bacteria as an effective tool for sustainable acid mine bioremediation. Front Microbiol 9:1986

Bertrand D (1972) Interactions entre éléments minéraux et microorganismes du sol

Beveridge TJ (1978) The response of cell walls of Bacillus subtilis to metals and to electron-microscopic stains. Can J Microbiol 24(2):89–104

Boeva N, Bortnikov N, Slukin A, Shipilova E, Makarova M, Melnikov P (2021) Biofilms and biominerals in the lateritic weathering crust as exemplified by the central bauxite deposit (Siberian platform, Russia). Minerals 11(11):1184

Bopp LH, Ehrlich HL (1988) Chromate resistance and reduction in Pseudomonas fluorescens strain LB300. Arch Microbiol 150:426–431

Bosecker K (1997) Bioleaching: metal solubilization by microorganisms. FEMS Microbiol Rev 20(3-4):591–604

Bradshaw AD, McNeilly TS, Gregory RPG (1965) Industrialization, evolution and the development of heavy metal tolerance in plants. Ecol Ind Soc 5:327–343

Brimhall GH, Alpers CN, Swayze GA (2005) Characterization of waste rock associated with acid drainage at the Penn Mine, California, by ground-based visible to short-wave infrared reflectance spectroscopy assisted by digital mapping. Chem Geol 215(1-4):453–472

Cao X, Alabresm A, Chen YP, Decho AW, Lead J (2020) Improved metal remediation using a combined bacterial and nanoscience approach. Sci Total Environ 704:135378

Chen J, Guo Y, Li F, Zheng Y, Xu D, Liu H, Bao Y (2020) Exploring the effects of volcanic eruption disturbances on the soil microbial communities in the montane meadow steppe. Environ Pollut 267:115600

Chen J, Liu Y, Diep P, Mahadevan R (2022) Harnessing synthetic biology for sustainable biomining with Fe/S-oxidizing microbes. Front Bioeng Biotechnol 10:920639

Corstjens PL, De Vrind JP, Westbroek P, De Vrind-Jong EW (1992) Enzymatic iron oxidation by Leptothrix discophora: identification of an iron-oxidizing protein. Appl Environ Microbiol 58(2):450–454

Courchesne B, Schindler M, Mykytczuk NC (2021) Relationships between the microbial composition and the geochemistry and mineralogy of the cobalt-bearing legacy mine tailings in Northeastern Ontario. Front Microbiol 12:660190

De Vrind-Jong EW, De Vrind JP (1997) Algal deposition of carbonates and silicates. Rev Mineral 35:267–308

De Vrind-Jong ED, Corstjens PLAM, Kempers ES, Westbroek P, De Vrind JPM (1990) Oxidation of manganese and iron by Leptothrix discophora: use of N, N, N′, N′-tetramethyl-p-phenylenediamine as an indicator of metal oxidation. Appl Environ Microbiol 56(11):3458–3462

DeLeo PC, Ehrlich HL (1994) Reduction of hexavalent chromium by Pseudomonas fluorescens LB300 in batch and continuous cultures. Appl Microbiol Biotechnol 40:756–759

Dexter-Dyer B, Kretzschmar M, Krumbein WE (1984) Possible microbial pathways in the formation of Precambrian ore deposits. J Geol Soc 141(2):251–262

Dong H, Huang L, Zhao L, Zeng Q, Liu X, Sheng Y, Chen H (2022) A critical review of mineral–microbe interaction and co-evolution: mechanisms and applications. Natl Sci Rev 9(10):128

Douglas S, Beveridge TJ (1998) Mineral formation by bacteria in natural microbial communities. FEMS Microbiol Ecol 26(2):79–88

Ehrlich HL (1999) Microbes as geologic agents: their role in mineral formation. Geomicrobiol J 16(2):135–153

Ferguson J, Bubela B (1974) The concentration of Cu (II), Pb (II) and Zn (II) from aqueous solutions by particulate algal matter. Chem Geol 13(3):163–186

Gadd GM (2010) Metals, minerals and microbes: geomicrobiology and bioremediation. Microbiology 156(3):609–643

Garcia-Lopez E, Ruiz-Blas F, Sanchez-Casanova S, Peña Perez S, Martin-Cerezo ML, Cid C (2022) Microbial communities in volcanic glacier ecosystems. Front Microbiol 13:825632

Gavrilescu M (2022) Microbial recovery of critical metals from secondary sources. Bioresour Technol 344:126208

Gerrard TL, Telford JN, Williams HH (1974) Detection of selenium deposits in Escherichia coli by electron microscopy. J Bacteriol 119(3):1057–1060

Ghiorse WC (1984) Biology of iron-and manganese-depositing bacteria. Ann Rev Microbiol 38:515–550

Ghiorse WC, Ehrlich HL (1992) Microbial biomineralization of iron and manganese. Health Environ Res Online 21:75–99

Giller KE, Witter E, Mcgrath SP (1998) Toxicity of heavy metals to microorganisms and microbial processes in agricultural soils: a review. Soil Biol Biochem 30(10-11):1389–1414

Gorby YA, Lovley DR (1992) Enzymic uranium precipitation. Environ Sci Technol 26(1):205–207

Gross M (2018) Mining the mineral microbiome. Curr Biol 28(23):R1325–R1328

Hallbeck L, Pedersen K (1991) Autotrophic and mixotrophic growth of Gallionella ferruginea. Microbiology 137(11):2657–2661

Hedrich S, Lünsdorf H, Kleeberg R, Heide G, Seifert J, Schlömann M (2011) Schwertmannite formation adjacent to bacterial cells in a mine water treatment plant and in pure cultures of Ferrovum myxofaciens. Environ Sci Technol 45(18):7685–7692

Heinrich CA, Candela PA (2014) Fluids and ore formation in the Earth's crust. In: Treatise on geochemistry, vol 13, 2nd edn. Elsevier, Amsterdam, pp 1–28

Hinojosa MB, Carreira JA, Rodríguez-Maroto JM, García-Ruíz R (2008) Effects of pyrite sludge pollution on soil enzyme activities: Ecological dose–response model. Sci Total Environ 396(2-3):89–99

Hofmann BA (2011) Ores, microbial precipitation and oxidation. In: Encyclopedia of geobiology. Springer, Cham

Horikoshi T, Nakajima A, Sakaguchi T (1981) Studies on the accumulation of heavy metal elements in biological systems: XIX. Accumulation of uranium by microorganisms. Eur J Appl Microbiol Biotechnol 12:90–96

Huang LN, Zhou WH, Hallberg KB, Wan CY, Li J, Shu WS (2011) Spatial and temporal analysis of the microbial community in the tailings of a Pb-Zn mine generating acidic drainage. Appl Environ Microbiol 77(15):5540–5544

Islam E, Paul D, Sar P (2014) Microbial diversity in Uranium deposits from Jaduguda and Bagjata Uranium mines, India as revealed by clone library and denaturing gradient gel electrophoresis analyses. Geomicrobiol J 31(10):862–874

Ivanova E, Chizhikova N (2014) Degradative crystal–chemical transformations of clay minerals under the influence of cyanobacterium-actinomycetal symbiotic associations. Eur J Soil Sci 3(2):116–122

Jalilvand N, Akhgar A, Alikhani HA, Rahmani HA, Rejali F (2020) Removal of heavy metals zinc, lead, and cadmium by biomineralization of urease-producing bacteria isolated from Iranian mine calcareous soils. J Soil Sci Plant Nutr 20:206–219

Jiang FZ, Wang YW (2005) Marine volcanic rocks and related metallic ore deposits. Metallurgical Industry Press, Beijing

Joulian C, Fonti V, Chapron S, Bryan CG, Guezennec AG (2020) Bioleaching of pyritic coal wastes: bioprospecting and efficiency of selected consortia. Res Microbiol 171(7):260–270

Kang JK, Song Y, Moon JW, Moon HS (2001) Water quality impact of mining in the Wolmyoung area of Korea, and its short-term changes. Water Air Soil Pollut 129:349–367

Klusman RW (1993) Soil gas and related methods for natural resource exploration. Wiley, Chichester

Korobushkina ED, Mineev GG, Praded GP (1976) Mechanism of microbiological dissolution of gold. Mikrobiologiia 45:535–538

Kraemer SM, Duckworth OW, Harrington JM, Schenkeveld WD (2015) Metallophores and trace metal biogeochemistry. Aquat Geochem 21:159–195

Kretzschmar M (1982) Fossil fungi in iron stromatolites from Warstein (Rhenish Massif, Northwestern Germany). Facies 7:237–259

Krumbein WE (1979) Photolithotropic and chemoorganotrophic activity of bacteria and algae as related to beachrock formation and degradation (Gulf of Aqaba, Sinai). Geomicrobiol J 1(2):139–203

Kumar D, Kumar D (2018) Sustainable management of coal preparation. Woodhead Publishing, Sawston

Lado LR, Hengl T, Reuter HI (2008) Heavy metals in European soils: a geostatistical analysis of the FOREGS geochemical database. Geoderma 148(2):189–199

Lepleux C, Turpault MP, Oger P, Frey-Klett P, Uroz S (2012) Correlation of the abundance of Betaproteobacteria on mineral surfaces with mineral weathering in forest soils. Appl Environ Microbiol 78(19):7114–7119

Levett A, Gleeson SA, Kallmeyer J (2021) From exploration to remediation: a microbial perspective for innovation in mining. Earth Sci Rev 216:103563

Liu T, Chen D, Li X, Li F (2019) Microbially mediated coupling of nitrate reduction and Fe (II) oxidation under anoxic conditions. FEMS Microbiol Ecol 95(4):fiz030

Logan BE, Rossi R, Ragab AA, Saikaly PE (2019) Electroactive microorganisms in bioelectrochemical systems. Nat Rev Microbiol 17(5):307–319

Londry KL, Sherriff BL (2005) Comparison of microbial biomass, biodiversity, and biogeochemistry in three contrasting gold mine tailings deposits. Geomicrobiol J 22(5):237–247

Lovley DR, Phillips EJ (1992a) Bioremediation of uranium contamination with enzymatic uranium reduction. Environ Sci Technol 26(11):2228–2234

Lovley DR, Phillips EJ (1992b) Reduction of uranium by Desulfovibrio desulfuricans. Appl Environ Microbiol 58(3):850–856

Lovley DR, Stolz JF, Nord GL Jr, Phillips EJ (1987) Anaerobic production of magnetite by a dissimilatory iron-reducing microorganism. Nature 330(6145):252–254

Lovley DR, Phillips EJ, Gorby YA, Landa ER (1991) Microbial reduction of uranium. Nature 350(6317):413–416

Lowenstam HA, Weiner S (1989) On biomineralization. Oxford University Press, New York

Lu Y, Liu H, Huang X, Xu L, Zhou J, Qian G, Chen X (2021) Nitrate removal during Fe (III) bioreduction in microbial-mediated iron redox cycling systems. Water Sci Technol 84(4):985–994

Lütters S, Hanert HH (1989) The ultrastructure of chemolithoautotrophic Gallionella ferruginea and Thiobacillus ferrooxidans as revealed by chemical fixation and freeze-etching. Arch Microbiol 151:245–251

Lybrand RA, Austin JC, Fedenko J, Gallery RE, Rooney E, Schroeder PA, Qafoku O (2019) A coupled microscopy approach to assess the nano-landscape of weathering. Sci Rep 9(1):5377

Macaskie LE, Dean ACR (1984) Cadmium accumulation by a Citrobacter sp. Microbiology 130(1):53–62

Macaskie LE, Empson RM, Cheetham AK, Grey CP, Skarnulis AJ (1992) Uranium bioaccumulation by a Citrobacter sp. as a result of enzymically mediated growth of polycrystalline HUO_2PO_4. Science 257(5071):782–784

Macqueen RW (1985) Symposium: Role of organisms and organic matter in ore deposition/Le rôle des organismes et de la matière organique dans la formation des gisements métallifères. Can J Earth Sci 22(12):1890–1892

Mann S (1995) Biomineralization and biomimetic materials chemistry. J Mater Chem 5(7):935–946

Margaryan A (2021) Diversity and application of heavy-metal resistant microbes. In: Microbes in microbial communities: ecological and applied perspectives. Springer, Singapore, pp 153–174

Montgomery DM, Dean AC, Wiffen P, Macaskie LE (1995) Phosphatase production and activity in Citrobacter freundii and a naturally occurring, heavy-metal-accumulating Citrobacter sp. Microbiology 141(10):2433–2441

Nagvenkar GS, Ramaiah N (2010) Arsenite tolerance and biotransformation potential in estuarine bacteria. Ecotoxicology 19:604–613

Nagy B (1981) The geochemistry of organic matter in ore deposits. Carnegie Inst Washington Geophys Lab 9:7–99

Nazari B, Jorjani E, Hani H, Manafi Z, Riahi A (2014) Formation of jarosite and its effect on important ions for Acidithiobacillus ferrooxidans bacteria. Trans Nonferrous Metals Soc China 24(4):1152–1160

Neubauer U, Furrer G, Kayser A, Schulin R (2000) Siderophores, NTA, and citrate: potential soil amendments to enhance heavy metal mobility in phytoremediation. Int J Phytoremediation 2(4):353–368

Newsome L, Falagán C (2021) The microbiology of metal mine waste: bioremediation applications and implications for planetary health. GeoHealth 5(10):e2020GH000380

Ohtake H, Fujii E, Toda K (1990) Reduction of toxic chromate in an industrial effluent by use of a chromate-reducing strain of Enterobacter cloacae. Environ Technol 11(7):663–668

Ortiz-Castillo JE, Mirazimi M, Mohammadi M, Dy E, Liu W (2021) The role of microorganisms in the formation, dissolution, and transformation of secondary minerals in mine rock and drainage: a review. Minerals 11(12):1349

Pande V, Pandey SC, Sati D, Bhatt P, Samant M (2022) Microbial interventions in bioremediation of heavy metal contaminants in agroecosystem. Front Microbiol 13:824084

Potter-McIntyre SL, McCollom TM (2018) Jarosite and alunite in ancient terrestrial sedimentary rocks: reinterpreting Martian depositional and diagenetic environmental conditions. Life 8(3):32

Rajkumar M, Ae N, Prasad MNV, Freitas H (2010) Potential of siderophore-producing bacteria for improving heavy metal phytoextraction. Trends Biotechnol 28(3):142–149

Rasheed MA, Lakshmi M, Srinu D, Dayal AM (2011) Bacteria as indicators for finding oil and gas reservoirs: a case study of the Bikaner-Nagaur Basin, Rajasthan, India. Pet Sci 8:264–268

Reimer TO (1975) The age of the witwatersrand system and other gold-uranium placers: implications on the origin of the mineralisation. N Jahrbuch Min Monatshaft 1975:79–98

Reith F, Rogers SL, McPhail DC, Webb D (2006) Biomineralization of gold: biofilms on bacterioform gold. Science 313(5784):233–236

Robles-Fernández A, Areias C, Daffonchio D, Vahrenkamp VC, Sánchez-Román M (2022) The role of microorganisms in the nucleation of carbonates, environmental implications and applications. Minerals 12(12):1562

Rohwerder T, Gehrke T, Kinzler K, Sand W (2003) Bioleaching review part A: progress in bioleaching: fundamentals and mechanisms of bacterial metal sulfide oxidation. Appl Microbiol Biotechnol 63:239–248

Römer R, Schwartz W (1965) Geomicrobiological investigations V. Utilization of sulfate minerals and heavy metal tolerance in desulfurizers. J Gen Microbiol 5(2):122–135

Sajjad W, Zheng G, Din G, Ma X, Rafiq M, Xu W (2019) Metals extraction from sulfide ores with microorganisms: the bioleaching technology and recent developments. Trans Indian Inst Metals 72:559–579

Sana S, Neelam D, Gupta V, Devki D, Rahi R (2021) An overview: application of microorganisms in bio-mining of metals. Int J Pharm Bio Sci 11(1):1–8

Sánchez-Castro I, Molina L, Prieto-Fernández MÁ, Segura A (2023) Past, present and future trends in the remediation of heavy-metal contaminated soil-remediation techniques applied in real soil-contamination events. Heliyon 9(6):e16692

Sarkodie EK, Jiang L, Li K, Yang J, Guo Z, Shi J, Liu X (2022) A review on the bioleaching of toxic metal (loid)s from contaminated soil: Insight into the mechanism of action and the role of influencing factors. Front Microbiol 13:1049277

Sati D, Pande V, Pandey SC, Samant M (2023) Recent advances in PGPR and molecular mechanisms involved in drought stress resistance. J Soil Sci Plant Nutr 23(1):106–124
Sayqal A, Ahmed OB (2021) Advances in heavy metal bioremediation: an overview. Appl Bionics Biomech 2021(1):1609149
Schippers A, Jozsa PG, Sand W, Kovacs ZM, Jelea M (2000) Microbiological pyrite oxidation in a mine tailings heap and its relevance to the death of vegetation. Geomicrobiol J 17(2):151–162
Sharma I (2020) Bioremediation techniques for polluted environment: concept, advantages, limitations, and prospects. In: Trace metals in the environment-new approaches and recent advances. IntechOpen, London
Sher S, Rehman A (2019) Use of heavy metals resistant bacteria—a strategy for arsenic bioremediation. Appl Microbiol Biotechnol 103:6007–6021
Sokolova GA, Karavaĭko GI, Halpern Y, Rabinovitz E (1968) Physiology and geochemical activity of Thiobacilli. (No Title)
Southam G, Saunders JA (2005) The geomicrobiology of ore deposits. Econ Geol 100(6):1067–1084
Sun W, Xiao E, Xiao T, Krumins V, Wang Q, Haggblom M, Liu W (2017) Response of soil microbial communities to elevated antimony and arsenic contamination indicates the relationship between the innate microbiota and contaminant fractions. Environ Sci Technol 51(16):9165–9175
Thatoi H, Das S, Mishra J, Rath BP, Das N (2014) Bacterial chromate reductase, a potential enzyme for bioremediation of hexavalent chromium: a review. J Environ Manag 146:383–399
Thompson JB (1998) Bacterial carbonate precipitation. Oral presentation at symposium 87/J: bacterial mineral precipitation. American Society for Microbiology, 98th General Meeting, Atlanta, GA
Thompson JB, Ferris FG (1990) Cyanobacterial precipitation of gypsum, calcite, and magnesite from natural alkaline lake water. Geology 18(10):995–998
Tischler JS, Wiacek C, Janneck E, Schlömann M (2013) Microbial abundance in the schwertmannite formed in a mine water treatment plant. Mine Water Environ 4(32):258–265
Tsezos M, Volesky B (1982) The mechanism of uranium biosorption by Rhizopus arrhizus. Biotechnol Bioeng 24(2):385–401
Tyler PA, Marshall KC (1967) Form and function in manganese-oxidizing bacteria. Arch Mikrobiol 56:344–353
van den Ende FP, van Gemerden H (1993) Sulfide oxidation under oxygen limitation by a Thiobacillus thioparus isolated from a marine microbial mat. FEMS Microbiol Ecol 13(1):69–77
Van Der Wal P, De Jong EW, Westbroek P, De Bruijn WC, Mulder-Stapel AA (1983) Polysaccharide localization, coccolith formation, and Golgi dynamics in the coccolithophorid Hymenomonas carterae. J Ultrastruct Res 85(2):139–158
van Gemerden H (1986) Production of elemental sulfur by green and purple sulfur bacteria. Arch Microbiol 146:52–56
van Iterson W, Leene W (1964) A cytochemical localization of reductive sites in a Gram-negative bacterium: tellurite reduction in Proteus vulgaris. J Cell Biol 20(3):377–387
Velásquez L, Dussan J (2009) Biosorption and bioaccumulation of heavy metals on dead and living biomass of Bacillus sphaericus. J Hazard Mater 167(1-3):713–716
Wang YT, Shen H (1995) Bacterial reduction of hexavalent chromium. J Ind Microbiol Biotechnol 14(2):159–163
Wang YT, Xiao C (1995) Factors affecting hexavalent chromium reduction in pure cultures of bacteria. Water Res 29(11):2467–2474
Wang C, Che Q, Luo B, Zhu Y, Liu J, Tang M (2024) The impacts of volcanic activity on microbial growth—a simulation experiment in the qiliao section in Shizhu County, Chongqing, China. Biology 13(11):895
Wray RA (2011) Alunite formation within silica stalactites from the Sydney Region, South-eastern Australia

Xie Y, Luo H, Du Z, Hu L, Fu J (2014) Identification of cadmium-resistant fungi related to Cd transportation in bermudagrass [Cynodon dactylon (L.) Pers.]. Chemosphere 117:786–792

Xie Y, Fan J, Zhu W, Amombo E, Lou Y, Chen L, Fu J (2016) Effect of heavy metals pollution on soil microbial diversity and bermudagrass genetic variation. Front Plant Sci 7:755

Xu Z, Lǚ B, Wu J, Zhou L, Lan Y (2013) Reduction of Cr (VI) facilitated by biogenetic jarosite and analysis of its influencing factors with response surface methodology. Mater Sci Eng C 33(7):3723–3729

Zhou P, Li Y, Shen Y, Lan Y, Zhou L (2012) Facilitating role of biogenetic schwertmannite in the reduction of Cr (VI) by sulfide and its mechanism. J Hazard Mater 237:194–198

Zhou H, Jiang L, Li K, Chen C, Lin X, Zhang C, Xie Q (2021) Enhanced bioremediation of diesel oil-contaminated seawater by a biochar-immobilized biosurfactant-producing bacteria Vibrio sp. LQ2 isolated from cold seep sediment. Sci Total Environ 793:148529

Harnessing the Ability of Cyanobacterium *Anabaena variabilis* for Bioremediation of Waste Generated from Oil Refineries

Sibange Paul, Barihun Thyrniang, Sumit Deb, and Samrat Adhikari

Contents

Abstract ETP and tank bottom sludge are two forms of effluents from oil refineries, generated more in quantity than the usable oil products. These are rich in long- and short-chain hydrocarbons and heavy metals defining a toxic composition competent to deplete the surrounding soil and water quality if discarded untreated. The unassisted decomposition of these types of sludge are time-consuming; thus, a catalyzed decomposition by virtue of natural decomposing agents could solve the matter. Features such as positive effect, low cost, lack of secondary pollution, and little environmental influence make bioremediation utilizing petroleum hydrocarbon-degrading organisms an efficient and eco-friendly technology. Cyanobacteria are recognized as biogenic contributors in oil formation and thus are supposed to have in fact a better ability to decompose oil and its derivatives. Thus, the chapter is

S. Paul · B. Thyrniang · S. Adhikari (✉)
Advanced Biotechnology Research Laboratory, St. Edmund's College, Shillong, Meghalaya, India
e-mail: researchhub1@sec.edu.in

S. Deb
Department of Chemistry, St. Edmund's College, Shillong, Meghalaya, India

S. R. Joshi et al. (eds.), *Mineral Transformation and Bioremediation by Geo-Microbes*, Interdisciplinary Biotechnological Advances,
https://doi.org/10.1007/978-981-96-3033-2_21

devoted to the application and assessment of a cyanobacterial species *Anabaena variabilis* for its potential capacity for bioremediation of petroleum hydrocarbons.

Keywords Bioremediation · GC-FID · Cyanobacteria · Crude oil · ETP · Sludge

Abbreviations

BaP	Benzo(a)pyrene
BTEX	Benzene, toluene, ethylbenzene, xylene
ETP	Effluent treatment plant
GC-FID	Gas chromatography flame-induced detector
IOCL	Indian Oil Corporation Ltd
LPG	Liquefied petroleum gas
NRL	Numaligarh Refinery Limited
PAHs	Polycyclic aromatic hydrocarbons
TPH	Total petroleum hydrocarbon

1 Introduction

Effluent treatment plant (ETP) residues, tank bottom sludge, and residual crude oil and byproducts resulting from the extraction of oil refineries often contain pollutants like heavy metals and hydrocarbons. Tank bottom sludge comprises sediments and residues settled at the bottom of storage tanks, containing a mix of hydrocarbons and impurities. Residual crude oil refers to unprocessed or partially processed oil remaining after refining operations, presenting challenges for disposal or reprocessing. The production of such effluents exceeds 28,220 tons from India alone as reported by Bhattacharyya and Shekdar (2003). These refinery wastes pose a significant threat to both land and water ecosystems, largely due to the extensive use of natural resources by increased population and energy demands. This contamination profoundly impacts the survival of flora and fauna, disrupting their natural habitats. The storage, extraction, and transportation of refinery waste produce processing water contaminated with hazardous hydrocarbons, endangering not only wildlife but also human health. In 1977, Colwell and Walker (Colwell et al. 1977) reported that petroleum products are mainly composed of asphaltenes, resins, saturated hydrocarbons, and aromatic hydrocarbons. Saturated hydrocarbons, characterized by the general formula C_nH_{2n+2}, are categorized as: linear straight-chain alkanes, branched alkanes, and cycloalkanes. The simplest and slightest in reactivity, alkanes, occur in gaseous form with less than five carbon atoms and in liquid form with more than five carbon atoms (Kissin 1987). Aromatic compounds, consisting of one or additional planar rings bonded covalently, are a class of unsaturated chemical compounds. Benzene, the most popular aromatic hydrocarbon, is mentioned by Carey

2024. Additionally, aromatic compounds like the BTEX compounds (benzene, toluene, ethylbenzene, and xylene), known for their toxicity, compose 60% of the petroleum light fraction (Heibati et al. 2017).

The crude oil viscosity is determined by the asphaltenes proportion present. These asphaltenes are toluene-soluble compounds but precipitate in alkanes like pentane or heptane (Ilyin et al. 2016). Asphaltenes, along with resins, constitute the crude oil's higher molecular weight composition (Pierre et al. 2004). Resins constituted of branched paraffin, polar compounds like sulfur, nitrogen, or oxygen, and fused rings of aromatic compounds are fractions heavier than saturate and aromatics (Demirbaş and Taylan 2016). Asphaltenes constitute compounds such as phenols and fatty acids, while resins consist of compounds like sulfoxides and amides.

Aromatic and saturated hydrocarbons are present as a chief proportion in all crude oils, with hetero compounds (asphaltenes and resins) in a lesser portion (Radwan and Al-Hasan 2007). The oily wastes generated during petroleum processing could be categorized into crude oil and sludge (Hu et al. 2013).

With petroleum being one of the main sources of energy large amount of fuel are processed in petroleum refineries. Petroleum refineries are mostly vast, and emission of toxic gases and other hazardous petroleum-based hydrocarbons is more compared to other processing industrial units (Al-Rubaye et al. 2023a). Petroleum refineries process crude oil into other products such as gasoline, kerosene, diesel, liquefied petroleum gas (LPG), and others. The production and processing of petroleum products across India itself amounted to over 254 million metric tons in 2022. Thus petroleum refineries are taken as a member of the leading springs of environmental pollution. Carbon dioxide, hydrogen sulfides, sulfur, and oxygen in addition to organic compounds are some of the emissions emitted during refining process. Treatment furnaces, steam boilers, and processing heater equipment usually found in refining industries are the main sources of these emissions (Vasović et al. 2015). Besides these emissions, transportation of petroleum-based products leads to leakage and spilling. Even though less than 0.001%of spillage has been reported, some of these spills could have a significant impact and would require extensive cleaning (Al-Rubaye et al. 2023b). About 825 operational refineries are reported around the globe with the United States being regarded as one of the biggest refiners. This number was estimated to grow by 23% in the upcoming years. This makes it one of the top causes of environmental pollution and needs urgent attention.

2 Methods for Refinery Waste Remediation

Remediation refers to the technique or process whereby recalcitrant compounds are completely broken into harmless or less toxic substances or in some cases complete removal of these toxic substances. Traditional methods for degrading petroleum hydrocarbons such as solvent extraction, chemical oxidation, and natural attenuation have been proposed though these processes are time-consuming and expensive (Xu et al. 2011). The United States Environmental Protection Agency devised remediation procedures that could be categorized into physical, chemical, thermal, and biological methods.

Physical methods involve the application of physical and mechanical blockade to isolate and separate oil components on the soil surface or water bodies (Ossai et al. 2020) followed by containment and recovery of oil. Aftermath the containment of oil, oil removal process begins which can be done by using techniques such as skimmers and sorbents. Skimmers are equipment that recover oil from spillage area either by suction or by using gravity-skimmed oil into underwater storage tanks. Sorbents use the principle of absorption and adsorption or merger of both as a mean to recover oil components from water surface (Cumo et al. 2007).

Chemical methods are the application of certain chemicals to facilitate oil spill clean-up. Dispersants are primarily used and are composed of solvents, surfactants, and stabilizers. Surfactants reduce surface tension between oil and water or land particles, thereby making it easier to disperse or wash oil away. Solvents, on the other hand, make it possible for the surfactant to penetrate oil particles for the formation of emulsion (Dewling and McCarthy 1980). Stabilizers prevent corrosion, regularize pH, and stabilize the emulsion so formed. Another approach using chemicals for remediation involves three stages: the use of certain chemicals to stimulate the activity of indigenous degrading microorganisms (augmentation), which is then succeeded by chemical leaching, which involves the application of water and other solvents to wash away the pollutants. The third stage makes use of oxidizing chemicals such as permanganate and hydrogen peroxide to render the hydrocarbon compounds into non-toxic substances (Igolima et al. 2022).

Thermal method (in situ burning) involves the use of heating system to bring about the volatilization of volatile components within crude oil. In this method, the water surface coated with oil is ignited for getting rid of low-molecular-weight hydrocarbons effectively (Idris 2014). Fire-resistant booms are also in use for effectively collecting and concentrating oil thick enough to burn (Cumo et al. 2007).

3 Bioremediation

Bioremediation denotes the process of employing microorganisms such as fungi, yeast, bacteria, in addition to plants to degrade and detoxify harmful pollutants present in the milieu, as an aftermath of leakage and spillage of oil, and such other environmental disaster due to human activities (Pari et al. 2019). According to a report, from the Deepwater Horizon Oil spill in Gulf of Mexico in 2010, a minimum of 200,000 tons of oil was removed by naturally occurring bacteria habituated in that area. A total number of over 200 species and 70 genera of microorganisms were reported with the potentiality of petroleum hydrocarbons degradation (Cai et al. 2021). The main degraders in the environment are indigenous bacteria, yeasts, algae, and cyanobacteria.

Bacteria represent one of the microorganisms at the front line for the role of degradation. Bacteria, degrading petroleum hydrocarbons, are used in bioremediation, generally recognized to be an effective and environmentally benign method. Occurrences of bacterial species of *Acinetobacter*, *Mycobacterium*, *Pseudomonas*, and *Rhodococcus* were reported at the locations of petroleum-contaminated as well as pristine soil sites indicating their capabilities to both survive and flourish in the regions (Margesin et al. 2003). In the research piloted by Yakimov et al. (2007), a consecutive flourishing of indigenous marine bacterial species, namely *Alcanivorax*, *Marinobacter*, and *Thallassolituus* to mention a few, was observed when introduced to oil components resulting in rapid degradation of these oil compounds. These bacterial species were usually unheeded because of their negligible presence and their heightened growth increase observed could be because of the additional carbon source supplemented by oil compounds. Bacterial species consisting of *Pandoraea*, *Enterobacter*, and *Burkholderia* strains sequestered from the site of waste disposition at the Indian Oil Corporation Ltd. (IOCL) oil refinery, Digboi, Assam, showed high degradation of dodecane with *Pandoraea* showing degradation percentage of 67% (Sarkar et al. 2017).

Two yeast strains isolated from wastewater collected from petrochemical industry which were later identified to be akin to *Candida tropicalis* and *Trichosporon asahii* showed influential crude oil degradation capacity. These yeast strains could degrade an extensive array of aromatic hydrocarbons such as phenanthrene and pyrene (Gargouri et al. 2015). Three yeast species, namely *Rhodotorula mucilaginosa*, *Rhodotorula ingeniosa*, and *Candida parapsilosis*, identified from soils polluted with Khafji oil fie (Saudi Arabia) were reported to show biodegradation potency (Al-Dhabaan 2022).

Two fungi of filamentous structure, specifically *Purpureocillium lilacinum* and *Aspergillus*, were reported to show good growth when grown on diesel, crude oil, and UE oil. It was observed that these two organisms showed considerable growth and efficient degradation capacity on engine oil and crude oil with crude oil being the most degraded component followed by engine oil. However minimal growth and degradation rate was reported in the case of UE oil. *P. lilacium* in particular showed degradation capacity of more than 40% in crude oil and 27% in diesel oil (Benguenab and Chibani 2021). Two *Penicillium* species strains sampled from the Rumaila field of oil (Iraq) evince high degradation rate of aliphatic hydrocarbons of carbon numbers 11–16. However, from C21 to C25, alkanes reduction in degradation capability was reported (Al-Hawash et al. 2018). Hamad et al. 2021 described about an excellent degradation of PAHs by two fungal species, *Penicillium chrysogenum* and *Alternaria alternate*.

The use of algae has been extensive in the treatment of wastewater; however, their potential regarding biodegradation of petroleum is a field which still needs to be explored properly (Tam et al. 2002). *Scenedesmus quadricauda*, *Chlorella vulgaris*, *Selenastrum capricornutum*, and *Scenedesmus platydiscus* are four species of algae which were trialed to assess their capability in the removal of two petroleum hydrocarbon compounds, namely pyrene and fluoranthene. These four algal species displayed their ability to degrade PAHs as degradation was observed of these compounds with *Selenastrum capricornutum* showing the highest PAHs degradation (Lei et al. 2007). *Selenastrum capricornutum* and *Scenedesmus acutus* have been reported to remove petroleum hydrocarbons, specifically Benzo(a)pyrene (BaP), by both degradation or sorption. The BaP removal was efficient and quick by both the microalgae, but the removal rate was swifter in the case of *S. capricornutum* which was explained by the fact that BaP must have been reserved by the cells for a short span, as sorption is more prevalent than degradation on exposure for a short period (Llasera et al. 2016). *Protheca zopfii*, a green achlorophyllous alga found ubiquitously, has also been reported to degrade crude oil. Walker et al. (1975) in their study determined that *P. zopfii* showed better growth in crude oil than in motor oil which can be concluded that the organisms degraded a major portion of aromatics in comparison to the saturated hydrocarbons since crude oil contains higher percentage of aromatic compounds (Walker et al. 1975). Seven microalgae species, namely *Chlorella miniata*, *Chlamydomonas* sp., *Chlorella vulgaris*, *Selenastrum capricornutum*, *Scenedesmus platydiscus*, *Scenedesmus quadricauda*, and *Synechosystis*, were also tested and observed for their pyrene removal efficiency. The results obtained were significant with *S. capricornutum* emerging out as the most effective algae from the seven, whose activity led to complete removal and transformation of pyrene with no traces of pyrene remaining toward the end of the study (Lei et al. 2002).

3.1 Role of Cyanobacteria as Potential Candidate for Bioremediation

As organotrophs, remediating polluted environments is a supposed attribute of cyanobacterial species. Many studies revealed that cyanobacterial species existing as an axenic culture or in a consortium with other organisms such as bacteria or algae have been widely in use for the clean-up process of oil-polluted environments. Hamouda et al. (2023) presented a study in which a consortium of microalgae *Chlorella vulgaris* and *Synechococcus* species (cyanobacterium) was involved, reporting the ability of this consortium for removing petroleum hydrocarbons for example kerosene from water. *Nostoc punctiforme* and *Spirulina platensis* were two cyanobacterial species selected by El-Sheekh and Hamouda (2013) for their petroleum hydrocarbons degradation efficiency. They reported that aliphatic hydrocarbons such as Tridecane, Tetradecane, Pentadecane, Hexadecane, Heptadecane, and Tricosane were degraded completely when treated with *S. platensis* with a minimal amount of tetracosane left behind when the crude oil concentration was taken at 0.5%. In the case of *Nostoc punctiforme*, the concentration of compounds like aliphatics was decomposed entirely (at 1% crude oil concentration) after an incubation of 15 days (El-Sheekh and Hamouda 2013). Two isolated strains of cyanobacteria (*Oscillatoria agardhii* and *Anabaena spharica*) were made use of as trial organisms to scrutinize their degradation capability on two crucial crude oil components, viz.

PAHs and n-alkanes. It was inferred that the n-alkanes removal percentage by *Oscillatoria* was 97% and 99% in *Anabaena* after a period of 5 and 9 weeks of exposure to crude oil by the former and the latter. PAHs removal efficiency was 70% and 90% in *Oscillatoria* and *Anabaena*, respectively, over a time period of 5 and 9 weeks (Gamila et al. 2003). Previous studies have reported a cyanobacterial consortium segregated from water surfaces of Bodo Creek (Ogoniland) polluted by oil. The cyanobacterial species later specified as *Chlorellakessleri*, *Anabaena*, and *Phormidium* strains were concluded as Cyanobacteria that could be using crude oil as energy and carbon source resulting in its decomposition (Ichor et al. 2016). Cyanobacteria are omnipresent especially in oil-polluted location. One such bacterial mat in a petroleum production site, Indonesia, showed high hydrocarbons (HCs) degradation ability (Chaillan et al. 2006). Another cyanobacterial species, *Anabaena variabilis* sequestered from Al-Dora refinery (Iraq), was also reported to exhibit good growth when treated with crude oil at 1%. After an incubation of 14 days, the cultures showed remarkable growth which could be concluded that the cyanobacteria is making use of the extra energy and carbon source utilized in its reproduction and growth. Reduction in the hydrocarbon composition, specifically saturated hydrocarbons, was reported (Ban and Al-Mayaly 2018). Study on the biodegradation ability of two cyanobacterial species isolated from Nile river shows that n-alkanes and PAHs (polyaromatic hydrocarbons) were degraded after these cyanobacterial species were exposed to crude oil for a period of 5 weeks (Gamila et al. 2003). Algae such as *Chlorella kessleri* have been reported to decompose Benzo(a) Pyrene (Takáčová et al. 2014).

4 Approach of Study

Cyanobacterial species *Anabaena variabilis* were sequestered from the raw samples of soil obtained from the Jaintia Hills, Meghalaya having obtained the accession code KR709104 (Jungai and Adhikari 2015). The strain was grown in standard $BG11_0$ medium and maintained following the described culturing conditions with continuous Fluorescent White illumination at a relatively low intensity (<500 lux) incubated at temperatures 25 °C (Rippka et al. 1979). The three petroleum hydrocarbon samples of crude oil, ETP hydrocarbon sludge, and tank bottom sludge were sourced from the Numaligarh Refinery Limited (NRL), Assam. The minimal inhibitory concentrations were determined for each of the samples by inoculating it in to the *A. variabilis* cultures thereby obtaining the suitable concentrations for the subsequent steps of the study. The sample inoculated cultures were assessed for their growth in context to their pigment content, chlorophyll A, chlorophyll B, carotenoids, and the phycobilisome proteins (Phycocyanin, Phycoerythrin, Allophycocyanin), according to the protocols described by Bennett and Bogorad (1973) and Bryant et al. (1979), followed by the GC-FID analysis.

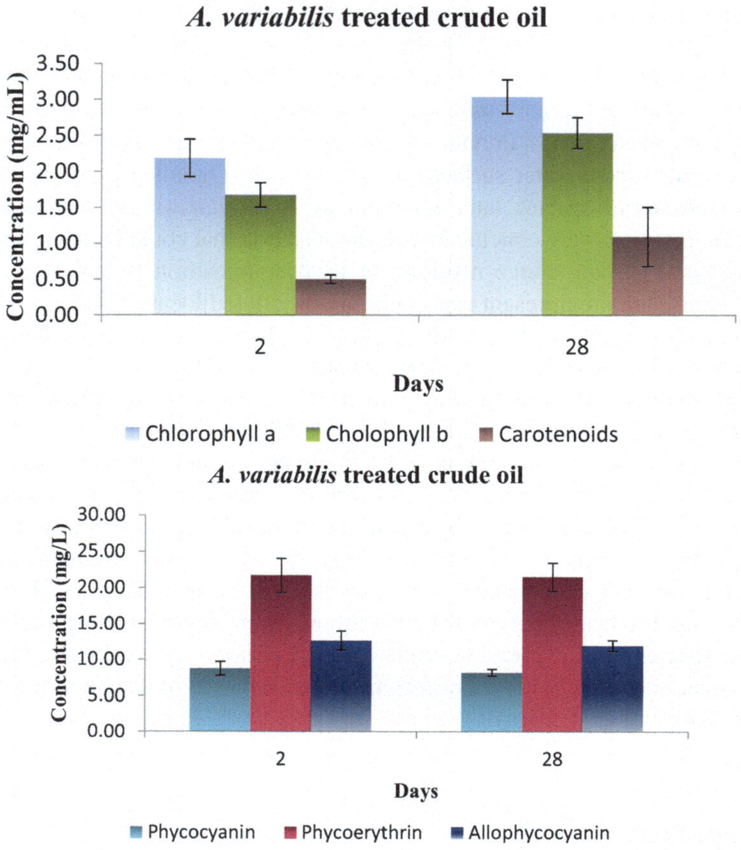

Fig. 1 (**a**) The chlorophyll A, chlorophyll B, and carotenoids content of *A. variabilis* treated crude oil culture. The values analyzed by one-way ANOVA were statistically significant (*p*-value <0.05). (**b**) The phycobilisome protein content of *A. variabilis* treated crude oil culture. The values analyzed by one-way ANOVA were statistically significant (*p*-value <0.05)

4.1 Biodegradation Efficacy in Context to the Pigment Content of the Cultures

The growth was evaluated on the basis of the chlorophyll, carotenoids, and phycobilisome protein content, and the generated pattern observed is represented graphically. The carotenoids, chlorophyll A, and chlorophyll B content of the *Anabaena variabilis* cultures in presence of crude oil on day 28 showed a positive increase (Fig. 1a). An about-constant concentration pattern was observed in regards to the phycobilisome proteins as well, lacking a decline in the concentrations due to the possible toxicity of crude oil (Fig. 1b). For the ETP hydrocarbon sludge treating cultures, a similar pattern was observed as a significant rise in the chlorophyll, the

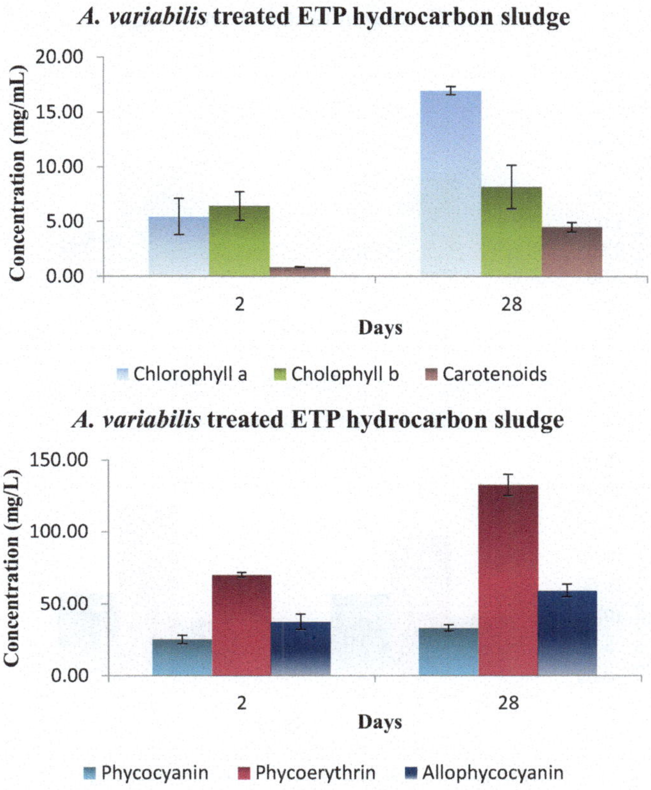

Fig. 2 (**a**) The chlorophyll A, chlorophyll B, and carotenoids content of *A. variabilis* treated ETP hydrocarbon sludge culture. The values analyzed by one-way ANOVA were statistically significant (p-value <0.05). (**b**) The phycobilisome protein content of *A. variabilis* treated ETP hydrocarbon sludge culture. The values analyzed by one-way ANOVA were statistically significant (p-value <0.05)

carotenoids, and the phycobilisome content on day 28 in relation to day 2 (Fig. 2a, b). In the culture degrading the tank bottom sludge, the chlorophyll A, chlorophyll B, and carotenoids content as well exhibited a sharp increase during the period of incubation (Fig. 3a). In context to the phycobilisome content, the phycoerythrin showed quite an increase alongside having the phycocyanin and allophycocyanin showing neither a toxicity induced decline in the concentrations nor a boost (Fig. 3b). Thus, the output suggests a stable and a potential growth capacity of the *A. variabilis* species, thereby indicating a strong biodegradation phenomenon toward the petroleum hydrocarbon substrates studied.

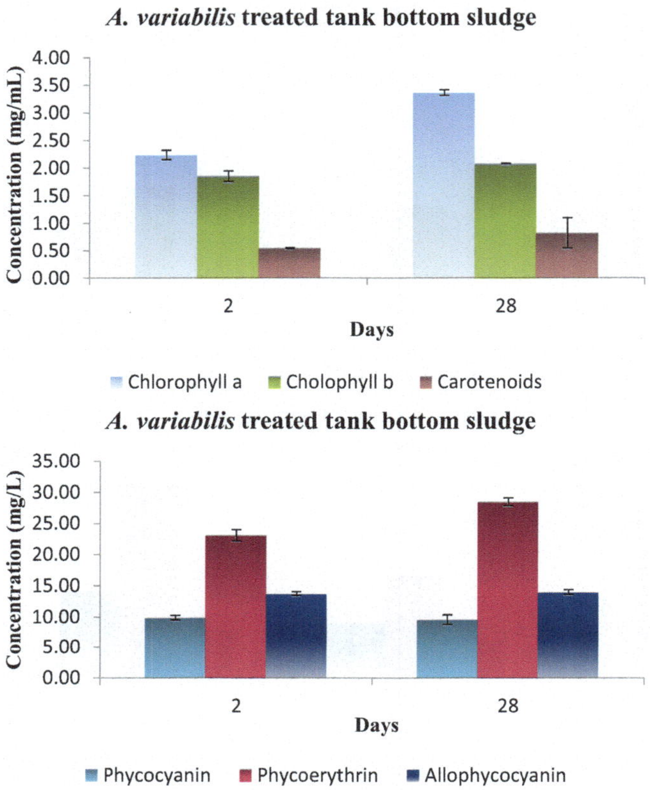

Fig. 3 (**a**) The chlorophyll A, chlorophyll B, and carotenoids content of *A. variabilis* treated tank bottom sludge culture. The values analyzed by one-way ANOVA were statistically significant (p-value <0.05). (**b**) The phycobilisome protein content of *A. variabilis* treated tank bottom sludge culture. The values analyzed by one-way ANOVA were statistically significant (p-value <0.05)

4.2 Using GC-FID to Assess the Biodegradation

It is recognized that hydrocarbon components are members of the family of organic contaminants that are cancer-causing agents and neurotoxic in nature (Das and Chandran 2010). Additionally, these are responsible for depleting the natural composition of water bodies and soil making it uninhabitable for the native microbiome in addition to other forms of life. It is through their essential enzymes and metabolic pathways that microbes are competent to degrade petroleum contaminants and remediate contaminated soils (Xu et al. 2023), by virtue of adaptation, either by utilizing them as an additional source of energy and carbon or feeding on them as their sole source of carbon. The degradation is assessed by approaches such as enzyme assays, GC-MS, GC-FID, and ICPOES. In this current study, the Flame–Ionization Detector equipped Gas Chromatograph (GC-FID) from the Thermofisher

TRACE 1600 Series having a TG-WAXMS GC COL capillary column with 30 m × 250 mm × 0.25 μm dimensions was opted for quantifying the biodegradation extent. The samples were processed in the system at 230 °C, 190 °C, and 250 °C under the inlet, oven, and the detector of the GC-FID, respectively, for 50 min in a splitless mode with a uniform flow of 1 ml min^{-1} of Hydrogen carrier gas.

4.3 Quantification of the Residual TPH Content

Table 1 summarizes the TPH (total petroleum hydrocarbon) content and the percentage of degradation of the petroleum hydrocarbon samples in each individual culture treated by *A. variabilis* on the 28th day of incubation. Figure 4a–f represents the GC-FID outputs depicting the hydrocarbon spectrum of the samples prior and after the treatment. The obtained data shows figures of highly significant reductions in the TPH content with the highest reduction in the crude oil sample with 99.98%, followed by ETP hydrocarbon sludge with 99.88%, and tank bottom sludge with 99.55%. The crude oil sample had completely degraded hydrocarbons from C12 to C30 having faint traces of C3 to C11 and C13 remaining. For the ETP hydrocarbon sludge, C7, C10, C11, and C14 to C29 were completely degraded in addition to the residual traces of C3 to C6 and C8, C9, and C12. In case of the tank bottom sludge, C4 to C6, C8, and C12 traces of hydrocarbon remained having C3 to C9, C13, C14, C18, C19, and C22 detected prior to the treatment was degraded completely.

Ulrici et al. (2000) and Perry (1984) ranked the susceptibility of hydrocarbons to microbe-mediated degradation in a linear order of linear alkanes followed by branched alkanes, small aromatics, and cyclic alkanes. Kim et al. (2018) reported removal of 92.1% of n-alkanes ranging from (C8) octadecane to (C32) dotriacontane, 19.0% of hentriacontane (C31) and 55.3% and 63.2% of aromatic compounds that are naphthalene and anthracene, by the strain *Gordonia iterans* Co17 after 7 days under culture conditions. They as well reported other bacterial species *Gordonia iterans* IFM 103 48T, *Rhodococcus wratislaviensis* N805T, *Rhodococcus pyridinivorans* PDB9T, *Shewanella aquimarina* SW-12 0T, and *Ochrobactrum anthropi* ATCC49188T with a crude oil R_E (removal efficiency) of 84.2%, 80.9%, 76.6%, 75.9%, and 72.3%. Lai et al. (2023) reported a consortium of heterotrophic oil-resistant bacteria (OG1 and OG2-*Erythrobacter citreus*) and a bacteria (AR3-*Pseudomonas pseudoalcaligenes*) that utilizes hydrocarbons, to have degraded the

Table 1 The total hydrocarbon content of crude inlet, ETP hydrocarbon sludge, and tank bottom sludge treated by *Anabaena variabilis* culture for 28 days

Treatment	TPH content (ppm)		
Sample	Untreated	Treated	Percentage of degradation (%)
Crude oil	400.30 × 10^3	17.11	99.98
ETP hydrocarbon sludge	495.63 × 10^3	572.32	99.88
Tank bottom sludge	250.20 × 10^3	1130.60	99.55

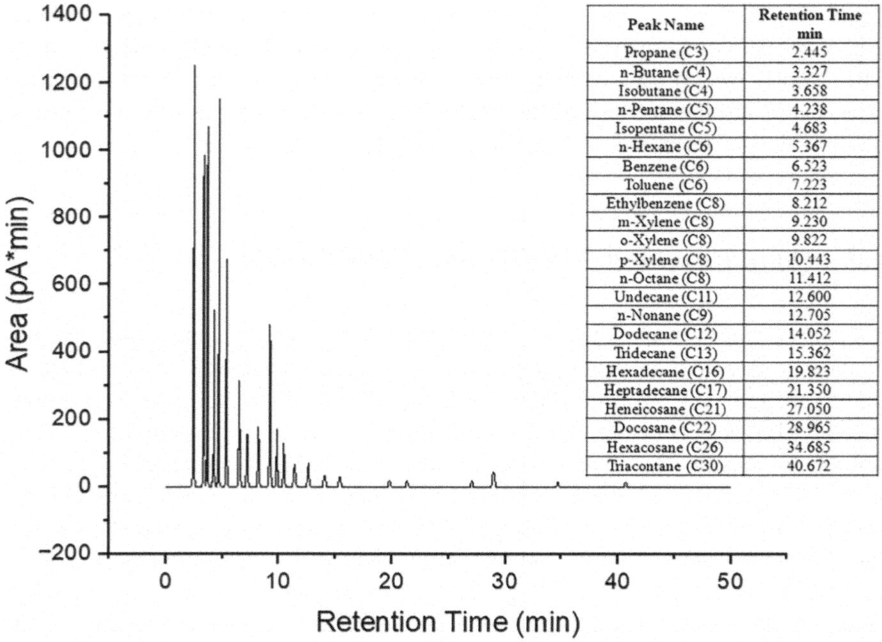

Peak Name	Retention Time min
Propane (C3)	2.445
n-Butane (C4)	3.327
Isobutane (C4)	3.658
n-Pentane (C5)	4.238
Isopentane (C5)	4.683
n-Hexane (C6)	5.367
Benzene (C6)	6.523
Toluene (C6)	7.223
Ethylbenzene (C8)	8.212
m-Xylene (C8)	9.230
o-Xylene (C8)	9.822
p-Xylene (C8)	10.443
n-Octane (C8)	11.412
Undecane (C11)	12.600
n-Nonane (C9)	12.705
Dodecane (C12)	14.052
Tridecane (C13)	15.362
Hexadecane (C16)	19.823
Heptadecane (C17)	21.350
Heneicosane (C21)	27.050
Docosane (C22)	28.965
Hexacosane (C26)	34.685
Triacontane (C30)	40.672

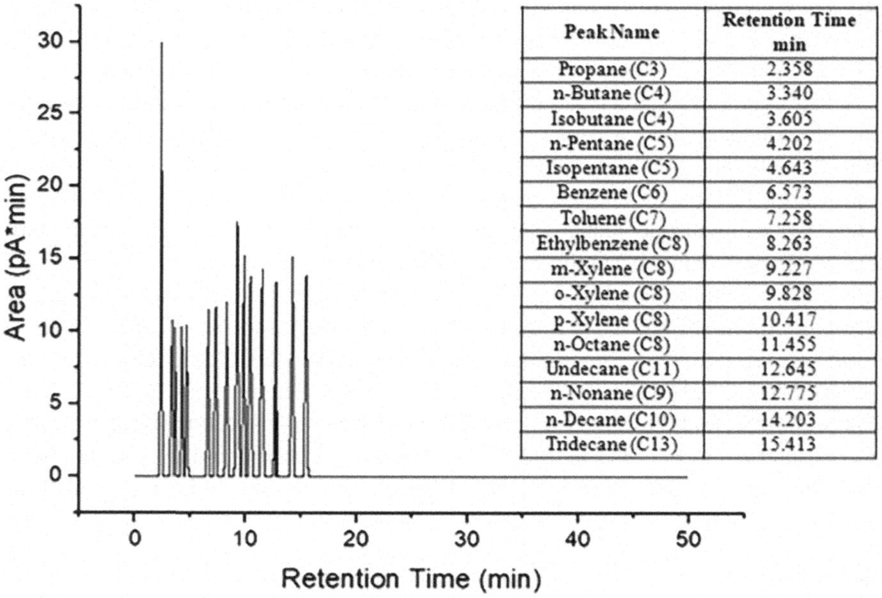

Peak Name	Retention Time min
Propane (C3)	2.358
n-Butane (C4)	3.340
Isobutane (C4)	3.605
n-Pentane (C5)	4.202
Isopentane (C5)	4.643
Benzene (C6)	6.573
Toluene (C7)	7.258
Ethylbenzene (C8)	8.263
m-Xylene (C8)	9.227
o-Xylene (C8)	9.828
p-Xylene (C8)	10.417
n-Octane (C8)	11.455
Undecane (C11)	12.645
n-Nonane (C9)	12.775
n-Decane (C10)	14.203
Tridecane (C13)	15.413

Fig. 4 (**a**) The GC-FID chromatogram for crude oil (standard) sample. (**b**) The GC-FID chromatogram for *Anabaena variabilis* culture treated crude oil. (**c**) The GC-FID chromatogram for ETP hydrocarbon sludge (standard) sample. (**d**) The GC-FID chromatogram for *Anabaena variabilis* culture treated ETP hydrocarbon sludge. (**e**) The GC-FID chromatogram for tank bottom sludge (standard) sample. (**f**) The GC-FID chromatogram for *Anabaena variabilis* culture treated tank bottom sludge

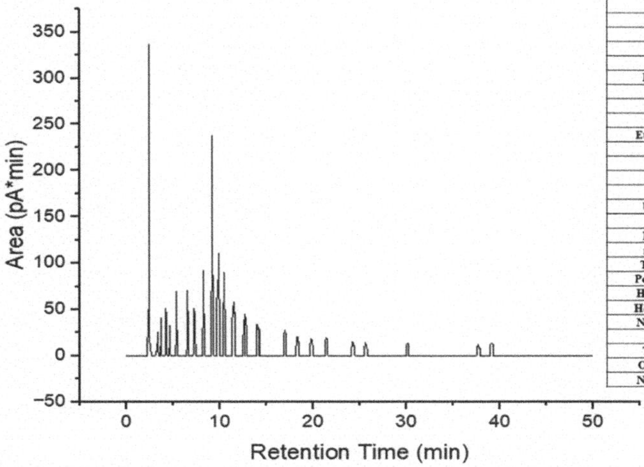

Peak Name	Retention Time min
Propane (C3)	2.430
n-Butane (C4)	3.308
Isobutane (C4)	3.678
n-Pentane (C5)	4.223
Isopentane (C5)	4.658
n-Hexane (C6)	5.348
Benzene (C6)	6.550
Toluene (C7)	7.245
Ethylbenzene (C8)	8.220
m-Xylene (C8)	9.215
o-Xylene (C8)	9.857
p-Xylene (C8)	10.428
n-Octane (C8)	11.495
Undecane (C11)	12.667
n-Nonane (C9)	12.753
Dodecane (C12)	14.010
n-Decane (C10)	14.183
Tetradecane (14)	16.967
Pentadecane (C15)	18.295
Hexadecane (C16)	19.802
Heptadecane (C17)	21.395
Nonadecane (C19)	24.230
Eicosane (C20)	25.617
Tricosane (C23)	30.105
Octacosane (C28)	37.690
Nonacosane (C29)	39.165

TPH of ETP hydrocarbon sludge treated by *Anabaena variabilis* culture

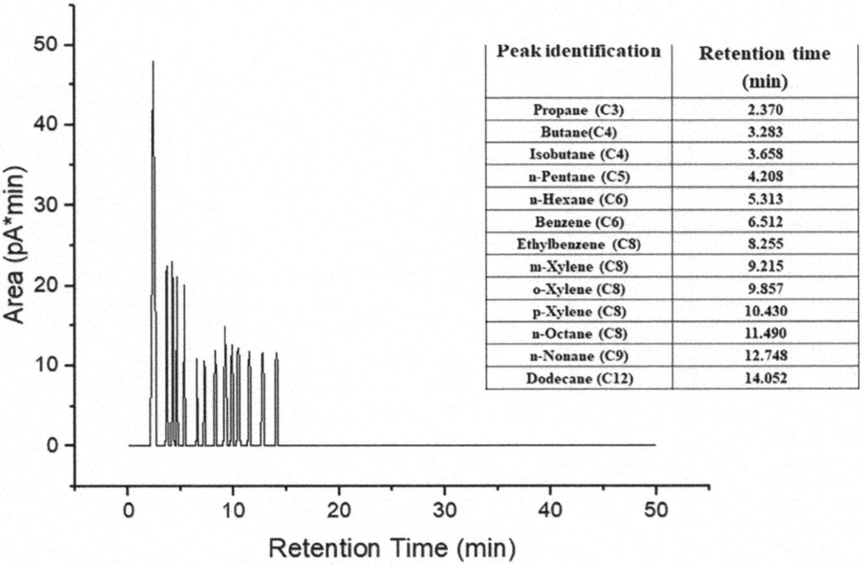

Peak identification	Retention time (min)
Propane (C3)	2.370
Butane(C4)	3.283
Isobutane (C4)	3.658
n-Pentane (C5)	4.208
n-Hexane (C6)	5.313
Benzene (C6)	6.512
Ethylbenzene (C8)	8.255
m-Xylene (C8)	9.215
o-Xylene (C8)	9.857
p-Xylene (C8)	10.430
n-Octane (C8)	11.490
n-Nonane (C9)	12.748
Dodecane (C12)	14.052

Fig. 4 (continued)

crude oil substrate to 9333.178 ppb (70.5%) of 16 aromatic hydrocarbons and 84,550.851 ppb (96.3%) of 16 aliphatic hydrocarbons in a designed bioreactor, with all suitable conditions provided. Raghukumar et al. (2001) stated the Bombay High crude oil containing 50% aliphatics, 14% aromatics, 31% waxes, bitumens etc, and 5% polar compounds, was degraded by three species of marine cyanobacteria namely *Oscillatoria salina* degrading 90.6%, 22.4%, 48.4%, *Plectonema terebrans*

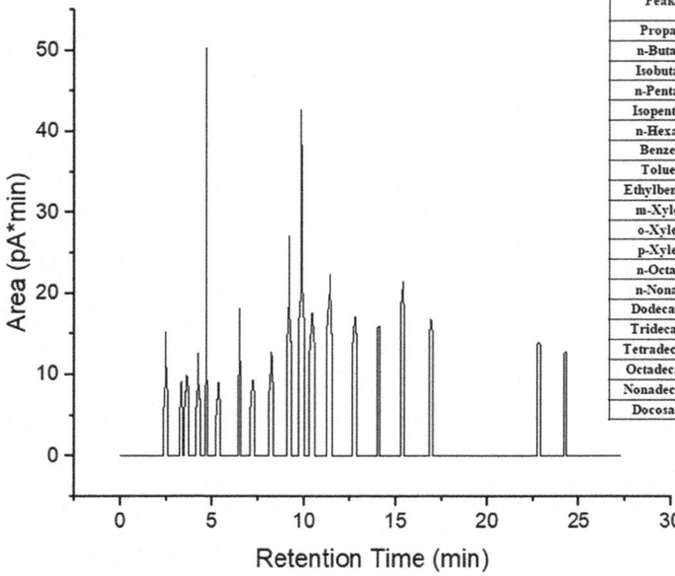

Peak Name	Retention Time (min)
Propane (C3)	2.492
n-Butane (C4)	3.317
Isobutane (C4)	3.628
n-Pentane (C5)	4.223
Isopentane (C5)	4.663
n-Hexane (C6)	5.327
Benzene (C6)	6.512
Toluene (C6)	7.222
Ethylbenzene (C8)	8.218
m-Xylene (C8)	9.220
o-Xylene (C8)	9.863
p-Xylene (C8)	10.443
n-Octane (C8)	11.412
n-Nonane (C9)	12.777
Dodecane (C12)	14.085
Tridecane (C13)	15.375
Tetradecane (C14)	16.920
Octadecane (C18)	22.780
Nonadecane (C19)	24.220
Docosane (C22)	28.885

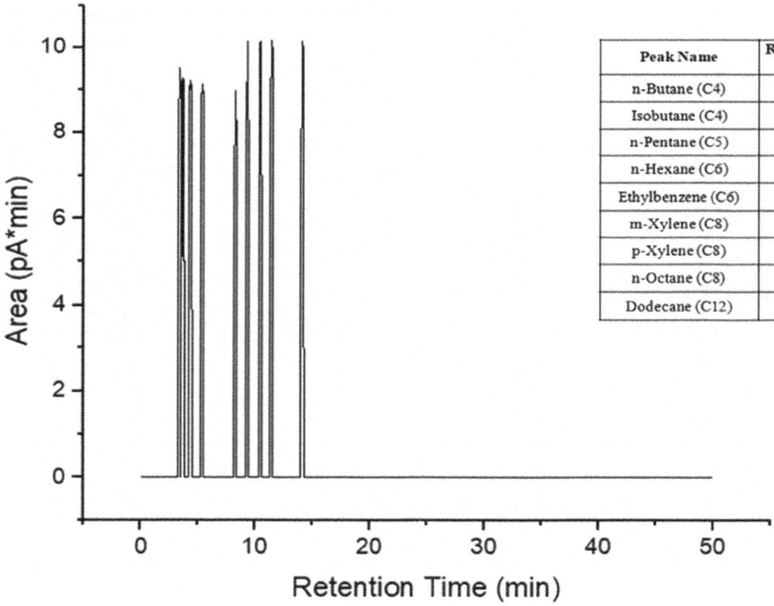

Peak Name	Retention Time (min)
n-Butane (C4)	3.368
Isobutane (C4)	3.633
n-Pentane (C5)	4.297
n-Hexane (C6)	5.327
Ethylbenzene (C6)	8.247
m-Xylene (C8)	9.277
p-Xylene (C8)	10.402
n-Octane (C8)	11.408
Dodecane (C12)	14.090

Fig. 4 (continued)

degrading 62.7%, 22.4%, 60.1% and *Aphanocapsa* sp. degrading 41.9%, 85.5%, 65.2% of Anthracene, Phenanthrene, and Hexadecane respectively. This approach of culturing the cyanobacterial species *Anabaena variabilis* to treat the three petroleum hydrocarbon samples individually exhibited a pattern of degradation in corroboration with the above-mentioned studies.

5 Conclusion

Petroleum oil is one of the main sources of energy. As the demand for energy increased, the processing and transport of petroleum products in different regions has resulted in contamination in various regions. The processing of these products results in huge quantities of toxic waste, be it liquid waste, solid waste, or semi-solid waste. Some of these toxic compounds are aromatic compounds. Be it low concentrations, reports describe aromatic compounds to interfere with chemoreception, which is essential for the feeding and mating signals of some marine organisms. Some petroleum hydrocarbons have carcinogenic properties. Proper disposal practices of these oil refinery-generated effluents are more than a concern for a safe environment. Remediation of these recalcitrant compounds is therefore of utmost importance. The use of microbial organisms to bioremediate these toxic pollutants presents a highly effectual approach because they occur in the environment naturally, thus reducing the risk of being or producing a latent threat to the environment and surrounding organisms.

Bacteria, yeast, and fungi are widely used to clean up petroleum hydrocarbon. Studies using algae and cyanobacteria in petroleum hydrocarbon degradation are relatively rare. However, there has been a surge in their usage in recent years following studies by Cerniglia et al. (1980) which meant to highlight their potential to metabolize naphthalene. There are reports for microbial species-mediated oil pollution remediation with tangible degradation activities with removal of hydrocarbon compounds evidenced by GC-FID technique. Microorganisms have been reported to decompose petroleum hydrocarbons on their own or in a consortium with other organisms or other genera and species of the same organisms. Microorganisms also produce biosurfactants which further accelerate the degradation process. Thus, in this context, biosurfactant production plays the catalytic role in the remediation process. Even the organisms that do not directly participate in the degradation process provide an indirect needed support by converting the original form of oil compounds into another form which is usable by other microbes in the consortium. *Mycobacterium vaccae* as reported by Beam and Perry (1974) practiced co-metabolism, wherein the bacteria convert propane into cyclohexane, which was then oxidized to cyclohexanol. Cyclohexanol was then used as a substrate by other bacteria in the consortium. According to several studies, the most widely studied catalyst enzymes contributing to the degradation of petroleum hydrocarbons include alkane oxygenases like alkB (di-iron alkane hydroxylases), Cytochrome P450, copper-containing methane monooxygenases, and di-iron methane soluble monooxygenases.

For promising execution regarding the rate of degradation of these petroleum compounds, researchers and experts from different areas of science are required to come together. Identification of many indigenous and exogenous organisms involving in petroleum hydrocarbons degradation has been widely assessed by microbiologists. Genes involving in the production of degradation enzymes are also characterized by molecular biologist. Genetic manipulation could be regularized for transformation of such genes in other microbial organisms to enhance the degradation capacity of these microbes. With the help of chemical and environmental engineers, the toxicity of by-products resulting from degradation could be further analyzed and the degradation strategy could be further improved. With updated technologies and research, an amalgam of microbial organisms and technology could be improvised to provide optimum conditions (temperature, salinity, pH, and limiting nutrients availability to name a few) for further acceleration of the degradation capacity of these microorganisms.

References

Al-Dhabaan FA (2022) Degradation of oil and hydrocarbons derivatives by isolated yeast strains from Ras Tanura in Saudi Arabia. J King Saud Univ Sci 34(7):102263. https://doi.org/10.1016/j.jksus.2022.102263

Al-Hawash AB, Zhang X, Ma F (2018) Removal and biodegradation of different petroleum hydrocarbons using the filamentous fungus Aspergillus sp. RFC-1. Microbiol Open 8(1):e00619. https://doi.org/10.1002/mbo3.619

Al-Rubaye AH, Jasim DJ, Jassam SA, Jasim HM, Ameen M, Al-Robai HA (2023a) The side effect of oil refineries on environment: as a mini review. IOP Conf Ser Earth Environ Sci 1262(2):022024. https://doi.org/10.1088/1755-1315/1262/2/022024

Al-Rubaye AH, Jasim DJ, Ameen HFM, Al-Robai HA, Al-Assal JR (2023b) The impacts of petroleum on environment. IOP Conf Ser Earth Environ Sci 1158:032014. https://doi.org/10.1088/1755-1315/1158/3/032014

Ban SA, Al-Mayaly IKA (2018) Biodegradation of crude oil by *Anabaena variabilis* isolated from al-dora refinery. IOSR J Pharm Biol Sci 13:51–56. https://doi.org/10.9790/3008-1303015156

Beam HW, Perry JJ (1974) Microbial degradation of cycloparaffinic hydrocarbons via cometabolism and commensalism. J Gen Microbiol 82(1):163–169

Benguenab A, Chibani A (2021) Biodegradation of petroleum hydrocarbons by filamentous fungi (*Aspergillus ustus* and *Purpureocillium lilacinum*) isolated from used engine oil contaminated soil. Acta Ecol Sin 41(5):416–423. https://doi.org/10.1016/j.chnaes.2020.10.008

Bennett A, Bogorad L (1973) Complementary chromatic adaptation in a filamentous blue-green alga. J Cell Biol 58(2):419–435. https://doi.org/10.1083/jcb.58.2.419

Bhattacharyya JK, Shekdar AV (2003) Treatment and disposal of refinery sludges: Indian scenario. Waste Manag Res 21(3):249–261. https://doi.org/10.1177/0734242x0302100309

Bryant DA, Guglielmi G, de Marsac NT, Castets AM, Cohen BG (1979) The structure of cyanobacterialphycobilisomes: a model. Arch Microbiol 123(2):113–127. https://doi.org/10.1007/bf00446810

Cai Y, Wang R, Rao P, Wu B, Yan L, Hu L, Park S, Ryu M, Zhou X (2021) Bioremediation of petroleum hydrocarbons using Acinetobacter sp. SCYY-5 isolated from contaminated oil sludge: strategy and effectiveness study. Int J Environ Res Public Health 18(2):819. https://doi.org/10.3390/ijerph18020819

Carey FA (2024) Aromatic compound. Encyclopedia Britannica. https://www.britannica.com/science/aromatic-compound. Accessed 29 August 2024

Cerniglia C, Gibson DT, Baalen CV (1980) Oxidation of naphthalene by cyanobacteria and microalgae. Microbiology 116:495–500. https://doi.org/10.1099/00221287-116-2-495

Chaillan F, Gugger M, Saliot A, Couté A, Oudot J (2006) Role of cyanobacteria in the biodegradation of crude oil by a tropical cyanobacterial mat. Chemosphere 62(10):1574–1582. https://doi.org/10.1016/j.chemosphere.2005.06.050

Colwell RR, Walker JD, Cooney JJ (1977) Ecological aspects of microbial degradation of petroleum in the marine environment. CRC Crit Rev Microbiol 5(4):423–445. https://doi.org/10.3109/10408417709102813

Cumo F, Gugliermetti F, Guidi G (2007) Best available techniques for oil spill containment and clean-up in the Mediterranean Sea. Water Resour Manag 103:527–535. https://doi.org/10.2495/wrm070491

Das N, Chandran P (2010) Microbial degradation of petroleum hydrocarbon contaminants: an overview. Biotechnol Res Int 2011:1–13. https://doi.org/10.4061/2011/941810

Demirbaş A, Taylan O (2016) Removing of resins from crude oils. Pet Sci Technol 34(8):771–777. https://doi.org/10.1080/10916466.2016.1163397

Dewling RT, McCarthy LT (1980) Chemical treatment of oil spills. Environ Int 3(2):155–162. https://doi.org/10.1016/0160-4120(80)90050-1

El-Sheekh MM, Hamouda RA (2013) Biodegradation of crude oil by some cyanobacteria under heterotrophic conditions. Desalin Water Treat 52(7-9):1448–1454. https://doi.org/10.1080/19443994.2013.794008

Gamila HA, Ibrahim MBM, El-Ghafar HHA (2003) The role of cyanobacterial isolated strains in the biodegradation of crude oil. Int J Environ Stud 60(5):435–444. https://doi.org/10.1080/0020723032000050367

Gargouri B, Mhiri N, Karray F, Aloui F, Sayadi S (2015) Isolation and characterization of hydrocarbon-degrading yeast strains from petroleum contaminated industrial wastewater. Biomed Res Int 2015:1–11. https://doi.org/10.1155/2015/929424

Hamad AA, Moubasher HA, Moustafa YM, Mohamed NH, Abd-elrhim EH (2021) Petroleum hydrocarbon bioremediation using native fungal isolates and consortia. Sci World J 2021:1–13. https://doi.org/10.1155/2021/6641533

Hamouda RA, Alhumairi AM, Saddiq AA (2023) Simultaneous bioremediation of petroleum hydrocarbons and production of biofuels by the micro-green alga, cyanobacteria, and its consortium. Heliyon 9(6):e16656. https://doi.org/10.1016/j.heliyon.2023.e16656

Heibati B, Pollitt KJG, Karimi A, Yazdani CJ, Ducatman A, Shokrzadeh M, Mohammadyan M (2017) BTEX exposure assessment and quantitative risk assessment among petroleum product distributors. Ecotoxicol Environ Saf 144:445–449. https://doi.org/10.1016/j.ecoenv.2017.06.055

Hu G, Li J, Zeng G (2013) Recent development in the treatment of oily sludge from petroleum industry: a review. J Hazard Mater 61:470–490. https://doi.org/10.1016/j.jhazmat.2013.07.069

Ichor T, Okerentugb PO, Okpokwasil GC (2016) Biodegradation of total petroleum hydrocarbon by a consortium of cyanobacteria isolated from crude oil polluted brackish waters of bodo creeks in Ogoniland, Rivers State. Res J Environ Toxicol 10(1):16–27. https://doi.org/10.3923/rjet.2016.16.27

Idris J (2014) Oil spills and sustainable cleanup approach. Aust J Basic Appl Sci 7(14):272–2280

Igolima MU, Abbey SJ, Ifelebuegu AO (2022) A systematic review on the effectiveness of remediation methods for oil contaminated soils. Environ Adv 9:100319. https://doi.org/10.1016/j.envadv.2022.100319

Ilyin S, Arinina M, Polyakova M, Bondarenko G, Konstantinov I, Kulichikhin V, Malkin A (2016) Asphaltenes in heavy crude oil: designation, precipitation, solutions, and effects on viscosity. J Pet Sci Eng 147:211–217. https://doi.org/10.1016/j.petrol.2016.06.020

Jungai N, Adhikari S (2015) Genetic diversity of free living filamentous cyanobacteria isolated from a variety of coal mining areas of Jaintia Hills District, Meghalaya, India. Int J Res Stud Biosci 3(12):26–34

Kim H, Dong K, Kim J, Lee S (2018) Characteristics of crude oil-degrading bacteria *Gordoniaiterans* isolated from marine coastal in Taean sediment. Microbiol Open 8(6):e00754. https://doi.org/10.1002/mbo3.754

Kissin YV (1987) Catagenesis and composition of petroleum: origin of n-alkanes and iso-alkanes in petroleum crudes. Geochim Cosmochim Acta 51(9):2445–2457. https://doi.org/10.1016/0016-7037(87)90296-1

Lai FC, Nawaz A, Daniel JD, Olakunle O, Habila MA, Woei YT, Misson M (2023) Investigating the crude oil biodegradation performance in membrane bioreactor by using a consortium of symbiotic bacteria. Chemosphere 337:139293–139293. https://doi.org/10.1016/j.chemosphere.2023.139293

Lei AP, Wong YS, Tam NFY (2002) Removal of pyrene by different microalgal species. Water Sci Technol 46(11-12):195–201. https://doi.org/10.2166/wst.2002.0738

Lei AP, Hu ZL, Wong YS, Tam NFY (2007) Removal of fluoranthene and pyrene by different microalgal species. Bioresour Technol 98(2):273–280. https://doi.org/10.1016/j.biortech.2006.01.012

Llasera G-MP, Olmos-Espejel J-J, Díaz-Flores G, Montaño-Montiel A (2016) Biodegradation of benzo(a)pyrene by two freshwater microalgae *Selenastrum capricornutum* and *Scenedesmus acutus*: a comparative study useful for bioremediation. Environ Sci Pollut Res 23(4):3365–3375. https://doi.org/10.1007/s11356-015-5576-2

Margesin R, Labbé D, Schinner F, Greer CW, Whyte LG (2003) Characterization of hydrocarbon-degrading microbial populations in contaminated and pristine alpine soils. Appl Environ Microbiol 69(6):3085–3092. https://doi.org/10.1128/aem.69.6.3085-3092.2003

Ossai IC, Ahmed A, Hassan A, Hamid FS (2020) Remediation of soil and water contaminated with petroleum hydrocarbon: a review. Environ Technol Innov 17:100526. https://doi.org/10.1016/j.eti.2019.100526

Pari P, Thillipan M, Queenie A (2019) Bioremediation of oil sludge using Pseudomonas aeruginosa, Bacillus subtilis and Brassica juncea. Int J Eng Res Technol 8:551–559

Perry JJ (1984) Microbial metabolism of cyclic alkanes, *Petroleum Microbiology*. Macmillan, New York, pp 61–98

Pierre C, Barré L, Pina A, Moan M (2004) Composition and heavy oil rheology. Oil Gas Sci Technol 59(5):489–501. https://doi.org/10.2516/ogst:2004034

Radwan SS, Al-Hasan RH (2007) Oil pollution and cyanobacteria. In: The ecology of cyanobacteria. Springer, Cham, pp 307–319. https://doi.org/10.1007/0-306-46855-7_11

Raghukumar C, Vipparty V, David JJ, Chandramohan D (2001) Degradation of crude oil by marine cyanobacteria. Appl Microbiol Biotechnol 57(3):433–436. https://doi.org/10.1007/s002530100784

Rippka R, Derueless J, Waterbury JB, Herdman M, Stanier RY (1979) Generic assignment, strain histories, and properties of pure cultures of cyanobacteria. J Gen Microbiol 111:1–61. https://doi.org/10.1099/00221287-111-1-1

Sarkar P, Roy A, Pal S, Mohapatra B, Kazy SK, Maiti MK, Sar P (2017) Enrichment and characterization of hydrocarbon-degrading bacteria from petroleum refinery waste as potent bioaugmentation agent for in situ bioremediation. Bioresour Technol 242:15–27. https://doi.org/10.1016/j.biortech.2017.05.010

Takáčová A, Smolinská M, Ryba J, Mackuľak T, Jokrllová J, Hronec P, Čík G (2014) Biodegradation of Benzo[a]Pyrene through the use of algae. Open Chem 12(11):1133–1143. https://doi.org/10.2478/s11532-014-0567-6

Tam NFY, Chong AMY, Wong YS (2002) Removal of tributyltin (TBT) by live and dead microalgal cells. Mar Pollut Bull 45(1-12):362–371. https://doi.org/10.1016/s0025-326x(02)00184-4

Ulrici W, Rehm HJ, Reed G (2000) Contaminated soil areas, different countries and contaminants, monitoring of contaminants. In: Biotechnology: environmental process II, 2nd edn. Wiley, Hoboken, p 507

Vasović D, Nikolić JM, Radosavljević J, Vukadinović A (2015) A brief overview of ippc/ied implementation. Zrenjanin, Serbia. In: V International Conference - Industrial Engineering and Environmental Protection (IIZS 2015), pp 95–99. http://www.tfzr.uns.ac.rs/iizs/files/zbornik%202015.pdf

Walker JD, Colwell RR, Petrakis L (1975) Degradation of petroleum by an alga, *Prototheca zopfii*. Appl Microbiol 30(1):79–81. https://doi.org/10.1128/am.30.1.79-81.1975

Xu J, Lei X, Huang T, Kim C (2011) Enhanced bioremediation of oil contaminated soil by graded modified Fenton oxidation. J Environ Sci 23(11):1873–1879. https://doi.org/10.1016/s1001-0742(10)60654-7

Xu C, Qaria MA, Qi X, Daochen Z (2023) The role of microorganisms in petroleum degradation: current development and prospects. Sci Total Environ 865:161112–161112. https://doi.org/10.1016/j.scitotenv.2022.161112

Yakimov MM, Timmis KN, Golyshin PN (2007) Obligate oil-degrading marine bacteria. Curr Opin Biotechnol 18(3):257–266. https://doi.org/10.1016/j.copbio.2007.04.006

Microbial Responses to Hydrocarbon Contamination and Its Bioremediation

Susmita Paul, Birson Ingti, Dipayan Das, and Santa Ram Joshi

Contents

S. Paul
Microbiology Laboratory, Department of Biotechnology and Bioinformatics, North-Eastern Hill University, Shillong, Meghalaya, India

Department of Microbiology, Royal School of Biosciences, The Assam Royal Global University, Guwahati, India

B. Ingti · D. Das
Department of Microbiology, Royal School of Biosciences, The Assam Royal Global University, Guwahati, India

S. R. Joshi (✉)
Microbiology Laboratory, Department of Biotechnology and Bioinformatics, North-Eastern Hill University, Shillong, Meghalaya, India

S. R. Joshi et al. (eds.), *Mineral Transformation and Bioremediation by Geo-Microbes*, Interdisciplinary Biotechnological Advances,
https://doi.org/10.1007/978-981-96-3033-2_22

Abstract Hydrocarbon contamination poses significant environmental challenges due to its persistence and adverse effects on ecosystems. Contamination of soil by petroleum hydrocarbons (PHCs) has attracted the attention of environmental scientists and engineers because of its threats to soil ecosystem services, groundwater quality, and human health. Thus, remediation of such pollutants from soil, water, etc. is extremely important. In recent years, systematic efforts have focused on developing feasible approaches for recovering PHC-polluted soils. Bioremediation through bio-stimulation (adding nutrients and electron acceptors to improve the proliferation and activity of native hydrocarbon-degrading microorganisms), as well as phytoremediation, was identified as a technically viable, profitable, and ecologically sustainable technology. Subsequently, studying microbial communities involved in soil bio-stimulation as well as phytoremediation known for its significant potential to eliminate hydrocarbons and other hazardous compounds in soil and water bodies is vital to recognize the precise microbial characteristics that regulate better-quality decontamination rates. Therefore, the field of geomicrobiology offers a promising alternative through bioremediation, utilizing microorganisms to degrade hydrocarbons. In this review, the dynamics of microbial communities (including bacteria, archaea, and fungi) intermediating the bioremediation of PHC-contaminated soils as well as water bodies through bio-stimulation in terms of activity, abundance, and taxonomic configuration will be critically analyzed, which is a geomicrobiological approach.

Keywords Petroleum hydrocarbons · Soil ecosystem · Bioremediation

Abbreviations

MNA	Monitored Natural Attenuation
PAHs	Polycyclic aromatic hydrocarbons
SRB	Sulfate-reducing bacteria
TCA	Tricarboxylic acid (TCA)
TPH	Total petroleum hydrocarbon

1 Introduction

Hydrocarbons are organic molecules primarily made up of hydrogen and carbon. They are widely used as fuels and are mostly found in nature as coal, natural gas, and petroleum. In the past few decades, there has been extensive use of these hydrocarbons in the form of diesel, mineral oil, gasoline, heavy oil, kerosene, lubricants, etc., leading to the increased incidence of hydrocarbon contamination worldwide (Xu et al. 2018).

These pollutants are mutagenic, carcinogenic, immunotoxic, and teratogenic in nature and are regularly released into the environment through various human activities. Their presence in the environment is a major concern worldwide due to their long-term effects on soil, water, and air, threatening all life forms ranging from microorganisms to human health (Varjani 2017; Spierings and Mulder 2017). They are persistent and are not easily degradable in natural environments, impacting the physicochemical and biochemical properties of soil and water. Their presence in soil affects plant growth by limiting access to vital nutrients and root penetration, resulting in the death of the plants or retardation of plant growth (Jyoti and Smita 2012). Additionally, it affects oxygen and light penetration in both fresh and marine water bodies, impacting the flora and fauna of aquatic ecosystems. Also, through the food chain, these pollutants can affect the health of both humans and animals. Therefore, it is crucial to remove or degrade these contaminants in order to protect the ecosystem. Numerous physical and chemical methods, including burying, evaporation, dispersion, and cremation, are available to remove polluted hydrocarbons from soil and water bodies. However, these technologies are expensive and may result in insufficient hydrocarbon removal. Therefore, finding an inexpensive and environmentally benign technique to remove hydrocarbons is crucial. Microorganism-based bioremediation is considered a safer and more economical method for treating hydrocarbon-contaminated sites. Since a wide range of microorganisms are used in this technique, they can grow in hydrocarbon-contaminated soil and water and convert toxic hydrocarbons into less hazardous than the original compounds. Numerous microbes, such as bacteria, actinomycetes, algae, and fungi, can be used in this process (Fowzia and Fakhruddin 2018). Thus, the purpose of this chapter is to give an overview of hydrocarbon contamination and its potential use of microorganisms in the elimination of hazardous hydrocarbons from the surroundings.

2 Sources of Soil and Water Hydrocarbon Pollution

Environmental pollution of hydrocarbons comes from a number of sources, including petroleum, insecticides, and other hazardous organic materials. These pollutants are released into the environment via various human activities, such as extensive use of pesticides and automobile oil, urban stormwater discharges, oil spillage during exploration, production, transportation, etc. Oil spillage also occurs due to leaking pipelines and subterranean oil storage tanks. Another cause of hydrocarbon pollution in water and soil is the release of automotive oil brought on by an increase in the number of people using automobiles. This kind of contamination happens when vehicle oil spills onto the ground, polluting the soil, and may also be carried into streams of water by runoff. Pesticides such as herbicides, fungicides, insecticides, etc. play an important role in polluting the environment. These pesticides are mostly hazardous and may be carried into the water streams by rainfall (Lee and Page 1997; Gros et al. 2014; Khan 1990). Further, increasing industrial activities on land also contributes to hydrocarbon contamination in soil and water. Rainfall carries these hydrocarbon- or toxic organic-compound-contaminated fields into streams, resulting in pollution. Urban stormwater runoff is another cause of hydrocarbon pollution. Roads and parking lots in metropolitan areas are regularly contaminated by the oil and gasoline used by cars, and these pollutants can taint streams of water when it rains (Gros et al. 2014; Khan 1990).

3 Microbial Communities in Hydrocarbon-Contaminated Environments

Microorganisms are highly skilled at modifying their environment to produce growth-promoting conditions. Many of these bacteria may convert and break down pollutants through their metabolic processes, using them as a food source. The synergistic proliferation of microbial communities observed in natural habitats produces a surprising mix of primary and secondary bioactive chemicals (metabolites), including hydrolytic and oxidative enzymes. These organic materials are essential for cell division and have been connected to the mineralization of different hydrocarbon fractions. Few microbes that are commonly isolated from hydrocarbon-contaminated soil include *Klebsiella* spp., *Rhodococcus* spp., *Alcanivorax* spp., *Pseudomonas putida, Pseudomonas aeruginosa, Corynebacterium* spp., *Staphylococcus* spp., *Micrococcus* spp., *Alcaligenes* spp., *Serratia* spp., *Arthrobacter* spp., *Nocardia* spp., *Flavobacterium* spp., *Escherichia* spp., *Acinetobacter* spp., *Proteus* spp., *Citrobacter* spp., *Rhizopus* spp., *Penicillium* spp., *Candida* spp., *Aspergillus* spp., *Chlorella vulgaris, Scenedesmus obliquus*, etc. (Nair et al. 2008; Xu et al. 2018).

4 Microbial Degradation of Hydrocarbon

Hydrocarbon degradation is a fundamental process where microorganisms break down hydrocarbon compounds, which are the primary constituents of petroleum and other fossil fuels. These hydrocarbons, if left unchecked, can lead to significant environmental pollution, particularly in soils, groundwater, and marine ecosystems. Understanding the mechanisms of hydrocarbon degradation is crucial for developing effective strategies to mitigate environmental contamination and restore ecological balance. The degradation of hydrocarbons by microorganisms can be broadly categorized into aerobic and anaerobic processes, each with distinct pathways and biochemical mechanisms.

4.1 Aerobic Degradation of Hydrocarbons

Aerobic degradation of hydrocarbons is a critical process in the bioremediation of contaminated environments. This process involves the breakdown of hydrocarbon molecules in the presence of oxygen by specialized microorganisms, leading to the detoxification and mineralization of these pollutants(Haddock 2010). The aptitude of microbes to degrade hydrocarbons aerobically is fundamental to the usual attenuation of petroleum and other fossil fuel-derived contaminants in the environment. Understanding the mechanisms of aerobic degradation is essential for developing effective strategies for environmental cleanup and mitigating the impact of hydrocarbon pollution (Liu et al. 2017).

Hydrocarbons, which are organic compounds consisting solely of carbon and hydrogen atoms, are prevalent in petroleum and its derivatives. These compounds are hydrophobic and relatively stable, making them challenging to degrade. However, certain microorganisms, including bacteria, fungi, and algae, have evolved the ability to exploit hydrocarbons as a source of carbon and energy (Das and Chandran 2010). Aerobic degradation of hydrocarbons occurs in the presence of oxygen, where oxygen serves as a fundamental electron acceptor and is involved in the initial activation and subsequent oxidation of the hydrocarbon molecules (Brzeszcz and Kaszycki 2018). The aerobic degradation pathway can be divided into three main stages: activation, oxidation, and assimilation into central metabolic pathways.

4.2 Activation of Hydrocarbons

The paramount step in aerobic hydrocarbon dilapidation is the activation of the hydrocarbon molecule. This activation is crucial because hydrocarbons are generally inert and hydrophobic, making them less accessible for microbial attack. The activation process comprises the integration of molecular oxygen into the hydrocarbon, which is catalyzed by a cluster of enzymes known as oxygenases (Haddock 2010).

Oxygenases are divided into two major classes: monooxygenases and dioxygenases. Monooxygenases incorporate one atom of oxygen into the hydrocarbon molecule while reducing the other atom to water. An example of a monooxygenase is alkane monooxygenase, which catalyzes the conversion of alkanes (saturated hydrocarbons) into alcohols (van Beilen and Funhoff 2007). This reaction can be represented as:

$$R-CH_3+O_2+NADH+H^+\rightarrow R-CH_2OH+NAD^++H_2O$$

In this reaction, the alkane is hydroxylated to form an alcohol, which is more reactive and can be further metabolized by the microorganism.

Dioxygenases, on the other hand, incorporate both atoms of oxygen into the hydrocarbon substrate, leading to the formation of a dihydroxylated intermediate. Dioxygenases are particularly important in the degradation of aromatic hydrocarbons, where they catalyze the initial ring cleavage, converting the aromatic ring into a more reactive, open-chain structure that can be further degraded (Seo et al. 2009).

4.3 Oxidation of Activated Hydrocarbons

Following activation, the hydroxylated intermediates undergo further oxidation. This step is crucial for converting the initially inert hydrocarbon into more polar, water-soluble compounds that can be easily assimilated into the microorganism's central metabolic pathways (van Beilen and Funhoff 2007). The oxidation process is typically catalyzed by dehydrogenases, which are enzymes that remove hydrogen atoms from the hydroxylated intermediates, converting alcohols into aldehydes and subsequently into carboxylic acids (Rojo 2009).

For instance, in the degradation of alkanes, the alcohols produced by monooxygenases are oxidized by alcohol dehydrogenases to form aldehydes:

$$R-CH_2OH+NAD^+\rightarrow R-CHO+NADH+H^+$$

The aldehyde is then further oxidized by aldehyde dehydrogenases to form a carboxylic acid:

$$R-CHO+NAD^++H_2O\rightarrow R-COOH+NADH+H^+$$

These carboxylic acids are more hydrophilic and can be readily transported into the cell, where they enter central metabolic pathways for further processing.

5 Assimilation into Central Metabolic Pathways

The final stage in aerobic hydrocarbon degradation is the assimilation of the oxidized intermediates into central metabolic pathways, such as the β-oxidation pathway and the tricarboxylic acid (TCA) cycle. These pathways are critical for the complete mineralization of hydrocarbons, converting them into carbon dioxide and water while generating energy for the microorganism. For alkanes, the β-oxidation pathway is the primary route for further degradation. In this pathway, the carboxylic acids produced from the oxidation of alkanes are activated by Coenzyme A to form acyl-CoA derivatives (Alvarez 2003). These derivatives undergo successive rounds of β-oxidation, where they are shortened by two carbon atoms in each cycle, releasing acetyl-CoA. The acetyl-CoA then enters the TCA cycle, where it is entirely oxidized to carbon dioxide, generating ATP through oxidative phosphorylation (Jimenez-Diaz et al. 2017).

In the case of aromatic hydrocarbons, the intermediates produced by dioxygenases are converted into catechols or other dihydroxylated compounds. These compounds are further degraded through ring cleavage pathways, producing intermediates that can enter the TCA cycle directly. The complete oxidation of these intermediates in the TCA cycle results in the production of carbon dioxide and water, with the concomitant generation of energy (Heider et al. 1998).

The aerobic degradation of hydrocarbons is not only crucial for the natural attenuation of hydrocarbon pollutants in the environment but also has significant applications in bioremediation. Bioremediation is a method that uses microbes to degrade or detoxify pollutants, and the efficiency of this process depends largely on the metabolic capabilities of the microbial communities involved (Timmis and Pieper 1999).

Aerobic bioremediation strategies often involve the stimulation of indigenous microbial populations by optimizing environmental conditions such as oxygen availability, nutrient levels, and temperature. This approach can enhance the natural degradation processes, leading to more effective and faster cleanup of contaminated sites. In some cases, bioaugmentation, which involves the introduction of specific hydrocarbon-degrading microorganisms, is employed to further accelerate the degradation practice (Pandolfo et al. 2023).

Aerobic degradation of hydrocarbons is a complex but highly efficient process that plays a vital role in the breakdown of hydrocarbon pollutants in the environment. This process involves the activation of hydrocarbons by oxygenases, the oxidation of these activated intermediates by dehydrogenases, and their subsequent assimilation into central metabolic pathways, leading to complete mineralization (Wang et al. 2023). The understanding of these mechanisms has important implications for bioremediation strategies, offering a sustainable and effective approach to mitigating the impact of hydrocarbon pollution on the environment. As research continues to advance, new insights into microbial metabolism and enzyme function will further enhance our ability to harness these processes for environmental cleanup and other biotechnological applications.

6 Mechanisms of Anaerobic Degradation of Hydrocarbons

Anaerobic degradation of hydrocarbons is a critical process in environments where oxygen is limited or absent, such as deep subsurface sediments, wetlands, and contaminated aquifers. Although less efficient than aerobic degradation, anaerobic degradation plays a vital role in the natural attenuation of hydrocarbon pollution in anoxic conditions (Laczi et al. 2020). Understanding the mechanisms underlying anaerobic degradation is essential for developing effective bioremediation strategies in environments where oxygen is scarce. This chapter explores the key biochemical trials and microbial processes involved in the anaerobic degradation of hydrocarbons.

6.1 Anaerobic Environments and the Importance of Anaerobic Degradation

Anaerobic environments are characterized by the absence of molecular oxygen, which creates unique challenges for the degradation of hydrocarbons. Oxygen, a potent electron acceptor, is absent in these environments, necessitating the use of alternative electron acceptors by microorganisms to drive the degradation process. Despite the slower rates of degradation compared to aerobic processes, anaerobic degradation is crucial for the long-term stability and health of ecosystems impacted by hydrocarbon pollution (Boll et al. 2020).

The significance of anaerobic degradation lies in its ability to occur in oxygen-depleted environments, which are often challenging to remediate through conventional means. In many contaminated sites, such as oil spills in marine sediments or hydrocarbon contamination in deep aquifers, anaerobic conditions prevail (Widdel and Rabus 2001). In such cases, aerobic degradation is limited, and anaerobic processes become the primary means of hydrocarbon breakdown.

7 Key Steps in Anaerobic Degradation of Hydrocarbons

Anaerobic degradation of hydrocarbons involves several distinct biochemical steps, each facilitated by specialized microorganisms adapted to thrive in anoxic conditions. The primary steps in this process include hydrocarbon activation, reduction, and the involvement of fermentative and methanogenic pathways.

7.1 Hydrocarbon Activation

The initial step in the anaerobic degradation of hydrocarbons is the activation of the hydrocarbon molecule. Unlike aerobic degradation, which relies on oxygenases to incorporate molecular oxygen into the hydrocarbon, anaerobic activation requires alternative mechanisms. In anaerobic environments, hydrocarbon activation is often initiated by the addition of functional groups, such as carboxyl or sulfate groups, making the hydrocarbon molecule more reactive and amenable to further degradation (Laczi et al. 2020).

One well-studied example of anaerobic hydrocarbon activation is the addition of fumarate to toluene by the enzyme benzylsuccinate synthase, resulting in the formation of benzylsuccinate. This reaction represents a typical activation mechanism in the anaerobic degradation of aromatic hydrocarbons. The addition of fumarate introduces a reactive functional group into the hydrocarbon structure, facilitating subsequent biochemical transformations (Heider and Fuchs 1997).

7.2 Reduction of Activated Hydrocarbons

Following activation, the hydrocarbons undergo reductive transformations. This step involves the removal of functional groups and the generation of reduced intermediates. The reduction process is facilitated by a series of enzymes that operate under anaerobic conditions, often involving electron transfer reactions. These enzymes exploit alternative electron acceptors, such as nitrate, sulfate, or carbon dioxide, to drive the reduction of activated hydrocarbons (Callaghan et al. 2006). For instance, in the anaerobic degradation of n-alkanes, the initial activation of the alkane by fumarate addition is followed by a series of reduction steps that produce reduced intermediates, such as alkylated succinates. These intermediates are further transformed through beta-oxidation-like pathways, ultimately yielding smaller molecules that can enter central metabolic pathways (Boll et al. 2020).

7.3 Fermentation and Methanogenesis

The reduced intermediates produced during the reduction step are further metabolized by a consortium of fermentative and methanogenic microorganisms. These microbial communities work together in a syntrophic relationship to completely degrade hydrocarbons under anaerobic conditions. Fermentative bacteria play a key role in breaking down the reduced intermediates into smaller molecules such as acetate, hydrogen, and carbon dioxide. These molecules serve as substrates for methanogenic archaea, which carry out the final step in anaerobic hydrocarbon degradation: methanogenesis. Methanogenesis is the process by which methane is produced from simple carbon compounds, such as acetate and carbon dioxide, in the absence of oxygen (Thauer et al. 1977).

Methanogenesis is particularly important in anaerobic environments, as it represents the terminal step in the degradation process. The production of methane, a simple and energy-rich molecule, allows for the complete breakdown of hydrocarbons in oxygen-depleted environments. Methane can then diffuse out of the anaerobic zone and be oxidized by methanotrophic bacteria in aerobic zones, further contributing to the complete mineralization of hydrocarbons (Thauer et al. 2008).

8 Microbial Communities Involved in Anaerobic Degradation

The anaerobic degradation of hydrocarbons is carried out by a diverse array of microorganisms, each adapted to specific niches within anaerobic environments. These microorganisms include fermentative bacteria, sulfate-reducing bacteria, nitrate-reducing bacteria, and methanogenic archaea. The interactions among these microbial groups are complex and often involve syntrophic relationships, where the metabolic activities of one group depend on the products generated by another. Sulfate-reducing bacteria (SRB) are among the most important microbial groups involved in anaerobic hydrocarbon degradation. These bacteria utilize sulfate as an electron acceptor to drive the reduction of hydrocarbons, leading to the production of hydrogen sulfide as a byproduct. SRB are particularly active in marine sediments and other environments with high sulfate concentrations (Kleindienst and Knittel 2020).

Nitrate-reducing bacteria also play a substantial role in anaerobic hydrocarbon degradation. These bacteria use nitrate as an electron acceptor to facilitate the reduction of hydrocarbons. Nitrate reduction often leads to the production of nitrogen gas, which diffuses out of the anaerobic zone, contributing to the removal of nitrogen from contaminated environments (Xu et al. 2018). Methanogenic archaea, as mentioned earlier, are the key players in the final step of anaerobic degradation. These microorganisms convert simple carbon compounds, such as acetate and carbon dioxide, into methane. Methanogenesis is a crucial process in anoxic environments, particularly in deep subsurface sediments and wetlands, where oxygen is entirely absent (Jiménez et al. 2016). While anaerobic degradation is essential for the natural attenuation of hydrocarbons in anoxic environments, it poses several challenges for bioremediation efforts. The slower rates of degradation, compared to aerobic processes, can limit the effectiveness of anaerobic bioremediation strategies. Additionally, the production of methane, a potent greenhouse gas, during anaerobic degradation raises concerns about the environmental impact of methane emissions (Widdel et al. 2006).

Despite these challenges, anaerobic degradation has important applications in bioremediation, particularly in environments where oxygen is scarce or absent.

Techniques such as biostimulation, which involves the addition of electron acceptors like nitrate or sulfate, can enhance anaerobic degradation processes and improve the effectiveness of bioremediation efforts (Mu et al. 2022). Additionally, the study of anaerobic microbial communities and their metabolic pathways has led to the discovery of novel enzymes and biochemical reactions that can be harnessed for biotechnological applications (Müller et al. 2012a, b). Anaerobic degradation of hydrocarbons is a complex and multifaceted process that plays a critical role in the natural attenuation of hydrocarbon pollution in oxygen-depleted environments. The key steps in anaerobic degradation, including hydrocarbon activation, reduction, and methanogenesis, involve a diverse array of microorganisms and biochemical pathways. Understanding these mechanisms is essential for developing effective bioremediation strategies and mitigating the environmental impact of hydrocarbon contamination in anoxic environments. As research in this field continues to advance, new insights into anaerobic degradation processes are likely to emerge, offering novel solutions for the remediation of contaminated sites and the sustainable management of hydrocarbon pollutants.

9 Factors Influencing Microbial Degradation During Microbial Responses to Hydrocarbon Contamination and Bioremediation

Microbial degradation of hydrocarbons is a crucial process in the natural attenuation of hydrocarbon contamination in the environment. This process is fundamental to bioremediation, a method that employs microorganisms to degrade or detoxify pollutants, restoring ecosystems impacted by petroleum spills, industrial discharges, and other sources of hydrocarbon pollution. The efficiency and success of microbial degradation during bioremediation efforts are prejudiced by a complex interplay of various features, including environmental conditions, the composition of microbial communities, the nature of the hydrocarbons, and the availability of nutrients and electron acceptors. This essay explores the key factors influencing microbial degradation in the context of hydrocarbon contamination and bioremediation.

9.1 Environmental Conditions

Environmental conditions play a critical role in determining the rate and efficiency of microbial degradation of hydrocarbons (Leahy and Colwell 1990a, b). Temperature, pH, oxygen availability, and moisture content are among the most significant environmental factors influencing microbial activity.

9.1.1 Temperature

Temperature directly affects microbial metabolism and enzyme activity, which in turn influence the rate of hydrocarbon degradation (Zekri and Chaalal 2005). Most hydrocarbon-degrading microorganisms are mesophilic, thriving at moderate temperatures (20–40 °C) (Parthipan et al. 2018). However, extremophiles, such as psychrophiles and thermophiles, can degrade hydrocarbons in extreme temperature conditions (Chia et al. 2024), such as in Arctic or deep-sea environments. Degradation rates typically increase with temperature up to an optimum point, beyond which enzyme denaturation and microbial inactivity occur (Kebede et al. 2021).

9.1.2 pH

The pH of the environment effects the microbial communal structure and enzyme function (Gomez et al. 2020). Most hydrocarbon-degrading microorganisms prefer neutral to somewhat alkaline conditions (pH 6–8) (Al-Hawash et al. 2018). Acidic or highly alkaline environments can inhibit microbial activity, reducing the efficiency of hydrocarbon degradation (Al-Hawash et al. 2018). The pH also affects the solubility and bioavailability of hydrocarbons, further influencing the degradation process (Heipieper and Martínez 2010).

9.1.3 Oxygen Availability

Oxygen is a critical factor in determining whether aerobic or anaerobic degradation pathways will dominate. Aerobic degradation is generally faster and more efficient due to the obtainability of oxygen as a potent electron acceptor (Pandolfo et al. 2023). In oxygen-limited or anoxic environments, anaerobic degradation prevails, which is slower and less efficient but essential in environments where oxygen is scarce, such as in deep subsurface sediments or wetlands (McGenity 2014).

9.1.4 Moisture Content

Water is essential for microbial life, and its availability affects the transport and bioavailability of hydrocarbons (Wang et al. 2021b). In soil environments, optimal moisture levels facilitate microbial activity by enabling the diffusion of nutrients and substrates (Trivedi et al. 2018). However, excessive moisture can lead to waterlogged conditions, reducing oxygen availability and shifting the degradation process from aerobic to anaerobic (Jia et al. 2020).

9.2 Composition of Microbial Communities

The structure and diversity of microbial populations are critical factors of the efficiency of hydrocarbon degradation. Microbial communities in contaminated environments are typically complex, consisting of various species with different metabolic capabilities. The presence of specific hydrocarbon-degrading microorganisms, such as bacteria, fungi, and archaea, is essential for effective degradation.

9.2.1 Bacterial Communities

Bacteria are the primary agents of hydrocarbon degradation, with species belonging to genera such as *Pseudomonas, Rhodococcus*, and *Alcanivorax* being particularly well-known for their hydrocarbon-degrading capabilities (Palleroni et al. 2010; Thi Mo et al. 2022; Chernikova et al. 2020). These bacteria possess enzymes that can degrade various hydrocarbons, from alkanes to complex aromatic compounds (Sorkhoh et al. 1990). The presence of a diverse bacterial community increases the likelihood of complete hydrocarbon degradation, as different species specialize in breaking down different hydrocarbon fractions (Santisi et al. 2015).

9.2.2 Fungal Communities

Fungi, particularly white-rot fungi, play a significant part in the degradation of complex, recalcitrant hydrocarbons, such as polycyclic aromatic hydrocarbons (PAHs) (Pozdnyakova 2012). Fungi can break down hydrocarbons through oxidative enzymes, such as laccases and peroxidases, which are capable of degrading highly stable aromatic rings (Diefenbach et al. 2024).

9.2.3 Archaeal Communities

In extreme environments, such as hydrothermal vents or highly saline conditions, archaea, particularly methanogens and halophiles, contribute to hydrocarbon degradation (Bacosa et al. 2022). Methanogens, in particular, are crucial in anaerobic degradation processes, converting hydrocarbons into methane through methanogenesis (Jiménez et al. 2016).

9.2.4 Community Interactions

The interactions between different microbial species, including syntrophic relationships, also play a significant role in hydrocarbon degradation. For instance, in anaerobic environments, fermentative bacteria break down hydrocarbons into reduced molecules, which are then exploited by methanogenic archaea to produce methane (Sherry et al. 2020). Such interactions enhance the overall efficiency of the degradation process.

9.3 Nature of the Hydrocarbons

The chemical structure and physical possessions of hydrocarbons significantly influence their susceptibility to microbial degradation. Hydrocarbons can be broadly categorized into four groups: alkanes, aromatics, asphaltenes, and resins, each with different degradation profiles.

9.3.1 Alkanes

Alkanes are the simplest hydrocarbons and are generally the most readily degraded by microorganisms. Short-chain alkanes are more bioavailable and degrade faster than long-chain alkanes (Xu et al. 2018). Microorganisms, such as alkane-degrading bacteria, possess specific enzymes, such as alkane hydroxylases, that initiate the breakdown of alkanes (Singh et al. 2011).

9.3.2 Aromatic Hydrocarbons

Aromatic hydrocarbons, which comprise one or more benzene rings, are more complex and less biodegradable than alkanes. Simple monoaromatics, such as benzene and toluene, are relatively easier to degrade, while polycyclic aromatic hydrocarbons (PAHs) are more resistant to microbial attack (Martirani-Von Abercron et al. 2016). The degradation of PAHs often requires specialized microorganisms with specific enzymes, such as dioxygenases, that can cleave aromatic rings (Bugg and Winfield 1998).

9.3.3 Asphaltenes and Resins

Asphaltenes and resins are the most complex and recalcitrant components of hydrocarbons, consisting of large, polycyclic structures with high molecular weights (Liao et al. 2009). These compounds are highly resistant to biodegradation and may persist in the environment for long periods. The degradation of asphaltenes and resins is typically slow and requires specialized microorganisms capable of breaking down these complex structures (Liao et al. 2009).

9.4 *Availability of Nutrients and Electron Acceptors*

The availability of nutrients and electron acceptors is crucial for sustaining microbial activity and facilitating hydrocarbon degradation. Key nutrients, such as nitrogen, phosphorus, and trace elements, are essential for microbial growth and enzyme function.

9.4.1 Nutrient Availability

Nitrogen and phosphorus are often limiting factors in hydrocarbon-contaminated environments, particularly in marine and terrestrial ecosystems (Sun et al. 2021). The addition of these nutrients, a process known as biostimulation, can enhance microbial activity and accelerate the degradation process. However, excessive nutrient levels can lead to nutrient imbalances and inhibit microbial activity(Roy et al. 2018a, b).

9.4.2 Electron Acceptors

In aerobic environments, oxygen serves as the principal electron acceptor, driving the oxidation of hydrocarbons. In anaerobic environments, alternate electron acceptors, such as nitrate, sulfate, and carbon dioxide, are used by microorganisms to facilitate the degradation process (Gupta et al. 2016). The obtainability and type of electron acceptors influence the pathways of hydrocarbon degradation and the overall efficacy of the process. For instance, nitrate-reducing bacteria can degrade hydrocarbons under anoxic conditions, producing nitrogen gas as a byproduct (Liu et al. 2020).

9.5 *Bioavailability of Hydrocarbons*

The bioavailability of hydrocarbons refers to the extent to which hydrocarbons are accessible to microorganisms for degradation. Several factors influence bioavailability, including the physical state of the hydrocarbons, the presence of sorbents, and the degree of weathering.

9.5.1 Physical State

Hydrocarbons can exist in different physical states, such as dissolved, emulsified, or adsorbed to solid surfaces. Dissolved hydrocarbons are more bioavailable and easier to degrade, while hydrocarbons adsorbed to soil particles or trapped in sediments are less accessible to microorganisms (Marschner and Kalbitz 2003).

9.5.2 Sorbents

The presence of organic or inorganic sorbents, such as clay, humic substances, or activated carbon, can reduce the bioavailability of hydrocarbons by adsorbing them and limiting their contact with microorganisms (Johnsen et al. 2005). Bioavailability can be enhanced by the use of surfactants or biosurfactants, which increase the solubility and dispersion of hydrocarbons, making them more accessible to microbial degradation (Mulligan et al. 2001).

9.5.3 Weathering

Weathering processes, such as evaporation, dissolution, and photo-oxidation, can alter the composition and bioavailability of hydrocarbons. Weathered hydrocarbons are often more resistant to biodegradation due to the loss of volatile and easily degradable components, leaving behind more recalcitrant compounds (Atlas and Bragg 2009).

The microbial degradation of hydrocarbons during bioremediation is a complex process influenced by various factors, including environmental conditions, microbial community composition, the nature of hydrocarbons, nutrient availability, and bioavailability. Understanding these factors is essential for optimizing bioremediation strategies and enhancing the efficiency of hydrocarbon degradation in contaminated environments. By managing these factors through techniques such as biostimulation, bioaugmentation, and the use of surfactants, it is possible to improve the outcomes of bioremediation efforts, leading to more effective and sustainable solutions for the cleanup of hydrocarbon pollution(Hazen et al. 2016).

9.5.4 Monitored Natural Attenuation (MNA)

Monitored Natural Attenuation (MNA) is a passive remediation strategy used for managing contaminated sites, where natural processes are relied upon to degrade, dilute, or stabilize pollutants over time. MNA leverages the inherent capabilities of the environment, including biological, chemical, and physical routes, to lessen the concentration and toxicity of impurities without active human intervention. This approach is increasingly being adopted as a cost-effective and sustainable alternative to more invasive remediation techniques, particularly for long-term management of contamination in soils and groundwater (Jørgensen et al. 2010).

9.5.5 Principles of Monitored Natural Attenuation

MNA is based on the principle that natural processes can effectively reduce contaminant levels in the environment to acceptable standards over time. These natural processes, mutually referred to as natural attenuation processes, include biodegradation, dilution, dispersion, sorption, volatilization, and chemical or biological stabilization.

MNA is not a "do nothing" approach; it requires comprehensive site characterization, ongoing monitoring, and risk assessment to ensure that natural processes are effectively controlling contaminant levels. The decision to employ MNA is based on site-specific factors, including the type and concentration of contaminants, site hydrogeology, and the presence of natural attenuation processes.

9.5.6 Processes Involved in Monitored Natural Attenuation

MNA relies on various natural processes to realize the decrease of contaminant concentrations. The effectiveness of these processes is contingent on the site conditions and the nature of the contaminants. The key processes involved in MNA are as follows:

9.5.7 Microbial Biodegradation

The most critical process in MNA, microbial biodegradation includes the itemization of organic contaminants by indigenous microorganisms. For example, petroleum hydrocarbons and chlorinated solvents can be dishonored by bacteria under aerobic or anaerobic conditions (Wang et al. 2021a). The degree and amount of biodegradation depend on the accessibility of electron acceptors (e.g., oxygen, nitrate, sulfate) and the presence of suitable microbial communities.

9.5.8 Abiotic Transformation

Some contaminants undergo chemical reactions in the environment that transform them into less harmful substances. For instance, certain metals can precipitate out of solution, reducing their mobility and toxicity. Abiotic processes such as hydrolysis, oxidation, and reduction can also subsidize the breakdown of organic contaminants (Kotthoff et al. 2019).

9.5.9 Physical Processes

Physical processes like dilution, dispersion, and volatilization play a role in reducing contaminant concentrations (Mulligan and Yong 2004). As contaminants move through groundwater or soil, they spread out, decreasing their concentration. Volatile contaminants can evaporate into the atmosphere, further reducing their presence in the subsurface environment (David and Niculescu 2021).

Sorption and Immobilization

Contaminants can adhere to soil particles or organic matter, reducing their mobility. This process is particularly important for metals and hydrophobic organic compounds. Sorption can prevent contaminants from reaching groundwater or being taken up by plants, thus reducing their ecological impact (Aumeier et al. 2023).

Benefits of Monitored Natural Attenuation

MNA offers several advantages over more active remediation methods, making it an attractive option for managing contaminated sites under certain conditions.

Cost-Effectiveness

MNA is often more cost-effective than active remediation methods, such as soil excavation or pump-and-treat systems. By relying on natural processes, MNA reduces the need for expensive equipment, energy, and labor, making it a financially viable option for large or complex sites(Jørgensen 2011).

Sustainability

MNA aligns with sustainable remediation principles by minimizing the environmental footprint of remediation activities. Since MNA does not require extensive infrastructure or energy inputs, it has a lower impact on the environment compared to active remediation methods (Banwart and Thornton 2010).

Minimal Disruption

MNA causes minimal disruption to the site and surrounding areas. Unlike invasive methods that may require excavation or drilling, MNA allows the site to remain undisturbed, preserving ecosystems and maintaining the integrity of the landscape (Declercq et al. 2012).

Long-Term Solution

MNA can provide a long-term solution to contamination, as it relies on ongoing natural processes that continue to operate as long as contaminants are present. With proper monitoring and management, MNA can ensure the gradual and sustained reduction of contaminant levels over time (Kouzuma and Watanabe 2011).

Limitations and Challenges of Monitored Natural Attenuation

Despite its benefits, MNA has several limitations and challenges that must be considered before it is implemented as a remediation strategy.

Time-Consuming

MNA is typically a slow process, and achieving regulatory cleanup standards can take many years or even decades. This extended timeframe may not be acceptable for sites requiring immediate risk reduction or where human health and environmental risks are high (Jørgensen 2011).

Site-Specific

The effectiveness of MNA is highly site-specific, depending on the presence and activity of natural attenuation processes. Not all contaminants or sites are suitable for MNA, and extensive site characterization is required to assess its feasibility (Mulligan and Yong 2004).

Uncertain Outcomes

Predicting the long-term success of MNA can be challenging, as natural processes are influenced by a variety of factors, including changes in environmental conditions, microbial activity, and contaminant characteristics. Unexpected changes in these factors can affect the rate and extent of natural attenuation (Ferrey 2006).

Regulatory Acceptance

MNA may face regulatory scrutiny due to its passive nature and the uncertainty associated with predicting long-term outcomes. Regulators may require robust monitoring and risk assessment to ensure that MNA is protective of human health and the environment.

Applications of Monitored Natural Attenuation

MNA is used in a variety of settings, particularly where the contamination is relatively stable, and the risks are manageable. Common applications of MNA include the following:

Petroleum Hydrocarbon Contamination

MNA is widely used for sites contaminated with petroleum hydrocarbons, such as gasoline, diesel, and crude oil. Biodegradation by indigenous microorganisms is a key process for reducing hydrocarbon concentrations in soil and groundwater (Truskewycz et al. 2019).

Chlorinated Solvent Contamination

Chlorinated solvents, such as trichloroethylene (TCE) and perchloroethylene (PCE), are common contaminants at industrial sites. MNA can be effective for these contaminants, particularly when biodegradation or abiotic degradation processes are active (Grandel and Dahmke 2004).

Metals and Radionuclides

MNA can be applied to sites contaminated with metals or radionuclides, where sorption, precipitation, and immobilization processes reduce the mobility and bioavailability of these contaminants (Rittmann et al. 2007).

Nitrate and Ammonium Contamination

MNA can be used to address nitrate and ammonium contamination in groundwater, where microbial denitrification and volatilization processes play a key role in reducing contaminant levels (Huno et al. 2018).

Monitored Natural Attenuation (MNA) is a valuable and increasingly popular approach to managing contaminated sites, leveraging natural processes to degrade, dilute, or stabilize pollutants over time. While MNA offers significant benefits, including cost-effectiveness, sustainability, and minimal disruption, it also presents challenges, such as extended timeframes, site-specificity, and uncertain outcomes. Successful implementation of MNA requires thorough site characterization, continuous monitoring, and risk assessment to ensure that natural attenuation processes are effectively managing contamination and protecting human fitness and the environment. As our understanding of natural attenuation processes continues to evolve, MNA is likely to play an even more prominent role in the future of environmental remediation, offering a sustainable solution to managing contamination in a wide range of settings.

10 Bioremediation Strategies

With rising standards of living and increased awareness of environmental issues, society has made the repair of polluted places a priority. There has been a lot of attention in the past few decades in creating methods for cleaning up environmental toxins. Biological instruments are a preferable alternative for cleaning up these persistent pollutants since they are less expensive than physicochemical strategies (Omokhagbor Adams et al. 2020). A few strategies are developed in the bioremediation process of environmental pollutants.

10.1 Bio-Stimulation

The application of nutrients, terminal electron acceptors, and additives, which are found in every site, is the basis of one of the most fundamental designed bioremediation approaches, known as bio-stimulation. This technique encourages the activity of native degrading bacteria. The frequency of biodegradation directly rises with the concentration of microbial cells. This approach closely addresses the control ensuing from a rough C:N:P (Carbon: Nitrogen: Phosphorus) ratio. A low biodegradation rate on a mass scale is evident when there is even a slight deficiency of P, which leads to a significant number of residual hydrocarbons. By adding N and P sources, such as ordinary fertilizers, this can be fixed (Ortega et al. 2018). Today, to achieve maximum biodegradation efficiency, bio-stimulation and bioaugmentation, another bioremediation technique, are frequently used in tandem (Roy et al. 2018a, b; Varjani and Upasani 2019). This strategy worked well most of the time, although there have also been accounts of unsuccessful attempts (Ramadass et al. 2018). Furthermore, several research suggest that bio-stimulation serves as the primary engine within the context of these integrated techniques (Roy et al. 2018a, b; Safdari et al. 2018). Based on a comparison of the two approaches using various reports, it is observed that bio-stimulation produces better results while bio-augmentation has a smaller impact (Wu et al. 2019). Introducing the bioaugmented species in an immobilized state by natural carriers that also function as nutrition sources is an intriguing concept (Tao et al. 2019). It should be noted, nevertheless, that the kind of alteration needs to be chosen carefully because certain nutrients may speed up the process while others have the opposite effect (Oualha et al. 2019).

The rate at which hydrocarbons biodegrade in soil can be restricted by a variety of parameters, such as the presence of contaminants, pH, temperature, moistness, oxygen, and nutrients (Atagana 2008). By altering the surroundings, bio-stimulation entails promoting already-existing bacteria that can perform bioremediation. It is possible to accomplish this by adding different types of electron acceptors and limiting nutrients, such as phosphorus, nitrogen, oxygen, or carbon (for example, in molasses), which are ordinarily present in amounts that are sufficient to limit microbial activity (Piehler et al. 1999). Perfumo et al. (2007) defined it as the concentration or the activity of microbes that are found naturally and can be used for bioremediation, and it entails supplementing the coordinated site with nutrients, oxygen, or other electron donors and acceptors. By maximizing conditions such as aeration, fertilizer addition, pH control, and temperature management, bio-stimulation is a type of natural remediation that can improve pollutant breakdown, according to Margesin and Schinner (2001). To ascertain whether bio-stimulation is an appropriate remediation technique for the removal of petroleum pollutants from soil, they suggested evaluating the intrinsic degradation capacities of the autochthonous microflora as well as the environmental factors involved in the kinetics of the in situ process (Omokhagbor Adams et al. 2020).

Native microorganisms that are already present and widely distributed across the subsurface will perform bioremediation because they are well adapted to the subsurface environment. This is the main benefit of bio-stimulation. The main difficulty is dispersing additives so that bacteria below the surface may readily access them, reliant on the local geology of the subsurface. Distributing additives over the impacted region is difficult when the underlying lithology is tight and impermeable (tight clays or added fine-grained material). The sub-surface's fractures produce preferential channels for the additives to follow, preventing the additives from dispersing evenly. Heterotrophic microbial growth, which does not naturally break down petroleum hydrocarbons, may be encouraged by the addition of nutrients, leading to competition among the local microflora (Adams et al. 2014).

10.2 Nutrient Amendments

It has been noted from a few research that organic nutrients may be helpful as bioremediation encouraging nutrients. Atagana (2008) used sewage sludge and wood chip compost to treat polluted soil with a total petroleum hydrocarbon (TPH) content of over 38,000 mg/kg, the petroleum pollutants. The moisture content, nutrient composition, and temperature commands in the compost systems were relevant to this study. In the control experiment, temperature variations among 12 °C and 30 °C were noted. A high initial microbial load was the reason for the temperature in the composting of sewage sludge to rise to 58 °C in the second month of the trial. As the experiment went on, he saw that the nitrogen content in the compost made from sewage sludge dropped more quickly than in the control group. He attributed this to an increased rate of decomposition and microbial activity in the compost system. According to the study, 68.8% of TPH had been broken down by the sewage sludge compost in 2 months, compared to 10% in the control group and at the conclusion of the 19-month period (Omokhagbor Adams et al. 2020). To bioremediate spent motor oil in soil, Abioye et al. (2012) used a variety of organic nutrient sources. The study examined the soil bioremediation after it was modified with 15% BSG, BS, and SMC and polluted with 5% and 15% w/w wasted oil (waste mushroom compost). The investigation was conducted for 84 days ex situ. Soil treated with brewery-used grain and polluted with 5% spent lubricating oil showed up to 92% biodegradation. In 15% of the utilized lubricating oil-contaminated soil, the same substrate only showed 55% biodegradation. The study found that the inclusion of nutrient sources significantly increased the amount of moisture in the matrix of soil, nutrients, and oil. This resulted from the increased moisture content of the nutrients—in particular, BSG—which reached 71.84%. The present study's results indicate a larger proportion of degradation compared to the findings of Adesodun and Mbagwu (2008), who discovered a 42% overall degradation of petroleum hydrocarbons after 3 months. Utilizing a variety of nutrition sources, however, would have been beneficial in the experimental group with a higher waste lubricating oil composition (15%), as the researchers found that BSG enhanced the microbial population, moisture level, and nutrient content. This is especially significant considering the findings of Nduka et al. (2008), who found that combined sources of nutrients increased microbial proliferation and enhanced the "diet" source for microorganisms, which in turn gave soils with high pollutant concentrations a high potential for biodegradation. Their study also showed that the degradation process was shorter and that a high percentage of degradation occurred within the first 15 days. Finally, it can be said that TPH breakdown in contaminated soil requires the proper amounts of nutrients; the degree of degradation is dependent on the concentration of pollutants and the accessibility of nutrients, which can be further enhanced by using nutritional sources in consortiums. Abioye et al. (2012) conducted a study that was comparable to this one. The potential for biowastes (potato skin, soy cake, and tea leaf) to restore soil damaged with diesel was noted by Agamuthu and Dadrasnia (2013). Their 126-day investigation revealed that the proportion of oil pollution affected the rate of remediation; test samples damaged by 10% oil showed reductions of up to 82% TPH, although test samples had an impact which was by 20% oil showed reductions of only 25%. The colony-forming unit count range for both test categories was 150 X 106–176 X 106 CFUg-1. The use of organic wastes in their study significantly increased the activity of the soil enzyme dehydrogenase. With soy cake exhibiting the maximum biodegradation rate constant of 0.153 day-1 at 10% oil pollution, diesel range carbons were specifically destroyed by applying biowastes. In addition, the control soil's nitrogen (0.8%) and phosphorus (0.6%) contents were lower than those of the treatment groups that included organic wastes. Due to the organic wastes' nutrient content, particularly in areas with 10% oil pollution, TPH decreased quickly in all treatment categories that were treated with organic wastes when compared to soil that was not altered. This supported the findings of Bartha (1986), who found that adding oil to soil at weight-based rates of 0.5–10% caused rapid degradation within 3 months. One study that employed chicken droppings as a source of nutrients for ex situ remediation (Chikere 2012) identified 56 hydrocarbon-utilizing bacterial isolates from contaminated soil and chicken manure, such as *Staphylococcus* spp., *Pseudomonas* spp., *Citrobacter* sp., *Klebsiella* sp., *Micrococcus* spp., *Corynebacterium* spp., *Bacillus* spp., *Rhodococcus* spp., *Alcanivorax* spp., *Alcaligenes* sp., *Serratia* spp., *Arthrobacter* spp., *Nocardia* spp., *Flavobacterium* sp., *Escherichia* sp., *Acinetobacter* sp., *Proteus* spp. One kilogram wet weight of sediments, 1 L of seawater, 20 mL of crude oil, 20 mg of anthracene, and 20 g of droppings from fowl were placed into a bioreactor vessel. Control 1 and Control 2 were an autoclaved bioreactor with no nutrients and another bioreactor vessel with sediment and no nutrients, respectively. Determining if native bacteria in sediments and seawater had a natural tendency to break down petroleum hydrocarbons was the main goal of this study. Abiotic factors also had a role in determining the amount of petroleum hydrocarbons lost. The results of their analysis showed that the total amount of hydrocarbon-consuming and heterotrophic bacteria was within the array of 105 cfu/g, suggesting that the bacterial population was able to use petroleum hydrocarbons. The initial Polycyclic Aromatic Hydrocarbons (PAH) content with in the petroleum pollutants (treatment categories and the control) was initially between 96 and 104.4 ppm. The baseline values of TPH and PAH were between 106 and 124.2 ppm in all treatments and the control. After the study was over, the amount of TPH and PAH in soil modified with chicken droppings had dropped to 95.35% and 98.92% respectively. The scientists noted that the ability of autochthonous microorganisms, specifically isolated gram-negative bacteria present in the sediments, to flourish without using amino acids and carbohydrates as growth substrates was remarkable, despite the observably high rate of deterioration. Gram-negative bacteria's ability to produce glucolipid biosurfactant allows them to almost entirely exploit hydrocarbons as carbon sources, according to reports by Head et al. (2006). Remarkably, several researchers (Abioye et al. 2012; Adams et al. 2014) have proposed that the inherent degradative capacity of microorganisms, especially gram-negative bacteria, is enhanced using organic nutrients, particularly nitrogen and phosphorus, as an amendment. The significance of abiotic elements like agitation and homogenization through stirring in the lowering of hydrocarbon concentrations was also demonstrated by this study (Chikere 2012).

10.3 Bio-Stimulation Using Inorganic Nutrients

Worldwide, the use of inorganic fertilizer sources as bio-stimulation agents has also been documented. In a 2010 study, Chorom et al. (2010) examined how well inorganic fertilizer (NPK) promoted microbial breakdown of petroleum hydrocarbons in soil. The results of gas chromatography revealed that in less than 10 weeks, standard paraffin as well as isoperoid (Phitane and Pristane) dropped in the range of 40–60% in all treatment categories. A boost in the test setup's CNP ratio from the addition of inorganic fertilizer eventually encouraged microbial decomposition. In an ex situ study, Agarry and Owabor (2011) investigated increased bioremediation of soil intentionally polluted with wasted motor oil. With TPH decrease as the primary dependent variable, the concentrations of non-ionic surfactant and inorganic NPK fertilizer were employed as independent bio-stimulation factors. There was a 67.20% decrease in TPH levels after 42 days. The scientists found that the ideal values for the bio-stimulation agent investigated utilizing a computational optimization method based on the desirability function, the amounts of NPK and nonionic surfactant required to accomplish 67.20% degradation of TPH were 4.22 g and 10.69ug/g, respectively. Moreover, total petroleum hydrocarbon degradation was obtained by (Agarry and Owabor 2011) utilizing kerosene as a source of TPH and inorganic NPK (4.30 g) as a basis of nutrients. Chromatographic results indicated that the kerosene-contaminated soil had lighter chains of hydrocarbons than the used engine oil-contaminated soil, which may have contributed to NPK's superior efficacy in lowering TPH in the latter soil. This supported past research (Venosa et al. 2002) that showed microorganisms broke down light-end hydrocarbons more quickly than heavy-end hydrocarbons, in the surface layer, non-saturated layer, and saturated layer of diesel-polluted soil (20,000 mg/kg), (Omokhagbor Adams et al. 2020) reported that according to a study researchers obtained 58%, 57%, and 32% TPH reduction by means of altered Fenton and NPK fertilizer. They disclosed that a bacterial population underwent specialization and differentiation soon following soil contamination, indicating improved degrading biological activity and an improvement in the degrading microbiota upon stimulation.

10.4 Surfactants

A family of structurally varied compounds made by various microbes is known as microbial surface-active chemicals. Their chemical makeup and microbiological origins are the main classification criteria. Their hydrophilic moiety is made up of mono-, di-, or polysaccharides; their hydrophobic moiety is made up of fatty acids; and their composition includes an acid, peptide anions, or cations. These structures have many different attributes, such as the capacity to reduce liquids' surface and interfacial tension and to create micelles and micro-emulsions between various phases. These materials can be broadly divided into two groups: bio-emulsifiers, which are polymers of polysaccharides, lipoproteins, or lipopolysaccharide proteins with a high molecular weight, and biosurfactants, which are low molecular weight (<1500 Da) substances such as proteins, glycolipids, and lipopeptides (Hamley 2015) (Fig. 1). The most studied class of microbial surfactants are glycolipids (Sar et al. 2023).

Among these are the most well-known compounds, namely rhamnolipids, trehalolipids, sophorolipids, and mannosylerythritol lipids (MELs) (Yu et al. 2007). Microbial metabolism inhibits the production of the hydrophilic moiety through a glycolytic pathway and the lipogenic pathway (lipid formation), as carbohydrates are the lone accessible basis of carbon for the synthesis of biopolymers. The glycolytic pathway produces intermediates like glucose 6-phosphate (G6P) when a range of enzymes break down a water-soluble substrate like glucose. In the hydrophilic zone of biopolymers, this G6P is one of the main pioneers of carbohydrates. Glycolysis first converts glucose to pyruvate, which is then turned to acetyl-CoA. AcetylCoA then combines with oxaloacetate to generate malonyl-CoA, which

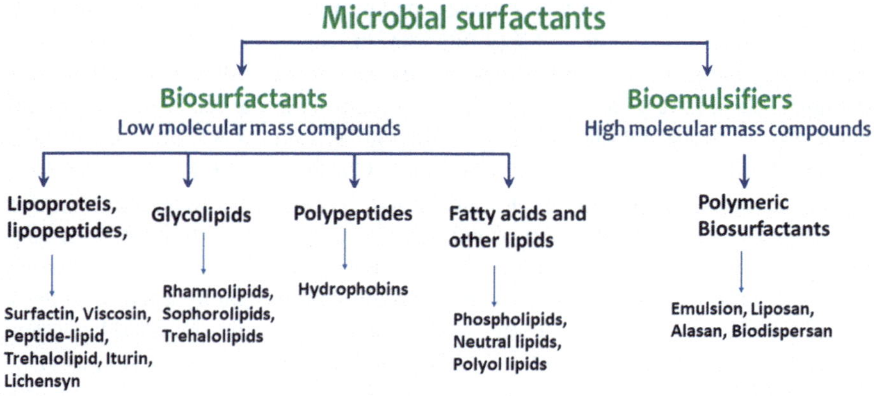

Fig. 1 Arrangement of several representative designs for microbial surfactants

is further processed into a lipid. Significant surface activities can be produced by a range of amphiphilic chemicals by environmental microbiological organisms. Biologically generated natural surfactants are known to have numerous uses across diverse fields due to their surfactant nature (Santos et al. 2016). Like synthetic surfactants, biosurfactants also produce micelle-like nano-aggregates in the 1–200 mg L − 1 concentration range. Recently, however, a higher CMC value of biosurfactants has been reported; for instance, the CMCs of glycolipids and glycoproteins generated by *Rhizopus arrhizus* (Pele et al. 2019) and *Acinetobacter baumannii* (Gupta et al. 2020) are 1200 and 1700 mg L − 1, respectively. As demonstrated for other biosurfactants, two well-known normal surfactants that reduce water's surface tension from 73 to 30 mN m − 1 are rhamnolipids and surfactin (Müller et al. 2012a, b). The uses of bio-derived surfactants are widely used, albeit not fully investigated. These include cosmetic, pharmaceutical, food, textile, painting, petroleum recovery, and many other industries. Using biosurfactants for remediation is a very profitable and vital way to eliminate contaminants from the atmosphere using the microbial degradation technique (Rangarajan and Narayanan 2018). Conventional remediation technologies typically encounter significant levels of resistance and difficulty in eliminating the contaminants, which are mainly organic pollutants. Furthermore, several poisonous heavy metals have a significant ability to interact with the soil matrix, creating a great deal of complexity that is necessary to eliminate the dangerous metals from the soil. Contaminated soil can be easily detoxified using the well-established bioremediation approach, which is aided by the existence of surfactants that are biologically active. The first issue that arises throughout the microbial bioremediation process is the contaminants' accessibility to the microorganism. Hydrophobic PAH molecules have a strong adhesion to the soil matrix due to their poorer solubility, which makes it difficult to remove them from soil medium. Natural surfactants produced by a variety of microorganisms show a noteworthy role in the biodegradation of organic hydrophobic compounds and polluted PAHs from soil. The degree of PAH biodegradation is significantly increased by biologically produced surfactants, as they increase the bioavailability of hydrophobic substrates. Since hydrophobic organic compounds are often difficult to access, biologically generated surfactants increase the hydrophobic compounds' desorption efficiency and accessibility, making organic contaminants more readily bio-accessible (Bezza and Nkhalambayausi-Chirwa 2015). Many researchers looked at the possible activity of a wide range of microbial species that are used to break down PAHs. According to some reports, the biosurfactant rhamnolipid shows promise in the breakdown of hydrocarbons. The removal of hexane is notably aided by bio-trickling saponin filters (Zhou et al. 2019). It has also been demonstrated that a biofilm harboring a *Pseudomonas* strain can function as a (bio) filter for volatile organic compounds, degrading hexane and producing biosurfactants. Because they can remove pollution and stop biomass from collecting, microorganisms that produce biosurfactants from hydrocarbons are essential to the biofiltering process (He et al. 2020). The initial soil washing by the biosurfactant is a helpful step that plays a major role in the breakdown of petroleum hydrocarbons from polluted soil (Olasanmi and Thring 2020). A dirt roll-up process and solubilization are the two mechanistic paths by which soil washing, a practical method for eliminating pollutants, works (Mao et al. 2015). Once it was discovered that as the biosurfactant concentration hit a threshold, PAH biodegradation accelerated. Since a larger dose of biosurfactant interferes with the microbial species' cellular membrane, the biosurfactant's breakdown effectiveness somewhat decreases beyond this threshold level. Biosurfactants regulate the microbial cell's surface hydrophobicity in addition to the solubility of organic pollutants. Once more, cell adhesion to hydrophobic chemical substances and cell-to-cell contacts are mediated by the hydrophobicity of the cell surface (Rabodonirina et al. 2019). Based on their molecular weight and concentration, the biosurfactants have three distinct modes of action when it comes to the bioremediation of hydrophobic chemicals from contaminated soil (Mańko-Jurkowska et al. 2019). Another method of the bioremediation process operates at lower concentrations and involves solubilizing hydrophobic organic compounds by means of the combined nanostructure of biosurfactants (above the CMC). The concentration of the used low-molecular-weight biosurfactant determines both mechanistic modalities of the biosurfactant-mediated remediation process. A high-molecular-weight biosurfactant's emulsifying capacity plays a mechanistic role in soil bioremediation, particularly when the biosurfactant is utilized to remove hydrocarbon-contaminated soil. After organic hydrophobic pollutants are solubilized and emulsified, a biodegradation process is facilitated by a biosurfactant with a larger molecular weight. In order to improve our understanding of the mechanisms by which biosurfactants break down hydrophobic pollutants, more investigation is required to fully understand how biosurfactants interact with cells (Karamchandani et al. 2022).

10.5 Bioaugmentation

Bioaugmentation, or the addition of oil-degrading microorganisms to boost native populations, has been suggested as a substitute method for bioremediation of oil-contaminated settings since the 1970s. The perceptive behind this strategy is that native microbial populations might be stressed from recent exposure to the spill or they might not be able to break down the broad diversity of possible substrates found in complex blends like petroleum (Leahy and Colwell 1990a, b). Bioaugmentation could also be used in situations where there is a low native population of hydrocarbon-degrading organisms, where the key factor is the rate of decontamination, or where seeding could condense the time, it takes for the bioremediation process to begin (Omokhagbor Adams et al. 2020). For the method to be successful in the field, the seed microorganisms used in this approach must be able to degrade the majority of petroleum components, maintain genetic stability and viability during storage, withstand harsh environments, effectively compete with native microorganisms, and pass through sediment pores to reach the contaminants (Atlas 1984). As it relates to the breakdown of oil molecules, different microbial species exhibit varying enzymatic capacities and preferences. Alkanes that are branching, cyclic, or linear can be broken down by some microbes. Some prefer mono- or polynuclear aromatics, whereas others degrade both alkanes and aromatics. It is feasible researching microorganisms in bioremediation systems, and it is possible to find those that can break down and create compounds with biotechnological uses in the petrochemical and oil industries. The usage of inocula made up of microbial strains or microbial groupings that have been specifically tailored to the site that has to be cleaned up is essential for the success of bioaugmentation therapies. The successful application of foreign microorganisms (inocula) is contingent upon their capacity to contend with native microorganisms, predators, and diverse abiotic variables. When looking for microorganisms to be applied, one should consider factors that affect the production of the microorganisms used for bioaugmentation, such as the chemical structure and concentration of pollutants, the availability of the pollutant to the microorganisms, the size and makeup of the microbial population, and the physical environment. Assuming and/or confirming that native organisms within the impacted site are unable to biodegrade petroleum hydrocarbon, bioaugmentation entails introducing microorganisms that have been isolated from the contaminated site, from a historical site, or that have been carefully chosen and genetically revised to support the remediation of petroleum hydrocarbon-polluted sites (Omokhagbor Adams et al. 2020).

10.6 Genetically Engineered Microorganisms (GEMs)

Recent developments in synthetic biology have shown to have a significant impact on environmental issues. Environmental biotechnology uses GEMs to remove pesticides, xenobiotics, and dangerous compounds from the environment (Pant et al. 2021). Understanding natural microbial communities is essential to the creation of a synthetic one. In a natural setting, determining which species are engaged in bioremediation can be difficult (Nwankwegu et al. 2022). In another study synthetic microbial community was used, as a model system for structural, functional, and ecological features. In which two microbial species were cultivated based on their interactions, bioremediation resulting to the formation of a non-natural microbiome with functionally specialized species (Anjum et al. 2022). These apparatuses have an impact on the community's performances and structure (Patel et al. 2022a). These metabolically driven outlines of microbial contact led to microbial community interaction (Saleh et al. 2022). Two conceivable outcomes from social interactions (such as mutualism, collaboration, and competition, for example) between two microbial populations are +/+ and/+ or +/. Collaboration is essential to the structure and operation of communities (Yue et al. 2022). The result of intentional collaboration is the synthetic community, which dictates how cooperation affects community dynamics. A variety of environmental manipulation techniques, including the insertion and removal of genes, can be used to engineer cooperation between two microbes (De Melo 2022). Furthermore, extra patterns of interaction have been examined with the artificial microbial species of the synthetic community. The use of intentional interaction is seen in bioremediation systems (Huang 2022). Synthetic biology holds the potential to improve the sustainability of huge populations of microorganisms. As a result, the application of non-natural microbial communities as bioremediation techniques for various contaminants, including pesticides, oil spills, acid drainage, etc. Engineered interactions between and across species can increase cellular activity and microbial consortia's capacity to create synthetic microbial communities in a variety of polluted situations (Abdi et al. 2022). A density-dependent phenomenon in bacterial signaling, quorum sensing combines population-level activity and cell-to-cell transmission. Chemical molecules are released and received by microbial organisms within a population to facilitate communication (Yin et al. 2022). This could be used by engineers to generate multicellular behavior and artificial communities with specialized roles. Synthetic models can be used to get a suitable strategy that might become less useful when competing with other species in the normal environment (Chavan et al. 2022). Over time, mutations in genomic elements and gene transfer could lead to the extinction of genetic circuits. The constancy of communities is defined in large part by species rivalry, as demonstrated by a recent study by Coyte and colleagues. For this reason, competition between species may have an adverse effect on the constancy of the artificial community. These dynamics need to speed up to build bioremediation synthetic communities with specific functionalities (Fan et al. 2022). The only technique that has been proven to be secure, economical, and long-lasting is bioremediation, which involves cleaning up contaminated areas (Dai et al. 2020). This makes the process of choosing the right microbial strain and then engineering it very slow. A few considerations need to be taken while designing and growing bacteria, such as whether they contain genes for the creation of chelators and metal absorption, as well as genes for existence in both normal and non-natural atmospheres.

Recombinant DNA technology allows for the transformation of any species into the chosen form (Sakshi and Haritash 2020). The desired gene can then be inserted into the vector's genome to create a compatible host gene in a different organism (such as phage, plasmid, or virus) (Patel et al. 2022b). This approach facilitates the bioremediation procedure. Numerous instruments are required, such as host enzymes, T4 polynucleotide kinase, alkaline phosphatase, reverse transcriptase, DNA ligase, as well as various DNA ligases (Pal et al. 2020). The function of metabolic and genetic engineering in bioremediation. Thanks to scientific and technological advancements, DNA molecule fragments holding one or more nucleotides can be exactly inserted, deleted, or swapped into an organism's genome's cells, enabling rational genetic engineering at the global (genome) or local (gene) level (Soutullo et al. 2017). Transcription activators like effector nucleases (TALEN) and clustered regularly interspaced short palindromic repeat (CRISPR)-related nucleases are among the most widely utilized tools for gene editing (Sharma and Shukla 2022). CRISPR-Cas is an effective and comprehensible gene editing tool. The application of these tools may help bioremediation. The sequence-specific DNA-binding module of TALEN is unique to the host genome (Priya et al. 2022). When TALEN attaches to DNA and leaves sticky ends for stability, double-stranded breaks (DSBs) are produced. Thirty amino acids make up ZFNs' DNA-binding domain. DSB is introduced at the intended place in the host genome by the Fok 1 cleavage domain. As a result of the molecular complexity, TALENs and ZFN nucleases, among other hybrid nucleases, have gained new insight (Gao et al. 2022). This allows the CRISPR-Cas system to precisely alter numerous genes at once at the same time. A bacterium called *Streptococcus pyogenes* provides immunity against the virus. In the CRISPR-Cas system, guide RNA (gRNA) connects crisper-derived RNA (crRNA) with trans-acting antisense RNA (trcRNA) (Sarma et al. 2021). The gRNA sequence identifies the Cas9 enzyme and directs it to induce a double-strand break (DSB) into the target DNA sequence. Researchers studying bioremediation are anticipating the release of these gene editing tools that produce knock-in and knock-out (Liu et al. 2021). Recent articles indicate that the CRISPR-Cas system has mostly been used by researchers in model organisms like as *Pseudomonas* or *Escherichia coli* (Okoli et al. 2022). With the progress of gRNA to direct function-specific genes crucial to remediation and new understandings of CRISPR tools, even non-model organisms like *Achromobacter* sp. HZ01 and *Comamonas testosteroni* may be exploited for bioremediation (Zhang et al. 2021). Since contaminant-inhabited bacteria are used to survive and find refuge in stressful environments caused by a variety of lethal, refractory, and non-degradable xenobiotics, they are outstanding contenders for metabolic engineering and genome editing (Patel et al. 2022b). The study of microbial bioremediation, for instance, the bioremediation of toxic pollutants by haloalkane dehalogenases and the decontamination of pyrethroid from soil via the fenpropathrin biodegradation mechanism discussed in *Bacillus* sp. DG-02, requires an understanding of metabolic pathways (Kurade et al. 2021). By altering the current course, metabolic engineering might be utilized to enhance the bioremediation process. This method thoroughly examines a range of enzymes, such as oxidases, esterases, monooxygenases, oxidoreductases, and phenoloxidases (Mehta et al. 2021). Since enzyme-based bioremediation is an environmentally favorable technique, it offers numerous benefits. Genetic engineering is more likely to produce recombinant enzymes. Studies have indicated that extracellular enzymes play a part in enzymatic bioremediation. The white rot fungus releases enzymes that degrade TNT (2,4,6-Trinitrotoluene), PCBs (polychlorinated biphenyls), and PAHs (polycyclic aromatic hydrocarbons). Simpler molecules are produced once the enzyme Esterase D combines by means of the insecticide endosulfan, an organochlorine. Hemoproteins encoded by LiPs from *Phanerochaete chrysosporium* break down PAHs (Gallo et al. 2021). Examples of contaminants that can be used up by microbes as an intermediate in biological pathways otherwise as a basis of energy even if they are only partially degraded include benzopyrene, which is broken down into three quinine compounds by the enzyme lignin peroxidase (LiP): 1,6-quinone, 6,12-quinine, and 3,6-quinine. MnP causes organic molecules to oxidize when Mn (II) is present (Marycz et al. 2022). Additional enzymes implicated in the biodegradation of refractory materials comprise cytochrome P450, laccase, glutathione S transferase, and varied function oxidases. *Pseudomonas* NP-6's extrinsic catechol 1,2-dioxidase converts catechol into muconate molecules. Enzyme immobilization significantly increases half-life, constancy, and enzyme activity. This strategy may lead to the effective and ecologically adequate enzymatic bioremediation of insistent xenobiotics (Wiener and LeFevre 2022). The limited production of these microbes has made it challenging to isolate and characterize those proficient in generating enzymes. Agricultural soil contains the primary constituents of insecticides, organophosphates (OP), and organochlorines (OC) in addition to the runoff from agriculture, which carries them to water sources. It has been demonstrated that genetically modified microbes are useful for bioremediating OC and OP (Sazanova et al. 2022). Additionally, tests were conducted on organophosphate and pyrethroid bioremediation by means of a genetically modified *P. putida* KT2440. Subsequently, with the progress of metabolic engineering, several persistent compounds were revealed to experience catabolism and destruction. The breakdown of methyl parathion and -hexachlorocyclohexane was demonstrated by *Pseudomonas* sp. WBC-3 and *Sphingobium japonicum* by bioremediation. One persistent ingredient in fumigants is 1, 2, 3-trichloropropane (Xu et al. 2022). Heterologous catabolism, a process in which the enzymes from two diverse microorganisms are united, is how *E. coli* releases this chemical into the environment. *P. putida* is created by introducing organophosphorus hydrolase (OPD) and the pnp operon, which encode the enzymes which catalyze the alteration of p-nitrophenol into adipate (Husna et al. 2022). The pobA and chcpca gene clusters of *Rhodococcus opacus* R7 aid in naphthenic acid bioremediation. An enzyme that breaks down lengthy chains of linear and non-alicyclic naphthenic acid is encoded by the aliA 1 gene. The aforementioned method uses microbes to partially or completely mineralize persistent compounds to lessen accumulation (Kharitonov et al. 2022). CRISPR is an influential technology that makes genome editing possible. It is easy for scientists to modify gene activity and DNA sequences in microorganisms; also this can be functional to bacteria that are proposed to remove organic contaminants. Several CRISPR-Cas systems are used in genome editing for bioremediation (types I, II, and III). Cas9, also known as CRISPR-associated, functions as a couple of molecular scissors that can cut DNA. A 30- to 40-nucleotide direct repeat sequence in CRISPR that is corresponding to foreign DNA is what makes up CrRNA. Guided RNA is another product of CRISPR. CrRNA and Cas protein are combined to create Cr ribonucleoproteins (RNPs) (Zhang et al. 2022). This crRNP in the invader initiates cleavage of DNA and RNA. The precision of the CRISPR mechanism is attributed to gRNA. To increase the microbial strain's degrading effectiveness, we can employ CRISPR-Cas9 to add or remove the desired gene. After deleting the yvmC gene in *Bacillus licheniformis* by means of the CRISPR-Cas9 technique, biotransformation proficiency increased to 100%. In the future, CRISPR-Cas9 technology might be applied to improve organism strains. It may be possible to create microbial strains using CRISPR-cas9 that can bioremediate soil pollutants (SPs) in the atmosphere (Qian et al. 2022).

11 Phytoremediation

Plants are utilized in phytoremediation, which purges contaminated zones. Several pollutants, for example, metals, pesticides, explosives, and oil, can be removed with the aid of plants. Nevertheless, large quantities may hinder plant growth and increase cleanup time; as a result, they thrive in low-contamination situations. Additionally, plants help prevent contaminants from being transported by wind, rain, and groundwater flow either further below or to neighboring places.

When certain plants' roots take up water and nutrients from polluted groundwater, sediment, or soil, the plants could remove or decompose dangerous substances from the environment (Saroj et al. 2022). As far down as their roots can go, plants can use natural mechanisms to help remove pollutants from the following areas:

- Contaminants should be stored in stems, roots, or leaves.
- Change them to low hazardous chemicals inside the plant or, more frequently, the root zone.
- Transform them into vapors that are released into the atmosphere.
- Sorb (stick) pollutants onto their roots, where "microbes," or little soil-dwelling creatures, including bacteria, convert the sorbed toxins into less toxic substances.

When it comes to removing contaminants, certain plants are more efficient than others. Plants must be able to withstand the types and concentrations of contaminants present for phytoremediation to be successful. Additionally, they need to be able to thrive and survive in the local climate. Another factor is the pollution level. Small plants, such as ferns, and grasses, have been employed in places where contamination is superficial. Because their roots grow deeper, trees like poplars and willows are used for hydraulic management, or to remove further soil contamination and contaminated groundwater.

Cleaning up a place with phytoremediation could take several years. Numerous variables will affect how long cleanup takes. Phytoremediation will take longer, for instance, where:

- The levels of contaminants are elevated.
- There is a lot of ground or depth contamination.
- Plants with extended growth periods are employed.
- The growth season is brief.

Phytoremediation includes rhizoremediation, phyto-volatilization, and phyto-stabilization (Davis 2001; Shahzad et al. 2016).

11.1 Rhizoremediation

Rhizoremediation comprises phytoremediation as well as bioremediation. Inoculating oily sludge-contaminated locations with a bacterial consortium that degrades hydrocarbons is known as bioremediation. Furthermore, phytoremediation is the process of breaking down the hydrocarbons in oily sludge-contaminated locations using plants that are tolerant to hydrocarbons, such as lucerne, soybean, perennial ryegrass, fescue, or Kaller grasses. The rhizosphere, or the environment created by plants, offers a special environment in which hydrocarbon-degrading microorganisms can proliferate and grow. The combined effect of these microorganisms and the release of root exudates, which include sugars, amino acids, and organic acids, among other things, accelerates the rate of hydrocarbon degradation compared to non-rhizosphere environments found in oily sludge-polluted sites. The usual reintegration procedure of oily sludge degradation with rhizoremediation proceeds more quickly than through phytoremediation or bioremediation. The complex and dynamic environment found in the rhizosphere of soils polluted with petroleum hydrocarbons allows a variety of plant-associated microbial genera that break down hydrocarbons into biomass, CO2, and H2O by β-oxidation (Siddiqui and Adams 2002). Different kinds of hydrocarbons break down in different ways inside microbial bodies. For example, isoalkanes break down into cyclic alcohols that are then dehydrogenated to ketone, while aromatics break down into diols that break down spontaneously to catechol (Jones et al. 2008). According to earlier research, there are several species of aerobic bacteria that live in rhizosphere environments. The most prevalent ones include *Arthrobacter, Rhodococcus, Bacillus, Pseudomonas putida, Pseudomonas aeruginosa, Corynebacterium* spp., and *Pseudomonas methanica* (Nair et al. 2008). *Acinetobacter baumannii*, which can break down straight-chain hydrocarbons, was discovered by Mishra et al. (2001). *Burkholderia cepacia* can speed up the breakdown of aromatic and NSO compounds in oily sludge, while *Pseudomonas* can extract asphaltene and alkane fractions from oily sludge. Four common bacterial strains were discovered by Das et al. (2018) from different soils that had gotten polluted with oily sludge through the previous 15–100 years: *Pseudomonas putida* MTCC 978, *Pseudomonas citronellolis* MTCC 1191, *Pseudomonas aeruginosa* MTCC 1034, and *Pseudomonas aeruginosa* MTCC 2642. Over the course of a year, they discovered that strains enhanced the rate of hydrocarbon breakdown, resulting in an overall reduction in hydrocarbons in oily sludge of 25–4%. After being inoculated with a hydrocarbon-degrading bacterium, the rate of aliphatic hydrocarbon degradation in oily sludge was shown to be faster than that of aromatic hydrocarbons (Shahzad et al. 2016).

11.2 Phytovolatilization and Phytostabilization

Phytoremediation is a useful technique for cleaning contaminated soils. This approach reduces pollutant toxicity in contaminated environments by utilizing plant relations at the physical, biological, chemical, biochemical, and microbiological stages. Numerous techniques are used in phytoremediation, depending on the kind and amount of pollution (Wei et al. 2021). Common techniques for eliminating contaminants like heavy metals include extraction, sequestration, and transformation. Organic pollutants, for instance, oils and chloro-compounds, can be effectively decayed, immobilized, undergo rhizoremediation, and evaporated by employing plants like alfalfa or willow (Odoh et al. 2019). In plants that act as phytoremediators, the tap root system or fibrous root system, penetration, toxicity levels, adaptableness to the harsh environmental situations of the pollutants, plant annual growth, direction, and, most importantly, the period required to influence the standard of cleanliness are all significant aspects. Additionally, the plant needs to fight insects and diseases (Chakrabartty and Rashid 2021). The elimination of contaminants from the roots and shoots is a crucial step in the phytoremediation process. Transpiration and partitioning are also necessary for the flow of water and nutrients (Ali et al. 2020). This process can be changed in relation to pollutants and plant nature. Most of the plants that are found in a contaminated environment can aid in phytoremediation. Inherent plants can be bioaugmented in contaminated areas by man-made, normal, or a blend of both types of plants. One of these is phytomining, the practice of using plants to eliminate precious metals from polluted areas (Nkrumah et al. 2019). Since many plants (more than 300) absorb Cu, Zn, and Ni optimally, they make better candidates for phytoremediation. Phytostabilization, also known as in situ immobilization or inactivation of heavy metals, lowers the metals' bioavailability as well as stops them from moving off-site. Metals are absorbed by the plant and restored at the roots. Several species exhibit Cd photostability, including *Virola surinamensis* and *Acanthus ilicifolius*. Five ornamental plants—*Ligustrum vicaryi*, *Euonymus japonicus*, *Osmanthus fragrans*, *Cinnamomum camphora*, and *Loropetalum chinense*—were selected because of their capability to phytostabilize CD (Zeng et al. 2018). Using bacterially aided phytoremediation, metal-contaminated water from different sources can be effectively treated. Coalitions of degrading, growth-promoting, and endophytic bacteria can employ the phytoremediation technique of metal reduction in wastewater by using plants (Capuana 2020).

12 Conclusion

Hydrocarbon contamination has a wide range of effects on ecosystems. Since the start of the industrial revolution and the use of fossil fuels, this problem has existed. Hydrocarbon pollution has greatly increased by a number of factors, including pipeline breaks, regular oil spills, widespread pesticide use, and unprecedented population growth. Together, these are endangering not only the natural microbial population but also the lives of marine, terrestrial, as well as airborne creatures. Therefore, environmental rehabilitation is the most important aspect of human survival. In addition to outlining the rationale, this chapter explains various biotechniques used for removing hydrocarbons from the environment, such as phytoremediation.

References

Abdi J, Sisi AJ, Hadipoor M, Khataee A (2022) State of the art on the ultrasonic-assisted removal of environmental pollutants using metal-organic frameworks. J Hazard Mater 424:127558. https://doi.org/10.1016/j.jhazmat.2021.127558

Abioye OP, Agamuthu P, Abdul Aziz AR (2012) Biodegradation of used motor oil in soil using organic waste amendments. Biotechnol Res Int 2012:587041. https://doi.org/10.1155/2012/587041

Adams G, Tawari-Fufeyin P, Ehinomen I (2014) Bioremediation of spent oil contaminated soils using poultry litter. Res J Eng Appl Sci 3:124

Adesodun JK, Mbagwu JSC (2008) Biodegradation of waste-lubricating petroleum oil in a tropical alfisol as mediated by animal droppings. Bioresour Technol 99:5659–5665. https://doi.org/10.1016/j.biortech.2007.10.031

Agamuthu P, Dadrasnia A (2013) Potential of biowastes to remediate diesel fuel contaminated soil. Glob Nest J 15:474–484. https://doi.org/10.30955/gnj.001031

Agarry SE, Owabor CN (2011) Anaerobic bioremediation of marine sediment artificially contaminated with anthracene and naphthalene. Environ Technol 32:1375–1381. https://doi.org/10.1080/09593330.2010.536788

Al-Hawash A, Dragh M, Li S, Alhujaily A, Abbood H, Zhang X, Ma F (2018) Principles of microbial degradation of petroleum hydrocarbons in the environment. Egypt J Aquat Res 44:71. https://doi.org/10.1016/j.ejar.2018.06.001

Ali S, Abbas Z, Rizwan M et al (2020) Application of floating aquatic plants in phytoremediation of heavy metals polluted water: a review. Sustainability 12:1927

Alvarez HM (2003) Relationship between β-oxidation pathway and the hydrocarbon-degrading profile in actinomycetes bacteria. Int Biodeterior Biodegrad 52:35–42. https://doi.org/10.1016/S0964-8305(02)00120-8

Anjum F, Yasin Naz M, Ghaffar A et al (2022) Sustainable insulating porous building materials for energy-saving perspective: stones to environmentally friendly bricks. Constr Build Mater 318:125930. https://doi.org/10.1016/j.conbuildmat.2021.125930

Atagana HI (2008) Compost bioremediation of hydrocarbon-contaminated soil inoculated with organic manure. Afr J Biotechnol 7:1516–1525. https://doi.org/10.5897/AJB08.193

Atlas RM (1984) Petroleum microbiology. MacMillan Pub. Co., New York, NY, United States

Atlas R, Bragg J (2009) Bioremediation of marine oil spills: when and when not—the Exxon Valdez experience. Microb Biotechnol 2:213–221. https://doi.org/10.1111/j.1751-7915.2008.00079.x

Aumeier BM, Georgi A, Saeidi N, Sigmund G (2023) Is sorption technology fit for the removal of persistent and mobile organic contaminants from water? Sci Total Environ 880:163343. https://doi.org/10.1016/j.scitotenv.2023.163343

Bacosa HP, Ancla SMB, Arcadio CGLA, Dalogdog JRA, Ellos DMC, Hayag HDA, Jarabe JGP, Karim AJT, Navarro CKP, Palma MPI, Romarate RA, Similatan KM, Tangkion JAB, Yurong SNA, Mabuhay-Omar JA, Inoue C, Adhikari PL (2022) From surface water to the Deep Sea: a review on factors affecting the biodegradation of spilled oil in marine environment. J Marine Sci Eng 10:426. https://doi.org/10.3390/jmse10030426

Banwart SA, Thornton SF (2010) Natural attenuation of hydrocarbon compounds in groundwater. In: Timmis KN (ed) Handbook of hydrocarbon and lipid microbiology. Springer, Berlin, pp 2473–2486. https://doi.org/10.1007/978-3-540-77587-4_182

Bezza FA, Nkhalambayausi-Chirwa EM (2015) Desorption kinetics of polycyclic aromatic hydrocarbons (Pahs) from contaminated soil and the effect of biosurfactant supplementation on the rapidly desorbing fractions. Biotechnol Biotechnol Equip 29:680–688. https://doi.org/10.1080/13102818.2015.1028444

Boll M, Estelmann S, Heider J (2020) Anaerobic degradation of hydrocarbons: mechanisms of hydrocarbon activation in the absence of oxygen. In: Boll M (ed) Anaerobic utilization of hydrocarbons, oils, and lipids. Springer, Cham, pp 3–29. https://doi.org/10.1007/978-3-319-50391-2_2

Brzeszcz J, Kaszycki P (2018) Aerobic bacteria degrading both n-alkanes and aromatic hydrocarbons: an undervalued strategy for metabolic diversity and flexibility. Biodegradation 29:359–407. https://doi.org/10.1007/s10532-018-9837-x

Bugg TDH, Winfield CJ (1998) Enzymatic cleavage of aromatic rings: mechanistic aspects of the catechol dioxygenases and later enzymes of bacterial oxidative cleavage pathways. Nat Prod Rep 15:513–530. https://doi.org/10.1039/A815513Y

Callaghan AV, Gieg LM, Kropp KG, Suflita JM, Young LY (2006) Comparison of mechanisms of alkane metabolism under sulfate-reducing conditions among two bacterial isolates and a bacterial consortium. Appl Environ Microbiol 72:4274–4282. https://doi.org/10.1128/AEM.02896-05

Capuana M (2020) A review of the performance of woody and herbaceous ornamental plants for phytoremediation in urban areas. iForest 13:139–151. https://doi.org/10.3832/ifor3242-013

Chakrabartty M, Rashid GMH-O (2021) Feasibility study of the soil remediation Technologies in the Natural Environment. Am J Civ Eng 9:91–98. https://doi.org/10.11648/j.ajce.20210904.11

Chavan D, Arya S, Kumar S (2022) In: Hussain C, Hait SBT-AOWM (eds) Open dumping of organic waste: associated fire, environmental pollution and health hazards. Elsevier, pp 15–31

Chernikova TN, Bargiela R, Toshchakov SV, Shivaraman V, Lunev EA, Yakimov MM, Thomas DN, Golyshin PN (2020) Hydrocarbon-degrading bacteria Alcanivorax and Marinobacter associated with microalgae Pavlova lutheri and Nannochloropsis oculata. Front Microbiol 11:572931. https://doi.org/10.3389/fmicb.2020.572931

Chia XK, Hadibarata T, Jusoh MNH, Sutiknowati LI, Tan IS, Foo HCY (2024) Role of extremophiles in biodegradation of emerging pollutants. Top Catal 2024:1–8. https://doi.org/10.1007/s11244-024-01919-7

Chikere C (2012) Culture-independent analysis of bacterial community composition during bioremediation of crude oil-polluted soil. Br Microbiol Res J 2:187–211. https://doi.org/10.9734/BMRJ/2012/1565

Chorom M, Sharifi HS, Motamedi H (2010) Bioremediation of a crude oil—polluted soil by application of fertilizers. Iran J Environ Heal Sci Eng 7:319–326

Dai X, Lv J, Yan G et al (2020) Bioremediation of intertidal zones polluted by heavy oil spilling using immobilized laccase-bacteria consortium. Bioresour Technol 309:123305

Das N, Chandran P (2010) Microbial degradation of petroleum hydrocarbon contaminants: an overview. Biotechnol Res Int 2011:941810. https://doi.org/10.4061/2011/941810

Das S, Kuppanan N, Channashettar VA, Lal B (2018) In: Adhya TK, Lal B, Mohapatra B et al (eds) Remediation of oily sludge-and oil-contaminated soil from petroleum industry: recent developments and future prospects BT—advances in soil microbiology: recent trends and future prospects: volume 1: soil-microbe interaction. Springer, Singapore, pp 165–177

David E, Niculescu V-C (2021) Volatile organic compounds (VOCs) as environmental pollutants: occurrence and mitigation using nanomaterials. Int J Environ Res Public Health 18:13147. https://doi.org/10.3390/ijerph182413147

Davis GG (2001) A Citizen ' s Guide to 95814:16

De Melo J (2022) Towards an Environmental Goods Agreement Style (EGAST) agenda to improve the regime complex for climate change. ECONSTOR, pp 202–219

Declercq I, Cappuyns V, Duclos Y (2012) Monitored natural attenuation (MNA) of contaminated soils: state of the art in Europe—a critical evaluation. Sci Total Environ 426:393–405. https://doi.org/10.1016/j.scitotenv.2012.03.040

Diefenbach T, Sumetzberger-Hasinger M, Braunschmid V, Konegger H, Heipieper HJ, Guebitz GM, Lackner M, Ribitsch D, Loibner AP (2024) Laccase-mediated degradation of petroleum hydrocarbons in historically contaminated soil. Chemosphere 348:140733. https://doi.org/10.1016/j.chemosphere.2023.140733

Fan H, Wang Y, Wang Y, Coyte PC (2022) The impact of environmental pollution on the physical health of middle-aged and older adults in China. Environ Sci Pollut Res Int 29:4219–4231. https://doi.org/10.1007/s11356-021-15832-z

Ferrey M (2006) Guidelines: natural attenuation of chlorinated solvents in ground water. Minnesota Pollution Control Agency

Fowzia A, Fakhruddin ANM (2018) A review on environmental contamination of petroleum hydrocarbons and its biodegradation. Int J Environ Sci Nat Resourc 11(3):555811. https://doi.org/10.19080/IJESNR.2018.11.555811

Gallo G, Puopolo R, Carbonaro M et al (2021) Extremophiles, a nifty tool to face environmental pollution: from exploitation of metabolism to genome engineering. Int J Environ Res Public Health 18:5228. https://doi.org/10.3390/ijerph18105228

Gao J-J, Wang B, Li Z-J et al (2022) Metabolic engineering of Oryza sativa for complete biodegradation of thiocyanate. Sci Total Environ 820:153283. https://doi.org/10.1016/j.scitotenv.2022.153283

Gomez EJ, Delgado JA, Gonzalez JM (2020) Environmental factors affect the response of microbial extracellular enzyme activity in soils when determined as a function of water availability and temperature. Ecol Evol 10:10105–10115. https://doi.org/10.1002/ece3.6672

Grandel S, Dahmke A (2004) Monitored natural attenuation of chlorinated solvents: assessment of potential and limitations. Biodegradation 15:371–386. https://doi.org/10.1023/B:BIOD.0000044683.44549.99

Gros J et al (2014) First day of an oil spill on the open sea: early mass transfers of hydrocarbons to air and water. Environ Sci Technol 48(16):9400–9411

Gupta A, Gupta R, Singh RL (2016) Microbes and environment. In: Principles and applications of environmental biotechnology for a sustainable future. Springer, Singapore, pp 43–84. https://doi.org/10.1007/978-981-10-1866-4_3

Gupta B, Puri S, Thakur IS, Kaur J (2020) Enhanced pyrene degradation by a biosurfactant producing Acinetobacter baumannii BJ5: growth kinetics, toxicity and substrate inhibition studies. Environ Technol Innov 19:100804. https://doi.org/10.1016/j.eti.2020.100804

Haddock JD (2010) Aerobic degradation of aromatic hydrocarbons: enzyme structures and catalytic mechanisms. In: Timmis KN (ed) Handbook of hydrocarbon and lipid microbiology. Springer, Berlin, Heidelberg, pp 1057–1069. https://doi.org/10.1007/978-3-540-77587-4_74

Hamley IW (2015) Lipopeptides: from self-assembly to bioactivity. Chem Commun 51:8574–8583. https://doi.org/10.1039/c5cc01535a

Hazen TC, Prince RC, Mahmoudi N (2016) Marine oil biodegradation. Environ Sci Technol 50:2121–2129. https://doi.org/10.1021/acs.est.5b03333

He S, Ni Y, Lu L et al (2020) Simultaneous degradation of n-hexane and production of biosurfactants by pseudomonas sp. strain NEE2 isolated from oil-contaminated soils. Chemosphere 242:125237. https://doi.org/10.1016/j.chemosphere.2019.125237

Head IM, Jones DM, Röling WFM (2006) Marine microorganisms make a meal of oil. Nat Rev Microbiol 4:173–182. https://doi.org/10.1038/nrmicro1348

Heider J, Fuchs G (1997) Anaerobic metabolism of aromatic compounds. Eur J Biochem 243:577–596. https://doi.org/10.1111/j.1432-1033.1997.00577.x

Heider J, Spormann AM, Beller HR, Widdel F (1998) Anaerobic bacterial metabolism of hydrocarbons. FEMS Microbiol Rev 22:459–473. https://doi.org/10.1111/j.1574-6976.1998.tb00381.x

Heipieper HJ, Martínez PM (2010) Toxicity of hydrocarbons to microorganisms. In: Timmis KN (ed) Handbook of hydrocarbon and lipid microbiology. Springer, Berlin, Heidelberg, pp 1563–1573. https://doi.org/10.1007/978-3-540-77587-4_108

Huang FH (2022) Influence of reduced air pollution source emission information on user Behavioural intention towards E-scooter products. Promet-Traffic Transp 34:53–67. https://doi.org/10.7307/ptt.v34i1.3762

Huno SKM, Rene ER, van Hullebusch ED, Annachhatre AP (2018) Nitrate removal from groundwater: a review of natural and engineered processes. J Water Supply Res Technol AQUA 67:885–902. https://doi.org/10.2166/aqua.2018.194

Husna HA, Shah M et al (2022) Heavy metal tolerant endophytic fungi aspergillus welwitschiae improves growth, ceasing metal uptake and strengthening antioxidant system in Glycine max L. Environ Sci Pollut Res Int 29:15501–15515. https://doi.org/10.1007/s11356-021-16640-1

Jia B, Niu Z, Wu Y, Kuzyakov Y, Li XG (2020) Waterlogging increases organic carbon decomposition in grassland soils. Soil Biol Biochem 148:107927. https://doi.org/10.1016/j.soilbio.2020.107927

Jiménez N, Richnow HH, Vogt C, Treude T, Krüger M (2016) Methanogenic hydrocarbon degradation: evidence from field and laboratory studies. J Mol Microbiol Biotechnol 26:227–242. https://doi.org/10.1159/000441679

Jimenez-Diaz L, Caballero A, Segura A (2017) Pathways for the degradation of fatty acids in bacteria. In: Rojo F (ed) Aerobic utilization of hydrocarbons, oils and lipids. Springer International Publishing, Cham, pp 1–23. https://doi.org/10.1007/978-3-319-39782-5_42-1

Johnsen AR, Wick LY, Harms H (2005) Principles of microbial PAH-degradation in soil. Environ Pollut 133:71–84. https://doi.org/10.1016/j.envpol.2004.04.015

Jones DM, Head IM, Gray ND et al (2008) Crude-oil biodegradation via methanogenesis in subsurface petroleum reservoirs. Nature 451:176–180. https://doi.org/10.1038/nature06484

Jørgensen KS (2011) In situ bioremediation. In: Moo-Young M (ed) Comprehensive biotechnology, 2nd edn. Academic Press, Burlington, pp 59–67. https://doi.org/10.1016/B978-0-08-088504-9.00372-X

Jørgensen KS, Salminen JM, Björklöf K (2010) Monitored natural attenuation. Methods Mol Biol 599:217–233. https://doi.org/10.1007/978-1-60761-439-5_14

Jyoti L, Smita C (2012) Effect of diesel fuel contamination on seed germination and growth of four agricultural crops. Univers J Environ Res Technol 2(4):311–317

Karamchandani BM, Pawar AA, Pawar SS et al (2022) Biosurfactants' multifarious functional potential for sustainable agricultural practices. Front Bioeng Biotechnol 10:1047279. https://doi.org/10.3389/fbioe.2022.1047279

Kebede G, Tafese T, Abda EM, Kamaraj M, Assefa F (2021) Factors influencing the bacterial bioremediation of hydrocarbon contaminants in the soil: mechanisms and impacts. J Chem 2021:9823362. https://doi.org/10.1155/2021/9823362

Khan RA (1990) Parasitism in marine fish after chronic exposure to petroleum hydrocarbons in the laboratory and to the Exxon Valdez oil spill. Bull Environ Contam Toxicol 44:59–763

Kharitonov S, Shchegolkova N, Alexandrova AV et al (2022) Taxonomic diversity of fungi and bacteria in Azoé-NP®vertical flow constructed wetlands. Water 14:698

Kleindienst S, Knittel K (2020) Anaerobic hydrocarbon-degrading sulfate-reducing bacteria at marine gas and oil seeps. In: Teske A, Carvalho V (eds) Marine hydrocarbon seeps: microbiology and biogeochemistry of a global marine habitat. Springer International Publishing, Cham, pp 21–41. https://doi.org/10.1007/978-3-030-34827-4_2

Kotthoff L, Keller J, Lörchner D, Mekonnen TF, Koch M (2019) Transformation products of organic contaminants and residues—overview of current simulation methods. Molecules 24:753. https://doi.org/10.3390/molecules24040753

Kouzuma A, Watanabe K (2011) Molecular approaches for the analysis of natural attenuation and bioremediation. In: Moo-Young M (ed) Comprehensive biotechnology, 2nd edn. Academic Press, Burlington, pp 25–36. https://doi.org/10.1016/B978-0-08-088504-9.00375-5

Kurade MB, Ha Y-H, Xiong J-Q et al (2021) Phytoremediation as a green biotechnology tool for emerging environmental pollution: a step forward towards sustainable rehabilitation of the environment. Chem Eng J 415:129040. https://doi.org/10.1016/j.cej.2021.129040

Laczi K, Erdeiné Kis Á, Szilágyi Á, Bounedjoum N, Bodor A, Vincze GE, Kovács T, Rákhely G, Perei K (2020) New Frontiers of anaerobic hydrocarbon biodegradation in the multi-omics era. Front Microbiol 11:590049. https://doi.org/10.3389/fmicb.2020.590049

Leahy JG, Colwell RR (1990a) Microbial degradation of hydrocarbons in the environment. Microbiol Rev 54:305–315. https://doi.org/10.1128/mr.54.3.305-315.1990

Leahy JG, Colwell RR (1990b) Microbial degradation of hydrocarbons in the environment. Microbiol Rev 54:305–315

Lee RF, Page DS (1997) Petroleum hydrocarbons and their effects in subtidal regions after major oil spills. Mar Pollut Bull 34(11):928–940

Liao Y, Geng A, Huang H (2009) The influence of biodegradation on resins and asphaltenes in the Liaohe Basin. Org Geochem 40:312–320. https://doi.org/10.1016/j.orggeochem.2008.12.006

Liu Q, Tang J, Gao K, Gurav R, Giesy JP (2017) Aerobic degradation of crude oil by microorganisms in soils from four geographic regions of China. Sci Rep 7:14856. https://doi.org/10.1038/s41598-017-14032-5

Liu X, Li Z, Zhang C, Tan X, Yang X, Wan C, Lee D-J (2020) Enhancement of anaerobic degradation of petroleum hydrocarbons by electron intermediate: performance and mechanism. Bioresour Technol 295:122305. https://doi.org/10.1016/j.biortech.2019.122305

Liu Y, Chang J, Yang C et al (2021) Genome-wide CRISPR-Cas9 screening in Bombyx mori reveals the toxicological mechanisms of environmental pollutants, fluoride and cadmium. J Hazard Mater 410:124666. https://doi.org/10.1016/j.jhazmat.2020.124666

Mańko-Jurkowska D, Głowacka R, Górska A, Ostrowska-Ligęza E (2019) The role of biosurfactants in soil remediation. Zesz Probl Postępów Nauk Rol 2019:33–43. https://doi.org/10.22630/ZPPNR.2019.596.4

Mao X, Jiang R, Xiao W, Yu J (2015) Use of surfactants for the remediation of contaminated soils: a review. J Hazard Mater 285:419–435. https://doi.org/10.1016/j.jhazmat.2014.12.009

Margesin R, Schinner F (2001) Bioremediation (natural attenuation and biostimulation) of diesel-oil-contaminated soil in an alpine glacier skiing area. Appl Environ Microbiol 67:3127–3133. https://doi.org/10.1128/AEM.67.7.3127-3133.2001

Marschner B, Kalbitz K (2003) Controls of bioavailability and biodegradability of dissolved organic matter in soils. Geoderma 113:211–235. https://doi.org/10.1016/S0016-7061(02)00362-2

Martirani-Von Abercron S-M, Pacheco D, Benito-Santano P, Marín P, Marqués S (2016) Polycyclic aromatic hydrocarbon-induced changes in bacterial community structure under anoxic nitrate reducing conditions. Front Microbiol 7:01775. https://doi.org/10.3389/fmicb.2016.01775

Marycz M, Brillowska-Dąbrowska A, Muñoz R, Gębicki J (2022) A state of the art review on the use of fungi in biofiltration to remove volatile hydrophobic pollutants. Rev Environ Sci BioTechnol 21:225–246

McGenity TJ (2014) Hydrocarbon biodegradation in intertidal wetland sediments. Curr Opin Biotechnol 27:46–54. https://doi.org/10.1016/j.copbio.2013.10.010

Mehta K, Shukla A, Saraf M (2021) Articulating the exuberant intricacies of bacterial exopolysaccharides to purge environmental pollutants. Heliyon 7:e08446. https://doi.org/10.1016/j.heliyon.2021.e08446

Mishra S, Jyot J, Kuhad RC, Lal B (2001) Evaluation of inoculum addition to stimulate in situ bioremediation of oily-sludge-contaminated soil. Appl Environ Microbiol 67:1675–1681. https://doi.org/10.1128/AEM.67.4.1675-1681.2001

Mu J, Chen Y, Song Z, Liu M, Zhu B, Tao H, Bao M, Chen Q (2022) Effect of terminal electron acceptors on the anaerobic biodegradation of PAHs in marine sediments. J Hazard Mater 438:129569. https://doi.org/10.1016/j.jhazmat.2022.129569

Müller MM, Kügler JH, Henkel M et al (2012a) Rhamnolipids—next generation surfactants? J Biotechnol 162:366–380. https://doi.org/10.1016/j.jbiotec.2012.05.022

Müller M, Mentel M, van Hellemond JJ, Henze K, Woehle C, Gould SB, Yu R-Y, van der Giezen M, Tielens AGM, Martin WF (2012b) Biochemistry and evolution of anaerobic energy metabolism in eukaryotes. Microbiol Mol Biol Rev 76:444–495. https://doi.org/10.1128/MMBR.05024-11

Mulligan CN, Yong RN (2004) Natural attenuation of contaminated soils. Environ Int 30:587–601. https://doi.org/10.1016/j.envint.2003.11.001

Mulligan C, Yong RN, Gibbs BF (2001) Surfactant-enhanced remediation of contaminated soil: a review. Eng Geol 60:371–380. https://doi.org/10.1016/S0013-7952(00)00117-4

Nair CI, Jayachandran K, Shashidhar S (2008) Biodegradation of phenol. Afr J Biotechnol 7:4951–4958. https://doi.org/10.4018/978-1-5225-8903-7.ch045

Nduka JK, Orisakwe OE, Ezenweke LO (2008) Some physicochemical parameters of potable water supply in Warri, Niger Delta area of Nigeria. Sci Res Essays 3:547–551

Nkrumah PN, Echevarria G, Erskine PD, van der Ent A (2019) Phytomining: using plants to extract valuable metals from mineralized wastes and uneconomic resources. In: Extracting innovations. CRC Press, pp 313–324. https://doi.org/10.1201/b22353-23

Nwankwegu AS, Zhang L, Xie D et al (2022) Bioaugmentation as a green technology for hydrocarbon pollution remediation. Problems and prospects. J Environ Manage 304:114313. https://doi.org/10.1016/j.jenvman.2021.114313

Odoh CK, Zabbey N, Sam K, Eze CN (2019) Status, progress and challenges of phytoremediation—an African scenario. J Environ Manag 237:365–378

Okoli AS, Blix T, Myhr AI et al (2022) Sustainable use of CRISPR/Cas in fish aquaculture: the biosafety perspective. Transgenic Res 31:1–21. https://doi.org/10.1007/s11248-021-00274-7

Olasanmi IO, Thring RW (2020) Evaluating rhamnolipid-enhanced washing as a first step in remediation of drill cuttings and petroleum-contaminated soils. J Adv Res 21:79–90. https://doi.org/10.1016/j.jare.2019.07.003

Omokhagbor Adams G, Tawari Fufeyin P, Eruke Okoro S, Ehinomen I (2020) Bioremediation, biostimulation and Bioaugmention: a review. Int J Environ Bioremediation Biodegrad 3:28–39. https://doi.org/10.12691/ijebb-3-1-5

Ortega MF, Guerrero DE, García-Martínez MJ et al (2018) Optimization of Landfarming amendments based on soil texture and crude oil concentration. Water Air Soil Pollut 229:234. https://doi.org/10.1007/s11270-018-3891-1

Oualha M, Al-Kaabi N, Al-Ghouti M, Zouari N (2019) Identification and overcome of limitations of weathered oil hydrocarbons bioremediation by an adapted Bacillus sorensis strain. J Environ Manag 250:109455. https://doi.org/10.1016/j.jenvman.2019.109455

Pal A, Singh J, Soni R et al (2020) The role of microorganism in bioremediation for sustainable environment management. Elsevier, Amsterdam

Palleroni NJ, Pieper DH, Moore ERB (2010) Microbiology of hydrocarbon-degrading pseudomonas. In: Timmis KN (ed) Handbook of hydrocarbon and lipid microbiology. Springer, Berlin, pp 1787–1798. https://doi.org/10.1007/978-3-540-77587-4_129

Pandolfo E, Barra Caracciolo A, Rolando L (2023) Recent advances in bacterial degradation of hydrocarbons. Water 15:375. https://doi.org/10.3390/w15020375

Pant G, Garlapati D, Agrawal U et al (2021) Biological approaches practised using genetically engineered microbes for a sustainable environment: a review. J Hazard Mater 405:124631. https://doi.org/10.1016/j.jhazmat.2020.124631
Parthipan P, Elumalai P, Ting YP, Rahman PKSM, Rajasekar A (2018) Characterization of hydrocarbon degrading bacteria isolated from Indian crude oil reservoir and their influence on biocorrosion of carbon steel API 5LX. Int Biodeterior Biodegradation 129:67–80. https://doi.org/10.1016/j.ibiod.2018.01.006
Patel AK, Singhania RR, Albarico FPJB et al (2022a) Organic wastes bioremediation and its changing prospects. Sci Total Environ 824:153889. https://doi.org/10.1016/j.scitotenv.2022.153889
Patel H, Shakhreliya S, Maurya R et al (2022b) In: Pandey VBT-AP (ed) CRISPR-assisted strategies for futuristic phytoremediation. Elsevier, Amsterdam, pp 203–220
Pele MA, Ribeaux DR, Vieira ER et al (2019) Conversion of renewable substrates for biosurfactant production by Rhizopus arrhizus UCP 1607 and enhancing the removal of diesel oil from marine soil. Electron J Biotechnol 38:40–48. https://doi.org/10.1016/j.ejbt.2018.12.003
Perfumo A, Banat IM, Marchant R, Vezzulli L (2007) Thermally enhanced approaches for bioremediation of hydrocarbon-contaminated soils. Chemosphere 66:179–184. https://doi.org/10.1016/j.chemosphere.2006.05.006
Piehler MF, Swistak JG, Pinckney JL, Paerl HW (1999) Stimulation of diesel fuel biodegradation by indigenous nitrogen fixing bacterial consortia. Microb Ecol 38:69–78. https://doi.org/10.1007/s002489900157
Pozdnyakova NN (2012) Involvement of the Ligninolytic system of white-rot and litter-decomposing fungi in the degradation of polycyclic aromatic hydrocarbons. Biotechnol Res Int 2012:243217. https://doi.org/10.1155/2012/243217
Priya A, Dutta K, Daverey A (2022) A comprehensive biotechnological and molecular insight into plastic degradation by microbial community. J Chem Technol Biotechnol 97:381–390. https://doi.org/10.1002/jctb.6675
Qian S, Chen Y, Xu X et al (2022) Advances in amplification-free detection of nucleic acid: CRISPR/Cas system as a powerful tool. Anal Biochem 643:114593. https://doi.org/10.1016/j.ab.2022.114593
Rabodonirina S, Rasolomampianina R, Krier F et al (2019) Degradation of fluorene and phenanthrene in PAHs-contaminated soil using Pseudomonas and Bacillus strains isolated from oil spill sites. J Environ Manag 232:1–7. https://doi.org/10.1016/j.jenvman.2018.11.005
Ramadass K, Megharaj M, Venkateswarlu K, Naidu R (2018) Bioavailability of weathered hydrocarbons in engine oil-contaminated soil: impact of bioaugmentation mediated by pseudomonas spp. on bioremediation. Sci Total Environ 636:968–974. https://doi.org/10.1016/j.scitotenv.2018.04.379
Rangarajan V, Narayanan M (2018) In: Adhya TK, Lal B, Mohapatra B et al (eds) Biosurfactants in soil bioremediation BT—advances in soil microbiology: recent trends and future prospects: volume 1: soil-microbe interaction. Springer Singapore, Singapore, pp 193–204
Rittmann B, Ford R, Wilkin R, Everett J, Kennedy L (2007) Monitored natural attenuation forum: MNA of metals and radionuclides. Remediat J 18:121–129. https://doi.org/10.1002/rem.20156
Rojo F (2009) Degradation of alkanes by bacteria. Environ Microbiol 11:2477–2490. https://doi.org/10.1111/j.1462-2920.2009.01948.x
Roy A, Dutta A, Pal S et al (2018a) Biostimulation and bioaugmentation of native microbial community accelerated bioremediation of oil refinery sludge. Bioresour Technol 253:22–32. https://doi.org/10.1016/j.biortech.2018.01.004
Roy A, Dutta A, Pal S, Gupta A, Sarkar J, Chatterjee A, Saha A, Sarkar P, Sar P, Kazy SK (2018b) Biostimulation and bioaugmentation of native microbial community accelerated bioremediation of oil refinery sludge. Bioresour Technol 253:22–32. https://doi.org/10.1016/j.biortech.2018.01.004
Safdari M-S, Kariminia H-R, Rahmati M et al (2018) Development of bioreactors for comparative study of natural attenuation, biostimulation, and bioaugmentation of petroleum-hydrocarbon contaminated soil. J Hazard Mater 342:270–278. https://doi.org/10.1016/j.jhazmat.2017.08.044

Sakshi, Haritash AK (2020) A comprehensive review of metabolic and genomic aspects of PAH-degradation. Arch Microbiol 202:2033–2058. https://doi.org/10.1007/s00203-020-01929-5

Saleh S, Sambakunsi H, Makina D et al (2022) "We threw away the stones": a mixed method evaluation of a simple cookstove intervention in Malawi. Wellcome Open Res 7:52. https://doi.org/10.12688/wellcomeopenres.17544.3

Santisi S, Cappello S, Catalfamo M, Mancini G, Hassanshahian M, Genovese L, Giuliano L, Yakimov MM (2015) Biodegradation of crude oil by individual bacterial strains and a mixed bacterial consortium. Braz J Microbiol 46:377–387. https://doi.org/10.1590/S1517-838246120131276

Santos DKF, Rufino RD, Luna JM et al (2016) Biosurfactants: multifunctional biomolecules of the 21st century. Int J Mol Sci 17:401. https://doi.org/10.3390/ijms17030401

Sar P, Kundu S, Ghosh A, Saha B (2023) Natural surfactant mediated bioremediation approaches for contaminated soil. RSC Adv 13:30586–30605. https://doi.org/10.1039/d3ra05062a

Sarma H, Islam NF, Prasad R et al (2021) Enhancing phytoremediation of hazardous metal(loid)s using genome engineering CRISPR–Cas9 technology. J Hazard Mater 414:125493. https://doi.org/10.1016/j.jhazmat.2021.125493

Saroj, Garg Bala S, Garg D et al (2022) Recent strategies for bioremediation of emerging pollutants: a review for a green and sustainable environment. Toxics 10:1–24

Sazanova KV, Zelenskaya MS, Vlasov AD et al (2022) Microorganisms in superficial deposits on the stone monuments in Saint Petersburg. Microorganisms 10:316. https://doi.org/10.3390/microorganisms10020316

Seo J-S, Keum Y-S, Li QX (2009) Bacterial degradation of aromatic compounds. Int J Environ Res Public Health 6:278–309. https://doi.org/10.3390/ijerph6010278

Shahzad A, Siddiqui S, Bano A (2016) Rhizoremediation of petroleum hydrocarbon, prospects and future. RSC Adv 6:108347–108361. https://doi.org/10.1039/c6ra12458e

Sharma B, Shukla P (2022) Futuristic avenues of metabolic engineering techniques in bioremediation. Biotechnol Appl Biochem 69:51–60. https://doi.org/10.1002/bab.2080

Sherry A, Grant RJ, Aitken CM, Jones M, Bowler BFJ, Larter SR, Head IM, Gray ND (2020) Methanogenic crude oil-degrading microbial consortia are not universally abundant in anoxic environments. Int Biodeterior Biodegradation 155:105085. https://doi.org/10.1016/j.ibiod.2020.105085

Siddiqui S, Adams WA (2002) The fate of diesel hydrocarbons in soils and their effect on the germination of perennial ryegrass. Environ Toxicol 17:49–62. https://doi.org/10.1002/tox.10032

Singh S, Kumari B, Mishra S (2011) Microbial degradation of alkanes. In: Microbial degradation of xenobiotics. Springer, Cham, pp 439–469. https://doi.org/10.1007/978-3-642-23789-8_17

Sorkhoh NA, Ghannoum MA, Ibrahim AS, Stretton RJ, Radwan SS (1990) Crude oil and hydrocarbon-degrading strains of Rhodococcus rhodochrous isolated from soil and marine environments in Kuwait. Environ Pollut 65:1–17. https://doi.org/10.1016/0269-7491(90)90162-6

Soutullo S, Sánchez MN, Enríquez R et al (2017) Bioclimatic vs conventional building: experimental quantification of the thermal improvements. Energy Procedia 122:823–828

Spierings ELH, Mulder MJHL (2017) Persistent orofacial muscle pain: its synonymous terminology and presentation. Cranio 35(5):304–307

Sun Y, Chen W, Wang Y, Guo J, Zhang H, Hu X (2021) Nutrient depletion is the main limiting factor in the crude oil bioaugmentation process. J Environ Sci 100:317–327. https://doi.org/10.1016/j.jes.2020.07.025

Tao K, Zhang X, Chen X et al (2019) Response of soil bacterial community to bioaugmentation with a plant residue-immobilized bacterial consortium for crude oil removal. Chemosphere 222:831–838. https://doi.org/10.1016/j.chemosphere.2019.01.133

Thauer RK, Jungermann K, Decker K (1977) Energy conservation in chemotrophic anaerobic bacteria. Bacteriol Rev 41:100–180. https://doi.org/10.1128/br.41.1.100-180.1977

Thauer RK, Kaster A-K, Seedorf H, Buckel W, Hedderich R (2008) Methanogenic archaea: ecologically relevant differences in energy conservation. Nat Rev Microbiol 6:579–591. https://doi.org/10.1038/nrmicro1931

Thi Mo L, Irina P, Natalia S, Irina N, Lenar A, Andrey F, Ekaterina A, Sergey A, Olga P (2022) Hydrocarbons biodegradation by Rhodococcus: assimilation of hexadecane in different aggregate states. Microorganisms 10:1594. https://doi.org/10.3390/microorganisms10081594

Timmis KN, Pieper DH (1999) Bacteria designed for bioremediation. Trends Biotechnol 17:200–204. https://doi.org/10.1016/s0167-7799(98)01295-5

Trivedi P, Wallenstein MD, Delgado-Baquerizo M, Singh BK (2018) Microbial modulators and mechanisms of soil carbon storage. In: Singh BK (ed) Soil carbon storage. Academic Press, Cambridge, pp 73–115. https://doi.org/10.1016/B978-0-12-812766-7.00003-2

Truskewycz A, Gundry TD, Khudur LS, Kolobaric A, Taha M, Aburto-Medina A, Ball AS, Shahsavari E (2019) Petroleum hydrocarbon contamination in terrestrial ecosystems—fate and microbial responses. Molecules 24:3400. https://doi.org/10.3390/molecules24183400

van Beilen JB, Funhoff EG (2007) Alkane hydroxylases involved in microbial alkane degradation. Appl Microbiol Biotechnol 74:13–21. https://doi.org/10.1007/s00253-006-0748-0

Varjani SJ (2017) Microbial degradation of petroleum hydrocarbons. Bioresour Technol 223:277–286. https://doi.org/10.1016/j.biortech.2016.10.037

Varjani S, Upasani VN (2019) Influence of abiotic factors, natural attenuation, bioaugmentation and nutrient supplementation on bioremediation of petroleum crude contaminated agricultural soil. J Environ Manag 245:358–366. https://doi.org/10.1016/j.jenvman.2019.05.070

Venosa AD, Lee K, Suidan MT et al (2002) Bioremediation and biorestoration of a Crube oil-contaminated freshwater wetland on the St.Lawrence river. Biorem J 6:261–281. https://doi.org/10.1080/10889860290777602

Wang B, Kuang S, Shao H, Wang L, Wang H (2021a) Anaerobic-petroleum degrading bacteria: diversity and biotechnological applications for improving coastal soil. Ecotoxicol Environ Saf 224:112646. https://doi.org/10.1016/j.ecoenv.2021.112646

Wang L, Cheng Y, Naidu R, Bowman M (2021b) The key factors for the fate and transport of petroleum hydrocarbons in soil with related in/ex situ measurement methods: an overview. Front Environ Sci 9:756404. https://doi.org/10.3389/fenvs.2021.756404

Wang M, Ding M, Yuan Y (2023) Bioengineering for the microbial degradation of petroleum hydrocarbon contaminants. Bioengineering (Basel) 10:347. https://doi.org/10.3390/bioengineering10030347

Wei Z, Van Le Q, Peng W et al (2021) A review on phytoremediation of contaminants in air, water and soil. J Hazard Mater 403:123658. https://doi.org/10.1016/j.jhazmat.2020.123658

Widdel F, Rabus R (2001) Anaerobic biodegradation of saturated and aromatic hydrocarbons. Curr Opin Biotechnol 12:259–276. https://doi.org/10.1016/s0958-1669(00)00209-3

Widdel F, Boetius A, Rabus R (2006) Anaerobic biodegradation of hydrocarbons including methane. In: Dworkin M, Falkow S, Rosenberg E, Schleifer K-H, Stackebrandt E (eds) The prokaryotes, Ecophysiology and biochemistry, vol 2. Springer, New York, NY, pp 1028–1049. https://doi.org/10.1007/0-387-30742-7_33

Wiener EA, LeFevre GH (2022) White rot fungi produce novel Tire Wear compound metabolites and reveal underappreciated amino acid conjugation pathways. Environ Sci Technol Lett 9:391–399. https://doi.org/10.1021/acs.estlett.2c00114

Wu M, Wu J, Zhang X, Ye X (2019) Effect of bioaugmentation and biostimulation on hydrocarbon degradation and microbial community composition in petroleum-contaminated loessal soil. Chemosphere 237:124456. https://doi.org/10.1016/j.chemosphere.2019.124456

Xu X, Liu W, Tian S, Wang W, Qi Q, Jiang P, Gao X, Li F, Li H, Yu H (2018) Petroleum hydrocarbon-degrading bacteria for the remediation of oil pollution under aerobic conditions: a perspective analysis. Front Microbiol 9:2885. https://doi.org/10.3389/fmicb.2018.02885

Xu M, Xue Z, Liu J et al (2022) Observation of few GR24 induced fungal-microalgal pellets performance for higher pollutants removal and biogas quality improvement. Energy 244:123171. https://doi.org/10.1016/j.energy.2022.123171

Yin C, Salmador MP, Li D, Lloria MB (2022) Green entrepreneurship and SME performance: the moderating effect of firm age. Int Entrep Manag J 18:255–275. https://doi.org/10.1007/s11365-021-00757-3

Yu J, Grossiord N, Koning CE, Loos J (2007) Controlling the dispersion of multi-wall carbon nanotubes in aqueous surfactant solution. Carbon N Y 45:618–623. https://doi.org/10.1016/j.carbon.2006.10.010

Yue F, Xie Z, Zhang Y et al (2022) Latitudinal distribution of gaseous elemental mercury in tropical Western Pacific: the role of the doldrums and the ITCZ. Environ Sci Technol 56:2968–2976. https://doi.org/10.1021/acs.est.1c07229

Zekri A, Chaalal O (2005) Effect of temperature on biodegradation of crude oil. Energy Sources 27:233–244. https://doi.org/10.1080/00908310490448299

Zeng P, Guo Z, Cao X et al (2018) Phytostabilization potential of ornamental plants grown in soil contaminated with cadmium. Int J Phytoremediation 20:311–320

Zhang J, Zhang H, Li S et al (2021) Increasing yield potential through manipulating of an ARE1 ortholog related to nitrogen use efficiency in wheat by CRISPR/Cas9. J Integr Plant Biol 63:1649–1663. https://doi.org/10.1111/jipb.13151

Zhang H, Zhang X, Geng A (2022) Construction of CRISPR-Cas9 genome editing platform for white-rot fungus Cerrena unicolor BBP6 and its effects on extracellular ligninolytic enzyme biosynthesis. Biochem Eng J 185:108527. https://doi.org/10.1016/j.bej.2022.108527

Zhou C, Huang D, Xu P et al (2019) Efficient visible light driven degradation of sulfamethazine and tetracycline by salicylic acid modified polymeric carbon nitride via charge transfer. Chem Eng J 370:1077–1086. https://doi.org/10.1016/j.cej.2019.03.279

Index

S. R. Joshi et al. (eds.), *Mineral Transformation and Bioremediation by Geo-Microbes*, Interdisciplinary Biotechnological Advances,
https://doi.org/10.1007/978-981-96-3033-2

The manufacturer's authorised representative in the EU is Springer Nature Customer Service Centre GmbH, Europaplatz 3, 69115 Heidelberg, Germany. If you have any concerns regarding our products, please contact ProductSafety@springernature.com

Printed and bound by CPI Group (UK) Ltd, Croydon, CR0 4YY
01/10/2026
02225913-0020